# Topics in Applied Physics

Volume 139

Topics in Applied Physics is a well-established series of review books, each of which presents a comprehensive survey of a selected topic within the domain of applied physics. Since 1973 it has served a broad readership across academia and industry, providing both newcomers and seasoned scholars easy but comprehensive access to the state of the art of a number of diverse research topics.

Edited and written by leading international scientists, each volume contains high-quality review contributions, extending from an introduction to the subject right up to the frontiers of contemporary research.

Topics in Applied Physics strives to provide its readership with a diverse and interdisciplinary collection of some of the most current topics across the full spectrum of applied physics research, including but not limited to:

- Quantum computation and information
- Photonics, optoelectronics and device physics
- Nanoscale science and technology
- Ultrafast physics
- Microscopy and advanced imaging
- Biomaterials and biophysics
- Liquids and soft matter
- Materials for energy
- Geophysics
- Computational physics and numerical methods
- Interdisciplinary physics and engineering

We welcome any suggestions for topics coming from the community of applied physicists, no matter what the field, and encourage prospective book editors to approach us with ideas. Potential authors who wish to submit a book proposal should contact Zach Evenson, Publishing Editor:

zachary.evenson@springer.com

Topics in Applied Physics is included in Web of Science (2019 Impact Factor: 0.633), and is indexed by Scopus.

More information about this series at http://www.springer.com/series/560

David J. Lockwood · Lorenzo Pavesi
Editors

# Silicon Photonics IV

## Innovative Frontiers

Springer

*Editors*
David J. Lockwood
National Research Council of Canada
Metrology Research Centre
Ottawa, ON, Canada

Lorenzo Pavesi
Department of Physics
University of Trento
Povo, Italy

ISSN 0303-4216 ISSN 1437-0859 (electronic)
Topics in Applied Physics
ISBN 978-3-030-68221-7 ISBN 978-3-030-68222-4 (eBook)
https://doi.org/10.1007/978-3-030-68222-4



This Springer imprint is published by the registered company Springer Nature Switzerland AG
The registered company address is: Gewerbestrasse 11, 6330 Cham, Switzerland

# Preface

After 20 years of development, silicon photonics has reached commercial maturity with many available products. From 2020 to 2025, predicted growth rates of the associated market are in excess of 20% for conservative estimates to more than 30% for more optimistic estimates. In addition to Datacom and Telecom, many other applications are entering the silicon photonic products market. Particularly promising are the applications in the healthcare sector and in the automotive sector where high-quality and low-cost silicon photonic sensors, laser ranging systems and detectors allow a pervasive penetration of the technology. From the very start of silicon photonics, we follow its blossoming development from basic science to engineering and technology via a series of books entitled Silicon Photonics, all published by Springer in their Topics in Applied Physics series. The original intention, when the first book was published, was to underline the update to silicon electronics through the use of photons, which avoids the bottlenecks created with electrons for information exchange (interconnection bottleneck). Silicon Photonics I was based on the underlying fundamental research that was needed to bring these concepts into reality. Volume II of the series, then, covered the burgeoning applications scenario, and Silicon Photonics III provided overviews of practical applications of this technology in industry. Luckily, the field of silicon photonics is still presenting many interesting developments spanning from fundamental research where the use of new materials enables novel functionalities in silicon to the development of new technologies and computation paradigms where silicon photonics is making the difference. We feel the time is right to discuss these recent progresses to encompass the widening interest in the use of silicon photonics.

This fourth book in the series on silicon photonics gathers in-depth discussions of recent advances that go beyond already established and applied concepts. The chapters provided in this book by experts in their fields cover not only new research into the highly desired goal of light production in Group IV materials, but also change in the principles driving the development of integrated circuits, novel measurement strategies and novel technologies. Finally, the new paradigms in information processing and telecommunication are covered with the claim that silicon photonics is a proper platform for their realization.

Chapters in the first part of the book on *Advances in Fundamental Research* review recent progresses in the enduring search for crystalline-silicon-compatible light emitters and innovative applications of silicon photonics. For two decades now, there has been intensive research into overcoming the difficulty of obtaining efficient light emission from Si and Ge due to their indirect band gaps. In the past, various methods of inducing a direct gap have been introduced and lasing action has been observed in both Si and Ge, but so far, the performance at room temperature has not been good enough to produce the desired CMOS compatible room temperature laser. Recent work has shown that strained nanostructures of Group IV materials are the most promising candidates for room temperature lasers, as shown for Si in Chap. 1 by Kateřina Dohnalová and Kateřina Kůsová, for disordered Ge in Chap. 2 by Moritz Brehm, for GeSn and SiGeSn alloys in Chap. 3 by Vincent Reboud et al., and crystalline Ge in Chap. 4 by Nelson L. Rowell and David J. Lockwood. In Chap. 5, Simone Rossi, Elisa Vitiello and Fabio Pezzoli demonstrate that Ge-based heterostructures offer innovative solutions to the challenges of integrating spin functionalities into silicon photonics.

Chapters in the second part of the book on *Advances in Integration Architectures* review novel concepts and recent progress in the design of integrated photonic circuits for, and diverse applications of, silicon photonics. In Chap. 6, Hon Ki Tsang et al. introduce some of the numerous applications of subwavelength grating structures employed within silicon photonics devices. Subwavelength gratings have attracted interest recently, as they provide a useful degree of freedom for the crafting of the effective refractive index of the material incorporated in photonic devices. The implementation of quantum mechanics has been the driving force of the most intriguing designs of photonic structures. Non-Hermitian physics, which breaks the conventional scope of quantum mechanics based on the Hermitian Hamiltonian, has been widely explored in the silicon photonics platform, with promising new designs involving the refractive index, modal coupling and gain–loss distribution. In Chap. 7, Changqing Wang, Zhoutian Fu and Lan Yang explore the merger of the non-hermitian physics and classical silicon platforms. Topological photonic insulators have attracted considerable attention due to their unique ability to transport light via topologically protected edge states that are immune to defect scattering so that engineering of robust photonic devices that are insensitive to fabrication imperfections is made possible. Chapter 8 comprises a review by Shirin Afzal, Tyler James Zimmerling and Vien Van of the key concepts of topological insulator systems in one and two dimensions and their practical realization using coupled microring resonator lattices. In Chap. 9, Scott Jordon presents a new parallel digital gradient search technique for rapid automated alignment of devices on silicon photonics integrated circuits. Such a production tool is badly needed now owing to the increasing complexity of silicon photonics integrated circuits.

Chapters in the third part of the book on *Advances in Computation Schemes* review paradigm changing architectures where recent progress in silicon photonics enables new computer technology. In Chap. 10, Bicky A. Marquez et al. report on neuromorphic architectures on silicon photonics platforms, which enable high-bandwidth, low-latency, low-energy applications. Neuromorphic photonics implements the machine learning methodology in integrated photonics. It exploits the advantages of optics, including the ease of analog processing and full parallelism achieved at the speed of light by employing multiple signals in a single waveguide. Due to its mature semiconductor fabrication processes and the capability to integrate large quantum photonic circuits on a single device, silicon quantum photonics is emerging as a promising platform to develop photonic quantum processor chips. In Chap. 11, Stefano Paesani and Anthony Laing review recent results in developing key building blocks for chip-scale photonic quantum devices and deliberate on progress made toward useful large-scale quantum photonic computers. Finally, in Chap. 12, Marco Fiorentino et al. discuss recent progress toward an open silicon photonics ecosystem targeted at computercom applications where there is now an urgent need for standardized industrial production systems.

We hope that the readers will share our enthusiasm about this book, which shows the healthy status of silicon photonics despite the difficulties given by the present situation. We thank all the authors of the present volume for their invaluable contributions, especially considering the extensive restrictions in their work arena resulting from the COVID-19 global pandemic. At the same time, we are grateful to the editorial and production staff of Springer Nature for their support, patience and professional editing. Last but not least, we are indebted to our co-workers who are the real driving force behind our work. They do share with us their research, passion and, mostly importantly, friendship.

Ottawa, Canada — David J. Lockwood
Trento, Italy — Lorenzo Pavesi
December 2020

# Contents

# Contributors

**Shirin Afzal** Department of Electrical and Computer Engineering, University of Alberta, Edmonton, AB, Canada

**M. Ashkan Seyedi** Hewlett Packard Laboratories, Milpitas, USA

**Raymond G. Beausoleil** Hewlett Packard Laboratories, Milpitas, USA

**Moritz Brehm** Institute of Semiconductor and Solid State Physics, Johannes Kepler University Linz, Linz, Austria

**D. Buca** Institute of Semiconductor Nanoelectronics, Peter Grünberg Institute 9 (PGI 9) and JARA Fundamentals of Future Information Technologies, Forschungszentrum Jülich, Jülich, Germany

**V. Calvo** Univ. Grenoble Alpes, CEA, IRIG-DePhy, Grenoble, France

**A. Chelnokov** Université Grenoble Alpes, CEA, LETI, Grenoble, France

**Xia Chen** Optoelectronics Research Centre, University of Southampton, Southampton, UK

**Zhenzhou Cheng** School of Precision Instruments and Optoelectronics Engineering, Tianjin University, Tianjin, China

**Kateřina Dohnalová** Institute of Physics, University of Amsterdam, Amsterdam, The Netherlands

**Marco Fiorentino** Hewlett Packard Laboratories, Milpitas, USA

**Zhoutian Fu** Department of Electrical and Systems Engineering, Washington University, St. Louis, MO, USA

**J. M. Hartmann** Université Grenoble Alpes, CEA, LETI, Grenoble, France

**Chaoran Huang** Department of Electrical Engineering, Princeton University, Princeton, NJ, USA

**Zhihong Huang** Hewlett Packard Laboratories, Milpitas, USA

**Z. Ikonic** School of Electronic and Electrical Engineering, Pollard Institute, University of Leeds, Leeds, UK

**Scott Jordan** Physik Instrumente (PI) GmbH & Co., California, USA

**Kateřina Kůsová** Institute of Physics, Czech Academy of Sciences, Prague 6, Czech Republic

**Anthony Laing** Quantum Engineering Technology Labs, H. H. Wills Physics Laboratory, University of Bristol, Bristol, UK;
Department of Electrical and Electronic Engineering, University of Bristol, Bristol, UK

**Di Liang** Hewlett Packard Laboratories, Milpitas, USA

**David J. Lockwood** Measurement Science and Standards, National Research Council Canada, Ottawa, ON, Canada

**Bicky A. Marquez** Department of Physics, Engineering Physics and Astronomy, Queen's University, Kingston, ON, Canada

**Sagi Mathai** Hewlett Packard Laboratories, Milpitas, USA

**Stefano Paesani** Quantum Engineering Technology Labs, H. H. Wills Physics Laboratory, University of Bristol, Bristol, UK;
Department of Electrical and Electronic Engineering, University of Bristol, Bristol, UK

**N. Pauc** Univ. Grenoble Alpes, CEA, IRIG-DePhy, Grenoble, France

**Fabio Pezzoli** L-NESS and Dipartimento di Scienza dei Materiali, Università degli Studi di Milano Bicocca, Milan, Italy

**Paul R. Prucnal** Department of Electrical Engineering, Princeton University, Princeton, NJ, USA

**V. Reboud** Université Grenoble Alpes, CEA, LETI, Grenoble, France

**P. Rodriguez** Université Grenoble Alpes, CEA, LETI, Grenoble, France

**Simone Rossi** L-NESS and Dipartimento di Scienza dei Materiali, Università degli Studi di Milano Bicocca, Milan, Italy

**Nelson L. Rowell** Measurement Science and Standards, National Research Council Canada, Ottawa, ON, Canada

**Bhavin J. Shastri** Department of Physics, Engineering Physics and Astronomy, Queen's University, Kingston, ON, Canada

**H. Sigg** Laboratory for Micro- and Nanotechnology, Paul Scherrer Institut, Villigen, Switzerland

**Yeyu Tong** Department of Electronic Engineering, The Chinese University of Hong Kong, Shatin NT, Hong Kong SAR, People's Republic of China

**Hon Ki Tsang** Department of Electronic Engineering, The Chinese University of Hong Kong, Shatin NT, Hong Kong SAR, People's Republic of China

**Vien Van** Department of Electrical and Computer Engineering, University of Alberta, Edmonton, AB, Canada

**Elisa Vitiello** L-NESS and Dipartimento di Scienza dei Materiali, Università degli Studi di Milano Bicocca, Milan, Italy

**Changqing Wang** Department of Electrical and Systems Engineering, Washington University, St. Louis, MO, USA

**Lan Yang** Department of Electrical and Systems Engineering, Washington University, St. Louis, MO, USA

**Wen Zhou** Department of Materials, University of Oxford, Oxford, UK

**Tyler James Zimmerling** Department of Electrical and Computer Engineering, University of Alberta, Edmonton, AB, Canada

# Part I
# Advances in Fundamental Research

# Chapter 1
# Optical Properties of Si Nanocrystals Enhanced by Ligands

**Kateřina Dohnalová and Kateřina Kůsová**

**Abstract** Compared to bulk silicon, silicon nanocrystals (Si-NCs) show modified properties, such as tunable emission and enhanced radiative rate, as a result of the quantum confinement, surface chemistry and environment. While the effect of quantum confinement is well understood and experimentally confirmed on the hydrogen-capped Si-NCs, the surface effects in Si-NC with other types of ligands can be very complex and hard to predict. In our work, we argue that the surface chemistry, be it ligands and/or shell, can be designed to further improve the radiative rate of the Si-NCs, beyond what is achievable by the quantum confinement alone. Our experimental work shows a number of effects that indicate that in many instances, the core and surface capping cannot be separated, and optical properties cannot be clearly interpreted as "extrinsic" (related to the surface capping agent) or "intrinsic" (related to the core only). To this end, we performed also a detailed theoretical analysis of a number of surface ligands, to identify the role of chemistry and how that improves the optical properties of Si-NCs. Based on these investigations and findings, we realized two main things. Firstly, we argue that one cannot derive a simple rule to predict which type of element or molecule will improve or deteriorate the optical properties, because every individual element added (covalently) to the surface of Si-NC contributes to the electronic density via several mutually dependent effects, such as (i) orbital displacement, (ii) direct contribution of surface species into the density of states close to the bandgap, (iii) charge transfer due to the relative polarity of the surface capping element and Si, or (iv) ligand/matrix induced strain. Secondly, we realized that the **k**-space projections of the molecular orbitals, i.e., the band structure

K. Dohnalová (✉)
Institute of Physics, University of Amsterdam,
Science Park 904, 1098XH Amsterdam, The Netherlands
e-mail: k.newell@uva.nl

K. Kůsová
Institute of Physics, Czech Academy of Sciences,
Cukrovarnická 10, 162 00 Prague 6, Czech Republic
e-mail: kusova@fzu.cz

 
D. J. Lockwood and L. Pavesi (eds.), *Silicon Photonics IV*,
Topics in Applied Physics 139, https://doi.org/10.1007/978-3-030-68222-4_1

of the nanocrystal, are an essential and critical tool for investigations of the electronic and optical properties in materials with an originally indirect bandgap, since the surface chemistry in our simulations affects strongly the whole band structure.

## 1.1 Introduction

Silicon is a cornerstone of the modern civilization. Thanks to all its superlatives, such as abundant resources, non-toxicity, bio-compatibility and biodegradability, chemical robustness, low cost production, naturally forming oxide, and many more advantageous properties, it is in fact desirable material for any application. Silicon makes up about 28% of the Earth's crust by mass. Bulk silicon dominates CMOS micro-electronics technologies and enabled digital technologies through transistors to the computing central processing unit (CPU). It also plays an important role in photovoltaics and the detector industry, despite its poor band-edge absorption, for which it compensates by a higher material thickness. Silicon is currently entering also battery applications, for its enormous capacity for Li ion intake [27]. Silicon was reported to be, even in its nanocrystalline form, non-toxic [3, 9, 56, 125] and bio-degradable [154], with superior photo- and pH-stability [82], which opens opportunities also in the traditionally high health risk areas such as cosmetics, agriculture or medicine, for example, as theranostic agents [170]. However, for optical applications in lighting, displays, lasers, and amplifiers, as well as thin film photovoltaics, bulk silicon is not best suited due to its indirect bandgap (Fig. 1.1a). Radiative recombinations of electrons and holes, as well as optical excitations of electrons across the bandgap, require the participation of phonons, which lowers the probability of such transitions. The resulting weak oscillator strength of the optical transitions leads to a low radiative rate and a slow absorption onset with a weak absorption at the band-edge.

Nevertheless, finding/designing a silicon form that can efficiently emit light is obviously highly desirable. A silicon light source would enable the realization of the long awaited on-chip-integrated silicon laser and hence also all-Si photonics (the main topic of this book series), and consequently also optical CPU architecture [1]. Moreover, such a light source could also be implemented in the on-chip integration of the light emitting diodes (LEDs) [146], desired for the light-weight and modular micro-LED displays and lighting. Efficient band-edge absorption would be beneficial for silicon-based thin solar cells, which would also enable a wide spread of solar energy for transportation and other areas where besides efficiency also portability or light-weightiness are essential. Also, because of its non-toxicity and bio-compatibility, silicon can play an essential role in the bio-imaging [56] and bio-integration (with human body) of optically driven micro-devices, sensors, and interfaces [101].

For the last two decades, we have been searching experimentally and theoretically for paths toward enhanced optical capabilities of silicon, especially the wavelength tunability and radiative rate, via combining the quantum confinement and surface engineering in ligand-capped silicon nanocrystals (Si-NCs) (We note that for simplicity, we use the "Si-NCs" term also for materials, where the crystallinity of the nanoparticle has not been proven or measured and should be better named silicon nanoparticles). Properties induced by the surface chemistry are measured experimentally on both single nanoparticle and ensemble levels. Such analysis is combined with theoretical simulations via semi-empirical tight binding and from first principles by the use of self-consistent ground state density functional theory (DFT).

### 1.1.1 Quantum Confinement

The most reliable route toward an enhanced oscillator strength, and hence also the enhanced radiative rate and absorption cross section, has been so far via the utilization of the quantum confinement in Si-NCs. Bright size-tunable emission from the Si-NCs was first reported in 1990 in a pioneering work of Canham et al. [19] on H- and oxidized porous silicon, followed by many more reports from differently prepared and capped Si nanostructures in the following three decades [8, 26, 29, 33, 43, 67, 78, 107, 140, 143, 144, 150, 179, 185].

Quantum confinement in Si-NCs has a considerable effect on their electronic structure, but only for radii comparable or smaller than the bulk silicon's excitonic Bohr radius of ~4.9 nm. In such a case, strong spatial confinement of carriers results in a shift of valence and conduction bands, leading to a bandgap "opening", where the bandgap energy increases with the decreasing Si-NC size (Fig. 1.1b, d–g). According to the simplest effective mass approximation (EMA) model, the optical bandgap $E_g$ scales with the NC diameter $d$ via an inverse parabolic dependence $E_g(d) \propto d^{-2}$ as a result of the quantum localization. We also need to add the Coulomb term, which scales with $\propto d^{-1}$ and small polarization terms [14]. In reality, experimentally investigated Si-NC samples show slightly different values of the exponent, reported in the literature between 1.3 and 2.0 [122, 214]. The bandgap energy determines the photoluminescence (PL) peak from the Si-NCs, which can be tuned in a wide spectral range, from ultraviolet (UV) at ~260 nm to near infrared (NIR) at 1100 nm [43, 179] (Fig. 1.1d–g). This broad spectral tunability has been experimentally and theoretically proven in the hydrogen-capped Si-NCs (Fig. 1.1b, d, g) [43]. Despite being an ideal model of an Si-NC for theoretical calculations, H-capped Si-NCs are only rarely studied experimentally due to their increased sensitivity to air and UV light, leading almost immediately to at least partial oxidation, as discussed in more detail below. This is why only a few reports are included in the collection of literature data plotting the experimentally reported PL maxima as a function of NC size, which yields an exponent of 1.33, but shows a low spread of data points around the fitted curve. In Si-NCs with other than H-capping, often a much narrower tunability range is observed [37, 46, 76, 124, 172, 207] due to either a presence of surface sites that act as a traps for the excited carriers [43, 211], or due to the presence of strain [116], or simply as a result of limitations of the preparation procedure. Moreover, chemically

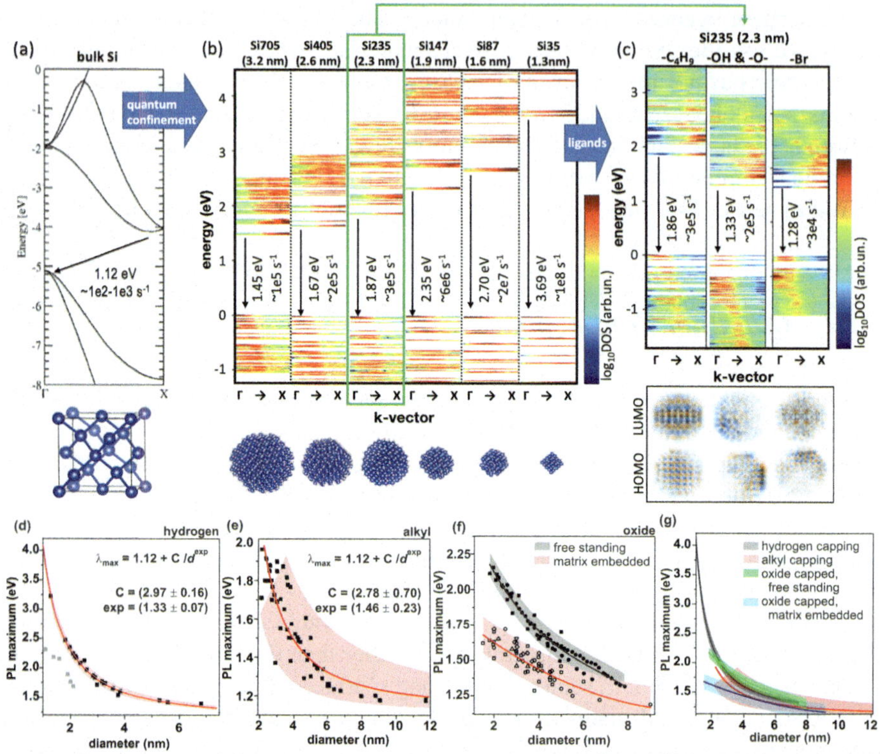


**Fig. 1.1** Quantum confinement effects in Si-NCs. **a** Bulk silicon band structure in the Γ–X direction, critical for the optical properties, where the bandgap can be identified. Bottom panel shows a silicon unit cell with the diamond crystalline structure. **b k**-space resolved density of states (DOS) for the H-capped Si-NCs of diameter from 1.3 to 3.2 nm, simulated by the DFT using cp2k code (full settings are discussed in [44]). Bottom panel shows the relative sizes and shapes of the simulated Si-NCs. **c** "Fuzzy" band structures of 2.3 nm Si-NC capped 50% by hydrogen and 50% by a butyl ligand $-C_4H_9$ (left), partly capped with –H and partly oxidized with –OH and -O- bonds (middle) and fully –Br capped surface (right). Below are shown real-space 2D cross sections of the highest occupied molecular orbital (HOMO) and the lowest unoccupied molecular orbital (LUMO) wavefunctions. The color scheme in (**b**, **c**) represents the DOS on a log-scale. Bandgap energy and phonon-less thermalized ($T = 300$ K) radiative rates are given in (**b**, **c**) in each respective band structure plot. **d–g** Comparison of photoluminescence (PL) tunability, experimentally reported by various sources: **d** Fully H-capped Si-NCs [45, 65, 211]. The gray data points [180] are possibly influenced by strain and are therefore not included in the fit. **e** Alkyl-capped Si-NCs [68, 79, 85, 96, 145, 162, 175, 207, 214]. **f** Oxidized Si-NCs [116]. Solid symbols in (**f**) denote Si-NCs prepared as free-standing (e.g., by wet etching or plasma synthesis followed by slow oxidation). Open symbols in (**f**) stand for matrix-embedded Si-NCs (prepared, e.g., as $SiO_x$/Si superlattices or by ion implantation or samples involving thermal oxidation). The fits are meant only as a description of the data and are PL max $= -0.0027d^3 + 0.056d^2 - 0.47d + 2.9$ for the free-standing and PL max $= 0.0058d^2 - 0.13d + 1.9$ for the matrix-embedded samples. Data from [116] are reused with permission from AIP Publishing. Panels (**d–f**) contain also fits of the data and the corresponding 95% confidence bands. **g** Comparison of fit curves from panels (**d–f**)

different surface ligands were themselves experimentally [22] and theoretically [215] shown to cause spectral shifts.

The effects of the quantum confinement on the bandgap energy is common to all types of semiconductor NCs. In an indirect bandgap semiconductor, such as silicon, additional effects occur due to the relaxation of the **k**-selection rule as a consequence of the larger spatial confinement of carriers. In silicon NCs, this in turn enhances the radiative rate $k_{rad}$, based on the theoretical calculations, proportionally to the inverse cubic function $k_{rad}(d) \propto d^{-3}$ [8]. In particular, there is about a 3–4 orders of magnitude increased phonon-less radiative rate in the smallest Si-NCs, when compared to the bulk Si, reaching $\sim 10^7$ s$^{-1}$ for the simulated ~1 nm H-capped Si-NCs [39, 44, 75, 159]. Despite such a considerable improvement, radiative rates for the Si-NCs emitting in the visible spectral range still remain much lower than those in the traditionally employed direct bandgap semiconductor NCs with the radiative rates exceeding $10^8$–$10^9$ s$^{-1}$, suggesting a persistently indirect nature of the bandgap in the Si-NCs and the importance of the $k$-space information in general.

An interesting development with respect to the enhanced radiative rate in Si-NCs has been published only very recently [180]; The authors study a series of H-capped Si-NCs prepared by a standard bottom-up sol-gel approach and observe a gradual size-induced PL shift from 700 to 530 nm. Whereas the larger Si-NCs exhibit the typical slow PL decay rates, at the size range below 1.7 nm, the PL decay rate is enhanced by about five orders of magnitude. (Argumentation is provided as to whether this PL decay rate is most likely the radiative rate.) This observation could be interpreted as the switch between indirect bandgap-like and direct bandgap-like emission in Si-NCs; the authors interpret their experimental data in terms of a switch between a bulk-like and molecular size regime. It is unknown why these particular very small Si-NCs exhibit fast PL decay, but the authors admit to the possibility that the small Si-NCs are strained.

Despite the generally limited success in the enhancement of the radiative rate for Si-NCs with emission in a visible spectral range and often observed limited tunability range of the bandgap in various ligand-capped Si-NCs, some of the best Si-NC materials with bright emission have already been implemented in various optoelectronic prototypes of light emitting devices [6, 43, 51, 55, 139, 163]. Nevertheless, such devices have not yet been commercialized and remain in a research and development stage, suggestive of the fact that a much needed improvement of their optical properties is still required.

### *1.1.2 Complex Role of Surface Chemistry*

The most frequently quoted property of a nanocrystalline semiconductor material is the above discussed "size-effect". One of its manifestations is the optical bandgap tunability via quantum confinement in the NCs smaller than the Bohr's excitonic radius. In most semiconductor NCs, such as the direct bandgap III–V or II–VI NCs, the size of the core offers a robust parametrization for the optical bandgap energy.

However, in the case of the Si-NCs, the size of the NC core is not enough to determine neither the optical bandgap energy nor the radiative rate, both being significantly affected by the surface chemistry, e.g., by organic coatings (Fig. 1.1e), oxide capping and strain (Fig. 1.1f).

The often observed strong influence of ligands on the optical properties of the Si-NCs is due to a complex interplay between the core and the covalent surface chemistry with other elements, and the relatively large percentage of atoms being located at the surface (a very useful tool giving insight into the basic possible arrangements of atoms and bonds can be found in [105]). This interplay manifests via charge transfer, strain, orbital displacement, and the direct contribution of the surface elements to the density of states near the bandgap. Silicon is a light element, and its valence electron states are relatively close to the elements that are often considered as capping agents, such as oxygen, nitrogen, or carbon. Silicon can covalently bond with many elements with a various degree of electron sharing/transfer, stability, opto-chemical robustness or resistance against oxidation (which is a strongly preferred reaction under normal conditions). The most robust and studied types of cappings in the literature are H–, silica oxide (O–), hydrocarbon (organic) C–, O– or N-linked ligands, and halides (–Cl, –Br, –I or –F), together with the P- and B- surface co-doped Si-NCs [182]. For the sake of completeness, we note that it is important to realize that the photo-chemical stability of a certain bond cannot be easily deduced from the Si-X bond dissociation energy (DE), derived for the diatomic species [132], because the covalent bond strength changes in the presence of other bonded species, or under strains [202]. This is a result of the different orbital displacement caused by the additional bonded species, which can weaken the bond. Hence, the silane $SiH_4$ molecule is extremely unstable, violently reacting with oxygen, but Si-NC capped with H can resist oxidation for a very long time. This is because the dissociation of the Si-H bond is easier when in the –Si-$H_2$ or even –Si-$H_3$ form, and therefore, the –$SiX_3$ species on the Si-NC surface are less stable than the –SiX species. Also, various facets on the Si-NC surface have different Si coordination and are therefore subject to different strains in the presence of ligands due to steric hindrance effects. The highly curved surface of small Si-NCs also results in distorted/strained bonds, which further decreases their stability and makes them more prone to reactions or disintegration. For these reasons, stability of, e.g., H-capping on the surface of a Si-NC will differ for a crystalline and an amorphous Si-NC, as well as for large or small Si-NCs, and possibly also for Si-NCs prepared by different methods. This further complicates the main question of this chapter on *how a certain element or molecule influences the optical properties of a Si-NC*, because one type of ligand can lead to very different results, when placed in different positions on the Si-NC surface.

The H-capping is the simplest and best understood ligand, and as such is also a great "clean" starting point for the subsequent surface treatments [149]. Despite the extreme reactivity of the $SiH_4$ with air, as discussed above, the H-terminated Si surfaces are on their own chemically very stable, because the Si-H bond is strong and nonpolar; however, a direct photo-desorption occurs under a UV-blue irradiation [104]. As a result of the native indirect bandgap, persisting also in small Si-NCs [80, 159], Si-NCs absorb efficiently only from the blue-UV range, rendering the H-capped Si-NCs surfaces photo-chemically unstable for the light-related applications. Nevertheless, despite their opto-chemical instability, H-capped Si-NCs are often presented in the literature as a reference material and a starting point for theoretical simulations, because their optical properties are purely core-driven.

#### 1.1.2.1 Oxide Capping

Silica oxide is naturally forming on the surface of silicon under exposure to air. Therefore, silica oxide-capped Si-NCs are one of the most accessible and analyzed types of Si-NCs. Capping with oxygen atoms can be done in several ways, as a double bond Si=O, as a bridging bond Si-O-Si, or via –OH groups [25]. The first two types have been reported to lead to limited bandgap tunability [74, 166, 167, 200, 211, 216] and slower radiative rates [40, 78]. The limiting effect of oxidation on spectral tunability has been first experimentally reported by Wolkin et al. [211] on porous silicon, prepared from bulk crystalline silicon by electro-chemical etching in solution composed of ethanol, water, and HF acid. This resulted in a H-passivated "nano-sponge" surface [83, 211], which contained small H-capped Si-NCs. As demonstrated by Wolkin et al. [211], the H-capped surface oxidized under exposure to air, when excited by UV light, which was accompanied by a change in color of the emitted PL.

For the larger oxide-capped Si-NCs, the PL peak was shown to be size tunable [10, 23, 52, 84, 122, 187, 211]; however, the tunability would often end in the red range close to 620–650 mm, indicating possible involvement of the surface oxide states [133]. This PL is always characterized by a slow $\sim$ μs decay, for which it has been named in literature as an "S-band" [17]. For the smaller Si-NCs, the limit to which the emission can be tuned varies greatly through the literature [46]. A detailed overview can be found in [116] with the data sets re-plotted in Fig. 1.1f, clearly illustrating the existence of two distinct classes of oxide-capped Si-NCs, characterized by a different size dependence of the PL maximum. These two classes of Si-NCs correlate with the way the samples were prepared—either as "free-standing" (e.g., by a wet etching) or as "matrix-embedded" (e.g., by an ion implantation). We have hypothesized that in the matrix-embedded samples, the matrix exerts a compressive strain on the crystalline Si-NC cores, which leads to the corresponding shift in the electronic bands and consequently to the reported red-shifted emission, when compared to the free-standing oxidized Si-NCs. Interestingly, an analogical influence of the matrix was also reported in $CdS_xSe_{1-x}$ NCs [176]. An opposite effect, i.e., the expansion of the crystalline lattice in oxidized Si-NCs, was demonstrated by doping Si-NCs with lithium [102], leading to a blue-shifted PL emission. Interestingly, the collection of the experimental data on the PL peak maximum as a function of the size of the NC $d$ from the literature in Fig. 1.1f could not be satisfactorily fitted using the EMA-like model $E_g^{NC} = E_g^{bulk} + C/d^{exp}$. Therefore, a polynomial, meant simply as a quantitative description, was used. The different shape of the size dependence of the bandgap might imply that an additional effect of possibly size-dependent strain is present in such oxidized samples. These oxide-capped Si-NCs exhibit typical stretched-exponential PL decay of the order of tens or hundreds of microseconds

[72], whose various properties [81, 197], including (the lack of) saturation of its power dependence [36, 73], were discussed in countless publications.

Interestingly, unlike other types of capping, a thin oxide shell was shown to lead to a very narrow luminescence linewidth [183], which could be very beneficial, for example, a high color definition required in application for displays. Oxidized Si-NCs can be also very interesting for application in the medical field. They are relatively photo-stable and were shown to be non-toxic [98, 155, 170, 177, 193]. In aqueous environments, oxide-capped Si-NCs slowly dissolve into the omnipresent, benign silicic acid [155], i.e., they are bio-degradable. The size of the lumminescent oxide-capped Si-NCs is always below ~5 nm, which is a limit for a safe excretion from a living organism via urine [126], but can be an issue for its possible entry through the blood-brain barrier [71]. Also, upon excitation, oxide-capped Si-NCs are a source of oxygen radicals that are toxic to the neighboring tissue, an effect suggested for possible use in photo-induced local cancer treatment [108, 156, 191], or as a carrier of radioactive isotopes [158]. The red and slow decaying emission offers also a great contrast to the often fast decaying blue-green emission from organic tissue [144]. For the optical applications, however, the oxidation of the Si-NC surface might not be desired, when a full spectral tunability through the whole visible range is required, as well as high radiative rates.

#### 1.1.2.2 Organic Capping

Another type of surface capping, well represented in the literature, but less understood theoretically, is via organic molecules, often long alkyl chains attached via hydrosilylation. Unlike in oxidized Si-NCs, the bandgap tunability curve for alkyl-capped SiNCs (Fig. 1.1e) can be fitted with an EMA-like model, albeit using a slightly different exponent than that in H-capped Si-NCs. Thus, this shape of the tunability curve suggests that in the alkyl-capped Si-NCs, the emitted PL resembles more that of the "ideal" H-capped Si-NCs system rather than the PL of the oxidized Si-NCs, in agreement with ab-initio studies [44, 171]. The large spread of the reported data around the fitted curve in Fig. 1.1e, also critically pointed out in [85] for this type of Si-NCs, is partly a result of the inclusion of many reports in this data set, where every measurement, especially the determination of size, is inevitably connected with an error [189, 190]. However, as some recent literature suggests, it can also signify some degree of influence of the alkyl chains on the emitted PL, be it their length, type of attachment or surface coverage [44]. Furthermore, synthesis of organically capped Si-NCs often involves heating of organic solvents, which has been reported to lead to the presence of other than Si-NC emitters, such as e.g. carbon dots, which could skew the reported PL properties if not carefully eliminated [17, 151, 209].

The comparison of the PL tunability curves in Fig. 1.1g confirms the presence of a limit around the end of the red spectral range for most of the experimentally investigated Si-NCs (except the H-capped ones) [37, 76, 180, 207]. It is very likely that most of the experimentally investigated organically capped Si-NCs are partly oxidized under UV-blue illumination, since due to the steric hindrance between the organic ligands, the surface cannot be 100% capped by organic molecules [44]. A possible culprit of the observed difference in the tunability limit of alkyl- and oxide-capped Si-NCs with respect to the H-capped ones might also be a "stability problem", where the surface of a too small Si-NC might be too reactive to be passivated after the core has been formed, since the attachment of ligands can simply lead to the disintegration of the whole Si-NC. Such a stability problem could in principle be circumvented by the various one-pot synthesis schemes of already-passivated Si-NCs [165].

The PL decay is for the larger, red-NIR emitting alkyl-capped Si-NCs, similarly slow as for the oxide-capped Si-NCs, albeit with much better quantum yields [85, 97, 175]. Moreover, there are certainly some new developments in this field of research: Ensemble-induced effects start to be discussed in these Si-NCs [92, 145], the discussion on the periodicity of the core and its influence on the PL has been opened [189, 190], and even a fast PL component has been reported within this red emission band [11]. Pressure-dependent PL studies of organically capped Si-NCs perfectly agree with bulk-like band shifts [79], pointing toward a possible core-related origin. Interestingly, very similar band shifts were reported also in oxide-capped Si-NCs [116]. Comparable PL changes in alkyl- and oxide-capped Si-NCs were also reported in temperature-resolved measurements [138]. On the contrary, alkyl- and oxide-capped SiNCs were observed to exhibit different phonon modes in Raman measurements [86], interpreted in terms of oxide-induced strain.

While origin of the slow decaying PL in organically capped Si-NCs is accepted as core-related, the vast majority of the bottom-up wet-chemically synthesized Si-NCs exhibit unusually fast, nanosecond decaying PL in the blue-green spectral range [41, 53, 85, 113, 173, 206, 210, 217], whose origin is still a subject of intense debate and will be discussed in the following sections.

#### 1.1.2.3 Direct Bandgap-Like Emission from Si-NCs: The "F-Band"

Quite controversially, a bright emission with direct bandgap-like radiative rates of $10^7$–$10^9$ $s^{-1}$, and often a blue-green spectrum, has been reported experimentally from various Si-NCs capped with oxide [5, 90, 109] (see the latest review in [17]), and also for organic ligands [41, 53, 85, 113, 173, 206, 210, 217]. Throughout the literature, to differentiate between the traditionally slow decaying Si emission and the often observed fast decaying emission bands, the two are often labeled as the "S-band" for the slow emission and the "F-band" for the fast one. The "S-band" PL has typically 1–100s microsecond lifetime and red-NIR tunable energy, as described above. The "F-band" is very often confined to the blue-green spectral region and has a fast 1–10 nanosecond PL decay. In the past, the "F-band" has been assigned mostly to an extrinsic origin, such as oxygen- [12, 58, 59, 66, 69, 77, 168, 192, 194, 212] or nitrogen-related surface sites [34]. Other literature sources gave evidence for its possible intrinsic origin [42, 99, 152, 164, 195, 205] connected with the core or a combination of intrinsic and surface-related emissive sites [30, 44, 64]. A large body of literature explores also the possibility that this emission is entirely unrelated to the Si-NCs and is caused by carbon dots or other organic impurities [18, 20, 21, 24, 35, 57, 121, 129, 130, 151, 201, 209]. With respect to the carbon impurities, in our recent critical study [209], we show that despite the confirmed presence of Si and NCs in the elemental and materials analysis of the sample, the emission still did not originate from the Si-NC, but from ill-defined organic impurities, possibly carbon dots. Hence, one must perform synthesis of a control sample, as well as a direct (preferably correlative) single-dot microscopy, to be able to confirm that the emissive nanoparticle is indeed the Si-NC, and not a carbon dot, which would also exhibit size-dependent properties.

At least some blue PL can be emitted by nearly any (mostly organic or oxide-related) compound, when detected with high enough sensitivity. Therefore, clearly, extreme care needs to be taken to rule out other species than Si-NCs as the source of the observed PL. However, if we exclude the possibly of an erroneous assignment, given the broad scope of the reported origins of the fast emission, it is highly improbable that a single, all-encompassing explanation exists. More probably, different physical phenomena are responsible for the fast emission in the different samples, and often, more than one of the proposed mechanisms might apply even in a single sample. Thus, we do not believe that the F-band label should apply to such a broad scope of phenomena and that several, possibly overlapping, sub-groups exist within this type of emission, as we have already suggested elsewhere [117]. Therefore, a detailed preparation history of any studied sample needs to be complemented by a thorough characterization to differentiate between the individual modes of emission. Additionally, taking into account the very small size of just a few crystallographic planes, sometimes even the concepts of, e.g., a "surface-site" might not be that easy to grasp. In most cases, the wavefunction of an electron or a hole inside a NC will be spread over a large part of the core [44], rather than just residing on a certain surface site or being homogeneously delocalized over the whole core volume. Such an effect is evident, e.g., from the real-space localization of the highest occupied molecular orbital (HOMO) and the lowest unoccupied MO (LUMO) wavefunctions in, e.g., Fig. 1.1c. Therefore, except for few extreme cases, a differentiation between a surface- or core-related origin can be blurred [110]. In fact, the reality of the light emission of the Si-NC is governed by a complex interplay of many intertwined phenomena, as we have recently discussed in [44], and parallel experimental and theoretical approaches are "a must" in order to achieve further progress.

Despite the described uncertainties, the silicon community is dedicated to uncover the origin of this fast decaying and often very efficient "F-band" emission, as it holds great promises for optical applications of silicon. In our experimental work, we researched the "F-band" emission using various methods and differently prepared Si-NCs, including reports of positive optical gain [49]. In our most recent work, we focus on the **k**-space projected density of states of Si-NCs with various ligands [48, 80, 115, 159], described in the following section, to uncover the possible evolution of the "direct bandgap-like" optical transitions that could stand behind the F-band emission.

### *1.1.3 The* **K***-Space Projections of the Density of States*

The physics of emission from the Si-NCs is in general challenging and not yet fully controlled due to a complex, almost organic-like covalent chemistry on the silicon surfaces [16], which can, moreover, depend also on the preparation technique of the Si-NC core [208]. This leads to a complex electronic interplay between the silicon and ligand states, where the surface bonded elements can strongly affect the whole electronic structure of the Si-NCs [44] (Fig. 1.1c), especially the real- and **k**-space density of states (DOS) profiles, and consequently also the bandgaps and radiative rates. In fact, the **k**-space resolved DOS in NCs is under-represented in most of the available theoretical literature and is not a standard option in the DFT simulations packages (unlike for bulk crystals). However, this might be slowly changing, as more recent work is adopting this approach as well [11, 162].

Band structure, or a **k**-space projected DOS, is a very useful formalism for the description of the electronic and optical properties in bulk crystalline materials. This often used approximation uses a simplified description of a quantum state of a solid based on single electron states. It assumes that the electrons travel in a static potential without dynamically interacting with the lattice vibrations, other electrons, etc., (or in other words the adiabatic and Hartree–Fock approximations are applied to the corresponding Schrödinger equation). This approach has proven very successful even beyond the scope of the initial approximation, as, for example, an exciton or a dopant/ impurity/ defect can be viewed as a small correction to the states within the band structure. One important assumption from which the band structure concept is derived is the requirement for a long-range translational symmetry in an ideally infinite crystalline material. In a macroscopic crystal, the lattice can be considered infinite without too much simplification, since 1 $cm^3$ of a crystalline material can contain some $10^{23}$ atoms. Similar simplification, however, would be far too crude for a small nanocrystal, whose crystalline core can contain as few as 500 atoms. Despite clearly breaking the requirement for being infinite and having a long-range translational symmetry, the band structure description has been used by the experimentalists to qualitatively and sometimes quantitatively describe some of the aspects of the behavior of semiconductor nanocrystals [60, 89].

In our tight binding simulations [48], we adopted this approach for the highest occupied molecular orbital (HOMO) and lowest unoccupied molecular orbital (LUMO) (see Supplementary information in [48]; Fig. 1.2a). To uncover the **k**-space profile of the density of states $|\Psi_i(\mathbf{k})|^2$, we performed a Fourier transform of the real-space molecular orbital $|\Psi_i(\mathbf{r})|$. The density of states $\rho_i\mathbf{k} = |\Psi_i\mathbf{r}|^2$ can be then plotted along a specific direction, e.g., $\Gamma-X$ and $\Gamma-L$. This allowed us to identify an enhancement in the DOS around the $\Gamma$ point for the LUMO, indicating direct bandgap-like transitions. This finding was accompanied by a $1000\times$ enhancement in the phonon-less radiative rates and enhanced absorption cross section (Supplementary information in [48]).

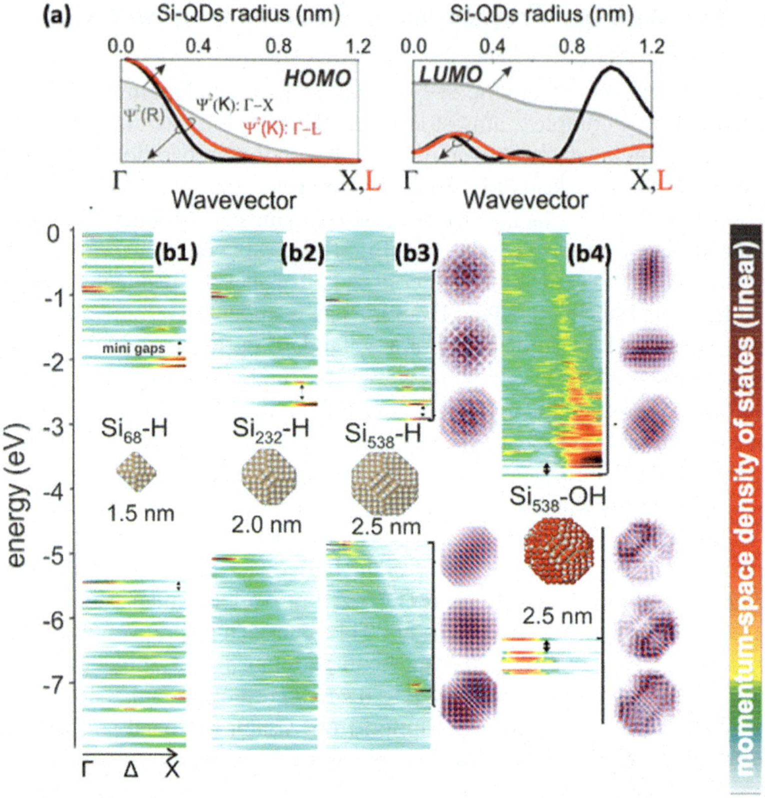


**Fig. 1.2** **a** Real-space (gray) and **k**-space projection in $\Gamma - X$ (black) and $\Gamma - L$ (red) directions of the HOMO and LUMO states. Data are simulated using the tight binding approach described in more detail in [159]. Data are reused from [48] with permission from Nature Springer. **b**1–**b**4 "Fuzzy" band structures of three sizes of H-capped (**b**1–**b**3) and one size of OH-capped (**b**4) Si-NCs. Data are simulated using DFT code "Fireball" [95]. On the right side of panels, (**b**3) and (**b**4) are shown the real-space cross sections of the wavefunctions of the three states closest to the bandgap. Modified after [80], reused with permission from APS Physics

The HOMO–LUMO **k**-space profile is very important for the emission from the Si-NC at $T = 0$ K, but for a more general understanding of the ligand effect and a better comparison with experiments, performed at $T > 0$ K, many more states around the Fermi energy need to be resolved in the **k**-space. To achieve this goal, a more suitable approach was demonstrated independently at that time by Hapala et al. [80]. In this approach, hundreds of molecular orbitals (MOs) in real space are Fourier transformed and projected along the specific line (direction) in the momentum space. This is done for a broad range of energy levels $E_i$, symmetrically around the Fermi energy. The most important message in [80], with respect to the earlier works that utilized Fourier transform for similar purposes, is the detailed discussion concerning the folding of the bands. This is not a trivial step, since the first Brillouin zone is almost empty and most of the momentum space density is located in the higher reciprocal cells. Also, not all the symmetry bulk crystal directions (such as $\Gamma-\text{X}$) in the small NCs are equivalent, as the symmetry is broken by the involvement of the surface and ligands. After folding all the contributions of the higher reciprocal cells into the first Brillouin zone, all the densities from the all the Γ-X directions are summed up to form the final "fuzzy" band structure, named after its blurred character in the **k**-direction.

Examples of such a "fuzzy" band structure for H- and OH-capped Si-NCs, simulated using DFT code "Fireball" [95], are shown in Fig. 1.2b1–b4. Panels (b1–b3) demonstrate the evolution of this **k**-space projection and the related radiative rates of an H-capped Si-NCs, depending on the size of the crystalline core. For the smallest size (1.5 nm) in Fig. 1.2b1, the **k**-space blur is very high, which makes it difficult for any trend to be discernible. Despite the blurring, the highest DOS seems to be close to the Γ point for the HOMO, and close to the $X$ point for the LUMO states, suggesting a persistently indirect bandgap in the HOMO–LUMO transitions. For the two larger H-capped Si-NCs in Fig. 1.2b2, b3, despite the extensive blurring in the **k**-direction, we can clearly recognize the shape of the original bulk Si bands in the $\Gamma-\text{X}$ direction, which makes the adoption of the band structure concept in this size range even more tenable. The original shape of the bulk silicon bands remains present even when the H-capping is replaced with various ligands: e.g., an –OH-capping in Fig. 1.2b4. More examples of the same type of simulations are shown also in the introductory Figs. 1.1b, c and 1.4b and later in Sect. 1.4, this time with a logarithmic DOS scale and simulated using DFT code cp2k [31, 44]. Nevertheless, despite this clear similarity in the overall band structure for all the different simulated Si-NC-ligand systems, there are also great differences, especially close to the bandgap. The comparison of the band structures for Si-NCs with various ligands presented here clearly indicates that *the surface capping plays a very complex role and influences the whole band structure*.

These results demonstrate that the band structure concept is still applicable even to a small NC with a relatively short-range translational symmetry. There are some limitations, though. First of all, the "blurring" in the **k**-space is a consequence of the Heisenberg's uncertainty relations: A higher spatial localization of carriers implies higher uncertainty in their **k**-vector; this effect had already been proposed a long time ago [89]. Secondly, bands become very sparse close to the bandgap, where the DOS is naturally the lowest, and become separated by so-called "mini-gaps", indicated by arrows in Fig. 1.2b1–b4. At a first glance, such "splitted" states could be mistaken for a "surface state", but this is most of the time not the case. In order to find out whether a state is introduced by the surface capping, one should analyze the so-called projected DOS (PDOS), where element (and/or orbital) contributions to the DOS is given. An example is given in Fig. 1.4d1–d4 and e1–e4 for the various oxide cappings. From the PDOS elemental fraction analysis, one can see that for all the levels, including the split states inside the bandgap, more than 90% of the DOS is contributed by the Si core. Yet, there are clearly large differences between the, e.g., HOMO–LUMO energies and radiative rates, as well as the overall shape and the DOS distribution. Surface chemistry might or might not contribute a surface-localized state close to the band-edge, and there is no simple way of predicting it. Thus, such states separated by a mini-gap can, but do not have to be connected with the surface of the NC. Rather than applying such a distinction, these states should be treated as a result of a much more complex interplay between the electronic states of the surface ligands and the Si core. This close cooperation between silicon and the ligands makes the distinction between a "core" and a "surface" state more complicated than traditionally thought, as we pointed out also in [110] and our latest DFT study [44]. Importantly, the band structure approach is not only applicable, but also highly relevant, since it provides an explanation for the persistently low radiative rates in such small Si-NCs. Thus, instead of the traditional, "atomistic" approach, assuming that the NC behaves solely as a sum of its parts (core, surface, interface), we propose that a more holistic approach needs too be adopted, as backed up by in-depth simulations of the band structure and projected density of states.

In this chapter, we describe parallel and often joint efforts of our two independent research teams toward the understanding the Si-NCs with such enhanced optical properties, specifically with focus on (i) the bright fast decaying F-band emission from Si-NCs, as well as (ii) the **k**-space representation of the effects of the surface chemistry on the directness of the radiative transitions in the Si-NCs. Our research of optical properties of Si-NCs covers parallel efforts in all three main directions of the research of nanomaterials, namely the technology of the preparation of samples, their experimental characterization as well as theoretical calculations. In the following text, we focus mainly on the promising yet still somewhat controversial topic of the fast radiative transitions detected in the various types of Si-NCs, and we discuss these transitions from both the experimental and theoretical point of view. We believe that experiment and theory need to go in this case hand in hand, especially given the number of issues with the precise material analysis and controversies surrounding the interpretations of the PL origin. To this date, we have employed the **k**-space DOS projection theoretical approach to uncover the separate effects of ligand-induced tensile strain by DFT code "Fireball" in [115], effects of electronegative ligand and environment by tight binding in [159], and the joint effects of ligand-induced charge transfer and strain by DFT code "cp2k" [31] in [44]. These topics will be further discussed in Sect. 1.4. The chapter is organized as follows. In the two following sections, Sects. 1.2 and 1.3, experimental results related to fast radiative rates in Si-NCs from our two teams are discussed in detail. In the next Sect. 1.4, we present theoretical calculations aiming at the explanation of fast radiative rates experimentally found in a particular class of organically capped Si-NCs and discuss the implications of these theoretical results in detail. The presented work covers a long time-span of more than a decade, which lets us gain perspective into this problem.

## 1.2 Fast Radiative Rate in Hydrogen- and Oxide-Capped Silicon Nanocrystals

### *1.2.1 Oxidation of Hydrogen-Terminated Silicon Nanocrystals*

To understand the underpinning mechanism of the emission in the oxide-capped Si-NCs, we resolved the effect of oxidation on PL in time [45], using a porous silicon, prepared by a similar process as described in [19, 211], but with the addition of hydrogen peroxide to achieve smaller Si-NCs, reported to yield blue emitting Si-NCs with fast emission rates [148]. The freshly etched porous silicon, with mostly H-capped Si-NCs, emitted in the green spectral range at around 525 nm with a fast, ∼ns decay (Fig. 1.3) [45]. Such an emission wavelength could be attributed to a band-to-band radiative recombination from a Si-NC with core of diameter of ∼2 nm (Fig. 1.1d), which agrees with our experimental size estimation from TEM and Raman spectroscopy for this sample [47]. We need to note here that in the absence of a PL quantum yield measurement, we cannot a priori link the experimentally detected fast rates $k_{m\mathrm{meas}}$ to fast radiative rates $k_{\mathrm{rad}}$, because the measured rates are determined by both the radiative and non-radiative processes [114] $k_{\mathrm{meas}} = k_{\mathrm{rad}} + k_{\mathrm{non-rad}}$ and the effect of non-radiative rates needs to be determined separately. However, the green PL is easily observable with the naked eye, see Fig. 1.3b, which suggests that the radiative rates play an important role in the observed emission and the corresponding radiative rates might actually be relatively high.

To oxidize such a porous silicon thin layer, we immersed the sample in an ethanol bath (containing air, as it has not been degassed) and exposed it for few minutes to laser UV irradiation at 355 nm, while simultaneously registering the PL spectra. During this time, we observed a small gradual shift in the fast green component, until it eventually completely disappeared Fig. [45]. This fast component was gradually replaced by a much slower, microsecond decaying component in a deep red range at ∼650 nm (Fig. 1.3b). At the time, we have interpreted this emergent band as a surface oxide-induced emission, due to its gradual appearance during the oxidation under the UV illumination. The gradual shift of the PL emission was also found to be in a good agreement with the available theoretical simulations [133]. In retrospect, however, we would probably use the wording "we observed emission characteristic of an oxidized Si-NC" instead of a "surface state" (see Sect. 1.2.2 later on. Hence, exposure of H-terminated freshly etched Si-NCs to oxygen and UV irradiation leads to oxidation, accompanied by an emergence of a typical S-band emission, very much like in [211].

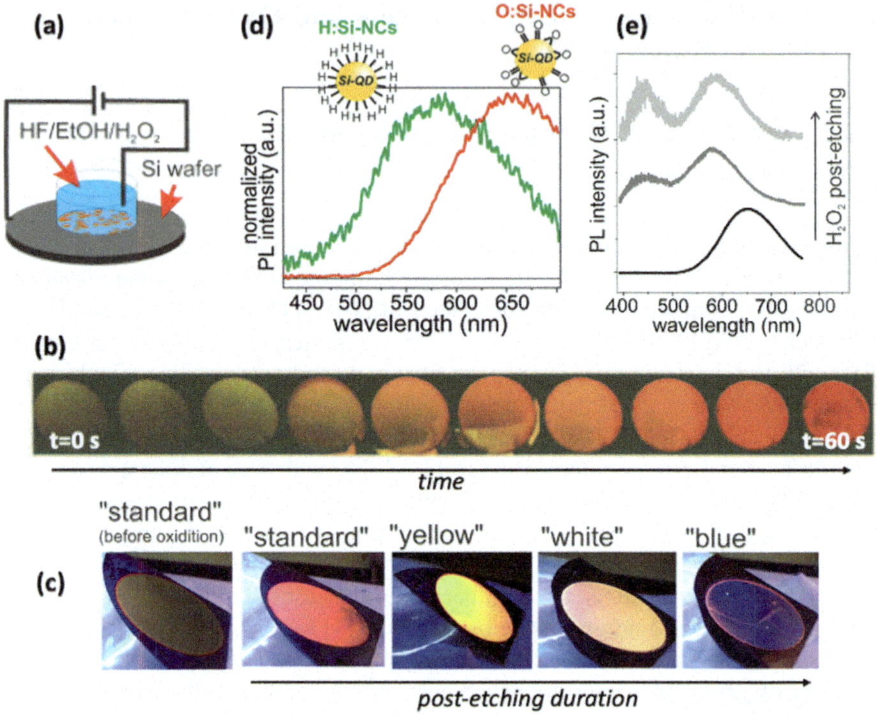


**Fig. 1.3** **a** Schematic of the electro-chemical etching procedure to obtain porous silicon with small H-capped Si-NCs. **b** Sequence of real-color photographs of the PL from freshly etched porous silicon surface (the circle is about 1 in. in diameter, as defined by the size of the etching teflon container). Freshly etched H-capped Si-NCs exhibit a green emission ($t = 0$ s), which after about 1 min of continuous UV irradiation turns red. **c** Real-color photo of PL of the "standard" porous silicon before and after oxidation, emitting green and consequently red; "yellow", "white", and "blue" samples are made by prolonged post-etching in a bath of hydrogen peroxide. **d** Steady-state PL spectrum of freshly etched and oxidized "standard" porous silicon. **e** PL spectra of "standard" porous silicon after prolonged post-etching in hydrogen peroxide. Reprinted from [45] and [46] with permission from AIP Publishing

## *1.2.2 Emergence of the F-Band in Oxidized Si-NC*

As discussed above in Sect. 1.1.2.1, oxygen can bond on the surface of silicon in several different ways, causing different spectral limitations. Using the "Firebal" DFT code, we have simulated the effect of the –OH-capping on a 2.5 nm Si-NC (Fig. 1.2b4) and found quite a profound effect of the –OH ligand onto the whole band structure, including the levels close to the band-edge. Nevertheless, a closer look suggests that the band-edge states are not related to a surface localization, but are general Si-NC core states. This can be clearly demonstrated on the spatial cross section of the three HOMO and LUMO states, depicted for the largest H-

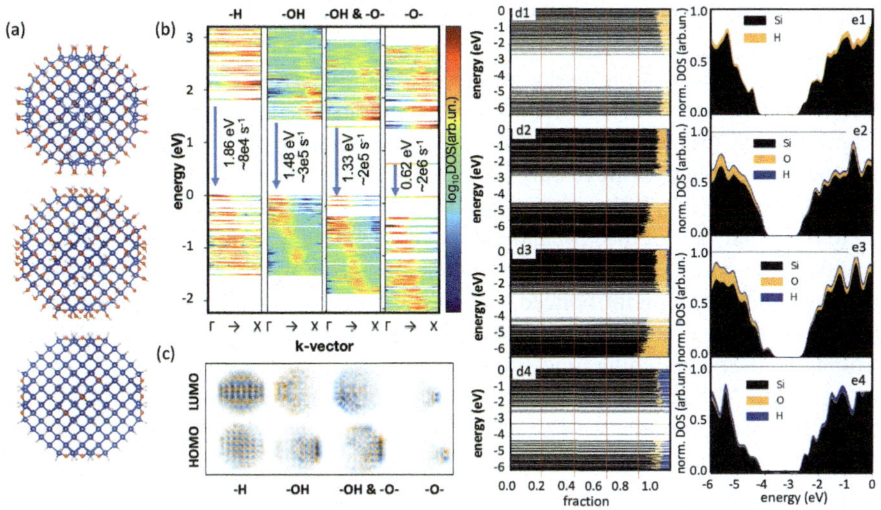

**Fig. 1.4** **a** From top down: 2.2 nmm Si-NC ($Si_{235}$) fully capped by –OH, –OH with Si-O-Si bridges on the edges and only the Si-O-Si bridges. **b** "Fuzzy" band structure of H-, –OH, –OH with Si-O-Si and Si-O-Si capping. Thermalized radiative rates at $T$ =300 K and HOMO–LUMO energies are indicated in the graphs. The color scheme depicts the DOS on a log-scale. **c** Real-space projection of the HOMO and LUMO wavefunctions. **d**1–**d**4, **e**1–**e**4 Element-color-coded PDOS fraction (color legend is in (**e**1–**e**4)) for (**d**1, **e**1) H-capped, (**d**2, **e**2) OH-capped, (**d**3, **e**3) OH- and O-bridge-, and (**d**4, **e**4) O-bridge-capped Si-NCs

and OH-capped Si-NC in Fig. 1.2b4, showing only a weak localization effect in the –OH-capped Si-NC.

A similar simulation for a slightly smaller 2.2 nm Si-NC by the DFT code cp2k, using settings described in [44], is shown Fig. 1.4, where we also for comparison simulate bridging Si-O-Si bonds, located on the edges of the Si-NC (Fig. 1.4a), and a combination of the two. Again we see, with respect to the H-capped Si-NC of the same size, a profound effect on the whole electronic structure, including the HOMO–LUMO energy (assumed to be an approximation to a bandgap energy), radiative rates and spatial localization of the carriers (Fig. 1.4c). Panels (d1–d4) and (e1–e4) show element-resolved PDOS, confirming that over 90% of the DOS for all the systems originate from the Si core, despite the fact that the real-space cross section of the HOMO and LUMO wavefunctions, shown in (Fig. 1.4c), is clearly localized close to the surface. This might come across as a surprise, considering that the two mid-gap states found for the Si-O-Si capping are typically ascribed to trapping on Si-O-related sites [140, 167, 211].

In order to introduce more –OH bonds onto the surface, which should have a good impact on the HOMO-LUMO energy and surface localization of the electron and hole (Fig. 1.4b,c), we introduced into our Si-NC preparation routine a hydrogen peroxide $H_2O_2$ post-etching procedure, assuming the following chemical reaction [46]:

$$2Si + 4H_2O_2 \rightarrow 2Si(OH)_4 \leftrightarrow Si(OH)_3 - O - Si(OH)_3 + H_2O.$$

The idea that including hydrogen peroxide in the production of porous silicon was introduced by Nayfeh et al. [148], who demonstrated production of ultra small, ~1 nm sized Si-NCs. In our case, when including the hydrogen peroxide in the etching bath, we were able to achieve only the ~2.2 nm green emitting H-capped Si-NC, as described before (Fig. 1.3b, c; "standard" sample). Adding the post-etching procedure led to a gradual shift of the "S-band" of the oxidized sample slightly beyond 650 nm, resulting in a porous silicon layer we labeled as "yellow" (Fig. 1.3c). Nevertheless, prolonged post-etching introduced a strong "F-band" in the blue spectral range at ~ 450 nm (Fig. 1.3e) with a nanosecond decay time [46]. At a certain point, with further prolonging the post-etching time, the "F-band" became in the steady-state PL equally strong as the "S-band", leading to the "white" emitting porous silicon (Fig. 1.3c, e), with CIE chart coordinates of (0.35, 0.34), close to the white "daylight" standard D65 with coordinates (0.313, 0.329) [12] (Fig. 1.5e). Eventually, the prolonged post-etching led to a drastic decrease in the "S-band" and further increase in the "F-band", resulting in a "blue" emitting porous silicon (Fig. 1.3c) with emissive properties similar to those observed by Nayfeh et al. [148]. In this way, we have demonstrated that by continuous tuning of the post-etching time, we can "continuously" shift the PL from the slow red (S-band) to a fast blue (F-band) spectral range, but the tunability obviously skipped the green spectral range, where we observed the original PL for the non-oxidized freshly etched H-capped porous silicon (Fig. 1.3b–e) [46].

Despite the fact that we observed the F-band in clearly oxidized Si-NCs, the origin of this emission cannot be straightforwardly assigned to an oxide surface state emission. In fact, in a collaborative research with Valenta et al., we found that the origin of the F-band in our material is likely of an intrinsic origin, from small and distorted Si-NCs [195], as inferred from the behavior of the excitation-dependence of the F-band. A similar conclusion was offered by the temporally resolved low-temperature PL measurements [152]. Nevertheless, a possible link to surface (silica-related) defects has not been completely ruled out. Interestingly, such an "accompanying" blue F-band is usually not reported in Si-NCs with long alkyl capping, but exceptions of the rule can be found also here. Fast blue emission has been recently reported in Si-NCs synthesized in plasma from liquid precursors and was interpreted as the "intrinsic" direct $\Gamma - \Gamma$ hot-electron emission [162].

### *1.2.3 Silica Defects*

Another possible emission channel in the oxidized Si-NCs is the silica oxide defects. Numerous literature sources confirm that radiation damaged silica and its emission bands that can occur almost anywhere in between the UV and NIR spectral ranges (see for example [12, 17, 153] and references therein), coinciding with the spectral tunability range of the Si-NCs core. It is possibly inevitable that a strained, curved

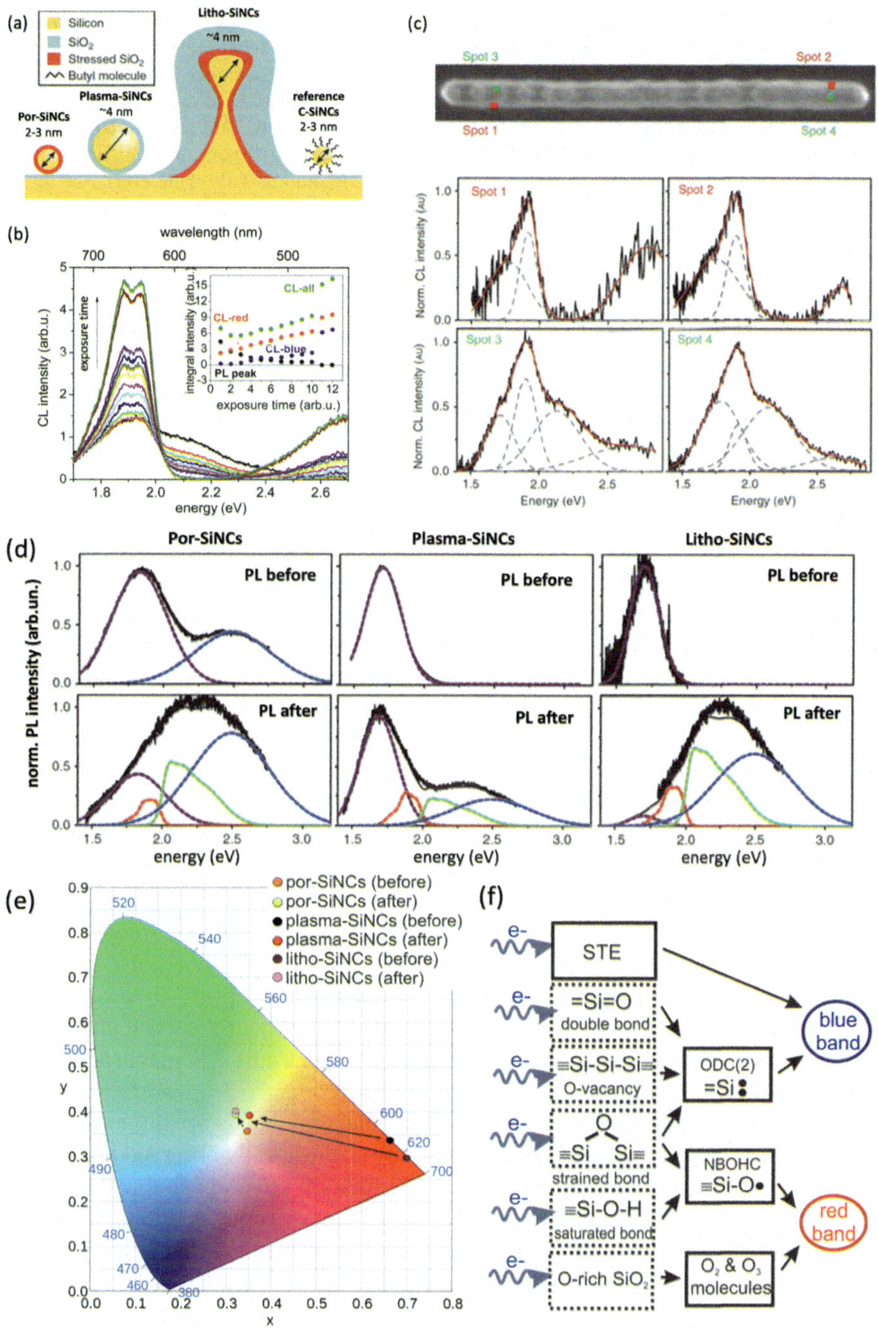
(a)
Silicon
SiO2
Stressed SiO2
Butyl molecule
Litho-SiNCs
~4 nm
Plasma-SiNCs
~4 nm
Por-SiNCs
2-3 nm
reference
C-SiNCs
2-3 nm
(b)
wavelength (nm)
CL intensity (arb.u.)
exposure time
integral intensity (arb.u.)
CL-all
CL-red
CL-blue
PL peak
exposure time (arb.u.)
energy (eV)
(c)
Spot 3
Spot 2
Spot 1
Spot 4
Norm. CL intensity (AU)
Energy (eV)
(d)
Por-SiNCs
Plasma-SiNCs
Litho-SiNCs
PL before
PL after
norm. PL intensity (arb.un.)
energy (eV)
(e)
por-SiNCs (before)
por-SiNCs (after)
plasma-SiNCs (before)
plasma-SiNCs (after)
litho-SiNCs (before)
litho-SiNCs (after)
x
y
(f)
e-
STE
=Si=O
double bond
≡Si-Si-Si≡
O-vacancy
≡Si Si≡
strained bond
≡Si-O-H
saturated bond
O-rich SiO2
ODC(2)
=Si:
NBOHC
≡Si-O•
O2 & O3
molecules
blue
band
red
band

◂**Fig. 1.5** **a** Schematics of the Si-NCs samples used for the e-beam irradiation experiment. **b** Cathodoluminescence (CL) upon prolonged e-beam irradiation from the porous silicon sample. **c** CL from the lithography-made Si-NCs—(top) Scanning electron microscopy (SEM) image of the structure and schematics of the cross section (bottom) PL spectrum of the strained areas (spot 3 and 4) contain also green CL band, while the unstrained areas (spots 1 and 2) do not. **d** PL before and after e-beam irradiation for all the oxide-capped samples. **e** PL in the chromaticity diagram CIE for all the oxide-capped Si-NCs, before and after e-beam irradiation. **f** Possible origins of the red and blue bands in the PL and CL of the radiation damaged silica. Figures after [12] are reprinted with permission from Springer Nature

silica shell on the surface of oxidized Si-NCs contains numerous defects, some of them emissive.

To this end, we assembled a series of oxidized Si-NCs prepared by various methods by different research groups with different type of oxide—naturally grown and thermally induced, and an organically capped Si-NC as a control sample (Fig. 1.5a) and analyzed their cathodoluminescence (CL) and PL before and after irradiation by the e-beam during the CL measurement [12] (Fig. 1.5b–d).

While virtually no effect of the electron beam irradiation was observed for the organically capped Si-NCs, all the oxide-capped Si-NCs have exhibited drastic changes to their PL after the CL measurement. In particular, during the electron beam irradiation in vacuum, all samples have shown a decreasing original PL peak (with different spectral position for the different samples) and a rise of several new emission bands. The most prominent new peaks were a narrow, strong red double peak at 650 nm, and a broad featureless blue band at 450–480 nm (Fig. 1.5b–d). For the strained parts of the large, lithography prepared Si-NCs, we have also observed emergence of a green emission band at ~560 nm (Fig. 1.5c), which did not show in the unstrained parts of the same sample (Fig. 1.5c), or the other oxide-capped Si-NCs.

After exposure to air, PL has stabilized for all the Si-NCs, independently of their origin, size, and original PL, into a broad, white PL composed of a mixture of the original PL band and the new blue, green, and red emission bands introduced by the electron beam irradiation (Fig. 1.5d). In fact, we realized that by the use of e-beam irradiation, we can modify the PL spectrum of any oxide-capped Si-NCs in such a way that it becomes very similar and white, independently of the particular Si-NCs synthesis method, size, shape, strain, or original PL spectrum (U.S. patent application number 16/331704) (Fig. 1.5d).

Due to the striking similarities between the emissive bands arising in the oxide-capped Si-NCs under electron beam irradiation and the usually observed blue F-band and red S-band in oxide-capped Si-NCs, we put forward a hypothesis that these bands are closely related to the well-known oxygen-deficiency centers in the Si-rich silica shell [12] (Fig. 1.5f). In radiation damaged silica, emission from such bands can be efficiently excited only over the wide silica bandgap (~9 eV). However, in the case of oxide-capped Si-NCs, the very same defects are excitable under blue/UV light through the Si-NC core. This is a great advantage that could help such silica defects to be utilized in white light emitting phosphors for applications in lighting (U.S. patent application number 16/331704). Thus, we have realized that silica defects could be an additional emissive channel and offer an additional degree of freedom to tune the PL in the Si-NCs. Their role can be very important in the F-band emission, where the electron beam irradiation induced a spectrally broad blue band, possibly related to the double dangling bond =Si: site (Fig. 1.5f). In poorly emitting samples, they can contribute even to the S-band emission, which can be inferred from the observation that the oxidized Si-NCs had kept (at least part) of their original emission after the electron beam irradiation. Generally speaking, the S-band, as observed in various samples, does not completely copy the behavior of a silica defect, because the position of the PL peak can be changed by preparation conditions and tuned throughout the red spectral region (see e.g. [187]). Nevertheless, the S-band spectral tunability usually ends close to the range 620–650 nm, where this characteristic silica defect band related to oxygen dangling bond =Si-O· resides (Fig. 1.5f) [12]. Consequently, this work again emphasizes the possibility of different emissive channels and even the coexistence of the individual channels in certain samples.

### *1.2.4 Role of Nitrogen*

The discussion over the origin of the blue "F-band" emission has been eventually stirred also toward a possible presence of nitrogen-based surface sites, inducing a fast blue emission even in the originally red-emitting Si-NCs [34]. To introduce nitrogen in our materials, we have treated our porous silicon with a non-thermal plasma in water [63, 64]. This treatment was shown to lead to the incorporation of nitrate-water complexes –O-$NO_2$ into the shell of the Si-NCs, which alleviates some of the strain in the surface oxide shell. We have investigated the time-resolved PL of both samples under different excitation wavelengths and applied a dedicated analysis [120]. We found that several components are present within the blue PL band Fig. 1.6a–d) The original oxide-capped Si-NCs exhibited a fast, single-exponential ($\tau_{PL}^{(2)} = 1.2$ ns) component, identified as related to a silica defect. Additionally, we also identified a slightly longer stretched-exponential ($\tau_{PL}^{(3)} \approx 10$ ns) emission-wavelength dependent component, assigned to non-thermalized, direct core-related PL. Interestingly, in the nitrate-containing Si-NCs, the silica-defect-related PL is eliminated, and only a very weak signal from the non-thermalized carriers remains. These results prove that whereas most of the blue PL emitted by our oxide-capped Si-NCs is connected with the silica-related defects, a small portion of this PL still originates in the core states. This finding is in agreement with and an extension of our previous results [152, 195], where we were not able to identify the silica-defect-related contribution to the PL due to a less complex analysis of data [152].

In parallel to the optical investigations, DFT calculations were also conducted for the nitrate-containing Si-NCs, shown in Fig. 1.6e [64]. Replacement of some of the –OH ligand groups with –O-$NO_2$ leads to the emergence of a few in-gap states with high surface real-space localization [the blue dashed line at the bottom of panel (e)]. However, these states, which are deep in the bandgap, would therefore act rather as non-radiative traps and would be inaccessible through the traditional PL mea-

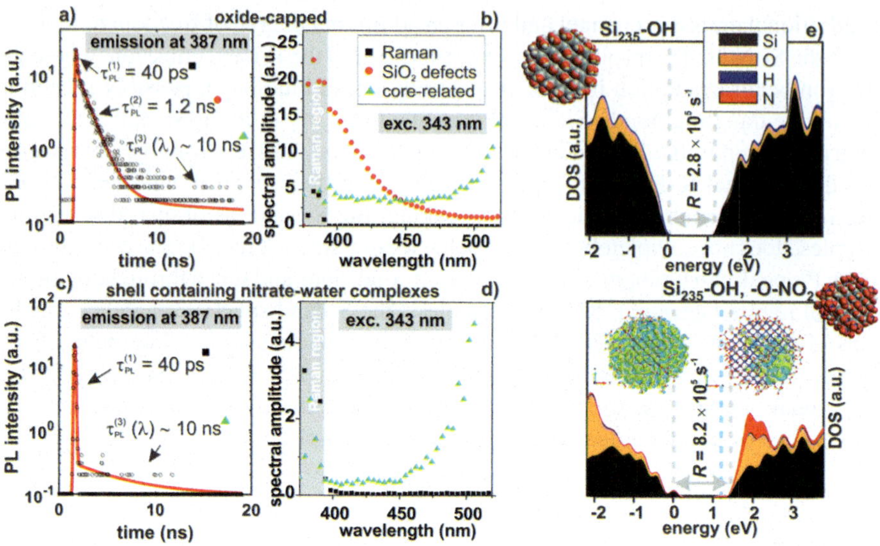


**Fig. 1.6** **a**, **c** Examples of the PL decays from **a** an O-capped Si-NCs and **c** a nitrogen-containing O-capped Si-NCs, containing **a** 3 and **c** 2 decay components. **b**, **d** The PL spectra of the PL decay components obtained by fitting, the corresponding symbols are also included in panels (**a**) and (**d**). The ultrafast ($\tau_{PL}^{(1)} = 40$ ps) component corresponds to the Raman signal of the solvent and therefore is not of interest for the study of Si-NCs. **e** The simulated (elementally resolved) PDOS in a Si-NC capped with –OH (simulating oxide surface, top) and a combination of –OH and –O-$NO_2$ (bottom). The bottom panel also shows the projection of the real-space wavefunction for the HOMO and LUMO states, highlighted by the gray dashed lines. The values of the corresponding computed radiative rates are also included. Reproduced from [64] with permission from The Royal Society of Chemistry

surements as a result of a phonon bottleneck. Therefore, the theoretical description relevant to the performed experiment compares the band-edge transitions in the two systems as indicated by the gray dashed lines in Fig. 1.6e. Clearly, for our O-linked nitrogen surface atoms, the change in the surface passivation leads to a small red-shift in PL, accompanied by a several-time enhancement in the radiative rate, in agreement with our experiment [63].

The presented results are not contradictory to the results obtained by the Veinot group [34], who reported the emergence of a large blue shift and a fast PL emission after the Si-NCs were exposed to nitrogen (and oxygen), because the type of incorporation of the nitrogen atoms into the Si-NC system can very well be different in the two cases (e.g., direct Si–N vs. our –O-$NO_2$ surface). However, they do illustrate the wealth of processes which can be observed in the PL of the Si-NCs and prove that the incorporation of nitrogen does not necessarily *have to* lead to a blue emission. A similar conclusion has been recently put forward also by an independent group [147].

## 1.3 Organically Coated Si-NCs with a Fast Emission Rate

The fast decaying blue emission, very similar to that observed in our oxide caped Si-NCs, has been routinely reported from number of organically capped Si-NCs. Therefore, we decided to steer our investigations also toward this promising and rapidly developing field. During our investigation of suspensions of oxide-capped Si-NCs, we developed a photo-chemical capping procedure, which resulted in the re-passivation of the oxide-capped Si-NCs by organic ligands. In particular, the oxide-capped Si-NCs (made from from porous silicon) were kept in a mixture of hydrocarbons containing xylene and were periodically (for a long time) irradiated at room temperature by a UV laser (325 nm, 2.5 mW) [112]. As a result, the original S-band PL at ~650 nm blue-shifted to an unusual yellow spectral region at ~570 nm (Fig. 1.7a, b). The corresponding PL decay dramatically shortened by several orders of magnitude without any notable decrease in the emission efficiency [112, 113] (Fig. 1.7a). The sample's surface chemistry was characterized by Fourier transform infrared (FTIR) spectroscopy in the attenuated total reflectance (ATR) mode and nuclear magnetic resonance (NMR) measurements (Fig. 1.7d). These analyses showed that the surface of the photo-chemically modified Si-NCs is likely methyl-passivated. The attachment of these short alkyl groups to the surface also agrees with the fact that after the photo-chemical procedure, the modified Si-NCs are well-dispersed in nonpolar solvents, such as xylene or chloroform. As the measured PL decay lifetime $\tau_{PL}$ is determined by the lifetime of both the radiative ($\tau_{\mathrm{rad}}$) and the

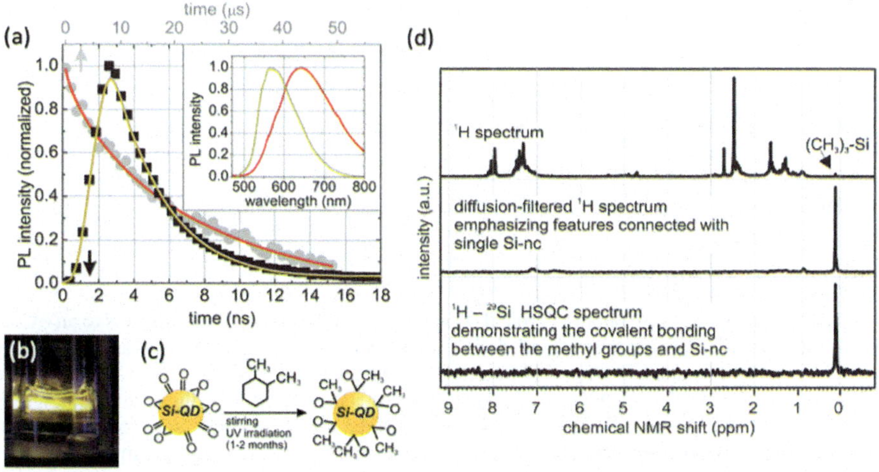


**Fig. 1.7** **a** Dramatic change in the PL decay and PL spectra of the oxide-capped Si-NCs, re-capped by organic methyl-based ligands upon UV irradiation in xylene-based solvent. **b**)Real-color photo of PL from the organically re-capped Si-NCs. **c** Schematic of the surface changes induced by the UV irradiation in xylene containing solvent. **d** Measured NMR spectra of the Si-NCs with methyl-based capping. Reprinted with permission from [113] Copyright (2010) American Chemical Society

non-radiative ($\tau_{rad}$) processes as $1/\tau_{PL} = 1/\tau_{rad} + 1/\tau_{nonrad}$, we verified also the change in the external quantum yield (QY) of the PL of the oxidized and photo-chemically modified samples and found an increase from 2–3%, for the O-capped original material, to 20% for the resulting organically capped Si-NCs. This allows us, assuming that no other external factors significantly affect the measured QY, to quantify the radiative lifetime of the photo-chemically modified sample to $\tau_{rad} = 10$ ns, which are values well comparable to those of direct bandgap semiconductors. This result is clearly quite controversial, since organic ligands were from the theoretical point of view expected to induce only minor changes to the electronic properties of the Si-NCs [103, 106, 123, 171].

Later, we performed a reverse experiment [48] with butyl-capped Si-NCs, prepared by a wet-chemical synthesis. This material was photo-chemically oxidized under intense illumination with a pulsed UV laser beam at 4.4 eV in an ethanol dispersion for different durations of time. Within the first 30 min, the intensity of the originally blue-green fast decaying PL band decreased and another PL band appeared in the red around 1.85 eV with a slow, 12 μs PL decay [48] (Fig. 1.8b). The PL spectrum before and after oxidation was compared with the theoretical and experimental values from [188, 211], as well as to the size distribution of the Si-NCs measured by

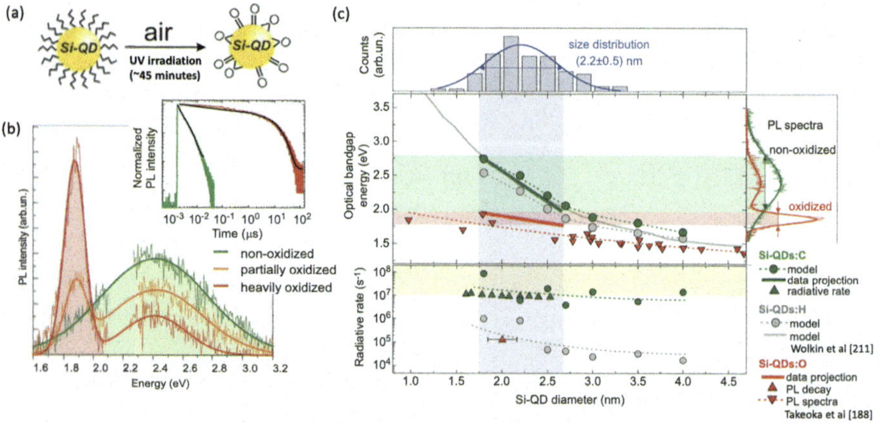


**Fig. 1.8** **a** Schematic of the photo-chemical recapping, from the original butyl ligands to an oxide-capped surface, via exposure to an intense 4.4 eV laser beam in ethanol. **b** PL spectra and PL lifetime (inset) of the butyl–capped Si-NCs for different UV exposure times. Upon oxidation, a red PL band appears with a slow, microsecond decay, replacing the fast, nanosecond decaying original blue-green PL band. **c** (top) TEM analysis of the size distribution of the butyl-capped Si-NCs. (middle left) PL peak for original butyl-capped Si-NCs (green thick line) and for the photo-chemically oxidized Si-NCs (red thick line) compared to the experimental results by Takeoka et al. [188] (=[23] in the figure) and Wolkin et al. [211] (=[5] in the figure). (middle right) PL spectra of the original (green) and the photo-chemically oxidized (red) Si-NCs. (bottom) PL lifetime of the original (green triangles) and the photo-chemically oxidized (red) Si-NCs, compared to our tight binding simulations for C-capped (green dots) and H-capped (gray) Si-NCs. Reprinted with permission from Springer Nature, after [48]

transmission electron microscopy (TEM) (Fig. 1.8c). We found a surprisingly excellent match for both the original PL band and the red-shifted narrower PL band after oxidation, and also, the PL decays in both cases (Fig. 1.8c), indicating a successful oxidation of the originally butyl-capped Si-NCs.

Hence, we confirmed that the replacement of an oxide capping with short alkyls and vice versa both lead to the corresponding changes in the light emission, with the fast yellow/green and slow red PL emission being characteristic for the organic and oxide types of capping, respectively [48, 113].

### *1.3.1 Single-Dot Spectroscopy of Si-NCs with Fast Radiative Rates*

The best possible experimental technique to uncover size-dependent optical qualities of Si-NCs is the (time-resolved) single-dot micro-spectroscopy PL (SPL) [111], which allows one to overcome the inhomogeneous broadening of the ensemble PL spectra. To better understand the origin of the controversial fast decaying PL from the organically capped Si-NCs, we performed SPL analysis. The fast PL rate is convenient for such an analysis, while the SPL in the oxide-capped Si-NCs with the slow PL rate is known to be notoriously difficult, because of the inherently low photon rates.

First, we analyzed SPL from the organically capped Si-NCs prepared by the wet-chemical synthesis [42], the same material that we oxidized by UV laser in [48]. SPL typically reveals PL structure related to the phonon replicas (Fig. 1.9a), which can help in interpretation of the emission origin. At room temperature, phonon replicas related to the surface vibrations of larger energies can be detected, such as Si–O or Si–C phonon replica with energies of ~130–140 meV and ~160–170 meV, respectively. Si-Si-related phonon vibrations at ~50 meV can be resolved at the low-temperature SPL measurements. In our sample, we analyzed a number of structured SPL spectra (Fig. 1.9b), over a broad spectral range, covering a wide color space (Fig. 1.9c; colored circles). When resolved in time, SPL spectra exhibited blinking with observable spectral shifts (Fig. 1.9d). To identify the origin of the phonon replica structure observed in our SPL spectra, we compared the histogram of the measured replica energies with the FTIR spectrum and found a match with a strong narrow signal at 16 meV, possibly related to Si–C vibration (Fig. 1.9e).

The ensemble PL peak was found to be tunable via excitation wavelength and exhibited a weakly size-dependent PL lifetime (Fig. 1.9f), both being an indication of a size-effect in a system of Si-NCs with a broad size distribution. Also, SPL decay was found to be of the order of a few ns (Fig. 1.9g), very similar to the ensemble PL lifetime. All the PL lifetimes exhibited double exponential behavior with a sub nanosecond component and a few nanosecond component. To avoid nonlinear effects, such as Auger non-radiative recombination, we analyzed all the SPL in the linear excitation regime, which was tested on several single Si-NCs (Fig. 1.9h). A similarly

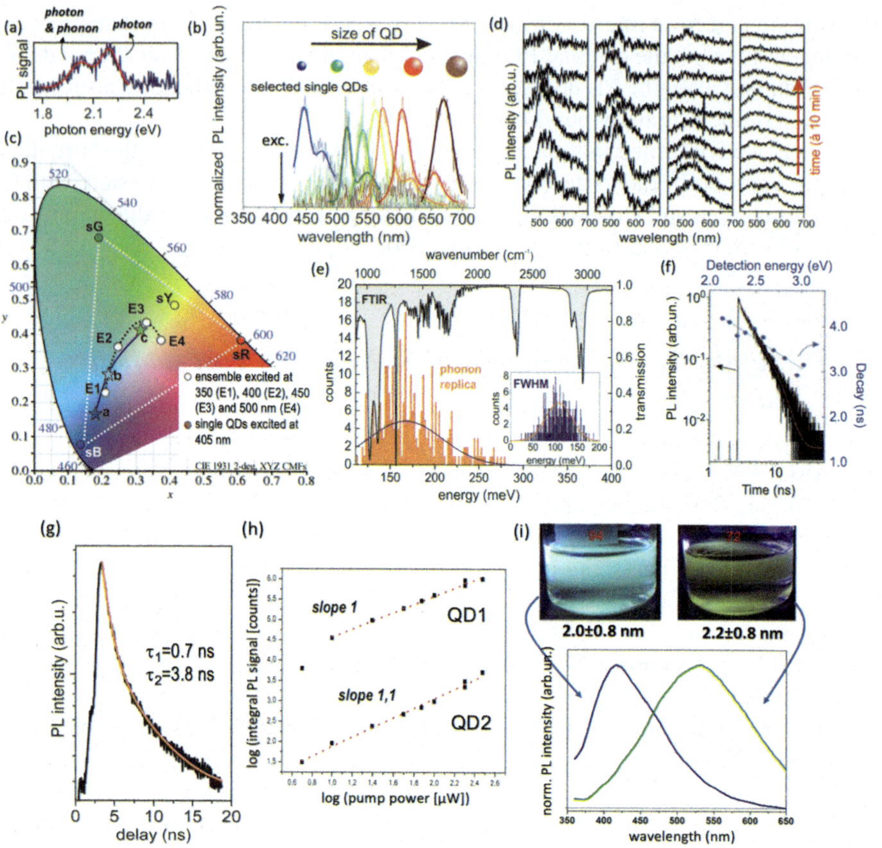


**Fig. 1.9** **a** Example of a single Si-NC PL spectrum with a structure related to a phonon replica. **b** SPL spectra of several selected Si-NCs with different sizes. **c** Chromaticity CIE chart with indicated points corresponding to the single Si-NCs (sG, sB and sR; colored circles), compared to the ensemble PL spectrum excited at different wavelengths (E1–E4; white circles) and the size selected ensembles (**a**–**c**; colored stars). **d** Time sequence of the SPL spectra for four different single Si-NCs, showing blinking and spectral shifts. **e** FTIR spectra compared to the statistics of the phonon replica energies extracted from fitting the observed SPL spectra. The inset shows FWHM statistics of the fitted SPL spectra. **f** Time resolved PL of the ensemble (black line) and PL lifetimes measured at the different PL wavelengths (blue circles), showing a size-dependent trend. **g** Time resolved SPL signal fitted with a bi-exponential function with an ultra fast (0.7 ns) and a fast (3.8 ns) decay components. **h** Plot of the integral PL intensity as a function of the excitation power (in $\mu$W) on a log-log scale for two different single NCs, showing a desired linear regime. **i** Size separated batches of butyl-capped Si-NCs—real-color photos (top) and the PL spectra (bottom) of the Si-NCs ensembles (courtesy of Dr. H. Nie and Dr. J. Paulusse, unpublished results). Reprinted with permission from John Wiley and Sons after [42] and Springer Nature [48]

**Table 1.1** Comparison of the room temperature (RT) SPL measurements of the Si-NCs from the literature. In the SPL results, the fraction of unstructured spectra (unstr.), the peak width ($w$), and the spacing ($\Delta$) within the structured spectra, are listed. The emission range denoted with an asterisk (*) corresponds to the same excitation wavelength as used for the SPL measurements. OT stands for octanethiol. The corresponding references are: Burrato [142], Valenta [196, 198], Linnros [174, 184, 186, 198], Cichos [141], Korgel [54], Dohnalová [42], Kůsová [113, 119]

| Reference | sample | | Single-dot photoluminescence | | | | | Macro PL | |
|---|---|---|---|---|---|---|---|---|---|
| | surface | size (nm) | unstr. | $w$ (meV) | $\Delta$ (meV) | Emission (nm) | Origin | Peak (nm) | Radiative lifetime |
| Burrato | oxide | 5–20 | 50% | 115 | 160 | 600–700 | Si–O ph. | $\approx$ 600 | 100 μs |
| Valenta | oxide | 3 | 100% | not observed | | | | 550–800 | 0.1–1 ms |
| Linnros | oxide | | 100 % | not observed | | | | | slow |
| Cichos | oxide | $5 \pm 2$ | some | 100 | 130–160 | 515–630 | Si–O ph. | 650–1000 | 50–900 μs |
| Korgel | OT | 3 | 50% | 150 | 133 | 525–700 | Si–O ph. | 550–800 | 100 ns |
| Dohnalová | $n$-butyl | $2.2 \pm 0.5$ | 80% | 110 | $167 \pm 40$ | 450–670 | Si–C ph. | 420–700* | 100 ns |
| Kůsová | methyl | 2.5 | 0% | 100 | $\approx$ 150 | | trion | 515–750 | 10 ns |

prepared sample was also size separated using size-exclusion chromatography into two batches: one with emission at ~410 nm and the other peaking at ~540 nm, with mean sizes of $2.0\pm0.8$ nm and $2.2\pm0.8$ nm (Fig. 1.9i), additionally confirming a size-effect and therefore an intrinsic origin of the PL from these organically coated Si-NCs (Table 1.1).

In Fig. 1.10 are shown SPL results on our methyl-capped Si-NCs prepared from the oxide-capped porous silicon by UV irradiation in xylene-based solvent [113]. Similarly to our butyl-capped Si-NCs, prepared by a wet-chemical synthesis [42], this sample also exhibited a high energy ~150 meV replica in nearly all the SPL spectra (Fig. 1.10). For this sample, we also analyzed a low-temperature SPL, which revealed much narrower phonon replicas with 50 meV spacing, attributed to the Si-Si phonon replicas. This further connects the observed PL with the intrinsic band-to-band radiative recombination. However, in contrast to the previous study [42], the broader structure seemed to be connected here with the discrete states inside the Si-NC and was interpreted as the PL from a radiative recombination of a trion quasi-particle [119]. A blinking study performed on this sample at room temperature [118] confirmed that the on-time distributions contain power-law distributed temporal regions, which are a typical feature of blinking of NCs, in contrast to a PL from defects or molecules.

The SPL spectral structure described above has been observed previously for oxide- and organically capped Si-NCs, but with often different replica energies and different interpretation of the results. In the literature, oxide-capped Si-NCs of various origin exhibited either partly structured SPL [54, 141, 142, 178] or completely unstructured SPL [184, 186, 196, 198]. The energy of the observed Si–O replica was typically observed in the range from 130 meV up till 160 meV. A thorough study

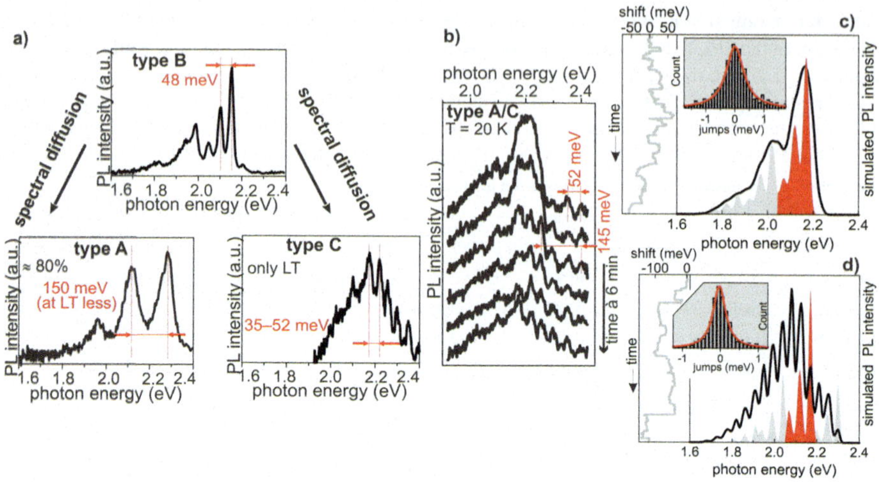


**Fig. 1.10** **a** Examples of the different types of SPL spectra measured from the same material, whose shape is determined by the spectral diffusion. **b** A measured series of SPL spectra at 20 K. The shape of the spectrum gradually changes due to the spectral diffusion. **c**, **d** Simulations of the influence of the spectral diffusion on the measured SPL. The SPL spectrum (the gray and red areas) undergoes random Lorentzian-distributed (the insets) shifts during the measurement due to the spectral diffusion. These randomly generated shifts form a time series shown as the gray curves on the left of both the panels. The sum of the emitted SPL spectra, undergoing the presented spectral shift during the measurement, is shown as the black curves, which represent a possible outcome of the measurement. Please note the good correspondence of the resulting spectrum in panel (**c**) and the type-A spectrum in panel (**a**), as well as the resulting SPL in panel (**d**) and the type-C spectrum in panel (**a**). Reprinted with permission from Springer Nature, after [119]

[178] revealed that the structured SPL spectra from octanethiol-capped Si-NCs with side peaks previously ascribed to Si–O phonon replicas [54] have nanosecond decay times on the single nanoparticle level. Moreover, in this follow-up study, unstructured broader spectra with a much longer ($\approx \mu$s) SPL lifetime also appeared. Drawing parallels with the SPL measurement of silica nanoparticles by the same group [28], they concluded that the fast structured SPL is purely from an Si-O-related defect, whereas the slow unstructured SPL comes from the intrinsic quantum-confined Si-NCs core, in agreement with the previous reports [196, 198]. The low-temperature SPL from the litography-defined Si-NCs with an oxidized surface [184, 186] revealed a mixture of structured and unstructured SPL spectra; however, the structure is completely different from the room temperature cases discussed above. The unstructured SPL spectra tended to be narrower, down to an ultra-narrow 400 $\mu$eV at 10 K, whereas the slightly broader SPL spectra exhibited a side band $\approx$60 meV apart from the main line, interpreted as a TO phonon in silicon. Due to the narrow linewidth, the unstructured SPL spectra were interpreted as an emission from a localized state while the SPL spectra with the TO phonon replica were considered to be from the Si-NCs core. A similar trend was observed also in the investigation of the formation of the Si-NCs from the nano-walls prepared by a self-limiting oxidation [13].

Additionally, we would like to point out a strong effect of spectra diffusion, which could lead to misinterpretation of the structured SPL. In general, the PL spectra of the single NCs are known to be influenced by a spectral diffusion [111] or random shifts of the whole emitted SPL spectrum, as a result of fluctuations in the surrounding environment. The extent, to which the measured SPL spectrum can change, is illustrated in Fig. 1.10b, measured at $T = 20$ K, where a series of SPL spectra collected over the same spot gradually changes its shape. A similar effect, with lower resolution and at room temperature, was also observed in our study of SPL from the wet-chemically synthesized Si-NCs (Fig. 1.9d). A simple simulations of the effect of spectral diffusion on the measured SPL spectrum in Fig. 1.10c, d confirm that spectral diffusion can indeed result in a complete change of the observed spectral shape integrated over time (Fig. 1.10). The "intrinsic" SPL spectrum can shift due to thermal fluctuations or changes in the electrostatic field of the environment, caused, e.g., by excitation of a nearby Si-NC. Therefore, even though several types of spectral shapes were observed, as shown in Fig. 1.10a, only one of them represents the "intrinsic" SPL spectrum of the Si-NC (type B), while the remaining other two solely result from the spectral diffusion (type A and C). Regardless of the origin of the observed PL, these measurements illustrate the importance of the careful interpretation of the results in the SPL spectroscopy.

Although we were able to confirm that the SPL originates in several cases from the Si-NCs, evidenced by the Si-Si phonon replica observed at low temperature, we also see that it is very difficult to deduce any detailed information about the origin of the SPL from the typical low resolution SPL measurements at room temperature due to the inevitable spectral diffusion occurring during the long acquisition times. Hence, SPL measurements, on their own, are again not a sufficient evidence for the PL origin and need to be combined with correlative material/optical analyses, such as Raman microscopy, TEM or scanning probe microscopy (SPM) techniques.

### *1.3.2 Enhanced Radiative Rate Measured by Drexhage Experiment*

The controversially high PL rate of the organically capped Si-NCs is a result of a combination of radiative and non-radiative rates $k_{PL} = k_{\mathrm{rad}} + k_{\mathrm{nrad}}$. These two cannot be separated by a PL decay analysis. Nevertheless, in the Drexhage-type experiments, where the local density of states (LDOS) is varied in a controlled way by a mirror placed in the vicinity of the emitter, the radiative and non-radiative rates can be separately identified. This is because the mirror affects the LDOS, but does not modify the direct environment of the emitter, so that the intrinsic non-radiative decay rate is not influenced. In our experiment, we adopt the implementation described in [131], where a spherical mirror is used in a combination with a fluorescence lifetime imaging microscopy (FLIM) [32] (Fig. 1.11a). A drop-casted thin film of butyl-capped Si-NCs is excited by a pulsed laser, and the PL decay rate is measured as a

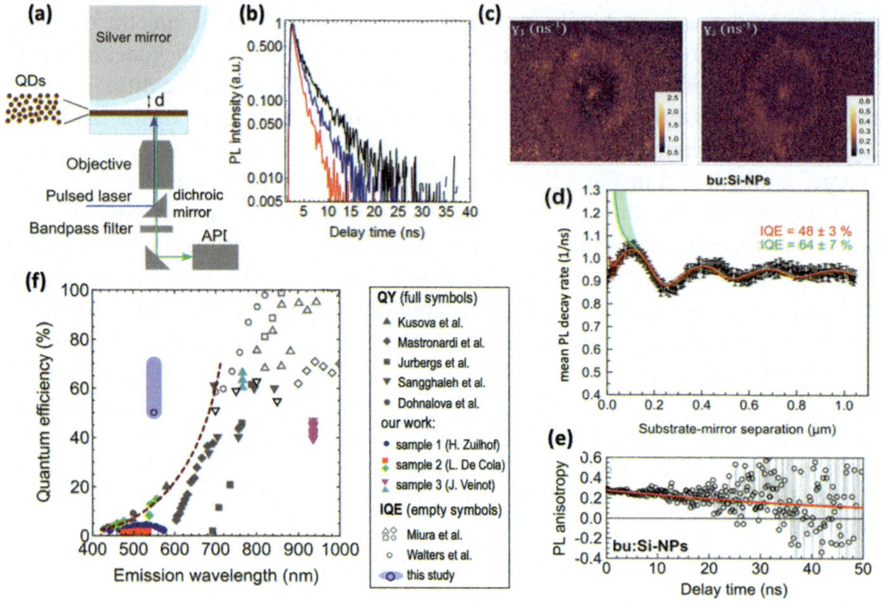


**Fig. 1.11** **a** Schematic of the Drexhage experiment. A spherical mirror is used to back-reflect the PL from the Si-NCs, which leads to an interference within the thin Si-NC layer and hence also to modulation of the local density of states (LDOS) that changes the radiative rate. **b** Time-resolved PL of the butyl-capped Si-NCs at different distances from the mirror, revealing changes in the emission rate. **c** PL lifetime maps of the fast (left) and the slower (right) PL lifetime components. **d** Measured PL decay as a function of the distance of the emitter (Si-NCs) from the mirror, fitted for an isotropic dipole orientation (green) and a parallel dipole orientation (red) LDOS. The respective fitted internal emission efficiencies are indicated in the same color in the graph. **e** Time resolved PL anisotropy of the butyl-capped Si-NCs, indicating a static dipole moment. **f** Internal quantum efficiency (IQE) and quantum yield (QY) of our butyl-capped Si-NC sample, in comparison to results from the literature (the respective references can be found in [32]). Our result is indicated in blue with a colored area representing the uncertainty in the final value. Reprinted with permission from APS Physics, after [32]

function of distance of the mirror from the Si-NC thin film (Fig. 1.11b, c), exhibiting oscillations and a baseline (Fig. 1.11d).

The oscillations are fitted by the LDOS for the relevant emitter dipole orientation—either parallel, perpendicular, or isotropic. We found that the Si-NCs have a fast radiative rate of the order of few nanoseconds, with a PL dipole orientation corresponding to that of a static parallel dipole orientation (Fig. 1.11d). The presence of a static dipole orientation might appear to be surprising, since in other semiconductor nanoparticles, such as CdSe-QDs, isotropic dipole orientation is observed, and a static dipole orientation is more expected for molecules. Nevertheless, we confirmed the static nature of the dipole orientation by an analysis of the time-resolved PL anisotropy (Fig. 1.11e). Our most recent work (Huang et al., to be published in 2021) indicates that a static dipole orientation is present in all various types of organically coated Si-NCs, prepared by various techniques, and is size dependent. In relation to our DFT simulations of the H- and organically coated Si-NCs, we came to a conclusion that this could be related to the observed highly non-degenerate nature of the optical transitions in Si-NCs [44], in contrast to the highly degenerate transitions in CdSe-QDs, because of the broken symmetry related to the randomly positioned covalently attached ligands, and inevitable Si-NC surface reconstructions and strains.

Thanks to the separated $k_{\mathrm{rad}}$ and $k_{\mathrm{nrad}}$, we can also evaluate the internal emission quantum efficiency IQE $= k_{\mathrm{rad}}/k_{\mathrm{PL}}$, which is found to be around 48% (Fig. 1.11f, blue hollow circle). This sets the upper limit for the PL quantum yield (QY), which is surprisingly high, considering the global emission efficiency trend (shown in Fig. 1.11f) with a rapid drop of emission efficiency below 20% for the PL emitting in the visible spectral range. The generally low PL QY in the visible spectral range is very unfortunate, considering that the organically coated Si-NCs are predicted to offer full tunability of the PL within the visible spectral range and generally show competitively direct bandgap-like radiative rates. Hence, this result gives a hope that the organically coated Si-NCs can have competitive emission efficiency. The global trend could be explained in terms of the presence of non-emissive impurities related to the synthesis, as well as possible presence of unsaturated Si bonds (dangling bonds) on the Si-NC surface. The latter being caused by the steric hindrance between the organic ligands that does not allow a full surface coverage by organic molecules, leaving part of the surface exposed to oxygen or left as dangling bonds, the main killer of the radiative transition in Si-NCs.

## 1.4 Theoretical Simulations of Si-NCs with Ligands

As already mentioned earlier, the direct bandgap-like radiative rates, observed in many organically capped Si-NCs, are still highly controversial, and the more conservative part of the Si-NCs community is convinced that the fast emission rate corresponds to an extrinsic radiative channel, such as oxide or carbon-related impurities. This is further supported by the fact that most of these fast decaying organically synthesized Si-NCs have PL confined to the blue-green spectral range. Moreover, one of the very few larger Si-NCs prepared by the bottom-up synthesis with organic coating from the sol-gel method [175] exhibited consistently only a very slow, $\mu$s PL decay, further supporting the idea that the blue-green emission might not be Si-related. To this end, we believe that theoretical modeling might offer some clarity, since one can remove the "distractions" of environmental effects and impurities, which are confusing the results in the experimental work.

### 1.4.1 Role of an Electronegative Ligand/Environment

Since most of the studied systems in practice contain more than 200–300 Si atoms, and a similar amount of atoms in the organic ligand capping, it has been till recently very difficult to simulate such a system by ab-initio self-consistent methods. Previous ab-initio and DF-TB calculations were performed only on very small Si-NCs and have predicted that alkyl capping should not alter dramatically the bandgap [106, 123, 171]. However, we could not find much information concerning the possible fast radiative rate or **k**-space projected DOS.

Atomistic, but empirical methods that are not self-consistent, such as tight binding (TB), allows us to get conceptual insights into the optical properties of Si-NCs. In our work, we performed TB simulations of an idealized system of a Si-NC with an electronegative ligand and an H-capped Si-NC in an electronegative environment. In this model, we expect to see a role of "charge transfer", which can be achieved either by a capping with polar ligands, or by placing the NC in a polar environment. Strongly electronegative elements, such as oxygen, chlorine or fluorine, have been already investigated theoretically in a great detail [91, 106, 134, 136, 167]. Oxygen was found to lead to a reduced bandgap and a surface trapping [91, 167, 211]. Konig et al. [106] found that highly polar –F, –OH or $-NH_2$ have a strong impact on bandgap, while the weakly polar $-CH_3$ does not, in agreement with the previous studies. Other studies have investigated, for example, –F [136], –Cl [134] or –Br, indicating strong changes to the bandgap.

In our work, we decided to try a different route and focus not on a specific element, but on a general undefined capping element with a broadly tunable electronegativity (i.e., ability to attract electrons from the Si-NC core), and a similarly functioning variant where the ligand is replaced by a general electronegative environment. For this purpose, we chose a convenient parametrization of the electronegativity $\chi_{\text{spec}}$, defined as a weighted average of the $s$- and $p$-orbital energies ($\epsilon_p$ and $\epsilon_s$, respectively), using equation [2]:

$$\chi_{\text{spec}} = (m\epsilon_p + n\epsilon_s)/(m+n),$$

where $m$ and $n$ are the numbers of electrons in the $p$- and $s$-orbital, respectively. In fact, we found that our results were not sensitive to the position of the $s$-level, which is therefore omitted in the further discussion, and we only specify the electronegativity by the shift of the $p$-state $\Delta_p$ (in eV). Using such a definition of electronegativity, we can say that an element is more electronegative than silicon, if its valence electron energy $p$-orbital energies are lower in energy by $\Delta_p$ (Fig. 1.12a).

In our pilot study [48], we analyzed a 2.5 nm Si-NC with –H and with a capping with a virtual element with $\Delta_p = -3$ eV, which is chosen to be similar to the $p$-level position of a carbon atom (Fig. 1.12c). The carbon-like atom is capped by three hydrogen atoms and hence plays a role of methyl-like capping. For this simulation, we used a $sp^3d^5s^*$ empirical 1st neighbor TB approach [93]. Our simulations revealed a surprising $\sim 10^3$ times enhancement in the phonon-less radiative rate, when compared to the H-capped Si-NC of the same size. Upon deeper inspection, we found this

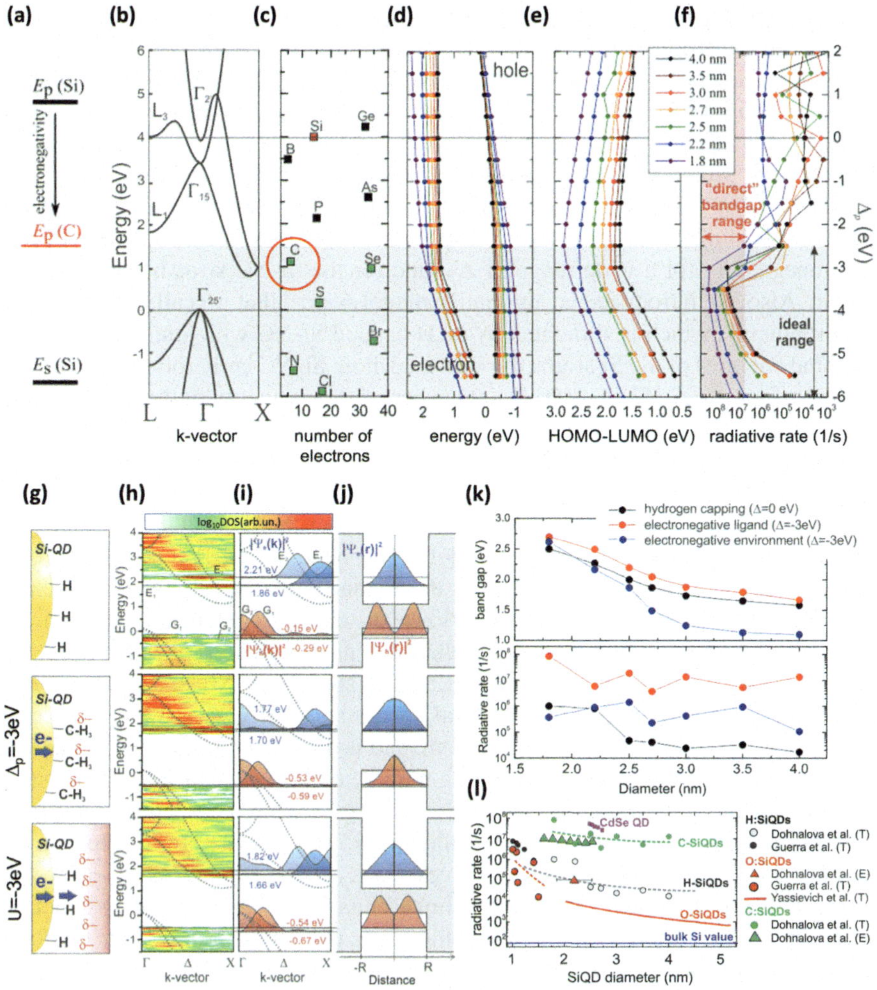

**Fig. 1.12** **a** Electronegative element simulated by a 3$p$-state shift by a $\Delta_p$. **b** Bulk silicon band structure. **c** Position of the atomic $p$-state of the different elements ($y$-axis on the left side, or relatively to silicon axis on the right side). **d** HOMO and LUMO energy for the different sizes of the Si-NCs. **e** HOMO–LUMO energy, i.e., the bandgap energy. **f** The phonon-less radiative rates. **g–j** Schematic sketches of the systems (**g**), band structures (**h**), DOS plotted for $HOMO_1$, $HOMO_2$ and $LUMO_1$ and $LUMO_2$ states with energies stated in the figure (**i**), real-space DOS for HOMO and LUMO state (**j**), for a 2.5 nm H-capped Si-NC (top row), a 2.5 nm Si-NC capped with elements with $\Delta_p = -3$ eV (middle row) and a 2.5 nm H-capped Si-NC in an electronegative environment with the $\Delta_p = -3$ eV (bottom row). **k** Bandgap energies and radiative rates obtained for the different sizes of Si-NCs for all three systems depicted in panels (**g–j**). **l** Comparison of theoretical (T) and experimental (E) data. Simulated radiative rates (Dohnalova et al. (T) [48, 159]) compared with simulations by Guerra et al. [75] and Yassievich et al. [213], and experimental results by Dohnalova et al. [48] and Driel et al. [50]. Reprinted with permission from American Physical Society and IOP Publishing, after [159] and [43]

enhancement correlates with a DOS $|\Psi_{LUMO}(\mathbf{k})|^2$ enhancement close to the $\Gamma$ point, making the transition effectively a direct bandgap one (Fig. 1.2a). At the same time, the real-space DOS $|\Psi_{LUMO}(\mathbf{r})|^2$ was found to be expanded toward the surface of the Si-NC (Fig. 1.2a), as we would expect for the electron-attractive electronegative surface chemistry.

This very promising result was further discussed in more detail in our follow-up TB study in [159], where we implemented the full band structure visualization and band folding, as introduced by Hapala et al. [80]. We analyzed the effect of electronegativity for a wide range of $\Delta_p$, and for the Si-NC sizes between 1.8 and 3.5 nm. Also, we introduced an alternative model system that we call electronegative "environment", which is simulated by an H-capped Si-NC with gradually changing $p$-orbital energies of the Si atoms through the whole Si-NC core, following a general equation $E_p(r) = E_p(\text{Si}) + U(r/R_{\max})^2$, where $E_p(r)$ is energy of the $p$-orbital level for Si atom on position $r$, $E_p(\text{Si})$ is energy of the $3p$-orbital in the bulk Si, $U$ is the variable factor defining magnitude of the electronegative effect (local field) and $R_{\max}$ is radius of the Si-NC.

We found a profound effect of the $\Delta_p$ energy shift on the HOMO–LUMO energies, bandgap and radiative rates (Fig. 1.12d–f). The bandgap energy peaks at around $\Delta_p = -2.5$ eV and then decreases for the higher electronegativity in agreement with literature [135, 137, 169, 204]. However, the most interesting result is the dramatic, three orders of magnitude enhancement in radiative rates for $-6$ eV $< \Delta_p < -2.7$ eV (Fig. 1.12f), present for all the studied Si-NC sizes. The highest radiative rate corresponds to the $p$-orbital energy shift of $\Delta_p = -3.5$ eV, when the $p$-orbital level is similar to the carbon atom, and is positioned within the Si bandgap (Fig. 1.12a–c). This means that the strong modification of the confined carrier states is a result of a resonant enhancement of the mixing between the electron states in the $X$-valley and the hole states in the $\Gamma$-valley [4, 15, 61, 62, 94, 160]. Interestingly, the opposite sign field, i.e., electro-positive (i.e., $\Delta_p > 0$), is found not to lead to such effect. This result indicates that carbon linked ligands might have a very beneficial influence on the radiative rate, while influencing the bandgap only very mildly, in agreement with previous ab initio studies on smaller systems [106, 123, 171].

The results are shown in detail for the mid-sized 2.5 nm Si-NC in Fig. 1.12g–j, which can be compared to our experimental work on 2–3 nm sized Si-NCs capped with butyl [32, 42, 48] and methyl [113]. The schematic sketch of the three studied systems—reference H-capped Si-NC, the system capped with an element with $\Delta_p = -3$ eV, and the H-capped Si-NC in an electronegative "environment" with $\Delta_p = -3$ eV—is shown in Fig. 1.12g (top, middle, and bottom row, respectively). The band structures with an overlaid bulk Si band structure (dotted line) are shown in Fig. 1.12h. The DOS resolved in the **k**-space for the first two HOMO and LUMO states is shown separately in Fig. 1.12i, next to the real-space DOS for the HOMO and LUMO states in Fig. 1.12j. From the comparison of columns (h–j) for all the three model systems, we see that there is a DOS enhancement close to the $\Gamma$ point for both, the "ligand" and the "environment" with $\Delta_p = -3$ eV. This is accompanied by a DOS expansion in the real space for the electron (LUMO) and compression for the hole (HOMO), as one would expect for the electronegative surface that is attractive to electrons and repulsive to holes.

In Fig. 1.12k, we show an overview of the results for the different Si-NC sizes. The bandgap energy with the $\Delta_p = -3$ eV "ligand" is very close to that of the reference H-capped Si-NC, in agreement with previous ab initio studies on small Si clusters [171]. The "environment", on the other hand, appears to reduce the bandgap for larger Si-NCs sizes. The radiative rate is enhanced by three orders of magnitude with respect to the H-capped Si-NCs, while this is less pronounced for the "environment". In comparison with literature in Fig. 1.12l, we see that the simulated radiative rate for the methyl-like "ligand" simulated here agrees well with the experimentally found emission rates [48] and is very close to those observed for the direct bandgap CdSe-QDs [50]. The radiative rates simulated by us for the reference H-capped Si-NCs system [159] are within the range expected for such a system [75]. For comparison, we also show in Fig. 1.12l data for the oxide-capped Si-NCs and bulk silicon radiative rate.

Such TB simulations give some insight into a possible trend that can occur as a result of charge transfer in the Si-NC-ligand/environment system. However, such TB simulations are only semi-empirical, are not self-consistent, and do not include realistic ligand and possible strain effects that could be present due to the full surface coverage capping. These effects are further explored in the following Sects. 1.4.2 and 1.4.3.

### *1.4.2 Role of Tensile Strain*

Another important effect that can enhance the radiative rate is a tensile strain. A simple phenomenological model explaining such an enhancement is based on a band structure engineering of bulk silicon, which can be described using the known deformation potentials. When tensile strain (lattice expansion) is applied to a bulk silicon, it causes an up-shift in energy of the indirect $\Delta_1$ conduction band minimum. At the same time, the band structure branch around the direct local conduction band $\Gamma_{15}$ maximum splits, and the local $\Gamma_{15}$ maximum transforms into a local minimum and down-shifts in energy [115]. This process brings the direct and indirect bandgaps closer in energies. Hence, if large enough tensile strain was applied, a switch between direct and indirect bandgap should be observed in bulk silicon. However, more detailed calculations [115] show that the amount of strain necessary for such a transformation would be far too high (over 6 % lattice constant expansion). This would inevitably lead to structural instabilities and formation of defects or trap states, which would inevitably quench any luminescence. Moreover, the experimental realization of tensile strain in a bulk crystal is, in contrast to a possible compression, not an easy task to achieve.

This is exactly where the Si-NCs have an advantage. First of all, the dramatic reduction in the size of the crystal gives rise to the quantum confinement, which on its own affects the band structure in the right way, since the influence of the quantum confinement acts on the band structure of Si-NCs in the same direction as the tensile strain does [10, 164]. On the one hand, it lifts up the indirect $\Delta_1$ conduction band minimum (otherwise also called for simplicity the X-point, for its close proximity to the X-point), known as the "bandgap opening" present in all the semiconductor NCs. On the other hand, as a result of the opposite local curvature of the conduction band in the vicinity of the direct $\Gamma_{15}$ local maximum, this branch of the band structure is downshifted as a result of quantum confinement [10, 164]. Consequently, the effect of the quantum confinement effectively *lowers* the amount of strain needed for the indirect-to-direct bandgap cross-over to occur. The cooperation of the quantum confinement and tensile strain is sketched in Fig. 1.13a. Another consequence of the quantum confinement is the delocalization of the electron and hole wavefunctions in the **k**-space. This property of nanosized objects implies a much higher level of (inter-valley) mixing of the various types of states, which implies that as a result of the energetic lift at the $\Delta_1$ point and the drop at the $\Gamma_{15}$, "intermediate" states having both the "indirect" and "direct" character will form. The enhanced "directness" of the HOMO–LUMO transition will then enhance the probability of radiative recombination. Thus, such **k**-space "blurring" can further decrease the amount of strain necessary to achieve the direct bandgap-like emission. An important factor is also the fact that Si-NCs can accommodate relatively high amounts of strain, which stands behind the great capacity of silicon for lithium intake, which makes it potentially a better candidate for batteries than carbon [128]. This property can also favorably contribute to the possibility of tensile strain-induced direct bandgap-like emission.

Naturally, a question arises as to how to realize tensile strain in a real-life Si-NC system. Whereas a compression can be relatively easy to achieve, using, for example, a diamond anvil cell [181], tensile strain is based on the *expansion* of a crystalline lattice, which is much more difficult to apply. Tensile-straining can be achieved, e.g., in films epitaxially grown on intentionally lattice-mismatched substrates, which was the case also for direct bandgap germanium [127], utilizing tensile strain in connection with $n^+$ doping. Another option is to grow silicon nanomaterials on a thin stretchable membrane [100]. Theoretically, tensile strain can also be achieved by using a high surface coverage of covalently bonded surface ligands, which is the path we explored in our simulations [115].

Figure 1.13b presents the calculated distribution of the Si–Si bond lengths for the three different types of capping, namely hydrogen, methyl $-CH_3$ and –OH (representing an oxide shell). Taking the H-capped Si-NC as the unstrained reference, we find the mean Si–Si bond length to be larger in the methyl-capped Si-QDs and smaller in the OH-capped Si-NCs. This calculation suggests a presence of tensile strain in the Si-NC systems with organic coating. Experimentally, this strain was confirmed through the shifts in the phonon energies in the SPL spectroscopy [115]).

In order to provide a theoretical confirmation of the model sketched in Fig. 1.13a, calculation of the band structure and oscillator strengths of the constructed Si-NC models were performed using the "Fireball" DFT code [95]. The simulation is performed on the local density approximation (LDA) level with a numerical atomic-like orbitals basis set including *s*,*p*, and *d* atomic orbitals [7]. The Si-NC core with 538 Si atoms ($Si_{538}$) was prepared by a cut of the bulk silicon structure along the (111) and (100) facets. All the available surface sites are passivated by the ligands—either –H,

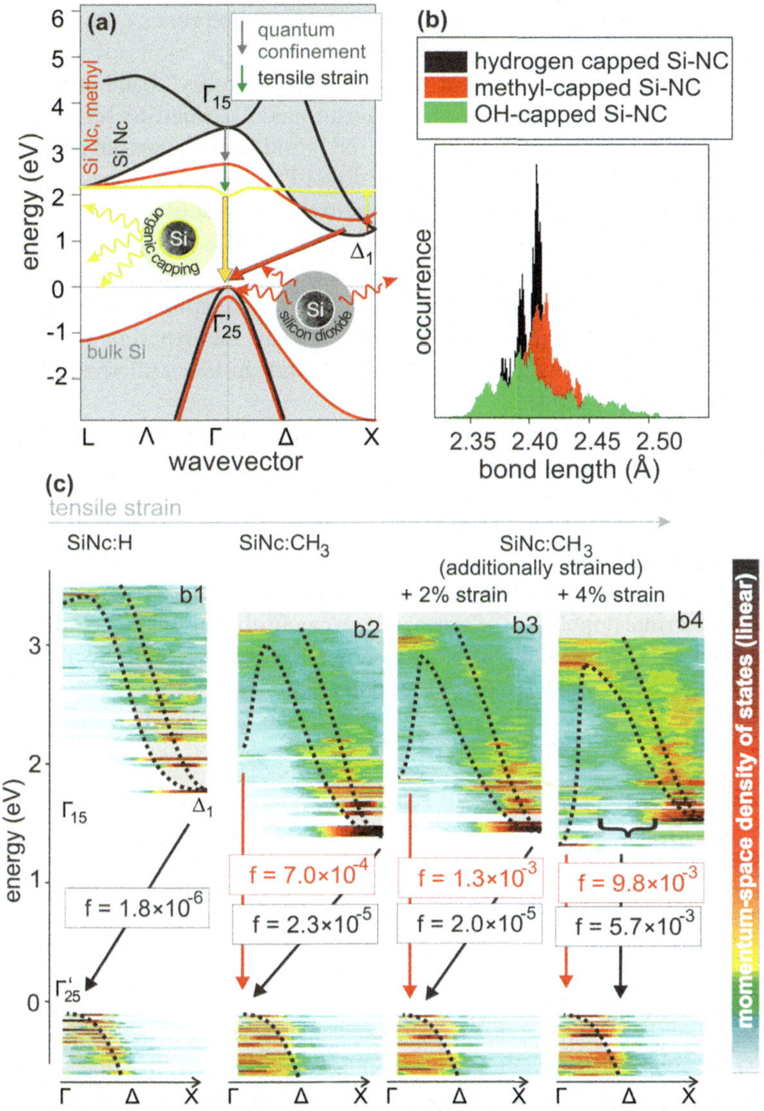


**Fig. 1.13** **a** Sketch of a phenomenological model explaining the transformation of the band structure of the Si-NCs by the cooperation of quantum confinement and tensile strain. **b** The calculated distributions of the Si–Si bond length for H-, methyl- and OH-capped 2.5 nm Si-NCs. **c** The calculated effect of a tensile strain on the band structure of the Si-NCs. Tensile strain increases from the strain-free hydrogen-capped Si-NCs to the naturally strained methyl-capped Si-NCs (with 100% ligand surface coverage). The indirect-to-direct bandgap cross-over occurs when an additional strain is applied. Reproduced with permission from John Wiley and Sons, after [115]

–$CH_3$, or –OH. The geometry is optimized with the convergence criteria of $10^{-6}$ eV and 0.05 eV/Åfor the total energy and forces.

The results of these calculations, in a form of the "fuzzy" band structures [80], are shown in Fig. 1.13c. Going from the unstrained H–caped Si-NCs (left) to the slightly naturally strained –$CH_3$-capped Si-NC (middle left), we already see a $\Gamma_{15}$-point descent in energy. The corresponding direct $\Gamma_{15}$-$\Gamma_{25'}$ transition is characterized by higher oscillator strength when compared to the indirect transition. However, the amount of the inherent tensile strain caused by the full surface coverage by the methyl groups is on its own not enough to induce the complete indirect-to-direct bandgap cross-over. Hence, we find that despite the tensile strain induced by the full surface coverage by methyl ligands, we cannot fully explain the experimentally observed results in [113], where we observed a direct bandgap-like radiative rates for the methyl-capped Si-NCs of similar sizes.

To analyze the additional amount of strain needed, we further artificially tensile-strained the methyl-capped Si-NCs with 2% and 4% extra strain (Fig. 1.13c, right). As expected, the $\Gamma_{15}$ further down-shifts with the increasing tensile strain, leading to a material with a fundamental direct bandgap at about over ∼4% of added tensile strain. These results imply that a homogeneously applied tensile strain of high enough levels can lead to a direct bandgap silicon. We need to add that this work did not discuss the thermal population of states, which can further affect the relation between the theory and the experiment, as will be shown later in the text in Sect. 1.4.4. Also, the radiative rates are calculated without phonon assistance.

### *1.4.3 Interplay Between the Charge Transfer and Tensile Strain*

In our previous studies of the electronegative ligands inducing a charge transfer [48, 159] and tensile strain [115], the picture was not complete, since both effects can interact in a complex way. In our recent work, using a self-consistent ground state DFT method, we analyze the role of these effects separately and together [44], to understand their respective relations to the enhancement of the radiative rates. The DFT calculations are performed using the cp2k code [31, 199] with a generalized gradient approximation (GGA) Perdew Burke Ernzerh (PBE) functional [157], core approximated by the Goedecker–Teter–Hutter (GTH) pseudopotential [70], a dual Gaussian and plane wave basis was implemented in Quickstep cp2k with the double zeta short-range Gaussian polarized basis set DVZP-MOLOPT [31] and a plane wave cutoff energy of 400 Ry. The geometry of all the structures was optimized to attain the lowest total energy state using the conjugated gradient (CG) algorithm, and the self-consistency convergence is set to $10^{-6}$ Ry. Our approach was validated by simulating a set of hydrogen-capped Si-NCs of sizes between 1 and 3 nm [44].

To be able to separate the roles of charge transfer and tensile strain, we designed a system of ∼2 nm Si-NC covered with a different surface coverage percentage by

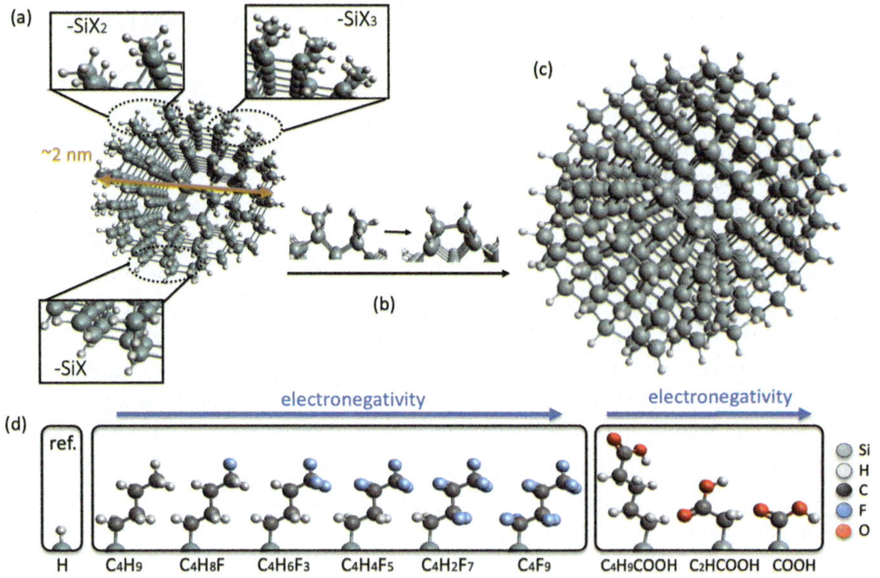


**Fig. 1.14** **a** 2-nm H-capped Si-NC with surface covered by hydrogen. Three types of surface species are present: the chemically unstable $-SiX_3$ (that are removed), the $-SiX_2$ species [that are reconstructed into the –SiX species (**b**)], and the -SiX species. **c** The 2-nm Si-NC geometry with a fully reconstructed surface, containing only –SiX species capped with hydrogen. **d** Different ligands used for surface capping: fluorinated hydrocarbons and –COOH terminated alkyls. **a**–**d** Reproduced with permission from the American Chemical Society, after [44]

ligands with increasing electronegativity. To simulate increasingly electronegative C-linked ligands, we start from butyl $-C_4H_9$ and substitute the H-atoms with an increasing number of Fluorine atoms in the alkyl backbone structure, i.e., $-C_4H_9 \rightarrow -C_3H_6CHF \rightarrow -C_3H_6CF_3 \rightarrow -C_2H_4C_2F_5 \rightarrow -CH_2C_3F_7 \rightarrow -C_4F_9$ (Fig. 1.14d). The C-linker is used because of the indications we obtained in our previous studies [48, 115, 159]. To make the system as realistic as possible, we removed all the unstable $-SiX_3$ sites. To unify the symmetry of the surface ligand bonding, we have reconstructed the edge (100) facets with the $-SiX_2$ sites, leaving us with only the –SiX bonding sites (Fig. 1.14a–c). For the unstrained systems, we used 50% surface coverage with the remaining surface sites capped by hydrogen. The (lack of) tensile strain is monitored by the analysis of the average Si–Si bond length of all the Si atoms of the geometrically optimized Si-NC. It is found that the 50% surface coverage does not induce any strain. Only a small bond shrinkage is observed for the most electronegative ligands, likely due to a reduced electronic density in the Si-NC core [44]. To prevent any changes in symmetry between the different ligands, we used the same surface sites for all the simulated systems. The ligand surface sites were randomly chosen from all the facets without any preference. For the tensile-strained systems, we used the 100% capping coverage.

The resulting band structures with indicated HOMO–LUMO energies and radiative rates at $T = 300$ K are shown in Fig. 1.15a. The ligands are sorted by their increasing electronegativity, described by the increased net charge transfer from the Si core to the ligand in the Mulliken population analysis (Fig. 1.15c). We see that more fluorine atoms on the carbon chain lead to a higher electronegativity of the ligand with respect to the Si core.

Increased electronegativity of the ligands seems to introduce only small variations in the DOS (Fig. 1.15a). The major difference is between the H-capped Si-NCs and all the other Si-NCs with C-linked ligand. Also, we see a gradual decrease in the Fermi energy [44] with the increasing ligand electronegativity. The real-space cross section of the HOMO and LUMO orbitals is shown in the bottom panel of Fig. 1.15a. We see only a weak localization for the Si-NCs with the more electronegative ligands, but no severe localization is observed. The HOMO–LUMO energy is separately shown in Fig. 1.15d, suggesting an almost constant bandgap energy for the fluorinated Si-NCs, with no obvious dependence on the electronegativity.

To evaluate the phonon-less radiative rates $k_{\mathrm{rad}}^{ji}$ at $T = 0$ K of a transition between $i$-th and $j$-th state, we use the Einstein's coefficient of the spontaneous emission $A_{ji}$ in atomic units:

$$A_{ji} = k_{\mathrm{rad}}^{ji} = \frac{1}{\tau_{\mathrm{rad}}^{ji}} = 2\frac{E_{ji}^2 f_{ji}}{c^3}, \tag{1.1}$$

where $E_{ji} = E_j - E_i > 0$ is energy of the transition, $c$ is speed of light and $f_{ji}$ is the oscillator strength given by the Fermi's "Golden Rule" in atomic units:

$$f_{ji} = \frac{2}{3} E_{ji} |d_{ji}|^2. \tag{1.2}$$

Here, $d_{ji} = \langle \Phi_j | \mathbf{r} | \Phi_i \rangle$ is the transition dipole moment matrix element and $\Phi_i$ and $\Phi_j$ are single-particle (doubly occupied) states of a hole and an electron (molecular orbitals), respectively. For a more reasonable comparison with experiment that is typically performed at $T > 0$ K, we evaluate also the thermally averaged radiative rates: [38]

$$\langle k_{\mathrm{rad}} \rangle (T) = \sum_{i,j} \frac{k_{\mathrm{rad}}^{ji} \mathrm{e}^{-\frac{E_{ji}}{k_B T}}}{\sum_{i,j} \mathrm{e}^{-\frac{E_{ji}}{k_B T}}}, \tag{1.3}$$

where $k_B$ is the Boltzmann constant and $T$ is temperature. This equation assumes that after the excitation, the thermalizations of the electron and the hole in the conduction and valence bands are more efficient than the radiative recombination, which is a very reasonable assumption [38, 161].

In Fig. 1.15e, we see the obtained radiative rates. Firstly, we see a considerable increase in the radiative rate between the H-capped Si-NC and the Si-NC with any of the C-linked ligands. Also, it seems that the higher electronegativity of the ligand

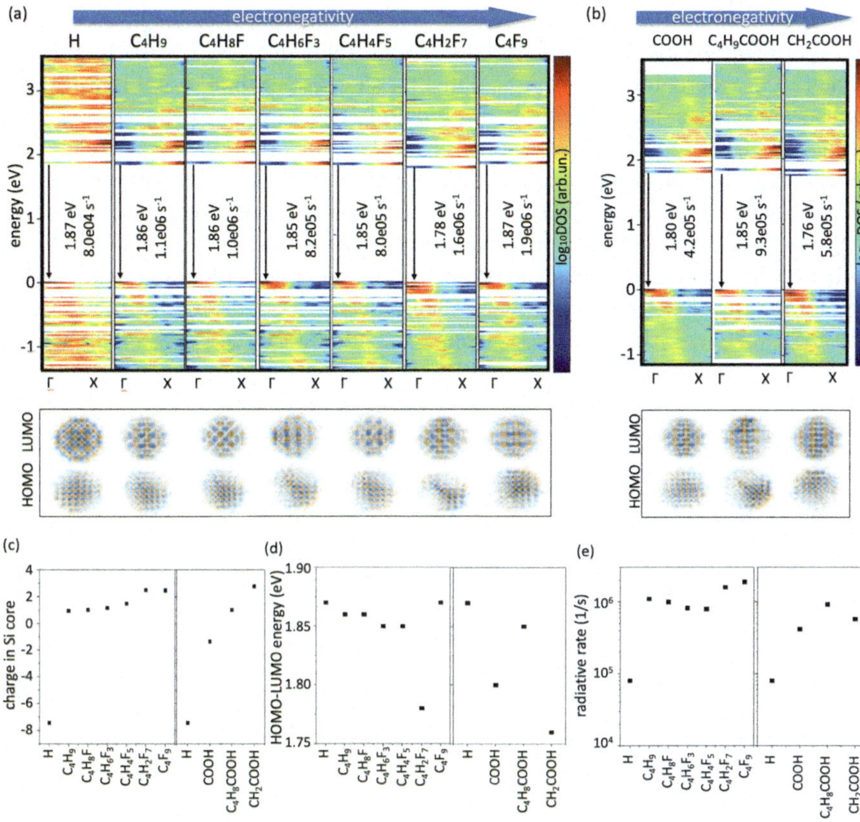


**Fig. 1.15** **a** Fuzzy band structures for 2 nm Si-NC capped 50% with –H, $–C_4H_9$, $–C_3H_6CHF$, $–C_3H_6CF_3$, $–C_2H_4C_2F_5$, $–CH_2C_3F_7$, and $–C_4F_9$. The real-space orbital projection (in x-z plane) is shown below for the HOMO and LUMO state. The HOMO–LUMO energy and thermalized radiative rate at 300 K is indicated in each panel. **b** The same as in (**a**) for cappings with –COOH, $–C_4H_8COOH$ and $–CH_2COOH$. **c** Charge in the Si-NC core. **d** HOMO–LUMO energy. **e** Phononless radiative rate at $T = 300$ K. The data connected with the fluorinated hydrocarbon ligands are reproduced with permission from the American Chemical Society, after [44]

improves the radiative rate further, but the trend is not as clear as what we obtained by the TB study in [48, 159]. The increase in the radiative rate is found to correlate in our study with an enhanced DOS contribution close to the $X$ point for the HOMO, leading to an enlarged overlap of the e–h wavefunction in the $X - X$ direct transition. This is qualitatively different output from our previous TB study [48, 159], where an enhanced DOS contribution was found for the LUMO state close to the Γ point instead. Nevertheless, the result is similar—the HOMO LUMO transition becomes more direct bandgap-like. Also, we find that with the increasing electronegativity, the fraction of the DOS on the ligand increases gradually (in agreement with the increasing electronegativity of the ligand and the resulting charge transfer). A similar effect has been used to qualitatively (i.e., without theoretical calculations) explain the experimentally observed enhanced radiative rates in Si-NCs with an N-linked ligand [203], where the authors argued that the increase of the Γ point LUMO DOS results from surface states. However, this is definitely not the case in the current study, because the LUMO DOS increase at Γ or *X* points clearly does not correlate with higher wavefunction spatial localization on the surface Fig. 1.15a (bottom panel).

Additionally, the same effect is studied also on the same geometry 2 nm Si-NC, using the same sites as in the previous study, but this time capped with the ligand (sorted by increasing electronegativity) $-COOH$, $-C_4H_8COOH$, and $-CH_2COOH$ (Fig. 1.15b–e); alkyl ligands of varying length (sorted again by electronegativity) $-C_4H_9$, $-C_{10}H_{21}$, $-C_2H_5$, and $-CH_3$ (Fig. 1.16a–e); various additional C-linked lig-

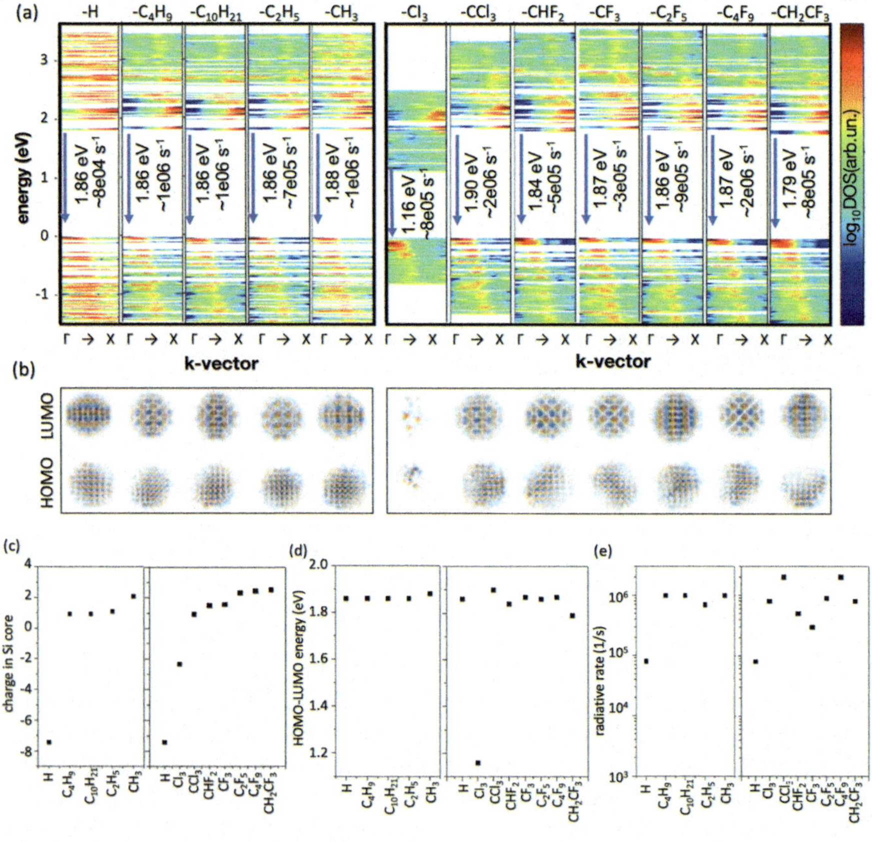


**Fig. 1.16** **a** Fuzzy band structures for 2 nm Si-NC capped with $-H$, $-C_4H_9$, $-C_{10}H_{21}$, $-C_2H_5$, $-CH_3$, and $-CI_3$, $-CCl_3$, $-CHF_2$, $-CF_3$, $-C_2F_5$, $-C_4F_9$, and $-CH_2CF_3$ (sorted by increasing electronegativity). The HOMO–LUMO energy and thermalized radiative rate at 300 K is indicated in each panel. **b** The real-space orbital projection (in x-z plane) is shown below for the HOMO and LUMO state. **c** Charge in the Si-NC core. **d** HOMO–LUMO energy. **e** Phonon-less radiative rate at $T = 300$ K

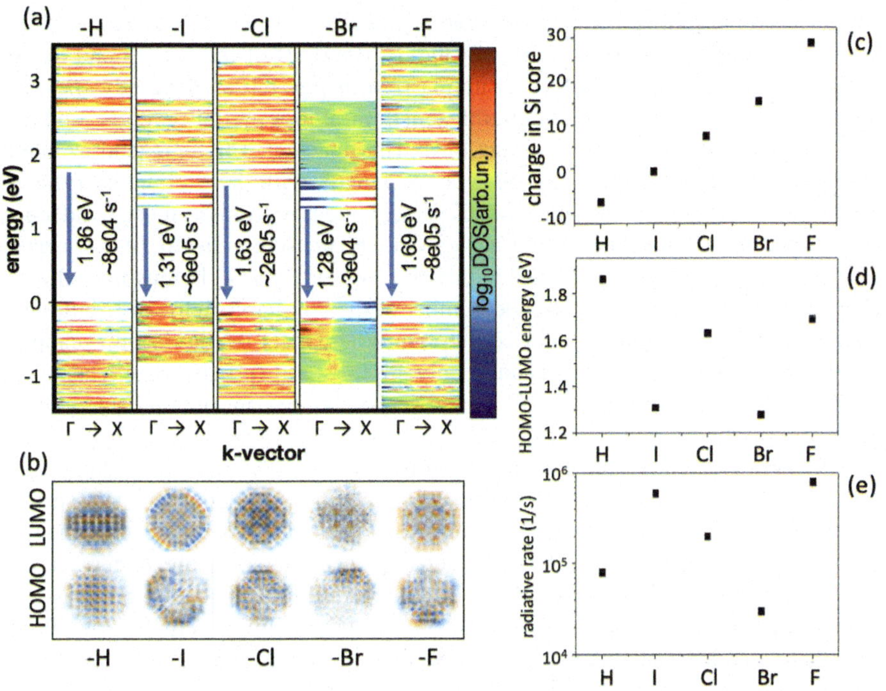


**Fig. 1.17** **a** Fuzzy band structures for 2 nm Si-NC capped with –H, –I, -Cl, –Br, and –F. HOMO–LUMO energy and thermalized radiative rate at 300 K is indicated in each panel. **b** The real-space orbital projection (in the $x - z$ plane) is shown below for the HOMO and LUMO state. **c** Charge in the Si-NC core. **d** HOMO–LUMO energy. **e** Phonon-less radiative rate at $T = 300$ K

ands (sorted by electronegativity) $-CI_3$, $-CCl_3$, $-CHF_2$, $-CF_3$, $-C_2F_5$, $-C_4F_9$, and $-CH_2CF_3$ (Fig. 1.16a–e); and halides (sorted by electronegativity) –I, –Cl, –Br, and –F (Fig. 1.17a–e). These additional studies reveal that there is, in fact, no explicit dependence between the electronegativity (i.e., charge transfer), as quantified by the Mulliken analysis, and the enhancement of the radiative rate. The main enhancement appears to be correlated with the presence of the C-linker, the first carbon atom in the ligand chain, which is slightly more electronegative than Si, but the further structure of the organic ligand is not adding any consistent effect. Nevertheless, the structure of the ligand beyond the linker element does play a role and affects strongly both the radiative rate and the HOMO–LUMO energy. We note that all the Si-NC studies shown here in Figs. 1.15, 1.16 and 1.17 are comparable between each other, due to the fact that in all cases we use an identical 2 nm Si-NC core and identical surface sites for the ligands.

To separately test the link between the surface capping and induced strain, we analyze the effect of low (50%) versus high (100%) surface coverage by ligands on the radiative rates. For this purpose, we use the same 2 nm Si-NC capped 50% or 100% by $-CH_3$, $-CF_3$, $-C_4H_9$, or $-C_4F_9$. The band structures, HOMO–LUMO energies and thermalized phonon-less radiative rates at $T = 0$ K, 100 K, and 300 K, are shown in Fig. 1.18. The analysis of the distributions of the Si–Si bond lengths in the differently capped Si-NC is summarized in Fig. 1.19a. Generally speaking, the attached alkyl chains exert negligible tensile strain for partial coverage, and only a small tensile strain at full coverage [44]. The fluorinated ligands tend to induce compression, or shrinkage of the Si–Si bond length, with the exception of the full $-C_4F_9$ capping, where the tensile strain is present, judging from the mean bond length [44]. However, the mean bond length should not be the sole indicator of how surface capping changes the strain balance within the NC core. As is obvious from the box chart in Fig. 1.19a, the width and the number of the "outliers" in the distributions of bond lengths change with the surface capping. Whereas for the partial alkyl capping, the widths of the distributions change only slightly, and full coverage leads to the widening of the distributions in all cases. These changes are then reflected also in the corresponding radiative rates.

The highest radiative rates are achieved for the partially covered $-C_4F_9$, $-C_4H_9$, and $-CH_3$ capped Si-NCs systems that are all slightly homogeneously tensile-strained with some positive net charge in the Si core. The lowest rates are obtained for the fully capped Si-NCs, especially for the $-CH_3$ capping at zero temperature. Comparing our results here with the previous study in [115], where full $-CH_3$ capping was used as a starting point for the enhanced radiative rate, we see that it is actually not only the amount of tensile strain, but also its homogeneity, which is important for the desired rate enhancement effect.

To sum up, the main consequence of the full coverage with ligands is the resulting inhomogeneously distributed strain within the NC, which appears to systematically lower the radiative rate of the transitions. Thus, in addition to the mean bond length, i.e., the applied tensile strain, also the shape of the histogram needs to be considered, i.e., the strain homogeneity. However, it should be noted here that the full ligand coverage is unrealistic and would be prevented by the steric hindrance effects in such small NC systems.

### *1.4.4 Thermal Population of States*

Thermal averaging of the radiative rates based on 1.3 is an important step in relating the computed radiative rates to an experiment. As evidenced by Fig. 1.19b, the thermal population of states can further enhance or decrease the corresponding radiative rate depending on the particular system, without any clear rule behind the calculated trends. Since the lowest-energy HOMO–LUMO transition is not necessarily the most efficient, it is interesting to examine in more detail what constitutes this averaged value at different temperatures. Therefore, we examine also the thermally averaged radiative rate resolved in energy for different temperatures:

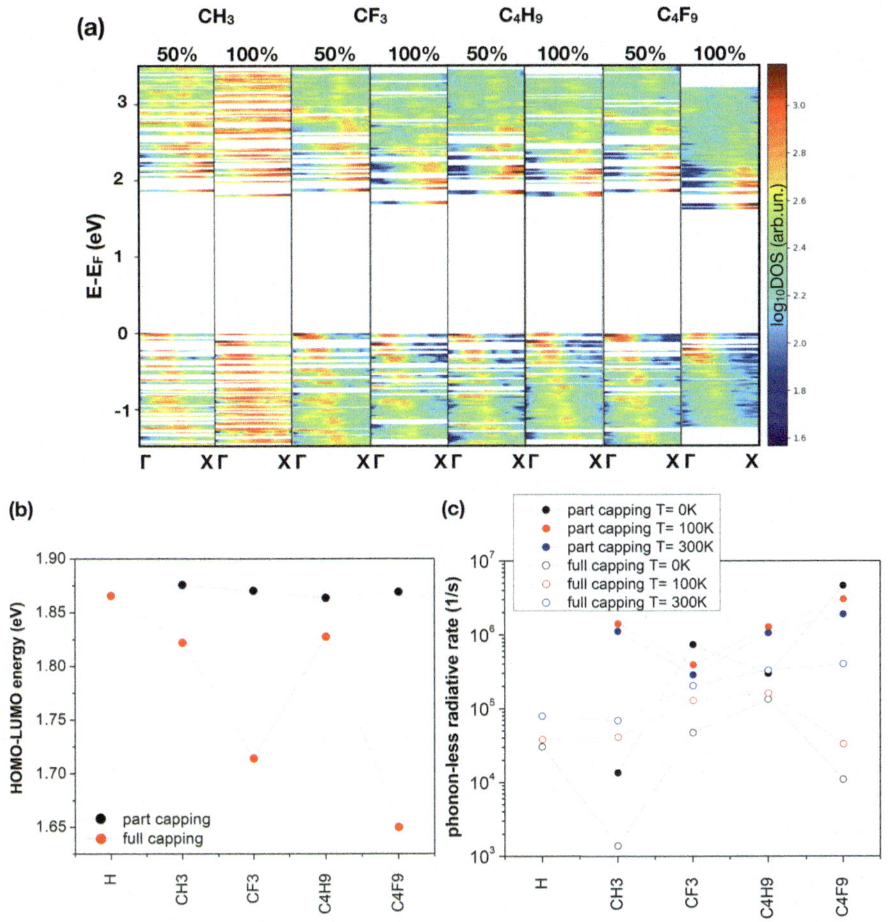


**Fig. 1.18** **a** Fuzzy band structures for 2 nm Si-NC capped 50% or 100% with $-CH_3$, $-CF_3$, $-C_4H_9$, or $-C_4F_9$ ligands. **b** HOMO–LUMO energy. **c** Thermalized phonon-less radiative rate at $T = 0$ K, 100 K, and 300 K. The figure is reprinted from [44] with permission from the American Chemical Society

$$k_{\text{rad}}^{ji}(T) = \frac{k_{\text{rad}}^{ji} e^{-\frac{E_{ji}}{k_B T}}}{\sum_{i,j} e^{-\frac{E_{ji}}{k_B T}}}. \tag{1.4}$$

This formula is a trivial decomposition of the sum in 1.3 to its constituents, where $\sum_{i,j} k_{\text{rad}}^{ji}(T) = \langle k_{\text{rad}} \rangle (T)$. In order to visualize these components of the thermally averaged radiative rates, we apply a "compressed-violin-plot" representation as explained in Fig. 1.20a. Violin plots [87] are typically used to represent a distribution of parameters as shaded areas, where the width of the area corresponds to the number of cases having that particular value similarly to a histogram. Often, as

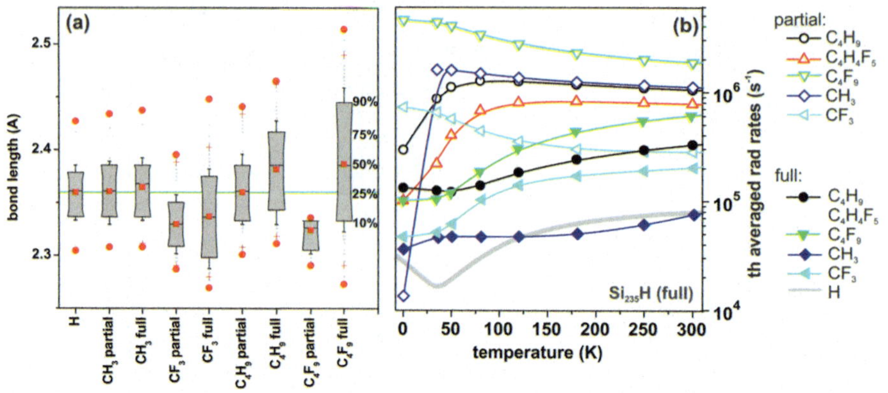


**Fig. 1.19** **a** Box chart summarizing the distribution of the Si–Si bond lengths for various calculated Si-NCs systems after relaxation. The result obtained for an H-capped Si-NC (full capping) is shown for comparison as an unstrained reference material; its median value is highlighted by the green line in the background. The gray filled areas show percentiles of the distribution as noted, the whiskers denote the 5–95 percentile. The red points correspond to the minimum/maximum (circles) and mean (square) of the distributions, the red crosses show 1–99 percentiles, and the gray diamonds are outlier data outside of the 5–95 percentile interval. **b** The overall comparison of all the calculated radiative rates for differently capped Si-NCs after thermal averaging at several temperatures. After [44], reprinted with permission

a result of broken symmetry, lowered degeneracy or even pure coincidence, several transitions $E_{ji}$ with different initial $|j\rangle$ and final $|i\rangle$ states can have a very similar transition energy $E_n \pm \delta E$ (see the color-coded examples in Fig. 1.20a). In some cases, the $\delta E$ is so small (1–2 meV) that we can think of such a group of states as a single "degenerate level". By plotting the radiative rates of such "levels" as violin plots, it is possible to visually summarize the information about the radiative transitions in a single column of data. This approach can then be used to demonstrate energy resolved changes in the radiative rates with temperature using 1.4 in a form of "compressed-violin-plots". More information about how exactly the plots are generated can be found in [44].

To illustrate the usefulness of this approach, two examples of the compressed-violin-plot representation of the butyl-capped ~2 nm Si-NCs are shown in Fig. 1.20b, c for partial (50%) and full (100%) capping, respectively. Clearly, for the partially capped Si-NC model, the dominant transition corresponds to the transition slightly energetically (~5 meV) above the lowest-energy HOMO–LUMO transition. As a result of its energetic position, the thermally averaged radiative rate of this level rises at low temperatures (below 120 K), but then it starts to decrease again. The situation is very different in the fully capped Si-NC, where the lowest-energy transition is the dominant one at low temperatures. However, its strength gradually decreases with increasing temperature, when the energetically higher (populated) levels dominate the radiative process. In the particular case of the fully butyl-capped Si-NC, multiple states at around 20–30 meV above the ground state dominate the emission at room

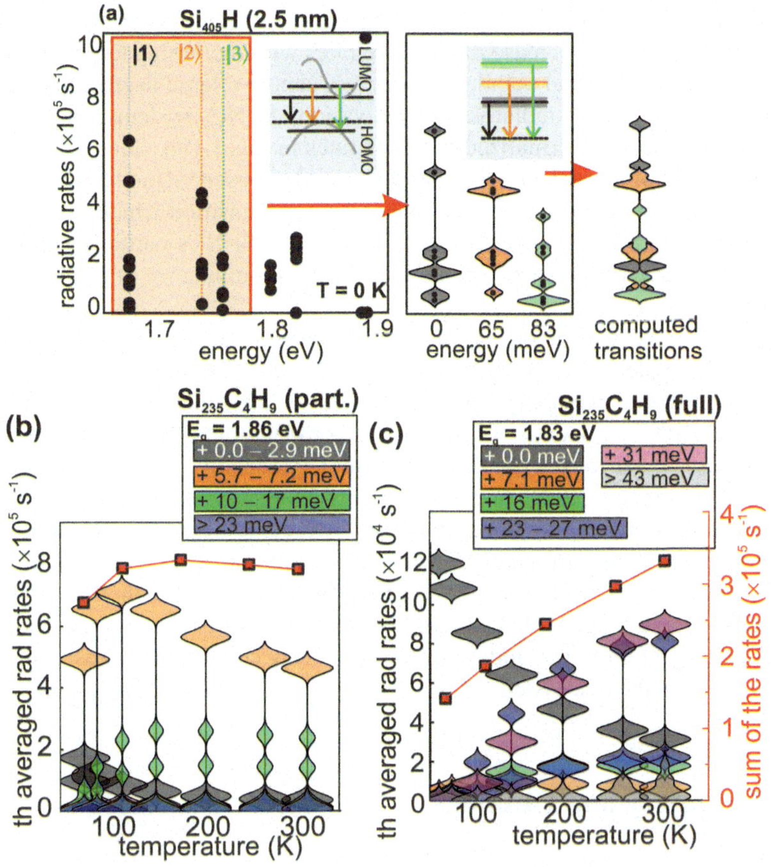


**Fig. 1.20** **a** Explanation of the compressed-violin-plot representation. We start with the calculated zero-temperature radiative rates $k_{\mathrm{rad}}^{ji}$ (black points). Clusters of different transitions $E_{ji}$ (for various $j$ and $i$) which have for any reason similar transition energy $E_n \pm \delta E$ can be represented by a single-colored line (levels) and indexed by $n$ (e.g., black for $n$ = 1, red for $n$ = 2, and green for $n$ = 3). The radiative rates corresponding to the transitions from these levels can be drawn as shaded areas with widths proportional to the number of transitions with the same radiative rate (a 90°-rotated histogram). The violin plot of the transitions from these levels, shown in the middle, is thus equivalent to the data in the red rectangle on the left, except for the energy axis, which is conveniently re-scaled to show all of the existing levels. The energy scale is kept in color-coding. All of the violin plots for the different levels then can be drawn for each temperature in a single column, as shown in the right part of panel (**a**), forming a compressed-violin-plot representation. **b**, **c** Compressed-violin-plot representation of the thermal evolution of the thermally averaged radiative rates $k_{\mathrm{rad}}^{ji}(T)$ resolved in energy from 1.4 (color-coded for the different levels) for partially and fully butyl-capped Si-NCs. The sum of the thermally averaged radiative rates $k_{\mathrm{rad}}^{ji}$ from 1.3 is also included as the red-and-black rectangles. Please note that in panel (**c**) the overall rates are connected with the red right vertical axis. After [44], reprinted with permission from the American Chemical Society

temperature and, on the whole disregarding the resolution in energy, the described trend translates into an increase of radiative rates with increasing temperature.

In this way, the thermal population of levels can, to a large degree, influence the experimentally observed radiative rates in various Si-NCs systems, explaining the large variations in the thermalized radiative rates in Fig. 1.19b. Generally speaking, the contribution of more than a single state was often observed, and especially in the smaller Si-NCs, it was often not the lowest-energy transition which was dominant at room temperature. An exception to this rule was Si-NCs partially capped with fluorinated hydrocarbon, where we observed a clear dominance of a single, strong, lowest-energy transition.

The significant (and often dominating) contribution of the transitions from the energetically higher states at elevated temperatures is generally an important observation. The thermal averaging from 1.3 and 1.4 is a product of the strength of the transition and the exponential Boltzmann factor describing the thermal population of the states, which makes the product extremely sensitive to a particular combination of the strength of the transition and its energetic position. Therefore, once a higher-lying transition is strong enough to overcome the Boltzmann factor from the thermal population of states in 1.4, for example, as a result of $\Gamma - X$ mixing, it will contribute or even dominate the emission. Therefore, the thermal population of the energetically higher states is clearly an important feature that determines the final efficiency of emission under typical laboratory conditions and as such cannot be omitted from the discussion.

## 1.5 Summary and Outlooks

Through the past two decades, we have focused our research on an in-depth analysis of the Si-NC emission, especially the controversial fast decaying F-band, and the theoretical possibilities of achieving an enhanced, direct bandgap-like radiative rates with Si-NCs.

In our experimental work, we have found evidence of the fast decaying F-band emission in the H-, oxide-, and organically capped Si-NCs. In H-capped porous silicon with $\sim$2 nm Si-NCs, we observed a nanosecond decaying (F-band) green emission [45], which turned into microsecond decaying red S-band after oxidation [45]. For the oxide-capped Si-NCs made from porous silicon, we have demonstrated that by decreasing the size of the Si-NCs, we can switch from the slow red S-band to the fast blue F-band, but this tunability is not continuous and skips the fast green PL band [46], where our sample originally emitted before oxidation [45]. Using an electron beam irradiation on several differently prepared oxide-capped Si-NCs, we created several brightly emissive surface sites, which could be related to the well-known oxygen vacancies in the damaged silica shell on the surface of oxide-capped Si-NCs [12], hinting at a possible relation between the S-band and F-band origins in the small oxidized Si-NCs based on the spectral similarities. We have also shown that the incorporation of nitrogen into the shell of Si-NCs does not necessarily induce blue F-band-like emission [64], as is sometimes suggested, but, on the contrary, a (silica-defect-related) part of the blue F-band emission is quenched.

In another two experiments, we tried to convert oxide capping into alkyl and vice versa. We have shown that the slow decaying red S-band PL can be transformed into the fast decaying yellow (F-band-like) emission when oxide-capped Si-NCs (from porous silicon) are re-capped by methyl groups using a photo-chemical reaction in organic solvent [113]. And for a butyl-capped Si-NC (synthesized wet-chemically), we have shown that it can be photo-chemically oxidized under exposure to UVC illumination, leading to a switch from the fast blue F-band to the slow red S-band emission [48]. In order to better understand the microscopic PL origins, we have performed room and low-temperature single-dot measurements. We have identified Si-C phonon replicas in the single-dot PL spectroscopy of the butyl-capped Si-NCs (prepared wet-chemically) with a fast decaying emission [42] and have size separated the sample into batches of smaller and larger Si-NCs that have shown blue and green PL, respectively (Fig. 1.9i), hinting toward an intrinsic PL origin. In methyl-capped Si-NCs (from porous silicon) with fast yellow emission, we have identified Si-Si phonon replicas in the low-temperature single-dot PL spectra, once again suggesting an intrinsic PL origin. However, a detailed low-temperature analysis of the phonon replica structures has revealed that for longer acquisition times, the broader structures in the single-dot PL spectra might not be possible interpret without error [119]. By a direct Drexhage-type experiment in our butyl-capped Si-NCs (prepared wet-chemically), we have confirmed the fast radiative rate of the F-band and estimated the internal quantum efficiency to be almost 50% [32].

In our theoretical work, we have developed a general model for a virtual electronegative capping/environment, simulated using the semi-empirical tight binding method [48, 159], showing a three orders of magnitude enhanced radiative rate for methyl-like surface ligand. For a deeper analysis of the ligand effect, we have analyzed the **k**-space profile of the HOMO and LUMO states by Fourier transform of the molecular orbitals in real space and their projection into the Γ–X direction [48] and later expanded this visualization by employing the method of zone folding for NCs, developed by Hapala et al. [80], which enabled us to plot the whole band structure of the Si-NC [159]. At the same time, we have also investigated, by self-consistent ground state DFT, the role of tensile strain and its possible realization via full surface capping by organic ligands [115], again using the band structure plotting method [80]. At this point, it has became clear that the **k**-space projected molecular orbitals in the most important direction Γ–X (we also investigated separately the Γ–L, but have not seen much added information), is an essential tool for understanding the changes to the radiative rate and/or bandgap energy, induced by the surface chemistry. It enabled us to see the whole extent to which a ligand placed at different surface sites, or at various densities, can influence the whole electronic structure and wavefunctions of the Si-NC system. The complex interplay of the surface and core states became even more apparent in our latest joint DFT project [44], where we have uncoupled the effects of charge transfer and tensile strain and have found that our previous assumptions about charge transfer and tensile strain were too simplistic. In particualr, we have found that beyond the first atom (linker atom), there is no consistent correlation between the electronegativity and enhancement in the radiative rate (see also some of the unpublished results presented in this chapter in Figs. 1.16 and 1.17). However, we have confirmed that the radiative rate in Si-NCs with C-linked organic ligands is enhanced with respect to H-capped Si-NCs. This was suggested in our older tight binding study [48, 159] to occur due to the right match of the C atomic orbital levels with respect to the silicon bandgap, resulting in a beneficial band mixing close to the bandgap. By separating the role of polar ligands and strain, we have found that it is not just simply the amount of tensile strain per se which enhances the radiative rate, but its homogeneity is also a very important factor [44]. In fact, ligands can often induce inhomogeneous tensile strain, especially when full ligand coverage is forced. Thus, even if homogeneous tensile strain is beneficial [115], high coverage of the surface capping is clearly detrimental and leads to lowered radiative rates. Nevertheless, full coverage with organic surface ligands is a situation unlikely to occur in realistic experiments.

The effect of the ligands on the Si-NC optical properties is very difficult to assess experimentally, especially if the emission is fast and located in the blue-green spectral region, due to a complex interplay between many different phenomena and a possible involvement of synthesis by-products, for example, hard-to-detect carbon-based impurities (probably carbon dots). In theoretical calculations, we see that the NC band structure is very sensitive not only to the first linking element, but also the more distant inner structure of the ligand, affecting both the bandgap energy and the radiative rate via the modifications to the real-space and *veck*-space profile of the density of states. This comes as as a surprise, because it has been widely accepted that the internal structure of the organic ligand should not have any impact on the optical properties. Also, the density of states is sensitive to where the ligand is attached (which facet and which bonding site), and with what surface coverage. Moreover, in only a very few isolated cases, we have found that it is really the surface capping which introduces trap states truly belonging to the surface element. In a vast majority of cases, even when such states deep inside the bandgap occur, they are (over 90%) Si (core) related, as evidenced by the projected real-space density of states analysis. Thus, we put forward a message that the $k$-space profile of the density of states together with the projected real-space density of states are essential tools for the analysis of the role of the ligands in an indirect bandgap NC. The importance of the **k**-space density of states projections is evident also from the fact that, in our theoretical calculations, even small H-capped Si-NCs have a persistently indirect bandgap character, evidenced by their slow radiative rates. We could not find a simple "recipe" for the radiative rate enhancement in the terms of charge transfer or simply the type of the ligand, but see a clear beneficial role in partial capping with C-linked ligands of any kind (as opposed to H– or oxide-capped Si-NCs).

The main message of our work can be summarized as follows. In general, we have found that the emission from Si-NCs cannot be simply discussed in terms of intrinsic (Si core) and extrinsic (such as surface sites related to the capping agent) channels, but rather it is a complex, indivisible interplay of both. This fact can be best evidenced using the real-space projected density of states. Also, we have found a profound and very complex influence of the ligands (not only the first linker atom, but also the farther structures) onto the density of states of the Si-NC-ligand system projected in the **k**-space. Based on our findings, we suggest that the $k$-space profile of the density of states together with the real-space projected density of states become essential tools for the analysis of the role of the ligands in an indirect bandgap NC, such as silicon, or germanium and diamond. Despite the uncertainties in the theoretical explanation of the emission rates in Si-NCs, our experimental work proves that the fast-emitting Si-NCs are a system stable under standard conditions, in contrast to just a theoretically calculated possibility of an arrangement of atoms. The experimental realization of the discussed fast-emitting system is a strong motivation for further theoretical research, which can bring deeper understanding of the system and thus provide further feedback to technology, potentially leading to efficient and non-toxic quantum-dot-based light sources.

**Acknowledgements** The authors thank several colleagues, whose contribution to this work has been essential. Firstly, we must acknowledge the pioneering approach of Dr. P. Hapala for plotting the NC band structures, which has been used throughout this chapter, as well as the important questions he asks to get to the bottom of things. Also, we warmly thank Dr. A. Poddubny for his critical contribution to the tight binding simulations presented here. Last but not least, we are very grateful to our experimental colleagues Prof. I. Pelant and Prof. J. Valenta, who have supervised most of our experimental projects in the past. In general, we truly appreciate the contribution of all our co-authors of the articles featured in this chapter, since without them we would have not achieved these results.
Authors acknowledge FOM Projectruimte No. 15PR3230 (KD) and MacGillavry Fellowship from University of Amsterdam (KD), grant No. 18-05552S (KK) and support by the Operational Programme Research, Development and Education (Project No. SOLID21 CZ.02.1.01/0.0/0.0/16_019/0000760) (KK).

## References

1. Silicon photonics as discussed at the web pages of intel (https://www.intel.com/content/www/us/en/architecture-and-technology/silicon-photonics/silicon-photonics-overview.html). Retrieved 21 Sept 2020
2. L.C. Allen, Electronegativity is the average one-electron energy of the valence-shell electrons in ground-state free atoms. J. Am. Chem. Soc. **111**, 9003–9014 (1989)
3. N.H. Alsharif, C.E. Berger, S.S. Varanasi, Y. Chao, B.R. Horrocks, H.K. Datta, Alkylcapped silicon nanocrystals lack cytotoxicity and have enhanced intracellular accumulation in malignant cells via cholesterol-dependent endocytosis. Small **5**, 221–228 (2009)
4. J.M. An, A. Franceschetti, S.V. Dudiy, A. Zunger, The peculiar electronic structure of pbse quantum dots. Nano Lett. **6**, 2728–2735 (2006)
5. A.V. Andrianov, D.I. Kovalev, V.B. Shuman, I.D. Yaroshetskir, Short-lived green band and time evolution of the photoluminescence spectrum of porous silicon. JETP Lett. **56**, 536 (1992)
6. P.O. Anikeeva, J.E. Halpert, M.G. Bawendi, V. Bulović, Quantum dot light-emitting devices with electroluminescence tunable over the entire visible spectrum. Nano Lett. **9**, 2532–2536 (2009)
7. M.A. Basanta, Y.J. Dappe, P. Jelinek, J. Ortega, Optimized atomic-like orbitals for first-principles tight-binding molecular dynamics. Comp. Mat. Sci. **39**, 759–766 (2006)

8. V.A. Belyakov, V.A. Burdov, R. Lockwood, A. Meldrum, Silicon nanocrystals: fundamental theory and implications for stimulated emission. Adv. Opt. Technol., pp.279502 (2008). https://doi.org/10.1155/2008/279502
9. S. Bhattacharjee, I. Rietjens, M. Singh, T.M. Atkins, T.K. Purkait, Z. Xu, S. Regli, A. Shukaliak, R.J. Clark, B.S. Mitchell, G.M. Alink, A.T. Marcelis, M.J. Fink, J.G. Veinot, S.M. Kauzlarich, H. Zuilhof, Cytotoxicity of surface-functionalized silicon and germanium nanoparticles: the dominant role of surface charges. Nanoscale **5**, 4870–4883 (2013)
10. W.D.A.M. de Boer, D. Timmerman, K. Dohnalova, I.N. Yassievich, H. Zhang, W.J. Buma, T. Gregorkiewicz, Red spectral shift and enhanced quantum efficiency in phonon-free photoluminescence from silicon nanocrystals. Nature Nanotech. **5**(12), 878–884 (2010)
11. S.L. Brown, J.B. Miller, R.J. Anthony, U.R. Kortshagen, A. Kryjevski, E.K. Hobbie, Abrupt size partitioning of multimodal photoluminescence relaxation in monodisperse silicon nanocrystals. ACS Nano **11**(2), 1597–1603 (2017). https://doi.org/10.1021/acsnano.6b07285
12. B. Bruhn, B.J. Brenny, S. Dekker, I. Doğan, P. Schall, K. Dohnalová, Multi-chromatic silicon nanocrystals. Light Sci. Appl. **6**, e17007 (2017). https://doi.org/10.1038/lsa.2017.7
13. B. Bruhn, J. Valenta, I. Sychugov, K. Mitsuishi, J. Linnros, Transition from silicon nanowires to isolated quantum dots: Optical and structural evolution. Phys. Rev. B **87**, 045404 (2013). https://doi.org/10.1103/PhysRevB.87.045404
14. L.E. Brus, Electron-electron and electron-hole interactions in small semiconductor crystallites: The size dependence of the lowest excited electronic state. J. Chem. Phys. **80**(9), 4403–4409 (1984)
15. C. Bulutay, Interband, intraband, and excited-state direct photon absorption of silicon and germanium nanocrystals embedded in a wide band-gap lattice. Phys. Rev. B **76**, 205321 (2007)
16. J.M. Buriak, Illuminating silicon surface hydrosilylation: An unexpected plurality of mechanisms. Chem. Mater. **26**(1), 763–772 (2014). https://doi.org/10.1021/cm402120f
17. L. Canham, Introductory lecture: origins and applications of efficient visible photoluminescence from silicon-based nanostructures. Faraday Discuss. **222**, 10–81 (2020). https://doi.org/10.1039/D0FD00018C
18. L. Canham, A. Loni, P. Calcott, A. Simons, C. Reeves, M. Houlton, J. Newey, K. Nash, T. Cox, On the origin of blue luminescence arising from atmospheric impregnation of oxidized porous silicon. Thin Solid Films **276**(1), 112–115 (1996). https://doi.org/10.1016/0040-6090(95)08072-4. URL http://www.sciencedirect.com/science/article/pii/0040609095080724. Papers presented at the European Materials Research Society 1995 Spring Conference, Symposium I: Porous Silicon: Material, Technology and Devices
19. L.T. Canham, Silicon quantum wire array fabrication by electrochemical and chemical dissolution of wafers. Appl. Phys. Lett. **57**(10), 1046–1048 (1990). https://doi.org/10.1063/1.103561
20. L.T. Canham, A.J. Groszek, Characterization of microporous si by flow calorimetry: Comparison with a hydrophobic $SiO_2$ molecular sieve. J. Appl. Phys. **72**(4), 1558–1565 (1992). https://doi.org/10.1063/1.351725
21. L.T. Canham, M.R. Houlton, W.Y. Leong, C. Pickering, J.M. Keen, Atmospheric impregnation of porous silicon at room temperature. J. Appl. Phys. **70**(1), 422–431 (1991). https://doi.org/10.1063/1.350293
22. G.M. Carroll, R. Limpens, N.R. Neale, Tuning confinement in colloidal silicon nanocrystals with saturated surface ligands. Nano Lett. **18**(5), 3118–3124 (2018). https://doi.org/10.1021/acs.nanolett.8b00680
23. M. Cazzanelli, D. Navarro-Urriós, F. Riboli, N. Daldosso, L. Pavesi, J. Heitmann, L.X. Yi, R. Scholz, M. Zacharias, U. Gösele, Optical gain in monodispersed silicon nanocrystals. J. Appl. Phys. **96**(6), 3164–3171 (2004)
24. H. Chen, Z. Zhen, W. Tang, T. Todd, Y. Chuang, L. Wang, Z. Pan, J. Xie, Label-free luminescent mesoporous silica nanoparticles for imaging and drug delivery. Theranostics **3**, 650–657 (2013). https://doi.org/10.7150/thno.6668

25. X. Chen, X. Pi, D. Yang, Bonding of oxygen at the oxide/nanocrystal interface of oxidized silicon nanocrystals: an ab initio study. J. Phys. Chem. C **114**(19), 8774–8781 (2010). https://doi.org/10.1021/jp100632u
26. X. Cheng, S.B. Lowe, P.J. Reece, J.J. Gooding, Colloidal silicon quantum dots: from preparation to the modification of self-assembled monolayers (sams) for bio-applications. Chem. Soc. Rev. **43**, 2680–2700 (2014). https://doi.org/10.1039/C3CS60353A
27. S. Chiluwal, N. Sapkota, A.M. Rao, R. Podila, Three-dimensional si anodes with fast diffusion, high capacity, high rate capability, and long cycle life. ACS Appl. Mater. Interf. **12**(31), 34763–34770 (2020). https://doi.org/10.1021/acsami.0c05888
28. A.M. Chizhik, A.I. Chizhik, R. Gutbrod, A.J. Meixner, T. Schmidt, J. Sommerfeld, F. Huisken, Imaging and spectroscopy of defect luminescence and electron-phonon coupling in single $SiO_2$ nanoparticles. Nano Lett. **9**(9), 3239–3244 (2009). https://doi.org/10.1021/nl901509k
29. R.J. Clark, M. Aghajamali, C.M. Gonzalez, L. Hadidi, M.A. Islam, M. Javadi, M.H. Mobarok, T.K. Purkait, C.J.T. Robidillo, R. Sinelnikov, A.N. Thiessen, J. Washington, H. Yu, J.G.C. Veinot, From hydrogen silsesquioxane to functionalized silicon nanocrystals. Chem. Mater. **29**(1), 80–89 (2017). https://doi.org/10.1021/acs.chemmater.6b02667
30. P.R. Coxon, Q. Wang, Y. Chao, An abrupt switch between the two photoluminescence bands within alkylated silicon nanocrystals. J. Phys. D: Appl. Phys. **44**(49), 495301 (2011)
31. CP2K-Foundation: Open source package for molecular dynamics simulations cp2k, retrieved 27 May 2020. URL https://www.cp2k.org
32. B. van Dam, C.I. Osorio, M.A. Hink, R. Muller, A.F. Koenderink, K. Dohnalova, High internal emission efficiency of silicon nanoparticles emitting in the visible range. ACS Photonics **5**(6), 2129–2136 (2018). https://doi.org/10.1021/acsphotonics.7b01624
33. M. Dasog, J. Kehrle, B. Rieger, J.G.C. Veinot, Silicon nanocrystals and silicon-polymer hybrids: synthesis, surface engineering, and applications. Angew. Chem. Int. Ed. **55**(7), 2322–2339 (2016). https://doi.org/10.1002/anie.201506065
34. M. Dasog, Z. Yang, S. Regli, T.M. Atkins, A. Faramus, M.P. Singh, E. Muthuswamy, S.M. Kauzlarich, R.D. Tilley, J.G.C. Veinot, Chemical insight into the origin of red and blue photoluminescence arising from freestanding silicon nanocrystals. ACS Nano **7**(3), 2676–2685 (2013). https://doi.org/10.1021/nn4000644
35. J.Z. Ddungu, S. Silvestrini, A. Tassoni, L. De Cola, Shedding light on the aqueous synthesis of silicon nanoparticles by reduction of silanes with citrates. Faraday Discuss. **222**, 350–361 (2020). https://doi.org/10.1039/C9FD00127A
36. E.M. De Jong, H. Rutjes, J. Valenta, M.T. Trinh, A.N. Poddubny, I.N. Yassievich, A. Capretti, T. Gregorkiewicz, Thermally stimulated exciton emission in Si nanocrystals. Light Sci. Appl. **7**(1), 17133–17137 (2018). https://doi.org/10.1038/lsa.2017.133
37. W.J.I. DeBenedetti, S.K. Chiu, C.M. Radlinger, R.J. Ellison, B.A. Manhat, J.Z. Zhang, J. Shi, A.M. Goforth, Conversion from red to blue photoluminescence in alcohol dispersions of alkyl-capped silicon nanoparticles: Insight into the origins of visible photoluminescence in colloidal nanocrystalline silicon. J. Phys. Chem. C **119**(17), 9595–9608 (2015). https://doi.org/10.1021/acs.jpcc.5b01137
38. C. Delerue, G. Allan, M. Lannoo, Theoretical aspects of the luminescence of porous silicon. Phys. Rev. B **48**, 11024–11036 (1993). https://doi.org/10.1103/PhysRevB.48.11024
39. C. Delerue, G. Allan, M. Lannoo, Theory of silicon nanocrystals, in *Towards the First Silicon Laser*, vol. 93, NATO Science Series (Series II: Mathematics, Physics and Chemistry), ed. by L. Pavesi, S. Gaponenko, L. Dal Negro (Springer, Dordrecht, 2003), pp. 243–260. https://doi.org/10.1007/978-94-010-0149-6_22
40. C. Delerue, G. Allan, C. Reynaud, O. Guillois, G. Ledoux, F. Huisken, Multiexponential photoluminescence decay in indirect-gap semiconductor nanocrystals. Phys. Rev. B **73**, 235318 (2006). https://link.aps.org/doi/10.1103/PhysRevB.73.235318
41. K. Dohnalová, A. Fučíková, C.P. Umesh, J. Humpolíčková, J.M.J. Paulusse, J. Valenta, H. Zuilhof, M. Hof, T. Gregorkiewicz, Microscopic origin of the fast blue-green luminescence of chemically synthesized non-oxidized silicon quantum dots. Small **8**, 3185–3191 (2012)

42. K. Dohnalová, A. Fučíková, C.P. Umesh, J. Humpolíčková, J.M.J. Paulusse, J. Valenta, H. Zuilhof, M. Hof, T. Gregorkiewicz, Microscopic origin of the fast blue-green luminescence of chemically synthesized non-oxidized silicon quantum dots. Small **8**(20), 3185–3191 (2012)
43. K. Dohnalová, T. Gregorkiewicz, K. Kůsová, Silicon quantum dots: surface matters. J Phys.: Condens. Matter **26**, 173201 (2014)
44. K. Dohnalová, P. Hapala, K. Kůsová, I. Infante, Electronic structure engineering achieved via organic ligands in silicon nanocrystals. Chem. Mater. **32**(15), 6326–6337 (2020)
45. K. Dohnalová, K. Kůsová, I. Pelant, Time-resolved photoluminescence spectroscopy of the initial oxidation stage of small silicon nanocrystals. Appl. Phys. Lett. **94**(21), 211903 (2009)
46. K. Dohnalová, L. Ondič, K. Kůsová, I. Pelant, J.L. Rehspringer, R.R. Mafouana, R.R.: White-emitting oxidized silicon nanocrystals: Discontinuity in spectral development with reducing size. J. Appl. Phys. **107**(5), 053102 (2010). https://doi.org/10.1063/1.3289719
47. K. Dohnalová, I. Pelant, K. Kůsová, P. Gilliot, M. Gallart, O. Crégut, J.L. Rehspringer, B. Hönerlage, T. Ostatnický, S. Bakardjieva, Closely packed luminescent silicon nanocrystals in a distributed-feedback laser cavity. New J. Phys. **10**, 063014 (2008). https://doi.org/10.1088/1367-2630/10/6/063014
48. K. Dohnalová, A.N. Poddubny, A.A. Prokofiev, W.D.A.M. de Boer, C.P. Umesh, J.M.J. Paulusse, H. Zuilhof, T. Gregorkiewicz, Surface brightens up Si quantum dots: direct bandgap-like size-tunable emission. Light Sci. Appl. **2**(1), e47 (2013). https://doi.org/10.1038/lsa.2013.3
49. K. Dohnalová, K. Žídek, L. Ondič, K. Kůsová, O. Cibulka, I. Pelant, Optical gain at the F-band of oxidized silicon nanocrystals. J. Phys. D: Appl. Phys. **42**, 135102 (2009). https://doi.org/10.1088/0022-3727/42/13/135102
50. A.F. van Driel, G. Allan, C. Delerue, P. Lodahl, W.L. Vos, D. Vanmaekelbergh, Frequency-dependent spontaneous emission rate from cdse and cdte nanocrystals: Influence of dark states. Phys. Rev. Lett. **95**, 236804 (2005). https://doi.org/10.1103/PhysRevLett.95.236804
51. Editorial: Nature Photon. **4**, 491–578 (2010)
52. M. Ehbrecht, B. Kohn, F. Huisken, M.A. Laguna, V. Paillard, Photoluminescence and resonant raman spectra of silicon films produced by size-selected cluster beam deposition. Phys. Rev. B **56**(11), 6958–6964 (1997)
53. D. English, L.P.Z. Yu, P. Barbara, B. Korgel, Size tunable visible luminescence from individual organic monolayer stabilized silicon nanocrystal quantum dots. Nano Lett. **2**, 681–685 (2002)
54. D.S. English, L.E. Pell, Z. Yu, P.F. Barbara, B.A. Korgel, Size tunable visible luminescence from individual organic monolayer stabilized silicon nanocrystal quantum dots. Nano Lett. **2**(7), 681–685 (2002)
55. T. Erdem, H.V. Demir, Semiconductor nanocrystals as rare-earth alternatives. Nat. Photonics **5**, 126 (2011)
56. F. Erogbogbo, K.T. Yong, I. Roy, R. Hu, W.C. Law, W. Zhao, H. Ding, F. Wu, R. Kumar, M.T. Swihart, P.N. Prasad, In vivo targeted cancer imaging, sentinel lymph node mapping and multi-channel imaging with biocompatible silicon nanocrystals. ACS Nano **5**(1), 413–423 (2011). https://doi.org/10.1021/nn1018945
57. J.B. Essner, J.A. Kist, L. Polo-Parada, G.A. Baker, Artifacts and errors associated with the ubiquitous presence of fluorescent impurities in carbon nanodots. Chem. Mater. **30**(6), 1878–1887 (2018). https://doi.org/10.1021/acs.chemmater.7b04446
58. J. Ewles, Luminescence of silica. Nature **165**, 812–813 (1950). https://doi.org/10.1038/165812a0
59. J. Ewles, R.F. Youell, Luminescence effects associated with the production of silicon monoxide and with oxygen deficit in silica. Trans. Faraday Soc. **47**, 1060–1064 (1951). https://doi.org/10.1039/TF9514701060
60. M.J. Fernée, P. Tamarat, B. Lounis, Spectroscopy of single nanocrystals. Chem. Soc. Rev. **43**(4), 1311–37 (2014). http://www.ncbi.nlm.nih.gov/pubmed/24121992
61. M. Friesen, S.N. Coppersmith, Theory of valley-orbit coupling in a si/sige quantum dot. Phys. Rev. B **81**, 115324 (2010)

62. Y. Fu, M. Willander, E.L. Ivchenko, A.A. Kiselev, Valley mixing in gaas/alas multilayer structures in the effective-mass method. Phys. Rev. B **47**, 13498–13507 (1993)
63. P. Galář, A. Fučiková, K. Newell, T. Popelář, I. Matulková, J. Valenta, V. Scholtz, K. Kůsová, Non-thermal pulsed plasma activated water: easy and cheap way for efficient surface modification of semiconductor nanoparticles. Green Chem. Under review
64. P. Galář, T. Popelář, J. Khun, I. Matulková, I. Němec, K. Newell, A. Michalcová, V. Scholtz, K. Kůsová, The red and blue luminescence in silicon nanocrystals with oxidized, nitrogen-containing shell. Faraday Discuss. **222**, 240–257 (2020). https://doi.org/10.1039/C9FD00092E
65. B. Garrido Fernandez, M. López, C. García, A. Pérez-Rodríguez, J.R. Morante, C. Bonafos, M. Carrada, A. Claverie, Influence of average size and interface passivation on the spectral emission of si nanocrystals embedded in $SiO_2$. J. Appl. Phys. **91**(2), 798–807 (2002). https://doi.org/10.1063/1.1423768
66. B. Gelloz, R. Mentek, N. Koshida, Specific blue light emission from nanocrystalline porous si treated by high-pressure water vapor annealing. Jpn. J. Appl. Phys. **48**(4), 04C119 (2009). https://doi.org/10.1143/Fjjap.48.04c119
67. B. Ghosh, N. Shirahata, Colloidal silicon quantum dots: synthesis and luminescence tuning from the near-UV to the near-IR range. Sci. Tech. Adv. Mater. **15**(1), 014207 (2014). https://doi.org/10.1088/1468-6996/15/1/014207
68. B. Ghosh, M. Takeguchi, J. Nakamura, Y. Nemoto, T. Hamaoka, S. Chandra, N. Shirahata, Origin of the photoluminescence quantum yields enhanced by alkane-termination of freestanding silicon nanocrystals: Temperature-dependence of optical properties. Sci. Rep. **6**, 36951 (2016). https://doi.org/10.1038/srep36951
69. Y. Glinka, A. Zyubin, A. Mebel, S. Lin, L. Hwang, Y. Chen, Photoluminescence from mesoporous silica akin to that from nanoscale silicon: the nature of light-emitters. Chem. Phys. Lett. **358**(3), 180–186 (2002). https://doi.org/10.1016/S0009-2614(02)00400-1. http://www.sciencedirect.com/science/article/pii/S0009261402004001
70. S. Goedecker, M. Teter, J. Hutter, Separable dual-space Gaussian pseudopotentials. Phys. Rev. B **54**, 1703–1710 (1996). https://doi.org/10.1103/PhysRevB.54.1703
71. D. Gonzalez-Carter, X. Liu, T.A. Tockary, A. Dirisala, K. Toh, Y. Anraku, K. Kataoka, Targeting nanoparticles to the brain by exploiting the blood-brain barrier impermeability to selectively label the brain endothelium. Proc. Natl. Acad. Sci. U.S.A. **117**(32), 19141–19150 (2020). https://doi.org/10.1073/pnas.2002016117. https://www.pnas.org/content/117/32/19141
72. M. Greben, P. Khoroshyy, I. Sychugov, J. Valenta, Non-exponential decay kinetics: correct assessment and description illustrated by slow luminescence of si nanostructures. Appl. Spectr. Rev. **54**(9), 758–801 (2019). https://doi.org/10.1080/05704928.2018.1517263
73. M. Greben, J. Valenta, Power-dependent photoluminescence decay kinetics of silicon nanocrystals under continuous and pulsed excitation. Faraday Discuss. **222**, 274–293 (2020). https://doi.org/10.1039/C9FD00100J
74. R. Guerra, F. Cigarini, S. Ossicini, Optical absorption and emission of silicon nanocrystals: From single to collective response. J. Appl. Phys. **113**(14), 143505 (2013). https://doi.org/10.1063/1.4799394
75. R. Guerra, S. Ossicini, High luminescence in small $Si/SiO_2$ nanocrystals: A theoretical study. Phys. Rev. B **81**(24), 245307 (2010). https://doi.org/10.1103/PhysRevB.81.245307
76. A. Gupta, M.T. Swihart, H. Wiggers, Luminescent colloidal dispersion of silicon quantum dots from microwave plasma synthesis: Exploring the photoluminescence behavior across the visible spectrum. Adv. Funct. Mater. **19**(5), 696–703 (2009). https://doi.org/10.1002/adfm.200801548
77. A. Gupta, H. Wiggers, Freestanding silicon quantum dots: origin of red and blue luminescence. Nanotechnology **22**(5), 055707 (2010). https://doi.org/10.1088/0957-4484/22/5/055707
78. O.B. Gusev, A.N. Poddubny, A.A. Prokofiev, I.N. Yassievich, Light emission from silicon nanocrystals. Semiconductors **47**(2), 183–202 (2013). https://doi.org/10.1134/S1063782613020103

79. D.C. Hannah, J. Yang, P. Podsiadlo, M.K.Y. Chan, A. Demortiére, D.J. Gosztola, V.B. Prakapenka, G.C. Schatz, U. Kortshagen, R.D. Schaller, On the origin of photoluminescence in silicon nanocrystals: pressure-dependent structural and optical studies. Nano Lett. **12**(8), 4200–4205 (2012). https://doi.org/10.1021/nl301787g
80. P. Hapala, K. Kůsová, I. Pelant, P. Jelínek, Theoretical analysis of electronic band structure of 2- to 3-nm si nanocrystals. Phys. Rev. B **87**(19), 195420 (2013). https://doi.org/10.1103/PhysRevB.87.195420
81. A.M. Hartel, S. Gutsch, D. Hiller, M. Zacharias, Fundamental temperature-dependent properties of the si nanocrystal band gap. Phys. Rev. B **85**(16), 165306 (2012). https://doi.org/10.1103/PhysRevB.85.165306
82. Y. He, Y. Zhong, F. Peng, X. Wei, Y. Su, Y. Lu, S. Su, W. Gu, L. Liao, S.T. Lee, One-pot microwave synthesis of water-dispersible, ultraphoto- and ph-stable, and highly fluorescent silicon quantum dots. J. Am. Chem. Soc. **133**, 14192–14195 (2011)
83. J.L. Heinrich, C.L. Curtis, G.M. Credo, M.J. Sailor, K.L. Kavanagh, Luminescent colloidal silicon suspensions from porous silicon. Science **255**(5040), 66–68 (1992). https://doi.org/10.1126/science.255.5040.66. https://science.sciencemag.org/content/255/5040/66
84. E.J. Henderson, J.A. Kelly, J.G.C. Veinot,Influence of $HSio_{1.5}$ sol-gel polymer structure and composition on the size and luminescent properties of silicon nanocrystals. Chem. Mater. **21**, 5426–5434 (2009)
85. C.M. Hessel, D. Reid, M.G. Panthani, M.R. Rasch, B.W. Goodfellow, J. Wei, H. Fujii, V. Akhavan, B.A. Korgel, Synthesis of ligand-stabilized silicon nanocrystals with size-dependent photoluminescence spanning visible to near-infrared wavelengths. Chem. Mater. **24**(2), 393–401 (2012). https://doi.org/10.1021/cm2032866
86. C.M. Hessel, J. Wei, D. Reid, H. Fujii, M.C. Downer, B.A. Korgel, Raman spectroscopy of oxide-embedded and ligand-stabilized silicon nanocrystals. J. Phys. Chem. Lett. **3**(9), 1089–1093 (2012). https://doi.org/10.1021/jz300309n
87. H. Hoffmann, Violin plot (https://www.mathworks.com/matlabcentral/fileexchange/45134-violin-plot), Matlab Central File Exchange. Retrieved May 27, 2020
88. F. Huisken, G. Ledoux, O. Guillois, C. Reynaud, Light-emitting silicon nanocrystals from laser pyrolysis. Adv. Mater. **14**(24), 1861–1865 (2002). https://doi.org/10.1002/adma.200290021
89. M.S. Hybertsen, Absorption and emission of light in nanoscale silicon structures. Phys. Rev. Lett. **72**(10), 1514–1517 (1994)
90. T. Ito, T. Ohta, A. Hiraki, Light emission from microcrystalline Si confined in $SiO_2$ matrix through partial oxidation of porous silicon. Jpn. J. Appl. Phys. **31**, L1 (1992)
91. I. Vasiliev, J.R. Chelikowsky, R.M. Martin, Surface oxidation effects on the optical properties of silicon nanocrystals. Phys. Rev. B **65**, 121302 (2002)
92. M. Jakob, M. Javadi, J.G.C. Veinot, A. Meldrum, A. Kartouzian, U. Heiz, Ensemble effects in the temperature-dependent photoluminescence of silicon nanocrystals. Chem. Eur. J. **25**(12), 3061–3067 (2019). https://doi.org/10.1002/chem.201804986. https://chemistry-europe.onlinelibrary.wiley.com/doi/abs/10.1002/chem.201804986
93. J.M. Jancu, R. Scholz, F. Beltram, F. Bassani, Empirical *spds*∗ tight-binding calculation for cubic semiconductors: General method and material parameters. Phys. Rev. B **57**, 6493–6507 (1998)
94. J.M. Jancu, R. Scholz, G.C.L. Rocca, E.A. de Andrada e Silva, P. Voisin, Giant spin splittings in gasb/alsb *L*-valley quantum wells. Phys. Rev. B **70**, 121306 (2004)
95. P. Jelínek, H. Wang, J. Lewis, O.F. Sankey, J. Ortega, Multicenter approach to the exchange-correlation interactions in ab initio tight-binding methods. Phys. Rev. B **71**, 235101 (2005)
96. X. Jia, P. Zhang, Z. Lin, R. Anthony, U. Kortshagen, S. Huang, B. Puthen-Veettil, G. Conibeer, I. Perez-Wurfl, Accurate determination of the size distribution of si nanocrystals from pl spectra. RSC Adv. **5**, 55119–55125 (2015). https://doi.org/10.1039/C5RA02805A
97. D. Jurbergs, E. Rogojina, L. Mangolini, U. Kortshagen, Silicon nanocrystals with ensemble quantum yields exceeding 60 %. Appl. Phys. Lett. **88**(23), 233116–233118 (2006). https://doi.org/10.1063/1.2210788

98. M.H. Kafshgari, N.H. Voelcker, F.J. Harding, Applications of zero-valent silicon nanostructures in biomedicine. Nanomedicine **10**(16), 2553–2571 (2015). https://doi.org/10.2217/nnm.15.91
99. Y. Kanemitsu, T. Futagi, T. Matsumoto, H. Mimura, Origin of the blue and red photoluminescence from oxidized porous silicon. Phys. Rev. B **49**(20), 14732–14735 (1994). https://doi.org/10.1103/PhysRevB.49.14732
100. A.K. Katiyar, K.Y. Thai, W.S. Yun, J. Lee, J.H. Ahn, Breaking the absorption limit of Si toward SWIR wavelength range via strain engineering. Sci. Adv. **6**(31), eabb0576 (2020). https://doi.org/10.1126/sciadv.abb0576. https://advances.sciencemag.org/content/6/31/eabb0576
101. M.S. Kim, *Flexible Si Nanomembrane Sensor for Human–Machine Interface* (Wiley & Sons, Ltd, 2016), pp. 169–202. 10.1002/9783527691005.ch7. URL https://onlinelibrary.wiley.com/doi/abs/10.1002/9783527691005.ch7
102. E. Klimcšová, K. Kůsová, J. Vacík, V. Holý, I. Pelant, I.: Tuning luminescence properties of silicon nanocrystals by lithium doping. J. Appl. Phys. **112**(6), 064322 (2012). https://doi.org/10.1063/1.4754518
103. V. Kocevski, Temperature dependence of radiative lifetimes, optical and electronic properties of silicon nanocrystals capped with various organic ligands. J. Chem. Phys. **149**(5), 054301 (2018). https://doi.org/10.1063/1.5039281
104. K. Kolasinski, Silicon surface photochemistry, in *Encyclopedia of Interfacial Chemistry*, ed. by K. Wandelt (Elsevier, Oxford, 2018), pp. 611 – 620. https://doi.org/10.1016/B978-0-12-409547-2.14046-6
105. D. König, Number series of atoms, interatomic bonds and interface bonds defining zinc-blende nanocrystals as function of size, shape and surface orientation: analytic tools to interpret solid state spectroscopy data. AIP Adv. **6**(8), 085306 (2016). https://doi.org/10.1063/1.4960994. http://scitation.aip.org/content/aip/journal/adva/6/8/10.1063/1.4960994
106. D. König, J. Rudd, M.A. Green, G. Conibeer, Role of the interface for the electronic structure of Si quantum dots. Phys. Rev. B **78**(3), 035339 (2008)
107. U.R. Kortshagen, R.M. Sankaran, R.N. Pereira, S.L. Girshick, J.J. Wu, E.S. Aydil, Nonthermal plasma synthesis of nanocrystals: fundamental principles, materials, and applications. Chem. Rev. **116**(18), 11061–11127 (2016). https://doi.org/10.1021/acs.chemrev.6b00039
108. D. Kovalev, M. Fujii, Silicon nanocrystals: photosensitizers for oxygen molecules. Adv. Mat. **17**(21), 2531–2544 (2005). http://doi.wiley.com/10.1002/adma.200500328
109. D.I. Kovalev, I.D. Yaroshetzkii, T. Muschik, V. Petrova-Koch, F. Koch, Fast and slow visible luminescence bands of oxidized porous si. Appl. Phys. Lett. **64**(2), 214–216 (1994). https://doi.org/10.1063/1.111508
110. K. Kůsová, Silicon nanocrystals: From indirect to direct bandgap. Phys. Stat. solidi (a) **215**(7), 1700718 (2018). https://doi.org/10.1002/pssa.201700718
111. K. Kůsová, Photoluminescence spectroscopy of single semiconductor quantum dots, in *21st Century Nanoscience—A Handbook: Advanced Analytic Methods and Instrumentation (Volume 3)*, ed. by K.D. Sattler (Taylor&Francis Publisher, CRC Press, Boca Raton, Florida, USA, 2020), pp. 21(1)–(25). https://doi.org/10.1201/9780429340420
112. K. Kůsová, O. Cibulka, K. Dohnalová, I. Pelant, A. Fučíková, J. Valenta, JYellow-emitting colloidal suspensions of silicon nanocrystals: Fabrication technology, luminescence performance and application prospects. Physica E **41**(6), 982–985 (2009). https://doi.org/10.1016/j.physe.2008.08.022
113. K. Kůsová, O. Cibulka, K. Dohnalová, I. Pelant, J. Valenta, A. Fučíková, K. Žídek, J. Lang, J. Englich, P. Matějka, P. Štěpánek, S. Bakardjieva, Brightly luminescent organically capped silicon nanocrystals fabricated at room temperature and atmospheric pressure. ACS Nano **4**(8), 4495–4504 (2010). https://doi.org/10.1021/nn1005182
114. K. Kůsová, K. Dohnalová, Organically capped silicon nanocrystals, in *Silicon Nanomaterials Sourcebook: Low-Dimensional Structures, Quantum Dots, and Nanowires*, vol. One, ed. by K.D. Sattler (Taylor&Francis Publisher, CRC Press, Boca Raton, Florida, USA, 2017), pp. 367–397

115. Kůsová, K., Hapala, P., Valenta, J., Jelínek, P., Cibulka, O., Ondič, L., Pelant, I.: Direct bandgap silicon: Tensile-strained silicon nanocrystals. Adv. Mater. Interfaces **1**(2), 1300042 (2014). https://doi.org/10.1002/admi.201300042
116. K. Kůsová, L. Ondič, E. Klimešová, K. Herynková, I. Pelant, S. Daniš, J. Valenta, M. Gallart, M. Ziegler, B. Hönerlage, P. Gilliot, Luminescence of free-standing versus matrix-embedded oxide-passivated silicon nanocrystals: the role of matrix-induced strain. Appl. Phys. Lett. **101**(14), 143101 (2012). https://doi.org/10.1063/1.4756696
117. Kůsová, K., Ondič, L., Pelant, I.: Comment on "ultrafast photoluminescence in quantum-confined silicon nanocrystals arises from an amorphous surface layer". ACS Photonics **2**(3), 454–455 (2015). https://doi.org/10.1021/ph500403j/
118. K. Kůsová, I. Pelant, J. Humpolíčková, M. Hof, Comprehensive description of blinking-dynamics regimes in single direct-band-gap silicon nanocrystals. Phys. Rev. B **93**(3), 035412 (2016). https://doi.org/10.1103/PhysRevB.93.035412
119. K. Kůsová, I. Pelant, J. Valenta, Bright trions in direct-bandgap silicon nanocrystals revealed by low-temperature single-nanocrystal spectroscopy. Light Sci Appl **4**, e336 (2015). https://doi.org/10.1038/lsa.2015.109
120. K. Kůsová, T. Popelář, On the importance of onset times and multiple-wavelength analysis of photoluminescence decays. J. Appl. Phys. **125**(19), 193103 (2019). https://doi.org/10.1063/1.5097065
121. C.W. Lai, Y.H. Hsiao, Y.K. Peng, P.T. Chou, Facile synthesis of highly emissive carbon dots from pyrolysis of glycerol; gram scale production of carbon dots/m$SiO_2$ for cell imaging and drug release. J. Mater. Chem. **22**, 14403–14409 (2012). https://doi.org/10.1039/C2JM32206D
122. G. Ledoux, J. Gong, F. Huisken, O. Guillois, C. Reynaud, Photoluminescence of size-separated silicon nanocrystals: confirmation of quantum confinement. Appl. Phys. Lett. **80**(25), 4834–4836 (2002). https://doi.org/10.1063/1.1485302
123. Q.S. Li, R.Q. Zhang, S.T. Lee, T.A. Niehaus, T. Frauenheim, Optimal surface functionalization of silicon quantum dots. J. Chem. Phys. **128**(24), 244714 (2008). https://doi.org/10.1063/1.2940735
124. X. Li, Y. He, S. Talukdar, M. Swihart, Process for preparing macroscopic quantities of brightly photoluminescent silicon nanoparticles with emission spanning the visible spectrum. Langmuir **19**, 8490–8496 (2003). https://doi.org/10.1021/la034487b
125. J. Liu, F. Erogbogbo, K.T. Yong, L. Ye, J. Liu, R. Hu, H. Chen, Y. Hu, Y. Yang, J. Yang, I. Roy, N.A. Karker, M.T. Swihart, P.N. Prasad, Assessing clinical prospects of silicon quantum dots: studies in mice and monkeys. ACS Nano **7**(8), 7303–7310 (2013)
126. J. Liu, F. Erogbogbo, K.T. Yong, L. Ye, J. Liu, R. Hu, H. Chen, Y. Hu, Y. Yang, J. Yang, I. Roy, N.A. Karker, M.T. Swihart, P.N. Prasad, Assessing clinical prospects of silicon quantum dots: studies in mice and monkeys. ACS Nano **7**(8), 7303–7310 (2013). https://doi.org/10.1021/nn4029234
127. J. Liu, X. Sun, R. Camacho-Aguilera, L.C. Kimerling, J. Michel, Ge-on-Si laser operating at room temperature. Opt. Lett. **35**(5), 679–681 (2010)
128. X.H. Liu, L. Zhong, S. Huang, S.X. Mao, T. Zhu, J.Y. Huang, Size-dependent fracture of silicon nanoparticles during lithiation. ACS Nano **6**(2), 1522–1531 (2012). https://doi.org/10.1021/nn204476h
129. A. Loni, A.J. Simons, P.D.J. Calcott, L.T. Canham, Blue photoluminescence from rapid thermally oxidized porous silicon following storage in ambient air. J. Appl. Phys. **77**(7), 3557–3559 (1995). https://doi.org/10.1063/1.358587
130. A. Loni, A.J. Simons, P.D.J. Calcott, J.P. Newey, T.I. Cox, L.T. Canham, Relationship between storage media and blue photoluminescence for oxidized porous silicon. Appl. Phys. Lett. **71**(1), 107–109 (1997). https://doi.org/10.1063/1.119442
131. P. Lunnemann, F.T. Rabouw, R.J.A. van Dijk-Moes, F. Pietra, D. Vanmaekelbergh, A.F. Koenderink, Calibrating and controlling the quantum efficiency distribution of inhomogeneously broadened quantum rods by using a mirror ball. ACS Nano **7**(7), 5984–5992 (2013). https://doi.org/10.1021/nn401683u

132. Y.R. Luo, *Comprehensive Handbook of Chemical Bond Energies* (CRC Press, Boca Raton, FL, USA, 2007)
133. M. Luppi, S. Ossicini, Ab initio study on oxidized silicon clusters and silicon nanocrystals embedded in $sio_2$: beyond the quantum confinement effect. Phys. Rev. B **71**, 035340 (2005)
134. Y. Ma, X. Chen, X. Pi, D. Yang, Theoretical study of chlorine for silicon nanocrystals. J. Phys. Chem. C **115**(26), 12822–12825 (2011). https://doi.org/10.1021/jp203064m
135. Y. Ma, X. Chen, X. Pi, D. Yang, Theoretical study of chlorine for silicon nanocrystals. J. Phys. Chem. C **115**, 12822–12825 (2011)
136. Y. Ma, X. Pi, D. Yang, Fluorine-passivated silicon nanocrystals: surface chemistry versus quantum confinement. J. Phys. Chem. C **116**(9), 5401–5406 (2012). https://doi.org/10.1021/jp211177d
137. Y. Ma, X. Pi, D. Yang, Fluorine-passivated silicon nanocrystals: surface chemistry versus quantum confinement. J. Phys. Chem. C **116**, 5401–5406 (2012)
138. F. Maier-Flaig, E.J. Henderson, S. Valouch, S. Klinkhammer, C. Kübel, G.A. Ozin, U. Lemmer, Photophysics of organically-capped silicon nanocrystals–a closer look into silicon nanocrystal luminescence using low temperature transient spectroscopy. Chem. Phys. **405**, 175–180 (2012). https://doi.org/10.1016/j.chemphys.2012.07.009
139. F. Maier-Flaig, J. Rinck, M. Stephan, T. Bocksrocker, M. Bruns, C. Kübel, A.K. Powell, G.A. Ozin, U. Lemmer, Multicolor silicon light-emitting diodes (sileds). Nano Lett. **13**, 475–480 (2013)
140. L. Mangolini, Synthesis, properties, and applications of silicon nanocrystals. J. Vac. Sci. Technol. **31**(2), 020801 (2013). https://doi.org/10.1116/1.4794789
141. J. Martin, F. Cichos, F. Huisken, C. von Borczyskowski, Electron-phonon coupling and localization of excitons in single silicon nanocrystals. Nano Lett. **8**(2), 656–660 (2008)
142. M.D. Mason, G.M. Credo, K.D. Weston, S.K. Buratto, Luminescence of individual porous si chromophores. Phys. Rev. Lett. **80**(24), 5405–5408 (1998). https://doi.org/10.1103/PhysRevLett.80.5405
143. R. Mazzaro, F. Romano, P. Ceroni, Long-lived luminescence of silicon nanocrystals: from principles to applications. Phys. Chem. Chem. Phys. **19**, 26507–26526 (2017). https://doi.org/10.1039/C7CP05208A
144. B.F.P. McVey, R.D. Tilley, Solution synthesis, optical properties, and bioimaging applications of silicon nanocrystals. Acc. Chem. Res. **47**(10), 3045–3051 (2014). https://doi.org/10.1021/ar500215v
145. J.B. Miller, N. Dandu, K.A. Velizhanin, R.J. Anthony, U.R. Kortshagen, D.M. Kroll, S. Kilina, E.K. Hobbie, Enhanced luminescent stability through particle interactions in silicon nanocrystal aggregates. ACS Nano **9**(10), 9772–9782 (2015). https://doi.org/10.1021/acsnano.5b02676
146. S. Morozova, M. Alikina, A. Vinogradov, M. Pagliaro, Silicon quantum dots: synthesis, encapsulation, and application in light-emitting diodes. Front. Chem. **8**, 191 (2020). https://doi.org/10.3389/fchem.2020.00191. https://www.frontiersin.org/article/10.3389/fchem.2020.00191
147. G. Morselli, F. Romano, P. Ceroni, Amine functionalised silicon nanocrystals with bright red and long-lived emission. Faraday Discuss. **222**, 108–121 (2020). https://doi.org/10.1039/C9FD00089E
148. M. Nayfeh, N. Barry, J. Therrien, O. Akcakir, E. Gratton, G. Belomoin, Stimulated blue emission in reconstituted films of ultrasmall silicon nanoparticles. Appl. Phys. Lett. **78**(8), 1131–1133 (2001)
149. M.H. Nayfeh, L. Mitas, Silicon nanoparticles: new photonic and electronic material at the transition between solid and molecule, in *Nanosilicon*, ed. by V. Kumar (Elsevier, Amsterdam, The Netherlands, 2008), pp. 1–78. https://doi.org/10.1016/B978-008044528-1.50002-6
150. Z. Ni, S. Zhou, S. Zhao, W. Peng, D. Yang, X. Pi, Silicon nanocrystals: unfading silicon materials for optoelectronics. Mater. Sci **138**, 8–117 (2019). https://doi.org/10.1016/j.mser.2019.06.001. http://www.sciencedirect.com/science/article/pii/S0927796X19300609
151. B.V. Oliinyk, D. Korytko, V. Lysenko, S. Alekseev, Are fluorescent silicon nanoparticles formed in a one-pot aqueous synthesis? Chem. Mater. **31**(18), 7167–7172 (2019). https://doi.org/10.1021/acs.chemmater.9b01067

152. L. Ondič, K. Kůsová, M. Ziegler, L. Fekete, V. Gärtnerová, V. Cháb, V. Holý, O. Cibulka, K. Herynková, M. Gallart, P. Gilliot, B. Hönerlage, I. Pelant, A complex study of the fast blue luminescence of oxidized silicon nanocrystals: the role of the core. Nanoscale **6**(7), 3837–3845 (2014). https://doi.org/10.1039/c3nr06454a
153. G. Pacchioni, L. Skuja, D. Griscom (eds.), *NATO Science Series II*, vol. 2 (Springer Netherlands, 2000). https://doi.org/10.1007/978-94-010-0944-7
154. J.H. Park, L. Gu, G. von Maltzahn, E. Ruoslahti, S.N. Bhatia, M.J. Sailor, Nat. Mater. **8**, 331–336 (2009)
155. J.H. Park, L. Gu, G. von Maltzahn, E. Ruoslahti, S.N. Bhatia, M.J. Sailor, Biodegradable luminescent porous silicon nanoparticles for in vivo applications. Nat. Mat. **8**(4), 331–336 (2009)
156. F. Peng, Z. Cao, X. Ji, B. Chu, Y. Su, Y. He, Silicon nanostructures for cancer diagnosis and therapy. Nanomedicine **10**(13), 2109–2123 (2015). https://doi.org/10.2217/nnm.15.53
157. J.P. Perdew, K. Burke, M. Ernzerhof, Generalized gradient approximation made simple. Phys. Rev. Lett. **77**, 3865–3868 (1996). https://doi.org/10.1103/PhysRevLett.77.3865
158. V.M. Petriev, V.K. Tischenko, A.A. Mikhailovskaya, A.A. Popov, G. Tselikov, I. Zelepukin, S.M. Deyev, A.D. Kaprin, S. Ivanov, V.Y. Timoshenko, P.N. Prasad, I.N. Zavestovskaya, A.V. Kabashin, Nuclear nanomedicine using si nanoparticles as safe and effective carriers of 188re radionuclide for cancer therapy. Sci. Rep. **9**, 2017 (2019). https://www.nature.com/articles/s41598-018-38474-7
159. A.N. Poddubny, K. Dohnalová, Direct band gap silicon quantum dots achieved via electronegative capping. Phys. Rev. B **90**, 245439 (2014). https://doi.org/10.1103/PhysRevB.90.245439
160. A.N. Poddubny, M.O. Nestoklon, S.V. Goupalov, Anomalous suppression of valley splittings in lead salt nanocrystals without inversion center. Phys. Rev. B **86**, 035324 (2012)
161. A.N. Poddubny, A.A. Prokofiev, I.N. Yassievich, Optical transitions and energy relaxation of hot carriers in si nanocrystals. Appl. Phys. Lett. **97**, 231116 (2010)
162. T.A. Pringle, K.I. Hunter, A. Brumberg, K.J. Anderson, J.A. Fagan, S.A. Thomas, R.J. Petersen, M. Sefannaser, Y. Han, S.L. Brown, D.S. Kilin, R.D. Schaller, U.R. Kortshagen, P.R. Boudjouk, E.K. Hobbie, Bright silicon nanocrystals from a liquid precursor: Quasi-direct recombination with high quantum yield. ACS Nano **14**(4), 3858–3867 (2020). https://doi.org/10.1021/acsnano.9b09614
163. F. Priolo, T. Gregorkiewicz, M. Galli, T.F. Krauss, Silicon nanostructures for photonics and photovoltaics. Nat. Nanotechnol. **9**, 19–32 (2014)
164. A.A. Prokofiev, A.S. Moskalenko, I.N. Yassievich, W.D.A.M. de Boer, D. Timmerman, H. Zhang, W.J. Buma, T. Gregorkiewicz, Direct bandgap optical transitions in Si nanocrystals. JETP Lett. **90**(12), 758–762 (2009)
165. S.P. Pujari, H. Driss, F. Bannani, B. van Lagen, H. Zuilhof, One-pot gram-scale synthesis of hydrogen-terminated silicon nanoparticles. Chem. Mater. **30**(18), 6503–6512 (2018). https://doi.org/10.1021/acs.chemmater.8b03113
166. A. Puzder, A.J. Williamson, J.C. Grossman, G. Galli, Surface chemistry of silicon nanoclusters. Phys. Rev. Lett. **88**, 097401 (2002). https://doi.org/10.1103/PhysRevLett.88.097401
167. A. Puzder, A.J. Williamson, J.C. Grossman, G. Galli, Surface control of optical properties in silicon nanoclusters. J. Chem. Phys. **117**(14), 6721–6729 (2002)
168. G.G. Qin, X.S. Liu, S.Y. Ma, J. Lin, G.Q. Yao, X.Y. Lin, K.X. Lin, Photoluminescence mechanism for blue-light-emitting porous silicon. Phys. Rev. B **55**, 12876–12879 (1997). https://doi.org/10.1103/PhysRevB.55.12876
169. E. Ramos, B.M. Monroy, J.C. Alonso, L.E. Sansores, R. Salcedo, A. Martinez, Theoretical study of chlorine for silicon nanocrystals. J. Phys. Chem. C **116**, 3988–3994 (2012)
170. Z. Ranjbar-Navazi, Y. Omidi, M. Eskandani, S. Davaran, Cadmium-free quantum dot-based theranostics. TrAC Trend Anal Chem **118**, 386–400 (2019). https://doi.org/10.1016/j.trac.2019.05.041. http://www.sciencedirect.com/science/article/pii/S0165993619300470
171. F.A. Reboredo, G. Galli, Theory of alkyl-terminated silicon quantum dots. J. Phys. Chem. B **109**(3), 1072–1078 (2005). https://doi.org/10.1021/jp0462254. http://pubs.acs.org/doi/abs/10.1021/jp0462254

172. J.R. Rodriguez Nunez, J.A. Kelly, E.J. Henderson, J.G.C. Veinot, J.R. Rodríguez Núñez, J.A. Kelly, E.J. Henderson, J.G.C. Veinot, Wavelength-controlled etching of silicon nanocrystals. Chem. Mater. **24**(2), 346–352 (2012). https://doi.org/10.1021/cm203085f. http://pubs.acs.org/doi/abs/10.1021/cm203085f
173. M. Rosso-Vasic, E. Spruijt, B. van Lagen, L. de Cola, H. Zuilhof, Alkyl-functionalized oxide-free silicon nanoparticles: synthesis and optical properties. Small **4**, 1835–1841 (2008)
174. F. Sangghaleh, B. Bruhn, T. Schmidt, J. Linnros, Exciton lifetime measurements on single silicon quantum dots. Nanotechnology **24**(22), 225204 (2013). https://doi.org/10.1088/0957-4484/24/22/225204. http://www.ncbi.nlm.nih.gov/pubmed/23644627
175. F. Sangghaleh, I. Sychugov, Z. Yang, J.G.C. Veinot, J. Linnros, Near-unity internal quantum efficiency of luminescent silicon nanocrystals with ligand passivation. ACS Nano **9**(7), 7097–7104 (2015). https://doi.org/10.1021/acsnano.5b01717
176. G. Scamarcio, M. Lugará, D. Manno, Size-dependent lattice contraction in $cds_{1-}xsex$ nanocrystals embedded in glass observed by Raman scattering. Phys. Rev. B **45**(23), 13792–13795 (1992). https://doi.org/10.1103/PhysRevB.45.13792
177. C. Schiattarella, R. Moretta, T. Defforge, G. Gautier, B. Della Ventura, M. Terracciano, C. Tortiglione, F. Fardella, P. Maddalena, L. De Stefano, R. Velotta, I. Rea, Time-gated luminescence imaging of positively charged poly-l-lysine-coated highly microporous silicon nanoparticles in living hydra polyp. J. Biophotonics **n/a**(n/a), e202000272 (2020). https://doi.org/10.1002/jbio.202000272. https://onlinelibrary.wiley.com/doi/abs/10.1002/jbio.202000272
178. T. Schmidt, A.I. Chizhik, A.M. Chizhik, K. Potrick, A.J. Meixner, F. Huisken, Radiative exciton recombination and defect luminescence observed in single silicon nanocrystals. Phys. Rev. B **86**, 125302 (2012)
179. N. Shirahata, Colloidal si nanocrystals: a controlled organic-inorganic interface and its implications of color-tuning and chemical design toward sophisticated architectures. Phys. Chem. Chem. Phys. **13**(16), 7284–7294 (2011)
180. N. Shirahata, J. Nakamura, J.I. Inoue, B. Ghosh, K. Nemoto, Y. Nemoto, M. Takeguchi, Y. Masuda, M. Tanaka, G.A. Ozin, Emerging atomic energy levels in zero-dimensional silicon quantum dots. Nano Lett. **20**(3), 1491–1498 (2020). https://doi.org/10.1021/acs.nanolett.9b03157
181. I.L. Spain, D.J. Dunstan, The technology of diamond anvil high-pressure cells: II. Operation and use. J. Phys. E: Sci. Instr. **22**(11), 923–933 (1989). https://doi.org/10.1088/0022-3735/22/11/005. http://iopscience.iop.org/0022-3735/22/11/005stacks.iop.org/0022-3735/22/i=11/a=005?key=crossref.effd33143aaa03ce650e3343fad74c27
182. H. Sugimoto, M. Fujii, K. Imakita, Synthesis of boron and phosphorus codoped all-inorganic colloidal silicon nanocrystals from hydrogen silsesquioxane. Nanoscale **6**, 12354–12359 (2014). https://doi.org/10.1039/C4NR03857F
183. I. Sychugov, A. Fucikova, F. Pevere, Z. Yang, J.G.C. Veinot, J. Linnros, Ultranarrow luminescence linewidth of silicon nanocrystals and influence of matrix. ACS Photonics **1**(10), 998–1005 (2014). https://doi.org/10.1021/ph500221z
184. I. Sychugov, R. Juhasz, J. Valenta, J. Linnros, Narrow luminescence linewidth of a silicon quantum dot. Phys. Rev. Lett. **94**, 087405 (2005). https://doi.org/10.1103/PhysRevLett.94.087405
185. I. Sychugov, J. Valenta, J. Linnros, Probing silicon quantum dots by single-dot techniques. Nanotechnology **28**(7), 072002 (2017). http://stacks.iop.org/0957-4484/28/i=7/a=072002
186. I. Sychugov, J. Valenta, K. Mitsuishi, M. Fujii, J. Linnros, Photoluminescence measurements of zero-phonon optical transitions in silicon nanocrystals. Phys. Rev. B **84**(12), 125326 (2011). https://doi.org/10.1103/PhysRevB.84.125326
187. H. Takagi, H. Ogawa, Y. Yamazaki, A. lshizaki, T. Nakagiri, Quantum size effects on photoluminescence in ultrafine Si particles. Appl. Phys. Lett. **56**(24), 2379 (2000)
188. S. Takeoka, M. Fujii, S. Hayashi, Size-dependent photoluminescence from surface-oxidized si nanocrystals in a weak confinement regime. Phys. Rev. B **62**(24), 16820–16825 (2000). https://doi.org/10.1103/PhysRevB.62.16820

189. A.N. Thiessen, M. Ha, R.W. Hooper, H. Yu, A.O. Oliynyk, J.G. Veinot, V.K. Michaelis, Silicon nanoparticles: are they crystalline from the core to the surface? Chem. Mater. **31**(3), 678–688 (2019). https://doi.org/10.1021/acs.chemmater.8b03074
190. A.N. Thiessen, L. Zhang, A.O. Oliynyk, H. Yu, K.M. O'Connor, A. Meldrum, J.G.C. Veinot, A tale of seemingly "Identical" silicon quantum dot families: structural insight into silicon quantum dot photoluminescence. Chemistry of Materials **32**(16), 6838–6846 (2020). https://doi.org/10.1021/acs.chemmater.0c00650
191. V. Timoshenko, A. Kudryavtsev, L. Osminkina, A. Vorontsov, Y. Ryabchikov, I. Belogorokhov, D. Kovalev, P. Kashkarov, Silicon nanocrystals as photosensitizers of active oxygen for biomedical applications. JETP Lett. **83**(9), 423–426 (2006)
192. L. Tsybeskov, J.V. Vandyshev, P.M. Fauchet, Blue emission in porous silicon: oxygen-related photoluminescence. Phys. Rev. B **49**, 7821–7824 (1994). https://doi.org/10.1103/PhysRevB.49.7821
193. C.C. Tu, K.P. Chen, T.A. Yang, M.Y. Chou, L.Y. Lin, Y.K. Li, Silicon quantum dot nanoparticles with antifouling coatings for immunostaining on live cancer cells. ACS Appl. Mater. Interfaces **8**(22), 13714–13723 (2016). https://doi.org/10.1021/acsami.6b02318
194. T. Uchino, N. Kurumoto, N. Sagawa, Structure and formation mechanism of blue-light-emitting centers in silicon and silica-based nanostructured materials. Phys. Rev. B **73**, 233203 (2006). https://doi.org/10.1103/PhysRevB.73.233203
195. J. Valenta, A. Fučíková, I. Pelant, K. Kůsová, K. Dohnalová, A. Aleknavičius, O. Cibulka, A. Fojtík, G. Kada, On the origin of the fast photoluminescence band in small silicon nanoparticles. New J. Phys. **10**, 073022 (2008). https://doi.org/10.1088/1367-2630/10/7/073022
196. J. Valenta, A. Fučíková, F. Vácha, F. Adamec, J. Humpolíčková, M. Hof, I. Pelant, K. Kůsová, K. Dohnalová, J. Linnros, Light-emission performance of silicon nanocrystals deduced from single quantum dot spectroscopy. Adv. Funct. Mater. **18**(18), 2666–2672 (2008). https://doi.org/10.1002/adfm.200800397
197. J. Valenta, M. Greben, Z. Remeš, S. Gutsch, D. Hiller, M. Zacharias, Determination of absorption cross-section of Si nanocrystals by two independent methods based on either absorption or luminescence. Appl. Phys. Lett. **108**(2), 023102 (2016). https://doi.org/10.1063/1.4939699. http://scitation.aip.org/content/aip/journal/apl/108/2/10.1063/1.4939699
198. J. Valenta, R. Juhasz, J. Linnros, Photoluminescence spectroscopy of single silicon quantum dots. Appl. Phys. Lett. **80**, 1070–1072 (2002)
199. J. VandeVondele, M. Krack, F. Mohamed, M. Parrinello, T. Chassaing, J. Hutter, Quickstep: Fast and accurate density functional calculations using a mixed gaussian and plane waves approach. Comput. Phys. Commun. **167**(2), 103–128 (2005). https://doi.org/10.1016/j.cpc.2004.12.014. http://www.sciencedirect.com/science/article/pii/S0010465505000615
200. I. Vasiliev, J.R. Chelikowsky, R.M. Martin, Surface oxidation effects on the optical properties of silicon nanocrystals. Phys. Rev. B **65**, 121302 (2002). https://doi.org/10.1103/PhysRevB.65.121302
201. N.C. Verma, A. Yadav, C.K. Nandi, Paving the path to the future of carbogenic nanodots. Nat. Commun. **10**, 2391 (2019). https://doi.org/10.1038/s41467-019-10394-8
202. R. Walsh, Bond dissociation energy values in silicon-containing compounds and some of their implications. Acc. Chem. Res. **14**(8), 246–252 (1981)
203. L. Wang, Q. Li, H.Y. Wang, J.C. Huang, R. Zhang, Q.D. Chen, H.L. Xu, W. Han, Z.Z. Shao, H.B. Sun, Ultrafast optical spectroscopy of surface-modified silicon quantum dots: unraveling the underlying mechanism of the ultrabright and color-tunable photoluminescence. Light Sci. Appl. **4**(1), e245 (2015)
204. R. Wang, X. Pi, D. Yang, Surface modification of chlorine-passivated silicon nanocrystals. Phys. Chem. Chem. Phys. **15**, 1815–1820 (2013)
205. X. Wang, R. Zhang, S. Lee, T. Niehaus, T. Frauenheim, Unusual size dependence of the optical emission gap in small hydrogenated silicon nanoparticles. Appl. Phys. Lett. **90**, 123116 (2007)
206. J.H. Warner, A. Hoshino, K. Yamamoto, R. Tilley, Water-soluble photoluminescent silicon quantum dots. Angew. Chem. Int. Ed. **44**, 4550–4554 (2005)

207. X. Wen, P. Zhang, T.A. Smith, R.J. Anthony, U.R. Kortshagen, P. Yu, Y. Feng, S. Shrestha, G. Coniber, S. Huang, Tunability Limit of Photoluminescence in Colloidal Silicon Nanocrystals. Sci. Rep. **5**, 12469 (2015). https://doi.org/10.1038/srep12469
208. L.M. Wheeler, N.C. Anderson, P.K.B. Palomaki, J.L. Blackburn, J.C. Johnson, N.R. Neale, Silyl radical abstraction in the functionalization of plasma-synthesized silicon nanocrystals. Chem. Mater. **27**(19), 6869–6878 (2015). https://doi.org/10.1021/acs.chemmater.5b03309
209. J. Wilbrink, C.C. Huang, K. Dohnalova, J.M.J. Paulusse, Critical assessment of wet-chemical oxidation synthesis of silicon quantum dots. Faraday Discuss. **222**, 149–165 (2020). https://doi.org/10.1039/C9FD00099B
210. J. Wilcoxon, G. Samara, P. Provencio, Optical and electronic properties of si nanoclusters synthesized in inverse micelles. Phys Rev B **60**, 2704–2714 (1999)
211. M.V. Wolkin, J. Jorne, P.M. Fauchet, G. Allan, C. Delerue, Electronic states and luminescence in porous silicon quantum dots: the role of oxygen. Phys. Rev. Lett. **82**, 197–200 (1999)
212. S. Yang, W. Li, B. Cao, H. Zeng, W. Cai, Origin of blue emission from silicon nanoparticles: direct transition and interface recombination. J. Phys. Chem. C **115**(43), 21056–21062 (2011). https://doi.org/10.1021/jp2075836
213. I.N. Yassievich, A.S. Moskalenko, A.A. Prokofiev, Confined electrons and holes in Si nanocrystals: theoretical modeling of the energy spectrum and radiative transitions. Mat. Sci. Eng. C **27**(5–8), 1386–1389 (2007)
214. Y. Yu, G. Fan, A. Fermi, R. Mazzaro, V. Morandi, P. Ceroni, D.M. Smilgies, B.A. Korgel, Size-dependent photoluminescence efficiency of silicon nanocrystal quantum dots. J. Phys. Chem. C **121**(41), 23240–23248 (2017). https://doi.org/10.1021/acs.jpcc.7b08054
215. T. Zhou, R.T. Anderson, H. Li, J. Bell, Y. Yang, B.P. Gorman, S. Pylypenko, M.T. Lusk, A. Sellinger, Bandgap tuning of silicon quantum dots by surface functionalization with conjugated organic groups. Nano Lett. **15**(6), 3657–3663 (2015). https://doi.org/10.1021/nl504051x
216. Z. Zhou, L. Brus, R. Friesner, Electronic structure and luminescence of 1.1- and 1.4-nm silicon nanocrystals: Oxide shell versus hydrogen passivation. Nano Lett. **3**(2), 163–167 (2003). https://doi.org/10.1021/nl025890q
217. J. Zou, R. Baldwin, K. Pettigrew, S. Kauzlarich, Solution synthesis of ultrastable luminescent siloxane-coated silicon nanoparticles. Nano Lett. **4**, 1181–1186 (2004)

# Chapter 2
# Light-Emission from Ion-Implanted Group-IV Nanostructures

**Moritz Brehm**

**Abstract** Silicon photonics is destined to revolutionize technological areas, such as short-distance data transfer and sensing applications by combining the benefits of integrated optics with the assertiveness of silicon-based microelectronics. However, the lack of practical and low-cost silicon-based monolithic light sources such as light-emitting diodes and, in particular, lasers remains the main bottleneck for silicon photonics to become the key technology of the twenty-first century. After briefly reviewing the state of the art regarding silicon-based light-emitters, we discuss the challenges and benefits of a highly flexible approach: The epitaxial incorporation of group-IV nanostructures into crystalline silicon. We argue that a paradigm change for group-IV quantum dots (QDs) can be achieved by the intentional incorporation of extended point defects inside the QDs upon low-energy ion implantation. The superior light-emission properties from such defect-enhanced quantum dots (DEQDs), our present understanding of their structural formation and light-emission mechanisms will be discussed. We will show that useful electrically driven devices, such as light-emitting diodes (LEDs), can be fabricated employing optically active DEQD material. These LEDs exhibit exceptional temperature stability of their light-emission properties even up to 100 °C, unprecedented for purely group-IV-based optoelectronic devices. Thereafter, we will assess the superior temperature stability of the structural properties of DEQDs upon thermal annealing, the scalability of the light-emission with the DEQD density and passivation schemes to further improve the optical properties. The chapter ends with a discussion of future research directions that will spark the development of this exciting field even further.

M. Brehm (✉)
Institute of Semiconductor and Solid State Physics, Johannes Kepler University Linz, Altenberger Strasse 69, 4040 Linz, Austria
e-mail: moritz.brehm@jku.at

 
D. J. Lockwood and L. Pavesi (eds.), *Silicon Photonics IV*,
Topics in Applied Physics 139, https://doi.org/10.1007/978-3-030-68222-4_2

## 2.1 Introduction to the Chapter

Silicon (Si), the dominating material for microelectronics [1], is generally a poor light-emitter which is mainly owed to the indirect nature of its energy bandgap. Nevertheless, silicon-based photonics (Si photonics) is rising to be a key technology for the twenty-first century and the implementation of practical and low-cost light sources based on group-IV nanostructures [2–9] and group-IV alloys [10–13] can ultimately be significant to this paradigm change. Si photonics is expected to solve challenges, such as, e.g., low-cost, high-bandwidth and CMOS-compatible optical interconnects with low-energy consumption [14, 15] or driving applications for sensing and lab-on-a-chip technologies [16, 17] with access to the manufacturing power of CMOS integrated circuits.

Given the fundamental limitations of Si as a light source, a wealth of strategies has been explored to develop material combinations capable of enhancing light-emission (see further discussion in Sects. 2.1.3 and 2.1.4). Among others, the introduction of germanium (Ge) nanostructures in Si (Ge/Si) sparked hopes and intensive research in various directions for the last three decades [2–9]. While self-assembled epitaxial Ge quantum dots (QDs) have been successfully used to study the rich physical phenomena leading to their nucleation [2, 6, 7, 9, 18], their performance as light-emitters has remained disappointingly low due to three key factors:

(i) Both Si and Ge are indirect semiconductors.
(ii) The band alignment of Ge on Si is of type-II, *i.e.*, only holes are confined inside the nanostructure, while electrons are weakly localized in the Si crystal around the QDs, leading to reduced matrix elements for radiative transitions.
(iii) The lattice mismatch between Ge and Si is relatively small (about 4.2%), and thus, the Ge/Si QD dimensions tend to be too large for efficient zero-dimensional (0D) quantum confinement.

Consequently, light-emission from Ge/Si nanostructures is typically weak and restricted to cryogenic sample temperatures only.

In this chapter, we discuss a path for efficient room temperature (RT) light-emission from Si that is based on QD material for which the optical properties are enhanced by ion implantation [19–22]. While such emitters have highly intriguing but to-date insufficiently well understood structural, electronic and optical properties, we argue that this approach has the intrinsic potential to be a solution to the missing bottleneck in Si photonics: It solves the above-addressed deficiencies of group-IV nanostructure light-emitters and enables efficient light-emission even up to RT and above, still keeping the beneficial group-IV material compatibility to integrated Si technology.

### *2.1.1 Background*

The elemental semiconductor Si and its oxide $SiO_2$ are by far the most important materials in integrated electronics as all integrated circuits are based on them. The ability to down-size active devices such as CMOS transistors led to a massive increase in computer performance over the last decades and has been thus the main driving force for the worldwide digital revolution. However, the ever-increasing amount of global data traffic has led to two fundamental problems: Bandwidth limitations and escalating power consumption.

A higher packing density of CMOS transistors requires a steadily increasing density of copper-based electrical interconnects. The small cross-section of down to $\sim 100 \times 100\,\text{nm}^2$ of the kilometers-long copper interconnects that carry the electronic signals in a single chip leads to substantial power dissipation due to Joule heating as well as parasitic capacitance and wiring delay, limiting the data transfer rates and bandwidth of integrated circuits. By using light instead of electrical current for data transfer, Si photonics can become a disruptive technology that aims at avoiding an explosion in energy consumption and at boosting data transfer rates at all levels of the interconnect hierarchy, from rack to rack to inter-chip and even intra-chip by creating low-cost, high-bandwidth optical interconnects.

Building a Si photonics platform or photonic integrated circuits (PIC) that are compatible with Si microelectronic technology requires several basic optical components such as waveguides [23–25], switches [26], filters [27], multiplexers [28], modulators [29, 30] and detectors [31, 32]. The development of these silicon-on-insulator (SOI)-based components is now, after more than a decade of extensive research, advancing rapidly and can be produced by Si-compatible technologies. However, the arguably most fundamental building block for such a universal platform remains elusive, due to several shortcomings of Si as a photonic material and mainly due to silicon's indirect bandgap: A practical, cost-efficient and electrically pumped group-IV laser source for monolithic integration on a Si chip and that can be easily coupled to standard Si waveguide technology and operated at room temperature (RT) and above has yet to be demonstrated. This ultimate bottleneck for Si photonics was already anticipated by Soref and Lorenzo in the mid-eighties as they wrote:

> From the foregoing information, we infer that single-crystal Si will be suitable for building directional couplers, filters, star couplers, optical switches, mode converters, polarizers, interferometers and modulators that operate at $\lambda = 1.3$ or $1.6\,\mu\text{m}$ (and beyond) essentially every integrated-optical component except an optical source. [33]

Further benefits of efficient light generation from Si would be that sensors, using Si-based near-infrared and especially mid-infrared (NIR and MIR) light-emitters merged with the power of Si microelectronics, could lead to many disruptive applications in, e.g., environmental and atmospheric monitoring [16, 17].

Propositions for overcoming the lack practical Si-based light-emitters include bulk Si approaches [34], using dislocation loops in crystalline bulk Si [35, 36], quantum confinement in group-IV nanocrystals [37–41], rare-earth doping [42–45],

Raman lasers [46, 47], SiGe lasers by the engineering of Ge layers toward a direct bandgap by applying strain [48–51] or alloying with Sn using SiGeSn layers [10–13, 52, 53], hexagonal Ge and SiGe alloys [54, 55] and last but not least, group-III-V lasers, bonded [56] or epitaxially grown on Si substrates [14, 15, 57, 58]. All of those approaches have both advantages and disadvantages, the latter mainly concerning the emission wavelength being above or below the telecommunication range, the temperature stability of the light-emission, especially in the temperature range from room temperature (RT) to 100 °C, the possibility of electrical injection, parasitic recombination due to innovative technology concepts, continuous-wave operation (cw), the threshold for lasing, state-of-the-art waveguide coupling and implementation into photonic circuits due to the necessary use of extensively thick buffer layers, CMOS process compatibility, requirements concerning substrate orientation and substrate size, device yield, power consumption, footprint and device lifetime. Probably, the most important point concerns the overall costs for large-scale integration of the lasers. This is mirrored in the long-term success of integrated Si technology. For almost all microelectronic devices, there exist counterparts, made from different materials than Si with higher single device performance. But there is a common saying that "*if it can be done with silicon, it will be done with silicon*," owing to the efficient large-scale integration possibilities of Si technology.

### *2.1.2 In a Nutshell: Potential Light Sources for Si Photonics*

At present, semiconductor light-emitting materials that can be easily implemented on Si platforms are understandably gaining attention as a potential game-changer in the previously mentioned efforts toward on-chip and inter-chip data communication. From this perspective, the last two decades have seen a surge of reports on various material combinations and designs, pushing the limits of what had previously been considered impossible. In Sects. 2.1.3 and 2.1.4, we will briefly review the respective advantages and disadvantages of different approaches for Si-compatible light-emission along with core challenges currently limiting their development and benefits of their practical deployment. As this is a highly active research field, red brick walls that were thought as being unsurmountable have been torn down within the last couple of years [59]. While, for the sake of conciseness, this certainly cannot be an exhaustive list, we believe that the brief summary will give the interested reader a valuable overview of the recent progress in the development of practical Si-based lasers and point out the ongoing need for a concerted effort to overcome present limitations.

### 2.1.3 All-Group-IV Approaches

- **Bulk silicon**: At the turn of the last century, emission from bulk Si at RT was demonstrated by reducing the density of non-radiative recombination centers, surface passivation and advanced doping profiles [34] and in another approach by limiting the carrier diffusion length by introducing dislocation loops that are formed upon ion implantation [35, 36, 60–63] or other defect centers in bulk silicon [64]. Besides emission yield considerations, the fundamental drawback of using bulk Si as the light-emitting material is the emission wavelength at ~1100 nm, causing problems regarding light absorption and losses in standard $Si/SiO_2$ waveguides.
- **Porous Si** [37, 65, 66] and **Si nanocrystal** [38–41, 67, 68] approaches aim to use spatial charge carrier confinement to enhance the radiative emission from Si through quantum confinement and increased wavefunction overlap. While optical gain was reported [40], the main issues are based on the performance of electrically driven devices and the light-emission energies well above the Si bandgap, the latter being problematic for light-coupling and propagation in $Si/SiO_2$ waveguides. The challenges of **epitaxial SiGe nanostructures on Si** and methods to overcome their limitations regarding light-emission will be discussed later in Sects. 2.2.2 and 2.2.3.
- In **rare-earth doping** approaches, rare-earth atoms such as erbium are introduced as optical recombination centers into Si, $SiO_2$ or Si nanocrystals [42–45]. While the emission wavelength of 1.55 μm is naturally perfectly matched to the telecommunication C-band, long radiative lifetimes in the ms range [69], limited electrical carrier injection due to the surrounding dielectric matrix, and high pumping thresholds make on-chip electrically driven lasing challenging [43].
- **Stimulated Raman scattering** in photonic crystal resonators (PhC) was used to demonstrate optically pumped lasing with low thresholds and small footprints [46, 47, 70]. However, Raman scattering-based lasers are intrinsically not ideal for on-chip applications as Raman scattering is always connected to optical pumping.
- **SiGe bandgap engineering** aims at lasing from group-IV structures. By applying tensile strain to the material, the small energy separation between the indirect and direct bandgap of bulk Ge can be reduced, or the band-ordering even reversed [48–51, 71–73]. Recently, lasing from tensile-strained micro-patterned Ge was reported [50] up to a sample temperature of about 100 K. Additionally, heavy n-type doping of Ge [74] can be used to induce electrically pumped lasing under pulsed excitation from bulk Ge structures [49]. In any case, the indirect transition remains close to the direct one and is, thus, always a concern of possible losses, especially at the mandatory operation temperatures for useful devices. Additionally, if the applied strain is too large, the emission wavelength shifts above the target wavelength of 1550 nm.
- By alloying **Ge with Sn** [10–13, 52, 53], another group-IV material, an optically pumped laser, was first demonstrated under pulsed excitation [10]. This group-IV laser marks a breakthrough because of its direct bandgap but still has severe

drawbacks concerning on-chip integration. As in the case of the Ge lasers above, the indirect bandgap is always in the energetic vicinity of the direct one. Nonetheless, recently optically pumped lasing up to a sample temperature of 270 K [75] and electrically pumped lasing up to 100 K was reported [76]. However, in all cases, strong emission quenching sets in already at cryogenic conditions, likely due to the energetically close indirect bandgap. The laser operation wavelength is generally larger than 2 μm [10, 12, 53, 75, 76] which makes this approach rather suitable for sensing in the mid-infrared wavelength region than for on-chip applications. A point of concern regarding applications of these metastable GeSn films is the efficient but detrimental GeSn phase separation [77–79] that can occur already at temperatures as low as 230 °C [77].

- The formation of a direct energy bandgap in group-IV materials has been predicted and was indeed observed in **hexagonal Ge and SiGe alloys** [54]. This approach makes use of Brillouin zone folding [80] by alternating two types of atomic stacking in Ge and in SiGe alloys in the form of a hexagonal crystal lattice. This can be achieved by the overgrowth of a group-III-V seed crystal that has the form of hexagonal GaAs wires, grown on GaAs (111)B substrates [55]. For those wires, the high surface-to-volume ratio enables the formation of metastable crystalline phases such as hexagonal Si or Ge. At present, the light-emission energies can be tuned from 0.35 to ~0.67 eV, i.e., well above the standard telecommunication wavelengths, making this approach probably more relevant for applications in chemical sensing by tracing the footprints of molecular vibration modes in this energy range. Additionally, arrays of such direct bandgap nanowires could be in principle envisioned as potential material for mid-infrared detectors, e.g., in lidar applications [54].

### *2.1.4 Group-III-V on Group-IV Approaches*

A straightforward corollary to avoid the intrinsic indirect bandgap nature of the group-IV elements Si, Ge and their alloys is to implement inherently bright light-emitters such as many of the compound semiconductors of the group-III-V or group-II-VI that exhibit a direct bandgap. Some of these hybrid group-III-V on Si prototype lasers represent the state of the art concerning their performance characteristics. In a way, also the above-mentioned hexagonal SiGe approach can belong to this section since group-III-V nanowires are needed to obtain optically active hexagonal group-IV material [54, 81].

- **Bonding III-V lasers onto Si substrates** by either wafer or die-bonding aims at combining the optical properties of group-III-V elements with the strength of Si electronics [82, 83]. Wafer bonding is fast but has several disadvantages: GaAs and InP wafers are typically of 100–150 mm in diameter, while standard Si and SOI wafers are 300 mm and, in future, possibly even 450 mm in diameter. This mismatch leads to a striking loss in material yield. Additionally, this approach is cost-extensive as group-III-V wafers are more expensive than Si and the bonding of whole wafers is an inefficient use of the group-III-V material.

- **Die-bonding** with die sizes >1 $mm^2$ circumvents some of those disadvantages. However, also the dicing step can result in small damage at the edges responsible for imperfect bonding. Additionally, this process is cost-intensive because it is more time-consuming than the aforementioned wafer bonding. Coupling to $Si/SiO_2$ waveguides is not straightforward but can be done if the optical mode is shared between the gain region and the waveguide region [56].
- **Transfer printing** of group-III-V materials on Si or SOI is more efficient than wafer bonding in terms of group-III-V material usage and it provides higher throughput and certain parallel fabrication possibilities as compared to die-bonding [84–86]. Impressively, photonic integrated circuits consisting of group-III-V material on Si have been demonstrated [86]. So far, limiting factors are the group-III-V coupon size that can be handled (<100 μm) while for larger coupon sizes, detrimental stress can be induced in the active devices due to the necessary underetching of the structures [85].
- **Monolithic integration of group-III-V semiconductors** on Si benefits from the excellent properties of the group-III-V constituents but faces issues with respect to the material implementation. Most prominent are uncontrolled and detrimental crystal defects originating from the different crystal structure and strain at the Si/Ge and group-III-V interface. Differences in thermal expansion coefficients of the group-III-V and group-IV constituents lead to additional defects once the substrate is cooled down to RT. Notably, all these defects can potentially contribute to device lifetime limitations. The yields of monolithically grown devices do not yet reach those of heterogeneous ones, and for all following approaches, compromises and disadvantages have to be accepted. In general, the major drawback of implementing group-III-V materials on a Si platform is that they do not offer the benefits and cost advantages inherent in monolithic Si integration and CMOS technology.
- One of the most promising group-III-V on Si approaches uses **InAs/GaAs QDs** as laser gain material [14, 15, 57, 58, 87, 88] for lasing at an emission wavelength of ~1.3 μm. The QDs are grown on several micrometer-thick group-III-V buffer layers to reduce the density of threading dislocations. However, the thick buffer layers impose a severe drawback concerning the optical coupling of the group-III-V device to the rest of the photonic integrated circuit, e.g., $Si/SiO_2$ waveguides. Additionally, dislocations can be present even in the QD layer region leading to non-radiative recombination of minority carriers and gain compression effects as observed in directly modulated Si-based QD lasers [89].
- **III-V nanowire lasers** grown on Si [90–93] show promise as the strain can be effectively accommodated due to the small footprint of the wire. Thus, group-III-V nanowires can be grown virtually defect-free on silicon [94] and electrical pumping of nanowire lasers in the visible wavelength range was reported in 2014 [95]. However, for obtaining vertical nanowires, (111)-type substrates are needed which are not employed in CMOS technology and, thus, routes toward efficient large-scale integration possibilities still have to be pursued.

- For **InP-based devices** [96, 97] RT lasing was demonstrated under pulsed optical excitation. Here, the group-III-V material is grown in defined nanotrenches for defect trapping. While this approach does not need micrometer-thick buffer layers and has been demonstrated to be compatible with 300 mm Si wafers [98], there exist some drawbacks. In general, the emission wavelength of 880–1040 nm is not in the telecom regime (leading to waveguide losses) which can be overcome by incorporating InGaAs QDs on the InP buffer layers, as reported recently [99]. Continuous-wave emission and electrical injection remain critical issues. While the latter was demonstrated for InGaAs on InP on Si lasers recently [100], the technology is to-date performance-wise less mature as compared to the aforementioned InGaAs on GaAs on Si lasers while in this case still suffering from the need of micrometers-thick group-III-V buffer layers [100].
- For **GaSb/Si and GaN/Si material systems** lasing was demonstrated [95, 101–104]. For the former, the main drawbacks concerning integrated photonics applications are high current densities, the long emission wavelength >2 μm and the need for micrometer-thick buffer layers [101, 102], for the latter, the short emission wavelength of 370–450 nm [95, 103, 104].
- **Template-assisted selective epitaxy** or TASE [105–108]. A flexible approach regarding device geometry that includes the incorporation of group-III-V bulk materials on Si was developed over the recent years by IBM [105–108]. Small seed openings are defined on the Si device layer of an SOI substrate which are directly overgrown by epitaxial group-III-V crystals. As for nanowires, the small footprint of the seed openings allows for defect confinement at the interface. Thus, TASE can avoid the need for thick SiGe or group-III-V virtual substrates to accommodate the accumulated strain caused by the epilayer-to-substrate lattice mismatch. The formation of $SiO_2$ shells of almost arbitrary shape in combination with selective group-III-V epitaxy by metal-organic chemical vapor deposition (MOCVD) allows for intrinsic adaptability regarding device shapes, e.g., for co-planar integration of group-III-V on Si devices [107], mandatory for the formation of photonic integrated circuits. Lasing action under picosecond-pulsed optical pumping was observed for GaAs microdisks on Si fabricated using TASE, exhibiting low thresholds for lasing of about 14 pJ/pulse [108]. Electrically injected lasing has yet to be demonstrated and is currently likely limited by the formation of defects at the original substrate interface. Further, bulk gain material has severe disadvantages as compared to QD lasers, especially concerning temperature stability and lasing threshold current [109]. It remains to be seen if group-III-V QDs with high optical quality can be formed by TASE for future Si photonics-based quantum dot lasers.

While the Si industry has been opened up regarding the diversity of elements of the periodic table from about 5% before the nineties (H, B, O, Al, Si and P) to about 45% nowadays, an aloofness and inertia toward severe changes comes naturally for such a large sector of industry (market share >400 billion US dollars [110]), as material innovations always significantly contribute to the complexity and thus to the costs. Thus, it remains to be seen if one of the before-described strategies to implement foreign group-III-V laser materials with Si microelectronics will be able to enter the mass market. This short review of the state of the art makes it very clear that the search for new, disruptive approaches to create electrically pumped group-IV light sources, operating above RT for practical monolithic integration on a Si platform needs to be continued.

## 2.2 Epitaxial Group-IV Nanostructures on Silicon

In the early nineties, defect-free group-IV epitaxial quantum dots embedded in crystalline Si were first discovered by Mo et al [111] and Eaglesham and Cerullo [112]. This sparked great hopes in the scientific community that quantum effects in low-dimensional nanostructures can be the game-changer to modify the inherently poor light-emitters Si and Ge into decent ones so that they can be utilized for applications compatible with Si integration technology [24]. The description—defect-free—was seen of particular importance since many kinds of defects, such as misfit dislocations, stacking faults can be electrically active, and, thus, leading to detrimental non-radiative recombination in low-dimensional group-IV material [113, 114]. Until today, the most research was devoted to the epitaxial growth of thin Ge layers on the technologically relevant Si(001) surface, although also other configurations such as Si epitaxy on Ge substrates [115–117] and SiGe and Ge on SOI [118–120] and Ge on Si(111) surfaces [121, 122] were heavily investigated. It can be foreseen that in the near future, significantly more research will be devoted to investigating also epitaxial Sn-based group-IV nanostructures on Ge and Si [123] due to the impact of GeSn bulk-based light-emission [10] and the aforementioned, group-III-V laser community-related considerations regarding the advantages of QD lasers over to bulk lasers [109].

The formation of epitaxial Ge/Si QDs follows the Stranski–Krastanow growth mode [124] and can be in short explained as follows: When thin films of Ge (crystal lattice constant 5.65 Å) are deposited on a Si(001) substrate surface (lattice constant ~5.43 Å), the first few monolayers of Ge form a pseudomorphically strained (~4.2%) two-dimensional wetting layer (WL), for which partial strain relief is provided to the formation of surface reconstruction [125, 126], *i.e.*, rebonding events of Ge atoms at the surface layer. As this provides only a limited amount of strain relief, the total strain energy is increasing almost linearly with the increasing film thickness. At a certain critical thickness (~3–5 monolayer (ML), depending on the Ge growth temperature), it is energetically favorable to expose a larger amount of surface atoms by the formation of three-dimensional objects—quantum dots—benefitting thereby from a significant lowering of the strain energy in the QDs, see Fig. 2.1.

Enormous efforts have been undertaken to optimize the fabrication schemes for defect-free epitaxial nanostructure growth, *i.e.*, strain-relieve occurring only elastically, as opposed to plastically by the introduction of dislocations. There along, the thin-film deposition of Ge on Si (along with the growth of InAs on GaAs) has

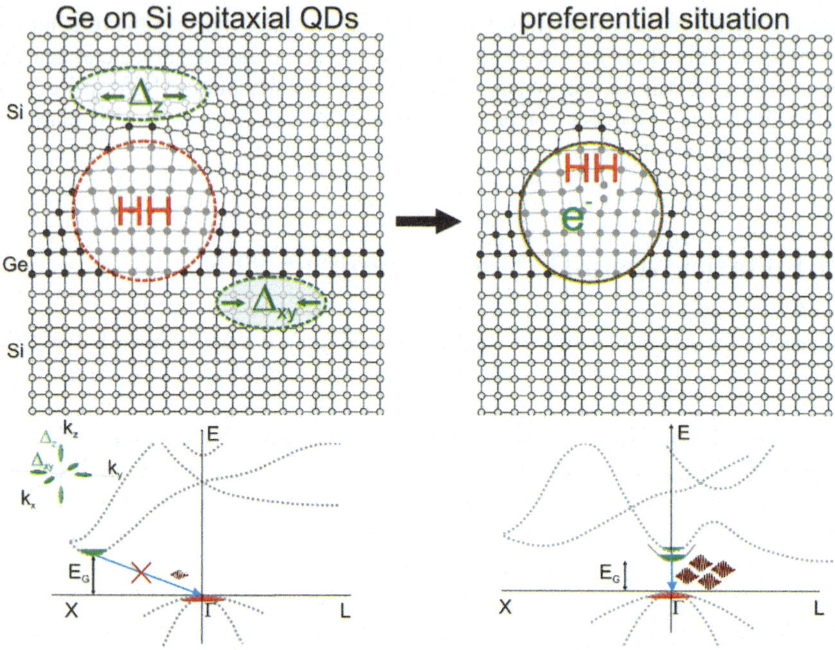


**Fig. 2.1** Up: Schematic representations of strain-relaxed Ge QDs (black) in a Si matrix (gray). Upper left: A defect-free QD. The respective locations of the ground states for the electrons ($\Delta_z$ and $\Delta_{xy}$) and heavy-hole states (HH) are indicated. Upper right: Preferential electronic configuration, achievable through the selective introduction of crystal defects. Bottom: Schematic band structure indicating optical transition paths (light blue arrows). Bottom, inset: splitting of the sixfold degeneracy of the electronic minima in the CB upon applied strain. Bottom left: Band structure indicating the weak radiative transition from Si-like electronic states at the $\Delta$-points to Ge-like hole states at the $\Gamma$-point for conventional QDs. Bottom right: Preferential optical recombination at the $\Gamma$-point. Radiative enhancement can be expected due to direct bandgap transitions and the large spatial overlap of the wavefunctions (type-I charge carrier confinement)

become the model system for strain-driven nanostructure formation in the Stranski–Krastanow growth system [127, 128]. This naïve picture would suggest that a suitable deposition of Ge on a Si surface leads to a formation of a nanometer-sized QD of pure Ge surrounded by Si. But it became clear very early that this description is by far too simple. Along the way, a myriad of growth-related phenomena were discovered which found their counterparts in other strain-driven thin-film growth systems [128] such as some group-II–VI systems like CdSe/ZnSe [129], nitride systems such as InN/GaN [130], metal systems like Au/Ni, or Au/Ag [131], to ice on platinum [132].

The Ge on Si(001) heteroepitaxial growth system is particularly rich when it comes to architectures of the QDs and islands. The different shapes are determined by low-index, low-energy crystal facets such as the {105}, {113}, {15 3 23}, {111} or {12 5 3}—facets [111, 133–135]. Depending on subtle variations in the growth parameters such as deposition temperature and—rate, the composition and deposited amount of material, different shapes were found to appear predominantly, such as unfaceted mounds [136], pyramids [111, 137], multi-facetted domes [133, 138] or structures with larger height to base aspect ratio such as barns [134] and cupolas [135], continuously approaching the shape of a half-sphere. The initial formation of the nanostructures occurs on a supersaturated wetting layer [138–140], *i.e.*, when pure Ge is grown on Si(001) by MBE at 700 °C and a low growth rate of 0.05 Å/s, the WL thickness initially exceeds its equilibrium thickness of roughly 3.2 ML by more than 30% (~1 ML) [138–140]. Then, counterintuitively, the first QD forming are not of the smaller, more shallow pyramids, but larger, but energetically more stable dome-like QDs with base diameters around 100 nm and heights of 20 nm [138]. Likely owed to the fact that the SiGe/Si is a binary system, as compared to ternary systems such as InGaAs/GaAs, a particularly rich theoretical and experimental insight was gained with respect to the physical aspects of QD formation phenomena. This includes the impact of two important structural parameters that crucially determine the electrical and optical properties of any heteroepitaxial system for which a lattice mismatch exists: The strain and the spatial chemical composition. Obviously, these two are linked to each other. As the QDs evolve, areas of Si and Ge start to mix through an efficient alloying mechanism that is triggered by surface diffusion. Such intermixing often leads to a Ge-lean QD base and a Ge-rich apex [141–143] or outer shell [144, 145]. The study of the chemical composition of individual SiGe QDs with sizes of only tens of nanometers in length and height was developed through two rivaling, yet complementary experimental methods: X-ray diffraction based on synchrotron radiation with sub-micrometer resolution [141] and nanotomography, a method of alternating SiGe composition sensitive selective chemical etching and atomic force microscopy [142]. Being much more than a mere note of spectacular material science, these local composition gradients strongly influence the localization and overlap of wavefunctions [143, 144] and thus directly influence recombination paths in the QDs [144] as Ge-rich inclusions at the nanostructure apex can lead to the formation of small and strongly quantum-confined QDs inside larger nanostructures [143, 146].

Furthermore, overgrowth of QDs with Si can influence the QD shape, internal SiGe composition [140, 147, 148], and thus the optical properties [140, 149, 150]. While overgrowth with Si at low temperatures (<~500 °C) preserves the QD properties to a large extent [140, 148], capping at higher temperatures (>~500 °C) can lead to flattening of larger QDs and even to a shape transformation of steeper domes to more shallow pyramids [140, 151]. Peculiar phenomena such as Ostwald ripening between QDs, coarsening and shape transformations by growing and shrinking nucleation cores [152] have been reported. Many more phenomena related to the growth of Ge and SiGe QDs on Si have been observed and can be found, *e.g.*, in the following reviews and the references therein. [6, 126, 128, 153]

As for the structural parameters, the optical and electrical parameters of Ge/Si QDs have also been subject to intense investigations [7, 8, 18, 24, 113, 138, 140, 143, 154]. The spatial location of the electron wavefunctions and hole wavefunctions is indicated in the sketch in Fig. 2.1a. In photoluminescence (PL), electron-hole pairs are excited in the crystalline Si matrix surrounding the QDs. Holes relax to the energetic minima in the QDs, where they are strongly confined by the large valence band (VB) offsets of ~700 meV between the Ge QD and the surrounding Si matrix. Thus, depending on the size and chemical composition of the QDs, the quantum confinement energy experienced by holes and the energy of the VB offsets can vary. For many Ge/Si QD structures, the thermal escape of holes from the VB potential to the surrounding Si is to be associated with high activation energies >300 meV. Thus, ionization of the QD through hole-escape is hardly possible at temperatures that are relevant for many optoelectronic device applications, *i.e.*, from about 0–100 °C. Therefore, for the VB alone, the electronic properties in Ge/Si QDs are very promising concerning light-emission and similar to those of the best semiconductor QD light-emitters such as InGaAs/GaAs QDs.

The real problem of these group-IV light-emitters originates from the properties of the conduction band (CB). The global energetic minima for the electrons are quantum-confined states in approximately triangular potentials. These are introduced by strain in the Si matrix that is induced by the elastic relaxation of the Ge QDs [155–157]. Thus, electron wavefunctions barely enter the QD and this spatial separation [157], a type-II band alignment, leads to a reduced wavefunction overlap between the electron states and the hole states. To make matters worse, these strain-induced potentials in the Si matrix are energetically very shallow. Depending on the QD size, an activation energy of only up to about 100 meV is obtained for the thermal escape of the electrons [143, 157, 158] which necessarily leads to pronounced PL quenching, already at cryogenic sample temperatures. Additionally, both Si and Ge are indirect semiconductors and optical transitions have to be assisted by phonons or scattering events. Thus, albeit all the efforts in understanding the physical nature of GeSi/Si QDs, the overall PL efficiency is rather poor, especially as compared to the best direct bandgap group-III-V nanostructures. Thus, over the past fifteen years, strategies have been employed to improve the overall light-emission yield from Ge/Si QDs. This includes the fabrication of site-controlled QDs with well-defined inter-QD distances [7, 18, 113, 143, 150, 159, 161] or by coupling the light-emission to photonic resonances of micro-resonators and plasmonic structures [162–168].

Significant improvement concerning light-emission was reported for randomly nucleated Ge/Si(001) QDs for which a subsequent photonic resonator was created using standard lithography and etching techniques. These resonator designs include, e.g., photonic crystals (162, 164, 165, 167) or microdisks [19, 163] and microrings [163]. An improvement of the optical properties can also be based on plasmonic effects, by coupling the QD layer beneath the Si cap to a plasmonic nanoparticle that is deposited on top of the Si capping layer [169, 170]. Naturally, for these resonator structures, no particular alignment to the QD positions is possible using self-assembled nanostructures. The resonance behavior in photonic microcavities is characterized by their cavity modes for which the location varies within the cavity. This, consequentially, leads to problems regarding the coupling between the cavity modes and randomly nucleated QDs for which no, or only minimal control over the microscopic, spatial nucleation position can be obtained. Therefore, after subsequent resonator formation, the QD positions and the positions of highest field enhancement will only coincidentally overlap and it is not possible to guarantee ideal QD cavity-mode overlap for ten thousands of optical devices that should be produced in photonic integrated circuits.

Group-IV nanostructures, and in particular, Ge/Si QDs, however, offer a valuable solution to this problem, namely deterministic site-control of QDs [7, 9, 113, 150, 159, 160]. Different approaches for defining QD nucleation sites have been investigated in the past two decades. These include, e.g., selective epitaxy using $SiO_2$ masks [11, 120, 171], scanning tunneling nanolithography [172], and local implantation by focused ion beams [173, 174]. Many works reported on the use of etched pits for defining preferential nucleation sites, during both molecular beam epitaxy (MBE) and chemical vapor deposition (CVD) growth [7, 18, 113, 143, 150, 159–161]. It was demonstrated that QD nucleation into periodic pit arrays can result in an improved morphological and chemical uniformity [7, 143, 144, 150, 175–177], which is associated with a substantially narrower ensemble photoluminescence (PL) emission line width [7, 113, 143]. Single SiGe QDs that are grown on pit-patterned Si substrates with wide inter-QD distances were used to study the optical response of individual QDs in this group-IV system [159, 178]. At the lowest excitation powers employed (100 nW), a full-width-at-half-maximum linewidth of only about 16 meV of the QD-related no-phonon peak emitted at 920 meV was reported, which is a record low for such epitaxial Ge/Si QDs [178].

Combining site-control of QDs and resonator formation allows for a deterministic placement of the QDs with respect to the resonator modes. Such a high degree of controllability was recently demonstrated by placing Ge/Si QDs at various position of L3 cavities of a photonic crystal resonator to increase the light-emission from defined resonator modes [167]. Additionally, in this way, an experimental mapping of the photon density of states of a resonator can be performed, allowing for a direct feedback loop between theoretical simulations and experimental realization of the devices [167].

In summary, the understanding of structural and optical properties of Ge/Si QDs and the coupling to microcavities brought significant progress in the field of CMOS-compatible light-emission using the flexible approach of Ge/Si heteroepitaxy. However, these methods of nano-optical light enhancement through photonic resonances seem to be a mere workaround that cannot answer the fundamental question: How to obtain light-emission from epitaxial group-IV quantum dots that can be comparable to that of group-III-V direct bandgap material, at least in the temperature window between 300 and 370 K. These QDs would be, through their structural properties, environmentally benign and, at the same time, chemically robust, perfect candidates for active material in optoelectronic devices due to their fabrication flexibility and intrinsic compatibility with Si microelectronic standards. In other words, what is really needed for epitaxial Ge/Si nanostructures is a paradigm change.

In the following, we will address a promising way to improve the intrinsic optical properties of Ge nanoclusters upon implantation of heavy ions. Ion implantation is one of the indispensable processes in microchip fabrication, as only implantation of donors such as phosphorus and acceptors such as boron allow us to control the electrical properties of Si and to make devices such as transistors. Almost all semiconductor manufacturing processes today make use of at least two but often many more ion implantation steps, *e.g.*, for almost all doping in silicon integrated circuits. Most commonly, B, As, P, Sb, In, Si, Ge N, He and H are implanted. Ion implantation offers high controllability and precision in incorporating the dopants at the designed location within the semiconductor device. For this, the implantation depth crucially depends on factors such as the mass and energy of the dopant ion, the mass of the target atoms, the implantation angle and the orientation of the target as well as the implantation temperature [179].

Concerning optical emission from Si, the implantation of ions, such as H, *etc* into crystalline bulk Si or other bulk group-IV layers was studied quite extensively within the last decades and led to the observation of interesting implantation-related light-emission phenomena from these structures [35, 36, 61, 180–182]. Further research was devoted to nanostructure formation driven by ion implantation [183–186]. Little, however, was reported about the direct implantation of ions into already existing, small, nanometer-sized nanostructures that are embedded in a crystalline host matrix. This is surprising, since both, nanostructure formation and ion implantation, have shown to provide benefits concerning optical and optoelectronic properties as compared to bulk Si. Following, we will discuss that what is needed to obtain efficient light from Si is interlacing these two approaches.

## 2.3 Ion Implantation into Ge Quantum Dots on Silicon

We will focus on the investigation and fostering of a novel material class, ion-implanted Ge/Si QDs that eventually might be the game-changer in providing a cost-efficient, practical and electrically driven laser, emitting in the telecom wavelength range at ~1.3 and 1.55 μm and operating at RT and above. The main point favoring such a technology would be that these lasers can be compatible with the standard SOI photonics platforms, *e.g.*, the SOI 220 nm device layer platform. Due to the absence of thick buffer layers, no complicated coupling to waveguides needs to be employed (see Fig. 2.2), in contrast to other material combinations (see Sects. 2.1.3 and 2.1.4). The need for thick buffer layers for the GeSn or group-III-V materials on Si (see Fig. 2.2a–c) is one of the bottlenecks for these approaches. While for the latter, the light sources themselves have excellent output characteristics, the simple and low-loss coupling of the light source to Si photonics waveguides remains an open issue. A scheme of the different layer stacks employed for the different material approaches is shown in Fig. 2.2. From this scheme, the flexibility and large-scale integration possibilities, as well as the benefit of employing Ge/Si-based light-emitters for on-chip or intra-chip data transmission, are evident as coupling to other photonic components such as photonic waveguides, filters, splitters, modulators, multipliers and detectors is rather straightforward (Fig. 2.2e).

Paradoxically, and as mentioned above, up to now, one of the main criteria for the fabrication of Ge/Si QDs and nanostructures, in general, was to control and

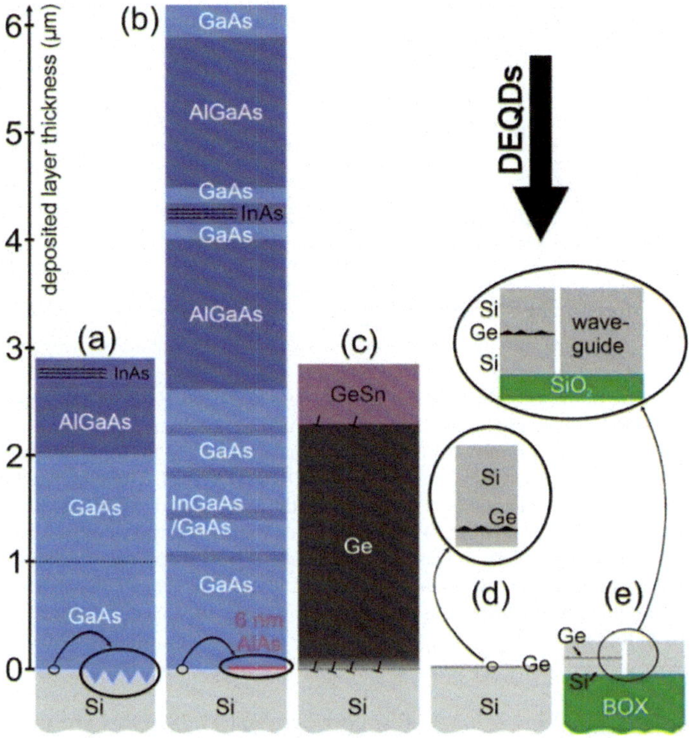


**Fig. 2.2** A comparison of possible monolithic implementations of semiconductor light sources on Si substrates. **a** and **b** Growth of multi-stacked InAs QDs on GaAs grown directly on Si [15, 58]. **c** Growth of bulk GeSn or GeSiSn heterostructures on top of a Ge virtual substrate grown on Si [10]. **d** Defect-enhanced GeSi quantum dots (DEQDs) grown directly on Si substrate [19, 20] or **e** on silicon-on-insulator (SOI) substrates. A scheme for evanescent coupling of DEQD light-emission to SOI waveguides is indicated in (e) image modified from [18] with permission from the authors

avoid defect formation. For course, this is particularly true for some defects, such as dislocations and stacking faults for which it was shown that their presence in the QD and within the surrounding Si matrix can fully suppress radiative recombination paths in the QDs [113]. But, as for bulk material, not all defects are detrimental for optical radiative recombination [35, 36]. The main idea behind the additional ion implantation into the nanostructures is to overcome the aforementioned shortcomings of conventional Ge/Si QDs (Fig. 2.1) by capturing holes and electrons in two fundamentally different but interlaced low-dimensional structures: epitaxial QDs and extended point defects. Such unique QD modification can result in exquisite and, in particular, superior structural and optical properties. Thus, these nanostructures are referred to as *defect-enhanced Ge quantum dots* or DEQDs.

First, we introduce the most striking benefits of DEQDs, before elaborating on the latter in Sect. 2.3. Already at cryogenic temperatures, *i.e.,* $T <\sim 200$ K, the luminescence yield of DEQDs is vastly enhanced as compared to conventional, defect-free

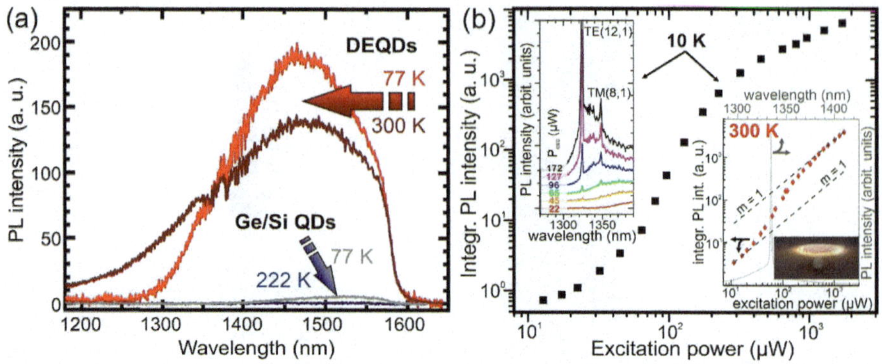


**Fig. 2.3** **a** Increase of the photoluminescence (PL) yield at 77 and 300 K of DEQDs as compared to conventional Ge/Si QDs, grown under the same growth conditions but without additional ion implantation. For those, the PL fully quenches already at ~200 K. **b** Signs for optically driven lasing at 1330 nm from microdisk cavities containing DEQD gain material. The integrated PL intensity versus excitation power is depicted on a double logarithmic scale and exhibits a clear s-shape and indicates lasing at 10 K (black squares) and 300 K (red-diamonds, right inset). Left inset in (**b**): Emergence of whispering gallery modes upon increased optical pumping. A schematic image of the underetched microdisk is shown in the bottom right inset in (**b**) Linewidth narrowing and mode competition from these resonator modes are discussed in [19]. (**b**) Image modified from [19] with permission from the authors

Ge/Si QDs [19], see Fig. 2.3a. At RT, conventional Ge/Si QDs are in essence not emitting any light, while DEQDs exhibit almost the same light-emission yield as at $T = 10$ K. For DEQDs, also clear signs for optically pumped lasing are present [19], *e.g.*, linewidth narrowing, threshold behavior and mode competition (Fig. 2.3b and [19]). The radiative carrier lifetimes in DEQDs can be very short, down to the range of less than 1 ns [19]. It is relatively straightforward to electrically contact DEQD layers. Consequently, DEQD light-emitting diodes (LEDs) that emit up to device temperature above 100 °C [21] have been reported, which is exceptional for group-IV light-emitters. Due to the small volume of the active DEQD nanostructure material, it is possible to enhance the active gain volume by vertical stacking of the DEQDs [22], therefore increasing the light-emission yield in a scalable way related to the DEQD density [21, 22]. Furthermore, we will discuss the surprising thermal stability of the structural properties of DEQDs which withstand thermal annealing at 600 °C for at least 2 h, or flash-lamp annealing at 800 °C for at least 20 ms [187, 188].

### 2.3.1 DEQD Fabrication Procedure

The current understanding of the DEQD formation process is depicted in Fig. 2.4. Si(001) or SOI (001) substrates are *ex-situ* chemically cleaned before their transfer

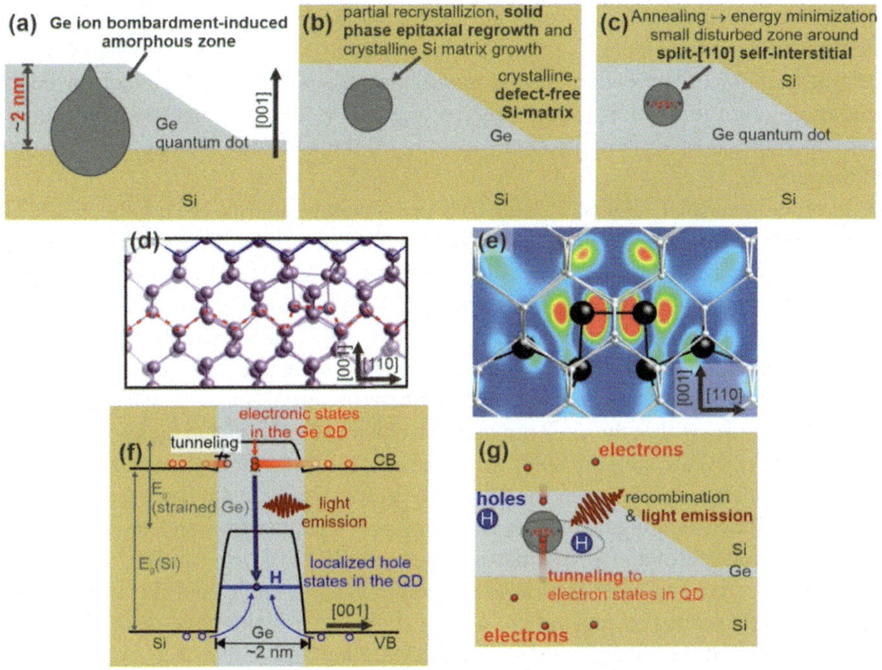


**Fig. 2.4** **a–c** Split-interstitial formation in Ge QDs upon Ge ion implantation and annealing. **d** Ground-state defect structure that results after computational crystallization of an amorphous state containing one extra Ge atom. The blue-solid and red-dashed line highlight the undisturbed lattice and the area containing the split—[110] self-interstitial. **e** DEQD crystal lattice including the split—[110] self-interstitial (black). The color code presents the highest (red) and lowest (blue) electronic orbital electron density of 0.00035 and 0.0 electrons/bohr$^3$, respectively. This indicates a strong modification of the electron states due to the defect site. **f**, **g** Proposed light-emission processes in DEQDs (f) as an energy level scheme and (g) in real space. Electrons tunnel to the defect-induced states in the QD, recombining with holes confined in the QD leading to direct recombination in real and reciprocal space. Image modified from [20] with permission from the authors

to a solid-source molecular beam epitaxy (MBE) chamber. *In-situ*, the samples are degassed, usually at temperatures of about 700 °C for ~15 min. Hereafter, a Si buffer layer of about 50–100 nm thickness is deposited to ensure a clean, epitaxially grown sample surface before the deposition of a thin Ge film. About 4.5–7 monolayers (ML) of Ge (*i.e.*, 6.3–9.9 Å) are deposited at typical growth rates of 0.03–0.1 Å/s and a growth temperatures $T_G$ typically between 400 and 700 °C. First, a two-dimensional wetting layer (WL) forms, followed by elastic relaxation of the strained layer that results in QD formation. Depending on the growth temperature, different QD shapes are obtained, as discussed above in Sect. 2.2. In the following, we will mainly focus on Ge growth at $T_G = 500$ °C which results in the formation of elongated pyramids, also called huts or hut clusters [111]. They are depicted in Fig. 2.4 as a light gray shape. At any point during the growth, a negative bias voltage between 0 V and up to 3 kV can be applied to the substrate. Since Ge and Si layers are evaporated from electron beam evaporators, Ge and Si ions are always present in the chamber as a certain fraction of the evaporated atoms again passes through the electron beam, leading to atom ionization. These positively charged ions can then be intentionally implanted into a specific epilayer, by turning on the substrate bias at a certain time of the growth.

In order to create DEQDs, the substrate bias is either applied during the whole Ge deposition or during the later stages of Ge QD growth, so that the low-energy Ge ions (<3 keV) impinge on the QD with a concentration that was estimated to be ~$10^4$ $\mu m^{-2}$. Thereby, the energy of the Ge ions is significantly higher than the displacement energy of Ge atoms in the crystal that is about 14 eV [189, 190]. Thus, even the first displaced atoms, so-called recoil atoms, have sufficient energy to displace further atoms and the resulting collision cascade is only stopped when the initial ion energy has dissipated [179]. Therefore, within these ~2 nm high hut clusters the heavy Ge ions with energies lower then 3 keV are causing shallow, about 1–2 nm large amorphized zones in the QD's crystal lattice, as displayed as dark-gray regions in Fig. 2.4a. During the ongoing Ge growth, the Si capping layer deposition and possible post-growth annealing, solid-phase epitaxial regrowth (SPER) partially recrystallizes the amorphized zone [191]. Note, that perfect recrystallization is impeded since an additional ion was brought into an already perfect crystal lattice, see Fig. 2.4b. Nevertheless, the recrystallization at the growth front allows for subsequent crystalline and defect-free overgrowth of the DEQDs with Si [19, 20] which is of particular importance regarding charge carrier confinement and light-emission and to avoid unwanted non-radiative recombination through surface states.

After the growth of a Si capping layer and, if applied, thermal annealing, a crystal configuration with low formation energy emerges out of the amorphized region in the QD, Fig. 2.4c. Theoretical calculations suggest that for a strain-relaxed Ge crystal that contains N+1 atoms in a volume for which N atoms would lead to a perfect crystal lattice, the minimum energy crystal configuration is in the form of a split—[110] Ge-Ge self-interstitial [20, 192–194]. The presence of such relatively large implanted ions must lead additionally to a pronounced deformation of the surrounding crystal lattice, involving about 45 neighboring atoms [20]. Figure 2.4d depicts the calculated ground-state defect structure of an amorphous state containing one extra Ge atom [20] that results after an extensive series of Monte–Carlo quench-and-anneal steps and subsequent geometry relaxation [20] using the Quantum Espresso DFT package [195]. The blue-solid and red-dashed line in Fig. 2.4d highlight the calculated undisturbed lattice and the area containing the resulting split—[110] self-interstitial. Figure 2.4e presents the core of the DEQD crystal lattice including the split—[110] self-interstitial (black) and the surrounding distorted crystal lattice (white). Overlaid in Fig. 2.4e is the electronic orbital isosurface cross-section indicating that the electron wavefunctions are influenced by the presence of the defect site (red color) due to electron states at the Γ-point ~70 meV below the Ge CB edge [20]. The wavefunctions of the holes seems to remain largely unaffected by the crystal distortion [20] and more details on the defect-induced band structure changes, especially band structure effects in the whole Brillouin zone, and the influence of strain will have to be evaluated in the future [196].

### 2.3.2 *Light-Emission from DEQDs*

Our present understanding of the light-emission dynamics from DEQDs is depicted in Fig. 2.4f, g as an energy band diagram (Fig. 2.4f) and in real space (Fig. 2.4g), respectively. Under non-resonant optical excitation or electrical excitation, free carriers are generated mostly in the Si matrix barriers surrounding the DEQDs from where they thermalize to the band edge minima of the Si conduction band (CB) and valence band (VB). Holes, indicated in blue in Fig. 2.4f, g, are efficiently trapped and confined in the DEQD VB potential as in conventional Ge/Si QDs (see Sect. 2.2). Note, that the average hut cluster height is only about 2 nm, due to the very shallow {105} facets of the QDs.

Thus, in the worst case, the split—[110] self-interstitial is located in the vertical center of the QD, *i.e.*, at most, ~1 nm away from the Si matrix layer. For such short distances, electrons, indicated in red in Fig. 2.4f, g, can tunnel from the Si layers through the ~1 nm thick QD-induced barrier to the QD with its defect site from where they can radiatively recombine with holes [20]. Since electrons and holes are confined within the DEQD and both carrier types can occupy states at the Γ-point [20], and the radiative transitions are direct in real and reciprocal space, in stark contrast to conventional QDs.

From the experimental evidence, we expect that defect-induced changes will lead to increased light-matter coupling, but the matrix elements for dipole-allowed transitions $\propto \left\langle f \middle| \hat{e} \cdot \vec{p} \middle| i \right\rangle$ (with $|i\rangle$ and $|f\rangle$ the initial and final electronic state, $\vec{p}$ the electron momentum, and $\hat{e}$ the light-polarization vector) were not reported so far for strained Ge [196]. In the following, we will discuss the experimentally observed optical properties of DEQDs that are in agreement with the above-described light-emission scheme.

Figure 2.5a, b depicts the influence of increasing PL excitation power ($P_{exc}$) on the light-emission from DEQDs. First, a pronounced blue-shift of the peak-PL emission from about 0.8 eV (1550 nm) to 0.925 eV (1350 nm) is observed for increasing $P_{exc}$. Interestingly, there exists an inner structure within the PL signal that can be separated into a lower-energy and a higher-energy signal, as can be seen in Fig. 2.5b. In a previous publication, we tentatively assigned the blue-shift to the filling of hole states with increasing $P_{exc}$ in the ensemble of measured QDs [19]. In Fig. 2.5b the integrated RT-PL intensities $I_{PL}$ of the total DEQD emission as well of the higher-energy and lower-energy peak are plotted versus $P_{exc}$. The lower-energy peak increases with a power factor m of about 0.6 according to $I_{PL} = P^m$. This power factor is commonly associated with Auger-recombination and is often observed in conventional GeSi QDs [144, 197, 198]. Noteworthy, the higher-energy peak rises with $P_{exc}$ according

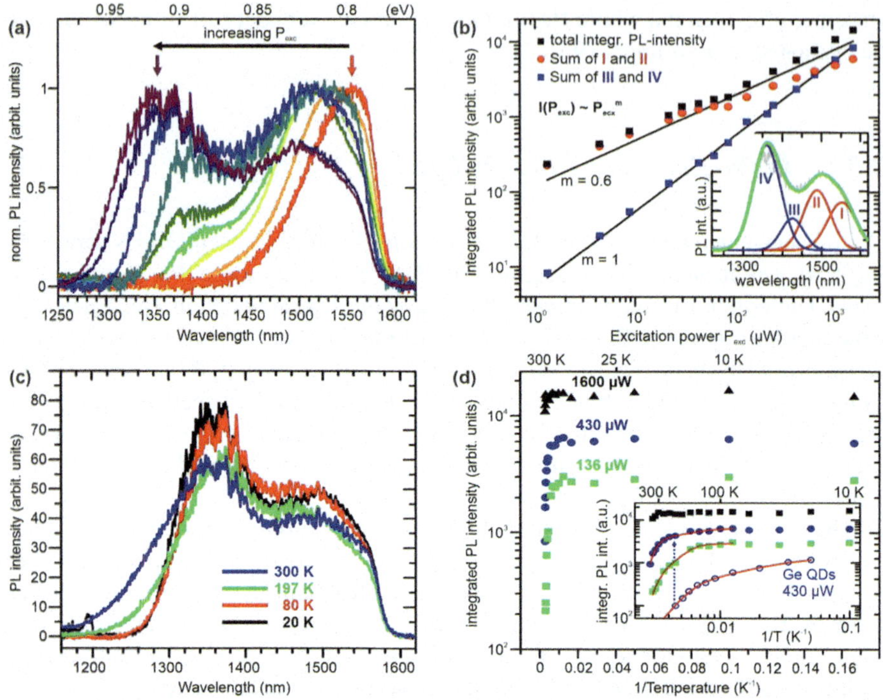


**Fig. 2.5** **a** Normalized PL spectra recorded at RT and for increasing laser excitation power ($P_{exc}$). A pronounced blue-shift with increasing $P_{exc}$ is observed **b** The integrated RT-PL intensity $I_{PL}$ versus $P_{exc}$ is plotted for the higher-energy peak (blue squares) the lower-energy peak (red spheres) and the total DEQD emission (black squares). The black solid lines represent power coefficients $m = 0.6$ and $m = 1$, according to $I_{PL} \sim P_{exc}^{m}$. **c** PL spectra of DEQDs for $T_{PL}$ of 20, 80, 197, and 300 K. The spectra are not normalized. **d** Full symbols: $I_{PL}$ of the DEQDs versus inverse $T_{PL}$ for $P_{exc} = 136$, 430, and 1600 μW. The red curves are Arrhenius fits to the data. The open blue circles show $I_{PL}$ of Ge QDs without ion implantation treatment versus $1/T_{PL}$ for $P_{exc} = 430$ μW. Image modified from [19] with permission from the authors

to a power factor $m$ of 1, which is assigned to optically direct recombination in semiconductors [199].

The probably most spectacular feature of the optical properties of DEQDs is depicted in Fig. 2.5c. The PL spectra were recorded at different sample temperatures $T_{PL}$ of 20, 80, 197 and 300 K. Note that these spectra are not normalized which immediately implies that temperature-induced PL quenching is basically absent in DEQD samples, at least in a temperature window from 4 K to RT and at elevated $P_{exc}$ [19]. To shed more light on this behavior, in Fig. 2.5d the detailed temperature dependence of the DEQD-PL signal is plotted in an Arrhenius-type graph with the PL intensity quenching at high $T_{PL}$ being fitted using Arrhenius fits, see solid red lines in Fig. 2.5d [19]. In the inset of Fig. 2.5d, the data are plotted on a double logarithmic scale to emphasize the PL quenching behavior at high $T_{PL}$ that is associated with activation energies $E_A$ of ~350 meV for both $P_{exc} = 136$ and 430 μW. This is in stark contrast to what was observed in conventional Ge QDs for which thermal quenching with $E_A \sim 60-80$ meV were reported [24, 200] (see inset of Fig. 2.5d). As discussed above, for conventional Ge/Si QDs, electrons condense at the global energy minima of the strain pockets in the Si matrix that are induced by the relaxation of the QD. These potentials are energetically shallow and, thus, already at cryogenic temperatures, light-emission quenching occurs, associated with the thermal escape of electrons to the surrounding Si.

In contrast, our present data suggest that thermal quenching in DEQD light-emission only sets in when holes can thermally escape from the QD valence band potential to the surrounding wetting layer ([188] and see Sect. 2.3.6). This process is associated with an activation energy of 300–400 meV [19, 188], making DEQDs suitable light-emitters even above RT. Additionally, the tunneling process for electrons from the Si matrix to the QDs is temperature–independent, and electron-hole recombination times have been reported to be very short [19]. Thus, the already long lifetime of electrons in the surrounding Si matrix [24] is further enlarged if holes are efficiently collected by the DEQD potentials. This could prevent detrimental electron-hole recombination in Si and ensure that electrons are always available to tunnel to the DEQD from where light-emission in the telecommunication regime occurs upon radiative carrier recombination. These processes make the DEQD system highly suitable even for device temperatures up to 100 °C, as will be seen in the following Sect. 2.3.4.

The carrier recombination scheme presented in Fig. 2.4f, g is consistent with the observed DEQD emission wavelength from 1300 to 1600 nm (Fig. 2.5), the high activation energies for thermal quenching (Fig. 2.5d), a power-coefficient of $m = 1$ (Fig. 2.5b), the negligible thermal PL quenching of the PL intensity at RT (Fig. 2.5c, d) and the dramatically shortened carrier lifetimes down to ~0.6 ns [19]. Our present understanding is that excitation of an ensemble of DEQDs with different sizes as well as filling of different defect-induced energy levels leads to the observed broad spectral emission range. Furthermore, in [19] it was found that the energetic difference between the calculated ground states and excited states for heavy holes in an ~2 nm high hut cluster is about 100 meV apart which would also agree with the energetic splitting between the high-energy and low-energy peak, observed for DEQD samples. Further research will be necessary to unambiguously track down the origin of multi-peak light-emission from DEQDs under high PL pumping power.

For DEQDs, a natural question concerns the role of the surrounding environment of the defects—the QD cages that host the defects. Can this approach also work for other structures with different dimensions? In a recent publication [20], we found that superior light-emission is only possible if the defects are introduced into QDs, as indicated in Fig. 2.6. There, the room temperature PL response of a DEQD sample (gray spectrum) is compared to two different quantum wells samples (QWs) (green and red spectra) for which Ge ion implantation was performed under the same experimental conditions as for the DEQDs. Pronounced light-emission is only observed for the DEQD sample while in the ion-implanted QW samples the optical properties at RT are dramatically reduced. We understand that when QDs, with their finite spatial boundaries in all three dimensions, are the host surrounding of the defects, then

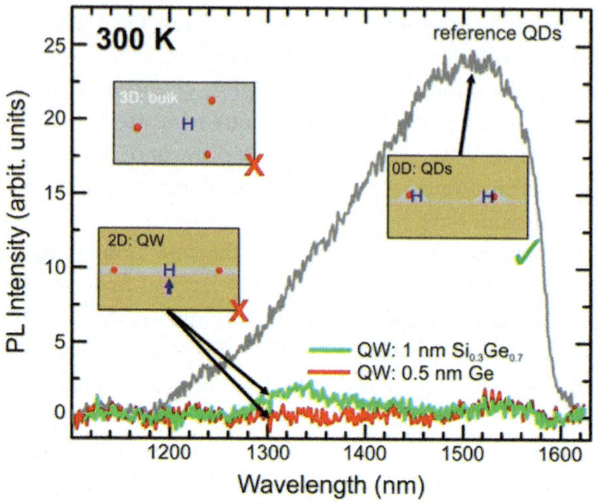


**Fig. 2.6** Room temperature PL spectra of DEQDs (gray spectrum) and two different quantum well samples (QWs) (green and red spectra) for which Ge ion implantation was performed under the same conditions as for the DEQDs. Only the DEQDs show strong PL enhancement. Inserts: Schematic views of DEQDs (right) versus ion-implanted QW (lower left) and a Ge bulk sample containing split—[110] interstitial defects. On the one hand, holes (blue H) are confined in DEQDs and thus overlap strongly with electrons trapped at defect sites (red dots). On the other hand, holes in an ion-implanted QW or in bulk Ge can diffuse away from the two adjacent point defects (blue arrow), leading to a reduced electron-hole wavefunction overlap. Image modified from [20] with permission from the authors

holes, indicated in blue color in the scheme in Fig. 2.6, are confined within the QDs because the QD potential imposes the global energy minimum in the valence band. Therefore, the hole wavefunctions are forced to overlap strongly with the electron wavefunctions that are attracted by the defect-induced changes in the conduction band at the QDs site (red dots), leading to direct light-emission from DEQDs. In contrast, in quantum wells and Ge bulk samples containing the same defects, we argue that even though the defect clearly seems to be attractive concerning electron recombination, the additional presence of strain around the defect site repels the holes from the QD site. In this way, holes can move to more strain-relaxed regions in between two such defects [20]. The resulting charge carrier separation reduces the spatial overlap of the electron-hole wavefunction and thus the transition matrix elements for optical transitions. Hence, enhanced PL yields at RT can neither be observed in control samples in which Ge ions were implanted into QW structures (Fig. 2.6), nor in bulk Ge [20], where split—[110] self-interstitials are well known for decades [192–194]. Thus, quantum dots are essential for DEQD light-emission and only the incorporation of the defect into the QD with nanometer precision in the growth direction makes the superior optical properties of DEQDs feasible. Therefore, for preferential defect formation, the defect must be created in the QD and, thus, implantation depth, and thus the ion energy, ion species and QD height have to be considered simultaneously [179].

### 2.3.3 Considerations Toward Large-Scale Integration Possibilities

Up to now, the DEQDs were grown by molecular beam epitaxy (MBE) combined with *in-situ* Ge ion implantation. For any industrial applications, where large-scale integration plays a role, a transfer of the growth of the active DEQD material to CMOS-compatible fabrication techniques has to be considered. This can be generally envisioned as Ge/Si QD growth can be achieved in high structural and optical quality also by chemical vapor deposition, CVD [2, 3]. Furthermore, ion implantation and rapid thermal annealing are standard processes in Si technology. To ensure defect formation inside the QD, implantation has to probably occur on uncapped samples or samples with a few nanometer-thick Si capping layer. Otherwise, it might be technologically challenging to ensure implantation of the ions into the QDs through several 100 nm thick layers due to the resulting smeared out implantation profiles. A transfer of the DEQD schemes to CVD methods seems to be feasible but still requires the necessary adaptation of machinery for CVD growth and in-situ transfer to the low-energy ion implantation to avoid detrimental contamination. Modern industrial implanters still face challenges concerning the productivity of the beam current at energies below 10 keV. But since the ongoing miniaturization of microelectronic devices demands decreasing ion implantation energies anyways, these challenges are likely to be solved by industry. Also, the implanter-induced cross-contamination has to be monitored and controlled as this can significantly influence the DEQD defect formation. Concerning the thermal budget, CMOS-compatibility for back-end implementation is guaranteed by the low DEQD growth temperatures of 400–500 °C [18–22].

### 2.3.4 Electrical Injection

Even though RT electroluminescence has been reported from conventional Ge/Si QDs [24, 201, 202], the reported light-emission has been subject to extensive thermal quenching and, thus, their potential as practical RT light-emitters is limited by all the aforementioned reasons. For DEQDs, the only significant change in the fabrication procedure concerns the additional ion implantation step into the QD layer. As in conventional Ge/SiQDs, DEQDs are surrounded by a crystalline and nominally defect-free Si matrix, which can be easily doped p-type or n-type using, *e.g.*, boron or phosphorous or antimony. Thus, the DEQD system is ideally suited for electrical injection, *e.g.*, by growing them into the intrinsic region of a p-i-n diode [21]. For such LED devices, the most essential observation is the outstanding temperature stability of the optical properties of DEQDs even under electrical pumping conditions. For the first generation of DEQD-LEDs, vertically stacked layers of QDs were employed that were placed in the middle of a 200 nm thick intrinsic region of a p-i-n silicon diode [21]. The p-type and n-type regions were doped with $2.5 \cdot 10^{19}$ cm$^{-3}$ of boron and about $1 \cdot 10^{19}$ cm$^{-3}$ of arsenic, respectively. To determine the scalability of the DEQD density on the LEDs, multi-DEQD layers of three and seven stacks were fabricated together with an empty Si diode, *i.e.*, without DEQDs. The device geometry of the $100 \times 100$ μm$^2$ sized mesa structure is depicted in the inset of Fig. 2.7a [21], where also the top Al metallization layer and the bonding wires are indicated. In Fig. 2.7a, the electrical characteristics of an LED with seven layers of DEQDs are presented for an injection current density of 10 kAcm$^{-2}$ and recorded at heat sink temperatures around 300 K and above [21]. No drastic change in the spectral emission characteristics was observed in this range of sample temperatures. Note the robust LED operation at heat sink temperatures as high as 100 °C (373 K) for a duty cycle up to 5%, which is remarkable for group-IV light-emitters. In the upper inset of Fig. 2.7a, the light-current characteristics are presented for the heat sink temperature range from 22 to 100 °C.

Noteworthy, the maximum values for the heat sink temperature and driving current were in [21] only limited by either the Peltier temperature controller or by the strength and number of bonding wires and the pulsed current source. No significant change in the shape of the curves with increasing current and only very weak

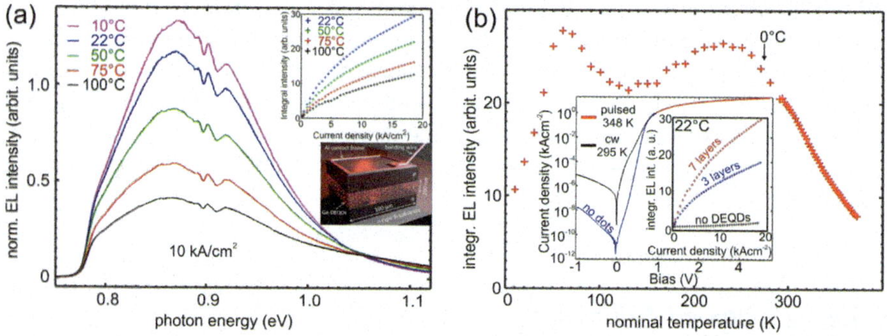


**Fig. 2.7** **a** Spectrally resolved evolution of the electroluminescence (EL) emission properties of an LED containing seven DEQD layers around and above RT in Peltier-controlled operation. The upper inset depicts the light-current characteristics for different high heat sink temperatures from 22 to 100 °C. The lower inset presents a scheme of the device geometry. **b** The integral intensity of the DEQD-LED (driving current density 10 kA/cm$^2$) versus the heat sink temperature. Emission quenching sets in above 230 K, but the LED performs well up to at least 100 °C (373 K). The first inset shows the T-dependence and I-V-characteristics of the DEQD-LEDs using continuous-wave (cw) (black and blue curves) and pulsed excitation (red curve). The second inset shows the low-current behavior of the DEQD-LEDs (blue and red) and the Si reference (black), respectively, measured under cw operation. Good scaling of the integrated emission intensity with the number of incorporated DEQD layers is demonstrated. Reprint (adapted) with permission from P. Rauter et al. ACS Photonics 5, 431 (2018). Copyright (2018) American Chemical Society

light-emission saturation is observed, implying that DEQDs allow for stable light-emission output under strong electrical pumping (at least 20 kAcm$^{-2}$) and very high sample temperatures.

In Fig. 2.7b, the integrated electroluminescence signal from the seven-layer DEQD-LED is plotted versus the heat sink temperature. The intensity remains at about 75% of the global maximum in a cryogenic temperature range from 60 to 230 K before an actual emission quenching sets in that is associated with the thermal escape of holes, as discussed earlier in Sect. 2.3.2. As a comparison, for optically pumped DEQD structures (Fig. 2.5), emission quenching at high pumping powers is delayed to sample temperatures higher than 300 K. The earlier onset of emission quenching from LEDs is likely related to the higher actual device temperature at a specific heat sink temperature due to ohmic heating induced by the driving current. Nevertheless, the integrated intensity still remains at about 30% of the global maximum for an impressive heat sink temperature of 100 °C (373 K), which is extremely late as compared to what is reported for other direct bandgap systems such as tensile-strained Ge of GeSn or SiGeSn materials [10, 12, 13, 50]. The superior temperature stability of the DEQD light-emission is of particular importance for electrically pumped laser resonators. These have to be fabricated on buried oxide layers to ensure optical confinement of the laser modes with respect to the gain material. In this case, the buried oxide permits efficient heat transfer between the laser and the heat sink, emphasizing the urgent need to employ temperature-stable gain material, such as DEQDs.

The first inset in Fig. 2.7b compares the current–voltage characteristics of a seven-layer DEQD-LED under low currents to the Si reference diode. Evidently, the reverse current is raised by about three orders of magnitude by the incorporation of Ge material in the form of DEQDs. However, in the relevant driving regime, *i.e.*, under forward bias and current densities >100 Acm$^{-2}$, the presence of DEQDs hardly influences the diode characteristics. Joule heating obviously plays a role when comparing the current–voltage characteristics obtained using cw excitation to pulsed operation at high driving currents and it can be seen that a heat sink temperature of 75 °C under pulsed excitation corresponds to a heat sink temperature of 22 °C under cw operation [21].

In the second inset of Fig. 2.7b, the light-current characteristics are plotted for the three-layer DEQD-LED and seven-layer DEQD-LED, as well as for the Si reference diode. The electroluminescence emission intensity is continuously increasing with the driving current up to the experimental limit at about 20 kAcm$^{-2}$, without any strong saturation effects being observed. Thus, it can be expected that higher driving currents will not impose a limiting factor for the future realization of an electrically pumped DEQD laser. The Si reference diode only shows very weak emission, originating from the optical interband transitions in bulk Si. Increasing the amount of DEQD multilayers from three to seven, *i.e.*, an increase in DEQD sheet-density by a factor of 2.3, increases the light-emission by a factor of 1.65 without changes in the shape of the light-current curve. The resulting emission power scaling by DEQD layer stacking in LED might turn out essential for the future development of electrically pumped laser sources with DEQD gain material. The slight deviation from a linear scaling behavior between light-emission and DEQD density is likely caused by charge carrier competition between the different DEQD layers, which can be mitigated by stronger electrical pumping of the devices and future optimization of the diode structure itself. No particular effort was invested by now to optimize doping concentrations and thickness of the contact and intrinsic regions to ensure ideal recombination currents in the DEQDs. Thus, improvements in this respect will be crucial for the fabrication of future DEQD laser devices.

In summary, electrical pumping of DEQD emitters is clearly feasible and excellent optical properties of these unique group-IV dots are maintained under electrical pumping. Comparing the PL and electroluminescence data from DEQD samples shows that for the same collection efficiency and under maximum available pumping conditions, the same time-averaged light-emission was obtained [21]. Considering the differences in the duty cycle between the results of Figs. 2.5 and 2.7, a 20-fold higher emission intensity from the DEQD-LEDs was obtained as compared to the DEQD emitters that show clear signs for optically pumped lasing (Fig. 2.2b) if they are embedded in microdisk resonators [19, 21]. Therefore, electrically pumped lasing from DEQD gain material in ridge-resonators or microdisk resonators is expected to be possible, even at application-relevant temperatures.

### *2.3.5 Scalability of DEQD Densities*

In the previous paragraph, we have seen that vertical DEQD stacking allows for very good scalability of the gain material volume which can be a significant factor concerning future laser applications. However, it has to be noted that the vertical stacking of DEQDs is intrinsically challenging due to the necessary ion implantation in each Ge layer. As noted above, the low-energy ion implantation during the DEQD process results in an amorphous zone in the QD (Fig. 2.8c) that partially recrystallizes via solid-phase epitaxial regrowth during the growth interrupts and the continued deposition of Si and Ge (Fig. 2.8d).

For stacked DEQD layers, the vertical distance between the QDs is preferably small so that high gain material volumes can be achieved at a certain layer thickness, which can be obtained by restricting the Si spacer thickness to about 10–30 nm. However, if the recrystallization is not leading for every QD in a crystalline growth front, then during the following Si overgrowth defects such as stacking faults and polycrystalline regions will be continued in the Si spacer layer above the DEQDs. These defects allow for a relaxation of the compressive strain that is built up by the elastic relaxation of the former QD layers and, thus, leads for the following Ge layer to a preferential nucleation site, at which a majority of the deposited Ge accumulates (Fig. 2.8f).

This, again, leads to non-crystalline overgrowth with Si which further feeds the Ge material accumulation cascade in the next-next Ge layers and leads to a massive increase of the defective region (Fig. 2.8g, h). This process can be avoided by using an elevated substrate temperature $T_G$ for the growth of the Si spacer layers. In Fig. 2.8a,

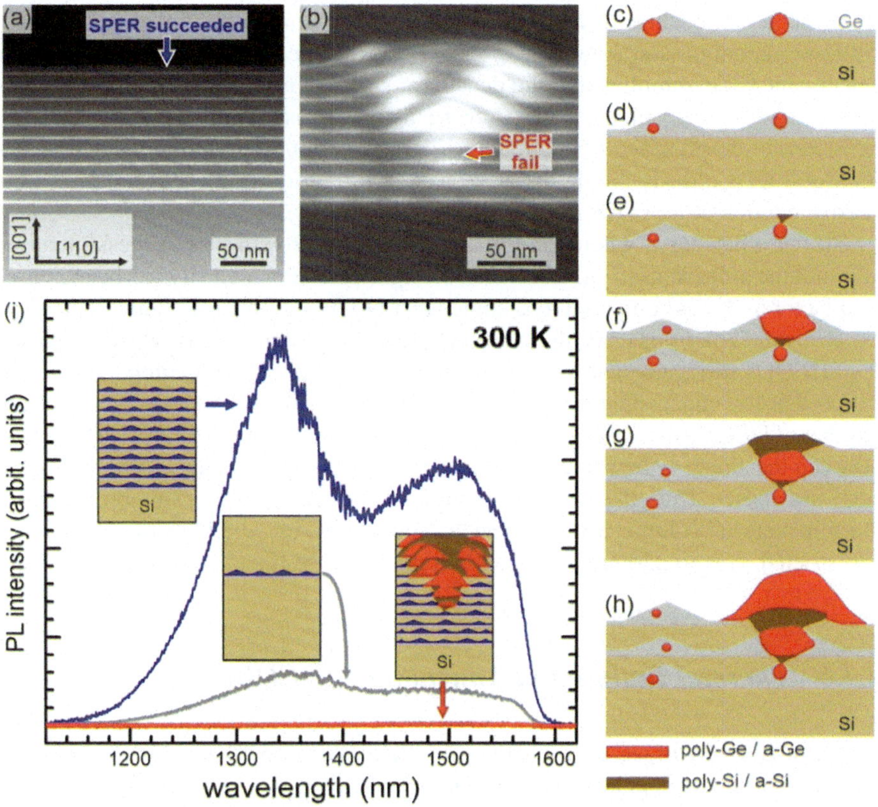


**Fig. 2.8** **a** HAADF-STEM images of an 11-fold stack of DEQDs (light gray) separated by 15 nm thick Si spacer layers grown at $T = 500$–$600$ °C. **b** HAADF-STEM image of a tenfold multi-stack of DEQDs where the 12 nm thick Si spacer layers were deposited at $T = 350$–$500$ °C. **c**–**h** Schematic evolution of the massive defects formed upon SPER-fail. Non-perfect recrystallization of the growth front **d** leads to defect formation in the Si spacer layer **e** leading to Ge accumulation in the subsequent DEQD layers **f**–**h**. **i** RT-PL spectra obtained from a single layer of DEQDs (gray) and from an 11-fold stack (blue) for which $T$ was 500–600 °C. For the multi-stack sample for which the growth parameters were not optimized ($T = 350$–$500$ °C), no PL signal was recorded (red). Image modified from [22] with permission from the authors

the cross-sectional transmission electron microscopy (TEM) image of an 11-fold multi-stack of DEQD layers is depicted for which the spacer layer $T_G$s were ramped up from 500 to 600 °C. On the contrary, if the spacer layer is grown at $T_G = 350$ to 500 °C, this leads to the formation of massive defect structures in the later periods of the multi-stack as can be seen in the cross-sectional TEM image in Fig. 2.8b.

The need to avoid such accumulating defect structures is immediately evident by evaluating the PL response from the successful and failed layer configuration depicted in the cross-sectional STEM images in Fig. 2.8a, b. Samples, where massive defects are present, do not show any PL signal at room temperature (red spectrum), even if the laser excitation and detection spots are several micrometers away from such a defect center. This implies that the density of states of non-radiative recombination centers must be very high in such defect clusters. Thus, the absence of DEQD-PL emission in the presence of the massive defects is not surprising considering that for Ge/Si QDs already single dislocations can fully quench the PL emission from nearby, but defect-free QDs [113].

On the other hand, when solid-phase epitaxial regrowth was successful for all layers of the multi-stack (Fig. 2.8a), then the light-emission intensity can be increased by a factor of 7.5 as compared to a single layer of DEQDs, measured under the same experimental conditions, see the blue spectrum in Fig. 2.8i. Likely, at the employed $P_{exc}$, DEQDs still compete for the generated electron-hole pairs leading to the observation that the enhancement factor is slightly smaller than the factor by which the number of DEQDs increases for different samples. However, the total layer thickness between the single-layer sample and the multilayer sample in Fig. 2.8 remained the same in all cases. Thus, the growth of multilayer DEQDs allows for suitable scaling of the gain material volume, even for device applications, for which the target thickness is restricted, *e.g.*, to 220 nm as in standard photonics SOI technology.

### 2.3.6 Thermal Budget and Annealing of DEQDs

A natural question targets the robustness whenever the properties of low-dimensional structures crucially depend on the formation and existence of point defects or defect complexes which consist only of a few atoms. For DEQDs, the question whether they are thermally robust enough to be employed in CMOS-compatible processes can be answered emphatically "yes," as it was shown that DEQDs retain their superior light-emission properties, even under thermal annealing at 600 °C for two hours [20, 188] and 800 °C for about 20 ms [187], which does not represent an upper boundary.

For the creation of CMOS-compatible photonic integrated circuits, monolithic light sources, such as DEQDs, have the distinct advantage of possible front-end integration, *i.e.*, placing the photonics layer in between the CMOS layer and the metallization layers. This integration possibility is favorable as compared to back-side or back-end integration since slow electrical driving of the integrated optics devices through vias (back-side) or the slowest metallization layers (back-end) can be avoided [59]. Of course, front-end integration requires that the thermal budget of the photonics layer must not deteriorate the underlying CMOS layer while being itself robust enough to withstand the fabrication of the subsequent metallization layers. The CMOS layer can typically sustain thermal budgets of about 1 h at 450 °C [203] or 0.5 h at 475 °C, while AlCu, or Cu metallization layers are typically deposited in a temperature window of at least 350–475 °C [203, 204].

Figure 2.9 depicts the spectral evolution of the DEQD low-temperature and RT-

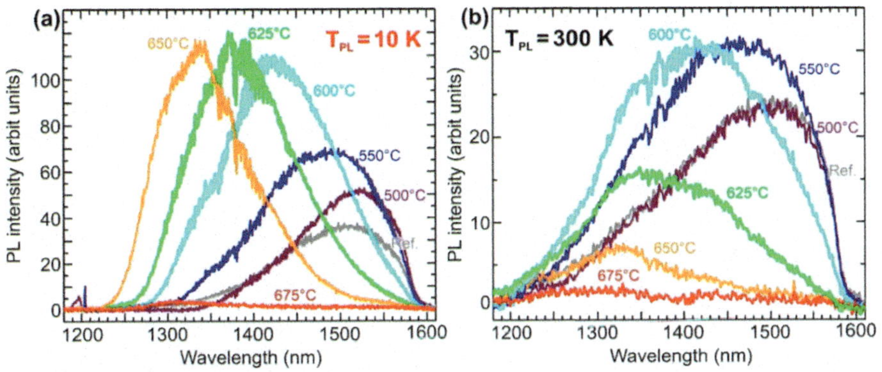


**Fig. 2.9** PL spectra of DEQDs, recorded at **a** $T_{PL} = 10$ K and **b** $T_{PL} = 300$ K. The spectrum of the DEQD reference sample is shown in gray. The other DEQD samples were in-situ annealed for 2 h at annealing temperature $T_A$ of 500 °C (violet spectra), 550 °C (dark blue), 600 °C (light blue) 625 °C (green), 650 °C (orange) and 675 °C (red). Image modified from [188] with permission from the authors

PL emission, following *in-situ* thermal annealing for 2 h at temperatures $T_A$, ranging from 500 to 675 °C. For $T_{PL} = 10$ K (Fig. 2.9a), a pronounced increase of the integrated PL emission is observed with increasing $T_A$, concomitant with a blue-shift of the peak emission. Only if the samples underwent annealing at $T_A = 675$ °C, the PL emission abruptly quenches (Fig. 2.9a). A similar, yet qualitatively different behavior is observed if the sample temperature is raised to 300 K, Fig. 2.9b. While the blue-shift with increasing $T_A$ is observed as well, the PL yield decreases already for $T_A > 600$ °C.

From Fig. 2.9, two main conclusions can be drawn. First, the abrupt decline in integrated PL intensities after thermal annealing at high $T_A$ can be associated with the thermally induced migration of the defect complex from the QD into the surrounding Si [20]. By evaluating an Arrhenius-fit of the integrated PL versus $T_A$ for $T_{PL} = 300$ K, a lower boundary for the activation energy for defect migration $E_A$ (migr) of 3.4 eV was found [188]. It was argued that the strong resilience of the thermally induced migration of the defect complex is based on the stabilizing effects of the strained regions around the split—[110] self-interstitial core [20, 188].

The second conclusion obtained from Fig. 2.9 concerns the changes in the Ge composition profiles of the DEQDs with increasing $T_A$ and the associated reason for the thermal quenching of the PL at high $T_{PL}$. The spectral shift of the DEQD-PL emission with increasing $T_A$ (Fig. 2.9) in combination with the activation energy $E_A$ for thermal quenching with increasing $T_{PL}$ allows for modeling of the composition profiles of the DEQDs and the surrounding wetting layer as a function of $T_A$. The results, which include a thermal degradation of the Ge profiles of DEQDs and the WL through bulk diffusion, are plotted in Fig. 2.10a, b. The actual QD ionization path at high sample temperatures was found to be caused by the thermal escape of holes, confined in QD (Fig. 2.10a) to the energetically higher quantum-confined states in

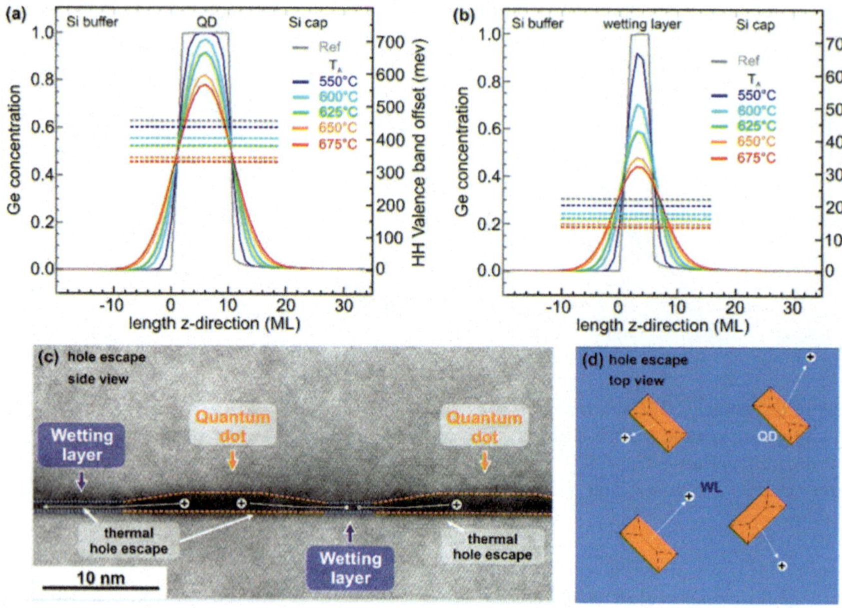


**Fig. 2.10** Heavy-hole ground-state energies (dashed lines), valence band offsets and simulated Ge composition profiles for different annealing temperatures $T_A$ for **a** a QD of 1.4 nm height, and **b** the surrounding wetting layer (WL) with a height of 0.7 nm, respectively. **c** Side-view and **d** top-view scheme depicting the thermal hole-escape from the QD VB potentials to higher-energy states in the surrounding WL potential, leading to spatial electron-hole-pair separation and thermally induced luminescence quenching. Image modified from [188] with permission from the authors

the surrounding wetting layer [188], Fig. 2.10b. This escape process is schematically presented in the cross-sectional TEM image and as a top-view scheme in Fig. 2.10c, d.

### 2.3.7 *Curing and Passivation of Non-radiative Recombination Centers*

In addition to the split—[110] self-interstitial, unwanted defects in DEQDs can arise from non-optimal sample cleaning before growth, growth of Ge and Si itself or the low-energy ion implantation conditions. For the latter, imperfect solid-phase epitaxial regrowth (Sect. 2.3.5) or an ion collision cascade that stops below the QDs can be detrimental to the optical properties since it might lead to the insertion of non-radiative recombination centers. These need to be avoided or, at least treated post-growth, so that their negative influence on the optical properties can be minimized. Furthermore, it was reported that the presence of extended crystal defects, such as misfit dislocations, can be associated with PL quenching in group-IV emitters [113], making the interpretation of the observed changes of the PL yield with respect to the performed *in-situ*/*ex-situ* treatments impossible.

In the previous Sect. 2.3.6, we have seen that post-growth thermal annealing can improve the optical quality of the DEQD- emitter by possibly recrystallizing such non-radiative recombination centers. However, this can come at a price of changed emission wavelengths (Fig. 2.9) originating from morphological and compositional changes due to the annealing (Fig. 2.10). As compared to in-situ thermal annealing or rapid thermal annealing, flash-lamp annealing provides the advantage of high temperatures, *e.g.*, for dopant activation or point-defect recrystallization while it avoids due to short exposure times severe structural changes that unavoidably also lead to changes in the optical properties, see Fig. 2.9 [188]. It was demonstrated that flash-lamp annealing at 800 °C for up to 20 ms, performed on DEQDs leads to only negligible modifications of the electronic band alignment [187]. For such accurately chosen parameters for the flash-lamp treatment, the PL emission yield can be increased by almost 50% while not changing the spectral shape of the DEQD emission. What is more, it was shown that flash-lamp annealing leads to improved stability of the PL emission intensity over a temperature range from 10 to 300 K [187]. Both, the increase in PL yield and the enhanced thermal stability is attributed to the thermal healing of unwanted and non-radiative defects in the Si matrix around the QDs. Overall, the results of Sects. 2.3.6 and 2.3.7 show that DEQDs can in principle be implemented into hybrid integrated circuits where high-temperature steps are necessary for the fabrication process.

In a different approach, hydrogen passivation can be employed to saturate the dangling bonds from unwanted defects in the Si matrix. This method is widely used in applications ranging from solar cells to CMOS technology to emission improvement of GeSi nanostructures [205–207]. In Fig. 2.11, we compare the PL emission for

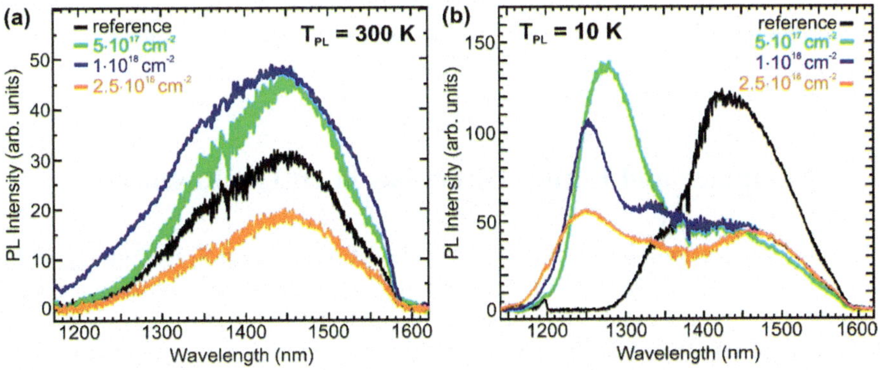


**Fig. 2.11** PL spectra recorded at **a** $T_{PL} = 300$ K and **b** $T_{PL} = 10$ K of a reference sample annealed at $T_A = 600$ °C. In addition to the untreated sample (black curves), samples have been investigated that underwent H-implantation with implantation dose of $5 \times 10$ cm$^{-2}$ (green curves), $1 \times 10^{18}$ cm$^{-2}$ (blue curves), and $2.5 \times 10$ cm$^{-2}$ (orange curves). Image modified from [188] with permission from the authors

a series of DEQD samples that underwent various low-energy proton irradiation treatment with different irradiation doses ($5 \times 10^{17}$ cm$^{-2}$, $1 \times 10^{18}$ cm$^{-2}$, and $5 \times 10^{18}$ cm$^{-2}$) to a DEQD reference sample (black spectra in Fig. 2.11). The spectra were recorded at sample temperatures of 300 K (Fig. 2.11a) and at $T_{PL} = 10$ K for Fig. 2.11b. The high-temperature spectra in Fig. 2.11 show that the DEQD emission yield can be increased by up to almost 90% for a low-energy proton irradiation dose of $1 \times 10^{18}$ cm$^{-2}$ (blue spectrum), as compared to the untreated reference sample (Fig. 2.11, black curve). On the other hand, a 2.5-fold higher H-implantation dose (orange spectra in Fig. 2.11) leads to a 63% decrease of the integrated PL emission (Fig. 2.11a). The DEQD-PL emission spectra recorded at 10 K reveal (Fig. 2.11b) that the proton implantation causes additional carrier recombination paths that can be observed by the appearing emission band between 1250 and 1300 nm. This additional peak exhibits a systematic shift to higher energies with increasing proton irradiation dose. Due to its thermal deactivation behavior that is associated with a rather low activation energy of about 40 meV, it can be assigned to so-called G-centers that are ascribed in the literature to the formation of Si-C vacancy clusters in Si [208]. The presence of the G-centers suggests that structural changes are induced in the Si capping layer upon low-energy proton irradiation, whereas the PL-results confirm that the DEQD layer is structurally not affected by H-treatment [188].

Moreover, the light-emission improvement observed for the H-treated samples measured at 300 K proves that H-treatment is a valid tool to be used for group-IV nanostructures since non-radiative recombination in the Si matrix layers can be successfully suppressed. Further investigations should be targeted at employing H-treatment methods that omit proton implantation techniques, thus providing the benefits of the passivation of unwanted dangling bonds while not causing structural harm to any part of the device layer. As such, H-treatment techniques as used in the photovoltaic industry should be excellent candidates to achieve this goal [206, 209]. There, H is supplied through the deposition of a H-rich SiN layer and then in a diffusion-driven process added to the Si crystal through advanced annealing techniques [209].

## 2.4 Summary and Future Directions in DEQD Research

Conventional wisdom, developed over decades of research on Si technology, holds that microelectronic-grade quality requires (apart from doping) ultralow concentrations of impurities and crystalline defects that act as electronic traps. To this end, and in retrospect, the bright PL in the telecommunication range from QDs containing split—[110] self-interstitial defects could have been discovered much earlier; the formation of dislocation-free QDs was demonstrated about 30 years ago [111] while the split—[110] self-interstitial, an easily forming extended point defect, has been discovered also decades ago [192–194]. All that really needed to be done was to combine these two approaches. However, this idea would have been hard to conceive of since both constituents, the split—[110] self-interstitial in bulk and the defect-free epitaxial Ge/Si QDs show only weak or even no luminescence emission at room temperature. In this respect, it seems to make quite some sense to revisit the combination of numerous group-IV nanostructure architectures and defect types, in terms of their light-emission properties, both on theoretical as well as on experimental grounds. Future research will need to be devoted to obtaining electrically driven lasing, and the microscopic understanding of the processes, leading to the enhanced light-emission from DEQDs, including the influence of SiGe composition and strain on the optical properties.

For DEQDs, efficient light-emission at room temperature was demonstrated, both under optical and electrical excitation. The band structure modifications induced by the implanted defects leads to optical direct bandgap transitions and type-I band alignment, thus overcoming the main drawbacks of epitaxial Ge/Si QDs. When placed into photonic resonators, clear signs for optically pumped lasing using DEQD light-emitters have been demonstrated. As a next step, the DEQD technology has to be transferred to an SOI platform, allowing also mode confinement in electrically pumped devices. This would allow achieving electrically pumped lasing, ideally at room temperature, using DEQD gain material. Such laser sources will form the basis for a monolithic optoelectronic platform enabling superior data transfer rates and novel optical sensing functionalities for a vast number of applications. Despite the truly exciting feature of DEQDs, we believe that the preliminary research on these just scratched the surface of what is possible using DEQDs as Si light sources.

**Acknowledgements** I would like to use this opportunity to gratefully thank the main mentors in my scientific career, Günther Bauer, Friedrich Schäffler, Thomas Fromherz and Armando Rastelli. This work was funded by the Austrian Science Fund (FWF): Y1238-N36, P29137-N36. Funding was also provided by the EU H2020 QuantERA ERA-NET via the Quantum Technologies project CUSPIDOR, which is co-funded by FWF(I3760) and the Linz Institute of Technology (LIT): LIT-2019-7-SEE-114.

## References

1. D.J. Paul, Semicond. Sci. Technol. **19**, R75 (2004)
2. G. Capellini, M. De Seta, F. Evangelisti, Appl. Phys. Lett. **78**, 303 (2001)
3. G. Capellini, M. De Seta, F. Evangelisti, J. Appl. Phys. **93**, 291 (2003)
4. J.L. Gray, S. Atha, R. Hull, J.A. Floro, Nano Lett. **4**, 2447 (2004)
5. I. Berbezier, A. Ronda, Surf. Sci. Rep. **64**, 47 (2009)
6. J.N. Aqua, I. Berbezier, L. Favre, T. Frisch, A. Ronda, Phys. Rep. **522**, 59 (2013)
7. D. Grützmacher et al., Nano Lett. **7**, 3150 (2007)
8. T. Tayagaki, S. Fukatsu, Y. Kanemitsu, Phys. Rev. B **79**, 041301 (2009)
9. J.M. Amatya, J.A. Floro, Appl. Phys. Lett. **109**, 193112 (2016)
10. S. Wirths et al., Nat. Photonics **9**, 88 (2015)
11. V. Schlykow et al., Nanotechnology **29**, 415702 (2018)
12. V. Reboud et al., Appl. Phys. Lett. **111**, 092101 (2017)
13. W. Dou et al., Opt. Lett. **43**, 4558–4561 (2018)
14. A.D. Lee, J. Qi, T. Mingchu, Z. Yunyan, A.J. Seeds, H. Liu, IEEE J. Sel. Top. Quantum Electron. **19**, 1701510 (2013)

15. S. Chen et al., Nat. Photonics **10**, 307 (2016)
16. G. Pellegrini et al., ACS Photonics **5**, 3601–3607 (2018)
17. R.W. Millar et al., Opt. Exp. **25**, 25374 (2017)
18. M. Brehm, M. Grydlik, Nanotechnology **28**, 392001 (2017)
19. M. Grydlik et al., ACS Photonics **3**, 298 (2016)
20. M. Grydlik et al., Nano Lett. **16**, 6802 (2016)
21. P. Rauter et al., ACS Photonics **5**, 431 (2018)
22. H. Groiss et al., Semicond. Sci. Technol. **32**, 02LT01 (2017)
23. R.A. Soref, J. Schmidtchen, K. Petermann, IEEE J. Quantum Electron. **27**, 1971 (1991)
24. L. Tsybeskov, D.J. Lockwood, Proc. IEEE **97**, 1284 (2009)
25. P. Chaisakul et al., Nat. Photonics **8**, 482 (2014)
26. B. Stern et al., Optica **2**, 530 (2015)
27. L. Zhou, A.W. Poon, Opt. Exp. **15**, 9194 (2007)
28. L.W. Luo et al., Nat. Commun. **5**, 3069 (2014)
29. G.T. Reed, G. Mashanovich, F.Y. Gardes, D.J. Thomson, Nat. Photonics **4**, 518 (2010)
30. D.C.S. Dumas et al., Opt. Exp. **22**, 19284 (2014)
31. J. Michel, J. Liu, L.C. Kimerling, Nat. Photonics **4**, 527 (2010)
32. L. Vivien et al., Opt. Exp. **20**, 1096 (2012)
33. R.A. Soref, J.P. Lorenzo, Quant. Electron. **22**, 873 (1986)
34. M.A. Green, J. Zhao, A. Wang, P.J. Reece, M. Gal, Nature **412**, 805 (2001)
35. E.Ö. Sveinbjörnsson, J. Weber Appl. Phys. Lett. **69**, 2686 (1996)
36. L. Ng, Nature **410**, 192 (2001)
37. L.T. Canham, Appl. Phys. Lett. **57**, 1046 (1990)
38. F. Priolo, T. Gregorkiewicz, M. Galli, T.F. Krauss, Nat. Nanotechn. **9**, 19 (2014)
39. S. Takeoka, M. Fujii, S. Hayashi, Phys. Rev. B **62**, 16820 (2000)
40. L. Pavesi, L. Dal Negro, C. Mazzoleni, G. Franzo, F. Priolo, Nature **408**, 440 (2000)
41. J. Valenta, R. Juhasz, J. Linnros, Appl. Phys. Lett. **80**, 1070 (2002)
42. H. Ennen et al., Appl. Phys. Lett. **46**, 381 (1985)
43. Z. Fang, Q.Y. Chen, C.Z. Zhao, Opt. Laser Techn. **46**, 103 (2013)
44. H. Krzyżanowska, K.S. Ni, Y. Fu, P.M. Fauchet, Mat. Sci. Eng. B **177**, 1547 (2012)
45. A. Tengattini et al., J. Lightwave Techn. **31**, 391 (2013)
46. Y. Takahashi et al., Nature **498**, 470 (2013)
47. H. Rong et al., Nature **433**, 292 (2005)
48. M.J. Süess et al., Nat. Photonics **7**, 466 (2013)
49. J. Liu, X. Sun, R. Camacho-Aguilera, L.C. Kimerling, J. Michel, Opt. Lett. **35**, 679 (2010)
50. F.T. Armand Pilon et al., Nat. Commun. **10**, 2724 (2019)
51. S. Bao et al., Nat. Commun. **8**, 1–7 (2017)
52. D. Stange et al., Optica **4**, 185 (2017)
53. A. Elbaz et al., Nat. Photonics **14**, 375–382 (2020)
54. E.M.T. Fadaly et al., Nature **580**, 205–209 (2020)
55. H.I.T. Hauge et al., Nano Lett. **15**, 5855–5860 (2015)
56. A.W. Fang et al., Opt. Exp. **14**, 9203 (2006)
57. M. Liao et al., IEEE J. Sel. Top. Quantum Electron. **23**, 1900910 (2017)
58. Y. Wan et al., Opt. Lett. **41**, 1664 (2016)
59. C. Cornet, Y. Léger, C. Robert, *Integrated Lasers on Silicon* (ISTE Press, London, UK, 2016)
60. V. Kveder et al., Appl. Phys. Lett. **84**, 2106 (2004)
61. A. Nikolskaya et al., Nucl. Instrum. Methods Phys. Res. Sect. B **472**, 32–35 (2020)
62. D.S. Korolev et al., J. Phys. Conf. Ser. **1410**, 012152 (2019)
63. A.A. Shklyaev, Y. Nakamura, F.N. Dultsev, M. Ichikawa, J. Appl. Phys. **105**, 063513 (2009)
64. S.M. Buckley et al., Opt. Exp. **28**, 16057–16072 (2020)
65. A.G. Nassiopoulos et al., Thin Solid Films **255**, 329–333 (1995)
66. O. Bisia, S. Ossicini, L. Pavesi, Surf. Sci. Rep. **38**, 1–126 (2000)
67. J. Valenta et al., Sci. Rep. **9**, 11214 (2019)
68. J. Heitmann, F. Müller, M. Zacharias, U. Gösele, Adv. Mater. **17**, 795–803 (2005)

69. A.J. Kenyon, Semicond. Sci. Technol. **20**, R65–R84 (2005)
70. P.A. Dmitriev et al., Nanoscale **8**, 9721–9726 (2016)
71. M. Virgilio, C.L. Manganelli, G. Grosso, T. Schroeder, G. Capellini, J. Appl. Phys. **114**, 243102 (2013)
72. M. El Kurdi et al., Appl. Phys. Lett. **108**, 091103 (2016)
73. X. Li, Z. Li, S. Li, L. Chrostowski, G. Xia, Semicond. Sci. Technol. **31**, 065015 (2016)
74. C. Xu, J. Kouvetakis, J. Menéndez, J. Appl. Phys. **125**, 085704 (2019)
75. Y. Zhou et al., ACS Photonics **6**, 1434–1441 (2019)
76. Y. Zhou et al., Optica **7**, 924–928 (2020)
77. H. Groiss et al., Sci. Rep. **7**, 16114 (2017)
78. P. Zaumseil et al., APL Mater. **6**, 076108 (2018)
79. J. Nicolas et al., Cryst. Growth Des. **20**, 3493 (2020)
80. U. Gnutzmann, K. Clausecker, Appl. Phys. **3**, 9–14 (1974)
81. Y. Qiu et al., Sci. Rep. **5**, 12692 (2015)
82. D. Liang, G. Roelkens, R. Baets, J. E. Bowers Mater. **3**, 1782 (2010)
83. K. Tanabe, K. Watanabe, Y. Arakawa, Sci. Rep. **2**, 349 (2012)
84. E. Menard, K.J. Lee, D.-Y. Khang, R.G. Nuzzo, J.A. Rogers, Appl. Phys. Lett. **84**, 5398–5400 (2004)
85. Z. Wang et al., Laser Photonics Rev. **11**, 1700063 (2017)
86. J. Zhang et al., APL Photonics **4**, 110803 (2020)
87. S. Pan et al., J. Semiconductors **40**, 101302 (2019)
88. Q. Feng et al., Appl. Sci. **9**, 385 (2019)
89. C. Hantschmann et al., J. Lightwave Technol. **38**, 4801 (2020)
90. R. Chen et al., Nat. Photonics **5**, 170 (2011)
91. G. Koblmüller, B. Mayer, T. Stettner, G. Abstreiter, J.J. Finley, Semicond. Sci. Technol. **32**, 053001 (2017)
92. T. Stettner et al., ACS Photonics **4**, 2537–2543 (2017)
93. J. Bissinger, D. Ruhstorfer, T. Stettner, G. Koblmüller, J.J. Finley, J. Appl. Phys. **125**, 243102 (2019)
94. S. Conesa-Boj et al., Nano Lett. **14**, 326–332 (2014)
95. T. Frost et al., Nano Lett. **14**, 4535–4541 (2014)
96. Z. Wang et al., Nat. Photonics **9**, 837 (2015)
97. J. Van Campenhout et al., Opt. Exp. **15**, 6744 (2007)
98. Y. Shi et al., Optica **4**, 1468–1473 (2017)
99. Y. Han et al., Optica **7**, 148 (2020)
100. Y. Xue et al., Opt. Exp. **28**, 18172 (2020)
101. J.R. Reboul, L. Cerutti, J.B. Rodriguez, P. Grech, E. Tournié, Appl. Phys. Lett. **99**, 121113 (2011)
102. M.R. Calvo et al., Optica **7**, 263–266 (2020)
103. Y. Sun et al., Light Sci. Appl. **7**, 13 (2018)
104. J. Wang et al., Photonics Res. **7**, B32–B35 (2019)
105. P.D. Kanungo et al., Nanotechnology **24**, 225304 (2013)
106. M. Borg et al., Nano Lett. **14**, 1914–1920 (2014)
107. H. Schmid et al., Appl. Phys. Lett. **106**, 233101 (2015)
108. B.F. Mayer et al., IEEE Photonics Technol. Lett. **31**, 1021–1024 (2019)
109. Z.I. Alferov, Rev. Mod. Phys. **73**, 767 (2001)
110. https://www.statista.com/statistics/266973/global-semiconductor-sales-since-1988/
111. Y.-W. Mo, D.E. Savage, B.S. Swartzentruber, M.G. Lagally, Phys. Rev. Lett. **65**, 1020 (1990)
112. D.J. Eaglesham, M. Cerullo, Phys. Rev. Lett. **64**, 1943 (1990)
113. F. Hackl et al., Nanotechnology **22**, 165302 (2011)
114. L. Vescan et al., Appl. Phys. A **71**, 423 (2000)
115. D. Pachinger et al., Appl. Phys. Lett. **91**, 23 (2007)
116. L. Persichetti et al., Surf. Sci. **683**, 31–37 (2019)
117. A.V. Novikov, M.V. Shaleev, D.V. Yurasov, P.A. Yunin, Semiconductors **50**, 1630 (2016)

118. M. Abbarchi et al., ACS Nano. **8**, 11181–11190 (2014)
119. V. Poborchii et al., Nanotechnology **31**, 195602 (2020)
120. G. Niu et al., Sci. Rep. **6**, 22709 (2016)
121. N. Motta et al., Mater. Sci. Eng. B **88**, 264–268 (2002)
122. C. Ishii, Y. Shigeta, Thin Solid Films **709**, 138007 (2020)
123. D. Rainko et al., Sci. Rep. **8**, 15557 (2018)
124. I.N. Stranski, L. Krastanow, Monatsh. Chem. **71**, 351–364 (1937)
125. B. Voigtländer, Surf. Sci. Rep. **43**, 127–254 (2001)
126. C. Teichert, Phys. Rep. **365**, 335–432 (2002)
127. G. Costantini et al., J. Cryst. Growth **278**, 38–45 (2005)
128. J. Zhang, M. Brehm, M. Grydlik, O.G. Schmidt, Chem. Soc. Rev. **44**, 26–39 (2015)
129. M. Strassburg et al., Appl. Phys. Lett. **76**, 685 (2000)
130. E. Dimakis et al., J. Appl. Phys. **97**, 113520 (2005)
131. S.G. Corcoran, G.S. Chakarova, K. Sieradzki, Phys. Rev. Lett. **71**, 1585 (1993)
132. S. Haq, J. Harnett, A. Hodgson, Surf. Sci. **505**, 171 (2002)
133. G. Medeiros-Ribeiro, A.M. Bratkovski, T.I. Kamins, D.A.A. Ohlberg, R.S. Williams, Science **279**, 353 (1998)
134. E. Sutter, P. Sutter, J.E. Bernard, Appl. Phys. Lett. **84**, 2262 (2004)
135. M. Brehm, H. Lichtenberger, T. Fromherz, G. Springholz, Nanoscale Res. Lett. **6**, 70 (2011)
136. A. Vailionis et al., Phys. Rev. Lett. **85**, 3672 (2000)
137. A. Rastelli, H. von Känel, Surf. Sci. **532–535**, 769–773 (2003)
138. M. Brehm et al., Phys. Rev. B **80**, 205321 (2009)
139. M.R. McKay, J.A. Venables, J. Drucker, Phys. Rev. Lett. **101**, 216104 (2008)
140. M. Brehm et al., J. Appl. Phys. **109**, 123505 (2011)
141. T.U. Schülli et al., Phys. Rev. Lett. **90**, 066105 (2003)
142. A. Rastelli et al., Nano Lett. **8**, 1404 (2008)
143. M. Brehm et al., Nanoscale Res. Lett. **5**, 1868 (2010)
144. F. Hackl et al., Ann. Phys. **531**, 1800259 (2019)
145. C. Georgiou, T. Leontiou, P.C. Kelires, AIP Adv. **4**, 077135 (2014)
146. P. Klenovský et al., Phys. Rev. B **86**, 115305 (2012)
147. G. Capellini, M. De Seta, L. Di Gaspare, F. Evangelisti, F. d'Acapito, J. Appl. Phys. **98**, 124901 (2005)
148. A. Rastelli, E. Müller, H. von Känel, Appl. Phys. Lett. **80**, 1438 (2002)
149. O.G. Schmidt, U. Denker, K. Eberl, O. Kienzle, F. Ernst, Appl. Phys. Lett. **77**, 2509 (2000)
150. M. Brehm et al., Nanotechnology **26**, 225202 (2015)
151. A. Rastelli, M. Kummer, H. von Känel, Phys. Rev. Lett. **87**, 256101 (2001)
152. F. Montalenti et al., Phys. Rev. Lett. **93**, 216102 (2004)
153. J. Stangl, V. Holý, G. Bauer, Rev. Mod. Phys. **76**, 725 (2004)
154. A.F. Zinovieva et al., Sci. Rep. **10**, 9308 (2020)
155. O.G. Schmidt, K. Eberl, Y. Rau, Phys. Rev. B **62**, 16715 (2000)
156. J.M. Baribeau, N.L. Rowell, D.J. Lockwood, *Nanoscale Materials* (Springer, New York, 2006), pp. 1–52
157. M. Brehm et al., New J. Phys. **11**, 063021 (2009)
158. R.O. Rezaev, S. Kiravittaya, V.M. Fomin, A. Rastelli, O.G. Schmidt, Phys. Rev. B **82**, 153306 (2010)
159. M. Grydlik, G. Langer, T. Fromherz, F. Schäffler, M. Brehm, Nanotechnology **24**, 105601 (2013)
160. A.V. Dvurechenskii et al., Phys. Stat. Sol. C **14**, 1700187 (2017)
161. M. Grydlik et al., Phys. Rev. B **88**, 115311 (2013)
162. J. Xia, Y. Ikegami, Y. Shiraki, N. Usami, Y. Nakata, Appl. Phys. Lett. **89**, 201102 (2006)
163. X. Xu et al., IEEE J. Sel. Top. Quantum Electron. **18**, 1830 (2012)
164. N. Hauke et al., New J. Phys. **14**, 083035 (2012)
165. T. Tsuboi et al., Appl. Phys. Exp. **5**, 052101 (2012)
166. J.S. Xia, Y. Takeda, N. Usami, T. Maruizumi, Y. Shiraki, Opt. Exp. **18**, 13945 (2010)

167. M. Schatzl et al., ACS Phot. **4**, 665 (2017)
168. A. Simbula et al., APL Photonics **2**, 056102 (2017)
169. A.I. Yakimov, V.V. Kirienko, A.V. Dvurechenskii, Opt. Mat. Exp. **8**, 3479 (2018)
170. V. Schlykow et al., Nanotechnology **31**, 345203 (2020)
171. T.I. Kamins, R.S. Williams, Appl. Phys. Lett. **71**, 1201 (1997)
172. P.D. Szkutnik, A. Sgarlata, S. Nufris, N. Motta, A. Balzarotti, Phys. Rev. B **69**, 201309 (2004)
173. J.K. Murphy et al., J. Vac. Sci. Technol. B **29**, 011029 (2011)
174. A. Karmous et al., Appl. Phys. Lett. **85**, 6401–6403 (2004)
175. Y.J. Ma et al., J. Phys. D Appl. Phys. **47**, 485303 (2014)
176. J.J. Zhang et al., Phys. Rev. Lett. **105**, 166102 (2010)
177. F. Pezzoli, M. Stoffel, T. Merdzhanova, A. Rastelli, O.G. Schmidt, Nanoscale Res. Lett. **4**, 1073 (2009)
178. M. Grydlik et al., Appl. Phys. Lett. **106**, 251904 (2015)
179. J. Ziegler, M. Ziegler, J. Biersack, Nucl. Instrum. Methods Phys. Res. Sect. B **268**, 1818 (2010)
180. G. Davies, Phys. Rep. **176**, 83–188 (1989)
181. L. Ouyang et al., Opt. Exp. **26**, 15899 (2018)
182. A. Shakoor et al., Physica B **407**, 4027–4031 (2012)
183. S. Facsko et al., Science **285**, 1551–1553 (1999)
184. T. Bobek et al., Phys. Rev. B **68**, 085324 (2003)
185. K.J. Kirkby, R.P. Webb, *Encyclopedia of Nanoscience and Nanotechnology*, edited by H.S. Nalwa, vol. 4. (American Scientific, Valecia, USA, 2004), pp. 283–291
186. N. Chekurov, K. Grigoras, A. Peltonen, S. Franssila, I. Tittonen, Nanotechnology **20**, 065307 (2009)
187. L. Spindlberger, S. Prucnal, J. Aberl, M. Brehm, Physica Status Solidi (a) **216**, 1900307 (2019)
188. L. Spindlberger et al., Crystals **10**, 351 (2020)
189. R.S. Averback, T.D. de la Rubia, Solid State Phys. **51**, 281–402 (1997)
190. G. Carter, W.A. Grant, *Ion Implantation of Semiconductors*. (London Edward Arnold Ltd. 1976), pp. 1976 1−214
191. J.S. Williams, Nucl. Instrum. Methods Phys. Res. **209**, 219–228 (1983)
192. P.E. Blöchl et al., Phys. Rev. Lett. **70**, 2435 (1993)
193. A.J.R. da Silva, A. Janotti, A. Fazzio, R.J. Baierle, R. Mota, Phys. Rev. B **62**, 9903 (2000)
194. M. Dionízio Moreira, R.H. Miwa, P. Venezuela, Phys. Rev. B **70**, 115215 (2004)
195. P. Giannozzi et al., J. Phys. Condens. Matter **21**, 395502 (2009)
196. F. Murphy-Armando et al., Phys. Rev. B **103**, 085310 (2021)
197. E.-K. Lee et al., Phys. Rev. B **80**, 049904 (2009)
198. B. Julsgaard, P. Balling, J.L. Hansen, A. Svane, A.N. Larsen, Nanotechnology **22**, 435401 (2011)
199. E.C. Le Ru, J. Fack, R. Murray, Phys. Rev. B **67**, 245318 (2003)
200. M.W. Dashiell, U. Denker, O.G. Schmidt, Appl. Phys. Lett. **79**, 2261 (2001)
201. T. Brunhes et al., Appl. Phys. Lett. **77**, 1822 (2000)
202. D.N. Lobanov et al., Semiconductors **46**, 1418–1422 (2012)
203. H. Takeuchi, A. Wung, X. Sun, R.T. Howe, T.-J. King, IEEE Trans. Electron. Dev. **52**, 2081–2086 (2005)
204. H. Xiao, *Introduction to Semiconductor Manufacturing Technology* (Prentice-Hall, Upper Saddle River, NJ, USA, 2001)
205. E. Cartier, J.H. Stathis, D.A. Buchanan, Appl. Phys. Lett. **63**, 1510 (1993)
206. S. Wang et al., Sol. Energ. Mater. Sol. Cells **193**, 403–410 (2019)
207. S. Fujimori, R. Nagai, M. Ikeda, K. Makihara, S. Miyazaki, Jpn. J. Appl. Phys. **58**, SIIA01 (2019)
208. G. Davies, Phys. Rep. **176**, 83–188 (1989)
209. B. Hallam et al., Phys. Status Solidi A **214**, 1700305 (2017)

# Chapter 3
# Lasing in Group-IV Materials

**V. Reboud, D. Buca, H. Sigg, J. M. Hartmann, Z. Ikonic, N. Pauc, V. Calvo, P. Rodriguez, and A. Chelnokov**

**Abstract** Silicon photonics in the near-IR, up to 1.6 μm, is already one of key technologies in optical data communications, particularly short range. It also is being prospected for applications in quantum computing, artificial intelligence, optical signal processing, where complex photonic integration is to be combined with a large-volume fabrication. However, silicon photonics does not yet cover a large portion of applications in the mid-IR. In the wavelength range of 2–5 μm, environmental sensing, life sensing and security, all rely on optical signatures of molecular vibrations to identify complex individual chemical species. The markets for such analysis are huge and constantly growing, with a push for sensitivity, specificity, compactness, low-power operation and low cost. An all-group-IV, CMOS-compatible mid-IR integrated photonic platform would be a key enabler in this wavelength range. As for other wavelengths, such a platform should be complete with low-loss guided interconnects, detectors, eventually modulators, and most important an efficient and integrated light sources. This chapter reviews the recent developments of mid-IR silicon-compatible optically and electrically pumped lasers, light emitting diodes and photodetectors based on Ge, GeSn and SiGeSn alloys. It contains insights into the fundamentals of these developments, including bandstructure modelling, material growth and processing techniques.

V. Reboud (✉) · J. M. Hartmann · P. Rodriguez · A. Chelnokov
Université Grenoble Alpes, CEA, LETI, 38054 Grenoble, France
e-mail: vincent.reboud@cea.fr

D. Buca
Institute of Semiconductor Nanoelectronics, Peter Grünberg Institute 9 (PGI 9) and JARA Fundamentals of Future Information Technologies, Forschungszentrum Jülich, 52074 Jülich, Germany

H. Sigg
Laboratory for Micro- and Nanotechnology, Paul Scherrer Institut, 5232 Villigen, Switzerland

Z. Ikonic
School of Electronic and Electrical Engineering, Pollard Institute, University of Leeds, Leeds, UK

N. Pauc · V. Calvo
Univ. Grenoble Alpes, CEA, IRIG-DePhy, 38054 Grenoble, France

 
D. J. Lockwood and L. Pavesi (eds.), *Silicon Photonics IV*,
Topics in Applied Physics 139, https://doi.org/10.1007/978-3-030-68222-4_3

## Abbreviations

| | |
|---|---|
| AFM | Atomic force microscopy |
| APT | Atom probe tomography |
| AP | Atmospheric pressure |
| APB | Anti-phase boundaries |
| ASE | Amplified spontaneous emission |
| CMOS | Complementary metal–oxide–semiconductor |
| CVD | Chemical vapour deposition |
| CW | Continuous wave |
| DBR | Distributed Bragg reflector |
| DFB | Distributed feedback |
| DHS | Double heterostructure |
| DOS | Density of states |
| EL | Electroluminescence |
| FCA | Free carrier absorption |
| FLP | Fermi-level pinning |
| F-P | Fabry–Perot |
| GBs | Grain boundaries |
| Ge | Germanium |
| GeOI | Germanium-on-insulator |
| GeSn | Germanium Tin |
| GR | Growth rate |
| HH | Heavy holes |
| IVBA | Inter-valence band absorption |
| LED | Light emitting device |
| LH | Light holes |
| L–L | Light-in–light-out |
| MD | Microdisc |
| MIGS | Metal-induced gap states |
| MPW | Multi-quantum well |
| MSM | Metal–semiconductor–metal |
| PAI | Pre-amorphized by implantation |
| PC | Photonic crystals |
| PD | Photodiode |
| PMMA | Polymethyl methacrylate |
| QDs | Quantum dots |
| rms | (Surface) root mean square (roughness) |
| RP-CVD | Reduced pressure–chemical vapour deposition |
| RSM | Reciprocal space map |
| Si | Silicon |
| SiGeSn | Silicon germanium tin |
| Sn | Tin |
| SO | Spin orbit |

| | |
|---|---|
| SRB | Strain-relaxed buffers |
| SSR | Solid-state reaction |
| TDD | Threading dislocations density |
| TEM | Surface root mean square |
| TMAH | Tetramethylammonium hydroxide |
| VLS | Variable stripe length |
| VS | Virtual substrate |
| WG | Waveguide |
| XRD | X-ray diffraction |

## 3.1 Introduction

Silicon photonics in the near-IR, up to 1.6 μm, is already one of key technologies in optical data communications, particularly short range. It also is being prospected for applications in quantum computing, artificial intelligence, optical signal processing, where complex photonic integration is to be combined with a large-volume fabrication. However, silicon photonics does not yet cover a large portion of applications in the mid-IR. In the wavelength range of 2–5 μm, environmental sensing, life sensing and security, all rely on optical signatures of molecular vibrations to identify complex individual chemical species. The markets for such analysis are huge and constantly growing, with a push for sensitivity, specificity, compactness, low-power operation and low cost.

An all-group-IV, CMOS-compatible mid-IR integrated photonic platform would be a key enabler in this wavelength range. As for other wavelengths, such a platform should be complete with low-loss guided interconnects, detectors, eventually modulators, and most important an efficient and integrated light sources. Since the first demonstration of lasing in a GeSn optical cavity in 2015 [1], remarkable progress has been achieved in the development of silicon germanium tin (SiGeSn) group-IV semiconductors as laser materials, including near room temperature laser operation [2], lasing at ultra-low threshold power densities [3] and near unity wall-plug efficiency [4]. Most recently, an electrically pumped germanium tin (GeSn) laser operating at temperatures up to 100 K was reported [5]. Our task here is to walk through the recent developments of optically pumped lasers and optoelectronic devices (including lasers, light emitting diodes and photodetectors). We will also present insights into the fundamentals of these developments, e.g. the material growth and processing techniques as well as the bandstructure modelling. We hope that this will give the reader the opportunity to learn from the past and find his own best strategy for the future.

Over the years, numerous concepts had been put forward such as SiGe superlattice zone-folding [6], confinement-induced reciprocal space engineering [7] as well as hexagonal core shell nanowires [8], to name a few. These are concepts that “by-pass” the problem of the indirect band structure of silicon and, to a lesser extent, germanium. Defect doping [9], as described in "stimulated emission in the near-infrared from disordered Ge dots" chapter of Moritz Brehm, Johannes Kepler University, Linz in this book, is another way of bypassing this limitation [10]. Here, instead of working around it, we solve the problem from scratch. Namely, a group-IV material is developed that, by its very nature, is a direct bandgap system. The community is currently exploring two directions. The first one consists in transforming the indirect Ge bandgap into a direct one using high amounts of tensile strain [4, 11]. The second one relies on the alloying of Ge with Sn and the control of the GeSn layer's strain [12, 13]. Both methods improve dramatically the Ge light emission properties in the short mid-infrared domain (in the 2–5 $\mu$m wavelength range) and, as shown in the following, enable the fabrication of lasing devices.

The current research is focussed on (i) the reduction of the laser threshold and (ii) the increase of the maximum lasing temperature. The lasing threshold density is closely related to loss mechanisms such as non-radiative recombination via defects at the surface, at the interfaces and also in the bulk of the optically active material. Such defects may be minimized by dedicated growth and etching strategies, as detailed below. Carrier confinement is another route to lower the laser threshold. This, however, requires appropriate barrier layers and the growth of hetero- or (multi)-quantum well structures. Maximal temperature operation is—as will be shown—in the first place related to the amount of Sn incorporated into the active layer. As expected, most of the strategies do come with trade-offs. For instance, the defect density is higher in higher Sn content, direct bandgap GeSn layers, resulting in non-radiative recombination channels which jeopardize the threshold density. Meanwhile, carrier confinement in wells induces quantization effects that reduce bandgap directness.

This review is meant to reflect such ongoing discussions related to material growth, device manufacturing and device concepts. As it happens often in a review, sometimes we had to choose between going into details but not cover all possible options, or the opposite, to mention all options but only cursorily. For the technical part, the growth and processing, we have indeed chosen the detailed approach, which necessarily reflects our personal experiences. This was to give the reader a sense of what is needed, but we of course acknowledge the many good works done elsewhere and thoroughly cite them. The review also gives an overview of recent experimental achievements.

Our review is organized as follows: Sect. 3.2 describes recent Ge and SiGeSn epitaxy developments and processes to fabricate devices in group-IV indirect and direct bandgap materials. The impact of strain and of Sn incorporation on the band structures of Ge, to transform it from an indirect into a direct bandgap material, is detailed in Sect. 3.3. In Sect. 3.4, the latest advances on lasing in Ge-based materials are reviewed. A state of the art of Ge-based optoelectronic devices that include light emitting devices (LEDs), photodetectors and lasers, is provided in Sect. 3.5. In Sect. 3.6, we will conclude by describing the opportunities Ge and GeSn offer to a future monolithic integration on a Si photonic platform and discuss future developments needed to reach this goal.

## 3.2 Fabrication: Ge-Based Epitaxy and Processes for Group-IV Indirect and Direct Bandgap Material

Germanium is a strategic material for microelectronic and photonic applications. It can be grown on Si(100): it is widely used for (i) near-infrared photodetectors thanks to its indirect bandgap of 0.77 eV at room temperature and (ii) metal–oxide–semiconductor field effect transistors (MOSFETs), with higher hole mobilities than in Si. We will review in this section Ge, GeSn and SiGeSn epitaxial growth and some of the processes, such as etching, which are specific to those materials. Metallization, in order to benefit from efficient contacts on GeSn-based layers and fabricate electrically pumped group-IV devices, will also be discussed, with a focus on thermal stability.

### *3.2.1 Germanium Growth*

Thick Ge strain-relaxed buffers (SRBs) on Si(001) can be used for a variety of purposes in microelectronics and optoelectronics. The top part of microns-thick Ge layers can be peeled-off from the Ge/Si stack underneath and bonded on oxidized Si using the SmartCutTM approach [14], resulting in germanium-on-insulator (GeOI) substrates. The latter can be used as templates for the fabrication of high mobility p-type metal–oxide–semiconductor field effect transistors [15]. The slight tensile strain present in thick Ge films on top of buried oxides can be exploited for the fabrication of highly strained suspended micro-bridges and micro-crosses [16–18]. Because of their lattice parameter, which is 4.2% higher than that of Si (5.65785 Å ⇔ 5.43105 Å) and close to that of GaAs (5.653 Å), Ge SRBs can be used as templates for the epitaxy of anti-phase boundaries-free, superior quality GaAs and InAlAs-based buffers [19, 20]. The later can be called upon for the fabrication of high electron mobility n-type MOSFETs on III-V-On-Insulator substrates [21]. Thick Ge layers grown selectively at the end of Silicon-On-Insulator (SOI) waveguides can act as the active cores of superior performance near-infrared photodetectors [18, 22, 23]. Large area, high-quality graphene 2D layers were recently deposited on Ge SRBs [24, 25]. Finally, Ge SRBs are handy for the deposition of high Sn content GeSn layers which are typically used for the fabrication of mid-IR optically pumped lasers [1, 2], as we will see in the following sections.

Such SRBs are typically grown on Si(001) using a low-temperature/high-temperature approach [26], with a short duration thermal cycling or anneal afterwards [27] to minimize the threading dislocations density (TDD). This TDD is typically around $10^7$ cm$^{-2}$ for 2.5-μm-thick Ge layers which are really smooth given the large lattice parameter mismatch between Si and Ge (surface root mean square roughness typically around 0.8 nm for 20 μm × 20 μm atomic force microscopy images). The TDD monotonously decreases with the Ge thickness, as shown for layers thicknesses up to 2.5 μm [28]. The universality of such a behaviour was conclusively demonstrated in the Ge/Si system and the GaAs/Si system in [29]. These findings were extended, for the Ge/Si system, to larger thicknesses (up to 5 μm) in [30] by CEA-LETI and recently confirmed by Leibniz-Institut für innovative Mikroelektronik (IHP) in [31].

The impact of thickness, in the 0.12–1.56 μm range, thermal cycling and substrate nature (nominal or slightly vicinal) on the properties of Ge layers grown on Si(001) was recently assessed in a 300 mm reduced pressure–chemical vapour deposition (RP-CVD) chamber from applied materials. The original results presented in the following subsections are typical of Ge SRBs grown in industrial epitaxy tools whatever their brand and the wafer diameters used.

#### 3.2.1.1 Ge Growth Protocol

The flow of $H_2$ carrier gas was set at a fixed value of a few tens of standard litres per minute. Germane ($GeH_4$) diluted at 2% in $H_2$ was used as the source of Ge. The $F(GeH_4)/F(H_2)$ mass–flow ratio was always equal to $10^{-3}$. The slightly p-type doped 300 mm Si(001) substrates used were either nominal (±0.25°) or slightly vicinal, with a 0.5° misorientation towards one of the ⟨110⟩ directions, to facilitate anti-phase boundaries (APB)-free GaAs growth on top [20]. During growth, the wafer laid horizontally on top of a circular SiC-coated susceptor plate that rotated to improve the spatial thickness uniformity of the films. It was heated by 76 tungsten–halogen lamps located above and below the susceptor assembly. Temperature monitoring and control was ensured through the lower pyrometer, i.e. the one looking at the backside of the susceptor plate on which the wafer laid.

The 0.12–1.56-μm-thick Ge layers were grown on nominal or 0.5° off-axis Si(001) substrates in three steps. After a 1100 °C, 2 min $H_2$ bake (to get rid of chemical oxide through the following reaction: $SiO_2(s) + 2H_2(g) \rightarrow Si(s) + 2H_2O(g)$), a 120-nm-thick Ge layer was grown in 460 s at 400 °C, 100 Torr, in order to start from a rather flat, nearly fully relaxed Ge "seed" layer. The Ge growth rate was then equal to 16 nm $min^{-1}$. In the second step, the temperature was ramped from 400 °C up to 750 °C (2.5 °C/s) and the growth pressure from 100 Torr down to 20 Torr while still having germane flowing into the growth chamber. Around 80 nm of Ge were deposited during the second step. In the third step, a Ge layer was grown at 750 °C, 20 Torr, with a 45 nm $min^{-1}$ growth rate, in order to obtain the desired thickness. A 3 × (875 °C, 10 s/750 °C, 10 s) thermal cycling under $H_2$ was used on the thickest Ge layers to minimize the threading dislocations density.

#### 3.2.1.2 Surface Morphology

2 μm × 2 μm and 20 μm × 20 μm atomic force microscopy (AFM) images of the surfaces of 1.56 μm thick, cyclically annealed Ge layers after growth on nominal and 0.5° off Si(001) substrates can be found in Fig. 3.1. They are characterized by the super-position of a cross-hatch along the ⟨110⟩ directions on top of holes and small hills bordered by bi-atomic step edges. This ⟨110⟩ cross-hatch is the surface

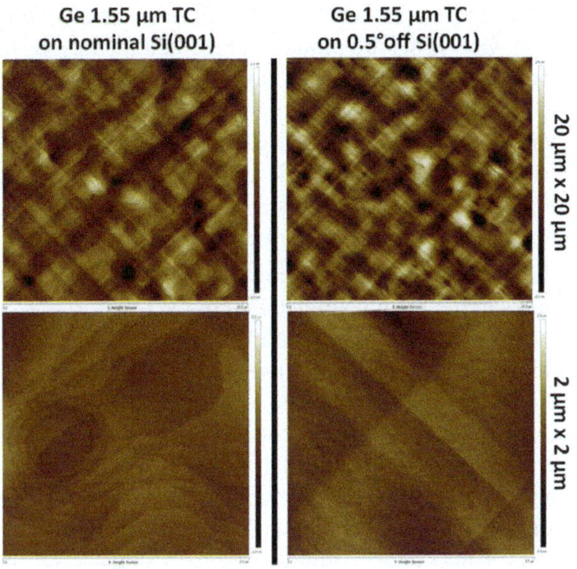


**Fig. 3.1** 20 µm × 20 µm AFM images and 2 µm × 2 µm close-ups of the surface of 1.56-µm-thick Ge layers grown on nominal and 0.5° off-axis Si(001) substrates (left and right images, respectively). Image sides are along the ⟨100⟩ directions, i.e. at 45° to the wafer notch and the cleaving directions

signature of the propagation on (111) planes of the threading arms of 60° misfit dislocations in the Ge layers, notably during thermal cycling. Those threading arms left in their wake ⟨110⟩ "plough" lines.

The main difference between the two types of surfaces is the bi-atomic step orientation and spacing. Those steps are randomly oriented, with a spacing between steps which fluctuates when starting from a nominal Si(001) substrate (Fig. 3.1 left column images). Meanwhile, they are parallel to one of the ⟨110⟩ directions (from the top left down to the bottom right of pictures) and are more closely spaced with a 0.5° off-axis Si(001) template (Fig. 3.1 right column images), which should be favourable for the growth of APB-free GaAs layers on top.

The surface root mean square (rms) roughness and the $Z$ ranges ($= Z_{\text{max.}} - Z_{\text{min.}}$) of those layers were extracted from those AFM images. They were rather small, with values for growth on a nominal surface 30% less, on average, than that for growth on a 0.5° off-axis substrate: 0.2 and 1.9 nm, to be compared with 0.3 and 2.5 nm (2 µm × 2 µm images) and 0.55 and 6.2 nm, to be compared with 0.7 and 7.1 nm (20 µm × 20 µm images).

#### 3.2.1.3 Macroscopic Degree of Strain Relaxation and Crystalline Quality from X-Ray Diffraction

High-resolution X-ray diffraction (XRD) was used to quantify the macroscopic degree of strain relaxation $R = (a_{\text{Ge}}^{//} - a_{\text{Si}})/(a_{\text{Ge}}^{//} - a_{\text{Si}})$ as a function of Ge layer thickness and cyclic anneal (or not). $a_{\text{Ge}}^{//}$, $a_{\text{Si}}$ and $a_{\text{Ge}}$ are the in-plane lattice parameter of the Ge layer, the lattice parameter of the Si substrate (5.43105 Å) and the lattice parameter of bulk Ge (5.65785 Å), respectively. Omega-2Theta profiles around the (004) XRD order are plotted in Fig. 3.2a for various thickness Ge layers on Si. Apart from the Si substrate peak (located at 34.564°), there is, at slightly more than 33°, the Ge layer peak, whose intensity drastically increases as the Ge layer thickness increases. The Ge peak is slightly asymmetric. The high incidence angle component is due to the interfacial GeSi alloy formed during the thermal cycling [27]. From the angular position $\omega_{Ge}$ of this peak, we can gain access to $a_{\mathrm{Ge}}^{\perp}$, the lattice parameter of the Ge layer in the growth direction, thanks to Bragg's Law: $2\left(a_{\mathrm{Ge}}^{\perp}/4\right)\sin\omega_{\mathrm{Ge}} = \lambda$, $\lambda$ being the Cu K$\alpha_1$ wavelength (1.5406 Å). $a_{\mathrm{Ge}}^{//}$ and then $R$ can then be determined thanks to the formalism detailed in [32].

Almost all Ge layers were slightly tensily-strained, see Fig. 3.2b, with R values which monotonously increased then stabilized (at a 104.5% value) as the thickness increased. Such a strain state was due to differences in thermal expansion coefficients between Ge and Si (see [32] for numerical values). The lattice parameter of Ge thick layers on Si substrates, which were almost fully relaxed at 750 °C (the upper growth temperature) or 875 °C (the upper temperature of thermal cycles), shrank differently from that of a bulk Ge crystal. $a_{\mathrm{Ge}}^{\perp}$ contracted with the thermal dilatation coefficient of Ge, whereas $a_{\mathrm{Ge}}^{//}$ shrank with the smaller one of the much thicker Si substrate underneath. We thus ended up with tensile-strained Ge layers as soon as there was a ramping-up to 750 °C or a thickening at 750 °C, with $R$ values in the 102.3–104.7% range. Using a short duration thermal cycling had no clear impact on R, which was exactly the same (104.7%) for a 1.56 μm cyclically annealed Ge layer than for the same thickness un-annealed one. The only layer which was still compressively strained was the 120-nm-thick Ge layer grown at 400 °C, with an R value, 94%, which was very close to 100% (full strain relaxation). Prestrain values as obtained by alternative methods are summarized in Sect. 3.3.1.5, Table 3.2.

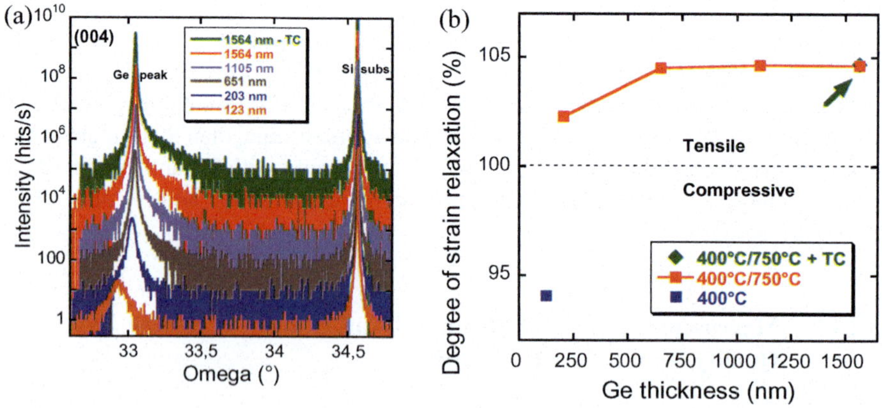


**Fig. 3.2** **a** Omega-2Theta scans around the (004) XRD order on various thickness Ge layers grown on Si(001). TC stands for thermal cycling, **b** macroscopic degree of strain relaxation as a function of Ge deposited thickness for layers grown at 400 °C, at 400 °C/750 °C or at 400 °C/750 °C followed by a short duration thermal cycling between 750 and 875 °C

#### 3.2.1.4 Threading Dislocations Density from X-Ray Diffraction

Ayers' theory [33] was then used in order to convert the full widths at half maximum $\beta$ of Ge peaks from omega scans around the (004) XRD order, such as the ones provided in Fig. 3.3a for various thickness Ge SRBs, into threading dislocation densities (TDDs). Ayers et al. have indeed shown that the TDD was related to $\beta$ by TDD $= (\beta/3b)^2$, $b$ being the length of Burger's vector if the rocking curve width is determined solely by threading dislocations. For pure Ge layers with 60° misfit dislocations, $b = 0.707a_{(\mathrm{Ge})}^{\mathrm{bulk}} = 4.00$ Å. The TDD (in $\mathrm{cm}^{-2}$) is then linked to $\beta$ (in arc seconds) by TDD $= 1632\beta^2$ for pure Ge.

TDDs coming from XRD and the use of Ayer's formula are plotted in Fig. 3.3b. Using thick Ge layers and high growth temperatures is advantageous: the TDD indeed drops from $3.6 \times 10^9$ $\mathrm{cm}^{-2}$ (123 nm @ 400 °C) down to $2.5 \times 10^7$ $\mathrm{cm}^{-2}$ (1564 nm @ 400 °C/750 °C + thermal cycling) as the Ge thickness increases.

One might wonder about the significance of such TDD values, which are indirect. We have shown in [34] that, for 2.5-μm-thick cyclically annealed Ge SRBs, we had after (i) the removal of one micron of Ge with Secco and chromium-free wet etching solutions and (ii) plane view transmission electron microscopy, a TDD of ~$10^7$ $\mathrm{cm}^{-2}$, to be compared with $8 \times 10^6$ $\mathrm{cm}^{-2}$ after HCl defect etching and $1.8 \times 10^7$ $\mathrm{cm}^{-2}$ from XRD in [30] layers. Figure 3.3b data thus seems trustworthy.

The reason as to why TDDs from HCl defect etching were on average 2.5 times lower than from XRD in [30] was tentatively due to the following: we had after HCl defect etching a TDD value which was a "surface" one, as only 130 nm of Ge was etched, while the XRD one came from the bulk of the Ge layer, with therefore a TDD mean value which might be higher than in reality. This might also be the reason why differences were deceptively small, in XRD, between as-grown and cyclically

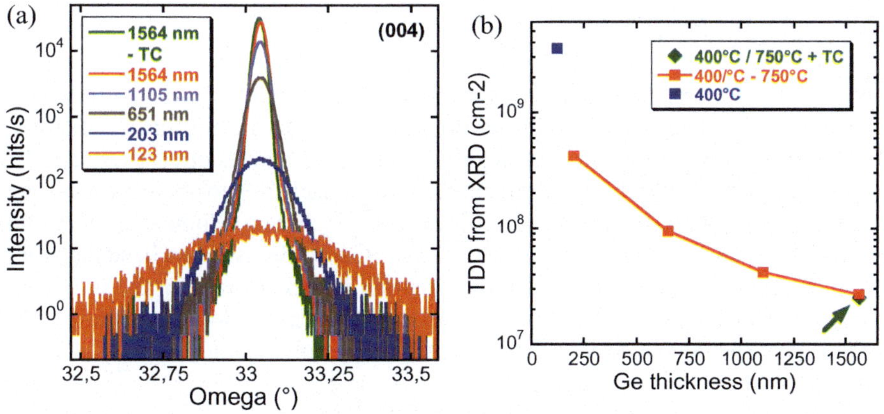


**Fig. 3.3** **a** Omega scans around the (004) XRD order on the various thickness Ge layers grown on Si(001). TC stands for thermal cycling, **b** threading dislocations densities ($\mathrm{cm}^{-2}$) inferred from Omega scans around the (004) Ge XRD peak and the use of Ayers' formula

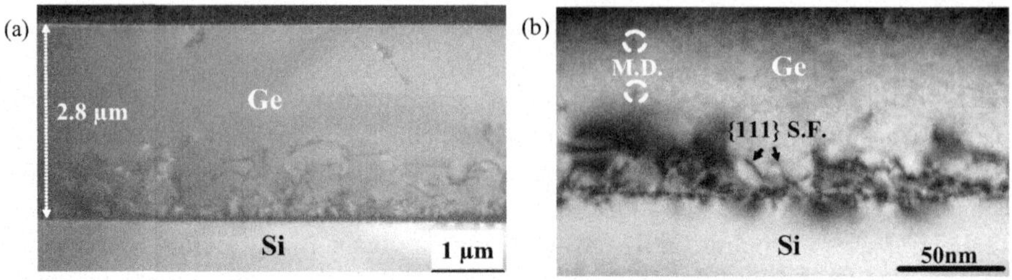


**Fig. 3.4** **a** Cross-sectional weak beam dark field image of a 2.8-μm-thick, annealed Ge layer grown on a nominal Si(001) substrate. **b** Dark field with close to Bragg conditions image (taken with $g = 220$) of the interface between Ge and Si. The {111} stacking faults (SF) observed in (**b**) are the result of the dissociation of perfect misfit dislocations (MD) into partial dislocations [35]

annealed 1.56-μm-thick Ge SRBs (Fig. 3.3b). We actually showed, for instance in Refs. [27, 28, 35], that the TDD, from Secco or HCl defect decoration, was 3–5 times lower in cyclically annealed than in as-grown Ge SRBs.

Finally, we have used cross-sectional transmission electron microscopy (TEM) to image, in [35], a 2.8-μm-thick cyclically annealed Ge layer. Although the volume probed in the TEM lamella was small, we showed that the 0.15-μm-thick zone close to the Ge/Si interface had a dislocation density of the order of $10^{10}$ cm$^{-2}$. The 1.3 μm Ge layer on top had a measured dislocation density around 1–2 × $10^9$ cm$^{-2}$. Finally, the topmost 1.35-μm-thick Ge layer was characterized by a dislocation density of the order of 3–6 × $10^7$ cm$^{-2}$ (see Fig. 3.4). Such a value is actually close to the ~2 × $10^7$ cm$^{-2}$ TDD from XRD in similar thickness Ge layers, validating our use of Ayers' theory.

### 3.2.2 *Germanium Tin Growth*

Silicon (Si) remains the backbone of the semiconductor industry due to the unique features offered by this semiconductor. Because of its indirect bandgap, it is, however, unable to efficiently emit light. The heterogeneous integration of direct bandgap III–V semiconductors on Si, which is the mainstream approach used to obtain near-infrared lasers onto Si [36, 37], has also some drawbacks, including a limited compatibility with CMOS fabrication techniques. Meanwhile, Ge, an almost direct bandgap group-IV semiconductor, gained considerable attention in recent years for the monolithic integration of lasers on Si platforms. Ge, which is fully CMOS-compatible, can indeed be turned into a direct bandgap material by applying high amounts of tensile stress [4, 16].

Germanium can otherwise be alloyed with tin [38], a semimetal, in order to fabricate optically pumped mid-infrared lasers [1–3, 39]. GeSn alloys can also be used in photo-diodes [40, 41], p-i-n light emitting devices and electrically pumped lasers [5, 42], Fin-type field effect transistors [43], vertical gate-all-around FETs [44] and so on. They indeed offer the possibility of engineering the energy bandgap by changing the Sn content and the built-in strain. They have a direct bandgap above a threshold Sn concentration around 8% (in unstrained layers). Their growth, which has to be conducted at low temperatures (because of Sn segregation and precipitation), can be an advantage for monolithic 3D integration. A GeSn alloy crystallizes in a diamond structure, enabling its epitaxial growth on Si(001) substrates. Its lattice parameter is much higher, however, making strain management complex ($a_{Sn} = 6.489$ Å $\Leftrightarrow a_{Si} = 5.43105$ Å and $a_{Ge} = 5.65785$ Å). These features triggered the interest of researchers and significant efforts have been made since 2010 to improve the epitaxial quality of GeSn, notably in chemical vapour deposition (CVD). However, the low solid solubility (<1%) of Sn in Ge and the low thermal stability of $Ge_{1-x}Sn_x$ alloys make their epitaxy quite challenging. This is particularly true for high Sn contents $x$. Thick Ge strain-relaxed buffers (SRBs) are typically used to mitigate the deleterious impact of the large lattice parameter mismatch between Si and $Ge_{1-x}Sn_x$. Growth conditions far from equilibrium must otherwise be used, with a combination of high growth rates and low growth temperatures (below 350 °C, typically). This can be achieved by selecting the proper CVD precursors and growth parameters.

B. Vincent et al. from imec were the first to use a mixture of digermane ($Ge_2H_6$) and tin tetrachloride ($SnCl_4$) to grow, at atmospheric pressure (760 Torr), pseudomorphic GeSn layers on Ge SRBs [45]. This chemistry was later on used by the ForschungsZentrum Juelich to build the first GeSn laser [1]. Since then, it has been adopted by, for instance, ForschungsZentrum Juelich, CEA-LETI or KTH Royal Institute of Technology in Stockholm to fabricate different types of devices. Typical growth features can be found in Refs. [46–50].

Digermane has, however, the reputation of being unstable, notably for high concentrations in the bottle (10%, typically). It is otherwise costly and difficult to order. This spurred ASM America and the University of Arkansas to evaluate mainstream germane ($GeH_4$). They reproductively showed that it yielded, together with $SnCl_4$, the same optical quality layers as $Ge_2H_6$ [51, 52]. They, together with other entities like imec or the University of Warwick, benchmarked $GeH_4$ and $Ge_2H_6$ for the growth of GeSn [53–55]. Process details were scarce, however, especially for $GeH_4$. Temperatures were provided but pressure data were at best vague or missing altogether. The only thorough comparison was that J. Margetis et al., from ASM America, who systematically explored the impact of the F($SnCl_4$)/F($GeH_4$) mass–flow ratio and the temperature on the GeSn growth rate and Sn content. There was no data on chamber pressure, however [55].

In the following, a one-to-one comparison of $GeH_4$ and $Ge_2H_6$ in a 200 mm RP-CVD cluster tool from applied materials, which can operate up to several hundreds of Torrs, will be presented. Growth temperature, pressure, precursor flows and $H_2$ carrier flows were changed in order to grow high-quality GeSn layers [56].

#### 3.2.2.1 GeSn Growth Protocol

The growth of Ge and $Ge_{1-x}Sn_x$ layers was performed on nominal Si(001) substrates. 2.5-μm-thick Ge strain-relaxed buffers were grown in a regular temperature epitaxy chamber to accommodate the lattice mismatch between $Ge_{1-x}Sn_x$ and Si. Germane was then used as a source of Ge. The Ge SRBs were grown using a low-temperature/high-temperature approach followed by a short duration thermal cycling in order to reduce the threading dislocations density, which was close to $10^7$ cm$^{-2}$. Samples were then kept at 20 Torr under ultra-pure $N_2$ in the load-locks of the tool to avoid surface oxidation. Afterwards, the wafers were loaded in the dedicated epitaxy chamber of the cluster tool equipped with low-temperature infrared pyrometers. Prior to $Ge_{1-x}Sn_x$ growth, the samples were annealed under $H_2$ at 800 °C for 2 min, i.e. a temperature significantly lower than the highest temperature used during the thermal cycling (875 °C). Surfaces had therefore, after such 800 °C bakes, the same cross-hatched morphology and roughness parameters as that of as-grown Ge buffers.

$GeH_4$ or $Ge_2H_6$ diluted at 2% in $H_2$ were used as Ge sources. As $SnCl_4$ is a liquid precursor, a bubbler was used to feed tin atoms into the growth chamber. The temperature and the pressure inside the bubbler were set to deliver $SnCl_4$ 1% in $H_2$ when hydrogen was flown in. In the following, the $GeH_4$ and the $Ge_2H_6$ flows were constant and such that $F(GeH_4) = 4 \times F(Ge_2H_6)$. This meant that the Ge atomic flow from germane was twice that from digermane. The growth pressures were always (i) 400 Torr for $GeH_4 + SnCl_4$ and (ii) 100 Torr for $Ge_2H_6 + SnCl_4$. The overall $H_2$ carrier flow was constant and the same for both chemistries. Finally, the $SnCl_4$ flow was either the same or half with $GeH_4$ than with $Ge_2H_6$.

#### 3.2.2.2 Impact of Temperature and Germanium Precursor on GeSn Growth Kinetics

As shown in the left part of Fig. 3.5, there was, over the 301–349 °C range, an exponential increase of the GeSn growth rate with the temperature, with similar activation energies for both chemistries. The activation energy increased with the $SnCl_4$ flow, from 8.6 kcal mol$^{-1}$ for $F(SnCl_4)/F(H_2) = 2 \times 10^{-5}$ up to 11.8 kcal mol$^{-1}$ for $F(SnCl_4)/F(H_2) = 4 \times 10^{-5}$. The latter is close indeed to the activation energy with $Ge_2H_6$ and the same $SnCl_4$ flow, 10.4 kcal mol$^{-1}$. Those values are very similar to those obtained a few years ago with $Ge_2H_6$ in the same epitaxy chamber: 10.4 kcal mol$^{-1}$ [48] and 10.6 kcal mol$^{-1}$ [50]. They are also close to those reported in [55] for $GeH_4 + SnCl_4$, although the chamber design was different and the pressure not disclosed: 9.0–12.0 kcal mol$^{-1}$.

As shown in the right part of Fig. 3.5, the Sn content decreased linearly as the growth temperature increased, with a slope that was less for $GeH_4$ (−1.0 or −1.1%/10 °C) than for $Ge_2H_6$ (−1.7%/10 °C). Those slopes are once again close to literature values. Slopes of −1.85%/10 °C [48] and −1.8%/10 °C [50] were indeed found for $Ge_2H_6$ in the same epitaxy chamber, in

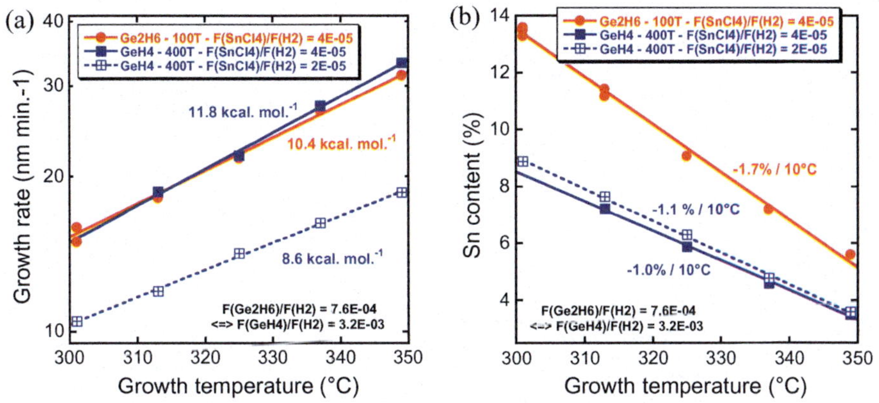


**Fig. 3.5** Evolution with the temperature of GeSn growth rate and Sn content in GeSn layers grown at **a** 400 Torr with $GeH_4 + SnCl_4$ and **b** 100 Torr with $Ge_2H_6 + SnCl_4$ [56]

full agreement with the −1.8%/10 °C value obtained, also for $Ge_2H_6$, in an ASM Epsilon 2000 epitaxy chamber [49]. Slopes of −1.1%/ 10 °C were associated, for $GeH_4$, with $F(SnCl_4)/F(GeH_4)$ MFRs of 0.0085 and 0.012 in [55]. The Sn content dropped with a −1.3%/10 °C slope in pseudomorphic GeSn layers grown with a $GeH_4 + SnCl_4 + H_2$ chemistry in a home-made low-pressure CVD reactor [57].

In order to gain more insight about the growth mechanisms, we have extracted from Fig. 3.5 data the elemental Sn and Ge growth rate components, by multiplying the GeSn Growth Rate (GR) by the Sn fraction $x$ or the Ge fraction $(1 - x)$ (i.e. Sn GR component = $x$ * GeSn GR and Ge GR component = $(1 - x)$ * GeSn GR)). Over the 301–349 °C range, the Ge growth rate component increased exponentially with the temperature, with similar activation energies for both chemistries (left part of Fig. 3.6). Ge GR components were almost the same for a given $SnCl_4$ flow, although the growth pressure was 4 times higher and the Ge atomic flow twice higher with $GeH_4$ than with $Ge_2H_6$. Halving, for the *same $GeH_4$ flow*, the $SnCl_4$ flow resulted in Ge GR components which were, *counterintuitively*, 40% lower.

Meanwhile, the Sn GR component monotonously decreased as the temperature increased, whatever the Ge precursor and the $SnCl_4$ flow probed (right part of Fig. 3.6). This was likely due to Sn sublimation. The Sn GR component was 50% higher, for the same $SnCl_4$ flow, with $Ge_2H_6$ than with $GeH_4$ (although the Ge atomic flow was half of it). Halving, for the same $GeH_4$ flow, the $SnCl_4$ flow otherwise resulted in Sn GR components 35% lower.

Complex interplays between Ge precursors and $SnCl_4$ thus governed the GeSn growth kinetics, with most likely the formation of reactive intermediaries. Margetis et al. [55] indeed suggested that, for the $GeH_4 + SnCl_4$ system, Ge incorporation mainly occurred through dissociative adsorption ($GeH_4(g) + 3 \rightarrow GeH_2 + 2H$, then $GeH_2 \rightarrow Ge + H_2(g)$). Meanwhile, Sn incorporation likely happened through the formation of dichloro-products in the gaseous phase (i.e. $GeH_4(g) + SnCl_4(g) \rightarrow GeH_2Cl_2(g) + SnH_2Cl_2(g)$, followed by $SnH_2Cl_2(g) \rightarrow SnHCl(g) + HCl(g)$,

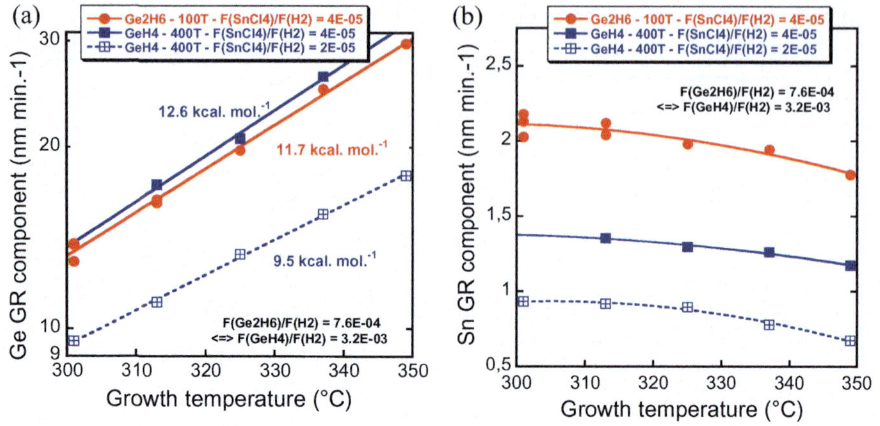


**Fig. 3.6** Evolution with the temperature of Ge and Sn growth rate components for GeSn layers grown at **a** 400 Torr with $GeH_4 + SnCl_4$ or **b** 100 Torr with $Ge_2H_6 + SnCl_4$ [56]

SnHCl(g) + → SnHCl and finally SnHCl → Sn + HCl(g)). This might be one of the reasons why Sn concentrations were, in Fig. 3.6, almost the same for $F(SnCl_4)/F(H_2) = 2 \times 10^{-5}$ and $4 \times 10^{-5}$: Sn incorporation might have been hampered by the $GeH_4$ flow, which was the same and not high enough to result in higher Sn contents for twice higher $SnCl_4$ flows.

#### 3.2.2.3 Structural Properties of Thin, Compressively Strained GeSn Layers

X-ray diffraction profiles highlighting the structural quality of thin, pseudomorphic GeSn layers grown at 400 Torr with $GeH_4 + SnCl_4$ in the 301–349 °C temperature range can be found in Fig. 3.7a. As the growth temperature decreased, the GeSn peak moved away from the Ge SRB peak, which was a clear sign that the Sn content increased. GeSn peaks were well defined and intense, with thickness fringes on both sides. They were properly reproduced by simulations, which was another sign that layers were fully monocrystalline.

The root mean square (rms) surface roughness and the *Z* range ($= Z_{max.} - Z_{min.}$) associated with 5 μm × 5 μm AFM images of those GeSn layers can be found in Fig. 3.7b. Surfaces were cross-hatched, as the Ge SRBs underneath, with little or no impact of the Ge precursor and growth temperature on roughness. The mean rms roughness and Z range of those pseudomorphic, tens of nm thick GeSn layers were equal to 0.54 nm and 4.84 nm, respectively.

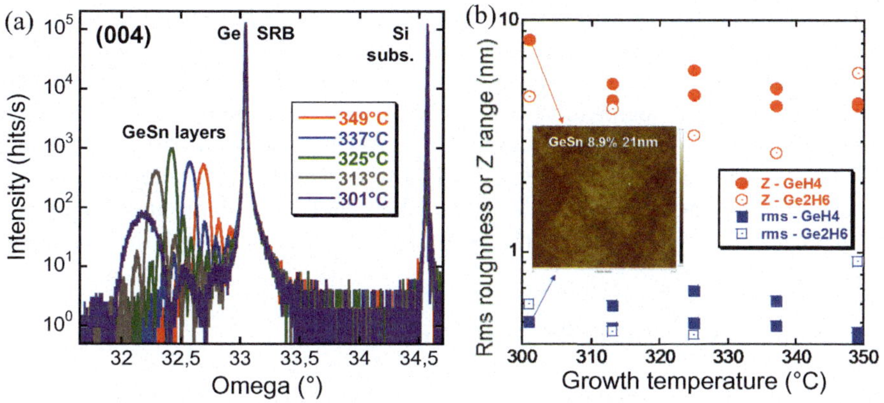


**Fig. 3.7** **a** $\omega$–$2\theta$ scans around the (004) XRD order associated with thin, pseudomorphic GeSn layers grown at 400 Torr with $GeH_4$ + $SnCl_4$ in the 301–349 °C temperature range, **b** rms roughness and $Z$ range associated with 5 μm × 5 μm images of the surface of GeSn layers grown with $GeH_4$ + $SnCl_4$ or $Ge_2H_6$ + $SnCl_4$ on Ge SRBs, over the 301–349 °C temperature range. ⟨100⟩ scan directions. Inset: AFM image of the highest Sn content layer grown with $GeH_4$ [56]

#### 3.2.2.4 SiGeSn Epitaxy: Some Specificities

Before ending Sect. 3.2.2 with a discussion on the properties of thick, relaxed GeSn layers on Ge SRBs which are typically used for optically pumped lasing, we will say a few words about SiGeSn epitaxy. Such ternary alloys can indeed be of use as energy barriers in SiGeSn/GeSn multi-quantum wells or double heterostructures [58]. Indeed, the energy bandgap of Si, 1.12 eV, is higher than that of Ge, 0.77 eV and Sn, which is actually a semimetal (−0.41 eV).

It was shown in [59] that, at 349 °C, 100 Torr and with a $Ge_2H_6$ + $SnCl_4$ + $Si_2H_6$ chemistry, there was, a *simultaneous increase of the Si and Sn contents with the* $Si_2H_6$ *flow* in thin, pseudomorphic SiGeSn layers on Ge SRBs, although the $SnCl_4$ flow was constant. Meanwhile, the SiGeSn growth rate decreased substantially with the $Si_2H_6$ flow, with some degradation, above a given threshold, of the surface morphology and the formation of Sn droplets. Those findings were confirmed at 337 and 325 °C in [50]. For given precursor flows, we also showed, in [50], that a growth temperature increase, in the 313–363 °C range, resulted in the following: (i) an exponential increase of the SiGeSn growth rate, with an activation energy, 9.5 kcal $mol^{-1}$, close to that of GeSn, 10.6 kcal $mol^{-1}$ (although growth rates were always lower for SiGeSn than for GeSn) and (ii) a Si content which increased, from 4 up to 12% (+1.31%/10 °C) and a Sn content which decreased, from 12 down to 4% (−1.25%/10 °C), *while the Ge content in the ternary alloy stayed constant at 84%.*

Results summarized above are definitely in line with ForschungsZentrum Juelich data in Refs. [60, 61], although the carrier gas and the precursor injection were different ($H_2$ and laminar flow above the wafer surface in Refs. [50, 59], to be compared with $N_2$ and a showerhead in Refs. [60, 61]), strengthening those findings.

The epitaxy of SiGeSn/GeSn stacks is not straightforward, because of requirements very much at odds with each other. High Sn content and thus definitely direct bandgap GeSn layers should, for instance, be grown at very low temperature to avoid Sn precipitation/surface segregation, while high Si contents and thus significant energy barriers would favour the high-temperature growth of SiGeSn barriers. A $SnCl_4$ flow which is lower than in binaries should also be selected to have, in the end, the targeted Sn content in SiGeSn. Indeed, the addition of $Si_2H_6$ to the gaseous mixture simultaneously increases the Si and Sn contents, reduces the growth rate and so on.

#### 3.2.2.5 Epitaxy of Thick, Relaxed GeSn Layers

As we will see in the following, Sn content and strain have a major impact on the bandgap directness of GeSn: the lower the compressive strain and the higher the Sn content in GeSn layers, the more direct their bandgap will be and the better their light emission properties will be. This is typically achieved, in GeSn laser devices, by growing GeSn layers on Ge SRB with a thickness well above the critical thickness for plastic relaxation. Layers will then relax, with the emission of vast amounts of strain-relieving dislocations.

We have explored, in [236], the impact of thickness on the properties of GeSn 10, 12 and 15% layers grown at 325, 313 and 301 °C with the same $Ge_2H_6$ and $SnCl_4$ mass-flows and growth pressure (100 Torr) as in Fig. 3.5. Above a given threshold that depended on the built-in strain and thus the Sn content, layers plastically relaxed, with, on top, the appearance of less heavily dislocated, higher Sn content layers, as shown in Fig. 3.8. This Sn enrichment was shown by ASM America and the University of Arkansas [62] or by Ecole Polytechnique de Montréal [63] to be due to a gradual dissipation of the compressive strain and therefore an increase of the GeSn

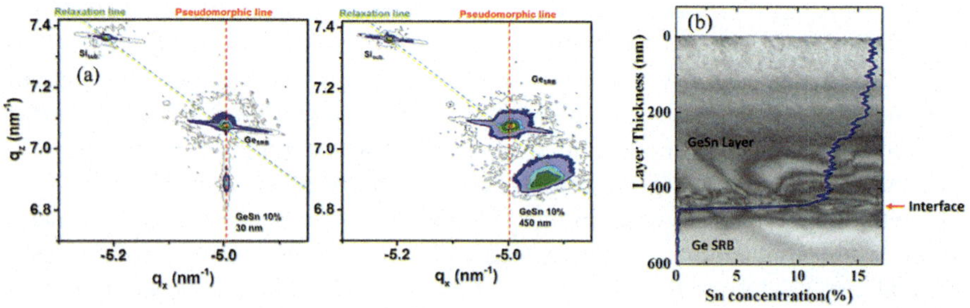


**Fig. 3.8** **a** Reciprocal space maps around the (224) XRD order showing the plastic relaxation occurring when switching from 30 to 450-nm-thick GeSn 10% layers grown at 325 °C. The thinner layer has the same in-plane lattice parameter and thus $q_X$ value as the Ge SRB underneath, while the top part of the thicker layer is 66% relaxed. **b** Sn concentration profile (from energy-dispersive X-ray analysis) superimposed on a Bright Field TEM image of a 465-nm-thick GeSn layer grown at 313 °C with nominally 12% of Sn [236]

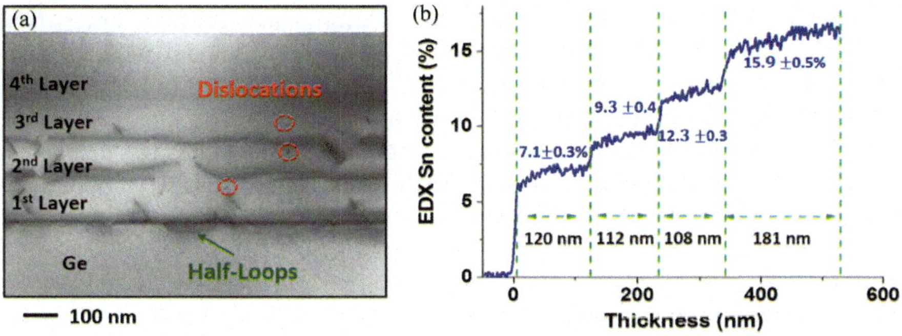


**Fig. 3.9** **a** Scanning transmission electron microscopy (STEM) images of a GeSn step-graded sample (top). Some dislocations are highlighted. **b** A Sn concentration profile extracted from TEM–EDX measurements is provided in the bottom part of the figure. Layers 1, 2, 3 and 4, with 7%, 9%, 12% and 16% of Sn were 95%, 92%, 78% and 64% relaxed, respectively [65]

in-plane constant, improving the incorporation of larger size Sn atoms. Such an involuntary Sn grading is actually advantageous for the fabrication of high performance lasers, as optical recombination will occur in the thick, higher Sn content and crystalline quality layers on top and not in the bottom, lower Sn content and therefore higher bandgap bottom layers, which are more defective [64].

The feasibility of growing GeSn step-graded heterostructures, in order to gradually relax the built-in compressive strain in a controlled manner and confine the misfit dislocations in the bottom layers, was evaluated in [65]. To that end, Fig. 3.5 data points were used, with a gradual decrease of the growth temperature from 349 down to 313 °C in 12 °C steps. We succeeded in obtaining, as shown in Fig. 3.9, superior quality stacks that were used to obtain optically pumped lasers in the mid-infrared with record high performances [2, 39, 241].

### *3.2.3 Ge-Based Materials Processing*

Lithographic steps used to fabricate Ge-GeSn optoelectronic devices must take into consideration the overall fabrication process. Indeed, many parameters such as the minimum pattern size, the layer thickness to be etched, the compatibility of some metals with the developer chemistry and so on have a direct impact on the lithography technique along with the resist type to be employed. For instance, diluting the resist with the proper solvent results in a thinner resist after spin coating, helping to reach smaller pattern sizes.

Optical lithography has been successfully used to fabricate waveguides [52]. However, electron beam lithography is the most versatile and spatially resolved tool found in most research papers. It enables to fabricate all sets of optical resonators, going from microdiscs to photonic crystals [39, 66, 67]. For processes employing the lift-off approach commonly used in academic research laboratories, polymethyl methacrylate (PMMA) is the resist of choice. Unfortunately, PMMA suffers from very poor selectivities when exposed to dry etching plasmas. Doing so requires resists which are more tolerant to plasma etching, such as the ZEP 520a (positive tone) or the ma-N 2400 series (negative tone). Both have a very good pattern resolution, particularly for infrared resonators where constraint on resolution is less demanding than for shorter wavelength resonators. These resists can be easily stripped in an oxygen plasma after the dry etching step.

#### 3.2.3.1 Anisotropic Etching

Etching is a critical step in the fabrication of semiconducting optical resonators. Control of the selectivity between the layer to be etched and the mask or the buried layer is of prime importance. In addition, the resonance properties of the light are strongly influenced by the shape of the etched sidewalls (verticality, roughness). Etching recipes must also address these aspects. Ge can be anisotropically dry etched in fluorine type plasmas ($SF_6$ [68], mainly, or a mixture of $C_4F_8$ and $SF_6$ [69]). $Cl_2$-based chemistries can be used for Ge [70] and GeSn [71]. The latter approach was shown to give very smooth etched surfaces and vertical sidewalls. Table 3.1 gives a representative recipe for the anisotropic etching of Ge and GeSn in an inductively coupled plasma reactive ion etching (ICP-RIE) reactor with a $Cl_2$ chemistry. Typical etching rates are of the order of 450–600 nm/min for Ge and GeSn, while the selectivity (semiconductor to resist etched thickness ratio) is in the range of 2.5–4 with ma-N 2400 or ZEP 520a resists.

Aluminium can be a very efficient hard mask for the anisotropic etching of Ge and GeSn. The selectivity is very high, reaching about 100 on Ge [71] with the recipe given in Table 3.1. A 10-nm-thick Al mask is thus enough to etch a 1-micron-thick Ge layer. The mask can be selectively removed afterwards in a hot and concentrated tetramethylammonium hydroxide (TMAH) solution, leaving the Ge and GeSn layers unattacked. The Al mask definition can also be done by locally etching the top 10-nm-thick Al layer through a soft mask, and with an oxygen-free Cl plasma [72]. After pattern transfer in the hard mask, oxygen is introduced in the recipe, leading to a continuous regeneration of the sputtered alumina via aluminium oxidation.

**Table 3.1** Anisotropic etching recipe of Ge and GeSn and isotropic etching recipe of Ge versus GeSn in an ICP-RIE reactor

| | Pressure (mT) | Gas flow (sccm) | | | Coil power (W) | Platen power (W) |
|---|---|---|---|---|---|---|
| Anisotropic etching of Ge and GeSn | 10 | $Cl_2$ | $N_2$ | $O_2$ | 100 | 200 |
| | | 100 | 25 | 10 | | |
| Isotropic etching of Ge versus GeSn | 50 | $CF_4$ | $N_2$ | $O_2$ | 500 | 0 |
| | | 30 | 40 | 50 | | |

#### 3.2.3.2 Isotropic Etching

Having GeSn layers grown on thick Ge strain relaxed buffers, as in most of the works, is very attractive. Indeed, under-etching the Ge SRB results in a better optical confinement and dissipates the (residual) compressive strain present in as-grown GeSn layers. This in turn improves the semiconductor directness. Gupta et al. noticed that low power $CF_4$ radio-frequency (RF) plasmas selectively etched Ge with respect to GeSn [73]. Such a selectivity was likely due to the formation of a thin surface passivation layer of tin fluoride on GeSn, because of Ge surface depletion and reaction of F with Sn atoms. To maximize the etching selectivity, the sputtering component of etching has then to be kept as low as possible to preserve the passivation layer. Working in an ICP reactor, in a pure «chemical» etching mode, therefore offers the highest selectivity. A typical recipe for the selective isotropic etching of Ge with respect to GeSn (etching rates around 500 nm/min) is given in Table 3.1. Figure 3.10c) shows a GeSn microdisc on a Ge pedestal (top) and a freestanding photonic crystal H4 hexagonal cavity (bottom) obtained with Table 3.1 recipes. A $SF_6$ chemistry can also be used, at zero RF power for the isotropic and selective etching of Ge with

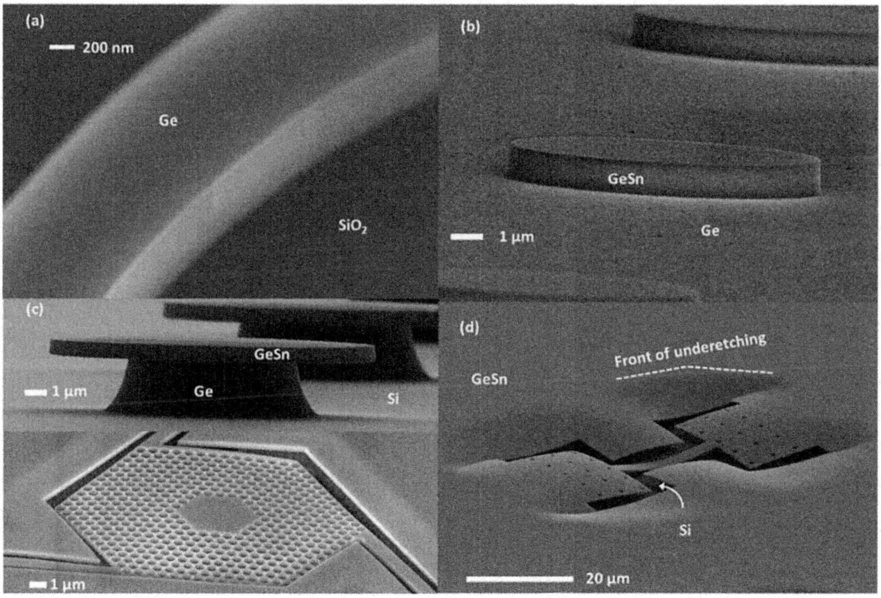


**Fig. 3.10** **a** Ge waveguide made from a germanium-on-insulator (GOI) substrate etched with the anisotropic etching recipe given in Table 3.1; **b** GeSn microdisc on the Ge buffer, etched with the same recipe as in (**a**); **c** GeSn microdisc (top) and H4 cavity (bottom) obtained after the anisotropic etching of GeSn and the isotropic and selective etching of Ge with respect to GeSn (see Table 3.1) and **d** released GeSn microstructure with two anchored sides, showing a dramatic membrane bending (note the presence of through holes in the GeSn membrane which can be used to initiate the under-etching reaction in areas far from the trench). Under-etched zones appear as bright areas on these SEM pictures

respect to GeSn. However, the selectivity is poorer than with $CF_4$. Indeed, etching of low Sn content layers will occur.

Under-etching a part or the whole resonator area can have a profound impact on the overall device since this step comes with a substantial displacement of matter at the micron scale (thick Ge films on Si are tensile strained and GeSn films on Ge are compressively strained). Special care must then be given to the total under-etching length or to the device design in order to benefit from strain release (see the strategies of strain redistribution in [4]), without generating undesired layer cracks or bending (see, for instance, Fig. 3.10d).

A wet chemical etching of GeSn was also used, for instance in [72]. Al-Kabi et al. developed a low-temperature (0 °C) etching of GeSn, with a mixture of $HCl:H_2O_2:H_2O$ 1:1:10. The authors reported the fabrication of smooth surfaces for GeSn waveguides on Ge, with lasing upon optical pumping. This section is far from being exhaustive and does not report all the different chemistries used for etching Ge and its alloys. Nevertheless, it seems at this stage interesting to note that hot TMAH does not attack Ge nor GeSn. This opens new perspectives and makes possible under-etching strategies involving Si as a sacrificial material.

#### 3.2.3.3 Surface Passivation of (SiGeSn) Based Materials

Parasitic and non-radiative recombination of free carriers in semiconductors is highly undesirable as it degrades the performances of light emitters or photodetectors (high threshold values and inherent heating problems in lasers, non-radiative "loss" of carriers in photodetectors, etc.). With the scaling down of components, the ratio of surface to volume atoms significantly increases. Recombination on surface traps might then influence the recombination dynamics of the whole device. Finding a way to keep the density of surface traps at reasonably low levels in Ge is the topic of recurring works (see, for instance, [74]). Nevertheless, the rise of SiGeSn alloys renewed research around surface passivation. Indeed, the growth of SiGeSn barriers is nowadays possible on GeSn [42]. Stange et al. demonstrated lasing in GeSn/SiGeSn multi-quantum wells with a threshold much lower than in bulk materials at cryogenic temperatures [58], indicating that these systems were suitable for carrier confinement and that the interface between the barriers and the wells were of good quality. Getting high barrier heights for both electrons and holes is, however, necessary for room temperature operation. Current barrier materials are limited to the low Sn and Si content range for $Ge_{1-x-y}Si_xSn_y$ compounds, requiring additional epitaxy developments to cover a more complete barrier materials spectrum. GeSn layers with a Sn content grading can also provide a way to confine carriers. J. Chrétien et al. demonstrated lasing in $Ge_{0.84}Sn_{0.16}$ layers embedded into $Ge_{0.87}Sn_{0.13}$ layers, with then a moderate electronic confinement [2].

Various wet cleaning and dielectric capping steps were evaluated in [75] for the passivation of GeSn layers. Mahjoub et al. quantified the density of interface traps $D_{it}$ in metal oxide semiconductor (MOS) capacitors with the capacitance voltage technique (an alumina layer was deposited beforehands thanks to atomic layer deposition) [76]. Sn droplets were found on the surface of $Ge_{0,9}Sn_{0,1}$ layers and removed by dipping in a HF:HCl mixture. The authors reported a retardation of the reoxidation prior to alumina deposition with the use of a $(NH4)_2S$-based wet treatment. They estimated the density of interface traps to be approximately $9 \times 10^{11}$ cm$^{-2}$ eV$^{-1}$.

### *3.2.4 Electrical Contacts on Ge-Based Materials: A Focus on GeSn*

Lasing, upon optical pumping, has been demonstrated in 2015 in GeSn alloys [1]. Electrically pumped GeSn-based photonics devices, such as photodetectors, LEDs or lasers remain elusive, however [5]. Fermi-level pinning at the metal/Ge(Sn) interface, doping and thermal stability of GeSn layers are indeed some of the technological hurdles that have to be overcome to obtain optimum electrical contacts.

The strong Fermi-level pinning (FLP) at the metal/germanium interface is well documented in the literature [77]; it is predominantly governed by metal-induced gap states (MIGS) and defects-induced gap states (DIGS). GeSn, in particular n-type doped layers, are not spared by the FLP issue. GeSn surface passivation by plasma treatment (typically $O_2$ treatment) or thin dielectric deposition ($Al_2O_3$) have been shown to alleviate FLP [78, 79] as well as the formation of stanogermanide or high doping in GeSn layers [80, 81].

Having high levels of electrically active dopants in GeSn layers, and therefore low contact resistivities, is challenging. Ion implantation and in situ doping have both been reported in the literature [82, 83]. One of the limiting factors for ion implantation is the activation step that can lead to a degradation of the junction and of the GeSn properties because of the poor thermal stability of GeSn layers (mainly due to Sn segregation [84]). The in situ Ga doping of GeSn layers that result in ultra-low p-type specific contact resistivity seems promising [85].

Beyond the above-mentioned topics, a comprehensive understanding of the solid-state reaction during the formation of stanogermanides is of utmost importance to precisely characterize the system and benefit from reliable contacts on GeSn. In this section, we will provide some data about the metallization of GeSn layers; more specifically, we will focus on Ni stanogermanide formation. In a second part, technological levers to enhance the thermal stability of stanogermanides will be proposed to improve GeSn devices performances.

#### 3.2.4.1 Metallization of GeSn Layers

A complete analysis of the literature concerning the metallization of GeSn layers showed that many metals were used to contact these layers. Al and TaN [78], Ti [86], Zr [87] and even Yb [80] were evaluated. Nevertheless, Ni [82] is by far the

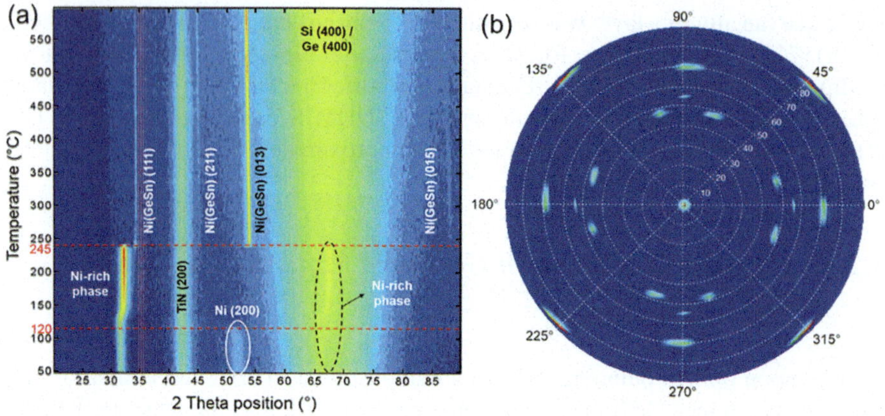


**Fig. 3.11** **a** In-situ X-ray diffraction measurement of the reaction between Ni and $Ge_{0.9}Sn_{0.1}$, adapted from [93], **b** in-plane pole figure recorded at $2\theta = 32°$ for Ni/$Ge_{0.9}Sn_{0.1}$ sample annealed at 180 °C, adapted from [93]

most used metal. Indeed, Ni enables to obtain ohmic contacts with low sheet and specific contact resistances at relatively low temperatures [88–91]. Several teams have investigated the solid-state reaction of the Ni/GeSn system. However, there were still open questions concerning the system evolution, the phases present, the impact of Sn content and the behaviour of Sn atoms during the solid-state reaction.

The Ni/$Ge_{0.9}Sn_{0.1}$ solid-state reaction can be monitored by in situ X-ray diffraction (XRD), in-plane reciprocal space map (RSM) measurements and in-plane pole figures. A sequential growth was evidenced using in situ XRD (Fig. 3.11a). The solid-state reaction between a Ni thin film and $Ge_{0.9}Sn_{0.1}$ layer can be summarized as follows. First, after the total consumption of Ni at 120 °C, a Ni-rich phase is formed. Then, at 245 °C, the Ni-rich phase disappears and the growth of the mono-stanogermanide phase Ni($Ge_{0.9}Sn_{0.1}$) is initiated. This phase is stable up to 600 °C [92]. Based on advanced in-plane RSM and pole figure measurements (Fig. 3.11b), it was demonstrated that the Ni-rich phase was actually the hexagonal ε-$Ni_5(Ge_{0.9}Sn_{0.1})_3$ metastable phase [92].

The role Sn might have during the Ni/GeSn solid-state reaction (SSR) was, until recently not fully understood. A comprehensive analysis focused on Sn segregation during the Ni/GeSn SSR was carried out. In situ X-ray diffraction and cross-sectional transmission electron microscopy measurements coupled with energy-dispersive X-ray spectrometry and electron energy-loss spectroscopy atomic mappings were performed to follow the phase sequence, Sn distribution and segregation [93, 94]. The results showed that, during the SSR, Sn incorporated into intermetallic phases. As the temperature increased, Sn segregation happened first around grain boundaries (GBs) and then towards the surface. At higher temperatures and when the Ni(GeSn) film agglomerated, Sn easily migrated towards the top surface (see Fig. 3.12).

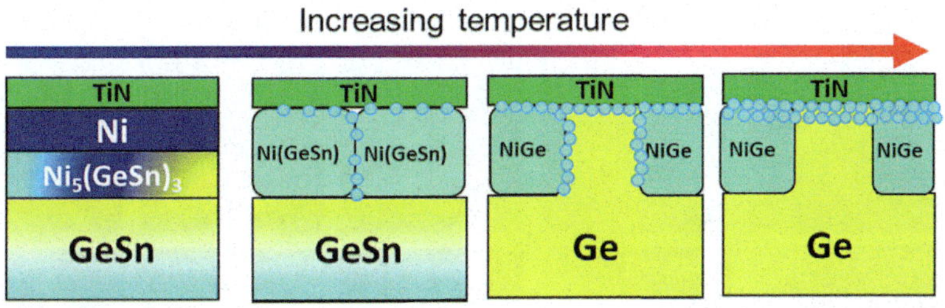


**Fig. 3.12** Schematics of Sn behaviour during Ni/GeSn solid-state reaction taken from [99]

Sn accumulation around GBs hampered atom diffusion, delaying the growth of the Ni(GeSn) phase. Higher thermal budgets will thus be mandatory for the formation of contacts in high-Sn-content photonic devices, which could be detrimental for thermal stability. This last point is crucial. Indeed, it has clearly been shown that the Ni/GeSn system suffered from a lack of thermal and electrical stability, mainly due to the agglomeration of the Ni(GeSn) phase and to Sn segregation. It was thus of utmost importance to propose various means of extending the thermal stability of Ni-based stanogermanides.

#### 3.2.4.2 Technological Levers to Extend the Thermal Stability of Ni-Based Stanogermanides

The metastability of GeSn alloys and surface Sn segregation phenomena led international research teams to propose alternative solutions to extend the thermal stability of Ni-based stanogermanides. Technological levers can be used at various steps of the contact module. The GeSn surface can be pre-amorphized by implantation (PAI) prior to Ni metallization. C implantation was successfully used to extend the thermal stability of Ni(GeSn) by about 100 °C [95]. The use of Pt interlayer between Ni and GeSn [96] or the co-sputtering of Pt and Ni [97] were shown to enhance the thermal stability of the Ni/GeSn system. More recently, the classical rapid thermal annealing was compared to the much shorter and thus more metastable pulsed laser thermal annealing. Results were promising [98].

Various pre-amorphization processes using Si, Ge, C or C + Ge implantation were implemented to prepare the GeSn surface prior to metallization. The use of C PAI not only extended the thermal stability of the Ni(GeSn) by delaying the agglomeration process as shown in Fig. 3.13, it also postponed Sn segregation towards higher temperatures (delay of 150–200 °C); the electrical properties were thus strongly improved [99]. Such results are very promising.

The impact of Pt or Co as alloying elements for Ni-based metallization of GeSn layers was also investigated [100]. As far as the solid-state reaction is concerned, the overall phase sequence is the same for all metallizations: at low temperature, a Ni-rich phase is obtained; it is then consumed to form the low resistivity mono-stanogermanide phase. Nevertheless, the addition of Pt or Co as alloying elements has

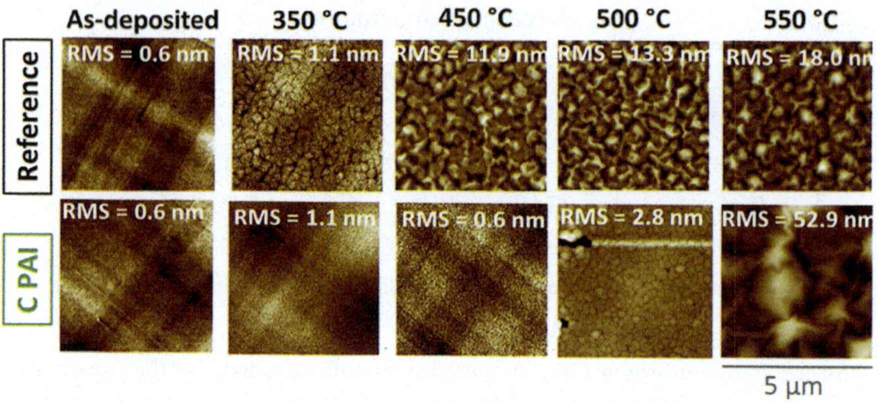


**Fig. 3.13** AFM images (scan 5 µm × 5 µm) of the surface of samples annealed at various temperatures for the Ni/GeSn system without PAI (top) and with C PAI (bottom) taken from [99]

an impact on Ni consumption, Ni-rich and mono-stanogermanide phases' formation temperatures. It also leads to the formation of $PtSn_x$ or $CoSn_y$ compounds. Moreover, the addition of Co or Pt positively influences Sn segregation by delaying this phenomenon. Co has a weak influence on morphological and electrical properties. On the other hand, Pt improves the surface morphology by delaying the Ni(GeSn) phase agglomeration and enhancing the process window in which the sheet resistance remains low [101, 102]. Figure 3.14 illustrates the impact of alloying elements on the electrical properties of the Ni/GeSn system.

The impact of annealing Ni/GeSn layers thanks to laser pulses (ultra-violet nanosecond laser or UV-NLA) was also analysed. The overall evolution of the phase sequence with the thermal budget was comparable to that in RTA samples: the $Ni_5(GeSn)_3$ phase was obtained first, followed by the mono-stanogermanide (Ni(GeSn)) phase. A texture change was, however, evidenced; indeed, the size of

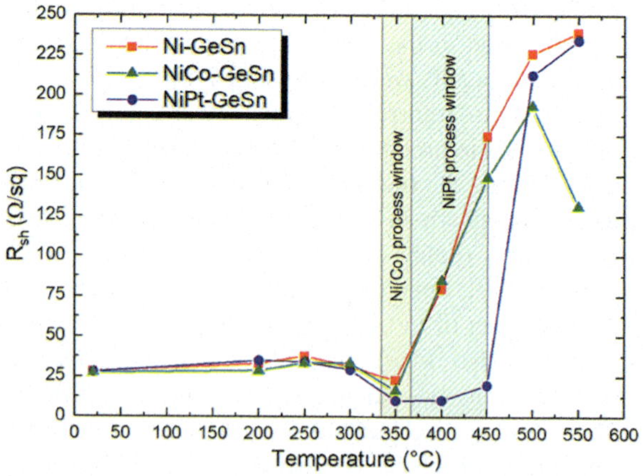


**Fig. 3.14** Evolution of the sheet resistance as a function of the rapid thermal annealing temperature for the Ni, $Ni_{0.9}Co_{0.1}$ and $Ni_{0.9}Pt_{0.1}/Ge_{0.9}Sn_{0.1}$ systems taken from [99]

crystallites in the Ni(GeSn) phase seemed to be reduced with UV-NLA, which could result in thermal stability improvements. Yet, agglomeration or Sn segregation (island formation on the surface) has not been noticed for laser-annealed samples, at variance with RTA samples. Laser annealing therefore seems to be promising to form, with low thermal budgets, optimum electrical performance Ni contacts on GeSn layers [99].

To conclude, the development of a complete contact module on GeSn is mandatory to have performant electrically activated devices. A better knowledge and understanding of the physico-chemical properties of the metal/GeSn systems was essential to propose technological and process levers to overcome the metastability of GeSn alloys and Sn segregation. The use of pre-amorphization by implantation, the addition of alloying elements such as Pt to Ni, the use of UV-nanosecond laser annealing or the combination of such technological levers are promising tracks towards a stable and reliable contact module on GeSn devices.

## 3.3 Effects of Strain and Sn for Bandstructure Manipulation

The direct bandgap in intrinsic Ge has to be carefully estimated as the deformation potentials model [103] under-estimates the crossover between the L- and Γ-valleys. A good experimental agreement of this crossover in intrinsic Ge has recently been achieved using the tight-binding model [104]. The crossing to the direct band is predicted to be close to 6% at 20 K for a [100] uniaxial stress [4] instead of previous predictions at around 4.4% [103, 105, 106]. The tight-binding model forecasts a crossover at around 2.1% at 20 K for (001) biaxial stress [4] instead of 1.8% predicted by the deformation potential model [107].

We present here the different stress induction schemes used in recent years to reach such high amounts of strain in Ge and GeSn layers. In the second part, we describe the principle of the bandstructure modifications due to strain, alloying, confinement and combinations of those. Where possible, the bandstructure effects are given in analytical form to express the general trends. However, for more detailed considerations we refer our readers to the more specialized literature.

### *3.3.1 Strain Control in Germanium-Based Materials*

#### 3.3.1.1 Strained Ge Growth

The growth of pseudomorphic Ge on larger lattice constant buffers such as InGaAs [108–112] or GeSn [113–115] can result in high biaxial tensile strain. Direct bandgap behaviour has been observed using photoluminescence analysis versus temperature in

Ge layers and Ge quantum dots [116]. Strains up, respectively, to 2.3% and 2.4% have been measured in Ge layers and Ge quantum dots, respectively [117]. This technique is limited to thin Ge layers below the critical thickness for plastic relaxation (typically around 50 nm [118, 119]). Similarly, III–V semiconductors—as virtual substrates [120]—are not acceptable in a Si-based platform. To the best of our knowledge, no tensily strained epitaxial GeSn layers have been reported yet in the literature.

#### 3.3.1.2 Thermal Strain

As presented in Part I—Fabrication: Ge-based epitaxy and processes for group-IV direct bandgap material—thick Ge layers epitaxially grown on Si are quasi-relaxed and present only slight tensile strains at room temperature. This residual strain is induced by the thermal expansion coefficients difference between Ge and Si coming into play during the cooling-down after epitaxy or annealing [32, 121]. The strain thus depends on the process temperature [122, 123] and may lead to thermally induced strains up to 0.3% [124] and 0.4% in patterned structures [125].

However, to achieve biaxial strain Ge with a direct bandgap, we need methods that provide much higher strain values than those given just above. Currently, the main techniques used are strain redistribution and external stress transducers or stressor layers. We are going to describe those technologies in the following. As source materials, we used GeOI wafers. Ge layers on top showed a homogeneously distributed residual strain of 0.16% [126] and were produced in a 200 mm wafer production line [127]. In the case of GeSn layers grown on thick Ge strain relaxed buffer (SRBs), Ge has once again a residual tensile strain of 0.16%, while GeSn is, because of the larger lattice constant parameter of Sn than of Ge, in a compressive strain state of −0.5%, typically.

#### 3.3.1.3 Micro-bridges and Membranes

In Germanium

The strain redistribution concept was first developed in Si using 0.6% pre-strained SOI [128]. The generalization of the method and its potential as photonic platform has been described in [129]. M. J. Suess et al. redistributed successfully the residual strain in Ge using micro-bridges up to uniaxial strain values of 3.1% [16]. Micro-bridges were patterned along the [100] crystallographic direction of Ge. The pre-strained Ge layer (Fig. 3.15a) grown on Si or SOI was first etched into a Ge micro-bridge (Fig. 3.15b). Then, the substrate or the buried oxide under the Ge layer was under-etched (Fig. 3.15c). The liberated structure relaxed, focusing the longitudinal strain of the two arms in the structure centre. The strain enhancement at the centre of the suspended micro-bridge can be analytically calculated from the four dimensions listed in d and the under-etch length below the Ge. The length *A* defines the central

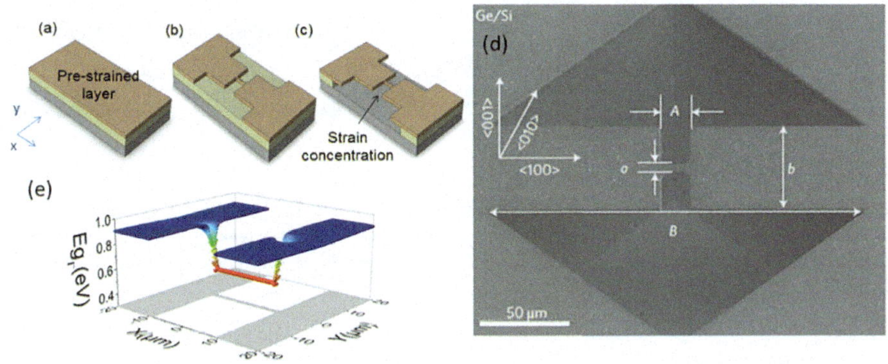


**Fig. 3.15** Schematic of the process steps used to fabricate strained Ge micro-bridges from **a** a residual tensile strained Ge layer, **b** which is patterned and **c** under-etched. **d** Tilted SEM image of a Ge/Si structure showing the critical dimensions A, a, B and b of the micro-bridge (figure from [16]). **e** Tight binding modelling of the bandgap energy of the strained micro-bridge. Note that strain is loaded uniaxially along ⟨100⟩, and the transition regime is narrow, i.e. less than 1 μm wide

micro-bridge length, $B$ is the overall length of the micro-bridge, $a$ the micro-bridge width and $b$ the micro-bridge arm width.

The strain enhancement factor $f = \varepsilon_{xx}/\varepsilon_0$ in the narrow constriction of the bridge is defined by the ratio of the $\varepsilon_{xx}$ micro-bridge strain divided by the $\varepsilon_0$ residual biaxial strain. It follows, for Ge deposited on a SOI substrate, that $f$ is approximately given by (1) where $E_i$ are Young's moduli in the [100] direction of Si and Ge and $t_i$ are the corresponding layer thicknesses,

$$f = \eta \frac{2L + B}{B}\left[1 + \frac{A}{B - A}\right] \Big/ \left[\frac{a}{b} + \frac{A}{B - A}\right] \quad \text{with } \eta = \frac{t_{Ge}E_{Ge}}{t_{Si}E_{Si} + t_{Ge}E_{Ge}}$$

Above 3.1% of uniaxial strain, corresponding to a 4.8 $cm^{-1}$ Raman frequency shift, the Ge micro-bridges grown on SOI broke due to the presence of misfit dislocations at the Ge/Si interface [16]. GeOI substrates with a significantly better crystallographic quality than Ge directly grown on Si, enabled the fabrication of highly strained micro-bridges. Record-breaking strain values of 4.9% at room temperature, corresponding to a Raman shift of 9.9 $cm^{-1}$, were reported [17]. Higher strains were achieved by lowering the temperature. At 20 K, strain values as large as 6% were obtained [4, 130].

#### In Germanium Tin

The situation is different for GeSn, as the GeSn layers grown on Ge SRBs on Si are compressively strained, as the Sn lattice constant is higher than the Ge one (6.489 Å, vs. 5.65785 Å). When the strategy described above is used on GeSn, the compressive

**Fig. 3.16** SEM images of the GeSn micro-bridge (left) before $SiO_2$ etching (GeSn micro-bridge bending downward due the release of its built-in compressive strain), (middle) at the beginning of the $SiO_2$ etching, (right) at the final stage of $SiO_2$ etching when the GeSn is tensile strained (figures from [2])

strain will inevitably lead to a bowing of the central part of the micro-bridge. A two-step lithography strategy has therefore been developed to use the residual strain in the Ge SRBs [2]. In Fig. 3.16, the two liberated Ge arms' length dictates the level of strain in the suspended GeSn micro-bridges. The residual tensile strain of the Ge SRBs is here the driving force to tensile strain GeSn micro-bridges, which are in a compressive strain state just after epitaxial growth.

The GeSn micro-bridge bends downward due the release of its built-in compressive strain after the etching of the GeSn/Ge/SOI and before the HF etching (Fig. 3.16 left). The $SiO_2$ etching allows the Ge arm to relax and slowly strains the GeSn micro-bridge (Fig. 3.16 middle). Using this method, GeSn micro-bridges with strain up to 2.2% were obtained [2] (Fig. 3.16 right).

#### 3.3.1.4 Biaxially Strained Micro-bridges

Suspended Membranes

Biaxial stress or, equivalently, strain in Ge and GeSn can be induced the same way as described previously by using a micro-cross design instead of a micro-bridge one. Strain relaxation in the large stretching arms during the $SiO_2$ under-etching results in a biaxial tensile strain in the central region. Figure 3.17a shows the strain in a Ge micro-cross-modelled by the finite element method (FEM) along $x = [110]$, $y = [110]$ and $z = [001]$ crystalline directions. The full strain tensor and the lattice orientation distribution in highly strained Ge micro-crosses were evaluated at the sub-micrometre scale using Laue micro-diffraction and simultaneous rainbow-filtered micro-diffraction [131]. The complete strain tensor maps extracted from standard Laue micro-diffraction measurements on micro-crosses showed a very good agreement with FEM modelling (Fig. 3.17a). This agreement is confirmed in Fig. 3.17b by comparing the strain profile measured along the $x = [110]$ direction with the modelled one.

The largest in-plane biaxial strain in Ge reported in the literature at room temperature is 1.9%, corresponding to a Raman shift of 8.1 $cm^{-1}$ [132]. This configuration suffers from shear stress at the edge of the stretching arms, which most likely is

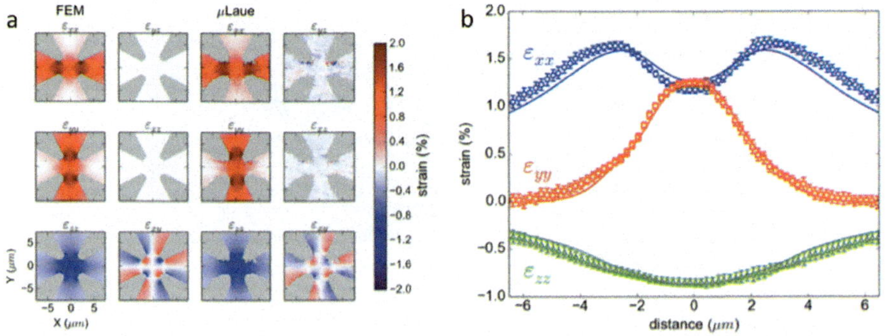


**Fig. 3.17** **a** Strain distribution in a Ge micro-cross along $x = [110]$, $y = [110]$ and $z = [001]$ crystalline directions from FEM modelling (left) and by rainbow-filtered Laue micro-diffraction measurements (right). **b** Profile of strain along the $x = [110]$ direction of the micro-cross calculated by FEM (solid line) and measured by Laue micro-diffraction (symbols) (figures from [131])

detrimental to carrier injection. Indeed, shear stress affects the band structure and in particular the alignment between the L- and Gamma-valleys.

Landed Membranes

Electrical contacts to suspended membranes could be very challenging. Specific strategies have therefore been developed to land on the substrate the entire under-etched structure after its initial release and the eventually obtained high strain levels. Strain redistribution can, for instance, be achieved by using a very thin sacrificial etch-layer. Jan Petykiewicz et al. adopted a 25-nm-thick layer of $Al_2O_3$ to release the strain. The strained structure was, immediately after the etching, re-attached to the substrate that consisted of Si with a 10-nm-thick conformal $Al_2O_3$ layer deposited by atomic layer deposition [133]. This thin layer can potentially passivate the Ge surface. Another strategy to land strained structures is by controlling the wet etching speed of the thick $SiO_2$ sacrificial layer (1 μm). Figure 3.18a shows cross-sectional schematics of a suspended strained structure using high-speed HF under-etching (bottom part) and of a landed strained structure using a low speed HF under-etching (upper part). Micro-crosses fabricated in a GeOI substrate with 1 μm of BOX have been landed on silicon (Fig. 3.18b) using a low-speed HF etching recipe. The stretching arms can subsequently be partially removed, provided that the bonding to the Si substrate is strong enough, with an elastic energy threshold of around 3 J/m$^2$, typically. Such a value is of the same order of magnitude as bonding energies in wafer bonding experiments [134]. In that case, the strain of a landed micro-bridge is kept after etching away the stretching arms, Fig. 3.18c and d.

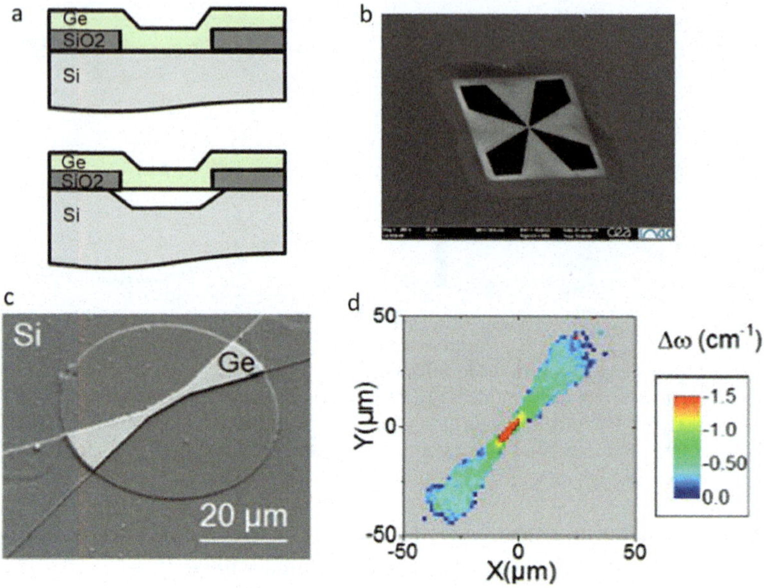


**Fig. 3.18** **a** Top (bottom) cross-sectional schematics of a landed (a suspended) strained structure obtained by slow (fast) etching. **b** Landed Ge micro-cross made from GeOI with a low speed HF etching recipe to remove $SiO_2$. **c** Landed Ge micro-bridge with a partial removal of stretching arms. **d** Raman spectroscopy mapping of the landed and strained Ge micro-bridge bonded on Si (figures from [134])

#### 3.3.1.5 Strain Induced by External Stress or by Stressor Layers

Tensile strain has also been applied to Ge using external force transponders [135–138] and flexible polyimide which could be set under pressure and on which Ge layers were deposited [139, 140].

Emission enhancement when reaching the Ge direct bandgap was observed for biaxial strain reaching 1.4% in [140]. A more promising technology for applications is strain-tensors deposited around Ge and/or GeSn layers. Several stressors have been studied including SiN [141], Cr [142], SiGe [143, 144], strained-Si [145] as well as GeSn [146, 147]. Ge photodetectors have, for instance, been tensile strained using tungsten layers with residual compressive stresses higher than 4 GPa [148]. However, SiN stressors are more often used, as their strain can be controlled from compressive to tensile through a proper selection of process parameters. In addition, SiN is transparent in the MIR and its electrical isolation properties as well as thermal properties are advantageous when combined with Ge or GeSn. Tensile strain has been induced using compressive SiN on Ge nanowires [71] and Ge membranes [133, 149–151], as well as on straight Ge waveguides [152–155], nanowires [156], microdiscs [69], micro-rings [157] and micro-pillars [158, 159].

However, inhomogeneous tensile strain is achieved when the SiN layer is deposited asymmetrically [69] around the host. Stressor layers in most of the cases

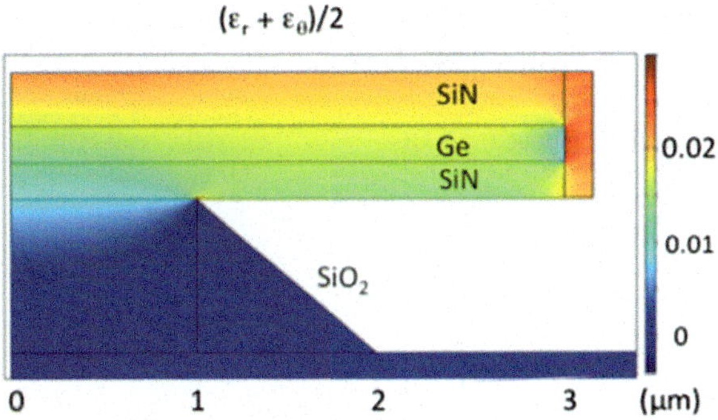


**Fig. 3.19** Strain distribution in a Ge microdisc with SiN all-around calculated by the finite elements method (figure from [160])

will thus lead to a poor overlap of, for instance, the optical mode with the strained media. To solve this issue, all-around Ge and GeSn structures were developed showing indeed a homogenous strain distribution in Ge and GeSn, Fig. 3.19. A direct bandgap behaviour was observed in Ge with up to 1.75% of biaxial tensile strain [160]. An all-around SiN strategy, with a SiN intrinsic stress of −1.9 GPa, was used to encapsulate GeSn, resulting in ultra-low lasing thresholds around 1 kW cm$^{-2}$ [3]. Table 3.2 summarizes the tensile strain induced in Ge and GeSn with the different approaches described previously.

## 3.3.2 *Band Structures and Band Alignment*

### 3.3.2.1 Band Structure of (Si)GeSn Alloys

Ge is an indirect bandgap semiconductor, but the lowest conduction band valley, at the L point, is just ~140 meV below the Γ point. Si has its conduction band minimum near the X point, while the Γ-valley is much higher than X. Alloying Si and Ge thus does not result in a direct gap material. On the other hand, Sn has a negative bandgap at the Γ point. In GeSn alloys, the Γ-valley energy $E_\Gamma$ decreases faster than the L-valley energy $E_L$, as the Sn content increases, which may result in a positive bandgap at Γ, and smaller than the indirect gap towards L, i.e. GeSn, can be a direct gap material. The transition into a fundamental direct bandgap semiconductor occurs at a Sn concentration of about 8 at.%. Adding Si, to have a ternary SiGeSn alloy, slows down the transition into a direct gap semiconductor.

The bandgap $E^{\Gamma,\mathrm{L},\mathrm{X}}$ of an unstrained (Si)GeSn alloy can been calculated from the concentrations of the three elements $x_{\mathrm{Ge}}$, $x_{\mathrm{Sn}}$, $x_{\mathrm{Si}}$ (where $x_{\mathrm{Ge}} + x_{\mathrm{Sn}} + x_{\mathrm{Si}} = 1$), as

$$\begin{aligned} E^{\Gamma,\mathrm{L},\mathrm{X}} = {} & E_{\mathrm{Ge}}^{\Gamma,\mathrm{L},\mathrm{X}} x_{\mathrm{Ge}} + E_{\mathrm{Sn}}^{\Gamma,\mathrm{L},\mathrm{X}} x_{\mathrm{Sn}} + E_{\mathrm{Si}}^{\Gamma,\mathrm{L},\mathrm{X}} x_{\mathrm{Si}} + b_{\mathrm{SiGe}}^{\Gamma,\mathrm{L},\mathrm{X}} x_{\mathrm{Si}} x_{\mathrm{Ge}} \\ & + b_{\mathrm{GeSn}}^{\Gamma,\mathrm{L},\mathrm{X}} x_{\mathrm{Ge}} x_{\mathrm{Sn}} + b_{\mathrm{SiSn}}^{\Gamma,\mathrm{L},\mathrm{X}} x_{\mathrm{Si}} x_{\mathrm{Sn}} \end{aligned}$$

**Table 3.2** Methods and reported tensile strain values in Ge and GeSn

| Straining method | Device form | Stress nature | Tensile strain (%) | Measurement | Ref. |
|---|---|---|---|---|---|
| Thermally induced strain | GeOI | Biaxial (001) | 0.16 | XRD, Raman, micro-Laue | [127, 131] |
| | | | 0.3 | XRD | [124] |
| | | | 0.4 | XRD | [125] |
| | Ge on Si | | | | |
| | Patterned GeOI | | | | |
| Strain by growth (pseudomorphic) | Ge on InGaAs | Biaxial (001) | 0.7 | XRD, Raman, PL | [110] |
| | | | 2.3 | XRD | [116] |
| | | | 2.4 | XRD, Raman, TEM | [117] |
| | Ge on InGaAs | | | | |
| | Ge quantum dots | | | | |
| External stress | Ge membrane | Biaxial (001) | 0.6 | Pressure sensor | [135] |
| | Ge membrane | Biaxial (001) | 2.0 | Raman | [139] |
| | Ge wire | Uniaxial ⟨111⟩ | 2.5 | Raman | [138] |
| Stressor layers | Ge strip | Biaxial (001) | 0.7 | Raman | [238] |
| | Ge membrane | Uniaxial ⟨100⟩ | 1 | Raman | [151] |
| | Ge pillar | Biaxial (001) | 1.3 | Raman | [239] |
| | Ge nanowire | Uniaxial ⟨111⟩ | 1.5 | Raman | [71] |
| | Ge microdisc | Biaxial (001) | 1.9 | Raman | [132] |
| | Ge micro-ring | Biaxial (001) | 2 | Raman | [157] |
| | GeSn microdisc | Biaxial (001) | 1.4 | Raman | [3] |
| Strain redistribution | Ge micro-bridge | Uniaxial ⟨100⟩ | 3.1 | Raman | [16] |
| | GeOI micro-bridge | Uniaxial ⟨100⟩ | 4.9 | Raman, micro-Laue | [4] |
| | GeSn micro-bridge | Uniaxial ⟨100⟩ | 2.2 | Raman | [2] |

(continued)

**Table 3.2** (continued)

| Straining method | Device form | Stress nature | Tensile strain (%) | Measurement | Ref. |
|---|---|---|---|---|---|
| | GeOI micro-cross | Biaxial (001) | 1.9 | Raman, Laue | [132] |

Methods include epitaxy as well as diverse routes based on a mechanical transfer of strain

where the first three terms are the weighted average (linear interpolation), and the other three terms give the correction (i.e. represent the quadratic interpolation). The bowing parameters $b_{\Gamma,L,X}$ are either derived from atomistic bandstructure calculations or are found experimentally. While either of these methods can be used to obtain even higher-order interpolation expressions, it is usually considered that the linear interpolation plus bowing is accurate enough for practical purposes. However, some recent publications cast doubt on whether the SiSn bowing for the Γ-valley can be described by a constant parameter $b^{\Gamma}_{\mathrm{SiSn}}$, for instance, because there is strong evidence that it depends on the Si and Sn concentration [61, 161].

Another way to tune the band structure is by applying strain. The six components of the strain tensor ($\varepsilon_{xx}$, $\varepsilon_{yy}$, $\varepsilon_{zz}$, $\varepsilon_{xy}$, $\varepsilon_{xz}$, $\varepsilon_{yz}$) may generally be independent and influence the band structure. However, the most interesting cases in real structures are the biaxial and uniaxial strains.

The biaxial strain is due to the lattice mismatch between the substrate and the overgrown thin layer of the active material or is set by a stressor layer (usually SiN) into which the laser active layer is embedded. In conventional, [001] grown structures, it has two equal in-plane components ($\varepsilon_{xx} = \varepsilon_{yy}$), while $\varepsilon_{zz}$ amounts to $\varepsilon_{zz} = -2C_{12}/C_{11}\varepsilon_{xx}$ (where $C_{11}$ and $C_{12}$ are the stiffness constants of the layer material). Off-diagonal components in the stress tensor are zero. Biaxial tensile strain corresponds to $\varepsilon_{xx} > 0$. If coming from the lattice mismatch, the biaxial strain is

$$\epsilon_{xx} = \epsilon_{yy} = \frac{a_{\mathrm{latt,subs}} - a_{\mathrm{latt}}}{a_{\mathrm{latt}}}$$

where the lattice constant depends on the alloy composition in a similar way as the bandgaps (weighted average plus bowing), i.e.

$$a_{\mathrm{latt}} = a_{\mathrm{Ge}}x_{\mathrm{Ge}} + a_{\mathrm{Sn}}x_{\mathrm{Sn}} + a_{\mathrm{Si}}x_{\mathrm{Si}} + b_{\mathrm{SiGe}}x_{\mathrm{Si}}x_{\mathrm{Ge}} + b_{\mathrm{GeSn}}x_{\mathrm{Ge}}x_{\mathrm{Sn}} + b_{\mathrm{SiSn}}x_{\mathrm{Si}}x_{\mathrm{Sn}}$$

The uniaxial strain usually appears in structures with external force applied to a thin rod-like semiconductor, e.g. in bridge-like structures. The value of one strain component, e.g. $\varepsilon_{zz}$, is there considered known, and the other two components are then found from $\varepsilon_{xx} = \varepsilon_{yy} = -C_{12}/(C_{11} + C_{12})\varepsilon_{zz}$, and the off-diagonal components are also zero. Uniaxial tensile strain corresponds to $\varepsilon_{zz} > 0$. The effect of strain on the band edge energies is described via deformation potentials, and the shifts of band energies (in cases of biaxial or uniaxial strains, as considered here) can be found from the 8-band k.p model for bands at Γ [162], for a zero wave vector, which gives the shifts of the conduction band at Γ and L points:

$$\Delta E_{\Gamma} = a_c(2\varepsilon_{xx} + \varepsilon_{zz}) \quad \Delta E_L = a_L(2\varepsilon_{xx} + \varepsilon_{zz})$$

and for the valence bands (taking zero energy as the valence band top for the no-strain case)

$$P_{\epsilon} = a_v(2\varepsilon_{xx} + \varepsilon_{zz}) \quad Q_{\epsilon} = -b_v(\varepsilon_{xx} - \varepsilon_{zz})$$

$$E_{\mathrm{HH}} = -P_{\epsilon} - Q_{\epsilon}$$

$$E_{\mathrm{LH,SO}} = -P_{\epsilon} + \frac{1}{2}\left(Q_{\epsilon} - \Delta_{\mathrm{SO}} \pm \sqrt{\Delta_{\mathrm{SO}}^2 + 2\Delta_{\mathrm{SO}} Q_{\epsilon} + 9Q_{\epsilon}^2}\right)$$

where $\Delta_{\mathrm{SO}}$ is the energy of the split-off band, below the heavy-hole and light-hole valence band (v.b.) degenerate top. The material parameters for Si, Ge and Sn, used in these calculations, and also their original sources, are given in [163].

Biaxial tensile strain decreases both the direct and indirect bandgap (the direct bandgap decreases faster, resulting in a conversion into a direct gap material). It also splits the heavy-hole (HH) and light-hole (LH) bands, degenerate at zero strain, so that LH comes above HH. Compressive biaxial strain has the opposite effect. Uniaxial tensile strain also decreases the direct gap faster than indirect, leading to a direct gap material, but in this case the HH band is the highest, and LH is below it.

In these considerations, it is important to note that in these strain-split valence band states, the heavy and light holes are defined (in terms of their character—the wave functions) with respect to the *z*-axis described above (normal to the biaxial strain plane, or along the uniaxial strain). Within the simple parabolic approximation, the HH band effective mass in the *z*-direction is $m_{\mathrm{HH}}^{z} = 1/(\gamma_1 - 2\gamma_2)$, while the in-plane mass is much smaller, $m_{\mathrm{HH}}^{||} = 1/(\gamma_1 + \gamma_2)$. The LH band masses are (approximately, without corrections for strain) $m_{\mathrm{LH}}^{z} = 1/(\gamma_1 + 2\gamma_2)$ and $m_{\mathrm{LH}}^{||} = 1/(\gamma_1 - \gamma_2)$. Therefore, the effective-mass-based labelling of HH and LH states applies only to their masses in the *z*-direction, if they are split by strain. This is important because the density of states of these bands is not the same as in unstrained bulk.

Also, the optical transition matrix elements (as known from k.p modelling) between the top valence band and the conduction band are larger for *z*-polarization in biaxial tensile strained materials (implying that the light propagation is in the strained layer plane). In the case of a compressive biaxial strain, the xy-polarization is dominant. On the other hand, for uniaxial tensile strained material (with HH being the topmost valence band), transitions for xy-polarized light are much stronger, i.e. for light propagation along the strain direction (which is a practically favourable set-up).

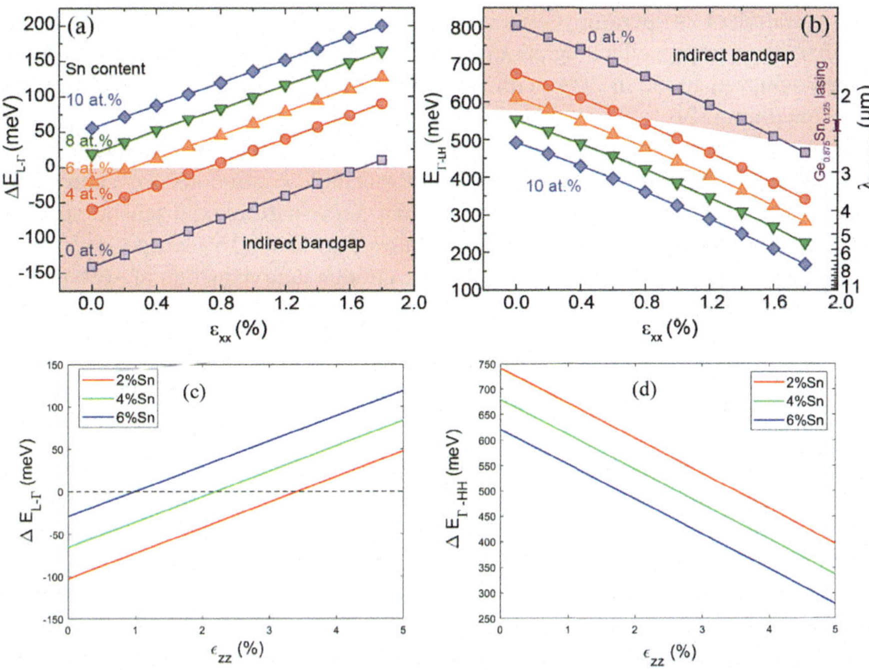


**Fig. 3.20** Directness $\Delta E_{\text{L-}\Gamma}$ (**a**), direct bandgap $E_{\Gamma\text{-LH}}$ and corresponding emission wavelengths (**b**) as functions of biaxial tensile strain for several Sn contents, at $T = 300$ K. For comparison, the lasing region of bulk $Ge_{0.875}Sn_{0.125}$ from [5] is marked on the wavelength scale (figure from [237]), **c** directness $\Delta E_{\text{L-}\Gamma}$ and **d** direct bandgap $\Delta E_{\Gamma\text{-HH}}$ functions of uniaxial tensile strain for several Sn contents, at $T = 300$ K. These results are obtained with parameters listed in [164], but some other methods, as discussed in [3], indicate that even larger strain is required for direct gap, e.g. 5–6% uniaxial strain for Ge

Combinations of tensile strain and Sn content yielding direct bandgap in GeSn alloys are shown in Fig. 3.20. Figure 3.20a shows the difference of indirect and direct bandgaps (“level of directness”), while Fig. 3.20b shows the direct bandgap, i.e. the approximate emission energy of a laser based on some material/strain combination. Given the much larger density of states in L-valleys than in the conduction band Γ-valleys, a directness of at least 100 meV is preferable for good lasing performance, where a good fraction of electrons will reside in the Γ-valley up to higher temperatures, c.f. our considerations given in Sect. 3.3.3. While pure Ge requires a rather large biaxial strain (~1.8–2.1%) to get zero directness (with other aspects of that further discussed in Sect. 3.3.2), alloys with over 4% Sn require much smaller strain levels. Figure 3.20c and d shows these two properties (directness and direct bandgap) for uniaxial tensile strain. Lower Sn content GeSn alloys may be interesting, in combination with high enough levels of tensile strain, for lasing. Indeed, the number of crystalline defects in such alloys is likely lower than in higher Sn content GeSn layers (see Sect. 3.2 of this chapter). Of course, alloys with large enough Sn contents can be direct even with some compressive strain, as was, e.g., the case in the first demonstration of an operating GeSn laser [1], and also in many other research results which showed improved lasing parameters. An operating temperature of 270 K was indeed achieved in GeSn 20% with a biaxial compressive strain of −0.5% in the active medium [52].

However, decreasing the compressive strain or, even better, introducing some tensile strain is generally beneficial. Ultralow threshold continuous wave and pulsed lasing were recently reported [3], using GeSn 5.4% with biaxial tensile strain of 1.4%. Lasing up to 273 K was recently achieved in GeSn 16% suspended micro-bridges with 2% of uniaxial strain [2]. A large enough uniaxial strain of ~6% yielded lasing in pure Ge [4].

#### 3.3.2.2 Band Alignment in Heterostructures

Using SiGeSn ternary alloys in GeSn-based lasers can be very attractive. While SiGeSn is not expected to be the laser active medium, it can be used for carrier confinement in GeSn, either in classical heterostructures or as the barrier material in quantum well structures, where GeSn is the well material. It is, therefore, important to consider the band alignment at GeSn/SiGeSn heterojunctions, i.e. to find the band discontinuities for electrons and holes. To find these discontinuities, we started from the average valence band energy (average of HH, LH and SO band energies), calculated according to the expression from [164]:

$$E_{v,\mathrm{av}} = -0.48\ \mathrm{eV} \cdot x_{\mathrm{Si}} + 0.69\,\mathrm{eV} \cdot x_{\mathrm{Sn}}$$

which gives the average valence band energy in SiGeSn alloys, the reference zero being this energy in Ge. The SO band energy in the same material is obtained by weighted average of SO energies in Si, Ge and Sn

$$E_{\mathrm{SO}} = 0.297\ \mathrm{eV} \cdot x_{\mathrm{Ge}} + 0.044\,\mathrm{eV} \cdot x_{\mathrm{Si}} + 0.8\,\mathrm{eV} \cdot x_{\mathrm{Sn}}$$

which then gives the HH, LL band top in unstrained material

$$E_v = E_{v,\mathrm{av}} + E_{\mathrm{SO}}/3$$

After that, by adding appropriate bandgaps, with applied strain, one can find energies of all bands of interest, all with the valence band top in Ge as the reference zero point. Calculating these for the two materials that make the heterojunction (with the strain in each one as exists in that heterojunction) will then deliver the discontinuity of any particular band at this heterojunction. There are also a couple of different expressions for the average valence band energies, instead of that in [164], e.g. [165, 166], which give somewhat different values for band discontinuities. A more detailed experimental investigation is necessary before concluding about the accuracy of those calculations.

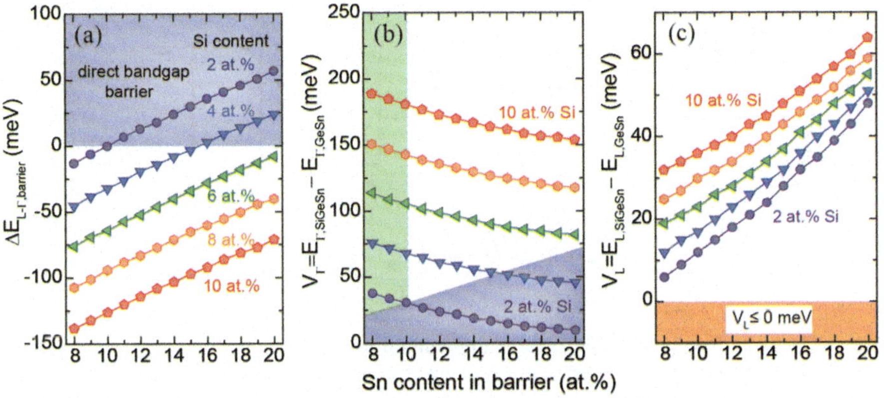


**Fig. 3.21** Directness of SiGeSn barrier (**a**), band discontinuity for Γ- (**b**) and L-valleys (**c**) for a $Ge_{0.92}Sn_{0.08}$/SiGeSn MQW. The coloured areas show regions where SiGeSn is a direct bandgap semiconductor (blue), regions where band discontinuities for L are negative (orange), and regions which correspond to realistically grown CVD layers (green) (figure from [163])

Band discontinuities in a GeSn/SiGeSn heterostructure, and also the level of directness of the barrier (which should preferentially be negative), are given as examples in Fig. 3.21.

#### 3.3.2.3 Quantized States in GeSn/SiGeSn Quantum Well Structures

The full 8-band k.p model can be used for the calculation of quantum states in quantum well structures. Usually, a simpler method based on the effective-mass approximation will deliver reasonably accurate results. Here, we briefly describe the latter method. The effective-mass Schrodinger equation, with the potential found from the band alignment calculated as described above, and with the effective mass as the $m^z_{HH}$, $m^z_{LH}$ component for HH, LH, or the effective mass for electrons, is solved e.g. by the finite-difference, or some other method to get the states of the system. For L-valley states, the projection of the effective mass in the z-direction is used. It is given by $m^z_L = 3m_l m_t/(2m_l + m_t)$, where $m_l$ and $m_t$ are the longitudinal and transverse components for L-valley ellipsoids. The Luttinger parameters necessary for the calculation of hole effective masses depend on the alloy composition. They are calculated from the fitting expressions obtained by the pseudopotential method [167, 168], which is more accurate than the simple weighted average of Luttinger parameters for elements. Within this effective-mass approximation, quantum states will have the density of states determined by the in-plane effective masses of the corresponding bands. This method in fact gives the same energies as k.p calculation does at subband edges (zero wave vector) and is reasonably accurate in cases where the quantum states of HH and LH are separated enough so that the HH-LH mixing is not large within the range of in-plane wave vectors of interest.

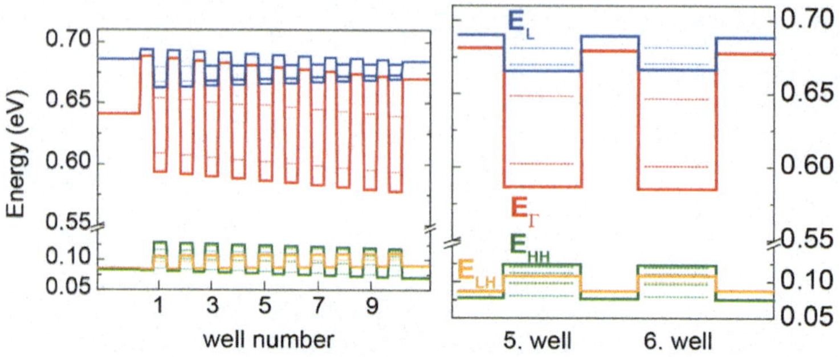


**Fig. 3.22** Band diagram of a MQW structure with 10 periods of $Si_{0.05}Ge_{0.82}Sn_{0.13}/Ge_{0.87}Sn_{0.13}$ with 22 nm thickness for each layer, with a gradient of strain (from −0.05% to −0.2%), and confined states derived from the strain distribution for the whole stack (left) and for the two central QW's (right). Inside the wells, the bulk band (solid) and quantized state energies (dotted lines) are shown for each of the valleys (figure from [58])

An example of bound states in a GeSn/SiGeSn multi-quantum well (MQW) structure, in which lasing was achieved, is shown in Fig. 3.22.

### *3.3.3 Gain Calculation of Strained Ge-Based Materials (Ge, GeSn, SiGeSn)*

The gain calculation is based on the 8-band k.p model, using the electron and hole states coming from the diagonalization of the 8 × 8 k.p matrix for a range of wave vectors $k$. The gain is given by

$$g(\hbar\omega) = \frac{e^2}{8\pi^2\omega n_r c\varepsilon_0 m_0^2} \sum_{c,v} |\hat{e}\cdot \boldsymbol{p}_{cv}|^2 \int \delta(E_c - E_v - \hbar\omega)(f_c(E_c(k)) - f_v(E_v(k)))\mathrm{d}^3k$$

where $n_r$ is the refractive index, $\omega$ the radiation frequency and $|\hat{e}\cdot \boldsymbol{p}_{cv}|^2$ is the interband momentum matrix element. $f_{c,v}$ are the Fermi–Dirac functions for electrons in the conduction and valence bands, with the quasi-Fermi levels $F_{c,v}$, and $E_c(k)$ and $E_v(k)$ coming from the 8-band k·p Hamiltonian. The summation is here made only over bands around the Γ point. However, the Fermi level for electrons should be calculated by including L-valleys. This equation does not account for any broadening of electron and hole states. This can be included by introducing the convolution of the gain spectrum calculated from the above formula with a Lorentzian function with appropriate linewidth. However, this is usually not essential for interband transitions in bulk, with very different dispersion of initial and final states.

Another important effect in semiconductor gain media is the inter-valence band absorption (IVBA), coming from transitions between different hole bands (HH, LH, SO). It is usually the transitions between HH, LH and SO bands which may result in considerable absorption at wavelengths close to interband transitions. This is calculated by an expression similar to that above for interband gain, except that both Fermi–Dirac functions are for the valence band states, and the summation should be done over all combinations of valence band states.

Finally, the calculated material gain has also to be corrected for free carrier absorption (FCA), which can be calculated using the second-order perturbation model described in [169], or using the Drude–Lorentz model, as in [170]. The acoustic phonon scattering, deformation potential scattering (L-valley), intervalley scattering, ionized impurity scattering and alloy scattering all contribute to and are included in the FCA calculation. After deducting the IVBA and FCA from the material gain, one gets the effective, net gain. Examples of the calculated gain achievable in biaxial tensile strained GeSn alloys and bandgap directness are given in Fig. 3.23.

In MQW structures, within the effective-mass approximation described above, used for state energies and dispersion, the gain is given [171]:

$$g(\hbar\omega) = \frac{\pi e^2}{n_r c \varepsilon_0 m_0^2 \omega} \sum_{n,m} \left| I_{\mathrm{hm}}^{\mathrm{en}} \right|^2 \int_0^{\infty} \rho_r^{2D} \left| \hat{e} \cdot \boldsymbol{p}_{cv} \right|^2 \frac{\gamma/\pi}{\left[ E_{\mathrm{hm}}^{\mathrm{en}} + E_t - \hbar\omega \right]^2 + \gamma^2} \times \left[ f_c^n(E_t) - f_v^m(E_t) \right] \mathrm{d}E_t$$

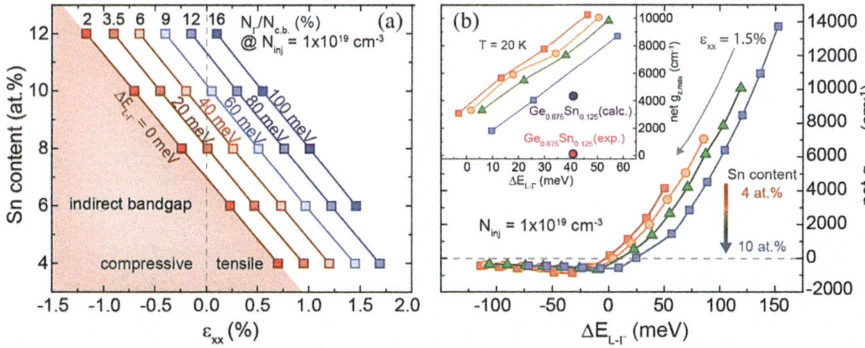


**Fig. 3.23** **a** Bandgap directness $\Delta E_{\mathrm{L-\Gamma}}$ as a function of biaxial strain and Sn content. **b** out-of-plane net gain maximum $g_{\mathrm{z,max}}$ for several Sn contents, as it depends on the directness, at an injection carrier density of $1 \times 10^{19}$ cm$^{-3}$ at 300 K. Highest values of $g_{\mathrm{z,max}}$ are achieved for strain values of 1.5% (figure from [237]). Clearly, the very high gain predicted for $1 \times 10^{19}$ cm$^{-3}$ carrier density means that even much lower densities can provide practically useful gain

where $|\hat{e} \cdot \boldsymbol{p}_{cv}|$ is the interband momentum matrix element, which depends on the in-plane wave vector (or $E_t$), as given in Sect. 9.5 of [171], for transitions between conduction band-HH and conduction band-LH quantized states. This expression also includes the transitions broadening, described by a Lorentzian function with a full width at half maximum of $2\gamma$. The density of states $\rho_r^{2D}$ is defined using the reduced in-plane effective mass $m_r^*$ as:

$$\rho_r^{2D} = \frac{m_r^*}{\pi \hbar^2 L_z}$$

where $L_z$ is the well width, and $I_{\text{hm}}^{\text{en}}$ is the overlap integral of envelope functions of electron and hole states $n$ and $m$. The Fermi-Dirac distributions $f_{c,v}^n$ for electrons and holes with the quasi-Fermi levels $F_{c,v}$ are here defined as:

$$f_{c,v}^n(E_t) = \frac{1}{1 + \exp\left[\left(E_g + E_{e,hn} + \frac{m_r^*}{m_{e,h}^*} E_t - F_{c,v}\right)/k_B T\right]}$$

Figure 3.24 shows the subbands in a MQW structure and the gain calculated by this method, as well as the influence of additional n-type doping on the net gain (this may improve the gain in a limited range of doping densities, but too large a doping degrades it).

We should note that the MQW gain calculations presented above are based on parabolic approximation within the k.p model for energy states, and that the SO band is not included. In some cases, this may be reasonably accurate, but in many other cases, the full k.p model should be used to include nonparabolic band dispersion and to calculate IVBA.

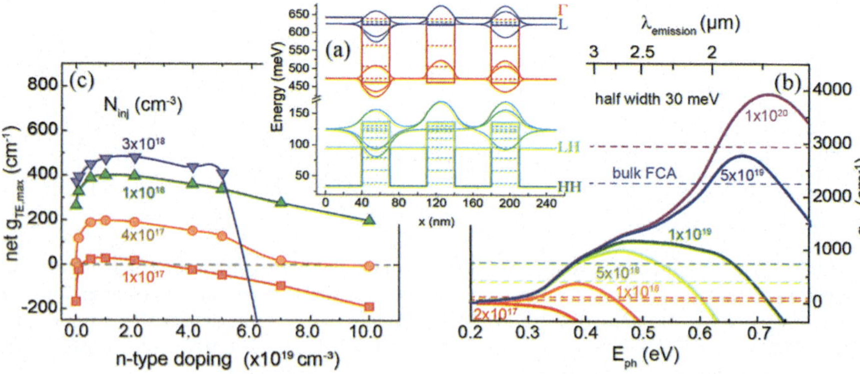


**Fig. 3.24** **a** Band diagram of an unstrained MQW with 3 × {30 nm $Ge_{0.84}Sn_{0.16}$/40 nm $Si_{0.1}Ge_{0.75}Sn_{0.15}$} stacks. **b** TE-mode interband gain at 300 K for different injected carrier densities. Dashed lines indicate FCA values for bulk $Ge_{0.84}Sn_{0.16}$. **c** Net gain dependence on n-doping density (with FCA subtracted; figure from [163])

Finally, it is worth noting that other methods of band structure and gain calculations have recently been presented for this material system (in bulk form), e.g. the empirical pseudopotential method (EPM) [172], which also enables accurate calculation of both interband gain and IVBA. As shown there for Ge with large uniaxial strain, good accuracy of these calculations is important for finding parameters leading to useful net gain (i.e. with IVBA definitely smaller than interband gain).

## 3.4 Group-IV Lasing

Lasing in silicon and Ge-based material—in addition to the impact it would make to traditional microelectronics—is an interesting topic for fundamental research. In this section, we review recent laser results, elaborate on the optical cavity designs used and review material developments and corresponding laser devices. An overview over the impressive achievements made by the community in just 5 years is given below.

Since the first demonstration in 2015 of lasing in GeSn layers [1], progress has been vast and continuous, as shown in Table 3.3 summarizing the recent achievements in chronological order. For example, threshold densities were reduced by more than 2 orders of magnitude. Similarly, the maximal operating temperature ($T_{max}$) was at the beginning around the boiling temperature of nitrogen. It now approaches room temperature (RT). This—as we will detail below—is directly correlated with progresses made in the epitaxial growth of high Sn content GeSn layers. Defect management is another method to improve lasing. Record low thresholds were obtained when interface defects were removed in transferred materials and an external stressor was used [3]. These developments are detailed in Sect. 3.4.3.2.

Two routes have been followed to obtain direct bandgap material, one based on mechanically induced tensile strain in Ge [16, 173, 174], the other based on the epitaxial growth of GeSn. Recently, these two approaches began to merge [2, 3]. Device structures and optical cavities are also different. In the case of suspended structures, the cavity design has to follow the strain enhancement rules. GeSn layers obtained by epitaxy on Ge SRBs allow the use of more traditional approaches such as Fabry–Perot waveguides or microdisc cavities. Most of the lasing reports quantify threshold excitation densities and $T_{max}$. Gain and quantum efficiency are, however, rarely given, most likely because of experimental difficulties. To stimulate such investigations, which enable to quantify non-radiative recombination and optical loss from, for instance, intervalence band [175] or free carrier absorption, a set of generic instructions how this could be done is presented in Sect. 3.4.1.2.

**Table 3.3** Group-IV lasing major milestones over the last 5 years

| Systems | Strain | Substrate | Cavity | L-Threshold | $T_{max}$ (K) | Gain ($cm^{-1}$) | *Q*-eff (%) | Ref | Year |
|---|---|---|---|---|---|---|---|---|---|
| GeSn (12.5%) | −0.6% biaxial | Ge-VS | Ridge WG F-P | 300 kW/cm$^2$ | 90 | <110 | ±1.5 | [1] | 2015 |
| GeSn (12.5%) | ±0% | Ge-VS | MD | 220 kW/cm$^2$ | 130 | – | – | [66] | 2016 |
| GeSn/SiGeS multi-QW (13%) | ±0% | Ge-VS | MD | 40 kW/cm$^2$ | 120 | – | – | [240] | 2018 |
| GeSn DHS (16%) | ±0% | Ge-VS graded | MD | 134 kW/cm$^2$ | 230 | – | – | [67, 241] | 2017, 2018 |
| GeSn (17.5%) | ±0% | Ge-VS | Ridge WG F-P | 170 kW/cm$^2$ | 180 | – | – | [242] | 2018 |
| Germanium micro-bridge | 1.6% uniaxial | GeOI bonded | MC DBR F-P | – | 180 | – | – | [245] | 2018 |
| GeSn (16%) Photonic crystal | ±0% | Ge-VS | PC membrane | 230 kW/cm$^2$ | <90 | | | [39] | 2018 |
| SiN stressor Germanium | 1.7% biaxial | Ge (SiN) bonded | MD | – | 180 | – | – | [246] | 2018 |
| GeSn (20%) step enhanced | −0.5% biaxial | Ge-VS | Large area F-P | 50 kW/cm$^2$ | 270 | – | – | [52] | 2019 |
| Germanium micro-bridge | +5.5% uniaxial | GeOI | MC CC F-P | 20 kW/cm$^2$ | 90 | >300 | >50 | [4] | 2019 |
| GeSn DHS (± 13%) | +2% uniaxial | GeOI | MC CC F-P | 10 kW/cm$^2$ | 273 | – | – | [2] | 2019 |

(continued)

**Table 3.3** (continued)

| Systems | Strain | Substrate | Cavity | L-Threshold | $T_{max}$ (K) | Gain ($cm^{-1}$) | *Q*-eff (%) | Ref | Year |
|---|---|---|---|---|---|---|---|---|---|
| SiN stressor GeSn (6%) | +1.5% biaxial | GeSn(SiN4) bonded | MD | 1.1 $kW/cm^2$ | 100 | – | – | [3] | 2019 |
| Electrically pumped GeSn p-i-n (11%) | ±0% | Ge-VS | Large area F-P | 600 $A/cm^2$ | 100 | – | 0.6 | [5] | 2020 |
| GeSn (7–10.5%) | ±0% | Ge-VS | MD | 10 $kW/cm^2$ | 90 | – | – | [243] | 2020 |

VS stands for virtual substrate, WG for waveguide, F-P for Fabry–Pérot, DBR for distributed Bragg reflector, MD for microdisc, MC for micro-cavity, CC for corner cube, PC for photonic crystal

### 3.4.1 Optical Cavity Design

The optical cavity or resonator is a major component of a laser; it provides the feedback of the laser light and thus eventually reinforces stimulated emission. The performance of a cavity is usually given by its quality factor ($Q$-factor), which is a measure of the number of oscillations cycles a photon takes before it decays, $Q = 2\pi \upsilon \tau_d$, where $\upsilon$ is the cavity resonance frequency and $\tau_d$ the photon decay time. Standard direct bandgap semiconductors stand out due to large gain and therefore—in principle—do not necessarily require cavities with ultra-high $Q$-factors. We expect this to be true for direct bandgap group-IV materials, by analogy. Fabrication concepts suitable for standard III–V-based semiconductor lasers should thus also suit group-IV lasers, as well as traditional characterization methods.

#### 3.4.1.1 Resonators

Resonator types used so far for group-IV lasers include (i) Fabry–Perot (F-P) cavities with partially reflecting end mirrors and single-mode ridge waveguides (WG) or multi-mode large area WG, (ii) micro-disc (MD) cavities, as well as (iii) F-P like micro-cavities (MC) with end-mirrors made of distributed feedback (DFB) reflectors or total reflection corner cubes (CC). The F-P MCs comprise mainly suspended structures. The above list is completed by (iv) photonic crystal designs for membranes and waveguide slaps. Their main properties are summarized in Table 3.4, and detailed below.

*Waveguide cavities* are conveniently used with GeSn layers epitaxially grown on a bulk or virtual substrate. They are fabricated with a planar technology and are typically 1 mm long. Thanks to an under-etching of the waveguides, the mode overlap with the active laser medium is improved and part of the transversal strain is relaxed. Waveguide F-P cavities are convenient for gain measurements, for example by investigating the laser threshold or light emission dependence on cavity length over which the waveguide is excited, see below. The latter method is named the variable stripe method [176]. Both of these methods were used to characterize the first group-IV lasing in GeSn [1]. The Hakki–Paoli or the Fourier transform methods [177], which can alternatively be used to measure the modal gain under subthreshold conditions, were not used so far for group-IV lasing. These methods are based on the mode spectra analysis at high resolution.

*Microdiscs (MD)* are the most used cavities. By design, the compressive strain of typically 0.5–0.8% in as-grown GeSn will then elastically relax [66]. Through the relaxation, the directness of the band structure improves and so does the laser performances, as detailed in the theory section. The characteristically high $Q$-factor of MD was for most of the investigations of less importance because—as said—the material gain is expected to be rather high. And indeed, the above-mentioned investigations performed on F-P cavities of GeSn layers with 12% of Sn [1] revealed gain values exceeding 100 $cm^{-1}$. MD is not only used with epitaxial GeSn layers

**Table 3.4** Laser cavities used to demonstrate lasing in group-IV semiconductors

| Cavity type | | Strain | Size | $Q$-factor, empty cavity | Filling factor | Benefit | Disadvantage | References |
|---|---|---|---|---|---|---|---|---|
| Waveguide ridge | 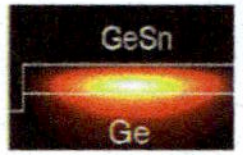  | Pre-served | ±1 mm | Typ. 10,000 | ca. 60% | Planar technology | Strain/mode leakages | [1] |
| Waveguide supported |   | Partially relaxed | ±1 mm | Typ. 10,000 | up to 90% | Planar technology | Mode quality | [244] |
| Microdisc supported |  | Nearly fully relaxed | ø ± 10 μm | >10,000 | ca. 90% | Strain relaxation | Heat management | [66] |
| Micro-bridges free standing |  | Tensily strained | ±10 μm | <2000 | ca. 20% | On wafer tunability | Low mode filling factor | [3, 173, 245] |
| Micro-cavity capped stressor |   | Tensily strained | ø ± 10 μm | >10,000 | ca. 90% | Defect removal | Complex fabrication | [246] |
| Photonic crystal suspended |  | Relaxed | ø ± 35 μm | >10,000 | ca. 40% | Single mode | Low fabrication tolerances | [39] |

but also in conjunction with GeSn layers which were transferred from the original epitaxial wafer and embedded in a SiN stressor layer to inject high amounts of tensile strain [3]. During the GeSn layer transfer, defects generated by plastic strain relaxation during the epitaxial growth on Ge SRBs were removed, making lasing possible under steady-state excitation [3], as detailed in Sect. 3.4.3.3.

*F-P micro-cavities* around strained micro-bridges were developed first with tensily strained Ge [4, 133, 178] and later on with GeSn grown on a Ge virtual substrate (VS) [2]. These designs have to comply with the specificities of strained bridges. The optical cavity should not, for instance, have too much of an impact on the amount of strain injected by anchors or pads in the optically active structures. Such constraints exclude, for example, photonic crystals with etched holes in the strained membrane or slab. The cavity should also match the small micrometre size of the gain medium. Designs used so far are of the Fabry–Perot type, with "end-mirrors" made of (i) a pair of distributed Bragg reflectors (DBR) on either side of the constriction [133] or (ii) two-corner cube-shaped surfaces where the light is confined via total reflections [2, 4]. Reported $Q$-factors range between 1000 and 4000, which is not particularly high given that the gain medium fills only 1/3 of the overall cavity. The corresponding optical loss can then quickly exceed 100 cm$^{-1}$ [4]. The principle of such cavities with mode calculation is shown in Fig. 3.25.

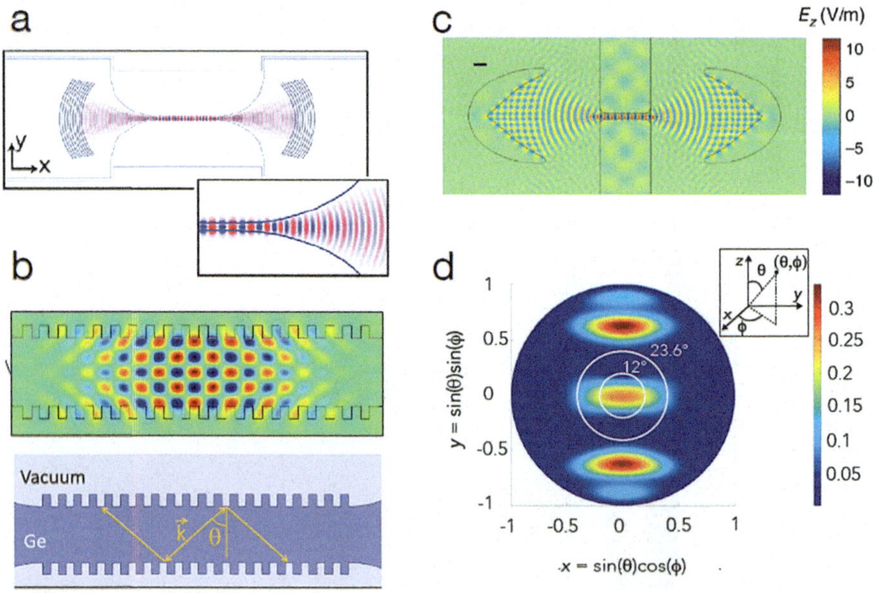


**Fig. 3.25** Microbrige laser cavities: **a** with DBR end mirror [173]. **b** Design for a higher-order mode DFB cavity. A high $Q$-factor (of >10k) is achieved by the strong distributed feedback which can exceed 0.5 per reflection unit [179]. **c** Mode pattern of micro-bridge cavity with parabolic end-mirrors forming a corner cube [4]. **d** Upwards scattered light of cavity (**c**), from S.I. of [4]

*Photonic crystal (PC)* designs yield laser cavities with high $Q$-factors which can operate in single mode. At the edge of the photonic Brillouin zone where the slope of the bands drops to zero, modes have a zero group velocity. These points of slow light are those where lasing can potentially take place. Such band edge modes were obtained in a 2D hexagonal photonic crystals membrane cavity obtained by under etching a $Ge_{0.84}Sn_{0.16}$ layer deposited on a step-graded GeSn buffer [39]. The performance of the laser thus obtained did not outperform the MD laser from the same wafer. This was partly due to the resonator frequency, which was not optimized for the wavelength red-shifted emission at elevated temperatures [39]. Indeed, the lasing wavelength in microdiscs adapts itself to gain when the temperature rises, not so in PC-based designs, where the resonance frequency is fixed by the periodicity of the hole—or corrugation—pattern.

Band edge modes with high $Q$ were also studied for use in suspended microstructures [179]. The laterally corrugated ridge structure, pictured in Fig. 3.25b, was found to maintain the imposed strain and may have offered $Q$-factors exceeding $10^4$ with suitable mode filling. A sketch of its working principle is shown in Fig. 3.25b, together with the mode pattern of a higher (5th-) order mode. In a semi-classical propagation picture, the strong localization of higher-order modes can be understood from resonant back reflection which may exceed 50% under certain angles. Key parameters for optimizations of the back reflection and thus the cavities $Q$-factor are the grating depth and period, the incidence angle—which is related to the mode order—and the duty cycle of the grating [179].

#### 3.4.1.2 Gain and Efficiency Measurements

A direct method to measure the modal gain is based on luminescence collected from one of the edges of a long waveguide structure that is excited over a variable length $L$ measured from that edge. If modal gain is present, the emission intensity increases exponential with the stripe length, $L$, and a substantial decrease in linewidth occurs. This is an indication of amplified spontaneous emission. The modal gain, $g$, is determined from:

$$I_{\mathrm{ASE}} = I_o + \frac{I_{\mathrm{SP}}}{g} \cdot \left(e^{gL} - 1\right),$$

where $I_{\mathrm{ASE}}$ and $I_{\mathrm{SP}}$ are the amplified and unamplified spontaneous emissions, respectively. $I_o$ contains higher-order modes as well as light scattered from the waveguide sidewalls. This method has successfully been used in [1], together with the threshold dependence on the cavity length. Similarly, the modal gain is obtained by following the lasing threshold change for a series of identically fabricated F-P waveguide cavities from the same material but with different lengths, l. The mirror—or extraction—loss $R$, expressed as gain needed to compensate this loss is:

$$\Delta g_{\text{mirror}} = -\frac{1}{2l}\ln(R_1 R_2)$$

where $R_j$ represents the reflection loss at the two waveguide edges. To give a simple example, by changing the length of a cavity from, for example, 300 to 600 μm, $\Delta g_{\text{mirror}}$ decreases from roughly 30 to 15 cm$^{-1}$ assuming $R_j = 0.4$. The thus obtained differential gain is obtained by dividing $\Delta g$ by the corresponding reduction of injection needed to reach threshold $\Delta I$. The latter is expressed in photon or current density for optical or electrical injection, respectively. This method has shown to give results in good agreement with the variable strip method [1]. Both methods gave a differential gain of approximately $\Delta g/\Delta I = 0.4$ cm kW$^{-1}$.

A measurement of a device's quantum efficiency is obtained from the measured (change of) optical output power normalized to the (change of) input power for conventional F-P lasers with end-mirrors made of waveguide facets. In such conventional cavity geometries, the total optical loss of the cavity corresponds indeed, and to a good approximation, to the reflection loss $R_j$ per facet. In closed cavities, for instance in micro-cavities around strained micro-bridges, the optical loss is primarily due to scattered light (in and out of plane). Its emission profile is a priori not known and thus has to be calculated using a 3D solver. This solver must contain the cavity structure and its environment. The detection window may be defined by a corresponding field of view, conventionally parametrized with a numerical aperture (NA).

In the case of the micro-bridge cavity shown in Fig. 3.25c, the light scatters out of the plane from two distinct places, (i) the intercept of the ellipsoids which form a discontinuity for the fields and (ii) the two filets of the micro-bridge. The amount of upwards scattered light from the latter depends on how narrow the bridge is and on how fast it narrows. The part which is collected within the NA is obtained by integrating the outwards scattered emission profile, Fig. 3.25d, over the corresponding solid angle $d\Omega = \sin\theta \cdot d\theta \cdot d\Phi$ normalized to the total intensity of the outwards scattered light $i^s_{\text{tot}}$:

$$\eta^s_{\text{col}} = \frac{i^s_{\text{col}}}{i^s_{\text{tot}}} = \frac{\int_{\text{NA}} i^s_{\text{out}} d\Omega}{i^s_{\text{tot}}}$$

$\eta^s_{\text{col}}$ represents thus the collection efficiency of the set-up. The external QE, $\eta^{\text{QE}}$, is defined as the ratio between scattered photons ($p^s_{\text{tot}}$) and the injected carriers. The latter is related to the number of absorbed photons ($p^{\text{abs}}$) or the injected current.

Quantifying the amount of scattered light (in terms of photons or power) requires a proper calibration of the set-up which includes all the loss channels starting from the light collection to transmission losses, coupling losses to the spectrometer and so on. The detector signal has also to be converted to an equivalent in power or photon flux. Since we are interested in the overall efficiency of the measurement system, its performance is conveniently calibrated by placing a test emitter at the sample position. A good choice for such an emitter is a hot filament black body emitter of known emissivity and temperature. Taking the emission profile into account which follows Lambert's cosine law, the power collected from the test emitter defined as $p'_{\mathrm{col}}$ is readily calculated. To account for the image blurring of the optical system, the size of the test emitter and laser device should be similar. The amount of scattered light from the lasing device follows from:

$$p^{s}_{\mathrm{tot}} = S_{\mathrm{det}} \cdot \left(\frac{p'_{\mathrm{col}}}{S'_{\mathrm{det}}}\right) \cdot \left(\eta^{s}_{\mathrm{col}}\right)^{-1}$$

where $S_{\mathrm{det}}$ and $S'_{\mathrm{det}}$ are the detectors signals as obtained from, respectively, the laser and test device. The above method is used to estimate the quantum efficiency of highly strained micro-bridge lasers under near resonant excitation [4].

### 3.4.2 Lasing in Ge

In comparison to GeSn, achievements in the field of strained germanium lasing might seem slower. One may therefore wonder about the significance of Ge laser experiments given the rapid developments underway with GeSn. Before answering this question, let us first define what we mean by a strained Ge laser. After all, the term "Ge laser" has already been used quite a bit, for instance for the inter-valence-band p-type Ge laser [180], the so-called pseudo-direct Ge structures obtained by band folding [6], confinement [7] and also for heavily n-type doped Ge [181].

Here, we mean by "strained Ge" those systems whose band structure is fundamentally modified by applying a high enough amount of tensile strain. We therefore often call it "direct bandgap strained Ge". This approach has the elegance of adding, through a careful tuning, the ability of pure Ge to lase to its already large application portfolio. One could, for instance, think of using its lasing capabilities for quantum technologies where germanium starts to play an important role [182]. Another good reason to continue this research is that strained Ge is the only semiconductor laser system that consists of only one element. Fundamental (ab initio) models—such as bandstructure calculations, phonon dispersion, charge carrier transport as well as fundamental laser modelling tools can simply be applied and tested on it.

However, the strained Ge approach inherently comes with disadvantages, notably when constructing a laser. Namely, semiconductor lasers are often built up of heterostructures, or quantum wells. For electrical carrier injection, some layers must be doped. In essence, a good laser will not consist of only a homogeneous layer such as Ge. Thus, for the moment, strained Ge may just be a powerful testbed helping the development of the final group-IV laser. Applications may nevertheless come, but in very specific areas.

In the following, we will highlight the impact research on Ge lasing has had so far. Firstly, we should mention the work performed by CEA-LETI on testing previously introduced bandstructure predictions, i.e. the deformation potential theory [183]. Similarly, the phonon dispersion under strain was investigated, with a pronounced bowing of the Raman strain tensor evidenced for any strain above 3%, notably for uniaxial loading [131]. For biaxial loading, nonlinear dispersion corrections are less pronounced. The bandstructure crossover from indirect to direct seems to match the deformation potential theory [183]. Obviously, this knowledge does have an impact on GeSn, in particular in a later stage when the fabrication and layer quality issues will have been resolved and accurate modelling will be needed for performance optimization.

What the lasing demonstration in strained Ge [4] has shown so far and what we can learn from it is the following:

(i) The direct bandgap concept works. Group-IV lasers of that kind behave very similarly to other semiconductor lasers.
(ii) Gain values as well as lasing efficiency are high, especially at low temperatures.
(iii) The latter is very prominent in gain clamping and mode competition, where one cavity mode prevails at the expense of all others.
(iv) A multi-mode spectrum—which is uniformly amplified as the excitation increases—does not mean lasing. Such a spectrum may well contain stimulated emission, but this can hardly be distinguished from transparency and is therefore not sufficient to evidence lasing. Instead, lasing with a high gain material such as Ge or GeSn will occur only on a few, possibly even only one mode(s), as Fig. 3.26, taken from [4], convincingly shows.

The above given features help to categorize previously published work, c.f. Table 3.5. We also mention, for comparison purposes, the most recent works on GeSn. Comparing the lasing experiments performed on biaxial strained Ge [191] and GeSn [3] is particularly instructive because the experiments were performed on practically identical set-ups. In the germanium case [191], only 2 of the 4 laser criteria were fulfilled. For GeSn [3], however, all 4 criteria were fulfilled. Biaxial strained germanium did not seem, in [191], to have reached the laser threshold, most probably because the tension (1.7%) was not high enough, then, to close the offset between Γ and L. Surprisingly, this possibility was not considered in [191].

Additional understanding to be gained from lasing experiments on strained Ge [4] concerns the valley scattering from Γ to L. This scattering is very fast in natural germanium, where the offset is about 140 meV. The situation, however, changes when the offset is approaching zero under high strain, c.f. Figure 3.26b. The scattering then depends on the energy the electron carries with respect to the L-band edge. The electron will hardly scatter as long its excess energy is less than the energy of phonons at the Brillouin zone boundary [4], Fig. 3.26c. Interestingly, the blocking of valley scattering as demonstrated here at low temperature reminds the original proposal by the MIT group [193] to block this scattering by filling up the L states at energy below the conduction band minimum at Γ through a heavy n-type doping of Ge with phosphorous. Seemingly, the blocking effect by doping is less effective and/or has drawbacks as level broadening.

For electrons with higher excess energy, valley scattering rates increases quickly. This effect is strongly dependent on temperature. Indeed, at elevated temperatures, (i) the probability electrons have of occupying states with energies definitely higher

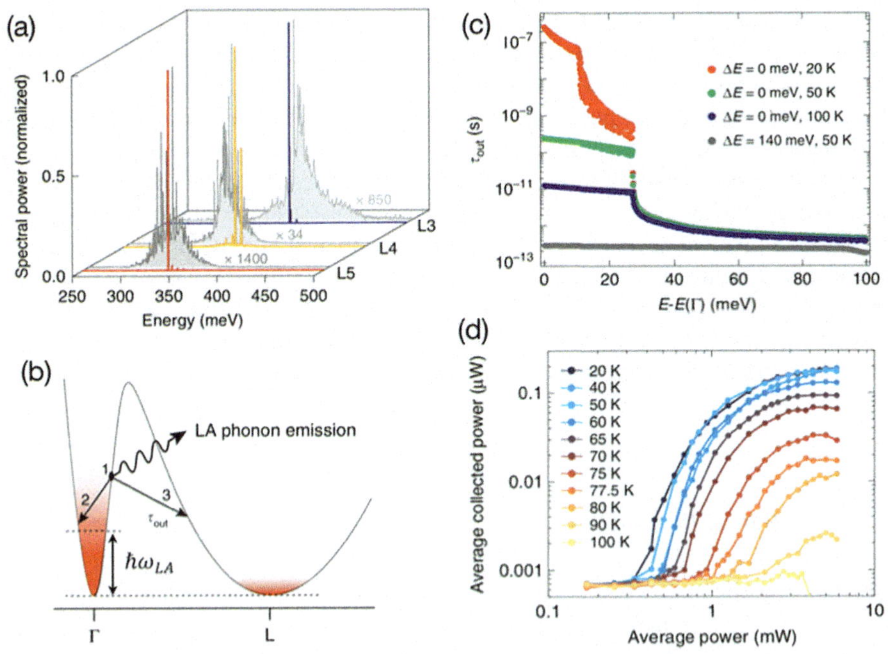


**Fig. 3.26** **a** Few mode to single mode behaviour of the strained Ge laser, typical of high gain semiconductor lasers. **b** and **c** At low temperature, the scattering of electrons from Γ- to the L-valley is blocked up to energies exceeding the energy of the zone boundary phonon $\hbar\omega_{\mathrm{LA}} \cong 28$ meV. **d** Light-in–light-out curve as a function of different temperatures. Images are adapted from [4]

**Table 3.5** Experiments performed on strained germanium and on GeSn (for comparison)

| | Cavity modes | Linewidth narrowing | Mode competition | Efficiency | References |
|---|---|---|---|---|---|
| n-doped Ge laser | No, only transversal | No | No | Low | [181] |
| Intermediate strained Ge | y | y | No | Low/not quantified, | [245] |
| Stressor induced Biaxial strain | y | y | No | Low/not quantified | [246] |
| Strained Ge laser | y | y | y | >50% | [4] |
| GeSn (microdisc) | y | y | y | High/not quantified | [3, 243] |

Examples given in the last two rows fulfil all four lasing criteria, longitudinal modes are resolved, linewidth narrowing, mode competition and high efficiency than the Fermi energy does increase, and so does (ii) the number of the zone boundary phonons. The latter may have the stronger impact, as is deduced from comparisons between experiments and model predictions, shown in Fig. 3.26d. Time resolved laser excitation experiments on strained Ge layers but also GeSn may be the best way to study these processes in more details. From such studies, we expect to gain a quantitative understanding of the impact of reduced scattering by phonon blocking for group-IV lasing. In [4], with nearly direct bandgap Ge and comparably short excitation, phonon blocking is indeed highly effective up to almost 100 K. For continuous wave excitation and under nanosecond long excitation, this effect is expected to be weaker, though not negligible. Indeed, in a vast amount of barely direct systems, lasing stops near $T < 100$ K [1, 3, 246], as shown in the following section. This peculiar conformity could come in part from the blocking effect vanishing at this temperature.

## *3.4.3 Lasing in GeSn*

The fabrication of an efficient, electrically pumped laser operating at room temperature based on the epitaxy of group-IV materials would have several advantages like an easy integration in the present day Si-photonics platform. In the subsection just above, Ge lasing was presented. Here, we will discuss lasing from GeSn alloys and GeSn/SiGeSn heterostructures. The bandstructure engineering lessons learned from Ge can also be applied here. There will however be specific issues coming from the epitaxy, on Ge SRBs, of thick GeSn layers or heterostructures.

### 3.4.3.1 From Bulk to Heterostructures

As detailed above, incorporating Sn atoms into the Ge lattice modifies the electronic band structure. The energy of the conduction band at the Γ-valley (direct transition) shrinks faster than the energy of the conduction band at the L-valley (indirect transition), resulting in a crossover to a fundamental direct bandgap semiconductor once a critical amount of Sn is reached. Photoluminescence (PL) experiments on a series of GeSn layers with Sn contents varying from 8 to 12.5 at.% proved that as early as in 2015 [1]. A definite change of the evolution of the integrated PL intensity with the measurement temperature highlighted a fundamental change in the semiconductor bandgap. Using the conduction band offset $\Delta E$ (energy difference between the Γ-valley and the L-valley) as a modelling parameter, an indirect bandgap with $\Delta E = -80$ meV was extracted for $Ge_{0.92}Sn_{0.08}$ and a direct bandgap with $\Delta E = +25$ meV for the $Ge_{0.875}Sn_{0.125}$ alloy. These values are in good agreement with k.p bandstructure calculations. The conduction band offset ΔE will be called, in the following, the bandgap "directness". It is an important parameter determining lasing properties. The above-mentioned PL experiments, together with strain effects

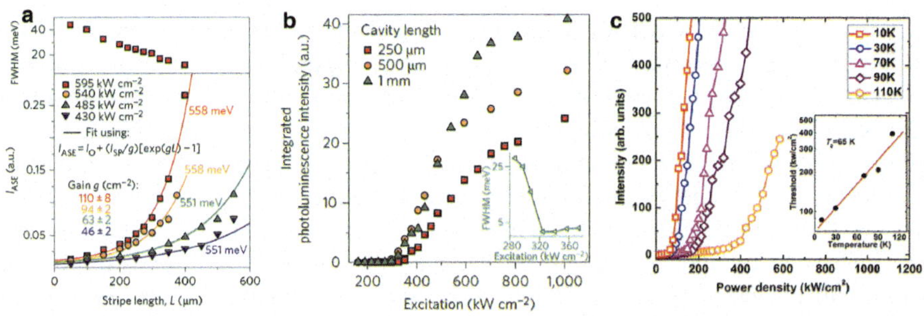


**Fig. 3.27** **a** Amplified spontaneous emission (ASE) spectra for different peak energies. The full line is the fit, using the equation given in the figure, to determine the modal gain [1], **b** integrated photoluminescence intensity as a function of optical excitation for 250 μm, 500 μm and 1 mm long waveguides. Inset: full width at half maximum of the peak around the lasing threshold for the 1-mm-long GeSn waveguide [51], **c** photoluminescence intensity of a 1100-μm-long waveguide between 10 and 110 K. Each curve shows lasing threshold characteristics. The temperature-dependent thresholds varied from 87 to 396 kW/cm$^2$. Inset: laser threshold versus temperature [51]

calculations, yielded an experimental Sn content of 8 at.% at which a fully relaxed, and therefore, cubic GeSn alloy became a direct bandgap semiconductor.

As shown in the theory section, a direct bandgap is a necessary condition for optical gain and lasing, in case the optical gain can compensate cavity and material losses. The modal optical gain—the remaining gain of a mode after subtracting the material losses—can be extracted experimentally by the variable stripe length (VSL) method as explained in Sect. 3.4.1.2. Using the VLS method, it was shown that a 500-nm-thick $Ge_{0.875}Sn_{0.125}$ alloy yielded, at the maximum emission energy of 558 meV, a modal gain of 110 cm$^{-1}$ (Fig. 3.27). Using the methods outlined in Sect. 3.4.1.2, experiments shown in Fig. 3.27a and b gave a differential modal gain of approximately 0.4 kW cm$^{-1}$.

With a direct bandgap semiconductor and the technology to fabricate Fabry–Perot resonators, the demonstration of lasing under optical pumping naturally followed. By optically pumping a 1-mm-long waveguide, lasing was unambiguously proved, with a distinct threshold in output intensity, as shown in Fig. 3.27. Despite the importance of being the first GeSn laser, performances were poor. The laser threshold at 20 K was about 350 kW/cm$^2$ and increased up to almost 1000 kW/cm$^2$ at the maximum operation temperature of 90 K.

Shortly after, these results were confirmed by the University of Arkansas group using a 760-nm-thick GeSn alloy with a Sn concentration of 11 at.% [51]. Lasing properties were similar, with a maximum lasing temperature of 110 K. The poor laser performances were attributed to different causes:

(i) a very low directness $\Delta E$;
(ii) a relatively high compressive strain in the layer;
(iii) a high density of misfit defects at the GeSn/Ge-substrate interface.

These drawbacks, which are interconnected and not easy to separate, are discussed in the following. A cubic GeSn alloy has a larger crystalline lattice than elemental Ge. During epitaxy, a large in-plane compressive strain will build up in GeSn alloys grown on Ge SRBs. According to theoretical bandstructure calculations, biaxial compressive strain results in an increase of the energy of the Γ-valley and therefore a reduction of the energy separation between Γ- and L-valleys, or directness, ΔE. This is equivalent to a lower "effective Sn content" in the GeSn alloys. The influence of tensile and compressive strain on bandgap directness is the opposite. Tensile strain will indeed increase the bandgap directness. For valence bands, an in-plane compressive strain will lift up the heavy-hole (HH) energy band while a tensile strain will lift up the light-hole (LH) energy band. In the following, we will first discuss the impact of biaxial compressive strain. Later on, we will elaborate on the role of uniaxial and biaxial tensile strain. The calculated electronic band structure of a $Ge_{0.875}Sn_{0.125}$ alloys as a function of compressive strain is shown in Fig. 3.28a. The solely strain-based indirect–direct bandgap transition is clearly seen for a biaxial compressive strain of −1.05%. The directness, $\Delta E$, of a $Ge_{0.875}Sn_{0.125}$ layer increases from 0 meV to 120 meV as the compressive strain decreases from −1.05% to 0%, respectively. As a consequence, the integrated PL intensity continuously increases as the compressive strain decreases [66].

The biaxial compressive strain build-up during the epitaxial growth can be relaxed by depositing thick layers, well above the critical thickness for strain relaxation. However, it was shown, in the well-known SiGe on Si system, that 100% strain

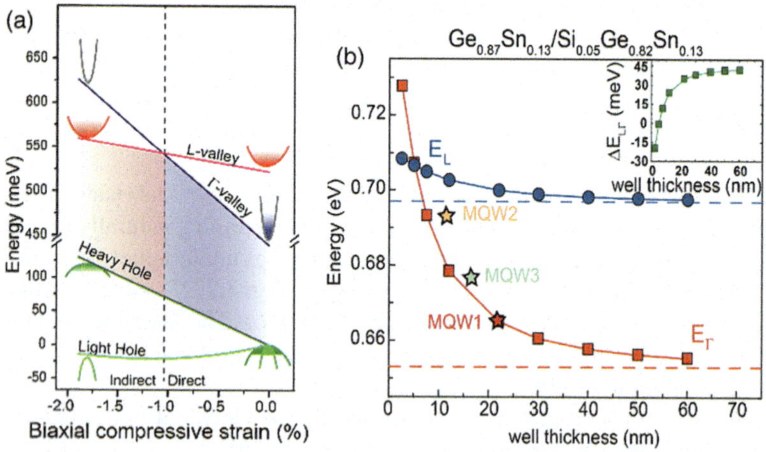


**Fig. 3.28 a** Calculated electronic band structure as a function of compressive strain in $Ge_{0.875}Sn_{0.125}$ alloys. The semiconductor has an indirect bandgap for compressive strains larger than −1.05%. It becomes a direct bandgap semiconductor for strains lower than −1.05%. **b** Energy value of Γ- and L-valleys as a function of the quantum well thickness. Inset: directness versus quantum well thickness

relaxation was hardly achievable in hundreds of nm thick layers [184]. In addition, a high density of crystalline defects density was present in those layers. Reduced crystallinity and point defects reduce the radiative emission by creating paths for non-radiative recombination.

Forming free standing regions where the lattice can elastically relax enabled to maximize relaxation in GeSn layers. Various strategies are described below. However, we would like first to discuss the possibility of improving laser performances thanks to the use of heterostructure and quantum wells into the active material. Growth aspects were discussed in a previous section.

Roughly an order of magnitude reduction in lasing threshold was achieved in III–V lasers each time the dimensionality of the gain material was reduced (from 3D to 2D to 1D and finally 0D, thanks to quantum wells, quantum wires and quantum dots, respectively). In order to go along this path with GeSn, a second group-IV semiconductor material with a larger bandgap than GeSn is required. The interest of using Ge in the barrier layers of p-i-n Ge/GeSn multi-quantum wells (MQW) LEDs was evaluated. Unfortunately, no advantages over p-i-n GeSn homo-junction LEDs [42] were evidenced. In 2017, efficient electro-luminescent GeSn/SiGeSn MQW [185] p-i-n LEDs were fabricated, showing the interest of Sn-based heterostructures for lasing.

The effective bandgap in a two dimensional biaxial compressive GeSn QW corresponds to the energy differences between the first quantized levels of heavy holes (HH) and Γ-valley electrons. The quantum state energies are inversely proportional to the effective masses of the conduction and valence bands. This has an impact on the directness of the material, $\Delta E$, which is considerably reduced in thinner wells since the electron effective mass at Γ ($m^*_\Gamma$ [100] $= 0.031 m_e$) is ~2.7 times smaller than at L ($m^*_L = 0.117 m_e$). In Fig. 3.28, the energy of the first quantized levels of Γ- and L-valleys as a function of the quantum well thickness is given for a $10 \times$ {$Si_{0.05}Ge_{0.82}Sn_{0.13}/Ge_{0.77}Sn_{0.13}$} MQW grown on a partially relaxed $Ge_{1-y}Sn_y$ buffer, itself on a Ge SRB. There is, for $Ge_{0.77}Sn_{0.13}$ wells less than 10 nm thick, a transition back to an indirect semiconductor.

A major way to improve GeSn lasing is reduced the amount of structural defects that result in non-radiative recombinations. For instance, there is, in thick plastically relaxed GeSn layers which are the most used for optically pumped lasing, a misfit dislocation network at the interface with the Ge SRB, i.e. very close to the optical mode in GeSn. To address this problem GeSn buffers with a grading [51, 67], or a lower Sn content buffer are grown first [186]. The optical active medium is then grown lattice matched on this buffer. Interestingly, ternary SiGeSn alloys offers a balance of lattice strain and mismatch reducing additional interfaces relaxation. A similar strategy was also used in GeSn/SiGeSn MQW lasers.

A 200-nm-thick strain relaxed $Ge_{0.9}Sn_{0.1}$ layer was first grown, as a separation layer, on a Ge SRB followed by a 10 periods {22 nm $Si_{0.05}Ge_{0.82}Sn_{0.13}$/22 nm $Ge_{0.77}Sn_{0.13}$} MQW (called MQW 1). Another MQW structure (called MQW2), with 12-nm-thick wells and 16-nm-thick barriers, as well as a double heterostructure (DHS) with a single 380-nm-thick $Ge_{0.855}Sn_{0.145}$ active layer cladded with

$Si_{0.055}Ge_{0.83}Sn_{0.115}$ were grown for comparison purposes. Details on the epitaxy of the heterostructures can be found in [186].

Undercut microdiscs were optically pumped using two different lasers: a Nd:YAG laser emitting at 1064 nm, with a pulse length of 5 ns and a repetition rate of 17 kHz and a fibre laser emitting at 1550 nm, with a pulse length of 800 ps and a repetition rate of 20 kHz [58]. Experimental conclusions, from comparing lasing data for these three heterostructures under two optical excitations, were the following:

- The maximum lasing temperatures were similar, for the DHS and the MQW1 heterostructures: 100 K and 120 K under 1064 nm and 1550 nm excitation, respectively, although the DHS directness, 82 meV, was almost 2.5 times higher than in the MQW1 structure (32 meV only).
- The MQW2 structure did not lase under 1064 nm but only under 1550 nm excitation and at very high power density above 1000 kW/cm$^2$. This is attributed to the very low directness of only 8 meV and the very low active volume of only 120 nm (compared with 220 nm for MQW1 and 380 nm for DHS), with a gain not able to compensate the optical losses.
- The most interesting comparison concerns the lasing threshold (Fig. 3.29). The laser threshold at 20 K for the DHS structure was found to be ~300 ± 25 kW/cm$^2$, to be compared with 40 ± 5 kW/cm$^2$ only for the MQW1 laser. However, the same laser threshold was obtained for temperatures above 50 K.

Those experimental results support the initial assumptions. Reduced non-radiative recombinations due to a *spatial separation from the dislocated interface* with the Ge-substrate by an intermediate strain relaxed buffer and an enhanced radiative recombination because of the *localization* of electrons and holes in multi-quantum wells result in the low directness MQW having a similar maximum lasing temperature than the larger directness DHS.

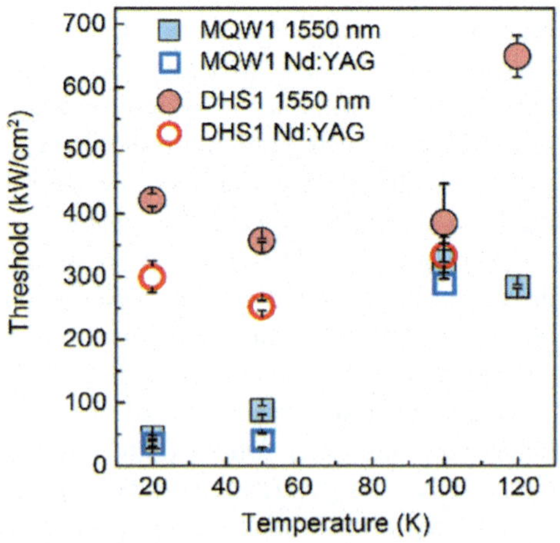


**Fig. 3.29** MQW and DHS laser threshold versus temperature for two different optical excitations

The lower threshold of MQW1 compared to DHS is mainly due to the reduced density of states (DOS), resulting in a decrease of the carrier concentration at transparency. Since the threshold reduction is more than 1 order of magnitude, it is reasonable to assume that the carrier lifetime is improved due to a localization of carriers in the QW. What is also striking is that, above 50 K, lasing thresholds for DHS and MQW become similar. This is explained based on bandstructure calculations which indicate that the conduction band offset is about 50 meV at 20 K and below 20 meV at room temperature. When pumped with high energy photons at high temperature, carriers easily can gain energy above the band offsets and escape the wells. Quantization disappears and the structure behaves as a DHS, then.

In conclusion, quantum well structures are viable solutions to strongly reduce the lasing threshold. Experimental data and theory tell us that, in a MQW, the thickness of the well should be such that there is some carrier confinement and the quantization just high enough to reduce the DOS. Large quantization effects will strongly and negatively influence the GeSn directness. MQW improvements would result from the epitaxy, if feasible, of higher Si content SiGeSn barriers: the QW to barrier energy band offset would be higher, then and carrier confinement still present at elevated temperatures.

Last but not least, optical excitation closer to the bandgap reduces the electron density and consequently the L-valley population, which contributes to losses by free carrier absorption. The effect was discussed, for the Ge laser, in a previous section.

#### 3.4.3.2 GeSn Optical Cavity Under-Etching

By deeply under-etching a GeSn epilayer grown on a Ge SRB, a significant fraction of the built-in strain will elastically relax. This compressive strain reduction will increase the bandgap directness and improve the gain. The increased refractive index contrast between air and semiconductor will also improve the mode filling. Last but not least, surface defects can be removed and/or passivated depending on the chemistry used. Those performance boosts are shown in the following discussion on undercut waveguide and microdisc lasers.

The fabrication process is based on standard CMOS processing. Section 3.2 gives some details about the selective etching used. The control samples (not under-etched) and the under-etched samples were passivated with 10 nm of $Al_2O_3$ dielectrics in order to passivate the GeSn surface and reduce surface carrier recombination. Moreover, under-etching will remove part of the highly defective region at the GeSn layer/Ge SRB interface, reducing non-radiative Shockley–Read–Hall recombinations.

The light-in–light-out (L–L) characteristics at 20 K of not-etched and under-etched, same dimensions $Ge_{0.875}Sn_{0.125}$ waveguides are shown in Fig. 3.30. It is clear that the under-etching definitely improves the lasing characteristics, as expected, due to the increased GeSn directness via strain relaxation (Fig. 3.30a).

The lasing threshold decreases from 430 down to 206 kW/cm$^2$, while the slope of amplified emission increases by a factor of 2.3. In the bulk state, the 560-nm-thick

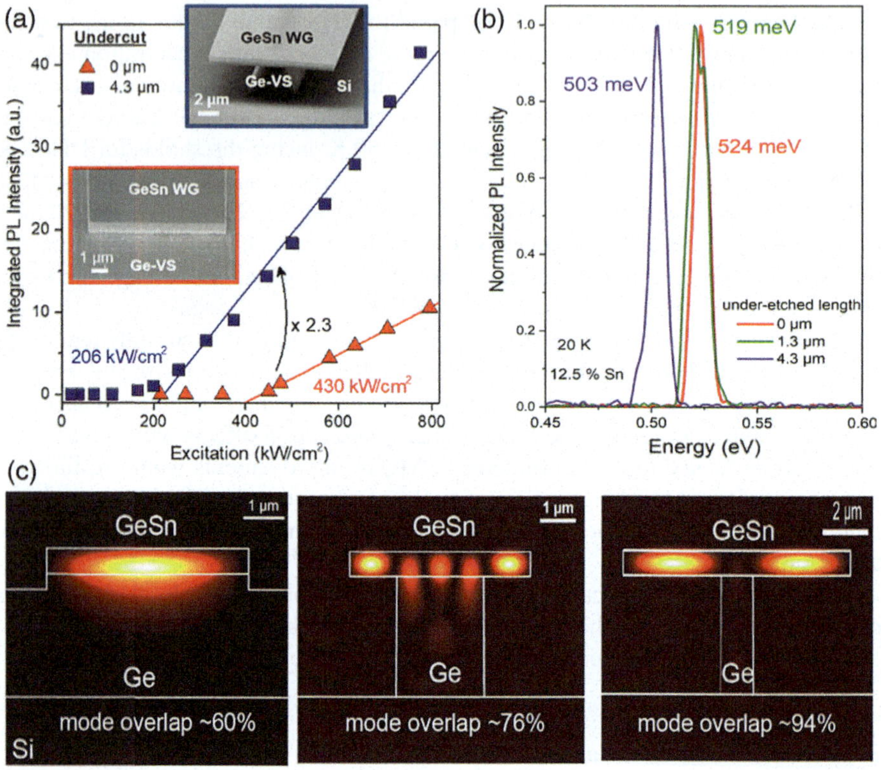


**Fig. 3.30** **a** L–L characteristics of non-under-etched (red plot) and 4.3 μm under-etched (blue plot) GeSn waveguides. SEM images of the two GeSn laser cavities are shown in the insets. **b** Laser emission from GeSn waveguides with different under-etch lengths. The shift of the laser emission energy is due to strain relaxation induced bandgap change. **c** Optical mode distribution in the GeSn waveguides with different under-etch lengths

$Ge_{0.875}Sn_{0.125}$ layer has a residual compressive strain of −0.6%, which translates into a $\Delta E = 25$ meV. According to strain relaxation simulations data used for band-structure calculations, a medium directness of about $\Delta E = 60$ meV at the edges of the GeSn "wings" is obtained in the under-etched structure. The bandgap change can be seen by the energy shift of the laser emission for waveguides with different under-etching lengths, as shown in Fig. 3.30b. The increased directness will result in an increase of the Γ-valley population, with less optical pumping required to reach the lasing threshold, as seen experimentally. Moreover, the Γ- to L-valley scattering is reduced for a higher bandgap directness, resulting in a higher lasing temperature. Indeed, the 4.3 μm under-etched waveguide lases up to 135 K, a temperature 45 K above that of the non–under-etched structure.

The additional advantage of suspended cavity designs is coming from a higher (GeSn/air) refractive index difference in the free standing regions, increasing the optical mode confinement in the gain material and reducing the bending losses. In the

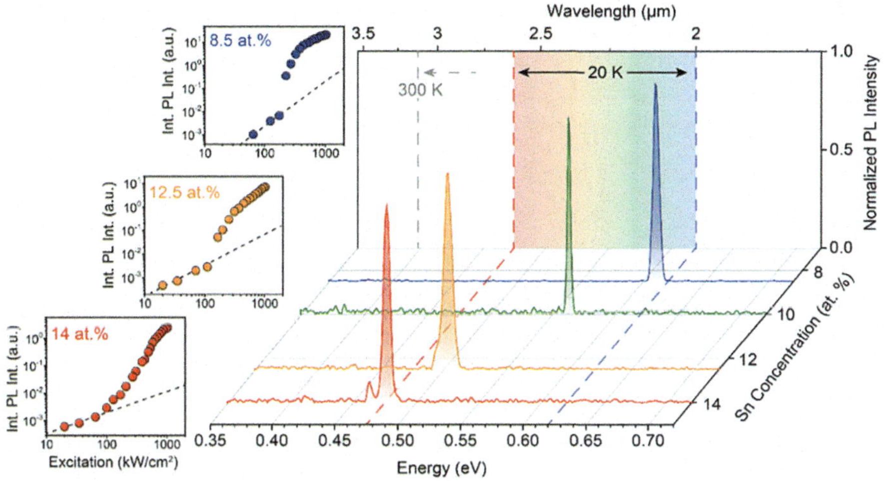


**Fig. 3.31** (Right side) Laser emissions of 8.5, 10, 12.5 and 14 at.% GeSn Fabry–Perot lasers with 4.3 μm undercuts. Bandgap changes induced by Sn contents which varied resulted in shifts of the laser emission energy. (left side) L–L characteristics at 20 K for 8.5, 12.5 and 14% GeSn Fabry–Perot lasers

present samples, the optical mode overlap is 64% in the bulk case (no under-etching) and increases up to 76% and 94% for 1.3 μm and 4.3 μm undercuts, respectively (see Fig. 3.30c). Such an effect is highly beneficial in order to increase the operating temperature and reduce the lasing threshold.

By fabricating such under-etched waveguides, it was possible for the first time to demonstrate multiple laser actions in the mid-IR region, or, in other words, tune the laser emission by changing the Sn atomic content. Using roughly 800-nm-thick layers with a large strain relaxation in the as-grown state together with under-etching waveguide cavities, laser emission from GeSn layers with Sn contents of 8.5, 10, 12.5 and 14 at.% was demonstrated. The laser peak energies and the L–L characteristics at 20 K for such optically pumped lasers are shown in Fig. 3.31.

The cavity fabrication technology described above was also used to fabricate micro-cavities with micrometres round trip lengths (microdisc resonators and micro-Fabry–Perot corner cube cavities). The advantage of such cavities is a simpler experimental set-up, with a homogeneous pumping of the complete cavity which is more difficult for long and wide waveguides. The importance of defects management on GeSn lasing characteristics is critical and can result, if done properly, in drastically reduced lasing thresholds in microdisc laser cavities [243].

#### 3.4.3.3 Towards Higher Operation Temperature

GeSn growth improvements, with, for instance, the use of GeSn graded buffers to gradually accommodate the lattice mismatch between GeSn and Ge, yielded good

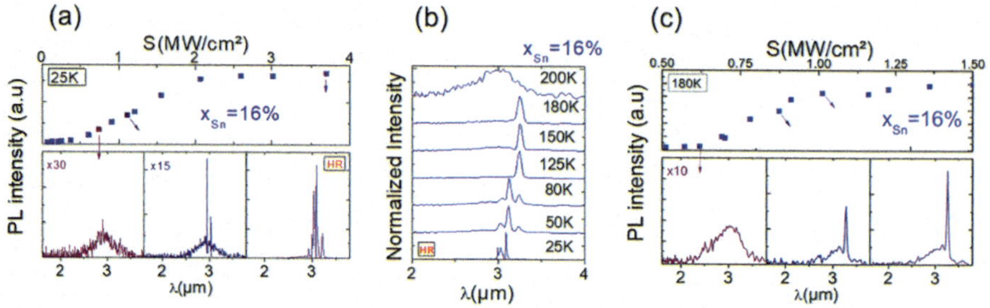


**Fig. 3.32** **a** L–L curve of a 8 μm microdisc fabricated using a $Ge_{0.84}Sn_{0.16}$ layer grown on a GeSn step-graded buffer at 25 K. **b** Evolution of the laser spectra as a function of temperature. **c** L–L curve at 180 K of the GeSn 16% microdisc [67]

optical quality GeSn layers with Sn contents up to 16 at.% [18]. Compared to the GeSn 12% layer discussed above, with a bandgap directness of only $\Delta E$ ~ 50 meV, the use of a graded GeSn buffer resulted in a large strain relaxation in the $Ge_{0.84}Sn_{0.16}$ as-grown layer on top, with defects confined in the bottom layers. The latter after the microdisc undercut became completely stress relaxed. The bandgap directness then increased to the 146 meV theoretical value.

Figure 3.32a shows the L–L characteristics at 25 K of a 8 μm diameter $Ge_{0.84}Sn_{0.16}$ microdisc together with emission spectra at three different pumping levels. The lasing threshold was 377 kW/cm$^2$ at 25 K. Figure 3.32b shows the evolution of the lasing spectrum as a function of temperature, with a mode switching as the temperature increased from 50 up to 125 K. The optical gain shifted rapidly with the temperature, while the cavity modes evolved at a much slower rate. At 200 K, a clear broadening of the emitted spectrum was observed compared to lower-temperature spectra, indicating no laser action anymore. Spectra were measured at 4.3 MW/cm$^2$. L–L curves at 180 K are shown in Fig. 3.32c. The increase of the lasing temperature, from 135 K up to 180 K, was attributed to the strongly increased bandgap directness, from 50 meV for $Ge_{0.875}Sn_{0.125}$ up to 146 meV for $Ge_{0.84}Sn_{0.16}$ samples. Increasing the Sn content, however, degraded the layer quality. This is believed to be responsible for the increased laser threshold at 25 K, from 220 kW/cm$^2$ for $Ge_{0.875}Sn_{0.125}$ sample up to 377 kW/cm$^2$ for $Ge_{0.84}Sn_{0.16}$.

With a thicker $Ge_{0.84}Sn_{0.16}$ active zone (418 nm) and a capping with a higher bandgap $Ge_{0.87}Sn_{0.13}$ layer, a maximum lasing temperature of 230 K was reached. There was also a decrease of the threshold, down to 134 kW/cm$^2$ at 15 K, because of a better overlap between the material gain and the optical modes [241].

Some conclusions can, at that point, be drawn regarding the evolution of the GeSn laser performance with the Sn content. Experiments described above show that an increase of the conduction band directness thanks to thick, higher Sn content GeSn alloys result in an increase of the maximum lasing temperature. However, a Sn content increase also seems to result in a lower crystalline quality of the material and thus to larger laser thresholds at high temperatures, where phonon scattering becomes important. For instance, the lasing threshold, for a Sn content of about 16 at.%, increases from approximately 100 kW/cm$^2$ at 20 K up to more than 3400 kW/cm$^2$ at 270 K [52]. A tentative solution to decrease the threshold while keeping high lasing temperatures could then be to grow SiGeSn/GeSn multi-quantum wells with high Sn contents in the wells. A decrease by one order of magnitude was indeed evidenced for a medium Sn content of 13% in the wells compared to bulk structures. It is then clear that the combination of two approaches seems the most reasonable way to address efficient room temperature GeSn-based laser.

All the GeSn lasing researches presented so far were based on the epitaxial growth of GeSn layers on Ge buffers, with very large thicknesses and/or under-etching to reduce the residual compressive strain. The main direct transition then took place between the conduction band Γ-valley and the heavy hole (HH) as the top valence sub-band. Theoretical estimations indicate that if some biaxial tensile strain is induced in GeSn, the light hole (LH) becomes the top valence band. Due to splitting of HH- and LH-bands, the majority of injected holes will be in the LH which is beneficial for lasing. The calculated maximum net gain, $g_{max}$, function of the GeSn directness for different Sn content GeSn alloys is shown in Fig. 3.33a. It can be seen that, for the same bandgap directness, the net gain is larger in low Sn content layers. This is why the optical transition matrix element for $z$-polarized radiation for the LH-G transition ($z$-polarized) is 30% larger than the matrix element for the HH-G ($x$-polarized) transition. The same directness can be obtained by injecting tensile strain in the GeSn material. Theoretical and experimental data [237] indicate that, under a biaxial tensile strain of +1.5%, a GeSn layer with 5 at.% of Sn has the same Γ- to L-valley energy difference as a 12.5 at.% GeSn layer under a biaxial compressive strain of −0.6% (as discussed above). The Sn content influences the

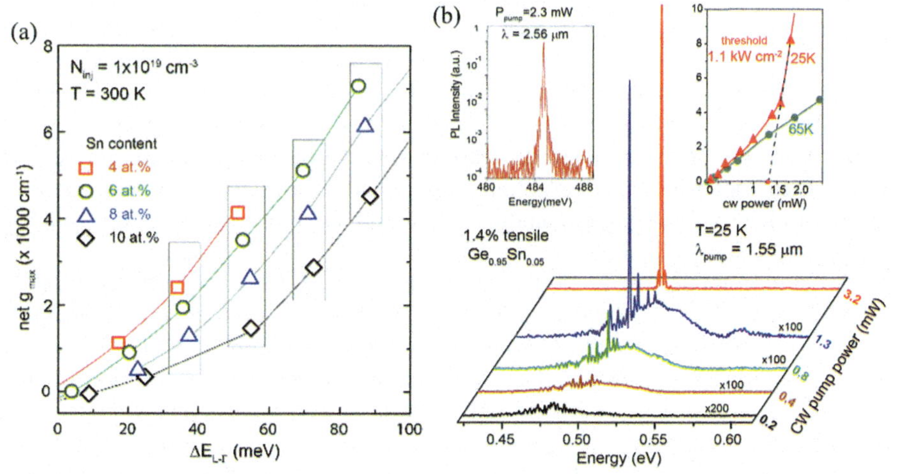


**Fig. 3.33** **a** Calculated net $g_{max}$ function of the GeSn directness for different Sn content GeSn alloys. A large directness is obtained by injecting high amounts of in-plane tensile strain [163]. **b** PL spectra of tensile strained $Ge_{0.95}Sn_{0.05}$ under CW excitation, insets: 2.55 μm emission peak intensity at 25 K and L–L curve detail taken close to the threshold

conduction band energy separation but *not* the LH and HH valence band separation. The tensile biaxial strain positively impacts both the conduction band directness and the LH-HH splitting. Biaxial tensile strain is the only type of strain that lifts the LH band above the HH band. The laser transition will then occur between two low DOS bands (Γ- and LH), with a large LH-HH energy separation. This should result in lower laser thresholds.

Based on the above considerations, a $Ge_{0.95}Sn_{0.05}$ layer was used as the optically active medium for laser emission. A $SiN_x$ stressor layer was deposited on the grown GeSn layer and subsequently transferred by wafer bonding onto a Si substrate covered with a thick Al layer [3]. The bottom part of the as-grown GeSn layer, which was defective with the presence of a regular array of edge dislocations, was at that stage on top of the stack. It was removed thanks by chemical etching. This step was crucial, as it eliminated one of the main sources of non-radiative recombination. This effect was also demonstrated in [243], where defects were removed at the bottom of microdiscs during the under-etching of 7–10.5% tin content alloys. Low-temperature lasing thresholds were then reduced to 10 kW/cm$^2$ upon optical pumping at 1.5 μm [25 MHz, 3.5 ns]. After microdisc patterning and under-etching, the GeSn optical cavity was supported by a large Al metallic post that acted as a heat sink. The final structure was covered with a second stressor $SiN_x$ layer, resulting in a GeSn layer embedded in a $SiN_x$ stressor layer and suspended on a large Al pillar. The strain injected into the nearly 300-nm-thick GeSn layer was 1.45%. It was high enough to transform the initially indirect bandgap $Ge_{0.95}Sn_{0.05}$ layer, with a directness of $\Delta E =$ − 60 meV, into a direct bandgap one, with $\Delta E = 70$ meV. The valence band splitting was $E_{LH} - E_{HH} = 170$ meV.

Lasing experiments were performed with a 1550 nm wavelength continuous wave pump laser focused on the sample surface into a 12 μm diameter spot. PL emission spectra from a 9 μm diameter disc, collected at various incident pump powers at 25 K, are shown in Fig. 3.33b. At low excitation levels, microdiscs exhibited a broad spontaneous emission background, attributed to Γ-LH direct transitions. For higher CW pump powers, whispering gallery modes (WGM) appeared. Higher excitation resulted in an exponential intensity increase, typical of amplified emission, of the main optical mode at 485 meV. At 2.3 mW pump power, the lasing emission was four orders of magnitude stronger than the background, as shown by the high-resolution spectrum shown in Fig. 3.33.

The emission intensity at 25 K of the lasing mode at 485 meV, corresponding to 2.55 μm wavelength, in the threshold region, is shown as a function of the incident pump power by the linear light-in–light-out (L–L) characteristic in Fig. 3.33-inset. The lasing threshold, as obtained by linear extrapolation the LL curve shown in the inset of Fig. 3.33b is 1.3 kW/cm$^2$ only. The laser emission is also characterized by a linewidth narrowing below the instrumental resolution of 58 μeV. This GeSn device was the first capable of continuous wave laser operation and had the smallest threshold and linewidth reported so far of any group-IV semiconductor lasers.

Finally, we focus on a device described in a previous section (Fig. 3.16), where an uniaxial stress was injected into a micro-bridge with a 400-nm-thick $Ge_{0.84}Sn_{0.16}$ layer on top of a step-grading [2]. The cavity was defined by two corner cube mirrors,

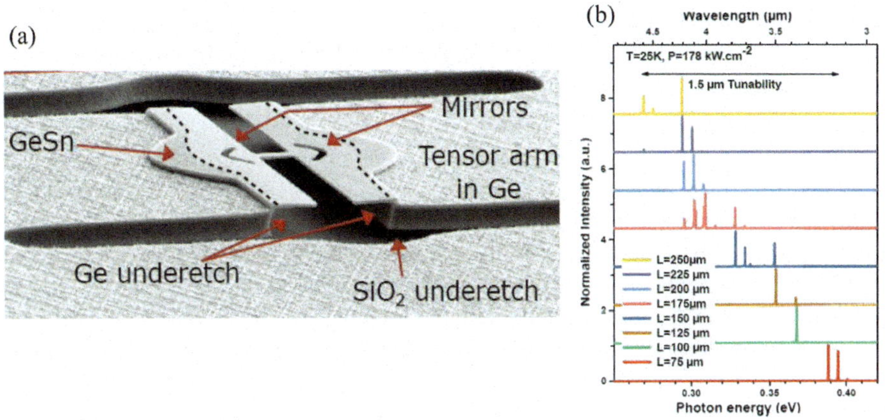


**Fig. 3.34** **a** Uniaxially tensile-strained $Ge_{0.84}Sn_{0.16}$ step-graded layer laser cavity, **b** laser emission spectra at 25 K versus arm length (the higher the arm length, the higher the uniaxial strain)

as shown in Fig. 3.34. The epitaxy of the step-graded stack occurred on a 5-µm-thick Ge SRB, itself on a silicon on insulator wafer. In contrast to pure germanium suspended membranes, the GeSn stack was in a compressive state. Etching of the membrane enabled to reverse the strain state thanks to pulling by the germanium arms, as discussed in Section "In Germanium Tin". The overall length $L$ of the Ge pulling arms governed the amount of tensile strain in the central part of the structure. At low temperature, a maximum of 2.2% uniaxial tensile strain was reached for $L = 250$ µm.

The uniaxial strain was evaluated from the photoluminescence spectral shift of the layer. Figure 3.34b shows the laser emission spectra are presented as a function of the arm's overall length (i.e. the tensile strain). A wide tunability of the laser wavelength, from 3.1 up to 4.6 µm, is obtained for a fixed 16% tin content in the active zone, at 25 K. Thresholds are around 10 kW/cm$^2$ for a pumping wavelength of 2.6 µm (82 MHz, 0.100 ns). As the built-in strain is created by thermal expansion coefficients differences, the strain decreases for higher temperature, and thus also the directness of the gain material. Laser emission is still observed up to 273 K, with a 2000 kW/cm$^2$ threshold for a pumping wavelength of 1.06 µm [50 kHz, 0.6 ns]. At 273 K, the strain is around 0.3%. So far, 273 K is the highest temperature reported for any GeSn laser operation. The race to reach room temperature lasing operation is still ongoing.

## 3.5 Optoelectronic Devices

A formidable interest has been building up for the last 10 years to develop and use photonic components operating in the mid-infrared, ranging from integrated biomedical and chemical sensors to short-range optical interconnections. A major roadblock is the current lack of a cost-efficient and complete MIR photonic platform with light sources and photodetectors fully compatible with the existing complementary metal–oxide–semiconductor (CMOS) platform. Current mid-IR technologies rely on bulky components and stand-alone laser sources operating in free space. Being able to monolithically integrate efficient and tunable MIR light sources and efficient group-IV photodetectors on Si would immediately give birth to new applications.

The continuous development of non-equilibrium growth schemes in industry-compatible reduced pressure chemical vapour deposition (CVD) systems have enabled the epitaxy of high-quality GeSn binary [46, 53, 187, 188] and SiGeSn ternary [50, 61, 189, 190] materials. Despite their metastable nature, alloys are thermally stable up to at least 300 °C (and possibly, depending on the Sn content and internal material strain, even above 600 °C) [191], which is high enough for back-end processing. Germanium and germanium tin (GeSn) have indeed shown great promises for the direct integration of group-IV semiconductor lasers and photodetectors. Light emission and photodetection will occur in the mid-infrared, thanks to the small bandgap of GeSn and tensile germanium. In the following, we will discuss the different strategies adopted in recent years to develop the required components to build an integrated circuit platform.

### 3.5.1 *Photodetectors*

Silicon and germanium, semiconductors from the group-IV column of the periodic table, are known for their pivotal role in microelectronics. The comparatively large elemental bandgaps of 1.12 eV (Si) and 0.80 eV (Ge, direct bandgap) at room temperature prevents their use as infrared detector materials, however. α-Sn is actually a group-IV semimetal with a negative bandgap of −0.41 eV. Alloying Ge with α-Sn and forming a binary with a diamond cubic crystal structure enable one to reach longer wavelengths in the infrared region. Even more control in designing material properties can be achieved thanks to SiGeSn ternaries. The incorporation of Si atoms increases the L-Γ-energy difference in the conduction band and might even reverse the Sn-induced direct bandgap transition. This may lead to the unique opportunity of simultaneously having a narrow and yet indirect bandgap material, not existing in III–V materials. The generally larger carrier lifetimes in indirect bandgap materials may result in larger generated photocurrents, which will increase the performance of photodiodes.

The wavelength tunability of SiGeSn alloys has been experimentally proved via transmission/ reflection measurements, as shown in Fig. 3.35. Reflection measure-

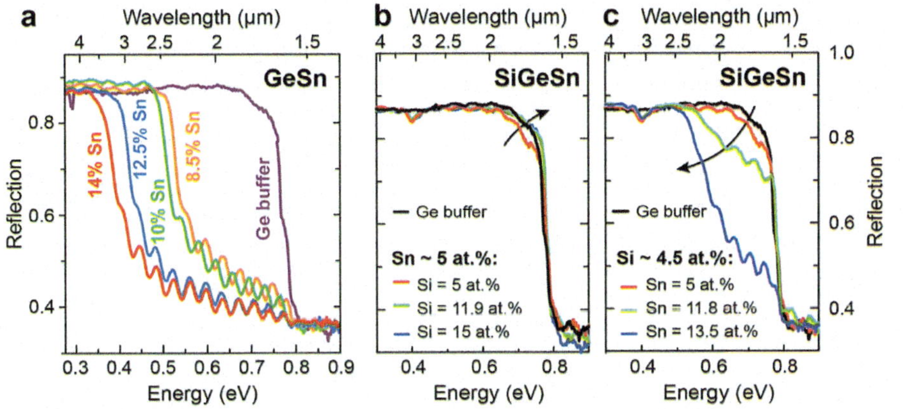


**Fig. 3.35** **a** Room temperature reflection spectra of 700–1000-nm-thick GeSn alloys, from [244]. SiGeSn ternaries enable to tune the absorption by changing the Si content (**b**) and also Sn content (**c**). Both from [61]

ments on 700–1000-nm-thick CVD-grown GeSn samples showed a strong absorption edge—linked to the direct bandgap transition in the Brillouin zone centre—which shifted into the short-wave Infrared as the Sn content increased (Fig. 3.35a). Being able to deposit SiGeSn ternaries yielded more freedom in designing the electronic structure. Individual tuning of incorporated Si atoms (Fig. 3.35b) and Sn atoms (Fig. 3.35c) led to decreased/increased absorption wavelengths.

In the following, a brief overview of the history of GeSn-based photodetectors is given. Several important works from that field and key figure-of-merits are summarized in Table 3.6. We will show how previous works could help to improve the performance of present-day detectors and stimulate new developments in this field.

Different types of GeSn-based photodetectors were developed since 2009, each with their specific advantages and disadvantages. Photoconductive devices made from GeSn, for example, rely on enhanced electrical conductivity due to generated photocurrents. They can be fabricated from intrinsic material without the need for any doping. In photodiodes (PD), the most common detector type, photo-generated carriers are separated by the internal electric field across a reverse-biased p-(i-)n junction. These structures can be combined with internal amplification mechanisms, e.g. through avalanche multiplication [192] or in a phototransistor geometry [193], to enable high-sensitivity photodetection. Metal–semiconductor–metal (MSM) structures were also explored for high-speed photodetection [194].

Ge photodetectors were already mature in 2009, with a monolithic integration into CMOS processes. They delivered acceptable performance up to the middle of the telecom C-band (roughly 1550 nm), where their direct absorption edge was positioned [195]. At higher wavelengths, however, the material's responsivity was low as only indirect bandgap absorption was available. Reducing the direct absorption edge by incorporating small amounts of Sn (~2 at.%) into Ge thus became an interesting option to enhance the responsivity. It also extended the spectral range of Ge

**Table 3.6** A brief overview of GeSn-based photodetectors history together with key figures-of-merit such as the responsivity at different wavelengths and the dark current

| Year | Sn (at.%) | Structure[a] | Responsivity (1.55 μm) @ −1 V (A/W) | Responsivity (> 1.55 μm) @ −1 V (A/W) | $I_{Dark}$ @ −1 V (A/cm$^2$) | Group |
|---|---|---|---|---|---|---|
| 2009 | 2 | PD | 0.05 @ −0.16 V | – | 1 | ASU [196] |
| 2011 | 3 | PD | 0.23 | 0.12 @ 1.64 μm | 1.8 | CAS [247] |
| 2012 | 9 | QW PC | 1 @ −5 V | 0.1 @ 2.2 μm | 5e−4 | Ghent [40] |
| 2013 | 3.9 | PD | 0.27 | 0.165 @ 1.6 μm | 0.4 | Taiwan [248] |
| 2013 | 3.6 | PD | – | 0.71 @1.8 μm (−3 V) | 6e−3 | CAS [249] |
| 2013 | 8 | MSM | 0.015 | – | 3e−6 | Stanford [194] |
| 2014 | 1.8 | WG PD | 0.18 @ 0 V | – | 5e−2 | Taiwan [205] |
| 2014 | 4.2 | PD | 0.22 @ 0 V | – | 8.9e−2 | Stuttgart [207] |
| 2014 | 7 | QW PD | 0.13 | – | 1.5e−1 | Stuttgart [250] |
| 2014 | 12 | PD | – | 0.03 @ 2 μm (−0.5 V, 100 K) | 5e-2 (-0.5 V) | Delaware [251] |
| 2015 | 5 | APD | 1.21 @ −9.8 V | 0.95 @ −9.8 V | 196 (-9.8 V) | NUS [192] |
| 2015 | 5 | PD | 0.18 | 0.06 @ 1.63 μm | 7.3e−2 | NUS [201] |
| 2016 | 10 | PC | 0.11 | – | – | UA [252] |
| 2016 | 2.5 | PD | 0.45 @ −0.1 V | 0.2/0.045 @ 1.71 μm, −0.1 V (front/backside illum.) | 1e−4 (−0.1 V) | NTU [209] |
| 2016 | 8 | PD | 0.07 @ 0 V | 0.093 @ 2 μm | 3 | CAS [253] |
| 2017 | 2.8 | MQW WG PD | 0.06 @ 0 V | – | 5.9e−2 | Taiwan [206] |
| 2017 | 3 | MQW PD | 0.07 @ 0 V | – | 4e−3 | Stanford [202] |
| 2017 | 10 | MQW PD | 0.2 | 0.023 @ 2 μm | 3e−2 | NUS [254] |
| 2017 | 6.5 | HPT | 1.8 | 0.043 @ 2 μm | 1.5e−1 | NUS [193] |
| 2018 | 2.5 | PD | 0.38 | – | 0.7 | Taiwan [204] |
| 2018 | 7 | Lateral PD | – | 0.016 @ 2 μm | – | NUS [208] |
| 2018 | 11 | PD | – | 0.25 @ 2 μm | 1.5 | UA [198] |

[a] *(A)PD* (Avalanche) photodiode; *PC* Photoconductor; *MSM* Metal–semiconductor–metal; *(M)QW* (Multi) quantum well; *WG* Waveguide; *HPT* Heterojunction phototransistor

photodetectors to span the complete telecommunication wavelength band between 1260 and 1675 nm [196]. Continuous progress in epitaxial growth techniques yielded a steady increase of Sn content in such devices, opening new optical I/O communication windows around a 2 μm wavelength [197]. A Sn content of 11 at.% led to an extension of the long-wavelength cutoff to roughly 2.6 μm (room temperature) in state-of-the-art GeSn photodetectors [198]. Even more freedom in shifting the wavelength regime could be obtained by integrating SiGeSn into SWIR photodetectors. However, only a limited amount of work has been devoted to that field up to now [199, 200].

Another important figure-of-merit for photodetectors is the dark current. In photodiodes, which are driven under reverse bias, this results in a serious source of noise, limiting the sensitivity of the device. Intrinsic factors, including the (thermal) generation of carriers in the depletion region, can partly be tackled by improving the crystalline quality of the material. In contrast to silicon, GeSn alloys do not have a single surface oxide, but, instead, several germanium and tin sub-oxides. The latter may even show some conductive behaviour. Leakage current from an insufficient surface passivation is therefore another important dark current source, which has previously been mitigated using low-temperature Si surface passivation [201] or ozone surface treatments [202].

Large material volumes are required to have high responsivities. However, the large lattice mismatch between GeSn and Ge/Si results in the formation of dislocations even in rather thin films. The critical thickness for strain relaxation of a $Ge_{0.9}Sn_{0.1}$ thin film grown on Ge is, for instance, around 90 nm [203]. Using GeSn wells separated by SiGeSn or Ge barriers with a lattice parameter close to or equal to that of the Ge SRBs, is an attractive way of working around the critical thickness for plastic relaxation issue and benefit from larger active volumes without the formation of lattice defects. Another way to mitigate the thickness limit in a vertically illuminated diode is to insert the absorbing layers in a resonant cavity. The responsivity of the detector can be enhanced by allowing not-absorbed photons to cross the active region multiple times [204]. Ultimately, the responsivity can completely be decoupled from the absorber thickness using a waveguide geometry. Light is then guided laterally into the detector material, yielding high-speed operation and high responsivity at the same time [205, 206].

Another advantage of heterojunction photodetectors is a decoupling of the active and contact material. Using a larger bandgap material for the top contact minimizes absorption losses and the deleterious impact of surface recombination. Furthermore, it acts as a low-pass filter to remove high energy photons and enhance the device's sensitivity in the targeted wavelength regime. While Ge has already been evaluated for such a purpose [40, 202, 206, 207], the use of SiGeSn ternaries, with similar or larger bandgaps, as shown in Fig. 3.35b, has not yet been explored.

The first steps towards a co-integration with electronics were undertaken by fabricating photodetectors on a GeSn-on-insulator platform [194, 208] or bonding a

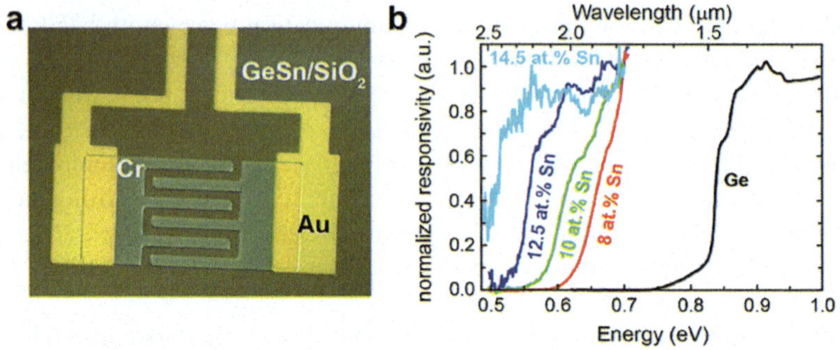


**Fig. 3.36** **a** GeSn-based photoconductive detector with interdigitated electrodes. **b** The normalized photo-response of several devices with different Sn contents shifted towards smaller energy/higher wavelengths as the Sn content increased

number of photodetectors to a customized readout circuitry for focal-plane arrays [209].

We have also evaluated the properties of photoconductive detectors, made from GeSn alloys with different Sn contents. Devices, shown in Fig. 3.36a, were built using a geometry with interdigitated electrodes, which helped to maximize the device area (minimize the RC delay) and keep the carrier drift region as short as possible. Using higher Sn contents shifted the photo-response towards higher wavelengths, as shown in Fig. 3.36b.

Besides intrinsic GeSn thin films, epitaxy of different types of photodiodes was performed [42, 61, 185]. Homo-junction p-i-n diodes were epitaxially grown on top of Ge-buffered Si(001) wafers, as schematically shown in Fig. 3.37a and b. In situ *p*- and *n*-type doped regions were separated by a 300-nm-thick intrinsic GeSn layer, which acted as the absorption region, as indicated by the SIMS depth profiles. Room temperature $I$–$V$-curves in Fig. 3.37c show distinct forward- and reverse bias regimes for different mesa structures. The dark current density at −0.5 V is roughly 1 A/cm$^2$, without any optimized surface passivation, however.

The combination of different group-IV materials for GeSn-based heterostructures has been also evaluated, including undoped material stacks [58, 240] and heterostructure diodes [42, 61]. Germanium was used as barrier material in GeSn/Ge MQW diodes, as shown by the TEM micrograph in Fig. 3.38a. When aiming for larger wavelengths—therefore higher Sn concentrations—Ge is not suitable anymore, since tensile strain from coherent growth on GeSn prevents them from being efficient barriers [42].

SiGeSn barriers can then be implemented in heterostructures, as shown by an atom probe tomography (APT) line scan of Si, Ge and Sn in a GeSn/SiGeSn MQW sample in Fig. 3.38b [240]. Furthermore, these material stacks were inserted in p-i-n light emitting diodes to fabricate room temperature short-wave infrared light emitters, as shown in Fig. 3.38c. Measured $I$–$V$ characteristics, shown in the inset of Fig. 3.38c, demonstrate that the diodes are generally also suited for applications under reverse bias, such as photodetection. To cancel absorption in contact layers, highly doped

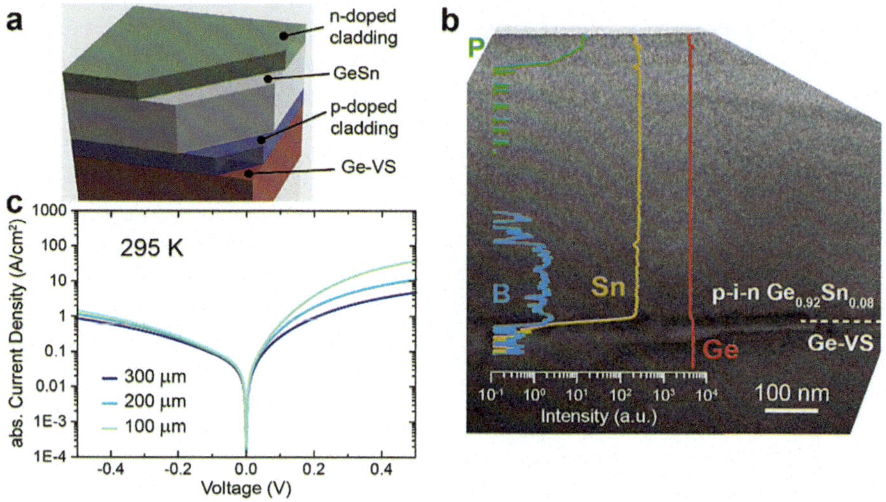


**Fig. 3.37** **a** A schematic of epitaxially grown GeSn p-i-n photodiode. **b** Cross-sectional TEM picture and SIMS depth profile of B, P, Sn and Ge in such photodetectors. Room temperature *I–V* curves for three fabricated devices with different mesa sizes. Adapted from [61]

larger-bandgap SiGeSn layers should be used as top layers. Low resistive p- and n-type SiGeSn layers were thus grown and evaluated, as shown by the resistivity measurements in Fig. 3.38d.

Additionally, the use of a double heterostructure design enables an electronic separation of the absorption region from the defective GeSn/(Ge/)Si interface. These defects are inherent to the GeSn material system and are due to the large lattice mismatch between GeSn and Ge/Si.

Table 3.6 provides the reader with an overview of various GeSn-based photodetectors.

### *3.5.2 Electrically Pumped Devices*

Interest has been growing in recent years to develop light sources for the mid-infrared. Such devices could be used for light detection and ranging [210], biomedical sensing [211], gas sensing [212] and MIR optical communication [213]. GeSn devices are good candidates, as the GeSn bandgap can be tuned by strain and the Sn content in the Ge matrix and because of their compatibility with existing CMOS platform. High-speed and efficient GeSn detectors have lately been demonstrated (see the photodetectors part of this chapter). We will review advances on GeSn light emitting diodes (LEDs) but start with earlier attempts using strained germanium and will conclude with the recent demonstration of an optically pumped GeSn laser.

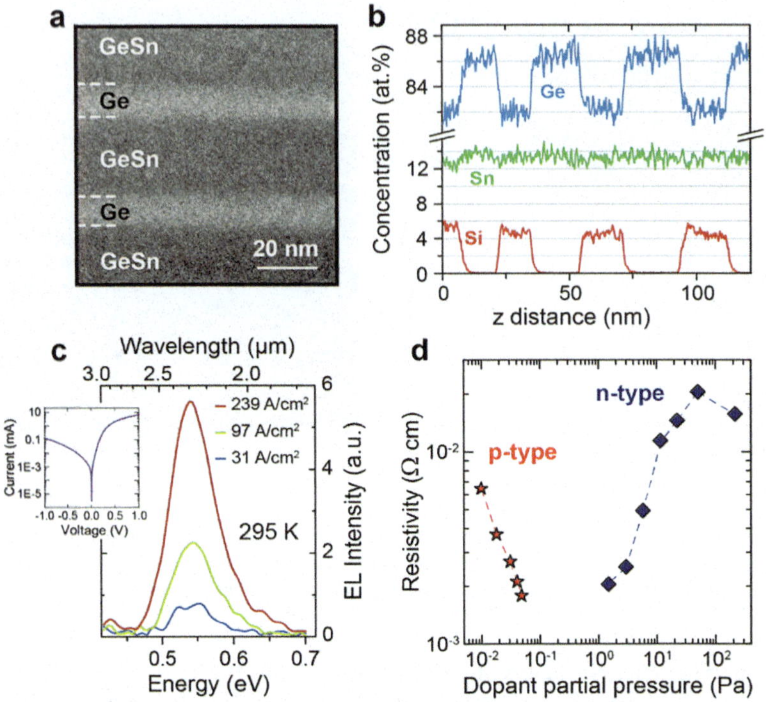


**Fig. 3.38** GeSn-based multi-quantum well heterostructures, with **a** Ge (TEM micrograph) or **b** SiGeSn barriers (APT line scan) [240]. **c** A GeSn/SiGeSn MQW diode shows room temperature electroluminescence and distinct forward and reverse bias regimes (inset). **d** n- and p-type doping of SiGeSn material yields low-resistive contact material

### 3.5.2.1 Light Emitting Diodes

We saw previously that the energy offset between the Γ- and the L-valley was reduced by injecting tensile strain in Ge. The electron occupancy was then increased in the Γ-valley, improving photon absorption and emission [214].

As early as 2012, an approach using external stressors technique was used to fabricate highly strained Ge LEDs. A tungsten stressor with a 4 GPa compressive stress was deposited at the bottom of the device (Fig. 3.39a) to take advantage of a biaxial tensile strain close to 0.75% in the active region of a vertical Ge p-n junction. *I–V* characteristic of the devices was improved at room temperature under stress by a factor of 3 in the forward current mode as compared to the as-grown device with 0.2% of strain. A redshift of 100 nm was measured for the strained LED with a light emission up to 1.7 μm [150].

Strain has been as well successfully transferred into Ge microdiscs using SiN stressors. Biaxially tensile strain up to 0.72% was locally achieved at room temperature in Ge LEDs [69]. A 250-nm n-doped Ge thick layer was grown on n-type GaAs substrates. The latter yielded electron injection and carrier confinement in the

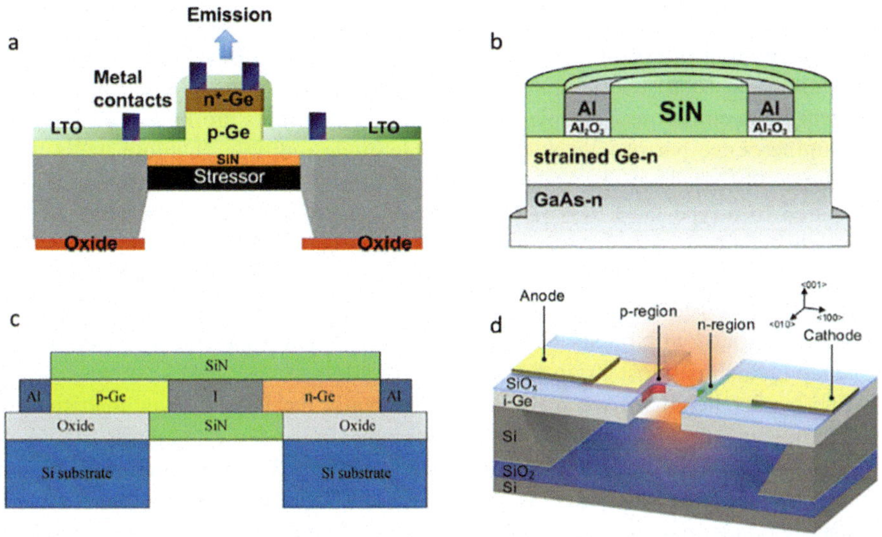


**Fig. 3.39** Schematic cross section of **a** a strained Ge LED with a tungsten stressor at the bottom of the device (figure from [148]), **b** a strained Ge LED with a silicon nitride stressor deposited on top of the Ge layer (figure from [69]), **c** a strained Ge LED with a silicon nitride stressor deposited on both sides of the Ge layer (figure from [215]), **d** schematic of a strained Ge micro-bridge LED with a lateral p-i-n configuration fabricated on a SOI substrate (figure from [216])

Ge layer thanks to the higher GaAs bandgap. A compressive SiN layer transferred locally tensile strain to Ge thanks to its mechanical relaxation (Fig. 3.39b). The electroluminescence maximal emission peak was shifted from 1580 nm for an unstrained LED up to 1680 nm for a LED with a 0.72% tensile strain, locally. A suspended Ge LED was bi-encapsulated in compressive SiN using GeOI wafers (Fig. 3.39c). A strain up to 1.92% in Ge was claimed, resulting in a 274 nm redshift of the emission spectrum [215]. Lateral p-i-n junction micro-bridges were used to reach 1.76% strain along the ⟨100⟩ direction [216]. The RP-CVD grown Ge devices were fabricated on SOI substrates with doped tensile arms made by implantation and activated by rapid thermal annealing (Fig. 3.39d). Electroluminescence up to 1885 nm was achieved at the centre of the bridge with a homogenous strain distribution. Strain in the Ge micro-bridges was limited due to dislocations between Ge and Si. A very different approach was explored using Ge-ion implanted Ge quantum dots (QDs) in a defect-free Si matrix. Extraordinary properties of those defect-enhanced Ge quantum dots were measured showing a short carrier lifetime and minor thermal quenching in the PL emission up to room temperature. Readers can refer to this work in the other chapter of this book or to [9, 217].

Bulk GeSn LEDs have been reported, with light emission between 1.8 and 2.2 $\mu$m. Low Sn content GeSn active layers were epitaxied on p-type Ge SRB and capped with n-type material. As expected, they did not show any direct bandgap behaviour [218–221]. Fabry–Perot modes were observed when such GeSn stacks were epitaxied on

SOI substrates [219]. Electroluminescence (EL) emission was reached with direct bandgap GeSn materials up to 3.4 μm for Sn contents up to 16% [222, 223]. Carrier confinement at room temperature seemed, however, to be insufficient to have an efficient light emission. Multiple quantum wells structures are known to improve the luminescence performances of LEDs. GeSn/Ge multiple quantum wells were investigated [224]. They showed a five time stronger electroluminescence emission compared to n-type Ge LEDs [225]. However, GeSn/Ge heterostructures seemed to poorly confine carriers at room temperature. SiGeSn carrier barriers have been introduced to overcome this limitation. These ternary alloys give more freedom to adjust the crystal lattice of the stack during the epitaxy. A type-I band alignment was demonstrated with binary GeSn alloys, with efficient light emission from direct bandgap layers in almost strain-free stacks [185].

The barrier height between the active layer and the barrier layer is critical to have efficient light emission at room temperature. In the last cited work, 10.5% of Si and 10.9% of Sn were incorporated in the SiGeSn barrier layers. The barrier height should increase as the Si/Sn ratio increases. Meanwhile, adding Sn results in an energy bandgap reduction. However, the exact band alignment between SiGeSn and GeSn is not yet know. Multiple SiGeSn/GeSn quantum well LEDs were investigated, with 10% of Si and 5% of Sn in the SiGeSn layers [226]. The EL intensity increased super-linearly with the applied voltage in those LEDs and was higher than in conventional GeSn/Ge heterojunction LEDS [227], showing that radiative recombination was efficient in such devices. Table 3.7 gives some of the specificities, at room temperature, of the GeSn-based LEDs discussed previously.

**Table 3.7** GeSn-based LED characteristics at room temperature

| LED structures | GeSn epitaxy | GeSn composition (%) | Biaxial strain | Wavelength emission | References |
|---|---|---|---|---|---|
| Ge/GeSn/Ge heterostructures | RP-CVD | 9.2 | −0.48% | 2.2 | [218] |
| | MBE | 4.5 | −0.66% | 1.9 | [219] |
| | MBE | 7.8 | – | 2.2 | [220] |
| | MBE | 9.7 | – | 2.25 | [221] |
| | RP-CVD | 8 | – | 2.06 | [227] |
| SiGeSn/GeSn/SiGeSn heterostructures | RP-CVD | 16 | −0.66% | 3.4 | [222, 223] |
| | RP-CVD | 13 | ~−0.5% | 2.8 | [223] |
| SiGeSn/GeSn/SiGeSn Multiple quantum wells | RP-CVD | 9.1 | −0.1% | 2.2 | [185] |
| | MBE | 8.5 | −1.6% | 2.06 | [226] |

*RP-CVD* Reduced pressure–chemical vapour deposition; *MBE* Molecular beam epitaxy

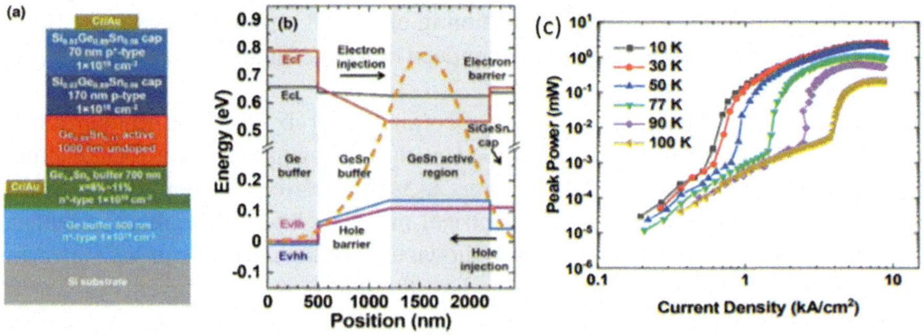


**Fig. 3.40** **a** Cross section of the electrically pumped laser stack, **b** band structures and optical mode profile of the type II GeSn heterostructure, **c** light output versus injection current of F-P cavity from 10 to 100 K. Figure from [5]

### 3.5.2.2 Lasers

Since the first claim of a laser effect in highly phosphorous-doped Ge [181, 228, 229], there is a growing effort around the world to fabricate an efficient CMOS-compatible Ge-based laser. No consensus has been reached yet on the physical interpretation of lasing by doping, and the impact of doping on the intervalence band absorption as shown in [175, 230]. To improve the performances of n-doped Ge layers for lasing, direct electron injection into the Γ conduction band of Ge was investigated in [231]. The radiative recombination of electrons injected by a forward biased p-i-n Zener diode with the holes in the indirect semiconductor Ge appeared to be more efficient than from a p-n junction with an n-doped active layer [231]. For both approaches, no further activities in the direction of lasing have been reported.

While the above-mentioned results did not result in any follow-up work, as far as we know, research on group-IV lasing continued, mostly on the development of direct bandgap GeSn. Such R&D efforts very recently resulted in an electrically pumped GeSn laser [5]. The there applied electrically pumping scheme is shown in Fig. 3.40a. A threshold close to 0.6 kA/cm$^2$ was obtained at 10 K, and lasing was observed up to 100 K (Fig. 3.40c). An external quantum efficiency of 0.3% and a peak power of 2.7 mW per laser facet were reported. Since the threshold was measured on a series of waveguides with various lengths, we may use, on data provided in Table SII of the supplementary information of [5], the differential gain analysis protocol described in Sect. 3.4.1.2. From a threshold that increases from $I = 0.77$ kA/cm$^2$ up to 1.2 kA/cm$^2$ for 0.6-mm and 0.3-mm-long waveguides, respectively, $\Delta g/\Delta I$ for the lasing device operated at 10 K is approximately 35 cm kA$^{-1}$. By comparison, the differential gain observed during the optical pumping of [1] devices was 0.4 cm kW$^{-1}$ only. The efficiency of the electrical injection thus seems to be orders of magnitude higher than the optical pumping, independent on the exact normalization one may apply to convert the photon injection into the injection by a current. This distinct difference may relate to the near resonance condition in electrical pumping. As shown for the strained Ge case, resonant pumping can indeed be very efficient, near unity as shown in [4] and detailed in Sect. 3.4.2.

Further lasing details are given in the upcoming report by the same group available on arXiv [232]. There, electrically pumped lasing is investigated for various thickness and Sn content optically active GeSn layers, as well as various contents in the SiGeSn barriers and indium doping for the electrical injection layers. The thickness of the capping layer above the gain region was shown to have a remarkable impact on lasing threshold. This was due to the mode overlap with the GeSn active layer (Fig. 3.40b). Performances as shown in the first report [5] were not exceeded, however. Subsequent developments may thus have to involve active layers with higher Sn contents and/or elastic-strain relaxation or p-i-n structures with tensile strained active layers. Single mode waveguide structures as well the use of shorter than 700 ns long pulses to avoid excessive heating of the device may also boost performances. Howsoever, these preliminary results encourages continuation of the research along that track and give good hope that group-IV semiconductor lasers once will perform on par with those of integrated group III–V devices.

## 3.6 Outlook and Conclusion

As in the entirety of this book, the starting point for this topic is the technological compatibility with silicon photonics. Silicon is a poor light emitter. Bringing light into a silicon photonic chip involves either a coupling to external light sources or a cumbersome integration of better-emitting semiconductors on the chip. For such integration, both direct epitaxy and bonding of III–V chips are intensely pursued today.

The idea of fabricating the entire Si photonic circuit in a standard CMOS foundry is obviously appealing. The only function lacking today is efficient and reliable light emission from group-IV semiconductors. A natural, straightforward approach would be to use a direct bandgap group-IV semiconductor. A by-product of such development of light emitters would be efficient photodiodes and modulators, all based on direct bandgap absorption by the same group-IV semiconductors.

To make a light emitter, particular attention was turned to Ge and its amendments for the following three reasons. (i) Although it is still indirect, Ge is much closer to a direct bandgap semiconductor than Si. (ii) Processing Ge in microelectronics fabrication lines is already an industrial reality; (iii) ditto the photonics developments, where Ge photodiode is a workhorse detector in Si photonic transceivers.

As shown, a global tensile stressing of pure Ge layers converts Ge into a direct semiconductor, and recently led to optically pumped Ge mid-IR lasers, although only up to mild cryogenic temperatures. Despite the fact that crystalline quality of Ge is very high and the straining technology very advanced, some intrinsic material properties, such as intra-valence band absorption and very high strain level requirements, make things difficult. Though technologically challenging, epitaxial schemes for electrical injection are already available. Strained bulk Ge layers enable in-depth fundamental investigations of light emission from group-IV semiconductors, as shown above in this chapter.

Modifying lattice parameters via tin-alloying is the second approach discussed here that yields a direct bandgap semiconductor. The good news for silicon photonics is that state-of-the-art GeSn layers and devices have been obtained with industry CMOS epitaxy tools and standard cleanroom technologies. This definitely favours GeSn alloys in material roadmaps. Still, many problems remain: GeSn has a large lattice mismatch with the Ge strain-relaxed buffers (on Si) typically used as templates. During the growth, the lattice mismatch is relaxed via a large number of interfacial defects. Moreover, an out-of-thermodynamic-equilibrium, low-temperature growth typically results in a large number of bulk point defects, which are responsible for a strong residual p-type doping and degrade the carrier mobility. All these crystalline defects have to be dealt with. To all those difficulties we can add many questions that still are to be answered: the band alignment at Ge/(Si)GeSn hetero-interfaces, predictive simulations of the band structure, etc.

However, even with an incomplete knowledge of the material parameters, by taking the best of the two approaches, external strain and alloying with Sn, several spectacular milestones were reached for GeSn lasers: low-threshold optically pumped strain- and dislocations-managed lasers [3], strain-wavelength-tuned optically pumped GeSn lasers operating at 273 K [2] and the first electrically pumped laser [5]. The GeSn lasers' characteristics are improving on a month-to-month basis. Outstanding challenges are not only to achieve continuous wave electrical pumping, but also to combine low lasing threshold and increased temperature lasing, as well as reaching reasonable wall-plug efficiencies.

The true goal, a room temperature electrically pumped laser, is not yet reached, but comes close as never before. Serious work is still to be done: optimizing the growth (residual doping, n-type and p-type doping, threading dislocations and point defect management), heterojunction band alignment within the (Si)GeSn quantum wells, designing optimized type-I heterostructures at room temperature, and finding strain-controlling cavities compatible with electrical injection.

While Ge is a already a legacy detecting material for datacom applications at 1.55 μm, strained Ge and/or GeSn may become equally relevant for applications in the mid-IR range, typically from 2 to 4.5 μm. LEDs and photodetectors emitting or detecting in the mid-IR range and fabricated with CMOS-relevant technologies will open to Si photonics a new large field of applications such as gas sensing [223] and mid-IR imaging [233].

As mentioned in the introduction, other approaches probed to have light emission from group-IV materials, in addition to those described here, include diamond under mechanical strain [234], defect levels in the Ge quantum dots [10], tunnel injection of electrons directly in the Γ-valley of Ge [231], hexagonal Ge and SiGe alloys [8], and others such as quantum cascade lasers for the THz [235]. Those approaches are recent, and their progress is certainly worth observing in the near future.

Since the demonstration of the first room temperature semiconductor laser in 1971, our understanding of semiconductors and nanostructures has dramatically improved and an exceptionally large set of tools became available: ab initio and quantum simulations, multiple epitaxial methods, nanofabrication and analysis of fine atomistic structure. Today, using a wide palette of such tools, we can tackle difficult solid-state physics problems, like modifying the band structure of an indirect semiconductor and converting it into a direct one.

Nobody knows yet which approach will be the most successful. Most probably, there will be some cross-links between some of them and that will certainly bring us spectacular developments in the near future.

**Acknowledgements** We would like to acknowledge our scientific collaborators we have been working with, in particular, Francesco T. Armand Pilon, Lara Casiez, Jérémie Chrétien, Marvin Frauenrath, Andrea Quintero, Quang M. Thai, Joris Aubin, Rami Khazaka, Alban Gassenq, Kevin Guilloy, Mathieu Bertrand, Richard Geiger, Thomas Zabel, Stephan Wirths, Daniela Stange, Yann-M. Niquet, Patrice Gergaud, Julie Widiez, Moustapha El Kurdi, Simone Assali, Oussama Moutanabbir, Guo-En Chang, Shui-Qing Yu and Jerome Faist, for our fruitful collaboration on group-IV lasers.

Some of the authors' work was partially supported by the CEA Phare Photonics project, the Swiss National Science Foundation (SNF), the Elegante ANR Project and the CEA Gelato Carnot Project.

## References

1. S. Wirths, R. Geiger, N. von den Driesch, G. Mussler, T. Stoica, S. Mantl, Z. Ikonic, M. Luysberg, S. Chiussi, J.M. Hartmann, H. Sigg, J. Faist, D. Buca, D. Grutzmacher, Lasing in direct bandgap GeSn alloy grown on Si (001). Nat. Photonics **9**, 88 (2015)
2. J. Chrétien, N. Pauc, F. Armand Pilon, M. Bertrand, Q.-M. Thai, L. Casiez, N. Bernier, H. Dansas, P. Gergaud, E. Delamadeleine, R. Khazaka, H. Sigg, J. Faist, A. Chelnokov, V. Reboud, J.-M. Hartmann, V. Calvo, GeSn lasers covering a wide wavelength range thanks to uniaxial tensile strain. ACS Photonics **6**(10), 2462–2469 (2019)
3. A. Elbaz, D. Buca, N. Von den Driesch, K. Pantzas, G. Patriarche, N. Zerounian, E. Herth, X. Checoury, S. Sauvage, I. Sagnes, A. Foti, R. Ossikovski, J.-M. Hartmann, F. Boeuf, Z. Ikonic, P. Boucaud, D. Grützmacher, M. El Kurdi, Ultra-low-threshold continuous-wave and pulsed lasing in tensile-strained GeSn alloys. Nat. Photonics **14**, 375–382 (2020)
4. F.T. Armand Pilon, A. Lyasota, Y.-M. Niquet, V. Reboud, V. Calvo, N. Pauc, J. Widiez, C. Bonzon, J.M. Hartmann, A. Chelnokov, J. Faist, H. Sigg, Lasing in strained germanium microbridges. Nat. Commun. **10**, 2724 (2019)
5. Y. Zhou, Y. Miao, S. Ojo, H. Tran, G. Abernathy, J.M. Grant, S. Amoah, G. Salamo, W. Du, J. Liu, J. Margetis, J. Tolle, Y.-H. Zhang, G. Sun, R.A. Soref, B. Li, S.-Q. Yu, Electrically injected GeSn lasers on Si operating up to 100 K. Optica **7**(8), 924–928 (2020)
6. U. Menczigar, G. Abstreiter, J. Olajos, H. Grimmeiss, H. Kibbel, H. Presting, E. Kasper, Enhanced band-gap luminescence in strain-symmetrized (Si)m/(Ge)n superlattices. Phys. Rev. B **47**(7), 4099–4102 (1993)
7. E.G. Barbagiovanni, D.J. Lockwood, P.J. Simpson, L.V. Goncharova, Quantum confinement in Si and Ge nanostructures: theory and experiment. Appl. Phys. Rev. **1**(1), 011302–011348 (2014)
8. E.M.T. Fadaly, A. Dijkstra, J.R. Suckert, D. Ziss, M.A.J. van Tilburg, C. Mao, Y. Ren, V.T. van Lange, K. Korzun, S. Kölling, M.A. Verheijen, D. Busse, C. Rödl, J. Furthmüller, F. Bechstedt, J. Stangl, J.J. Finley, S. Botti, J.E.M. Haverkort, E.P.A.M. Bakkers, Direct-bandgap emission from hexagonal Ge and SiGe alloys. Nature **580**(7802), 205–209 (2020)

9. M. Grydlik, F. Hackl, H. Groiss, M. Glaser, A. Halilovic, T. Fromherz, W. Jantsch, F. Schäffler, M. Brehm, Lasing from glassy Ge quantum dots in crystalline Si. ACS Photonics **3**(2), 298–303 (2016)
10. M. Brehm, Chapter X in Silicon Phonics IV (2021)
11. B. Dutt, D.S. Sukhdeo, D. Nam, B.M. Vulovic, Ze Yuan, K.C. Saraswat, Roadmap to an efficient germanium-on-silicon laser: strain vs. n-type doping. IEEE Photonics J. **4**, 2002–2009 (2012). https://doi.org/10.1109/jphot.2012.2221692
12. G.E. Chang, S.W. Chang, S.L. Chuang, Strain-balanced multiple-quantum-well lasers. IEEE J. Quantum Electron. **46**, 1813–1820 (2010). https://doi.org/10.1109/JQE.2010.2059000
13. S. Gupta, B. Magyari-Köpe, Y. Nishi, K.C. Saraswat, Achieving direct band gap in germanium through integration of Sn alloying and external strain. J. Appl. Phys. **113**, 073707 (2013). https://doi.org/10.1063/1.4792649
14. J. Widiez, J.M. Hartmann, F. Mazen, S. Sollier, C. Veytizou, Y. Bogumilowicz, E. Augendre, M. Martin, F. Gonzatti, M.C. Roure, J. Duvernay, V. Loup, C. Euvrard-Colnat, A. Seignard, T. Baron, R. Cipro, F. Bassani, A.M. Papon, C. Guedj, I. Huyet, M. Rivoire, P. Besson, C. Figuet, W. Schwarzenbach, D. Delprat, T. Signamarcheix, SOI-type bonded structures for advanced technology nodes. ECS Trans. **64**(5), 35 (2014)
15. L. Hutin, C. Le Royer, J.F. Damlencourt, J.M. Hartmann, H. Grampeix, V. Mazzocchi, C. Tabone, B. Prévitali, A. Pouydebasque, M. Vinet, O. Faynot, GeOI pMOSFETs scaled down to 30-nm gate length with record off-state current. IEEE Electron. Dev. Lett. **31**, 234 (2010)
16. M.J. Süess, R. Geiger, R.A. Minamisawa, G. Schiefler, J. Frigerio, D. Chrastina, G. Isella, R. Spolenak, J. Faist, H. Sigg, Analysis of enhanced light emission from highly strained germanium microbridges. Nat. Photonics **7**, 466–472 (2013)
17. A. Gassenq, S. Tardif, K. Guilloy, G. Osvaldo Dias, N. Pauc, I. Duchemin, D. Rouchon, J.M. Hartmann, J. Widiez, J. Escalante, Y.M. Niquet, R. Geiger, T. Zabel, H. Sigg, J. Faist, A. Chelnokov, F. Rieutord, V. Reboud, V. Calvo, Accurate strain measurements in highly strained Ge microbridges. Appl. Phys. Lett. **108**, 241902 (2016)
18. V. Reboud, A. Gassenq, J.M. Hartmann, J. Widiez, L. Virot, J. Aubin, K. Guilloy, S. Tardif, J.M. Fédéli, N. Pauc, A. Chelnokov, V. Calvo, Germanium based photonic components toward a full silicon/germanium photonic platform. Prog. Cryst. Growth Charact. Mater. **63**, 1 (2017)
19. N. Daix, E. Uccelli, L. Czornomaz, D. Caimi, C. Rossel, M. Sousa, H. Siegwart, C. Marchiori, J.M. Hartmann, K.T. Shiu, C.-W. Cheng, M. Krishnan, M. Lofaro, M. Kobayashi, D. Sadana, J. Fompeyrine, Towards large size substrates for III–V co-integration made by direct wafer bonding on Si. APL Mater. **2**, 086104 (2014)
20. Y. Bogumilowicz, J.M. Hartmann, R. Cipro, R. Alcotte, M. Martin, F. Bassani, J. Moeyaert, T. Baron, J.B. Pin, X. Bao, Z. Ye, E. Sanchez, Anti-phase boundaries-Free GaAs epilayers on "quasi-nominal" Ge-buffered silicon substrates. Appl. Phys. Lett. **107**, 212105 (2015)
21. V. Djara, V. Deshpande, E. Uccelli, N. Daix, D. Caimi, C. Rossel, M. Sousa, H. Siegwart, C. Marchiori, J.M. Hartmann, K.-T. Shiu, C.-W. Weng, M. Krishnan, M. Lofaro, R. Steiner, D. Sadana, D. Lubyshev, A. Liu, L. Czornomaz, J. Fompeyrine, An InGaAs on Si platform for CMOS with 200 mm InGaAs-OI substrate, gate-first, replacement gate planar and FinFETs down to 120 nm contact pitch, in *Proceedings of the 2015 VLSI Technology Symposium*, Kyoto (Japan) (2015), p. 176
22. L. Virot, L. Vivien, J.M. Fédéli, Y. Bogumilowicz, J.M. Hartmann, F. Boeuf, P. Crozat, D. Marris-Morini, E. Cassan, High-performance waveguide-integrated germanium PIN photodiodes for optical communication applications [invited]. Photonics Res. **1**, 140 (2013)
23. L. Virot, D. Benedikovic, B. Szelag, C. Alonso-Ramos, B. Karakus, J.M. Hartmann, X. Le Roux, P. Crozat, E. Cassan, D. Marris-Morini, C. Baudot, F. Boeuf, J.M. Fédéli, C. Kopp, L. Vivien, Integrated waveguide PIN photodiodes exploiting lateral Si/Ge/Si heterojunction. Opt. Express **25**, 19487 (2017)
24. M. Lukosius, J. Dabrowski, J. Kitzmann, O. Fursenko, F. Akhtar, M. Lisker, G. Lippert, S. Schulze, Y. Yamamoto, M.A. Schubert, H.M. Krause, A. Wolff, A. Mai, T. Schroeder, G. Lupina, Metal-free CVD graphene synthesis on 200 mm Ge/Si(001) substrates. ACS Appl. Mater. Interfaces **8**, 33786 (2016)

25. S. Brems, K. Verguts, N. Vrancken, B. Vermeulen, C. Porret, L. Peters, C. Han Wu, C. Huyghebaert, K. Schouteden, C. Van Haesendonck, S. De Gendt, Graphene synthesis and transfer improvements for applications in the semiconductor industry. ECS Trans. **77**(2), 3 (2017)
26. J.M. Hartmann, Chapter 3: Epitaxy of Ge layers on blanket and patterned Si(001) for nanoelectronics and optoelectronics, in *Photonics and Electronics with Germanium*, ed. by K. Wada, L.C. Kimerling, pp. 49–76 (Wiley, Germany, 2015). http://onlinelibrary.wiley.com/doi/10.1002/9783527650200.ch3/pdf
27. J.M. Hartmann, A. Abbadie, J.P. Barnes, J.M. Fédéli, T. Billon, L. Vivien, Impact of the $H_2$ anneal on the structural and optical properties of thin and thick Ge layers on Si; low temperature surface passivation of Ge by Si. J. Cryst. Growth **312**, 532 (2010)
28. J.M. Hartmann, J.F. Damlencourt, Y. Bogumilowicz, P. Holliger, G. Rolland, T. Billon, Reduced pressure-chemical vapor deposition of intrinsic and doped Ge layers on Si(0 0 1) for microelectronics and optoelectronics purposes. J. Cryst. Growth **274**, 90 (2005)
29. G. Wang, R. Loo, E. Simoen, L. Souriau, M. Caymax, M.M. Heyns, B. Blanpain, A model of threading dislocation density in strain-relaxed Ge and GaAs epitaxial films on Si (100). Appl. Phys. Lett. **94**, 102115 (2009)
30. J.M. Hartmann, J. Aubin, Assessment of the growth/etch back technique for the production of Ge strain-relaxed buffers on Si. J. Cryst. Growth **488**, 43 (2018)
31. O. Skibitzki, M.H. Zoellner, F. Rovaris, M.A. Schubert, Y. Yamamoto, L. Persichetti, L. Di Gaspare, M. De Seta, R. Gatti, F. Montalenti, G. Capellini, Reduction of threading dislocation density beyond the saturation limit by optimized reverse grading. Phys. Rev. Mater. **4**, 103403 (2020)
32. J.M. Hartmann, A.M. Papon, V. Destefanis, T. Billon, Reduced pressure chemical vapor deposition of Ge thick layers on Si(001), Si(011) and Si(111). J. Cryst. Growth **310**, 5287–5296 (2008)
33. J.E. Ayers, The measurement of threading dislocation densities in semiconductor crystals by X-ray diffraction. J. Cryst. Growth **135**, 71 (1994)
34. A. Abbadie, J.M. Hartmann, C. Deguet, L. Sanchez, F. Brunier, F. Letertre, Evaluation of different etching techniques in order to reveal dislocations in thick Ge layers. ECS Trans. **6**(4), 263 (2007)
35. J.M. Hartmann, A. Abbadie, N. Cherkashin, H. Grampeix, L. Clavelier, Epitaxial growth of Ge thick layers on nominal and 6° off Si(0 0 1); Ge surface passivation by Si. Semicond. Sci. Technol. **24**, 055002 (2009)
36. D. Thomson, A. Zilkie, J.E Bowers, T. Komljenovic, G.T. Reed, L. Vivien, D. Marris-Morini, E. Cassan, L. Virot, J.M. Fédéli, J.M. Hartmann, J.H. Schmid, D.X. Xu, F. Boeuf, P. O'Brien, G.Z. Mashanovich, M. Nedeljkovic, Roadmap on silicon photonics. J. Opt. **18**, 073003 (2016)
37. J. Durel, B.B. Bakir, C. Jany, S. Cremer, B. Szelag, T. Bria, V. Larrey, L. Sanchez, P. Brianceau, J.-A. Dallery, R. Guiavarch, T. Card, R. Thibon, J.E. Broquin, F. Boeuf, First demonstration of a back-side integrated heterogeneous hybrid III-V/Si DBR lasers for Si-photonics applications, in *Proceedings of the 2016 IEEE International Electron Devices Meeting (IEDM)* (2016), pp. 584–587
38. N. von den Driesch, D. Stange, S. Wirths, G. Mussler, B. Holländer, Z. Ikonic, J.-M. Hartmann, T. Stoica, S. Mantl, D. Grützmacher, D. Buca, Direct bandgap group IV epitaxy on Si for laser applications. Chem. Mater. **27**(13), 4693–4702 (2015)
39. Q.M. Thai, N. Pauc, J. Aubin, M. Bertrand, J. Chrétien, A. Chelnokov, J.M. Hartmann, V. Reboud, V. Calvo, 2D hexagonal photonic crystal GeSn laser with 16% Sn content. Appl. Phys. Lett. **113**(5), 051104–051105 (2018)
40. A. Gassenq, F. Gencarelli, J. Van Campenhout, Y. Shimura, R. Loo, G. Narcy, B. Vincent, G. Roelkens, GeSn/Ge heterostructure short-wave infrared photodetectors on silicon. Opt. Express **20**, 27297 (2012)
41. S. Xu, K. Han, Y.C. Huang, K.H. Lee, Y. Kang, S. Masudy-Panah, Y. Wu, D. Lei, Y. Zhao, H. Wang, C.S. Tan, X. Gong, Y.C. Yeo, Integrating GeSn photodiode on a 200 mm Ge-on-insulator photonics platform with Ge CMOS devices for advanced OEIC operating at 2 μm band. Opt. Express **27**, 26924 (2019)

42. D. Stange, N. von den Driesch, D. Rainko, C. Schulte-Braucks, S. Wirths, G. Mussler, A.T. Tiedemann, T. Stoica, J.M. Hartmann, Z. Ikonic, S. Mantl, D. Grützmacher, D. Buca, Study of GeSn based heterostructures: towards optimized group IV MQW LEDs. Opt. Express **24**, 1358 (2016)
43. D. Lei, K.H. Lee, Y.C. Huang, W. Wang, S. Masudy-Panah, S. Yadav, A. Kumar, Y. Dong, Y. Kang, S. Xu, Y. Wu, C.S. Tan, X. Gong, Y.C. Yeo, Germanium-Tin (GeSn) P-channel fin field-effect transistor fabricated on a novel GeSn-on-insulator substrate. IEEE Trans. Electron Dev. **65**, 3754 (2018)
44. M. Liu et al., First demonstration of vertical $Ge_{0.92}Sn_{0.08}$/Ge and Ge GAA nanowire nMOSFETs with low SS of 66 mV/dec and small DIBL of 35 mV/V, in *Proceedings of the 2019 IEEE International Electron Devices Meeting (IEDM)* (2019), pp. 693–696
45. B. Vincent, F. Gencarelli, H. Bender, C. Merckling, B. Douhard, D.H. Petersen, O. Hansen, H.H. Henrichsen, J. Meersschaut, W. Vandervorst, M. Heyns, R. Loo, M. Caymax, Undoped and in-situ B doped GeSn epitaxial growth on Ge by atmospheric pressure-chemical vapor deposition. Appl. Phys. Lett. **99**, 152103 (2011)
46. N. von den Driesch, D. Stange, S. Wirths, G. Mussler, B. Holländer, Z. Ikonic, J.M. Hartmann, T. Stoica, S. Mantl, D. Grützmacher, D. Buca, Direct bandgap group IV epitaxy on Si for laser applications. ACS Chem. Mater. **27**, 4693 (2015)
47. J. Aubin, J.M. Hartmann, J.P. Barnes, J.B. Pin, M. Bauer, Very low temperature epitaxy of heavily in situ phosphorous doped Ge layers and high Sn content GeSn layers. ECS J. Sol. State Sci. Technol. **6**, 21 (2017)
48. J. Aubin, J.M. Hartmann, GeSn growth kinetics in reduced pressure chemical vapor deposition from $Ge_2H_6$ and $SnCl_4$. J. Cryst. Growth **482**, 30 (2018)
49. A. Abedin et al., Heteroepitaxial growth of GeSn films on Ge, in *Proceedings of the ISTDM/ICSI 2018 Conference*, Potsdam, Germany, May 2018 (2018), pp. 273–274
50. R. Khazaka, E. Nolot, J. Aubin, J.M. Hartmann, Growth and characterization of SiGeSn pseudomorphic layers on 200 mm Ge virtual substrates. Semicond. Sci. Technol. **33**, 124011 (2018)
51. S. Al Kabi, S.A. Ghetmiri, J. Margetis, T. Pham, Y. Zhou, W. Dou, B. Collier, R. Quinde, W. Du, A. Mosleh, J. Liu, G. Sun, R.A. Soref, J. Tolle, B. Li, M. Mortazavi, H.A. Naseem, S.Q. Yu, An optically pumped 2.5 μm GeSn laser on Si operating at 110 K. Appl. Phys. Lett. **109**, 171105 (2016)
52. Y. Zhou, W. Dou, W. Du, S. Ojo, H. Tran, S.A. Ghetmiri, J. Liu, G. Sun, R. Soref, J. Margetis, J. Tolle, B. Li, Z. Chen, M. Mortazavi, S.-Q. Yu, Optically pumped GeSn lasers operating at 270 K with broad waveguide structures on Si. ACS Photonics **6**, 1434 (2019)
53. R. Loo, Y. Shimura, S. Ike, A. Vohra, T. Stoica, D. Stange, D. Buca, D. Kohen, J. Margetis, J. Tolle, Epitaxial GeSn: impact of process conditions on material quality. Semicond. Sci. Technol. **33**, 114010 (2018)
54. P. Jahandar, D. Weisshaupt, G. Colston, P. Allred, J. Schulze, M. Myronov, The effect of Ge precursor on the heteroepitaxy of $Ge_{1-x}Sn_x$ epilayers on a Si (001) substrate. Semicond. Sci. Technol. **33**, 034003 (2018)
55. J. Margetis, S.Q. Yu, B. Li, J. Tolle, Chemistry and kinetics governing hydride/chloride chemical vapor deposition of epitaxial $Ge_{1-x}Sn_x$. J. Vac. Sci. Technol. A **37**, 021508 (2019)
56. J.M. Hartmann, M. Frauenrath, J. Richy, Epitaxy of pseudomorphic GeSn layers with germane or digermane as Ge precursors and tin tetrachloride as Sn precursor. ECS Trans. **98**(5), 225 (2020)
57. E. Bouthillier, S. Assali, J. Nicolas, O. Moutanabbir, Decoupling the effects of composition and strain on the vibrational modes of GeSn. https://arxiv.org/ftp/arxiv/papers/1901/1901.00436.pdf
58. D. Stange, N. von den Driesch, T. Zabel, F. Armand-Pilon, D. Rainko, B. Marzban, P. Zaumseil, J.-M. Hartmann, Z. Ikonic, G. Capellini, S. Mantl, H. Sigg, J. Witzens, D. Grutzmacher, D. Buca, GeSn/SiGeSn heterostructure and multi quantum well lasers. ACS Photonics **5**, 4628 (2018)

59. R. Khazaka, J. Aubin, E. Nolot, J.M. Hartmann, Investigation of the growth of Si-Ge-Sn pseudomorphic layers on 200 mm Ge virtual substrates: impact of growth pressure, HCl and $Si_2H_6$ flows. ECS Trans. **86**(7), 207 (2018)
60. S. Wirths, D. Buca, Z. Ikonic, P. Harrison, A.T. Tiedemann, B. Holländer, T. Stoica, G. Mussler, U. Breuer, J.M. Hartmann, D. Grützmacher, S. Mantl, SiGeSn growth studies using reduced pressure chemical vapor deposition towards optoelectronic applications. Thin Sol. Films **557**, 183 (2014)
61. N. von den Driesch, D. Stange, S. Wirths, D. Rainko, I. Povstugar, A. Savenko, U. Breuer, R. Geiger, H. Sigg, Z. Ikonic, J.-M. Hartmann, D. Grutzmacher, S. Mantl, D. Buca, SiGeSn ternaries for efficient group IV heterostructure light emitters. Small **13**, 1603321 (2017)
62. W. Dou, M. Benamara, A. Mosleh, J. Margetis, P. Grant, Y. Zhou, S. Al-Kabi, W. Du, J. Tolle, B. Li, M. Mortazavi, S.Q. Yu, Investigation of GeSn strain relaxation and spontaneous composition gradient for low-defect and high-Sn alloy growth. Sci. Rep. **8**, 5640 (2018)
63. S. Assali, J. Nicolas, O. Moutanabbir, Enhanced Sn incorporation in GeSn epitaxial semiconductors via strain relaxation. J. Appl. Phys. **125**, 025304 (2019)
64. W. Du, Q.M. Thai, J. Chrétien, M. Bertrand, L. Casiez, Y. Zhou, J. Margetis, N. Pauc, A. Chelnokov, V. Reboud, V. Calvo, J. Tolle, B. Li, S.Q. Yu, Study of Si-based GeSn optically pumped lasers with micro-disk and ridge waveguide structures. Front. Phys. **7**, 147 (2019)
65. J. Aubin, J.M. Hartmann, A. Gassenq, J.L. Rouviere, E. Robin, V. Delaye, D. Cooper, N. Mollard, V. Reboud, V. Calvo, Growth and structural properties of step-graded, high Sn content GeSn layers on Ge. Semicond. Sci. Technol. **32**, 094006 (2017)
66. D. Stange, S. Wirths, R. Geiger, C. Schulte-Braucks, B. Marzban, N. von den Driesch, G. Mussler, T. Zabel, T. Stoica, J.-M. Hartmann, S. Mantl, Z. Ikonic, D. Grützmacher, H. Sigg, J. Witzens, D. Buca, Optically pumped GeSn microdisk lasers on Si. ACS Photonics **3**, 1279–1285 (2016)
67. V. Reboud, A. Gassenq, N. Pauc, J. Aubin, L. Milord, Q.M. Thai, M. Bertrand, K. Guilloy, D. Rouchon, J. Rothman, T. Zabel, F. Armand Pilon, H. Sigg, A. Chelnokov, J.M. Hartmann, V. Calvo, Optically pumped GeSn micro-disks with 16% Sn lasing at 3.1 $\mu$m up to 180 K. Appl. Phys. Lett. **111**, 092101 (2017)
68. K.-H. Shim, Y.-H. Kil, H.D. Yang, B.K. Park, J.-H. Yang, S. Kang, T.S. Jeong, T.-S. Kim, Characteristics of germanium dry etching using inductively coupled $SF_6$ plasma. Mater. Sci. Semicond. Process. **15**, 364–370 (2012)
69. A. Ghrib, M. El Kurdi, M. de Kersauson, M. Prost, S. Sauvage, X. Checoury, G. Beaudoin, I. Sagnes, P. Boucaud, Tensile-strained germanium microdisks. Appl. Phys. Lett. **102**, 221112 (2013)
70. M. Martin, S. Avertin, T. Chevolleau, F. Dhalluin, M. Ollivier, T. Baron, O. Joubert, J.M. Hartmann, Fabrication of high-density Si and $Si_xGe_{1-x}$ nanowire arrays based on the single step plasma etching process. JVST B **31**, 041806 (2013)
71. K. Guilloy, N. Pauc, A. Gassenq, P. Gentile, S. Tardif, F. Rieutord, V. Calvo, Tensile strained germanium nanowires measured by photocurrent spectroscopy and X-ray microdiffraction. Nano Lett. **15**, 2429–2433 (2015)
72. S. Al-Kabi, S.A. Ghetmiri, J. Margetis, T. Pham, Y. Zhou, W. Dou, B. Collier, R. Quinde, W. Du, A. Mosleh, J. Liu, G. Sun, R.A. Soref, J. Tolle, B. Li, M. Mortazavi, H.A. Naseem, S.Q. Yu, An optically pumped 2.5 $\mu$m GeSn laser on Si operating at 110 K. Appl. Phys. Lett. **109**, 171105 (2016)
73. S. Gupta, R. Chen, Y.C. Huang, Y. Kim, E. Sanchez, J.S. Harris, K.C. Saraswat, Highly selective dry etching of germanium over germanium-tin ($Ge_{1-x}Sn_x$): a novel route for $Ge_{1-x}Sn_x$ nanostructure fabrication. Nano Lett. **13**, 3783–3790 (2013)
74. Q. Xie, S. Deng, M. Schaekers, D. Lin, M. Caymax, A. Delabie, X.P. Qu, Y.L. Jiang, D. Deduytsche, C. Detavernier, Germanium surface passivation and atomic layer deposition of high-k dielectrics—a tutorial review on Ge-based MOS capacitors. Semicond. Sci. Technol. **27**, 074012 (2012)

75. S. Wirths, D. Stange, M.A. Pampillon, A.T. Tiedemann, G. Mussler, A. Fox, U. Breuer, B. Baert, E. San Andres, N.D. Nguyen, J.M. Hartmann, Z. Ikonic, S. Mantl, D. Buca, High-k gate stacks on low bandgap tensile strained Ge and GeSn alloys for field-effect transistors. ACS Appl. Mater. Interfaces, **7**, 62–67 (2015)
76. M.A. Mahjoub, T. Haffner, S. Labau, E. Eustache, J. Aubin, J.M. Hartmann, G. Ghibaudo, B. Pelissier, F. Bassani, B. Salem, Impact of wet treatments on the electrical performance of $Ge_{0.9}Sn_{0.1}$-based p-MOS capacitors. ACS Appl. Electron. Mater. **1**(2), 260–268 (2019)
77. T. Nishimura, K. Kita, A. Toriumi, Evidence for strong Fermi-level pinning due to metal-induced gap states. Appl. Phys. Lett. **91**, 123123 (2007)
78. K.Y. Chen, C.C. Su, C.P. Chou, Y.H. Wu, Formation of ohmic contact with low contact resistance on n-GeSn by fermi level depinning using plasma treatment. IEEE Electron Device Lett. **37**, 827–830 (2016)
79. B.K. Son, Y. Lin, W. Li, K.H. Lee, J. Margetis, D. Kohen, J. Tolle, L. Zhang, T.G. Xin, H. Wang, C.S. Tan, Metal-semiconductor-metal photodetectors on a GeSn-on-insulator platform, in *Optical Components and Materials XV* (2019)
80. S.C. Teng, C.C. Su, K.Y. Chen, C.P. Chou, Y.H. Wu, Fermi level depinning on n-epitaxial GeSn by Yb stanogermanide formation with low-contact resistivity. IEEE Electron Device Lett. **37**, 1207–1210 (2016)
81. Y. Zhang, J. Zheng, Z. Liu, C. Xue, C. Li, Y. Zuo, B. Cheng, Q. Wang, Electrical characteristics of ohmic contact on n-type in situ doped GeSiSn. Jpn. J. Appl. Phys. **57**, 106504 (2018)
82. G. Han, S. Su, Q. Zhou, P. Guo, Y. Yang, C. Zhan, L. Wang, W. Wang, Q. Wang, C. Xue, B. Cheng, Y.-C. Yeo, Dopant segregation and nickel stanogermanide contact formation on $p^+Ge_{0.947}Sn_{0.053}$ source/drain. IEEE Electron Device Lett. **33**, 634–636 (2012)
83. D. Kohen, A. Vohra, R. Loo, W. Vandervorst, N. Bhargava, J. Margetis, J. Tolle, Enhanced B doping in CVD-grown GeSn: B using B delta-doping layers. J. Cryst. Growth **483**, 285–290 (2018)
84. B.-Y. Tsui, H.-H. Liao, Y.-J. Chen, Degradation mechanism of Ge $N^+$-P shallow junction with thin GeSn surface layer. IEEE Trans. Electron Devices **67**, 1120–1125 (2020)
85. Y. Wu, S. Luo, W. Wang, S. Masudy-Panah, D. Lei, G. Liang, X. Gong, Y.-C. Yeo, Ultra-low specific contact resistivity ($1.4 \times 10^{-9}$ cm$^2$) for metal contacts on in-situ Ga-doped $Ge_{0.95}Sn_{0.05}$ film. J. Appl. Phys. **122**, 224503 (2017)
86. Y. Wu, W. Wang, S. Masudy-Panah, Y. Li, K. Han, L. He, Z. Zhang, D. Lei, S. Xu, Y. Kang, X. Gong, Y. Yeo, Sub-$10^{-9}$ ohm.cm$^2$ specific contact resistivity (down to $4.4 \times 10^{-10}$ ohm.cm$^2$) for metal contact on Ga and Sn surface-segregated GeSn film. IEEE Trans. Electron Devices **65**, 5275–5281 (2018)
87. A. Suzuki, O. Nakatsuka, M. Sakashita, S. Zaima, Alleviation of Fermi level pinning at metal/n-Ge interface with lattice-matched $Si_xGe_{1-x-y}Sn_y$ ternary alloy interlayer on Ge. Jpn. J. Appl. Phys. **57**, 060304 (2018)
88. T. Nishimura, O. Nakatsuka, Y. Shimura, S. Takeuchi, B. Vincent, A. Vantomme, J. Dekoster, M. Caymax, R. Loo, S. Zaima, Formation of $Ni(Ge_{1-x}Sn_x)$ layers with solid-phase reaction in $Ni/Ge_{1-x}Sn_x/Ge$ systems. Solid-State Electron. **60**, 46–52 (2011)
89. S. Wirths, R. Troitsch, G. Mussler, J.-M. Hartmann, P. Zaumseil, T. Schroeder, S. Mantl, D. Bucal, Ternary and quaternary Ni(Si)Ge(Sn) contact formation for highly strained Ge p- and n-MOSFETs. Semicond. Sci. Technol. **30**, 055003 (2015)
90. C. Schulte-Braucks, S. Glass, E. Hofmann, D. Stange, N.v.d. Driesch, J.M. Hartmann, Z. Ikonic, Q.T. Zhao, D. Buca, S. Mantl, Process modules for GeSn nanoelectronics with high Sn-contents. Solid-State Electron. **128**, 54–59 (2017)
91. J. Zheng, Y. Zhang, Z. Liu, Y. Zuo, C. Li, C. Xue, B. Cheng, Q. Wang, Fabrication of low-resistance Ni ohmic contacts on $n^+$-$Ge_{1-x}Sn_x$. IEEE Trans. Electron Devices **65**, 4971–4974 (2018)
92. A. Quintero, P. Gergaud, J. Aubin, J.-M. Hartmann, V. Reboud, P. Rodriguez, Ni/GeSn solid-state reaction monitored by combined X-ray diffraction analyses: focus on the Ni-rich phase. J. Appl. Cryst. **51**, 1133–1140 (2018)

93. A. Quintero, P. Gergaud, J.-M. Hartmann, V. Delaye, V. Reboud, E. Cassan, P. Rodriguez, Impact and behavior of Sn during the Ni/GeSn solid-state reaction. J. Appl. Cryst. **53**, 605–613 (2020)
94. A. Quintero, P. Gergaud, J.-M. Hartmann, V. Delaye, N. Bernier, D. Cooper, Z. Saghi, V. Reboud, E. Cassan, P. Rodriguez, Analysis of Sn behavior during Ni/GeSn solid-state reaction by correlated X-ray diffraction, atomic force microscopy, and ex-situ/in-situ transmission electron microscopy. ECS Trans. **98**, 365–375 (2020)
95. Q. Liu, W. Geilei, Y. Guo, X. Ke, H. Radamson, H. Liu, C. Zhao, J. Luo, Improvement of the thermal stability of nickel stanogermanide by carbon pre-stanogermanidation implant into GeSn substrate. ECS J. Solid State Sci. Technol. **4**, P67–P70 (2015)
96. W.-J. Wan, W. Ren, X.-R. Meng, Y.-X. Ping, X. Wei, Z.-Y. Xue, W. Yu, M. Zhang, Z.-F. Di, B. Zhang, Improvement of nickel-stanogermanide contact properties by platinum interlayer. Chin. Phys. Lett. **35**, 056802 (2018)
97. L. Wang, G. Han, S. Su, Q. Zhou, Y. Yang, P. Guo, W. Wang, Y. Tong, P.S.Y. Lim, B. Liu, E.Y.-J. Kong, C. Xue, Q. Wang, B. Cheng, Y.-C. Yeo, Thermally stable multi-phase nickel-platinum stanogermanide contacts for germanium-tin channel MOSFETs. Electrochem. Solid-State Lett. **15**, H179–H181 (2012)
98. E. Galluccio, G. Mirabelli, A. Harvey, M. Conroy, E. Napolitani, R. Duffy, Cell formation in stanogermanides using pulsed laser thermal anneal on $Ge_{0.91}Sn_{0.09}$. Mater. Sci. Semicond. Process. **121**, 105399 (2021)
99. A. Quintero, Development and electrical characterization of ohmic contacts on GeSn alloys, PhD Thesis, Université Paris-Saclay, 2020
100. A. Quintero, P. Gergaud, J.-M. Hartmann, V. Reboud, E. Cassan, P. Rodriguez, Impact of alloying elements (Co, Pt) on nickel stanogermanide formation. Mater. Sci. Semicond. Process. **108**, 104890 (2020)
101. A. Quintero, P. Gergaud, J. Aubin, J.-M. Hartmann, N. Chevalier, J.-P. Barnes, V. Loup, V. Reboud, F. Nemouchi, P. Rodriguez, Impact of Pt on the phase formation sequence, morphology and electrical properties of Ni(Pt)/$Ge_{0.9}Sn_{0.1}$ system during solid-state reaction. J. Appl. Phys. **124**, 085305 (2018)
102. A. Quintero, P. Gergaud, J. Aubin, J.-M. Hartmann, N. Chevalier, V. Reboud, E. Cassan, P. Rodriguez, Comparative analysis of Ni- and $Ni_{0.9}Pt_{0.1}$-$Ge_{0.9}Sn_{0.1}$ solid-state reaction by combined characterizations methods. ECS Trans. **86**, 299–307 (2018)
103. C.G. Van de Walle, Band lineups and deformation potentials in the model-solid theory. Phys. Rev. B **39**, 1871 (1989)
104. Y.M. Niquet, D. Rideau, C. Tavernier, H. Jaouen, X. Blase, Onsite matrix elements of the tight-binding Hamiltonian of a strained crystal: application to silicon, germanium, and their alloys. Phys. Rev. B **79**, 245201 (2009)
105. D.S. Sukhdeo, D. Nam, J.-H. Kang, M.L. Brongersma, K.C. Saraswat, Direct bandgap germanium-on-silicon inferred from 5.7% ⟨100⟩ uniaxial tensile strain. Photonics Res. **2**, A8 (2014). https://doi.org/10.1364/prj.2.0000a8
106. O. Aldaghri, Z. Ikonić, R.W. Kelsall, Optimum strain configurations for carrier injection in near infrared Ge lasers. J. Appl. Phys. **111**, 053106 (2012). https://doi.org/10.1063/1.3691790
107. D.S. Sukhdeo, D. Nam, J.-H. Kang, M.L. Brongersma, K.C. Saraswat, Bandgap-customizable germanium using lithographically determined biaxial tensile strain for silicon-compatible optoelectronics. Opt. Express **23**, 16740 (2015). https://doi.org/10.1364/OE.23.016740
108. Y. Bai, K.E. Lee, C. Cheng, M.L. Lee, E.A. Fitzgerald, Growth of highly tensile-strained Ge on relaxed $In_xGa_{1-x}As$ by metal-organic chemical vapor deposition. J. Appl. Phys. **104**, 084518 (2008). https://doi.org/10.1063/1.3005886
109. R. Jakomin, M. de Kersauson, M. El Kurdi, L. Largeau, O. Mauguin, G. Beaudoin, S. Sauvage, R. Ossikovski, G. Ndong, M. Chaigneau, I. Sagnes, P. Boucaud, High quality tensile-strained n-doped germanium thin films grown on InGaAs buffer layers by metal-organic chemical vapor deposition. Appl. Phys. Lett. **98**, 091901 (2011). https://doi.org/10.1063/1.3559231

110. M. de Kersauson, M. Prost, A. Ghrib, M. El Kurdi, S. Sauvage, G. Beaudoin, L. Largeau, O. Mauguin, R. Jakomin, I. Sagnes, G. Ndong, M. Chaigneau, R. Ossikovski, P. Boucaud, Effect of increasing thickness on tensile-strained germanium grown on InGaAs buffer layers. J. Appl. Phys. **113**, 183508 (2013). https://doi.org/10.1063/1.4804266
111. V.A. Volodin, L.V. Sokolov, M.A. Pytyato, N.I. Petikov, M. Stoffel, H. Rinnert, M. Vergnat, Optical properties of tensile-strained and relaxed Ge films grown on InGaAs buffer. J. Appl. Phys. **115**, 053518 (2014). https://doi.org/10.1063/1.4864466
112. V.A. Volodin, L.V. Sokolov, Redshift of the absorption edge in tensile-strained germanium layers. JETP Lett. **101**, 419–421 (2015). https://doi.org/10.1134/S0021364015060120
113. Y.-Y. Fang, J. Tolle, R. Roucka, A.V.G. Chizmeshya, J. Kouvetakis, V.R. D'Costa, J.B. Menendez, Perfectly tetragonal, tensile-strained Ge on $Ge_{1-y}Sn_y$ buffered Si(100). Appl. Phys. Lett. **90**, 061915 (2007). https://doi.org/10.1063/1.2472273
114. S. Takeuchi, Y. Shimura, O. Nakatsuka, S. Zaima, M. Ogawa, A. Sakai, Growth of highly strain-relaxed $Ge_{1-x}Sn_x$/virtual Ge by a Sn precipitation controlled compositionally step-graded method. Appl. Phys. Lett. **92**, 231916 (2008). https://doi.org/10.1063/1.2945629
115. Y.-Y. Fang, J. Tolle, J. Tice, A.V.G. Chizmeshya, J. Kouvetakis, V.R. D'Costa, J. Menéndez, Epitaxy-driven synthesis of elemental Ge/Si strain-engineered materials and device structures via designer molecular chemistry. Chem. Mater. **19**, 5910–5925 (2007). https://doi.org/10.1021/cm071581v
116. Y. Huo, H. Lin, R. Chen, M. Makarova, Y. Rong, M. Li, T.I. Kamins, J. Vuckovic, J.S. Harris, Strong enhancement of direct transition photoluminescence with highly tensile-strained Ge grown by molecular beam epitaxy. Appl. Phys. Lett. **98**, 011111 (2011). https://doi.org/10.1063/1.3534785
117. Y. Huo, H. Lin, R. Chen, Y. Rong, T.I. Kamins, J.S. Harris, MBE growth of tensile-strained Ge quantum wells and quantum dots. Front. Optoelectron. **5**, 112–116 (2012). https://doi.org/10.1007/s12200-012-0193-x
118. J.W. Matthews, A.E. Blakeslee, Defects in epitaxial multilayers. J. Cryst. Growth **27**, 118–125 (1974)
119. P.M.J. Marée, J.C. Barbour, J.F. van der Veen, K.L. Kavanagh, C.W.T. Bulle-Lieuwma, M.P.A. Viegers, Generation of misfit dislocations in semiconductors. J. Appl. Phys. **62**, 4413 (1987). https://doi.org/10.1063/1.339078
120. D. Jung, J. Faucher, S. Mukherjee, A. Akey, D.J. Ironside, M. Cabra, X. Sang, J. Lebeau, S.R. Bank, T. Buonassisi, O. Moutanabbir, M. Larry Lee, Highly tensile-strained Ge/InAlAs nanocomposites. Nat. Commun. **8**, 14204 (2017). https://doi.org/10.1038/ncomms14204
121. D.D. Cannon, J. Liu, Y. Ishikawa, K. Wada, D.T. Danielson, S. Jongthammanurak, J. Michel, L.C. Kimerling, Tensile strained epitaxial Ge films on Si(100) substrates with potential application in L-band telecommunications. Appl. Phys. Lett. **84**, 906 (2004). https://doi.org/10.1063/1.1645677
122. G. Capellini, M. De Seta, P. Zaumseil, G. Kozlowski, T. Schroeder, High temperature x ray diffraction measurements on Ge/Si(001) heterostructures: a study on the residual tensile strain. J. Appl. Phys. **111**, 073518 (2012). https://doi.org/10.1063/1.3702443
123. K. Oda, K. Tani, S. Saito, T. Ido, Improvement of crystallinity by post-annealing and regrowth of Ge layers on Si substrates. Thin Solid Films **550**, 509–514 (2014). https://doi.org/10.1016/j.tsf.2013.10.136
124. T.K.P. Luong, M.T. Dau, M.A. Zrir, M. Stoffel, V. Le Thanh, M. Petit, A. Ghrib, M. El Kurdi, P. Boucaud, H. Rinnert, J. Murota, Control of tensile strain and interdiffusion in Ge/Si(001) epilayers grown by molecular-beam epitaxy. J. Appl. Phys. **114**, 083504 (2013). https://doi.org/10.1063/1.4818945
125. M. Matsue, Y. Yasutake, S. Fukatsu, T. Hosoi, T. Shimura, H. Watanabe, Strain-induced direct band gap shrinkage in local Ge-on-insulator structures fabricated by lateral liquid-phase epitaxy. Appl. Phys. Lett. **104**, 031106 (2014). https://doi.org/10.1063/1.4862890
126. J.R. Jain, D.-S. Ly-Gagnon, K.C. Balram, J.S. White, M.L. Brongersma, D.A. Miller, R.T. Howe, Tensile-strained germanium-on-insulator substrate fabrication for silicon-compatible optoelectronics. Opt. Mater. Express **1**, 1121–1126 (2011)

127. V. Reboud, A. Gassenq, K. Guilloy, G. Osvaldo Dias, J.M. Escalante, S. Tardif, N. Pauc, J.M. Hartmann, J. Widiez, E. Gomez, E. Bellet Amalric, D. Fowler, D. Rouchon, I. Duchemin, Y.M. Niquet, F. Rieutord, J. Faist, R. Geiger, T. Zabel, E. Marin, H. Sigg, A. Chelnokov, V. Calvo, Ultra-high amplified strain on 200 mm optical Germanium-On-Insulator (GeOI) substrates: towards CMOS compatible Ge lasers, in *Silicon Photonics*, ed. by G.T. Reed, A.P. Knights, 97520F (2016)
128. R.A. Minamisawa, M.J. Süess, R. Spolenak, J. Faist, C. David, J. Gobrecht, K.K. Bourdelle, H. Sigg, Top-down fabricated silicon nanowires under tensile elastic strain up to 4.5%. Nat. Commun. **3**, 1096 (2012)
129. T. Zabel, R. Geiger, E. Marin, E. Müller, A. Diaz, C. Bonzon, M.J. Süess, R. Spolenak, J. Faist, H. Sigg, Top-down method to introduce ultra-high elastic strain. J. Mater. Res. **32**(4), 726–736 (2017). https://doi.org/10.1557/jmr.2017.31
130. R. Geiger, T. Zabel, E. Marin, A. Gassenq, J.-M. Hartmann, J. Widiez, J. Escalante, K. Guilloy, N. Pauc, D. Rouchon, G.O. Diaz, S. Tardif, F. Rieutord, I. Duchemin, Y.-M. Niquet, V. Reboud, V. Calvo, A. Chelnokov, J. Faist, H. Sigg, Uniaxially stressed germanium with fundamental direct band gap. Condens. Matter Physicsphys. (2015). http://arxiv.org/abs/1603.03454
131. S. Tardif, A. Gassenq, K. Guilloy, N. Pauc, G. Osvaldo Dias, J.-M. Hartmann, J. Widiez, T. Zabel, E. Marin, H. Sigg, J. Faist, A. Chelnokov, V. Reboud, V. Calvo, J.-S. Micha, O. Robach, F. Rieutord, Lattice strain and tilt mapping in stressed Ge microstructures using X-ray Laue micro-diffraction and rainbow filtering. J. Appl. Crystallogr. **49**(5), 1402–1411 (2016). https://doi.org/10.1107/s1600576716010347
132. A. Gassenq, K. Guilloy, G. Osvaldo Dias, N. Pauc, D. Rouchon, J.-M. Hartmann, J. Widiez, S. Tardif, F. Rieutord, J. Escalante, I. Duchemin, Y.-M. Niquet, R. Geiger, T. Zabel, H. Sigg, J. Faist, A. Chelnokov, V. Reboud, V. Calvo, 1.9% bi-axial tensile strain in thick germanium suspended membranes fabricated in optical germanium-on-insulator substrates for laser applications. Appl. Phys. Lett. **107**, 191904 (2015). http://dx.doi.org/10.1063/1.4935590
133. J. Petykiewicz, D. Nam, D.S. Sukhdeo, S. Gupta, S. Buckley, A.Y. Piggott, J. Vučković, K.C. Saraswat, Direct bandgap light emission from strained germanium nanowires coupled with high-Q nanophotonic cavities. Nano Lett. **16**, 2168–2173 (2016). https://doi.org/10.1021/acs.nanolett.5b03976
134. A. Gassenq, K. Guilloy, N. Pauc, D. Rouchon, J. Widiez, J. Rothman, J.-M. Hartmann, A. Chelnokov, V. Reboud, V. Calvo, Highly strained Ge micro-blocks bonded on Si platform for mid-infrared photonic applications. MRS Commun. **7**, 691–694 (2017). https://doi.org/10.1557/mrc.2017.53
135. M. El Kurdi, H. Bertin, E. Martincic, M. de Kersauson, G. Fishman, S. Sauvage, A. Bosseboeuf, P. Boucaud, Control of direct band gap emission of bulk germanium by mechanical tensile strain. Appl. Phys. Lett. **96**, 041909 (2010). https://doi.org/10.1063/1.3297883
136. C.-Y. Peng, C.-F. Huang, Y.-C. Fu, Y.-H. Yang, C.-Y. Lai, S.-T. Chang, C.W. Liu, Comprehensive study of the Raman shifts of strained silicon and germanium. J. Appl. Phys. **105**, 083537 (2009). https://doi.org/10.1063/1.3110184
137. T.-H. Cheng, K.-L. Peng, C.-Y. Ko, C.-Y. Chen, H.-S. Lan, Y.-R. Wu, C.W. Liu, H.-H. Tseng, Strain-enhanced photoluminescence from Ge direct transition. Appl. Phys. Lett. **96**, 211108 (2010). https://doi.org/10.1063/1.3429085
138. J. Greil, A. Lugstein, C. Zeiner, G. Strasser, E. Bertagnolli, Tuning the electro-optical properties of germanium nanowires by tensile strain. Nano Lett. **12**, 6230–6234 (2012). https://doi.org/10.1021/nl303288g
139. J.R. Sánchez-Pérez, C. Boztug, F. Chen, F.F. Sudradjat, D.M. Paskiewicz, R.B. Jacobson, M.G. Lagally, R. Paiella, Direct-bandgap light-emitting germanium in tensilely strained nanomembranes. Proc. Natl. Acad. Sci. **108**, 18893–18898 (2011)
140. C. Boztug, J.R. Sánchez-Pérez, F.F. Sudradjat, R. Jacobson, D.M. Paskiewicz, M.G. Lagally, R. Paiella, Tensilely strained germanium nanomembranes as infrared optical gain media. Small **9**, 622–630 (2013). https://doi.org/10.1002/smll.201201090
141. R. Kuroyanagi, L.M. Nguyen, T. Tsuchizawa, Y. Ishikawa, K. Yamada, K. Wada, Local bandgap control of germanium by silicon nitride stressor. Opt. Express **21**, 18553 (2013). https://doi.org/10.1364/OE.21.018553

142. Q. Guo, M. Zhang, Z. Xue, J. Zhang, G. Wang, D. Chen, Z. Mu, G. Huang, Y. Mei, Z. Di, X. Wang, Uniaxial and tensile strained germanium nanomembranes in rolled-up geometry by polarized Raman scattering spectroscopy. AIP Adv. **5**, 037115 (2015). https://doi.org/10.1063/1.4914916
143. M. Bollani, D. Chrastina, L. Gagliano, L. Rossetto, D. Scopece, M. Barget, V. Mondiali, J. Frigerio, M. Lodari, F. Pezzoli, F. Montalenti, E. Bonera, Local uniaxial tensile strain in germanium of up to 4% induced by SiGe epitaxial nanostructures. Appl. Phys. Lett. **107**, 083101 (2015). https://doi.org/10.1063/1.4928981
144. T. Mizutani, O. Nakatsuka, A. Sakai, H. Kondo, M. Ogawa, S. Zaima, Novel method to introduce uniaxial tensile strain in Ge by microfabrication of Ge/$Si_{1-x}Ge_x$ structures on Si(001) substrates. Solid-State Electron. **53**, 1198–1201 (2009). https://doi.org/10.1016/j.sse.2009.08.001
145. M. Keplinger, R. Grifone, J. Greil, D. Kriegner, J. Persson, A. Lugstein, T. Schülli, J. Stangl, Strain distribution in single, suspended germanium nanowires studied using nanofocused X-rays. Nanotechnology **27**, 055705 (2016). https://doi.org/10.1088/0957-4484/27/5/055705
146. S. Ike, O. Nakatsuka, Y. Moriyama, M. Kurosawa, N. Taoka, Y. Imai, S. Kimura, T. Tezuka, S. Zaima, Characterization of locally strained $Ge_{1-x}Sn_x$/Ge fine structures by synchrotron X-ray microdiffraction. Appl. Phys. Lett. **106**, 182104 (2015). https://doi.org/10.1063/1.4921010
147. Z.-M. Huang, W.-Q. Huang, S.-R. Liu, T.-G. Dong, G. Wang, X.-K. Wu, C.-J. Qin, Emission of direct-gap band in germanium with Ge-GeSn layers on one-dimensional structure. Sci. Rep. **6**, 24802 (2016). https://doi.org/10.1038/srep24802
148. D. Nam, D. Sukhdeo, A. Roy, K. Balram, S.-L. Cheng, K.C.-Y. Huang, Z. Yuan, M. Brongersma, Y. Nishi, D. Miller, others, Strained germanium thin film membrane on silicon substrate for optoelectronics. Opt. Express **19**, 25866–25872 (2011)
149. S. Gupta, D. Nam, J. Petykiewicz, D. Sukhdeo, J. Vuckovic, K. Saraswat, A novel, highly-strained structure with an integrated optical cavity for a low threshold germanium laser, in *CLEO: Science and Innovations* (Optical Society of America, 2015), p. SM2F-3. https://www.osapublishing.org/abstract.cfm?uri=CLEO_SI-2015-SM2F.3
150. D. Nam, D. Sukhdeo, S.-L. Cheng, A. Roy, K. Chih-Yao Huang, M. Brongersma, Y. Nishi, K. Saraswat, Electroluminescence from strained germanium membranes and implications for an efficient Si-compatible laser. Appl. Phys. Lett. **100**, 131112 (2012). https://doi.org/10.1063/1.3699224
151. J.R. Jain, A. Hryciw, T.M. Baer, D.A.B. Miller, M.L. Brongersma, R.T. Howe, A micromachining-based technology for enhancing germanium light emission via tensile strain. Nat. Photonics **6**, 398–405 (2012). https://doi.org/10.1038/nphoton.2012.111
152. G. Capellini, C. Reich, S. Guha, Y. Yamamoto, M. Lisker, M. Virgilio, A. Ghrib, M. El Kurdi, P. Boucaud, B. Tillack, T. Schroeder, Tensile Ge microstructures for lasing fabricated by means of a silicon complementary metal-oxide-semiconductor process. Opt. Express **22**, 399 (2014). https://doi.org/10.1364/OE.22.000399
153. G. Capellini, G. Kozlowski, Y. Yamamoto, M. Lisker, C. Wenger, G. Niu, P. Zaumseil, B. Tillack, A. Ghrib, M. de Kersauson, M. El Kurdi, P. Boucaud, T. Schroeder, Strain analysis in SiN/Ge microstructures obtained via Si-complementary metal oxide semiconductor compatible approach. J. Appl. Phys. **113**, 013513 (2013). https://doi.org/10.1063/1.4772781
154. M. Qi, W.A. O'Brien, C.A. Stephenson, N. Cao, B.J. Thibeault, M.A. Wistey, Stability of tensile-strained ge studied by transmission electron microscopy, in *2012 International Silicon-Germanium Technology and Device Meeting (ISTDM)*, IEEE, 2012, pp. 1–2. http://ieeexplore.ieee.org/xpls/abs_all.jsp?arnumber=6222505
155. M. de Kersauson, M.E. Kurdi, S. David, X. Checoury, G. Fishman, S. Sauvage, R. Jakomin, G. Beaudoin, I. Sagnes, P. Boucaud, Optical gain in single tensile-strained germanium photonic wire. Opt. Express **19**, 17925 (2011). https://doi.org/10.1364/OE.19.017925
156. L. Dupré, D. Buttard, P. Gentile, Q. Benoit à la Guillaume, T. Gorisse, H. Renevier, Strain control in germanium nanowires: the use of a silicon nitride shell: strain control in germanium nanowires: the use of a silicon nitride shell. Phys. Status Solidi RRL—Rapid Res. Lett. **8**, 317–320 (2014). https://doi.org/10.1002/pssr.201409050

157. R.W. Millar, K. Gallacher, J. Frigerio, A. Ballabio, A. Bashir, I. MacLaren, G. Isella, D.J. Paul, Analysis of Ge micro-cavities with in-plane tensile strains above 2%. Opt. Express **24**, 4365 (2016). https://doi.org/10.1364/OE.24.004365
158. P. Velha, D.C. Dumas, K. Gallacher, R. Millar, M. Myronov, D.R. Leadley, D.J. Paul, Strained germanium nanostructures on silicon emitting at >2.2 μm wavelength, in *10th International Conference on Group IV Photonics* (2013), pp. 142–143. https://doi.org/10.1109/group4.2013.6644411
159. P. Velha, K. Gallacher, D. Dumas, D. Paul, M. Myronov, D.R. Leadley, Long wavelength >1.9 um germanium for optoelectronics using process induced strain. ECS Trans. **50**, 779–782 (2012)
160. M. El Kurdi, M. Prost, A. Ghrib, S. Sauvage, X. Checoury, G. Beaudoin, I. Sagnes, G. Picardi, R. Ossikovski, P. Boucaud, Direct band gap germanium microdisks obtained with silicon nitride stressor layers. ACS Photonics **3**, 443–448 (2016). https://doi.org/10.1021/acsphotonics.5b00632
161. I.A. Fischer, A. Berrier, F. Hornung, M. Oehme, P. Zaumseil, G. Capellini, N. von den Driesch, D. Buca, J. Schulze, Optical critical points of $Si_xGe_{1-x-y}Sn_y$ alloys with high Si content. Semicond. Sci. Technol. **32**, 124004 (2017)
162. T.B. Bahder, Eight-band k.p model of strained zinc-blende crystals. Phys. Rev. B **41**, 11992 (1990). Erratum: Phys. Rev. B **46**, 9913 (1992)
163. D. Rainko, Z. Ikonic, N. Vukmirovic, D. Stange, N. von den Driesch, D. Grutzmacher, D. Buca, Investigation of carrier confinement in direct bandgap GeSn/SiGeSn 2D and 0D heterostructures. Sci. Rep. **8**, 15557 (2018)
164. M. Jaros, Simple analytic model for heterojunction band offsets. Phys. Rev. B **37**, 7112 (1988)
165. C.L. Senaratne, P.M. Wallace, J.D. Gallagher, P.E. Sims, J. Kouvetakis, J. Menendez, Direct gap $Ge_{1-y}Sn_y$ alloys: fabrication and design of mid-IR photodiodes. J. Appl. Phys. **120**, 025701 (2016)
166. H.S. Lan, C.W. Liu, Band alignments at strained $Ge_{1-x}Sn_x$/relaxed $Ge_{1-y}Sn_y$ heterointerfaces. J. Phys. D: Appl. Phys. **50**, 13LT02 (2017)
167. K. Low, Y. Yang, G. Han, W. Fan, Y. Yeo, Electronic band structure and effective mass parameters of $Ge_{1-x}Sn_x$ alloys. J. Appl. Phys. **112**, 103715 (2012)
168. S.-Q. Liu, S.-T. Yen, Extraction of eight-band k.p parameters from empirical pseudopotentials for GeSn. J. Appl. Phys. **125**, 245701 (2019)
169. C.-Y. Tsai, C.-Y. Tsai, C.-H. Chen, T.-L. Sung, T.-Y. Wu, F.-P. Shih, Theoretical model for intravalley and intervalley free-carrier absorption in semiconductor lasers: Beyond the classical Drude model. IEEE J. Quantum Electron. **34**, 552 (1998)
170. Q.M. Thai, J. Chretien, M. Bertrand, L. Casiez, A. Chelnokov, V. Reboud, N. Pauc, V. Calvo, Phys. Rev. B **102**, 155203 (2020)
171. S.L. Chuang, *Phys. Photonic Dev.*, 2nd edn. (Wiley, USA, 2009)
172. S. Gupta, D. Nam, J. Vučković, K. Saraswat, Room temperature lasing unraveled by a strong resonance between gain and parasitic absorption in uniaxially strained germanium. Phys. Rev. B **97**(15), 155127 (2018)
173. D. Nam, D.S. Sukhdeo, J.-H. Kang, J. Petykiewicz, J.H. Lee, W.S. Jung, J. Vučković, M.L. Brongersma, K.C. Saraswat, Strain-induced pseudoheterostructure nanowires confining carriers at room temperature with nanoscale-tunable band profiles. Nano Lett. **13**(7), 3118–3123 (2013)
174. A. Ghrib, M. de Kersauson, M. El Kurdi, R. Jakomin, G. Beaudoin, S. Sauvage, G. Fishman, G. Ndong, M. Chaigneau, R. Ossikovski, I. Sagnes, P. Boucaud, Control of tensile strain in germanium waveguides through silicon nitride layers. Appl. Phys. Lett. **100**(20), 201104 (2012)
175. L. Carroll, P. Friedli, S. Neuenschwander, H. Sigg, S. Cecchi, F. Isa, D. Chrastina, G. Isella, Y. Fedoryshyn, J. Faist, Direct-gap gain and optical absorption in germanium correlated to the density of photoexcited carriers, doping, and strain. Phys. Rev. Lett. **109** (2012). https://doi.org/10.1103/physrevlett.109.057402

176. K.L. Shaklee, R.F. Leheny, Direct determination of optical gain in semiconductor crystals. Appl. Phys. Lett. **18**(11), 475–477 (1971)
177. D. Hofstetter, J. Faist, Measurement of semiconductor laser gain and dispersion curves utilizing Fourier transforms of the emission spectra. IEEE Photonics Technol. Lett. **11**(11), 1372–1374 (1999)
178. A. Gassenq, G. Osvaldo Dias, K. Guiloy, S. Tardif, N. Pauc, D. Rouchon, J. Widiez, J.-M. Hartmann, D. Fowler, A. Chelnokov, J. Escalante, I. Duchemin, Y.-M. Niquet, R. Geiger, T. Zabel, H. Sigg, J. Faist, F. Rieutord, V. Reboud, V. Calvo, DBR based cavities in strained Ge microbridge on 200 mm Germanium-On-Insulator (GeOI) substrates: towards CMOS compatible laser applications, in *Proceedings 2015 European Conference on Lasers and Electro-Optics—European Quantum Electronics Conference, CLEO/Europe-EQEC* (2015), p. 124364
179. C.B.P. Bonzon, Phase and mode control of structured semiconductor lasers. Dissertation, ETHZ (2016). https://doi.org/10.3929/ethz-a-010656750
180. A.A. Andronov, I.V. Zverev, V.A. Kozlov, YuN Nozdrin, S.A. Pavlov, V.N. Shastin, Stimulated emission in the long-wavelength lR region from hot holes in Ge in crossed electric and magnetic fields. Pis'ma Zh. Eksp. Teor. Fiz. JETP Lett. **40**, 804 (1984)
181. R.J. Liu, X. Sun, R. Camacho-Aguilera, L.C. Kimerling, J. Michel, Ge-on-Si laser operating at room temperature. Opt. Lett. **35**, 679–681 (2010)
182. G. Scappucci, C. Kloeffel, F.A. Zwanenburg, D. Loss, M. Myronov, J.-J. Zhang, S. De Franceschi, G. Katsaros, M. Veldhorst, The germanium quantum information route (2020)
183. K. Guilloy, N. Pauc, A. Gassenq, Y.-M. Niquet, J.-M. Escalante, I. Duchemin, S. Tardif, G.O. Dias, D. Rouchon, J. Widiez, J.-M. Hartmann, R. Geiger, T. Zabel, H. Sigg, J. Faist, A. Chelnokov, V. Reboud, V. Calvo, Germanium under high tensile stress: nonlinear dependence of direct band gap vs strain. ACS Photonics **3**(10), 1907–1911 (2016)
184. J.M. Hartmann, A. Abbadie, S. Favier, Critical thickness for plastic relaxation of SiGe on Si(001) revisited. J. Appl. Phys. **110**, 083529 (2011)
185. D. Stange, N. von den Driesch, D. Rainko, S. Roesgaard, I. Povstugar, J.-M. Hartmann, T. Stoica, Z. Ikonic, S. Mantl, D. Grützmacher, D. Buca, Short-wave infrared LEDs from GeSn/SiGeSn multiple quantum wells. Optica **4** (2017). https://doi.org/10.1364/OPTICA.4.000185.5
186. N. von den Driesch, D. Stange D. Rainko I. Povstugar, P. Zaumseil, G. Capellini, T. Schröder, T. Denneulin, Z. Ikonic, J.-M. Hartmann, H. Sigg, S. Mantl, D. Grützmacher, D. Buca, Advanced GeSn/SiGeSn group IV heterostructure lasers. Adv. Sci. **5**(6), 1700955 (2018)
187. J. Margetis, A. Mosleh, S. Al-Kabi, S.A. Ghetmiri, W. Du, W. Dou, M. Benamara, B. Li, M. Mortazavi, H.A. Naseem, S.-Q. Yu, J. Tolle, Study of low-defect and strain-relaxed GeSn growth via reduced pressure CVD in $H_2$ and $N_2$ carrier gas. J. Cryst. Growth **463**, 128 (2017)
188. R. Chen, Y.-C. Huang, S. Gupta, A.C. Lin, E. Sanchez, Y. Kim, K.C. Saraswat, T.I. Kamins, J.S. Harris, Material characterization of high Sn-content, compressively-strained GeSn epitaxial films after rapid thermal processing. J. Cryst. Growth **365**, 29 (2013)
189. C. Xu, R.T. Beeler, L. Jiang, J.D. Gallagher, R. Favaro, J. Menéndez, J. Kouvetakis, Synthesis and optical properties of Sn-rich $Ge_{1-x-y}Si_xSn_y$ materials and devices. Thin Solid Films **557**, 177 (2014)
190. T. Yamaha, O. Nakatsuka, S. Takeuchi, W. Takeuchi, N. Taoka, K. Araki, K. Izunome, S. Zaima, Growth and characterization of heteroepitaxial layers of GeSiSn ternary alloy. ECS Trans. **50**, 907 (2013)
191. P. Zaumseil, Y. Hou, M.A. Schubert, N. von den Driesch, D. Stange, D. Rainko, M. Virgilio, D. Buca, G. Capellini, The thermal stability of epitaxial GeSn layers. APL Mater. **6**, 076108 (2018)
192. Y. Dong, W. Wang, X. Xu, X. Gong, D. Lei, Q. Zhou, Z. Xu, W. Khai Loke, S.-F. Yoon, G. Liang, Y.-C. Yeo, Germanium-Tin on Si avalanche photodiode: device design and technology demonstration. IEEE Trans. Electron Dev. **62**, 128 (2015)
193. W. Wang, Y. Dong, S.-Y. Lee, W.-K. Loke, D. Lei, S.-F. Yoon, G. Liang, X. Gong, Y.-C. Yeo, Floating-base germanium-tin heterojunction phototransistor for high-efficiency photodetection in short-wave infrared range. Opt. Express **25**, 18502 (2017)

194. J.J. Lin, S. Gupta, Y. Huang, Y. Kim, M. Jin, E. Sanchez, R. Chen, K. Balram, D. Miller, J.S. Harris, K. Saraswat, Fabrication of GeSn-On-Insulator (GSOI) to enable monolithic 3D co-integration of logic and photonics, in *Digest of Technical Papers—Symposium on VLSI Technology*, No. 6576674 (2013), p. 32
195. J. Michel, J. Liu, L.C. Kimerling, High-performance Ge-on-Si photodetectors. Nat. Photonics **4**, 527 (2010)
196. J. Mathews, R. Roucka, J. Xie, S.-Q. Yu, J. Menéndez, J. Kouvetakis, Extended performance GeSn/Si(100) p-i-n photodetectors for full spectral range telecommunication applications. Appl. Phys. Lett. **95**, 133506 (2009)
197. R.A. Soref, Group IV photonics: enabling 2 $\mu$m communications. Nat. Photonics **9**, 358 (2015)
198. H. Tran, T. Pham, W. Du, Y. Zhang, P.C. Grant, J.M. Grant, G. Sun, R.A. Soref, J. Margetis, J. Tolle, B. Li, M. Mortazavi, S.-Q. Yu, High performance $Ge_{0.89}Sn_{0.11}$ photodiodes for low-cost shortwave infrared imaging. J. Appl. Phys. **124**, 013101 (2018)
199. R.T. Beeler, C. Xu, D.J. Smith, G.J. Grzybowski, J. Menéndez, J. Kouvetakis, Compositional dependence of the absorption edge and dark currents in $Ge_{1-x-y}Si_xSn_y$/Ge(100) photodetectors grown via ultra-low-temperature epitaxy of $Ge_4H_{10}$, $Si_4H_{10}$, and $SnD_4$. Appl. Phys. Lett. **101**, 221111 (2012)
200. I.A. Fischer, T. Wendav, L. Augel, S. Jitpakdeebodin, F. Oliveira, A. Benedetti, S. Stefanov, S. Chiussi, G. Capellini, K. Busch, J. Schulze, Growth and characterization of SiGeSn quantum well photodiodes. Opt. Express **23**, 25048 (2015)
201. Y. Dong, W. Wang, D. Lei, X. Gong, Q. Zhou, S.Y. Lee, W.K. Loke, S.-F. Yoon, E.S. Tok, G. Liang, Y.-C. Yeo, Suppression of dark current in germanium-tin on silicon p-i-n photodiode by a silicon surface passivation technique. Opt. Express **23**, 18611 (2015)
202. M. Morea, C.E. Brendel, K. Zang, J. Suh, C.S. Fenrich, Y.-C. Huang, H. Chung, Y. Huo, T.I. Kamins, K.C. Saraswat, J.S. Harris, Passivation of multiple-quantum-well $Ge_{0.97}Sn_{0.03}$/Ge p-i-n photodetectors. Appl. Phys. Lett. **110**, 091109 (2017)
203. W. Wang, Q. Zhou, Y. Dong, E.S. Tok, Y.-C. Yeo, Critical thickness for strain relaxation of $Ge_{1-x}Sn_x$ ($x \leq 0.17$) grown by molecular beam epitaxy on Ge(001). Appl. Phys. Lett. **106**, 232106 (2015)
204. B.-J. Huang, J.-H. Lin, H.H. Cheng, G.-E. Chang, GeSn resonant-cavity-enhanced photodetectors on silicon-on-insulator platforms. Opt. Lett. **43**, 1215 (2018)
205. Y.-H. Peng, H.H. Cheng, V.I. Mashanov, G.-E. Chang, GeSn p-i-n waveguide photodetectors on silicon substrates. Appl. Phys. Lett. **105**, 231109 (2014)
206. Y.-H. Huang, G.-E. Chang, H. Li, H.H. Cheng, Sn-based waveguide p-i-n photodetector with strained GeSn/Ge multiple-quantum-well active layer. Opt. Lett. **42**, 1652 (2017)
207. M. Oehme, K. Kostecki, K. Ye, S. Bechler, K. Ulbricht, M. Schmid, M. Kaschel, M. Gollhofer, R. Körner, W. Zhang, E. Kasper, J. Schulze, GeSn-on-Si normal incidence photodetectors with bandwidths more than 40 GHz. Opt. Express **22**, 839 (2014)
208. S. Xu, Y.-C. Huang, K.H. Lee, W. Wang, Y. Dong, D. Lei, S. Masudy-Panah, C.S. Tan, X. Gong, Y.-C. Yeo, GeSn lateral pin photodetector on insulating substrate. Opt. Express **26**, 17312 (2018)
209. C. Chang, H. Li, C.-T. Ku, S.-G. Yang, H.H. Cheng, J. Hendrickson, R.A. Soref, G. Sun, $Ge_{0.975}Sn_{0.025}$ 320 $\times$ 256 imager chip for 1.6–1.9 $\mu$m infrared vision. Appl. Opt. **55**, 10170 (2016)
210. P.F. Ambrico, A. Amodeo, P. Di Girolamo, N. Spinelli, Sensitivity analysis of, differential absorption lidar measurements in the mid-infrared region. Appl. Opt. **39**, 6847–6865 (2000). https://doi.org/10.1364/AO.39.006847
211. J. Dong, D. Xiong, Applications of light emitting diodes in health care. Ann. Biomed. Eng. **45**, 2509–2523 (2017). https://doi.org/10.1007/s10439-017-1930-5
212. W. Zeller, L. Naehle, P. Fuchs, F. Gerschuetz, L. Hildebrandt, J. Koeth, DFB lasers between 760 nm and 16 $\mu$m for sensing applications. Sensors **10**, 2492–2510 (2010). https://doi.org/10.3390/s100402492

213. R. Wang, M. Muneeb, S. Sprengel, G. Boehm, A. Malik, R. Baets, M.-C. Amann, G. Roelkens, III-V-on-silicon 2-μm-wavelength-range wavelength demultiplexers with heterogeneously integrated InP-based type-II photodetectors. Opt. Express **24**(8), 8480–8490 (2016). https://doi.org/10.1364/OE.24.008480
214. M.V. Fischetti, S.E. Laux, Band structure, deformation potentials, and carrier mobility in strained Si, Ge, and SiGe alloys. J. Appl. Phys. **80**, 2234–2252 (1996)
215. J. Chen, B. Shu, J. Wu, L. Fan, H. Zhang, H. Hu, R. Xuan, J. Song, Enhanced electroluminescence from a free-standing tensilely strained germanium nanomembrane light-emitting diode. J. Semicond. **36**, 104004 (2015). https://doi.org/10.1088/1674-4926/36/10/104004
216. J. Jiang, M. Xue, C.-Y. Lu, C.S. Fenrich, M. Morea, K. Zang, J. Gao, M. Cheng, Y. Zhang, T.I. Kamins, J.S. Harris, J. Sun, Strain-induced enhancement of electroluminescence from highly strained germanium light-emitting diodes. ACS Photonics **6**, 915–923 (2019). https://doi.org/10.1021/acsphotonics.8b01553
217. P. Rauter, L. Spindlberger, F. Schäffler, T. Fromherz, J. Freund, M. Brehm, Room-temperature group-IV led based on defect-enhanced Ge quantum dots. ACS Photonics **5**, 431–438 (2018). https://doi.org/10.1021/acsphotonics.7b00888
218. Y. Zhou, W. Dou, W. Du, T. Pham, S.A. Ghetmiri, S. Al-Kabi, A. Mosleh, M. Alher, J. Margetis, J. Tolle, G. Sun, R. Soref, B. Li, M. Mortazavi, H. Naseem, S.-Q. Yu, Systematic study of GeSn heterostructure-based light-emitting diodes towards midinfrared applications. J. Appl. Phys. **120**, 023102 (2016). https://doi.org/10.1063/1.4958337
219. B.-J. Huang, C.-Y. Chang, Y.-D. Hsieh, R.A. Soref, G. Sun, H.-H. Cheng, G.-E. Chang, Electrically injected GeSn vertical-cavity surface emitters on silicon-on-insulator platforms. ACS Photonics **6**, 1931–1938 (2019). https://doi.org/10.1021/acsphotonics.8b01678
220. H.H. Tseng, K.Y. Wu, H. Li, V. Mashanov, H.H. Cheng, G. Sun, R.A. Soref, Mid-infrared electroluminescence from a Ge/$Ge_{0.922}Sn_{0.078}$/Ge double heterostructure p-i-n diode on a Si substrate. Appl. Phys. Lett. **102**, 182106 (2013). https://doi.org/10.1063/1.4804675
221. L.-F. Zhang, B. Shu, R.-H. Jiang, Y.-L. Gao, H.-M. Zhang, R.-X. Xuan, H.-Y. Hu, Short wave infrared LEDs based on strained GeSn. Microwave Opt. Technol. Lett. **60**(5), 1151–1154 (2018). https://doi.org/10.1002/mop.31131
222. M. Bertrand, N. Pauc, Q.M. Thai, J. Chrétien, L. Casiez, A. Quintero, Ph. Rodriguez, R. Khazaka, J. Aubin, J.-M. Hartmann, A. Chelnokov, V. Calvo, V. Reboud, Mid-infrared GeSn-based LEDs with Sn content up to 16%, in *IEEE 16th International Conference on Group IV Photonics, GFP-2019* (2019). https://doi.org/10.1109/group4.2019.8853926
223. L. Casiez, M. Bertrand, J. Chretien, A. Quintero, Q.M. Thai, M. Frauenrath, O. Lartigue, P. Barritault, N. Bernier, P. Rodriguez, A. Chelnokov, J.-M. Hartmann, N. Pauc, V. Calvo, V. Reboud, GeSn heterostructures LEDs for gas detection, in *IEEE 17th International Conference on Group IV Photonics* (2021)
224. B. Schwartz, M. Oehme, K. Kostecki, D. Widmann, M. Gollhofer, R. Koerner, S. Bechler, I.A. Fischer, T. Wendav, E. Kasper, J. Schulze, M. Kittler, Electroluminescence of GeSn/Ge MQW LEDs on Si substrate. Opt. Lett. **40**, 3209–3212 (2015). https://doi.org/10.1364/OL.40.003209
225. M. Oehme, M. Gollhofer, D. Widmann, M. Schmid, M. Kaschel, E. Kasper, J. Schulze, Direct bandgap narrowing in Ge LED's on Si substrates. Opt. Express **21**(2), 2206–2211 (2013). https://doi.org/10.1364/OE.21.002206
226. L. Peng, X. Li, J. Zheng, X. Liu, M. Li, Z. Liu, C. Xue, Y. Zuo, B. Cheng, Room-temperature direct-bandgap electroluminescence from type-I GeSn/SiGeSn multiple quantum wells for 2 μm LEDs. J. Lumin. **228**, 117539 (2020). https://doi.org/10.1016/j.jlumin.2020.117539
227. W. Du, Y. Zhou, S.A. Ghetmiri, A. Mosleh, B.R. Conley, A. Nazzal, R.A. Soref, G. Sun, J. Tolle, J. Margetis, H.A. Naseem, S.-Q. Yu, Room-temperature electroluminescence from Ge/$Ge_{1-x}Sn_x$/Ge diodes on Si substrates. Appl. Phys. Lett. **104**, 241110 (2014). https://doi.org/10.1063/1.4884380
228. R.E. Camacho-Aguilera, Y. Cai, N. Patel, J.T. Bessette, M. Romagnoli, L.C. Kimerling, J. Michel, An electrically pumped germanium laser. Opt. Express **20**, 11316–11320 (2012)

229. R. Koerner, M. Oehme, M. Gollhofer, M. Schmid, K. Kostecki, S. Bechler, D. Widmann, E. Kasper, J. Schulze, Electrically pumped lasing from Ge Fabry-Perot resonators on Si. Opt. Express **23**, 14815 (2015). https://doi.org/10.1364/OE.23.014815
230. S.A. Srinivasan, M. Pantouvaki, P. Verheyen, G. Lepage, P. Absil, J. Van Campenhout, D. Van Thourhout, Extraction of carrier lifetime in Ge waveguides using pump probe spectroscopy. Appl. Phys. Lett. **108**, 211101 (2016). https://doi.org/10.1063/1.4952432
231. R. Koerner, I.A. Fischer, D. Schwarz, C.J. Clausen, N. Hoppe, J. Schulze, Engineering of germanium tunnel junctions for optical applications. IEEE Photonics J. **10**(2), 2200912 (2018). https://doi.org/10.1109/JPHOT.2018.2818662
232. Y. Zhou, S. Ojo, Y. Miao, H. Tran, J.M. Grant, G. Abernathy, S. Amoah, J. Bass, G. Salamo, W. Du, J. Liu, J. Margetis, J. Tolle, Y.-H. Zhang, G. Sun, R.A. Soref, B. Li, S.-Q. Yu, Electrically injected GeSn lasers with peak wavelength up to 2.7 micrometer at 90 K. arXiv:2009.12229 (2020)
233. https://sigesn-muri.uark.edu/
234. Z. Shia, M. Daoa et al., Metallization of diamond. PNAS **117**(40), 24634–24639 (2020)
235. https://www.flash-project.eu/
236. J. Aubin, J.M. Hartmann, A. Gassenq, L. Milord, N. Pauc, V. Reboud, V. Calvo, Impact of thickness on the structural properties of high tin content GeSn layers. J. Cryst. Growth **473**, 20 (2017)
237. D. Rainko, Z. Ikonic, A. Elbaz, N. von den Driesch, D. Stange, E. Herth, P. Boucaud, M. El Kurdi, D. Grutzmacher, D. Buca, Impact of tensile strain on low Sn content GeSn lasing. Sci. Rep. **9**, 259 (2019)
238. R. Virgilio, T. Schroeder, Y. Yamamoto, G. Capellini, Radiative and non-radiative recombinations in tensile strained Ge microstrips: photoluminescence experiments and modeling. J. Appl. Phys. **118**, 233110 (2015)
239. R.W. Millar, K. Gallacher, A. Samarelli, J. Frigerio, D. Chrastina, T. Dieing, G. Isella, D.J. Paul, Expanding the Ge emission wavelength to 2.25 $\mu$m with $Si_xN_y$ strain engineering. Thin Solid Films **602**, 60–63 (2016). http://dx.doi.org/10.1016/j.tsf.2015.07.017
240. N. von den Driesch, D. Stange, D. Rainko, I. Povstugar, P. Zaumseil, G. Capellini, T. Schröder, T. Denneulin, Z. Ikonic, J.-M. Hartmann, H. Sigg, S. Mantl, D. Grützmacher, D. Buca, Advanced GeSn/SiGeSn group IV Heterostructure Lasers. Adv. Sci. **5**(6), 1700955–1700957 (2018)
241. Q.M. Thai, N. Pauc, J. Aubin, M. Bertrand, J. Chrétien, V. Delaye, A. Chelnokov, J.-M. Hartmann, V. Reboud, V. Calvo, GeSn heterostructure micro-disk laser operating at 230 K. Opt. Express **26**(25), 32500–32508 (2018)
242. J. Margetis, S. Al-Kabi, W. Du, W. Dou, Y. Zhou, T. Pham, P. Grant, S. Ghetmiri, A. Mosleh, B. Li, J. Liu, G. Sun, R. Soref, J. Tolle, M. Mortazavi, S.-Q. Yu, Si-based GeSn lasers with wavelength coverage of 2–3 $\mu$m and operating temperatures up to 180 K. ACS Photonics **5**(3), 827–833 (2017)
243. A. Elbaz, R. Arefin, E. Sakat, B. Wang, E. Herth, G. Patriarche, A. Foti, R. Ossikovski, S. Sauvage, X. Checoury, K. Pantzas, I. Sagnes, J. Chrétien, L. Casiez, M. Bertrand, V. Calvo, N. Pauc, A. Chelnokov, P. Boucaud, F. Boeuf, V. Reboud, J.-M. Hartmann, M. El Kurdi, Reduced lasing thresholds in GeSn microdisk cavities with defect management of the optically active region. ACS Photonics (2020)
244. Nils von den Driesch, Epitaxy of group IV Si-Ge-Sn alloys for advanced heterostructure light emitters. Dissertation, RWTH Aachen University (2018). https://doi.org/10.18154/RWTH-2018-221225
245. S. Bao, D. Kim, C. Onwukaeme, S. Gupta, K. Saraswat, K.H. Lee, Y. Kim, D. Min, Y. Jung, H. Qiu, H. Wang, E.A. Fitzgerald, C.S. Tan, D. Nam, Low-threshold optically pumped lasing in highly strained germanium nanowires. Nat. Commun. **8**(1), 195–197 (2017)
246. A. Elbaz, M. El Kurdi, A. Aassime, S. Sauvage, X. Checoury, I. Sagnes, C. Baudot, F. Boeuf, P. Boucaud, Germanium microlasers on metallic pedestals. APL Photonics **3**(10), 106102 (2018)

247. S. Su, B. Cheng, C. Xue, W. Wang, Q. Cao, H. Xue, W. Hu, G. Zhang, Y. Zuo, Q. Wang, GeSn p-i-n photodetector for all telecommunication bands detection. Opt. Express **19**, 6400 (2011)
248. H.H. Tseng, H. Li, V.I. Mashanov, Y.J. Yang, H.H. Cheng, G.E. Chang, R.A. Soref, G. Sun, GeSn-based p-i-n photodiodes with strained active layer on a Si wafer. Appl. Phys. Lett. **103**, 231907 (2013)
249. D. Zhang, C. Xue, B. Cheng, S. Su, Z. Liu, X. Zhang, G. Zhang, C. Li, Q. Wang, High-responsivity GeSn short-wave infrared p-i-n photodetectors. Appl. Phys. Lett. **102**, 141111 (2013)
250. M. Oehme, D. Widmann, K. Kostecki, P. Zaumseil, B. Schwartz, M. Gollhofer, R. Koerner, S. Bechler, M. Kittler, E. Kasper, J. Schulze, GeSn/Ge multiquantum well photodetectors on Si substrates. Opt. Lett. **39**, 4711 (2014)
251. S. Kim, N. Bhargava, J. Gupta, M. Coppinger, J. Kolodzey, Infrared photoresponse of GeSn/n-Ge heterojunctions grown by molecular beam epitaxy. Opt. Express **22**, 11029 (2014)
252. T. Pham, W. Du, B.R. Conley, J. Margetis, G. Sun, R.A. Soref, J. Tolle, B. Li, S.-Q. Yu, Si-based $Ge_{0.9}Sn_{0.1}$ photodetector with peak responsivity of 2.85 A/W and longwave cutoff at 2.4 $\mu$m. Electron. Lett. **51**, 854 (2015)
253. H. Cong, C. Xue, J. Zheng, F. Yang, K. Yu, Z. Liu, X. Zhang, B. Cheng, Q. Wang, Silicon based GeSn pin photodetector for SWIR detection. IEEE Photonics J. **8**, 1 (2016)
254. Y. Dong, W. Wang, S. Xu, D. Lei, X. Gong, X. Guo, H. Wang, S.-Y. Lee, W.-K. Loke, S.-F. Yoon, Y.-C. Yeo, Two-micron-wavelength germanium-tin photodiodes with low dark current and gigahertz bandwidth. Opt. Express **25**, 15818 (2017)

# Chapter 4
# Light Emission from Germanium Nanostructures

**Nelson L. Rowell and David J. Lockwood**

**Abstract** This Chapter discusses the phenomena associated with germanium (Ge) nanocrystals emitting near infrared radiation under optical excitation. We describe how the emission properties are influenced by various effects, including those of strain and particle confinement, as well as excitation mechanisms. Two example systems are discussed, namely one of isolated Ge quantum dots and another of Ge nanocrystals coherently imbedded in SiGe alloy layers, where both systems were grown by molecular beam epitaxy (MBE) on Si substrates. For the Ge dot ensembles, we show how particle size information can be derived from the emission spectrum. For the Ge nanocrystals, the emission spectra are analyzed for the effects of strain and particle confinement over a wide range of Ge fractions in the surrounding SiGe medium. This analysis provided significant insight into the properties of the Ge nanocrystals, including their size and shape, which were a 1.4 nm thickness in the MBE growth direction and a 7 nm lateral dimension. We also discuss the mechanisms leading to the high quantum efficiency observed for emission from the Ge nanocrystals at low temperatures.

## 4.1 Introduction

A silicon-compatible group IV coherent light source remains the missing link in Si-based photonics, despite the great variety of approaches explored such as, carrier localization, rare-earth doping, and stimulated inelastic light scattering. Although achieving efficient light emission from group IV semiconducting materials–a subject of intense research activity–has proven elusive, carrier localization methods have led to significantly enhanced optical emission from indirect gap materials and, notably, at higher temperatures. For example, room-temperature visible photoluminescence

N. L. Rowell (✉) · D. J. Lockwood
Measurement Science and Standards, National Research Council Canada, Ottawa, ON K1A 0R6, Canada
e-mail: nelson.rowell@nrc-cnrc.gc.ca

D. J. Lockwood
e-mail: David.Lockwood@nrc-cnrc.gc.ca

 
D. J. Lockwood and L. Pavesi (eds.), *Silicon Photonics IV*,
Topics in Applied Physics 139, https://doi.org/10.1007/978-3-030-68222-4_4

(PL) has been observed from porous Si [1] and strong infrared PL from various island or quantum dot systems [2–9]. A particular focus has been the study of self-organized structures containing nanometer sized Si [10] and Ge crystals [11], which being compatible with present Si technology have important capabilities for device engineering. Applications include those in optoelectronics [12], tunneling devices [13–15], and nanocrystal memories [10, 16]. Recently, lasing or very high efficiency photoluminescence (PL) has been revealed for Ge, a material with a relatively small indirect to direct bandgap (BG) difference, which can be overcome with tensile strain and/or high doping [17–21], or by alloying with tin [22].

## 4.2 Optical Properties of Bulk Ge

Germanium crystallizes in the diamond structure [23], which consists of two inter-penetrating face-centered cubic lattices displaced from each other by one quarter of the body diagonal. In zinc-blende semiconductors such as GaAs, the Ga and As atoms lie on separate sublattices, and thus the inversion symmetry of Ge is lost in polar III-V binary compounds. The difference in the crystal structures underlies the disparate electronic and optical properties of Ge and GaAs and results in the slow optical response of Ge together with its small electro-optic coefficient.

### *4.2.1 Band Structure*

The energy band structure in semiconductors is derived from the relationship between the energy and momentum of a charge carrier, which depends not only on the crystal structure but also on the bonding between atoms, the respective bond lengths, and the chemical species. The band structure is often quite complex and can only be calculated empirically [24].

Optical transitions conserve both energy and momentum. In direct gap GaAs, an excited electron at the bottom of the conduction band can relax spontaneously back into a hole in the valence band by emitting a photon at the band gap energy with momentum conservation (the photon momentum is negligible with respect to the electronic one). This electron-hole radiative recombination process can only occur in Ge with the assistance of a further process to conserve the momentum. This, in pure Ge, occurs via the transfer of the electron momentum to a phonon that is emitted or absorbed with equal and opposite wave vector to that of the initial state in the conduction band. Such a three-body process is quite inefficient compared with direct gap recombination [25]. Thus, the probability of spontaneous emission is very low for Ge and high for GaAs. Alternatively, by relating the recombination probability to a characteristic recombination lifetime, the spontaneous emission or radiative lifetime in Ge is very long (millisecond range) while in direct gap III-V semiconductors it is short (nanosecond range). In general, when excited carriers relax, non-radiative recombination pathways compete with radiative recombination. This competition is strongly thrown off balance for normal Ge, because of the long radiative lifetime, and hence non-radiative recombinations are dominant which in turn cause the very low luminescence efficiency. The indirect band gap of Ge is 0.66 eV at room temperature. The weak band-to-band emission at this energy in the near infrared was first observed [26] using visible light excitation. A review of the work on intrinsic and extrinsic radiative recombination in Ge is contained in [27, 28].

### 4.2.2 Absorption

The optical absorption spectrum of Ge at room temperature at energies close to the indirect and direct band gaps is shown in Fig. 4.1. The optical absorption coefficient

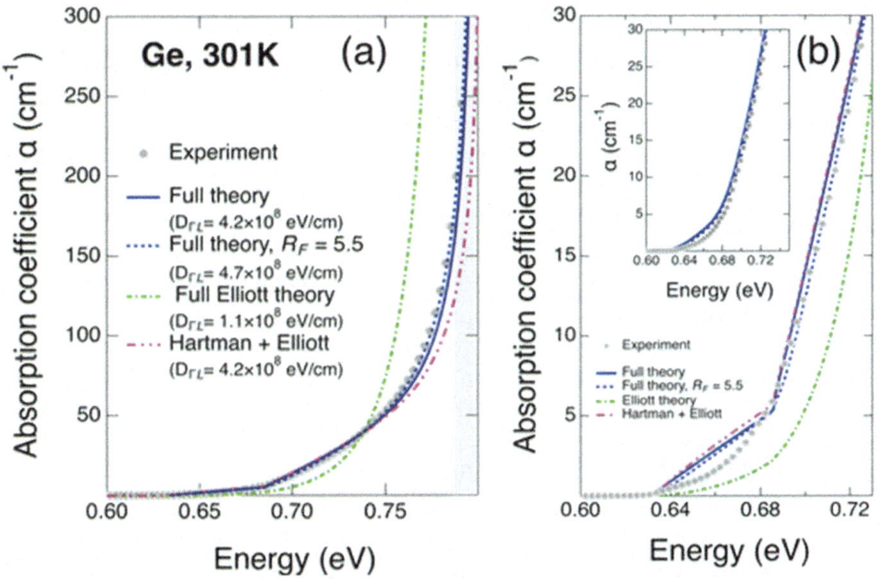


**Fig. 4.1** Ge absorption coefficient versus energy. **a** The circles indicate the experimentally determined absorption coefficient of Ge in the spectral range between the indirect and direct band gaps. The solid blue line is the fit with the theory using the average deformation potential $D_{\Gamma L}$ as the sole adjustable parameter. The dotted blue line is the same as the solid blue line but using the effective mass ratio $R_F$ as an additional adjustable parameter. The dash-dotted green line is the fit with the full Elliot model that neglects the energy-dependence of the intermediate states. The dash-double-dotted purple line is the fit with a hybrid model that combines Hartman's theory with Elliott's excitonic enhancement model. In all cases the fitted value of the deformation potential is indicated in the legend. **b** The same as for part (**a**) but showing the details of the experimental data and theoretical fits at the onset of the absorption. The inset shows the calculated solid blue line of the main panel convoluted with a Gaussian with a full-width-half-maximum of 0.015 eV to simulate lifetime broadening. After [30]

$\alpha$ begins to rise steeply from zero near the indirect band gap energy (0.66 eV) as the incident photon energy approaches that of the direct band gap of 0.8 eV.

Theoretically predicting the optical absorption properties of Ge is not trivial [29, 30]. First principles calculations have been carried out for Si that are able to reproduce the general absorption behavior, but these calculations are much more difficult for Ge due to the required energy resolution (see [29, 30]). Several analytical theories have been developed, with the most successful being those reported by Elliott [31] and Hartman [32], but until recently there has been no analytical approach that has been able to quantitatively reproduce the experimental results over a broad energy range (see [29, 30], and references therein). In [29, 30], a more complete theory of resonant indirect optical absorption is introduced to explain the experimental results. The main ingredient of the theoretical approach is a realistic account of the energy dependence of the intermediate states, which in Ge, unlike the case of Si, cannot be neglected. The resulting theory using independently determined values for most parameters agrees closely with the experimental results over the entire spectral range of the measurement, as can be seen in Fig. 4.1. The only adjustable parameter used in the fit was the average deformation potential $D_{\Gamma L}$ whose fitted value of $4.2 \times 10^8$ eV/cm was found to be equal to the independently determined value of this parameter. Thus, this parameter could be fixed such that the theory would predict indirect absorption without adjustable parameters. The relative uncertainty in the theoretical prediction is small over two orders of magnitude in the value of the absorption, with the largest discrepancy occurring near the indirect band gap energy, where excitonic effects at low temperature are not explained in the model. The small remaining discrepancy at higher energies can be further reduced by considering the effective mass ratio $R_F$ as an adjustable parameter. This factor is used to correct the bound exciton absorption, and it is more difficult to justify than the approximations used to deal with the continuum contribution. The result of freeing this parameter in the fits is shown as a blue dotted line in Fig. 4.1. The fitted deformation potential value of $4.7 \times 10^8$ eV/cm is still reasonable, but the mass ratio $R_F = 5.5$ has decreased by almost a factor of 3 relative to the predetermined value of 15. It is unclear if one can assign a definite physical meaning to this reduced value of $R_F$ [29].

### *4.2.3 Temperature Dependence—Ge Energy Gap*

Varshni [33] has determined that the experimental temperature dependence of the energy gap in semiconductors can be reproduced by

$$E_g = E_0 - \alpha T^2/(T + \beta) \tag{4.1}$$

where $E_g$ is the energy gap, which may be the direct or indirect one, $E_0$ is its value at 0 K, and $\alpha$ and $\beta$ are constants. Varshni found that this expression worked well for diamond, Si, Ge, 6H–SiC, GaAs, InP, and InAs. The energy gap variation with temperature is believed to arise mainly from two mechanisms [33]. At lower temperatures, the shift in the relative position of the valence and conduction bands is due to the temperature-dependent dilatation of the lattice as well as a shift in the relative position of the valence and conduction bands due to the temperature-dependent electron-lattice interaction. From fits of (4.1) to the experimental data for Ge [34], Varshni [33] determined the following parameter values for the direct (indirect) energy gap: $E_0 = 0.8893$ (0.7412) eV, $\alpha = 6.842 \times 10^{-4}$ ($4.561 \times 10^{-4}$) eV/K, and $\beta = 398$ (210) K.

### *4.2.4 Stress*

There has been considerable interest recently regarding the effects of strain on Ge, as it has been determined that tensile strain can lower the direct gap energy below that of the indirect gap and thus transforms Ge into a direct gap semiconductor. Such a transformation would then greatly improve the prospects for the use of Ge as a laser in Si photonics, particularly as Si is not amenable to this crossover due to the much wider energy spacing between the two energy gaps (see, for example, the review of this field in [35] and references therein).

Although crystalline Ge can be readily compressed hydrostatically, it is difficult to apply hydrostatic extension. Both theory and experiment have been used to explore the effects of strain on Ge that is lattice matched to some substrate, including Si and SiGe alloy. By considering biaxial strain, theory has shown that tensile strain can reverse the order of the direct and indirect band gaps [36]. A more recent theoretical result that demonstrates this reversal is shown in [37] (see references in this review paper for previous work). The calculation was made for a thin film of Ge that was released from its native substrate (i.e., a membrane of nanoscale thickness), and the direct energy gap was lowered from 0.8 to 0.48 eV, which means it now lies below that of the L point energy gap.

Regarding the application of one-dimensional tensile strain [38] to Ge, notable is the recent work of Guilloy et al. [39] who have used a micro-bridge geometry to uniaxially stress Ge along the [100] direction. These authors performed electro-absorption spectroscopy to measure the energy gap between the conduction band at the Γ point and the light- and heavy-hole valence bands. The results shown in Fig. 4.2 agree with the standard linear deformation potential theory for up to 2% strain. However, a significantly nonlinear behavior is observed at higher strains that is well described by a tight-binding model. The cross-over between the direct and indirect bands is now predicted to occur at 5.6% tensile strain [42].

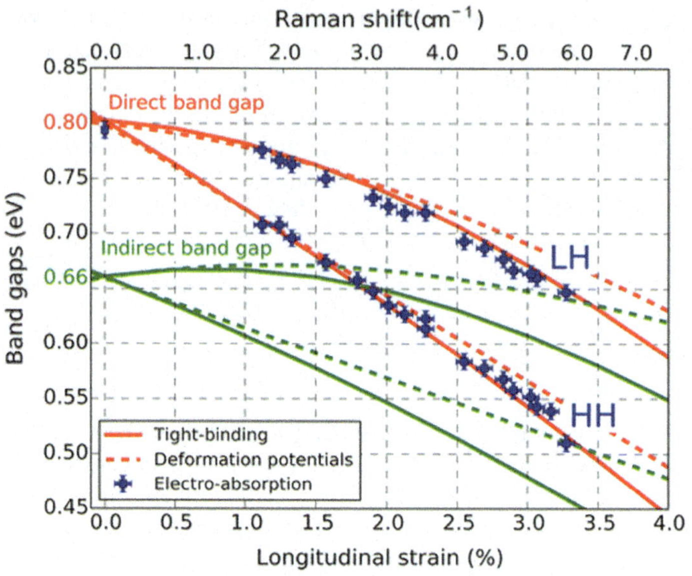


**Fig. 4.2** Direct band gap energy of Ge at 293 K measured by electro-absorption (blue squares) as a function of longitudinal strain measured by Raman spectroscopy. These experimental points are compared to theoretical band gap values computed with deformation potentials (dashed red lines) and with a tight-binding model (solid lines). The upper curve corresponds to the transition with the light hole (LH) band while the lower one is with the heavy hole (HH) band. After [39]

## *4.2.5 Optical Emission from Bulk Ge*

Recent measurements of the optical emission spectrum of intrinsic (low impurity concentration) bulk Ge as a function of temperature are shown in Fig. 4.3 [40]. The spectral range of the PL covers emission from both the indirect (phonon assisted) and direct gaps (see Sect. 2.1). At low temperatures, the contributions from the four types of phonons, i.e., including both acoustic and optic phonons, are readily apparent, but as the temperature increases the distinction blurs: longitudinal acoustic (LA) phonons dominate at 10 K, there are almost equal contributions from all but the transverse optic (TO) phonons at 160 K, and transverse acoustic (TA) phonons dominate at 295 K. For theory, the challenge is how to model this extreme variation in line shape and the overall intensity behavior as a function of temperature.

Menéndez et al. [40] have developed a theory to account for the experimental results using analytical expressions for the contributions from LA, TO, longitudinal optic (LO), and TA phonons. The coupling of electronic and phonon states at exactly the Γ or L Brillouin-zone point is forbidden by symmetry for the latter two phonon modes, but becomes allowed for nearby states and can be accounted for by using wave-vector dependent deformation potentials. The calculations require a knowledge of the optical absorption as a function of energy [29, 30] and need to consider the diffusion of excitons. With this theory, agreement was obtained between predicted

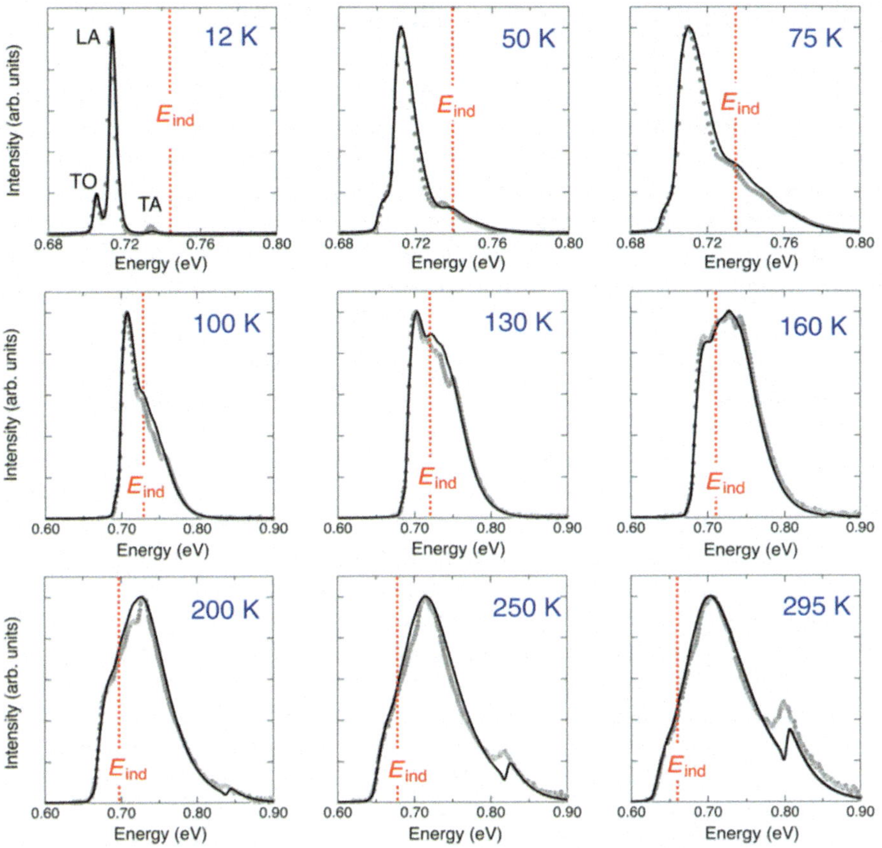


**Fig. 4.3** Photoluminescence from bulk Ge. Experimental (circles) and theoretically predicted (solid lines) PL spectra for Ge at nine selected temperatures. All spectra have been normalized to the same intensity for clarity. The vertical dotted lines indicate the position of the indirect gap at each temperature. Peak assignments are shown in the 12 K spectrum. The LO contribution in this spectrum appears as a weak shoulder in higher-resolution spectra. After [40]

and observed PL line shapes, as can be seen in Fig. 4.3. A decomposition of the predicted PL in terms of the different phonon contributions implies that for temperatures near room temperature indirect optical absorption and emission are dominated by "forbidden" processes, and that the deformation potentials for allowed processes are smaller than previously assumed. The theory [40] used to predict the PL line shape was not as successful in predicting the temperature dependence of the PL intensity.

## 4.3 Developments with Ge for Photonic Emitters

Germanium has recently been of considerable interest in photonics because of some notable properties, including chemical and crystallographic compatibility with silicon, a near infrared bandgap in the communications wavelengths, and that its indirect bandgap energy is only slightly smaller than its direct bandgap. As described below the various demonstrations of stimulated emission from Ge fall into four general categories, which are (i) heavily doped *n*-type Ge [17, 18], (ii) Ge under tensile strain [21, 41, 42], (iii) Ge alloyed with Sn [22, 43], and (iv) hexagonal Ge [44].

- With sufficient *n*-type doping in Ge, its direct gap becomes the lowest energy available for electronic transitions. Using this approach, two groups [17, 18] have reported stimulated emission and lasing in waveguide structures that contained heavily *n*-type doped ($5 \times 10^{19}$ cm$^{-3}$) Ge layers under 0.2% uniaxial tensile strain. In such structures, the doping effectively filled the lower indirect states in the conduction band, thus enabling direct gap emission under optical and electrical excitation. There have however been reports that such a process might have limited efficiency due to various inherent loss mechanisms [45–48].
- Tensile strain reduces the energy of both the indirect and direct gaps, but with the indirect energy declining more rapidly such that the direct gap energy becomes smaller than that of the indirect gap for sufficient strain. The first reports of strong PL or direct gap emission were on uniaxially strained (1.6%) nanowire samples or biaxially strained disc samples [21, 41]. In the nanowire work, extremely high optical excitation densities permitted direct gap emission by saturating the indirect gap transitions. Actual direct gap Ge was obtained more recently [42] with the material under sufficient tensile strain (~5.6%) in a microbridge configuration. In that work, low threshold, optically excited lasing was obtained at energies near 350 meV for temperatures up to 100 K. Tunability to higher photon energies would be necessary for this development to be useful in practice.
- Manipulation of the material's band structure by alloying Ge with a suitable group IV material can permit direct gap behavior, as for $Ge_{1-x}Sn_x$ alloys. In the original work [22] it was predicted that for a tin fraction greater than 0.08 the unstrained alloy would have a direct gap, as demonstrated by stimulated emission observed at 540 meV from $Ge_{0.87}Sn_{0.13}$. However, the laser threshold under optical excitation was relatively high. More recently [43], ultralow threshold lasing was obtained for tensile strained $Ge_{0.946}Sn_{0.054}$.
- The band structure of Ge can be altered by changing its crystal structure, as shown for example by Ge and $Si_{1-x}Ge_x$ being grown in a hexagonal lattice configuration [44]. For Ge fractions greater than 0.65, direct gap behavior is predicted by energy band calculations, with the bandgap varying from 670 meV ($x = 0.65$) to 350 meV ($x = 1$). An experimental verification of this analysis was provided through photoluminescence (PL) measurements in which the PL peak shifted as expected versus the Ge fraction.

## 4.4 Optical Emission from Ge Nanocrystals

### 4.4.1 *Outline*

This section begins with a general discussion of PL methodology, but the section mostly contains a detailed review of our recent results and analysis regarding Ge nanocrystals in two distinctly different environments.

In Sect. 4.3 we discuss the first system, which consisted of Ge quantum dots formed by thermal annealing of Ge layers on oxidized Si substrates. Such dots were electrically isolated from each other and from the substrate. Although the PL efficiency for such systems was relatively low, it was possible to analyse the emission spectrum to access the dot size distributions, for which the PL derived sizes were compared with those obtained with more direct microscopy methods including atomic force and electron microscopies. These comparisons have provided a useful check on the validity of the band gap models [52] for small Ge quantum dots as well as giving the PL relative efficiency versus dot size.

In Sect. 4.4 we discuss a second system consisting of platelet shaped Ge nanocrystals (NCs) imbedded as clusters within $Si_{1-x}Ge_x$ epitaxial layers grown directly on Si substrates. These NCs emit broad band PL with apparent high efficiency, due to the fact that the NCs act as potential wells trapping excitons created by the optical excitation. The factors to be evaluated in calculating the peak PL energy include: the Ge bandgap's variation with strain, the effect of Ge-fraction in the host $Si_{1-x}Ge_x$ epilayer on its bandgap, how the nanoscale dimensions affect the confinement shifts for both the nanocrystals and epilayers. Modeling these effects provided a NC height distribution over a range in strain from compressive to tensile. From an analysis of the NC PL linewidth, the NC lateral size was estimated to be about five times the NC height. Imaging experiments showed that part of the PL emission was confined to the waveguide formed by the $Si_{1-x}Ge_x$ layers on Si and that the emission was significantly collimated laterally due to processes in the waveguide.

### 4.4.2 *Photoluminescence (PL) Measurements*

#### 4.4.2.1 PL—Indirect Gap Materials

The PL emission process for bulk indirect gap materials is extremely inefficient [49]. In materials such as silicon and germanium, PL occurs after the generation of excitons. Excitons are electrically neutral quasiparticles each comprised of an electron and hole attracted to each other by the electrostatic Coulomb force. Such excitons eventually self-annihilate thus generating photons near the material's bandgap energy with a quantum efficiency of about one in ten thousand [49]. This weakness of the radiative process is due to the many competitive non-radiative processes active in such a semiconductor under excitation at photon energies above bandgap energy.

Of the several possible weak processes leading to PL, the one leading to bound exciton (BE) recombination is useful for the characterization of epitaxial material with respect to dopant, defect, and/or impurity concentrations [50].

#### 4.4.2.2 PL Enhancement by Exciton Diffusion/Concentration

The PL in indirect materials such as Si and Ge is often the result of electron-hole recombination within the excitons that are formed upon optical excitation. As will be discussed in further detail in Sect. 4.4.13, the PL intensity can be enhanced in nanostructures when the excitons created in a sample diffuse into a lower bandgap region, where they are trapped due to the energy difference illustrated in Fig. 4.4. In that figure, excitons are shown as trapped in one dimension within a graded SiGe layer on a silicon substrate. The excitons photo-excited in the substrate, diffuse in all directions such that a substantial proportion of them become trapped in the lower bandgap, thin SiGe layer. There, in relatively high concentrations, the excitons self-annihilate, leading to significant PL intensities. In this case the alloy fluctuations in the SiGe layer also result in the SiGe's no phonon line to be of similar intensity to the phonon replica lines. The PL intensity for the SiGe layer in systems such as those of Fig. 4.4 is approximately 100 times larger than for an isolated SiGe layer. This increase in intensity is due to the fact that, on the Si substrate, the SiGe layer

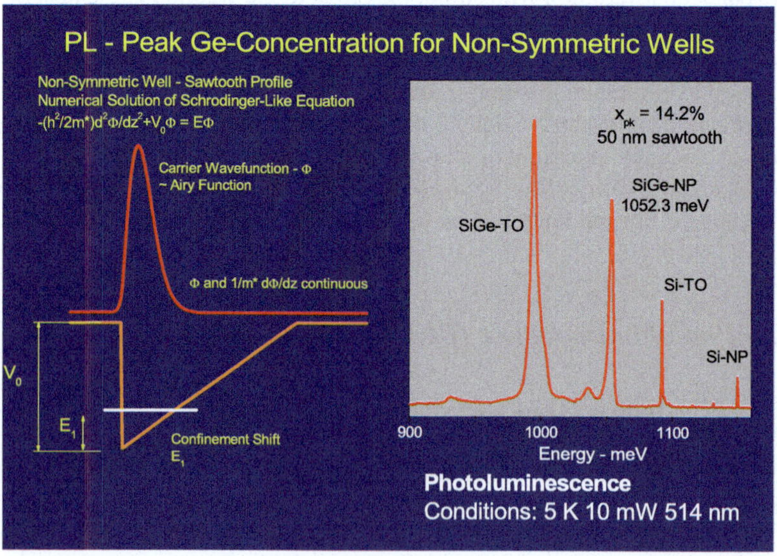


**Fig. 4.4** Exciton trapping in one dimension for a graded SiGe layer on a silicon substrate. Photo-excited excitons diffuse and are confined in the lower bandgap SiGe layer, where in relatively high concentrations they self-annihilate, leading to comparatively high PL intensities. Alloy fluctuations in the SiGe layer result in the layer's no phonon line being of similar intensity to the phonon replica lines

draws in significant quantities of excitons from the substrate for which the excitation depth is ~1000 nm as opposed to the ~10 nm SiGe layer thickness. Since the alloy bandgap varies with Ge concentration, the position of the no phonon line in the PL spectrum can be used to obtain the Ge concentration/distribution. This line's energy also depends on the layer thickness through the confinement effect. Exciton diffusion to smaller bandgap layers is prevented if there are relatively insulating intermediate barriers, such as oxide layers in the vertical direction. In such cases the emission is generally weak as it results only from the direct excitation of a layer, where the relatively small volume of the layer is the controlling factor.

#### 4.4.2.3 PL Method

In the following discussion the experimental method of choice was photoluminescence (PL). A PL apparatus as illustrated in Fig. 4.5 was used, where the PL spectra were obtained for the Ge quantum dot and the Ge nanocrystal samples with a FTIR spectrometer with the samples at liquid helium temperatures, excited with 0.02–400 mW of 405, 457.9, 488.0, or 514.5 nm laser light. Varying the wavelength of the exciting light affected the depth of penetration for the excitation in the sample. There are significant advantages to the use of a Fourier transform spectrometer for PL studies [51]. A radiometric calibration was performed to obtain the relative spectral

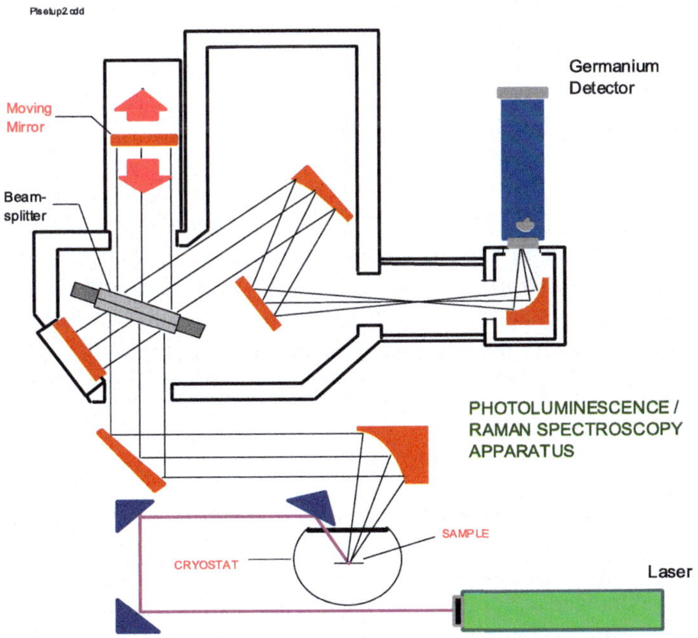


**Fig. 4.5** Fourier transform photoluminescence spectrometer system schematic

responsivity of the system from 500 to 1400 meV, which was used to normalize the raw spectral data. Enabling components for these measurements were ultra-sensitive optical detectors, namely, the Applied Detectors Corporation Ge PIN photodiode for the range from 700 to 1200 meV and the cold-filtered Cincinnati Electronics InSb photodiode from 500 to 1000 meV.

## *4.4.3 Ge Quantum Dot Ensembles*

### 4.4.3.1 Electrically Isolated Spherical Ge Quantum Dots

In this example, ensembles of isolated Ge quantum dots were formed by in situ thermal annealing of thin amorphous Ge layers grown by MBE on oxidized Si (001) substrates. These layers were deposited either directly on the $SiO_2$ layer or on a thin porous $TiO_2$ layer on the $SiO_2$ layer. After annealing the dots formed had diameters ranging from 3 to 35 nm, for which the photoluminescence (PL) consisted of a wide near-infrared band peaked near 800 meV. The peak energy of the PL band reflected the average dot size in the dot ensemble with the PL band shape depending on the size distribution within the dot ensemble. Here the QD PL was excited by the incident photons directly absorbed in the QDs. This PL was not affected by the excitons created in the substrate as their diffusion to the dots was blocked by the insulating layer between the substrate and the QDs. This barrier effect was responsible for the relatively weak PL from the QDs.

### 4.4.3.2 Models for the Ge Dot Bandgap

Two practical models that have been developed for the bandgap of spherical Ge quantum dots are the tight binding (TB) and k.p approximations [52] (see Fig. 4.6), where the assumption is made in both models that the dots are independent, with uniform and insulating surrounding media, which means the model dots are essentially non-interacting, either with each other, with the substrate, and/or the surrounding medium. As the TB model has proven in the present case to be more representative of the observed results, this model will be used in the subsequent discussion of QDs.

### 4.4.3.3 Dot Size Distribution from the PL Spectrum

In the case of independent Ge quantum dots, the peak energy of a PL band is reflective of the average dot size and the PL band shape depends on the dot size distribution. (see Fig. 4.7) Using the tight binding model results, the PL spectra were analyzed in terms of the dot size distribution. The PL spectra versus dot size were calculated for Ge/$SiO_2$ by assuming the PL emission occurred near the dot's TB bandgap energy.

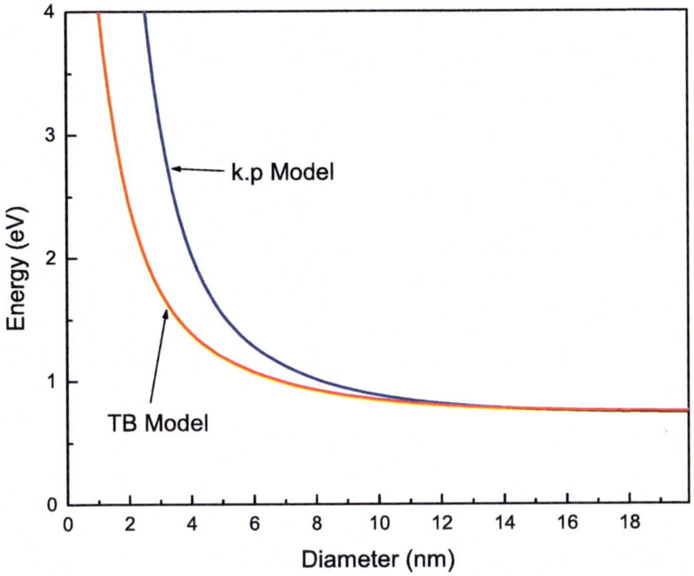


**Fig. 4.6** Variation of the bandgap with Ge dot diameter for the TB and k.p band models

The dot size distributions were measured from TEM (Fig. 4.7) and AFM images, from which best-fit Gaussian curves were obtained for the distributions (Fig. 4.8). The observed size distribution determined from TEM allowed the deduction that the PL efficiency increased non-linearly with decreasing dot diameter. This PL efficiency was derived by normalizing the PL intensity by the Gaussian size distribution for Ge dot samples on $SiO_2$. A dot diameter could also be obtained from the PL energy by inverting the TB model curve shown in Fig. 4.6.

#### 4.4.3.4 Comparison of Dot Size Measurements

The calculated efficiency curves were satisfactorily fitted by straight lines on a log-log plot, with essentially the same slope for all samples. This behavior demonstrated that there is a general power-law increase in PL efficiency with decreasing dot size. Knowing this generic PL efficiency, it was possible to evaluate the size distribution of Ge dots directly from the PL energy dependence. Since the fitted PL efficiency curve for Ge dots grown on $SiO_2$, although empirical, was generic for Ge dots, the dot size distributions were derived for the samples on $TiO_2$ from the PL data and then compared with those obtained more directly, for example, from AFM (see Fig. 4.8). In this way it was possible to obtain distributions simultaneously for several such samples. The size range obtained from PL was limited to dot diameters below 20 nm by the PL detector cutoff, since the bandgap sensitivity to particle size was low for larger diameters. By fitting Gaussians to these distributions, they were extrapolated

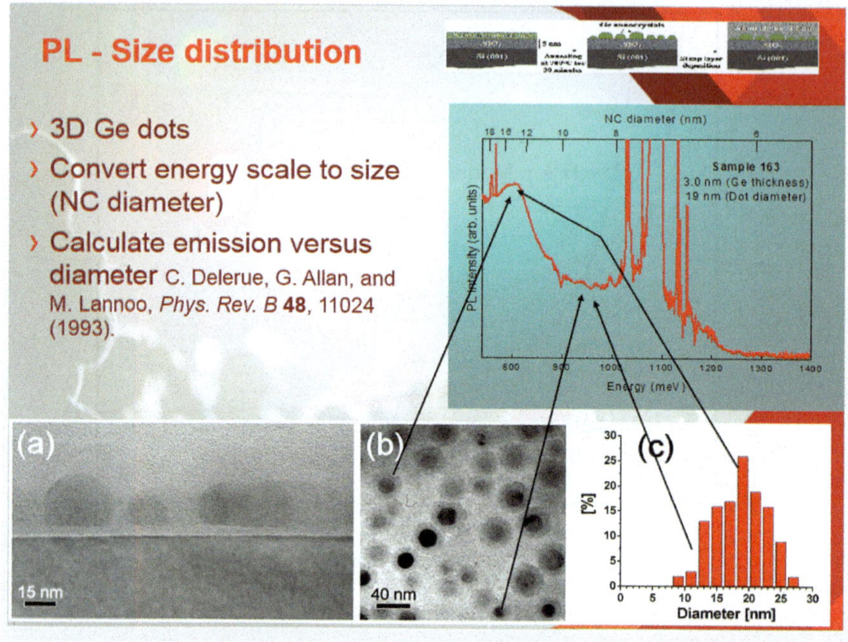


**Fig. 4.7** Cross sectional (**a**) and plan view (**b**) HRTEM image of a Ge dot sample, which shows a nearly symmetric dot size distribution (**c**) with a maximum at 19 nm, giving a theoretical bandgap of 0.75 eV. After the PL spectrum of this sample was corrected for the instrumental response, a dominant peak is in fact observed corresponding to dot diameters in the range 12–19 nm (the instrumental response cut-off at ~740 meV prevents observing PL to lower energy) consistent with theory and the measured sized distribution. The PL continuum becomes weaker to higher energy and extends to ~1200 meV, also consistent with the measured size distribution

for the dot density distributions at larger diameters and were then compared with the distributions obtained from the AFM images. These results are plotted in Fig. 4.9 along with the size distribution histograms obtained from AFM data with the best-fit Gaussian curves (dashed lines). For sample M01, the PL-derived distribution appears to be quite Gaussian, although only the lower half of the curve can be obtained with PL. Apparently for sample P01, the PL data predicts a somewhat non-Gaussian size distribution, which matches reasonably well the small-diameter onset of the distribution although the PL-derived distribution is limited for the reasons described above. It is to be noted that this procedure had the advantage of being immune to the relatively intense PL arising from the substrate, as shown by the well behaved curve below 9 nm. The results for the size distribution from the PL method agreed within experimental uncertainties with those of the AFM measurements in that the widths of the distributions are generally the same as are the positions of the distribution maxima.

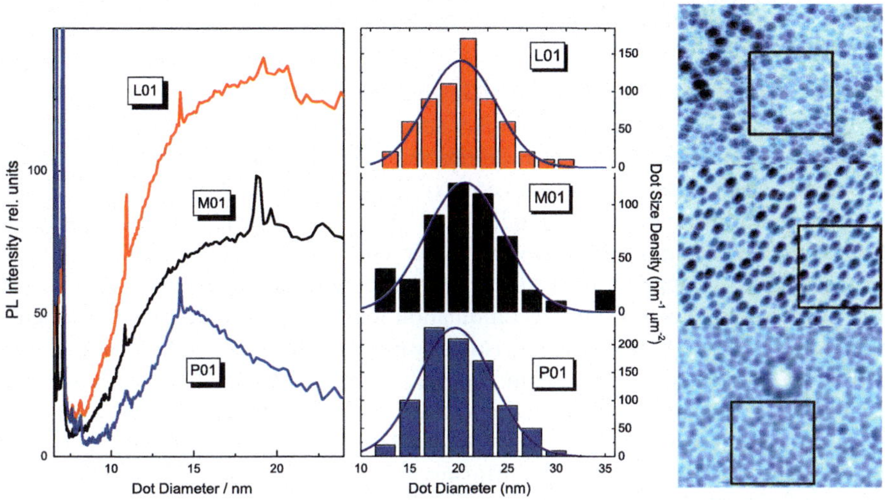


**Fig. 4.8** PL and dot density variation with dot diameter of self-organized Ge dots grown on a porous $TiO_2$ membrane for samples L01, M01, and P01. In the left panel are the PL spectra versus dot size as calculated by taking the PL emission to occur near the dot bandgap. The strong PL emission lines below 7 nm are due to the Si substrate. In the middle panel are shown the dot size distributions obtained from AFM measurements with best-fit Gaussian functions superimposed on the distributions: these AFM images are displayed on the right. The 200 nm wide squares in the images show the areas sampled for the size distributions in the middle panel

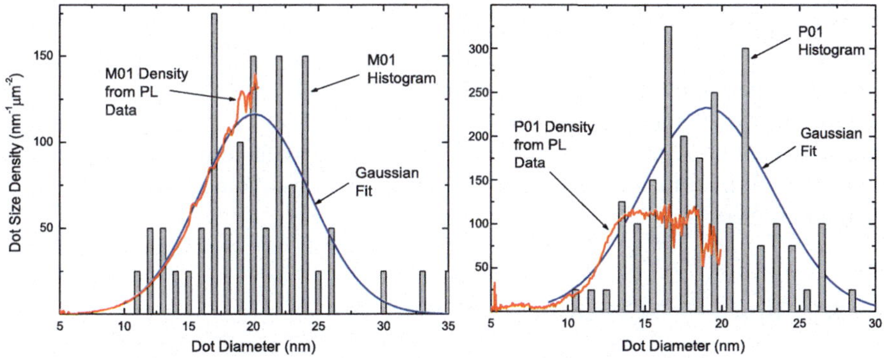


**Fig. 4.9** Size distribution derived from the PL data for samples M01 and P01. The size densities were calculated by taking the ratios of the PL spectra with the PL emission efficiency for sample L01. These results are compared with the size distribution histograms (with 1 nm bin sizes) obtained from AFM data and with the best-fit Gaussian functions for the histograms

## 4.4.4 *Self-organized Ge Nanocrystals (NCs)*

### 4.4.4.1 Ge NCs in MBE Grown SiGe Epitaxial Layers

Intense, low temperature PL has been observed [53] in many samples consisting of coherently strained molecular beam epitaxy (MBE)-grown $Si_{1-x}Ge_x$ epitaxial layers with Ge fractions ranging from 0.05 to 0.53. An effective quantum efficiency in the 5% range was obtained. This PL was neither defect nor dislocation related, but the enhancement was apparently due to exciton diffusion and congregation effects. The site of the actual electron-hole recombination was unknown until recently, when it was shown that the PL was characteristic of strained, self-assembled Ge nanocrystals (NCs) coherently contained within $Si_{1-x}Ge_x$ alloy epilayers. This conclusion was enabled by extensive experimental and theoretical investigation of the host alloy concentration dependence in the PL spectra. These NCs provide low bandgap sites that trap the neutral free excitons diffusing within the SiGe layers and Si substrate. The radiative efficiency was enhanced by as much as two orders of magnitude for an exciton trapped in a NC [53], because there was a low probability of an impurity atom or crystal defect being present in a NC [49].

### 4.4.4.2 Growth and Morphology

A brief description follows as to how the samples in question were grown and their PL [53] measured. The $Si_{1-x}Ge_x$ material was grown without intentional doping by conventional, solid source MBE in an ultrahigh vacuum deposition chamber. The growth temperature was in the range 350–900 °C. The Si growth rate was 1 Å/s and the Ge growth rate was set to achieve a given value of $x$. The single or multiple quantum well structures were designed to be stable with respect to misfit strain relaxation [54]. Further details regarding similar growths of other MBE samples can be found in Sect. 4.4.5.

### 4.4.4.3 Broad Ge NC Signature in PL Spectra

The PL spectra of the Ge NC samples were measured as described in Sect. 4.2.3. PL spectra containing broad peaks for more than 60 separate MBE growths were examined. Of these samples 30 had both broad and phonon resolved (PR) peaks in their spectra. On the right of Fig. 4.10 are the PL spectra for three $Si_{1-x}Ge_x$ multiple quantum well samples, each sample consisting of ten wells separated by Si spacers lattice matched in the $x$–$y$ plane on Si (001) $p$-type substrates, with increasing well thickness from trace (a) to trace (c). The Ge fraction varied slightly from 0.15 for the two thinner well samples to 0.19 for the thicker well sample. On the left are the plan-view TEM images for two of the samples, one with the thinnest wells and the other with the thickest wells. The TEM images indicate the presence of small platelets

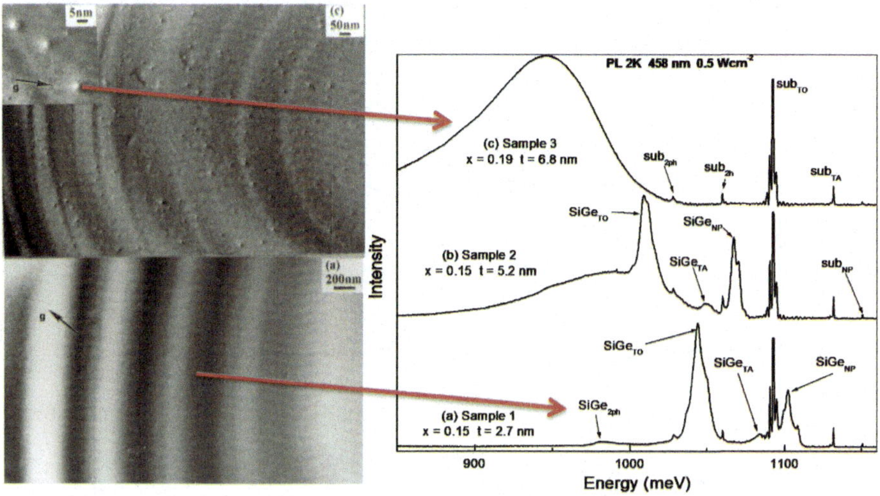


**Fig. 4.10** PL spectra for three $Si_{1-x}Ge_x$ multiple quantum well samples with TEM images. The phenomenology of the transition from PR to broad PL for three MBE $Si_{1-x}Ge_x$ multiple quantum well samples is shown. The box on the right depicts the PL obtained from the three samples with essentially the same Ge fraction at 2 K, with the TEM pictures obtained from samples 1 and 3 shown on the left

of elevated Ge content for the thicker quantum wells. For trace (a), only phonon resolved PL lines are seen, both for the quantum wells and for the substrate, with the broader lines being those associated with the quantum wells. The substrate lines are the NP line at 1150 meV, the TA PR line at 1130 meV, and the transverse optic (TO) PR around 1092 meV. The phonon resolved SiGe PL also consists primarily of a NP line and the TO replica, separated by the same Si–Si [69] TO phonon frequency (58 meV) as for Si, although the NP line is relatively much stronger and wider, due to alloy disorder. The position of the SiGe NP line depends on the Ge fraction and hence on the quantum well strain, and on the quantum well thickness through the confinement effect. When the well thickness is increased from 2.7 nm to 5.2 nm without changing the Ge fraction, the SiGe PR lines shift in tandem due to a reduction in the confinement shift for the thicker wells. There is also a rising broad background which peaks approximately 80 meV below the SiGe NP peak. For the thickest wells (c), the PR peaks are no longer present and an intense broad PL to lower photon energies dominates the spectrum.

#### 4.4.4.4 Alloy Layer Composition Dependence

As shown in Fig. 4.11, the PL attributed to Ge NCs consists of a broad peak with an asymmetry to low photon energies. This peak displays little variation in shape with Ge fraction and tracks the BG variation, but is ~00 meV below the indirect SiGe BG. If the material were Si or SiGe, the width of the peak at ~50 meV would be too small

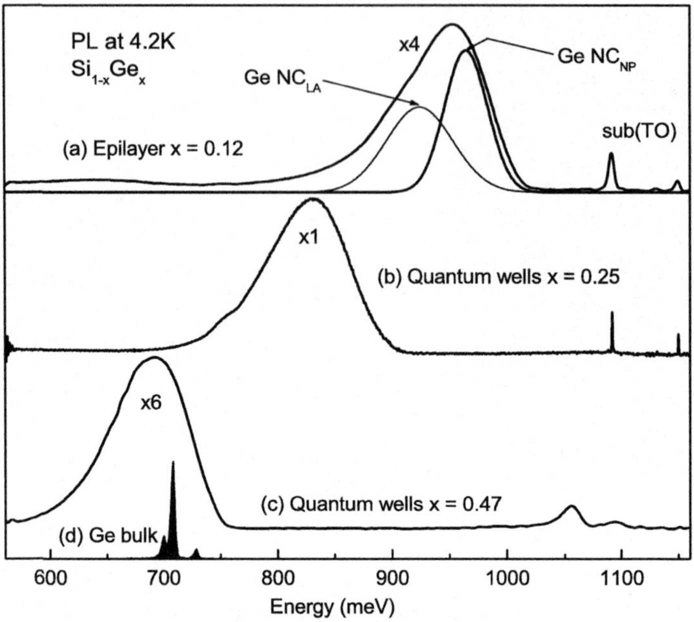


**Fig. 4.11** Low temperature PL from three SiGe samples and bulk Ge. For the spectrum in trace (a), the result is also illustrated with curve fitting using two Gaussian peaks for the broad PL peak

for it to be due to a NP line with its TO PR, as the NP-TO spacing is about 58 meV for those materials. Figure 4.11 shows that for higher Ge fractions the PL is emitted at energies significantly below those for bulk Ge, with its indirect BG of 744 meV at low temperatures. The origin of the broad, intense PL peak had remained unexplained, until recently [55]. In Fig. 4.11 the low temperature PL is displayed from three SiGe samples and bulk Ge. For the spectrum in trace (a), the result is also illustrated with curve fitting using two Gaussian peaks for the broad PL peak. The broad NC PL weakens with increasing temperature [53] but persists to approximately 100 K. The decay with temperature is nearly exponential, but the effective NC well depth is not sufficient to give emission at room temperature. As was shown in Fig. 4.11a, the broad peak's width and asymmetric shape can be curve resolved into two symmetric peaks, separated by ~35 meV, i.e., very near the momentum conserving TO phonon energy for Ge. The NP peak is wide (25–45 meV) due to NC confinement variations arising from size variability and to alloy disorder broadening in the SiGe. In Ge NCs, two main PL peaks are expected, a NP line and a LA replica line, separated by about 28 meV. So the NP-PR energies provided by curve resolving are relatively close to that (28 meV) for the most intense (LA) PR for Ge, but differ very significantly from the corresponding energy of the strongest (TO) PR for Si and SiGe, which, in both cases, is 58 meV.

#### 4.4.4.5 Background Information—Ge NC PL

A more detailed description of the PL properties observed with Ge NCs can be found in [9]. Carrier confinement, in principle, permits PL to be observed at higher temperature from NCs than in epilayers. For Sample (a) of Fig. 4.11, the NC—related SiGe broad peak persisted to about 100 K, but weakened with increasing temperature. Exciton diffusion and binding effects are seen in the PL through both the temperature dependence and excitation power variation. Although the NCs are small, they are efficient at capturing excitons [56]. Exciton diffusion effects are apparent in the temperature variation of PL spectra like those in Fig. 4.12, where the PR PL disappears by 20 K while the broad PL intensity was still apparent to higher temperatures. The broad PL displayed decay times (see below) and radiative efficiency [49] that were relatively large compared to the PR PL. Saturation with excitation was seen in the PL when the exciton and NC concentrations become similar.

The NC peak, present for power densities from 3 to 1.5 W/cm$^2$, increased sub-linearly in intensity with excitation density, and shifted slightly to higher energy with increasing excitation power. The variation with well location for a mixed PR/NC PL, three-well sample (68 Å wells) is illustrated in Fig. 4.13, which shows two PL spectra. The lower spectrum (a), taken from an as-deposited sample displays both the PR and NC peaks. Spectrum (b), from a sample of the same material after a Schimmel etch,

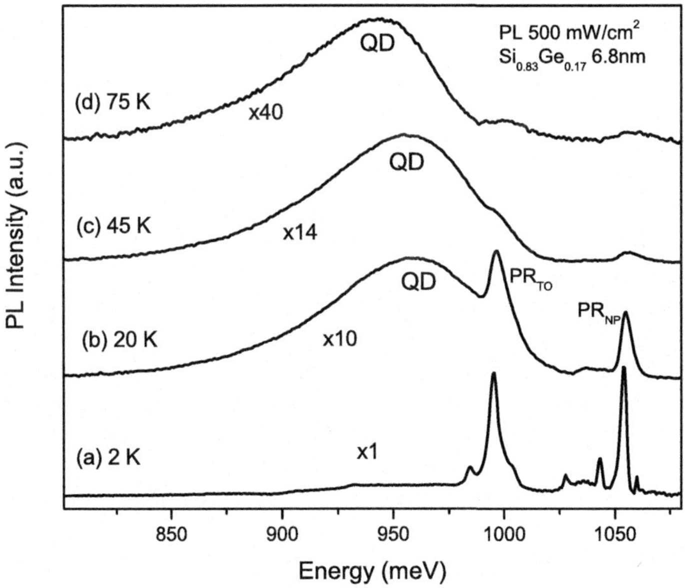


**Fig. 4.12** Temperature dependence of the PL for a MQW consisting of three 6.8 nm wells of $Si_{0.83}Ge_{0.17}$ with 20 nm Si spacer layers. As the temperature increases the bound exciton NP peak changes to a free exciton NP peak and the NC PL peak grows in intensity

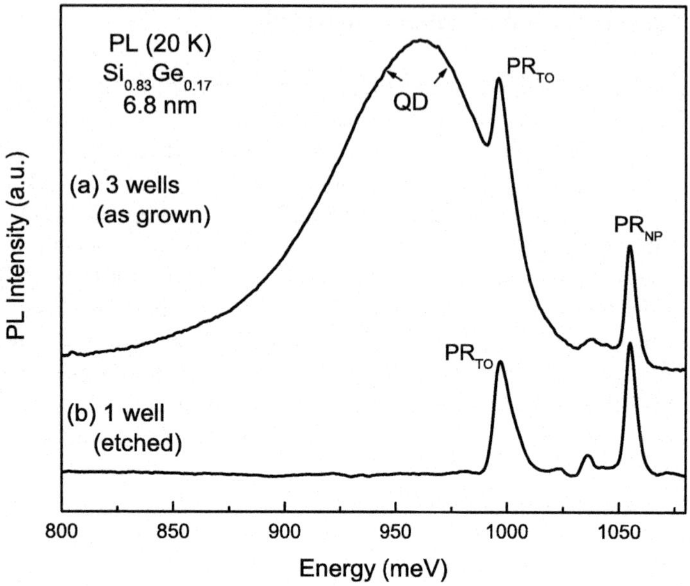


**Fig. 4.13** Variation of the PL spectrum with well position for a MQW sample consisting of three 6.8 nm wells of $Si_{0.83}Ge_{0.17}$ with 20 nm Si spacer layers. Spectrum (a), from an as-deposited sample, contains both the PR and NC PL peaks. In Spectrum (b) only the PR peaks are present as this spectrum is from a version of sample that has been processed such that only the first QW remains

had only one quantum well left. After etching, the NC PL peak was not present consistent with the nucleation of NCs during epitaxy only where the strain variations were higher, i.e., farthest from the substrate.

The PL time decay is shown in Fig. 4.14 Low temperature (5 K) PL spectra are shown for a 3.3 nm thick $Si_{0.55}Ge_{0.45}$ ten-layer structure grown at 675 °C. The PL spectra shown were obtained at various times beginning with the point when the exciting laser was turned off. The subsequent PL from the Ge NCs shown in Fig. 4.14 weakened with essentially the same lineshape but shifted towards lower photon energy. The decay took place in approximately compared with the substrate PL that disappeared after a few microseconds. The NC PL peak shifted to lower energy with increasing decay time (see Fig. 4.14). Similar results to those shown in Fig. 4.14 were observed for several other MBE-grown SiGe samples [9].

There have been relatively few reports of electroluminescence (EL) from Si/Ge NCs [57–59]. In the first observation of EL from NC's in MBE grown SiGe [3], the *p-n* $Si_{0.82}Ge_{0.18}$/Si heterostructure shown in Fig. 4.19 was used to fabricate light-emitting diodes. The structure grown by MBE was doped in situ [60–62]. Further details of the device fabrication can be found in [3]. The emission intensity from the diodes was proportional to the forward current density with a peak energy that varied with the Ge fraction. An EL spectrum is shown in Fig. 4.19, along with the corresponding PL spectrum. A substrate peak was seen at 1070 meV only in PL.

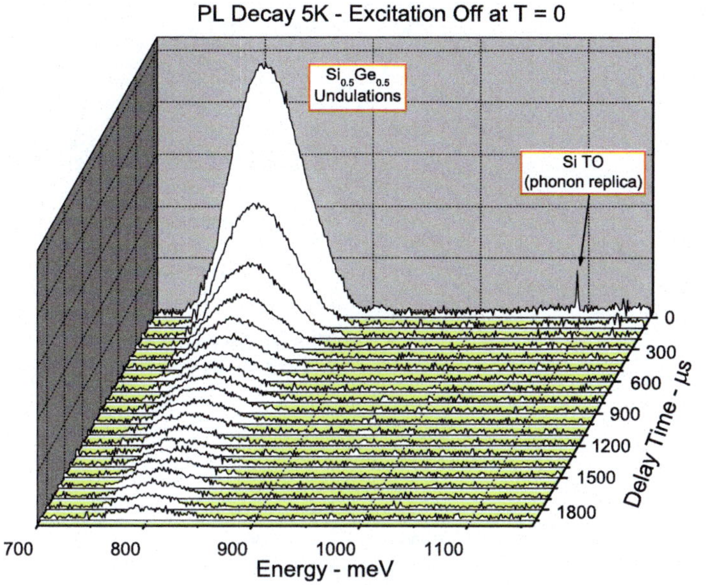


**Fig. 4.14** Time decay of the low temperature PL for a 3.3 nm thick $Si_{0.55}Ge_{0.45}$ ten-layer structure grown at 675 °C. The spectra were obtained after the excitation was turned off. The PL from the Ge NCs decayed towards lower photon energy in ~200 μs unlike the rapid substrate (labeled Si TO) PL decay

By comparing the EL and PL intensities, the EL quantum efficiency was seen to be 0.3 times that of the PL (see Sect. 4.4.13). Otherwise, the EL and PL spectra were similar in most respects, i.e., the same peak energies, a persistence to ~80 K, and with comparable internal quantum efficiencies. Such similarities point strongly to a common luminescence mechanism, i.e., excitonic recombination in the Ge NCs (Fig. 4.15).

#### 4.4.4.6 Phonon Replica Analysis

As was shown in Fig. 4.11a, the broad PL peak's width and asymmetric shape can be curve resolved into two symmetric peaks, separated by ~5 meV, i.e., very near the momentum conserving TO phonon energy for Ge. The NP peak is wide (25–45 meV) due to NC confinement variations arising from size variability and to alloy disorder broadening in the SiGe. In Ge NCs, two main PL peaks are expected, a NP line and a LA replica line, separated by about 28 meV. So the NP-PR energies provided by curve resolving are relatively close to that (28 meV) for the most intense (LA) PR for Ge, but differ very significantly to the corresponding energy of the strongest (TO) PR for Si and SiGe, which, in both cases, is 58 meV. Shown in Fig. 4.16 is a summary of the results from curve resolving two Gaussian peaks within the envelope of the broad PL peak. The number of samples with peak separations within 5 meV of each other

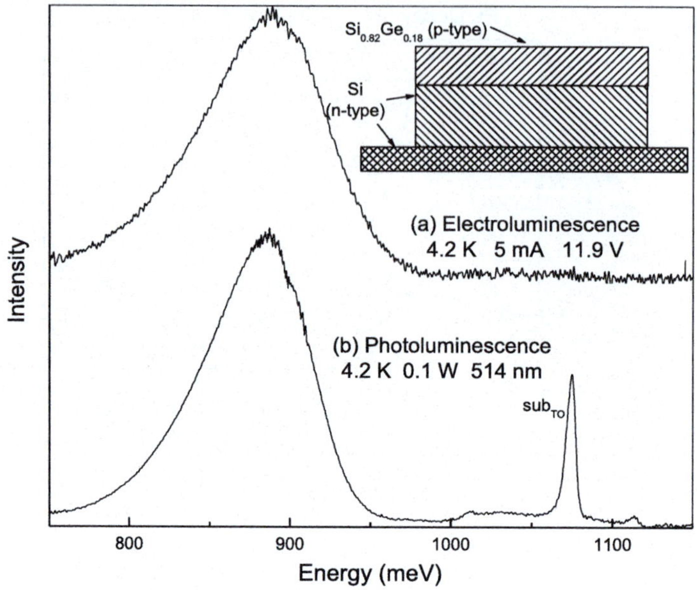


**Fig. 4.15** Electroluminescence (**a**) and PL (**b**) spectra at 4.2 K for the $Si_{0.82}Ge_{0.18}$ *p-n* diode structure. Inset: Schematic of the heterostructure processed into light-emitting diodes

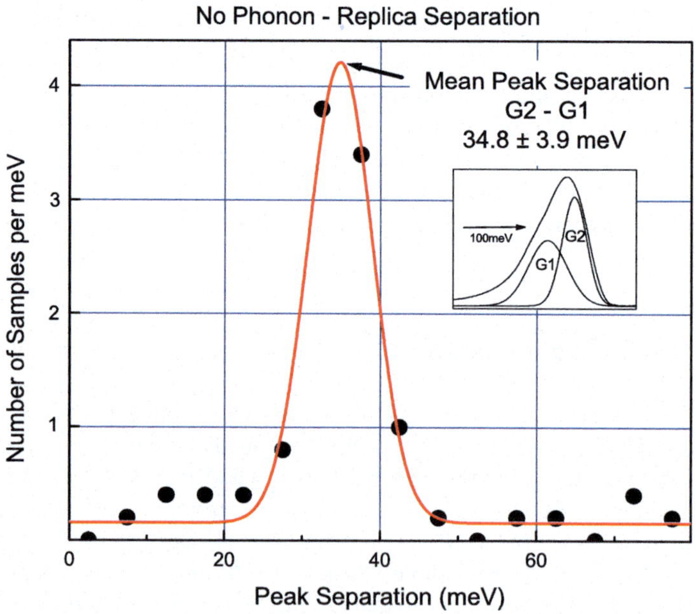


**Fig. 4.16** Number density for the PL samples versus the energy separation of the no phonon and phonon replica peaks, as evaluated using the curve fitting of two gaussian peaks within the envelopes of the broad PL peaks (see inset)

versus peak separation is shown. For the samples studied here, the mean separation is 34.8 ± 3.9 meV, which is between the LA phonon energy in bulk $Si_{1-x}Ge_x$ (30 meV for $x > 0.85$) and the same material's $TO_{Ge-Ge}$ energy (36 meV for $x > 0.85$) [69]. In bulk Si, the dopant type has a strong effect on the relative strengths of the NP and PR lines. In NCs, dopant effects are not significant with the NP- PR ratios dominated by localization effects that allow phonon-less, electron-hole recombination.

#### 4.4.4.7 NC Properties Resulting from the Conditions for Epitaxy

Referring to Fig. 4.17, the fact of epitaxial growth implies that the SiGe epilayers and the Ge NCs are lattice matched to Si (001) in the *x–y* plane, so that both the SiGe and NCs are under compression in that plane. However, the growth of SiGe is unconstrained in the vertical (*z*) direction. Therefore, the SiGe epilayer is under tensile strain in this direction, leading to a vertical lattice constant that steadily increases with Ge fraction and is larger than that for unstrained SiGe. Here the volume of the unit cell for the epitaxial SiGe layer is assumed to be the same as that for unstrained cubic (bulk) SiGe. The next important assumption is that the lattice constant of the Ge NCs matches that of the SiGe in all three directions, i.e., including the growth direction. For relatively dilute SiGe, the Ge NCs are under compression vertically, but for increasing Ge content in the SiGe epilayer the vertical lattice constant of the strained SiGe eventually exceeds that of bulk Ge. The point where the vertical strain in the Ge NC first becomes tensile occurs when the Ge fraction in the SiGe material is 0.36 (Fig. 4.18). With these constraints it is possible to write an expression for the strain in the *z* direction within the Ge NCs as:

$$\varepsilon_z = \frac{l_c^{\mathrm{Ge}} - l_c^{\mathrm{Si\,Gez}}}{l_c^{\mathrm{Ge}}} \tag{4.2}$$

where $l_c^{\mathrm{Ge}}$ is the bulk Ge lattice constant (5.658 Å) and $l_c^{\mathrm{SiGez}}$ is the strained lattice constant in the vertical direction for the tetragonally distorted SiGe material, which

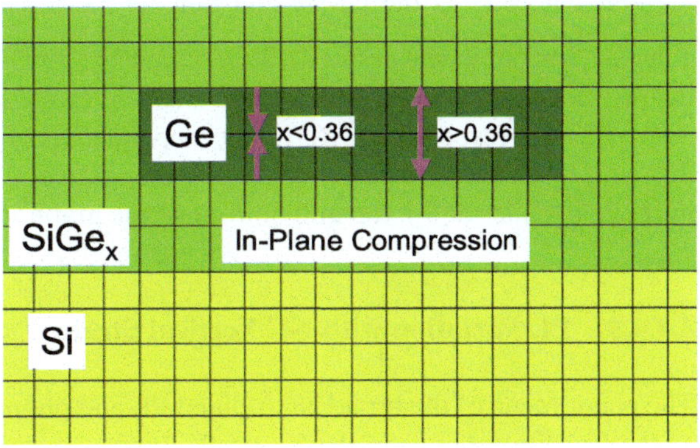


**Fig. 4.17** Cross sectional schematic for the *x–z* plane of lattice matched Ge NCs within SiGe on Si(001)

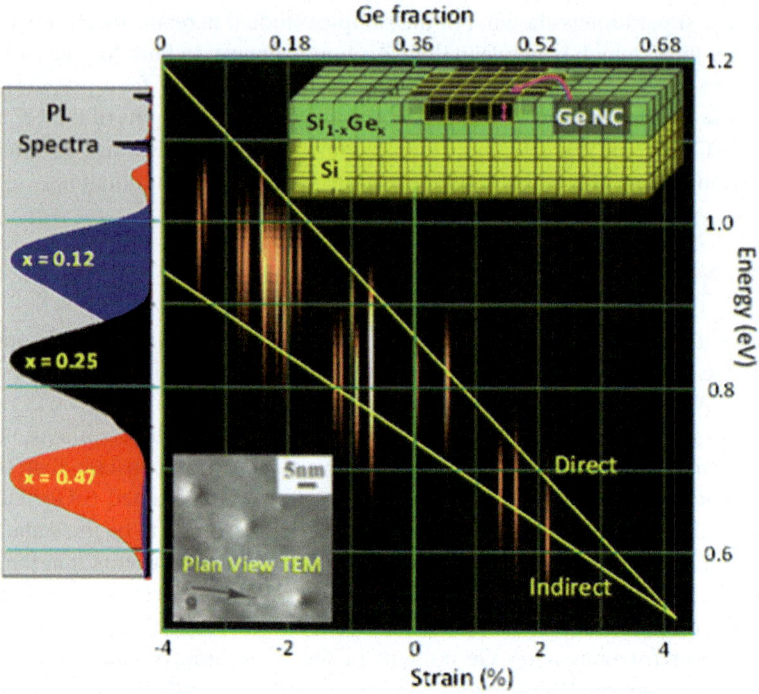


**Fig. 4.18** PL energy distributions for the broad PL versus the uniaxial vertical strain in the Ge NCs present in the $Si_{1-x}Ge_x$ layers, as depicted in the upper inset and shown in the plan view TEM image inset. The upper and lower solid lines are the direct and indirect bandgap energies for Ge calculated using deformation theory for uniaxial strain. The top horizontal scale is Ge fraction ($x$) from which the bottom scale strain values are derived using 4.2. On the left are shown three low-temperature NC PL spectra for different epilayer Ge concentrations

is lattice matched to Si in the horizontal plane. Note that the strain value is negative for compression and positive for tension. With increasing Ge fraction, $x$, this lattice constant varies from 5.431 to 6.141 Å as given by the following expression: $l_c^{\mathrm{SiGez}} = \frac{\left(l_c^{\mathrm{Si\,Geu}}\right)^3}{\left(l_c^{\mathrm{Si}}\right)^2}$, where $l_c^{\mathrm{Si}}$ is the lattice constant for Si (5.431 Å). Here $l_c^{\mathrm{SiGeu}}$ is the lattice constant for cubic (unstrained) SiGe given by: $l_c^{\mathrm{Si\,Geu}} = l_c^{\mathrm{Si}} + c_1 x + c_2 x^2$. In this expression, $x$ is the Ge fraction and the constants, $c_1$ and $c_2$, are equal to 0.2 Å and 0.027 Å, respectively [63]. The dependence of strain on the Ge-fraction for this simple model is readily derived for the vertical direction, as represented by the horizontal scales at the top and bottom of the graph in Fig. 4.18.

#### 4.4.4.8 PL Variation with NC Vertical Strain

The phenomena of the broad NC-related PL and phonon-resolved SiGe PL—as well as the transition between these two PL types—was illustrated in Fig. 4.10, where for thicker epilayers the TEM images show that Ge NCs had formed, causing the broad PL line. The box on the right depicts the PL obtained from three samples with essentially the same Ge fraction, with the plan-view TEM pictures obtained from samples 1 and 3 shown on the left. While in such Ge NCs the in-plane strain is constant, the vertical (epilayer growth direction) strain varies from compressive to tensile with increasing epilayer Ge-fraction.

There are several factors that must be evaluated to predict the Ge NC peak PL energy:

(a) Ge bandgap variation with vertical strain.
(b) Effect of the Ge-fraction in the host SiGe epilayer on the vertical lattice constant.
(c) Effect of strain on the SiGe bandgap energy.
(d) Confinement shifts in the Ge NCs.
(e) Confinement shifts in the SiGe epilayers.

Models incorporating these effects were developed that effectively explained the PL energy dependence over a wide range of SiGe samples with a multitude of separate epilayer configurations, for which the experimental data is shown in Fig. 4.18 [55, 64]. There, the PL energy distributions are shown for the Ge-NCs versus the vertical strain for the Ge NCs embedded within the $Si_{1-x}Ge_x$ layers. The upper and lower solid lines are the direct and indirect bandgap energies for Ge calculated using deformation theory for uniaxial strain. The top horizontal scale is Ge-fraction in the host $Si_{1-x}Ge_x$ layers. The upper inset in Fig. 4.18 depicts a Ge-NC within a SiGe layer and the lower inset contains a plan view TEM image of the Ge-NCs. On the left side of the figure, the PL spectra are shown for three Ge-NC samples with different Ge-fractions in the host $Si_{1-x}Ge_x$ layers.

#### 4.4.4.9 Models for Bandgap Energy Versus Strain

Strain affects the material bandgap energy, with compression generally increasing this energy while tension reduces it. A simple method for calculating this effect is to use the deformation potential (DP) model with linear relationships between strain and the direct and indirect bandgaps:

$$E_{\mathrm{BG}} = A + B\varepsilon_z \tag{4.3}$$

Here, for strain in per cent, the intercept and slope in (4.3) are 880 and −79.3 meV for the direct bandgap and 740 and −45.5 meV for the indirect gap at low temperature [39]. For example, for a tensile strain of 2%, the direct BG energy is 721 meV and the indirect BG is 649 meV.

In the DP model, as the strain becomes more strongly tensile both the direct and indirect gap energies decline, with the direct energy decreasing more rapidly than the indirect one. As shown in Fig. 4.19, the direct gap energy in the DP model crosses over the indirect gap energy at a tensile uniaxial strain of 4%, resulting in

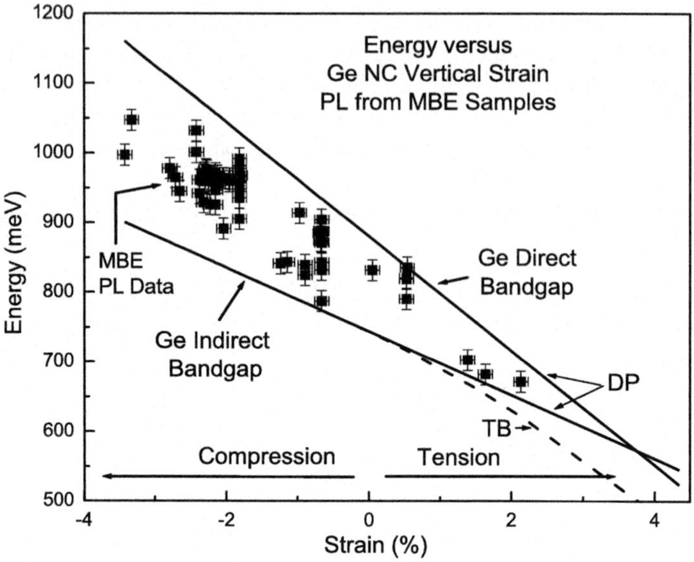


**Fig. 4.19** NP PL peak energy and bandgap (direct and indirect) energies of Ge versus Ge NC vertical strain. For the indirect gap under tensile strain, the energy has been calculated using both the DP and TB theories

Ge becoming a direct gap semiconductor. Due to inadequacies of the DP model, the actual crossover point is for tensile strains above 5%, as demonstrated recently [42]. Nonetheless, the present PL results point the way to this transition point, with the maximum vertical tensile strain for the present Ge NCs being under 2%. However, the fact that PL is seen at energies below the indirect bulk Ge BG is explained by the Ge NCs being under vertical tension, reducing their BG. As discussed later, there are other non-linear models for the bandgap energy such as the tight binding (TB) model [39], for which Fig. 4.19 contains the indirect bandgap calculation under tensile strain.

For the bandgap curves calculated versus strain in Fig. 4.19 the effects of quantum confinement have not been included. The data points were obtained by curve resolving the PL of the MBE-grown SiGe samples. In Fig. 4.19 the points are Ge NC NP PL peak energies and the solid lines are the direct and indirect NC BGs (4.3). The NP experimental peak energies have been obtained from the broad PL distributions depicted in Fig. 4.18 by fitting Gaussian functions to the observed PL spectra. In Fig. 4.18 and in Fig. 4.19 it is noted that the data falls generally above the indirect energy BG for uniaxially strained Ge, a differential which is most likely due to quantum confinement, as will be discussed next. This is a difference that appears to decrease as compression decreases, although there are only a few data points for the larger tensile strains. The energy uncertainties in the data points are those obtained from curve resolving the PL peaks. The strain uncertainties for these points are derived from the x-ray diffraction measurements of the composition and thickness of the SiGe layers. Figure 4.18 shows the color-weighted actual distributions in energy for the broad NC PL peaks, versus vertical NC strain. For a number of the strain values, there are several samples for which the PL spectra overlap giving rise to composite distributions that appear to be relatively intense.

#### 4.4.4.10 Calculation of Bound Exciton Energy for Ge NCs

To test the NC hypothesis further, the emission energy was calculated in a simple numerical model that accounts for the effects of strain on the Ge BG and of the confinement blue shifts in the exciton PL spectra for both the SiGe layers and the imbedded NCs, with the latter of a single thickness taken to be 2.5 nm high in the growth direction. In this model, the vertical direction is assumed to provide the lowest energy BG. A double well structure was used, consisting of both Ge and SiGe wells, with the smaller Ge wells completely imbedded in SiGe, as illustrated in Fig. 4.17. In such a system there are related confinement shifts in both the Ge NCs and the SiGe material. While the SiGe layer thicknesses were obtained from structural measurements, as discussed previously, the NC height is not known and can be treated as somewhat of a free variable. In the formulation, the effect of vertical strain on the NC BG was incorporated, first using a DP approach and later with TB theory.

Figure 4.20 depicts the shooting method [65] used in calculating the bound exciton energy in the double well Ge-SiGe/Si layered system. The shooting method is used for solving one-dimension Schrodinger-type equations for the bound energy with arbitrary potentials. Here an initial exciton energy was chosen and, using finite differences, the wave function was calculated numerically throughout a layered system. This wave function calculation was extended to distances far from the starting point. If the correct energy had been chosen, the wave function would have gone to zero for large distances, otherwise it would have diverged and another energy would have had to been chosen. The left panel in Fig. 4.20 shows the result for a $Si_{0.85}Ge_{0.15}$ quantum

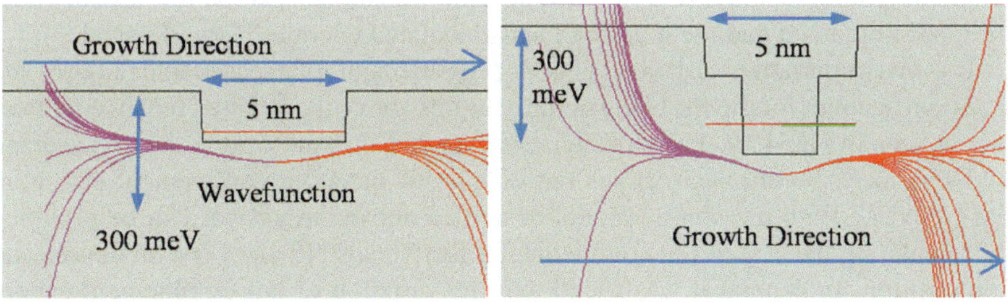


**Fig. 4.20** Shooting method results for calculating the bound exciton energy in a single well (left) SiGe/Si layered system and a double well (right) Ge-SiGe/Si layered system. The left panel depicts the result for a $Si_{0.85}Ge_{0.15}$ quantum well 5 nm thick. On the right, a 2.5 nm thick Ge NC has been added in the middle of the original SiGe well, which resulted in a reduction of the lowest bound exciton energy by 130 meV compared with the single well case

well 5 nm thick where the exciton energy obtained is approximately 1054 meV. On the right, a 2.5 nm thick strained Ge NC has been added. The lowest bound exciton energy in that case is 924 meV. The background energy is that of substrate bound excitons in silicon (1150 meV). The well structure shown illustrates the reduction in that energy due to the presence of Ge in the SiGe alloy layer and/or in the Ge NC. The horizontal lines in the quantum wells represent the energy solutions, obtained when the wavefunction is well behaved (goes to zero) far away from the quantum well in the growth direction. The exciton energy is of primary importance as it is the quantity that PL measures. The goal here is to compare this calculated no-phonon energy, which contains the confinement and strain BG shifts as well as exciton binding energies, to the observed PL energy.

#### 4.4.4.11 Comparison of Calculated Ge NC Energy with Measured PL Energy

The calculated energies for the samples of varying composition are compared with the measured PL energies in Fig. 4.21. Here for the solid square points a constant NC vertical size of 2.5 nm was used. In Fig. 4.21, the best-fit straight line for this size (shown by the dashed line) to the data is expected to have a slope near unity (the solid line in the figure) but is somewhat different from that value due to what are possibly lower order influences, such as variations in the NC size and deviations from the linear DP model for the strained BG. However, there is general agreement between the computed emission energies and those observed in PL, which provides reasonable validation for the hypothesis that imbedded Ge NCs have given rise to the intense, broad PL observed in MBE grown SiGe.

Figure 4.21 contains a comparison of the theoretical and experimental energies for Ge NC PL from SiGe samples with Ge fractions from 0.05 to 0.53. The dashed straight line is a linear least-squares fit to the results for the 2.5 nm NCs with the DP model and the dot-dash line is the similar fit for the TB model. It is noted that the two bandgap models do not give very different results in this respect for the range of strains available in the sample set. The solid line represents the result that would have been obtained had the measured and calculated energies been identical.

It seems important to test whether or not confinement for a reasonable range of NC sizes can account for the PL being at the energies above the strained bulk Ge indirect BG, as seen in Fig. 4.19. This difference is in the 50–150 meV range with an average of 103.4 meV. So are such values reasonable for quantum confinement effects in the Ge NCs? With the simple model described above, an answer can be provided by calculating the size of the confinement effect versus NC size, QW thickness and composition. In general it was found that the variation of NC confinement energy with thickness of the host SiGe was not that large (<5 meV) and the variation with Ge-fraction although larger was not a major contributor to the difference. For example, the NC confinement shift for a 1.5 nm NC in a 5 nm thick well was 110 meV for a Ge-fraction of 0.15 and 130 meV for a fraction of 0.50, everything else being equal. So it is seen that NC sizes in the range from 1 to 3 nm—corresponding to confinement

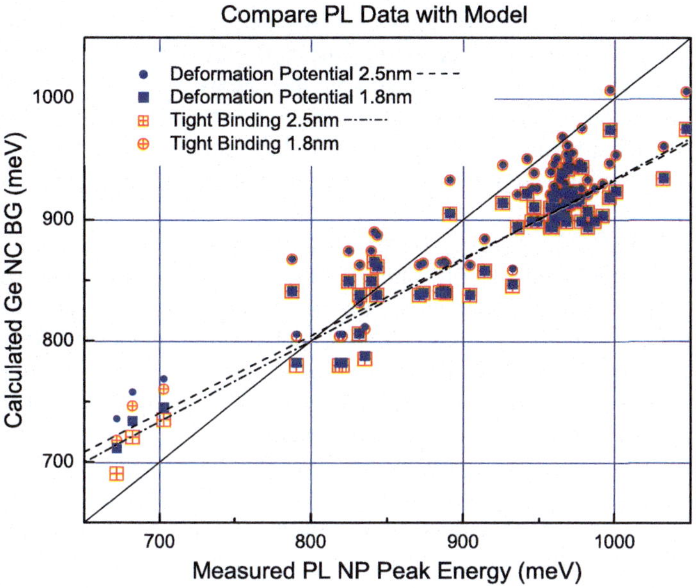


**Fig. 4.21** Theory versus experiment for the Ge NC PL from SiGe samples with Ge fractions from 0.05 to 0.53. Calculated energies for two BG models (DP and TB) for all samples and two NC sizes (1.8 and 2.5 nm). These results are plotted versus the measured NP energies obtained from the least squares fits of Gaussian peaks to the experimental PL data. Note that, for samples in which the Ge NCs are under compression vertically, the DP theory and TB model give identical results for the strained bandgap. The solid line has a slope of unity indicating where there would be exact agreement between experimental and calculated energies. The dashed line represents the best linear fit to the DP results for 2.5 nm NCs and the dot-dash line is the fit for the TB case

energies from 120 to 50 meV—can indeed account for all the blue shifts from the predicted Ge NC BG that is seen in the PL energies. This means that confinement shift provides an explanation for the difference seen between the measured PL energy and the NC bandgap. But it appears from the distribution of points in Fig. 4.21 that the average NC thickness of 2.5 nm that was employed might have been an overestimate. A value nearer 1.5 nm would appear to be more reasonable, leading to the question as to what influences and/or limits this size. Growth conditions should of course be critical in this respect.

There are other assumptions in the present simple model that must be examined, including whether or not the DP model describes adequately the strain effect. Other authors have indicated that a TB model could prove to be more appropriate [39], where for high tensile strain (>1%) the latter model provides significantly lower indirect BGs in Ge than does the DP approach. In that regard, we now discuss the results obtained for the two models and two NC sizes (1.8 and 2.5 nm) for all the samples, as illustrated in Fig. 4.21.

After completing the calculation for two models (DPs and non-linear TB) for all the samples and the two NC sizes (1.8 and 2.5 nm), the results in Fig. 4.21 were obtained for the calculated (vertical axis) versus measured (horizontal axis) energies. The solid line, which is for reference only, has a slope of unity and goes through the origin. In all cases the models overestimate the energy at the low end of the scale (corresponds to tensile strain) and underestimate it at the high end (compression). It appears the tensile-strained NCs are larger in the *z* direction than the nominal size, while the more compressed NCs are relatively smaller, with this trend skewing the curve in Fig. 4.21.

As seen in Fig. 4.22, the NC size trends with the *z*-direction strain, leading to a variation in the confinement shift. This effect is calculated to see if the sizes needed to explain the measured-model energy difference are reasonable. As is well known the confinement shift changes with NC size such that the smaller the NC, the larger the confinement shift, etc. The information needed to compute the variation in energy shift can be obtained as follows. For a given model with two NC sizes, the confinement shift per nm can be estimated in this range, which averages out to be approximately 39 meV/nm and is approximately constant across the energy scale. Now if it is known, for example, that a sample with a NP PL energy at 672 meV and a larger calculated BE energy of 718 meV (TB model for a NC thickness of 1.8 nm) would have a

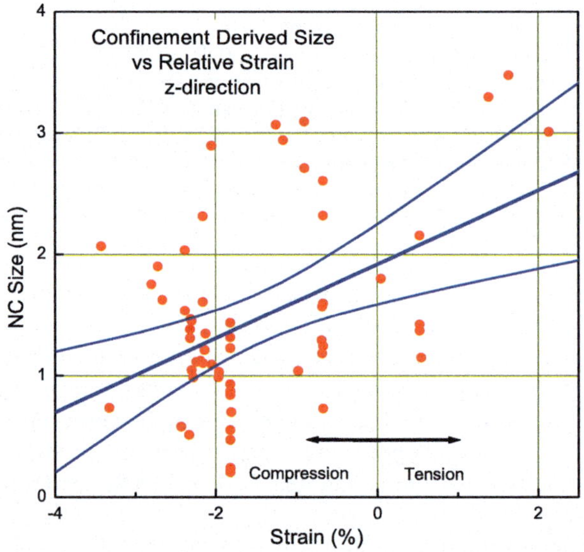


**Fig. 4.22** Confinement adjusted Ge NC size is plotted versus the relative strain in the *z*-direction. Negative abscissa values correspond to compression in the *z*-direction. The size is deduced from the amount of confinement shift needed to bring the calculated exciton energy in line with the observed NP energy. The DP model results shown in Fig. 4.21 have been used to obtain the size variation of the confinement energy (~39 meV/nm). By dividing this value into the difference between the calculated energy for a NC size of 1.8 nm and the observed NP peak energy, an amount is obtained to be added to the nominal NC size (1.8 nm) to deduce the NC size for which the theoretical energy equals that of the NP peak. The upper and lower curved blue lines correspond to the 95% upper and lower confidence bands for the linear fit

thickness dimension of 3.0 nm that is larger than the nominal one of 1.8 nm. On the other hand, another typical sample with a measured PL energy of 1047 meV but with a smaller calculated one of 1006 meV (for 1.8 nm NC thickness) would at 0.8 nm thickness be smaller than the nominal one of 1.8 nm: And so on. This procedure can be applied to all the samples to make the model energies equal to the measured ones. The NC sizes obtained in this manner are plotted versus relative strain in Fig. 4.22. This process results in an amplification of the scatter but the trend can still be seen. When this variation is fitted with a linear function, the NC size is seen to vary from about 1 nm for a compressive strain of 4% to nearly 3 nm for a tensile strain of 2%.

#### 4.4.4.12 Ge NCs Strain Compensated with C

Although there have been significant efforts regarding Ge nanostructures [4, 66], the most comparable example to the one described above is that of the Stuttgart group [7, 67]. In their work [67], fifty layers of C-layer-induced Ge NCs separated by 9.6 nm thick layers of Si were studied via PL. Each dot layer consisted of a 0.2 monolayer (ml) of C and 2.4 mls of Ge, resulting in C-induced Ge NCs 10–15 nm in diameter and 1–2 nm in height. The carbon layer strain compensated [68] the germanium so that the Ge NCs were unstrained. In the later work [7] a similar structure was used, although the Ge was thinner at 2.2 mls. For 2.4 mls, the confinement shift amounts to 220 meV and for 2.2 mls 240 meV. In Fig. 4.23 the Stuttgart PL data [67] is

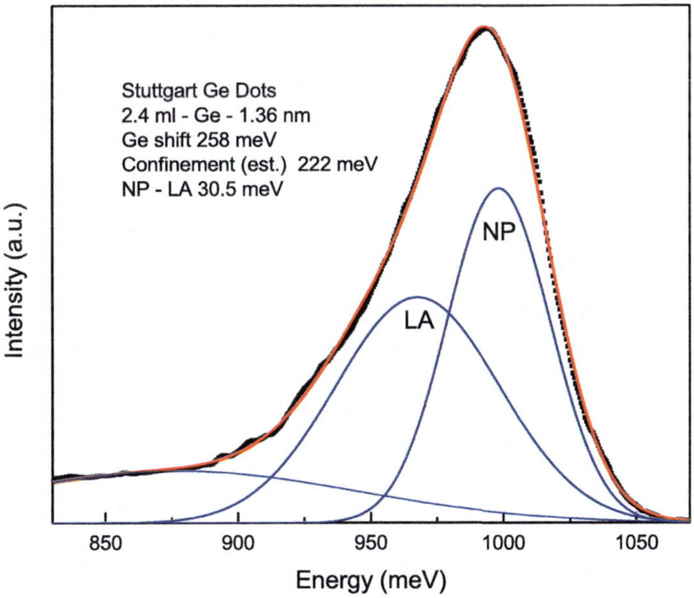


**Fig. 4.23** Curve resolving the Ge dot PL data from Stuttgart using three Gaussian functions

plotted; it has been curve resolved into three Gaussian peaks. Two of the fitted peaks are associated with the NP and primarily the LA replica lines, centered at 998.2 and 967.7 meV with widths of 38.2 and 61.7 meV, respectively. The NP-PR separation here is 30.5 meV, i.e., similar to Ge's bulk LA phonon value (28 meV) [69].

The spectral shape from the Stuttgart dots is similar to that of the NCs (compare with Fig. 4.10). The NP-PR separation (30.5 meV) is comparable to the value (34.8 meV) obtained for the NCs. The height of the Stuttgart dots, ~1.3 nm, is consistent with a NC height of about 1.6 nm for zero strain, as shown in Fig. 4.22. The lateral diameter of the Stuttgart dots, 10–15 nm, is somewhat larger than that of the present NCs (see Sect. 4.4.15). For such a range in diameter the lateral confinement effect would be relatively small, being ~10 meV. The material comprising the Stuttgart NCs does not appear to be the GeC alloy, which would have a carbon fraction of 0.11 [70] with a bandgap of approximately 1300 meV, which is 300 meV larger than the observed emission energies.

#### 4.4.4.13 Efficiency and Stability of NC Emission

Many of the Ge NC samples exhibited a very bright luminescence, with internal quantum efficiencies as high as 3%. The highest efficiency sample exhibited the PL spectra shown in Fig. 4.11b. This sample was comprised of a SiGe/Si multiple quantum well (MQW) structure, in which there were 40 epilayers of 7.6 nm thick $Si_{0.7}Ge_{0.3}$ separated by 20 nm of Si with the MQW structure grown on a Si (001) substrate. As the excitation wavelength was decreased (and thus the penetration depth) the PL peak shifted to higher energy, which suggested the smaller NCs were more concentrated in the layers near the upper surface of the MQW. This peak shift with excitation wavelength was monotonic with the peak energy given by 880–$0.12\lambda_{pk}$ meV for excitation wavelengths in nanometers. Comparing a recent PL spectrum of this sample with the spectrum taken at a much earlier date [2], the more recent data shows a spectrum that was remarkably stable in time, indicating that the NCs were unchanged structurally and enclosed in a benign environment (i.e., the SiGe eplayers). Higher excitation intensities shifted the PL peak to higher energy due to band bending.

It is of industrial interest to ascertain whether or not the sample's quantum efficiency for emission has changed significantly over the relatively long time since the sample growth. To this end a recent PL spectrum of this sample was compared with a Si sample with a large substrate PL signal. The Si signal from this sample was used as an intensity reference, since the NC Si signal was weak. In the NC spectrum, the integrated NC PL signal intensity per mW of excitation power was approximately 500 times as large as the substrate signal for the PR sample. Given a Si bound exciton emission efficiency of approximately 0.01% [49], an effective NC PL internal quantum efficiency is indicated of a few percent. This result in turn suggests the NC PL strength is constant with time.

#### 4.4.4.14 PL Linewidth Calculations—Exciton Penetration into Surrounding Medium

For the PL of bulk $Si_{1-x}Ge_x$ alloys, a simple expression [69] has been employed that describes the BE line broadening based on a Bernouilli concentration distribution [71] due to statistical alloy fluctuations within the exciton volume, as follows:

$$\Delta E_{BE} = 2.36\frac{dE_{gx}}{dx}\left[\frac{x(1-x)}{\frac{4}{3}\pi r_B^3 N}\right]^{1/2} \tag{4.4}$$

In (4.4), the derivative of the band gap energy versus $x$, $\frac{dE_{gx}}{dx}$, is 1.27 eV for $x > 0.85$, the Bohr radius ($r_B$) is that of germanium (11 nm) instead of that of silicon (3.8 nm), and $N$ is the density of lattice sites. The maximum linewidth for the germanium-rich material is approximately 6 meV ($\Delta E_B$) at $x \sim 0.94$ [69]. If in the case of Ge NCs the excitons are considered to be tightly confined to the NCs such that the exciton volume is reduced to essentially that of the NCs, it is possible to estimate a total NC volume—hence their size—from the BE NP linewidth. The many Ge NC samples in our collection yield an average BE PL linewidth of 65 meV ($\Delta E_{NC}$), practically independent of Ge fraction in the material surrounding the Ge NCs. In (4.4), above, the spherical exciton volume ($\frac{4}{3}\pi r_B^3$) is replaced by that ($\pi r_{NC}^2 h_{NC}$) of the NCs, assuming a cylindrical (platelet-shaped) NC with a height from confinement of 1.2 nm ($h_{NC}$). With these many assumptions and all other factors being equal, the expression for the NC lateral radius ($r_{NC}$) is

$$r_{NC} = \left[\frac{4}{3} r_B^3 h_{NC}^{-1}\right]^{1/2} \frac{\Delta E_B}{\Delta E_{NC}} \tag{4.5}$$

Using (4.5), the NC diameter is calculated to be 7.1 nm. This value is similar to the 5 nm seen in the plan-view TEM image (shown in Fig. 4.10), although this simple approach does not allow for cluster effects, which would only change the conclusion about the NC lateral size somewhat [72]. The number of germanium atoms, $N_{Ge}$, clustered in a NC is estimated to be approximately 260 using (4.6) and the Ge lattice constant ($a_{Ge} = 0.5658$ nm).

$$N_{Ge} = \pi \frac{r_{NC}^2 h_{NC}}{a_{Ge}^3} \tag{4.6}$$

#### 4.4.4.15 Optical Emission Images

The broad band PL was strong enough to be recorded with an uncooled infrared video camera from the NC sample for which data is shown in Fig. 4.11b. One such image is displayed in Fig. 4.24. In this case the laser excitation at normal incidence

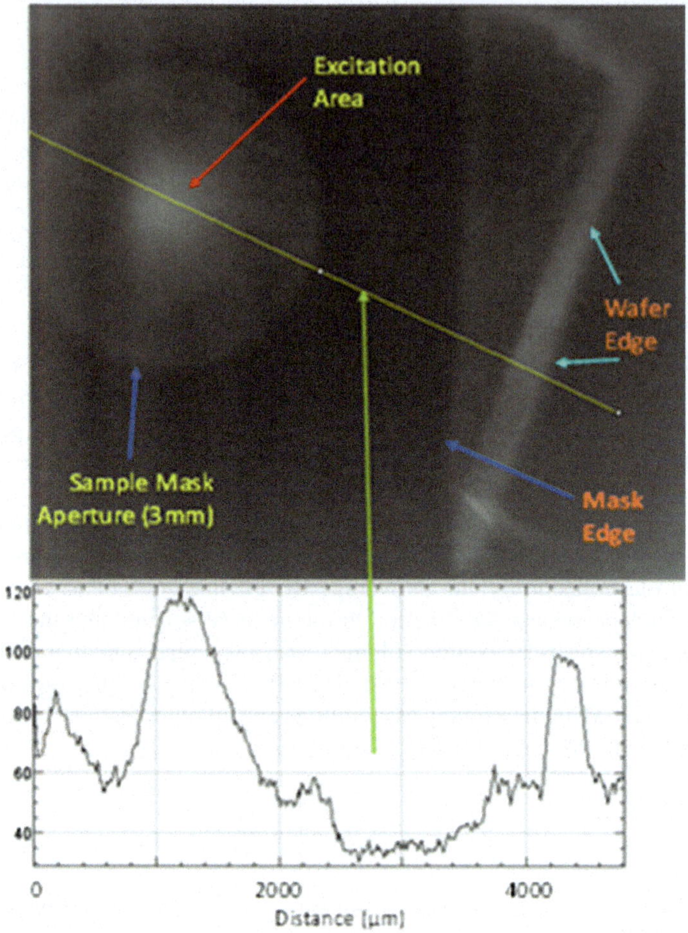


**Fig. 4.24** Broadband (1200–1900 nm) image obtained from the PL radiation, as excited with 100 mW of 514.5 nm laser light, 1 mm in diameter at the center of the aperture in the sample mask. An intensity profile from the excitation aperture to the sample edge is shown below the PL image

(at a wavelength of 514.5 nm with 100 mW power in a 1 mm diameter beam) was at the center of the 3 mm diameter sample mask aperture. The exciting radiation was excluded from the camera aperture with an optical filter. On the right of the sample mask in the image, the sample extends with its (110) edge protruding at a small angle (~20°) with respect to the edge of the sample mask. The PL was apparently brightest at the point of excitation, but was also quite intense at the edge of the sample, notably at the right hand edge. The trace below the image is the PL intensity profile along a line, which indicates an intensity at the edge of the sample comparable to that at the excitation location. In this experiment the incident excitation light was polarized roughly parallel to the wafer edge.

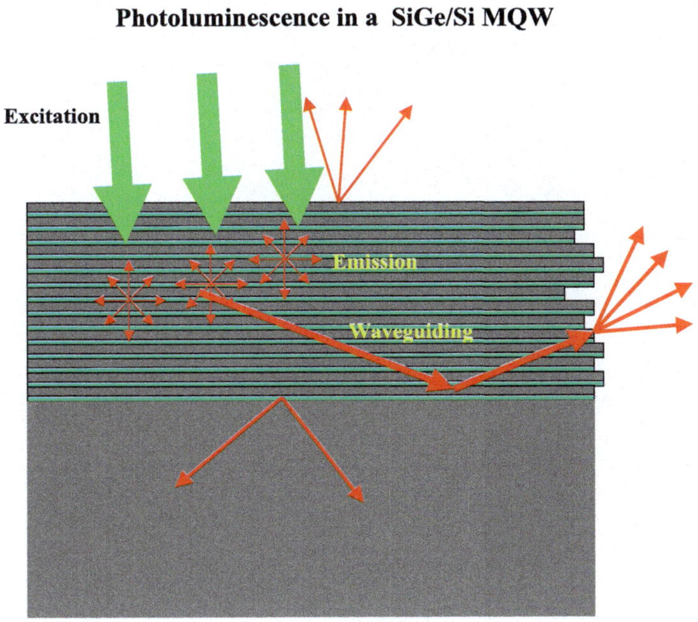


**Fig. 4.25** Schematic of waveguided PL in a SiGe MQW system

In the sample imaged, the MQW system was thick enough and had a high enough effective refractive index to act as a thin film waveguide [73] in the infrared. In addition to the waveguiding in the thin film structure, some bulk waveguiding was expected in the silicon substrate (see Fig. 4.25). The luminescence along the sample edge was brightest in a direction at ninety degrees to the excitation polarization, an elliptical intensity distribution also observed in fluorescence anisotropy [74].

### 4.4.4.16 Imaging with Focussed Excitation

A broadband PL image is shown in Fig. 4.26 for the sample excited with 100 mW of 514.5 nm laser light, focused to a spot 0.1 mm in diameter near the edge of the sample, where a second bright PL spot is also seen. Two intensity profiles are shown for the PL, one through both the excitation and edge spots and the other along the wafer edge. Edge brightening is observed to have also occurred, although the overall PL intensity is not greatly increased on focussing, possible due to NC saturation effects, which can occur when more than one exciton is captured by a Ge NC. Where and how such saturation occurs is related to the in-plane areal density of Ge NCs in the system. However, it is seen from Fig. 4.26 that the edge emission was in a spot about the same size as the emission at the excitation spot. This observation suggests that the emission was significantly collimated in the MQW waveguide, much more so than

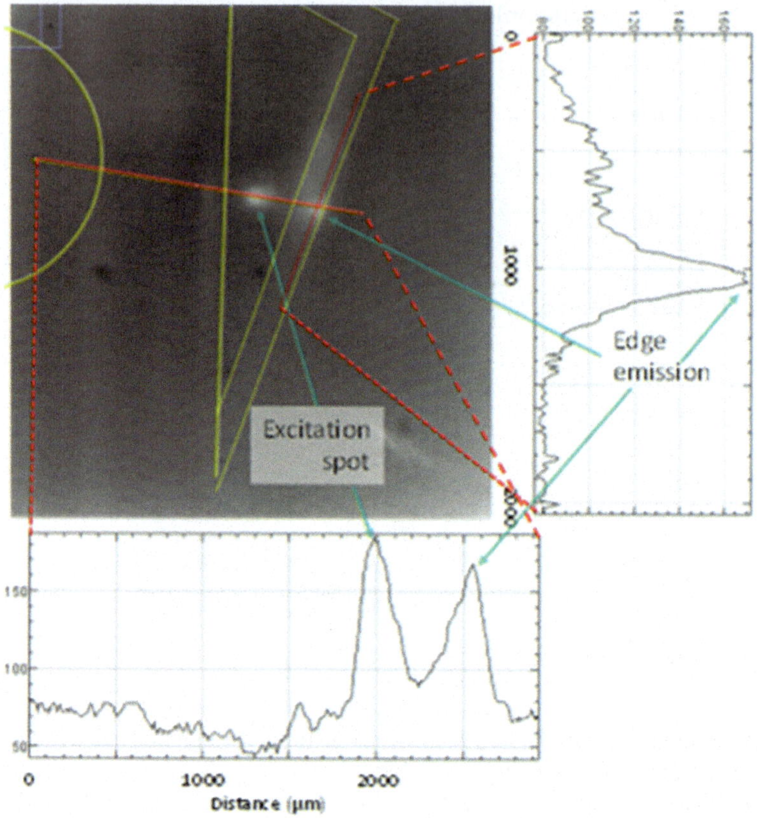


**Fig. 4.26** PL image excited with 100 mW of 514.5 nm laser light, focussed to a spot 0.1 mm in diameter near the edge of the sample, where a second bright PL spot is seen. Two intensity profiles are shown for the PL, one through both the excitation and edge spots and the other along the wafer edge

would have been expected from the simple directional effects based on excitation-emission polarization that are normally seen for bulk luminescence [74]. This lack of divergence between the excitation and the edge might be indicative of a resonant excitation process occurring in the MQW waveguide.

## 4.5 Prospects for CMOS Compatible Devices

Any mechanism that enhances emission from indirect gap materials is of interest, but clearly a process that depends on exciton diffusion and concentration can only be used in low-temperature devices. Fortunately higher temperature lasing and high efficiency PL has been achieved with Ge under tensile strain and/or with high doping

[17–21, 42] or by alloying with tin [22]. A practical Ge laser on silicon is considered such an important development to be included in various photonics roadmaps [75, 76]. In strained Ge, sufficient tensile strain reduces the relatively small indirect to direct bandgap (BG) difference to zero, making such Ge a direct gap material, albeit at a relatively low energy [42]. The use of alloying or novel crystalline symmetries possibly including confinement blue shifts have the potential to remedy this shortcoming. Our work on GE NCs indicates that it could now be possible with further appropriate device engineering to produce a stable room-temperature laser operating in the telecommunication band wavelength region from Si/Ge devices that could be readily produced in a CMOS factory.

## References


1. L.T. Canham, Appl. Phys. Lett. **57**, 1046 (1990)
2. J.-P. Noël, N.L. Rowell, D.C. Houghton, D.D. Perovic, Appl. Phys. Lett. **57**, 1037 (1990)
3. N.L. Rowell, J.-P. Noël, D.C. Houghton, M. Buchanan, Appl. Phys. Lett. **58**, 957 (1991)
4. Y.S. Tang, C.M. Sotomayor Torres, W.-X. Ni, G.V. Hansson, Superlattices Microstruct. **20**, 505 (1996)
5. S. Fukatsu, H. Sunamura, Y. Shiraki, S. Komiyama, Thin Solid Films **321**, 65 (1998)
6. L. Tsybeskov, K.D. Hirschman, S.P. Duttagupta, M. Zacharias, P.M. Fauchet, J.P. McCaffrey, D.J. Lockwood, Appl. Phys. Lett. **72**, 43 (1998)
7. K. Eberl, O.G. Schmidt, O. Kienzle, F. Ernst, Thin Solid Films **373**, 164 (2000)
8. D.N. Lobanov, A.V. Novikov, N.V. Vostokov, Y.N. Drozdov, A.N. Yablonskiy, Z.F. Krasilnik, M. Stoffel, U. Denker, O.G. Schmidt, Opt. Mater. **27**, 818 (2005)
9. J.-M. Baribeau, X. Wu, N.L. Rowell, D.J. Lockwood, J. Phys. Condens. Matter **18**, R139 (2006)
10. A. Karmous, I. Berbezier, A. Ronda, Phys. Rev. B **73**, 075323 (2006)
11. N.L. Rowell, D.J. Lockwood, G. Amiard, L. Favre, A. Ronda, I. Berbezier, M. Faustini, D. Grosso, J. Nanosci. Nanotechnol. **11**, 9190 (2011)
12. M.H. Nayfeh, S. Rao, N. Barry, J. Therrien, G. Belomoin, A. Smith, S. Chaieb, Appl. Phys. Lett. **80**, 121 (2002)
13. L. Tsybeskov, G.F. Grom, R. Krishnan, L. Montes, P.M. Fauchet, D. Kovalev, J. Diener, V. Timoshenko, F. Koch, J.P. McCaffrey, J.-M. Baribeau, G.I. Sproule, D.J. Lockwood, Y.M. Niquet, C. Delerue, G. Allan, Europhys. Lett. **55**, 552 (2001)
14. Q. Wan, T.H. Wang, M. Zhu, C.L. Lin, Appl. Phys. Lett. **81**, 538 (2002)
15. B.V. Kamenev, G.F. Grom, D.J. Lockwood, J.P. McCaffrey, B. Laikhtman, L. Tsybeskov, Phys. Rev. B **69**, 235306 (2004)
16. Y.C. King, T.-J. King, C. Hu, IEEE Trans. Electron Devices **48**, 696 (2001)
17. R.E. Camacho-Aguilera, Y. Cai, N. Patel, J.T. Bessette, M. Romagnoli, L.C. Kimerling, J. Michel, Kimerling. Opt. Express **20**, 11316 (2012)
18. R. Koerner, M. Oehme, M. Gollhofer, M. Schmid, K. Kostecki, S. Bechler, D. Widmann, E. Kasper, J. Schulze, Opt. Express **23**, 14815 (2015)
19. J.R. Sánchez-Pérez, C. Boztug, F. Chen, F.F. Sudradjat, D.M. Paskiewicz, R.B. Jacobson, M.G. Lagally, R. Paiella, PNAS **108**, 18893–18898 (2011)
20. J. Petykiewicz, D. Nam, D.S. Sukhdeo, S. Gupta, S. Buckley, A.Y. Piggott, J. Vučković, K.C. Saraswat, Nano Lett. **16**, 2168–2173 (2016)
21. K. Sawano, X. Xu, S. Konoshima, N. Shitara, T. Ohno, T. Maruizumi, ECS Trans. **75**(4), 191–197 (2016). https://doi.org/10.1149/07504.0191ecst
22. S. Wirths, R. Geiger, N. von den Driesch, G. Mussler, T. Stoica, S. Mantl, Z. Ikonic, M. Luysberg, S. Chiussi, J.M. Hartmann, H. Sigg, J. Faist, D. Buca, D. Grützmacher, Nat. Photonics **9**, 88 (2015)

23. S. Adachi, *Properties of Group-IV, III-V and II-VI Semiconductors* (Wiley, New York, 2005)
24. J.R. Chelikowsky, M.L. Cohen, Phys. Rev. B **14**, 556 (1976)
25. J.I. Pankove, *Optical Processes in Semiconductors* (Dover, New York, 1971)
26. E. Mayer, E.C. Lightowlers, J. Phys. C: Solid State Phys. **12**, L539 (1979); R.W. Martin, Solid State Commun. **14**, 369 (1974); E.F. Gross, B.V. Novikov, N.S. Sokolov, Fiz. Tverd. Tela **14**, 443 (1972); Sov. Phys. Solid State (Engl. Transl.) **14**, 368 (1972)
27. M. Allardt, VI. Kolkovsky, S. Kolodinski, N.V. Abrosimov, K. Irmscher, P. Clauws, J. Weber, K. Zuber, Solid State Phenomena Online: 2011-08-16, Vol. 178–179, pp 289–294. ISSN: 1662-9779. https://doi.org/10.4028/www.scientific.net/SSP.178-179.289
28. A.D Yoffe, Advan. Phys. **42**, 173 (1993)
29. J. Menéndez, M. Noel, J.C. Zwinkels, D.J. Lockwood, Phys. Rev. B **96**(12), 121201(R) (2017)
30. J. Menéndez, D.J. Lockwood, J.C. Zwinkels, M. Noël, Phys. Rev. B **98**(16), 165207 (2018)
31. R.J. Elliott, Phys. Rev. **108**, 1384 (1957)
32. R.L. Hartman, Phys. Rev. **127**, 765 (1962)
33. Y.P. Varshni, Physica **34**, 149–154 (1967)
34. T.P. McLean, Prog. in *Semiconductors 3* (Heywood & Co. Ltd., London, 1960), p. 53
35. S. Saito, A.Z. Al-Attili, K. Oda, Y. Ishikawa, Semicond. Sci. Technol. **31**, 043002 (2016)
36. M.V. Fischetti, S.E. Laux, J. Appl. Phys. **80**, 2234–2252 (1996)
37. Cicek Boztug, José R. Sánchez-Pérez, Francesca Cavallo, Max G. Lagally, Roberto Paiella, ACS Nano **8**, 3136–3151 (2014)
38. A. Gassenq, S. Tardif, K. Guilloy, I. Duchemin, N. Pauc, J.M. Hartmann, D. Rouchon, J. Widiez, Y.M. Niquet, L. Milord, T. Zabel, H. Sigg, J. Faist, A. Chelnokov, F. Rieutord, V. Reboud, V. Calvo, J. Appl. Phys. **121**, 055702 (2017)
39. K. Guilloy, N. Pauc, A. Gassenq, Y.M. Niquet, J.M. Escalante, I. Duchemin, S. Tardif, G. Osvaldo Dias, D. Rouchon, J. Widiez, J.M. Hartmann, R. Geiger, T. Zabel, H. Sigg, J. Faist, A. Chelnokov, V. Reboud, V. Calvo, ACS Photonics **3**, 1907 (2016)
40. J. Menéndez, C.D. Poweleit, S.E. Tilton, Phys. Rev. B **101**, 195204 (2020)
41. S. Bao, D. Kim, C. Onwukaeme, S. Gupta, K. Saraswat, K.H. Lee, Y. Kim, D. Min, Y. Jung, H. Qiu, H. Wang, E.A. Fitzgerald, C.S Tan, D. Nam, Nat. Commun. **8**, 1845 (2017)
42. F.T. Armand Pilon, A. Lyasota, Y.-M. Niquet, V. Reboud, V. Calvo, N. Pauc, J. Widiez, C. Bonzon, J.M. Hartmann, A. Chelnokov, J. Faist, H. Sigg, Nat Commun. **10**, 2724 (2019)
43. A. Elbaz, D. Buca, N. von den Driesch, K. Pantzas, G. Patriarche, N. Zerounian, E. Herth, X. Checoury, S. Sauvage, I. Sagnes, A. Foti, R. Ossikovski, J.M. Hartmann, F. Boeuf, Z. Ikonic, P. Boucaud, D. Grützmacher, M. El Kurdi, Nat. Photonics **14**, 375–382 (2020)
44. E.M.T. Fadaly, A. Dijkstra, J.R. Suckert, D. Ziss, M.A.J. van Tilburg, C. Mao, Y. Ren, V.T. van Lange, K. Korzun, S. Kölling, M.A. Verheijen, D. Busse, C. Rödl, J. Furthmüller, F. Bechstedt, J. Stangl, J.J. Finley, S. Botti, J.E.M. Haverkort, E.P.A.M. Bakkers, Nature **580**, 205 (2020)
45. S. Hall, AIP Scilight (2017). https://doi.org/10.1063/1.4991019
46. M.R. Barget, M. Virgilio, G. Capelini, Y. Yamamato, T. Schroeder, J. Appl. Phys. **121**, 245701 (2017)
47. L. Carroll, P. Friedli, S. Neuenschwander, H. Sigg, S. Cecchi, F. Isa, D. Chrastina, G. Isella, Y. Fedoryshyn, J. Faist, Phys. Rev. Lett. **109**, 057402 (2012)
48. S.A. Srinivasan, C. Porret, M. Pantouvaki, Y. Shimura, P. Geiregat, R. Loo, J. Van Campenhout, D. Van Thourhout, Appl. Phys. Lett. **113**, 161101 (2018)
49. L.C. Lenchyshyn, M.L.W. Thewalt, J.C. Sturm, P.V. Schwartz, E.J. Prinz, N.L. Rowell, J.-P. Noël, D.C. Houghton, Appl. Phys. Lett. **60**, 3174–3176 (1992)
50. J.-P. Noël, N.L. Rowell, J.E. Greene, J. Appl. Phys. **77**, 4623–4631 (1995)
51. N.L. Rowell, in *Proceedings of International Conference on Raman and Luminescence Spectroscopy in Technology*, SPIE, San Diego, vol. 822, pp. 161–171 (1987)
52. Y.M. Niquet, G. Allan, C. Delerue, M. Lannoo, Appl. Phys. Lett. **77**, 1182 (2000)
53. N.L. Rowell, J.-P. Noël, D.C. Houghton, A. Wang, L.C. Lenchyshyn, M.L.W. Thewalt, D.D. Perovic, J. Appl. Phys. **74**, 2790–2805 (1993)
54. D.C. Houghton, J. Appl. Phys. **70**, 2136 (1991)

55. N.L. Rowell, D.J. Lockwood, D.C. Houghton, J.P. Noël, J.M. Baribeau, ECS J. Solid State Sci. Technol. **7**(12), R195–R205 (2018)
56. P.L. Gourley, J.P. Wolfe, Phys. Rev. B **20**, 3319 (1979)
57. O. Chrétien, T. Stoica, D. Dentel, E. Mateeva, L. Vescan, Semicond. Sci. Technol. **15**, 920 (2000)
58. L. Vescan, O. Chretien, T. Stoica, E. Mateeva, A. Mück, Mater. Sci. Semicond. Process. **3**, 383 (2000)
59. Z. Pei, P.S. Chen, S.W. Lee, L.S. Lai, S.C. Lu, M.-J. Tsai, W.H. Chang, W.Y. Chen, A.T. Chou, T.M. Hsu, Appl. Surf. Sci. **224**, 165 (2004)
60. J.-P. Noël, J.E. Greene, N.L. Rowell, S. Kechang, D.C. Houghton, Appl. Phys. Lett. **55**, 1525 (1989)
61. J.-P. Noël, J.E. Greene, N.L. Rowell, D.C. Houghton, Appl. Phys. Lett. **56**, 265 (1990)
62. D.C. Houghton, Appl. Phys. Lett. **57**, 1434 (1990)
63. J.P. Dismukes, L. Ekstrom, R.J. Paff, J. Phys. Chem. **68**, 3021 (1964)
64. N.L. Rowell, D.J. Lockwood, ECS Trans. **97**, 3 (2020)
65. J.R. Merrill, p. 191, Using Computers in Physics, Houghton Miffin, Boston, ISBN 0-395-21411-4 (1976); W.H. Press, S.A. Teukolsky, W.T. Vetterling, B.P. Flannery, Section 18.1, Numerical Recipes: The Art of Scientific Computing (3rd ed.), (Cambridge University Press, New York, 2007). ISBN 978-0-521-88068-8
66. N.L. Rowell, D.J. Lockwood, I. Berbezier, P.D. Szkutnik, A. Ronda, J. Electrochem. Soc. **156**, H913 (2009)
67. O.G. Schmidt, K. Eberl, S. Schieker, N.Y. Jin-Phillipp, F. Phillipp, J. Auerswald, P. Lamperter, Mat. Res. Soc. Symp. Proc. **533**, 171 (1998)
68. D.C. Houghton, G.C. Aers, N.L. Rowell, K. Brunner, W. Winter, K. Eberl, Phys. Rev. Lett. **78**, 2441 (1997)
69. J. Weber, M.I. Alonso, Phys. Rev. B **40**, 5683 (1989)
70. J. Kolodzey, P.R. Berger, B.A. Orner, D. Hits, F. Chen, A. Shao, M.M. Waite, S. Ismat Shah, C.P. Swann, K.M. Unruh, J. Cryst. Growth **157**, 386 (1995)
71. E.F. Schubert, E.O. Gobel, Y. Horikoshi, K. Ploog, H.J. Queisser, Phys. Rev. B **30**, 813 (1984)
72. V.I. Litvinov, Appl. Phys. Lett. **77**, 2210 (2000)
73. N.L. Rowell, G.I. Stegeman, Phys. Rev. Lett. **41**, 970 (1978)
74. E.D. Cehelnik, I.D. Mielenz, R.A. Velapoldi, J. Res. Natl. Bureau Stan. A. Phys. Chem. **79A**, 1–15 (1975)
75. J.-M. Baribeau, D.J. Lockwood, M.W.C. Dharma-wardana, N.L. Rowell, J.P. McCaffrey, Thin Solid Films **183**, 17–24 (1989)
76. M. Glick, L.C. Kimmerling, R.C. Pfahl, Optics and Photonics News, Mar (2018)

# Chapter 5
# Optical Spin Orientation in Ge-Based Heterostructures

**Simone Rossi, Elisa Vitiello, and Fabio Pezzoli**

**Abstract** The paradigm shift introduced by Si photonics has had a tremendous impact on data transfer and processing speed. Beyond communication, many exciting opportunities for value creation across diverse disciplines, applications, and industries have been rapidly introduced throughout the years. Novel phenomena merging photonics and spin-based electronics are just beginning to emerge. Ge-based heterostructures can tremendously expand our capabilities offering creative and unprecedented solutions to the challenges of integrating spin functionalities onto Si photonics. In this chapter, we will review the latest advancements in this field, paying particular attention to the unique coexistence of notable optical properties and superior spin-dependent characteristics in Ge and novel GeSn alloys.

## 5.1 Introduction

### *5.1.1 A Minimal Guide to Optical Spin Orientation*

In nonmagnetic materials, such as group IV semiconductors, the generation of an out-of-equilibrium spin polarization can be achieved by leveraging light-matter interaction. This phenomenon, termed optical spin orientation, is a direct consequence of the conservation of angular momentum when circularly polarized light is absorbed by the medium. In this process, the momentum of light is transferred through spin-orbit interaction to the photogenerated electron and hole pairs, which eventually become spin polarized. Foundational works on optical spin pumping were initially conducted

S. Rossi · E. Vitiello · F. Pezzoli (✉)
L-NESS and Dipartimento di Scienza dei Materiali, Università degli Studi di Milano Bicocca, via Roberto Cozzi 55, 20125 Milan, Italy
e-mail: fabio.pezzoli@unimib.it

S. Rossi
e-mail: s.rossi72@campus.unimib.it

E. Vitiello
e-mail: elisa.vitiello@unimib.it

 
D. J. Lockwood and L. Pavesi (eds.), *Silicon Photonics IV*,
Topics in Applied Physics 139, https://doi.org/10.1007/978-3-030-68222-4_5

by Kastler on atomic gases [1], and subsequently extended to semiconductors by the pioneering demonstration of Lampel [2]. Since then, optical spin orientation has enabled the exploration of spin-dependent phenomena in solid-state materials and established as an effective tool primarily in the study of III–V compounds. We will not dwell here into such remarkable advancements, which have been subject of influential reviews in the field of spin physics [3–5]. In the following, we will rather focus on the exciting investigations opened by the recent application of optical spin orientation specifically to group IV semiconductors.

The fast-growing interest in this area of research was lately spurred by the unique properties of Ge. The Ge conduction band (CB) features a relatively small (~140 meV) energy difference between the local minimum at the center of the Brillouin zone (Γ-point) and the absolute minimum at the zone-edge (*L*-point). Such a characteristic endows Ge with a quasi-direct behavior. The resulting photonic properties are unmatched by Si, arguably the leading material in the group IV arena and the cornerstone of modern information technology. At a first glance, the Ge zone center resembles the one of GaAs. To intuitively capture the physics of the optical spin orientation process, we can therefore extend to Ge a toy model originally developed in the context of III-V compounds. This description, albeit useful, strictly holds at the Γ-point because it does not consider the inherently mixed orbital character of the wavefunctions of carriers having nonzero momentum. The next sections of this chapter will be thus devoted to take such subtleties into account by refining the minimal picture of optical spin orientation at the zone center herein introduced. Above all, we will show how a distinctive property of Ge, namely the multivalley nature of its CB offers extraordinary access to a very rich spin dynamics that has no counterparts in direct gap semiconductors.

We can thus start considering the projection $L_z = \pm\hbar$ of the angular momentum carried by circularly polarized light along the propagation direction, henceforth defined by the *z*-axis. Here, the plus and minus signs correspond to a right- and left-handed helicity of the light, respectively. The absorption of photons with energies above the $E_\Gamma$ bandgap generates an out-of-equilibrium distribution of spin-polarized carriers according to the selection rules and the relative oscillator strength of the optical transitions.

The valence band (VB) is *p*-like with a two-fold degenerate level at the Γ point and consists of heavy-holes (HH) and light-holes levels (LH). Spin-orbit interaction leads to an additional VB state, namely the split-off (SO) band, which lies at a lower energy with respect to the VB edge. The CB can be described by a single degenerate *s*-like level. All the possible $|j, m_j\rangle$ Bloch states are labeled by the quantum numbers $j$ and $m_j$ associated with the total angular momentum operator $\hat{J}$ and to its projection, $\hat{J}_z$, along the light propagation direction, respectively.

Owing to the correspondence principle, a quantum transition between an initial state $i$ located in the VB with energy $E_i$ and a final state $f$ in the CB with energy $E_f$ finds the classical limit in an electric dipole with frequency $\omega_{fi} = (E_f - E_i)/\hbar$. Each optical transition is then associated with a matrix element $\langle f|\hat{\boldsymbol{p}}|i\rangle$. Here, $\hat{\boldsymbol{p}} = -e\hat{\boldsymbol{r}}$ is the electric dipole momentum operator, where $e$ is the electron charge and $\hat{\boldsymbol{r}}$ is a unit vector along the direction of observation [3, 6]. Under circularly polarized excitation, the allowed transitions are those connecting the initial and final states that (i) are optically coupled via vertical transitions, i.e., occurring at the same $k = 0$ point, and (ii) satisfy the angular momentum conservation, namely $\Delta m = m_{jf} - m_{ji} = \pm 1$. The corresponding transition probabilities can be obtained as follows [6]:

$$\left|\langle j+1, m_j+1|\sigma^+|j, m_j\rangle\right|^2 = \tfrac{1}{2}(j+m_j+1)(j+m_j+2)B \tag{5.1.1}$$

$$\left|\left\langle j, m_j+1|\sigma^+|j, m_j^2\right\rangle\right| = \tfrac{1}{2}(j-m_j)(j+m_j+1)A \tag{5.1.2}$$

$$\left|\langle j-1, m_j+1|\sigma^+|j, m_j\rangle\right|^2 = \tfrac{1}{2}(j-m_j)(j-m_j-1)C \tag{5.1.3}$$

$$\left|\langle j+1, m_j-1|\sigma^-|j, m_j\rangle\right|^2 = \tfrac{1}{2}(j-m_j+1)(j-m_j+2)B \tag{5.1.4}$$

$$\left|\langle j, m_j-1|\sigma^-|j, m_j\rangle\right|^2 = \tfrac{1}{2}(j+m_j)(j-m_j+1)A \tag{5.1.5}$$

$$\left|\langle j-1, m_j-1|\sigma^-|j, m_j\rangle\right|^2 = \tfrac{1}{2}(j+m_j)(j+m_j-1)C \tag{5.1.6}$$

where expressing $\hat{\boldsymbol{r}}$ in terms of its cartesian components as $\hat{\boldsymbol{r}} = x\hat{\boldsymbol{i}} + y\hat{\boldsymbol{j}} + z\hat{\boldsymbol{k}}$ and $\sigma^{\pm} = (x \pm iy)/\sqrt{2}$, then the *A*, *B*, *C* constants are defined by:

$$A = \frac{|\langle j|p|j\rangle|^2}{j(j+1)} \tag{5.1.7}$$

$$B = \frac{|\langle j+1|p|j\rangle|^2}{(j+1)(2j+1)} \tag{5.1.8}$$

$$C = \frac{|\langle j-1|p|j\rangle|^2}{j(2j+1)} \tag{5.1.9}$$

Finally, the average spin polarization of the electrons promoted to CB levels is defined as:

$$S_0 = \frac{n_\uparrow - n_\downarrow}{n_\uparrow + n_\downarrow} \tag{5.2}$$

where $n_\uparrow\,(n_\downarrow)$ stands for the density of spin-up (down) polarized electrons.

Unless otherwise stated, we restrict ourselves to right-handed $\sigma^+$ circular polarization, whose induced transitions are summarized in Fig. 5.1a.

When the excitation energy matches the direct gap threshold ($\hbar\omega \cong E_\Gamma$), both LH and HH bands contribute to the absorption process. A direct comparison between the relevant matrix elements demonstrates that HH transitions possess an oscillator

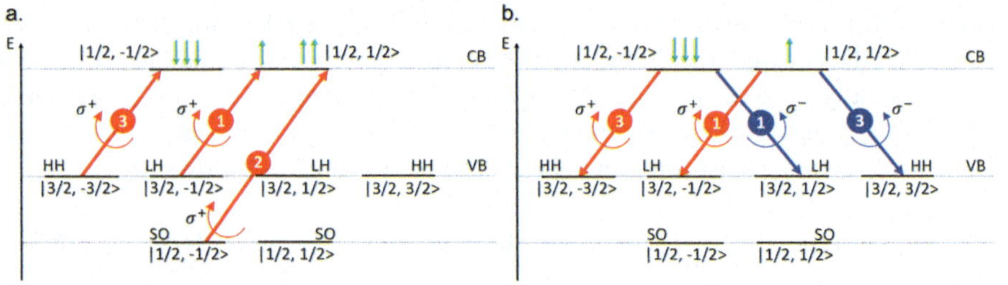


**Fig. 5.1** Schematic representation of states at the valence (VB) and conduction bands (CB) involved in the optical spin **a** excitation and **b** recombination processes near the Γ-point. Arrows in **a** indicate the allowed transitions upon $\sigma^+$ laser excitation, while in **b** arrows indicate the recombination of electrons in the CB with unpolarized holes in HH or LH states. The oscillator strengths for each transition are indicated in the circles

strength that is three times larger than those involving LHs. As a result, $n_\uparrow{:}n_\downarrow = 1{:}3$ and a net electron spin polarization $S_0 = -1/2$ is established. We can safely conclude that circularly polarized light impinging on a bulk material can induce a maximum polarization of 50%.

It should be noted that a fully polarized ensemble of electron spins can be effectively initialized upon resonance excitation with the fundamental transition once the Γ-point degeneracy at the VB edge has been removed. In the absence of external magnetic fields, convenient methods to induce the splitting of the HH and LH states at null wavevector rely on quantum confinement or on the reduction of the cubic symmetry of the crystal through the application of strain. As shown in the following sections of this chapter, such means proved effective in Ge, specifically when heteroepitaxial growth on Si is introduced.

If the photon energy is then continuously increased up to the SO onset ($\hbar\omega \geq E_\Gamma + \Delta_{SO}$), the polarization turns out to be washed out. Under such condition, the $\left|\frac{1}{2}, -\frac{1}{2}\right\rangle$ electron population that is optically coupled to the HH states is fully compensated by the counterpolarized $\left|\frac{1}{2}, \frac{1}{2}\right\rangle$ subset that originates from the combined action of SO and LH transitions. The former has a two-fold higher oscillator strength than the latter.

The optically induced nonequilibrium spins can eventually decay due to both relaxation and carrier recombination. The rate of change of the spin polarization is given by the following continuity equation for the average spin vector $S$ [4]:

$$\frac{\mathrm{d}S}{\mathrm{d}t} = -\frac{S}{\tau_S} - \frac{S - S_0}{\tau} \tag{5.3}$$

where the first term describes the spin relaxation occurring on the characteristic time scale $\tau_S$, the second one describes the generation through optical excitation and recombination, and $\tau$ is the carrier lifetime.

The solution of Eq. 5.3 in stationary conditions ($\mathrm{d}S/\mathrm{d}t = 0$) yields the average spin of thermalized electrons:

$$S = \frac{S_0}{1+\frac{\tau}{\tau_S}} \tag{5.4}$$

This leads to the important conclusion that, whenever the spin relaxation time of the photogenerated carriers exceeds the carrier lifetime, namely $\tau_S > \tau$, radiative recombination events would lead to the spontaneous emission of circularly polarized light, as guaranteed by the time-reversal symmetry of the optical selection rules. The observation of the photoluminescence (PL) helicity thus provides an ultimate compelling proof of the attained optical spin injection in a semiconductor. The PL helicity can be measured as:

$$\rho = \frac{I_+ - I_-}{I_+ + I_-} \tag{5.5}$$

where $I_+(I_-)$ represents the intensity of $\sigma^+(\sigma^-)$ polarized components of the emission.

Owning to the strong spin-orbit coupling in the VB, the depolarization of holes in bulk materials turns out to be very fast compared to the one of the electrons. Consequently, within the momentum relaxation time all the spin states of the VB are equally occupied by holes and available for the subsequent recombination process. Under this condition, only spin-polarized electrons contribute to the polarization of the PL (see Fig. 5.1b), and $I_+(I_-)$ is defined by the spin sublevels in the CB and their relative transition probabilities. Following an excitation with energy below the SO threshold in a bulk semiconductor:

$$\rho_0 = \frac{(n_\uparrow+3n_\downarrow)-(3n_\uparrow+n_\downarrow)}{(n_\uparrow+3n_\downarrow)+(3n_\uparrow+n_\downarrow)} \tag{5.6}$$

so that the initial 50% electron spin polarization is encoded in a maximum polarization $\rho_0 = -1/4$ of the PL. It is worth noting that, as shown in Fig. 5.1b, in this minimal $\Gamma$-point model the emitted light is expected to be copolarized with respect to the absorbed light.

The time evolution of the electron spin polarization introduced via Eqs. 5.3 and 5.4 reflects itself in the circular polarization degree of the PL measured under steady-state conditions as:

$$\rho = \frac{\rho_0}{1+\frac{\tau}{\tau_S}} \tag{5.7}$$

Interestingly, Eqs. 5.5 and 5.7 suggest that when a circularly polarized laser pulse is used to initialize the spin orientation, the subsequent decay of the spin ensemble toward the thermodynamic equilibrium maps out into the transient of the PL polarization, hence $\rho(t) \sim \rho_0 e^{-t/\tau_S}$. Polarization and time-resolved PL therefore provides a direct and convenient means to measure the spin relaxation time, offering insights into spin-dependent kinetics under the influence of different factors like lattice temperature, doping level and external perturbing fields [4].

By virtue of this notable capability, PL spectroscopy quickly established itself as a sensitive and practical method to gather insights into the dynamics of the out-of-equilibrium spin population, particularly in direct gap materials. Notably, the exceedingly long carrier lifetime and the strong electron-phonon coupling sustained by indirect gap semiconductors have been regarded as overwhelming research obstacles in the application of optical detection to indirect group IV materials. The fascinating spin physics of these semiconductors therefore remained completely overlooked, despite the first seminal demonstration of optical spin orientation in bulk crystals was elegantly accomplished in Si [2]. Exemplarily, also in this crucial experiment, optical detection via PL was not utilized. Rather, the occurrence of optical spin orientation of conduction band electrons was probed through their hyperfine interaction with $^{29}$Si nuclei in dedicated nuclear magnetic resonance measurements. As we will discuss, the combined use of optical spin orientation and luminescence polarization proved eventually successful in Ge due to its unique band structure and the inherently long spin lifetime.

#### 5.1.1.1 Spin Relaxation

Group IV semiconductors feature favorable and unmatched spin-dependent properties, which make them prime candidates for hosting future spin-based information and communication technologies. In these materials, spin-orbit coupling is inherently weak also because their centrosymmetric crystal structure prevents contributions arising from bulk inversion asymmetry. This guarantees an exceedingly long spin lifetime and manifests itself, specifically in Si, as a negligible deviation of the Landé $g$-factor from the isotropic value of free electrons [7, 8]. Owning to the lattice inversion symmetry, the seemingly cause of depolarization in elemental Si and Ge is the so-called Elliott-Yafet mechanism. Intuitively, this process stems from the interplay between momentum relaxation and lattice-induced spin-orbit coupling. The electron wave function is hardly in a pure spin eigenstate but rather composed of a spin-up ($|\uparrow\rangle$) and spin-down ($|\downarrow\rangle$) mixture, so that:

$$\Psi_{k,n\uparrow}(\boldsymbol{r}) = \left[a_{\boldsymbol{k},n}(\boldsymbol{r})|\uparrow\rangle + b_{\boldsymbol{k},n}(\boldsymbol{r})|\downarrow\rangle\right]e^{i\boldsymbol{k}\cdot\boldsymbol{r}} \tag{5.8.1}$$

$$\Psi_{k,n\downarrow}(\boldsymbol{r}) = \left[a^{*}_{-\boldsymbol{k},n}(\boldsymbol{r})|\downarrow\rangle - b^{*}_{-\boldsymbol{k},n}(\boldsymbol{r})|\uparrow\rangle\right]e^{i\boldsymbol{k}\cdot\boldsymbol{r}} \tag{5.8.2}$$

where $a$ and $b$ are complex coefficients, $\boldsymbol{k}$ the lattice momentum and $n$ the band index. Scattering events with phonons (high temperatures) or impurities (low temperatures) result in wavefunction renormalization and eventually spin-flip. About $10^5$ scattering centers are commonly needed before the carrier loses its pristine spin orientation [9] and, obviously, the faster the momentum scattering rate, the faster the occurrence of spin-flip events.

Theoretical works from Li and Tang [10, 11] elucidated the nature of intravalley and intervalley spin relaxation processes in Si and Ge due to electron-phonon interactions. Importantly, most of the intrinsic spin relaxation rate of conduction electrons can be explained by the coupling of the lowest conduction band to the upper conduction bands, rather than to the upper valence bands as in most of the semiconductors. Inspection of the electronic states reveal that the intraband spin-flip coupling is much smaller than the interband coupling between conduction bands [12].

In particular, Ge has a larger atomic mass than Si and sustains a stronger spin-orbit coupling. It notably exhibits a highly anisotropic electron *g*-factor [13], whilst demonstrates surprisingly long-lived spin states. A specific theory of spin-flip processes due to electron-phonon scattering in Ge was developed by Li et al. [10, 12]. In unstrained bulk Ge, the intervalley spin relaxation rate was shown to be two orders of magnitude faster than the intravalley one at $T > 20$ K. Therefore, for the whole temperature range between 20 and 300 K, intervalley scattering dominates and limits the spin relaxation (reaching ~1 ns at 300 K). On the contrary, below 20 K only intravalley scatterings dictates the spin dynamics leading to long spin relaxation time. However, as demonstrated in the next sections of the chapter, different effects can alter the spin lifetime, reducing especially at cryogenic temperatures the experimentally attainable values. Intriguingly, scattering processes can have a notable spin orientation dependence [10, 14]. Such relaxation anisotropy was experimentally confirmed along with a spin lifetime approaching ~1 μs at 30 K [14].

A strong dependence on strain direction was also predicted [10, 11]. When strain is applied along the [100] direction, no effect is to be expected on the scattering rate as the four $L$ valleys remain degenerate. On the contrary, for relatively large strain values (~1%) applied along [111] direction, the degeneracy is lifted and the energy separation can be sufficiently large to ensure suppression of the intervalley relaxation processes. However, the intravalley scattering still plays a key role in the overall spin dynamics. Tang et al. [11] pointed out that the spin lifetime can drastically increase from 1 ns up to the range of 100 ns with a suitable application of strain.

Solid solutions of group IV materials, for example, SiGe, GeSn, and SiGeSn, can break crystal inversion symmetry. This is expected to prompt the emergence of the Dyakonov-Perel mechanism, a leading cause of spin relaxation in III–V compounds. In noncentrosymmetric crystals the degeneracy of the spin sublevels is lifted. It can be regarded as an effective $\boldsymbol{k}$-dependent magnetic field $\boldsymbol{B}_i(\boldsymbol{k})$ around which electron spins can rotate. During momentum relaxation, the electron experiences a magnetic field varying both in intensity and in direction. The spin thus changes randomly, resulting in a loss of polarization. It is worth noting that, given the little work done on this topic, at present, there are no experimental or theoretical evidences of the actual conditions at which such a mechanism becomes dominant in group IV alloys.

Another potential source of spin relaxation is given by the electron–hole exchange interaction or Bir-Aronov-Pikus mechanism [4, 9, 15]. Such process has been pointed out to explain recent values of spin relaxation time observed at low temperature in substitutional GeSn alloys and sustained by the residual *p*-type background doping of the epitaxial layers [16]. Furthermore, electron–hole exchange interaction has been demonstrated to be relevant in optical spin orientation experiments in Ge/SiGe quantum wells, where the observation of such phenomenon was unveiled by high optical pumping excitation and facilitated by the significant overlap between electron and hole wavefunctions due to the quantum confinement.

Compared with other semiconductors, like III–V compounds, group IV materials also possess another particularly advantageous property, namely the preeminence of spin-less isotopes. The spin-bearing isotopes are $^{29}$Si and $^{73}$Ge, being solely 5% and 7.8% of the nuclei present in natural Si and Ge crystals, respectively. Such minimal abundance can be further reduced via isotopic enrichment and provides a viable solution to fully suppress hyperfine interactions that couple electron and nuclear spins. This is a crucial aid in the quest for a practical implementation of quantum computing schemes, which require long spin coherence times [17, 18].

We notice that under optical spin orientation of electrons, hyperfine interaction leads to dynamic polarization of the lattice nuclei. The steady-state value of this polarization is reached during the relaxation time of nuclei, which is typically longer than the electron lifetime. This nuclear field acts on the electron spins in the same way as an external magnetic field and depolarization can occur. Such effect is enhanced for electrons confined on impurity levels or trapped in quantum dots, since for itinerant electrons the nuclear spin averages out via motional narrowing.

#### 5.1.1.2 Magneto-Optics

The coupling between the spin states and magnetic fields increases substantially our capacity of investigating inherent relaxation processes. Such an approach is particularly convenient in semiconductors when utilized in conjunction with the optical generation of an out-of-equilibrium spin ensemble. The present section is therefore intended to provide very basic information related to the magneto-optical effects and the various physical process that can emerge under the influence of the external field.

A magnetic field applied perpendicular to the quantization axis (transverse magnetic field or Voigt configuration) induces a Larmor precession of the spin polarization. The time evolution of the average spin $S$ can be therefore obtained by the following general Bloch equation, which ignoring spin diffusion reads:

$$\frac{\partial S}{\partial t} = \begin{pmatrix} \Omega_x \\ \Omega_y \\ \Omega_z \end{pmatrix} \wedge \begin{pmatrix} S_x \\ S_y \\ S_z \end{pmatrix} - \begin{pmatrix} \frac{S_x}{T_2} \\ \frac{S_y}{T_2} \\ \frac{S_z - S_{0z}}{\tau_S} \end{pmatrix} \tag{5.9}$$

where $\Omega_i = g\mu_b B_i/\hbar$ is the precession frequency, $\mu_B$ is the Bohr magneton $g$ is the electron $g$-factor and $S_{0z}$ is the equilibrium value of the spin, which is null in group IV materials. The spin relaxation time $\tau_S$ describes the time required for spins along the longitudinal field to reach equilibrium. It is also termed spin-lattice time and commonly referred to as $T_1$. On the other hand, $T_2$ is the spin decoherence or dephasing time and is defined as the time required for transverse spins, initially precessing in phase about a longitudinal field, to lose their phase. Another related parameter that is more easily accessible in the experiments is $T_2^*$. This is the decoherence time of the spin ensemble [19] and takes into account inhomogeneous broadening of the precessing frequencies. It provides a lower limit for $T_2$.

Under continuous illumination, the spin precession together with the random character of carrier generation or diffusion leads to the spin relaxation. This phenomenon is known as Hanle effect [20], being the solid-state analog of the depolarization process of the resonance fluorescence of gases.

The steady-state spin polarization of the precessing electron spin can be calculated by solving the above-mentioned equation of motion including a term to describe recombination and generation through optical spin orientation, that is [4]:

$$\frac{dS}{dt} = \Omega \times S - \frac{S}{\tau_S} - \frac{S - S_0}{\tau} \tag{5.10}$$

Eventually this is the generalization of Eq. 5.3. Its solution describes the field-induced Hanle depolarization, namely

$$S(B) = S(B = 0)\frac{1}{1 + (\Omega_L T_s)^2} \tag{5.11}$$

where $T_S$ is the so-called spin lifetime given by

$$\frac{1}{T_s} = \frac{1}{\tau} + \frac{1}{\tau_s} \tag{5.12}$$

This precession leads to a decrease of the average projection of the electron spin on the direction of observation, which defines the degree of circular polarization of the luminescence. Consequently, the PL polarization decreases as a function of the transverse magnetic field so that $\rho = \rho(B)$, and the same relation of Eq. 5.11 can be applied also to the circular polarization of emission. By providing access to the spin lifetime, the steady-state Hanle effect turns out to be a powerful tool to investigate carrier dynamics in semiconductors through PL spectroscopy. As shown in the next sections, the Hanle effect was also fruitfully utilized, specifically in Ge, through the measurement of the inductive coupling of the optically induced magnetization of the samples.

It should be noted that the Larmor precession causes spontaneous oscillations in the average spin polarization along the light propagation direction. This coherent effect naturally manifests itself in a modulation, i.e., beats, of the polarized component of the luminescence as a function of time. This phenomenon was reported in GaAs quantum well heterostructures by Heberle et al. in a seminal work published in 1994 [21] and leads to the introduction of the technique of quantum beat spectroscopy.

The magnetic field amplitude affects both the period of the oscillations along with their damping time. First, by measuring the oscillation frequency of $\rho(t)$, that is the Larmor frequency as a function of magnetic field, a linear relationship is expected due to the induced Zeeman splitting of the spin sublevels. Quantum beat spectroscopy therefore allows a straightforward determination of the effective Landé $g$-factor.

Moreover, the damping component of $\rho(t)$ is ultimately related to the spin dephasing mechanisms. This technique therefore enables also the additional measurement of the characteristic $T_2^*$ via the observation of the quenching of the field-induced PL polarization. As shown in Sect. 5.4, quantum beat spectroscopy has been applied very recently to determine spin kinetics in novel GeSn solid solutions [16]. Similarly, coherent spin precession was notably observed in Ge via transmission pump and probe measurements, disclosing pivotal information on the spin-dependent properties of this material (see Ref. [13] and Sect. 5.2.2).

## 5.2 Bulk Ge

The investigation of spin-dependent phenomena in group IV materials lagged well-behind the advancements made in III–V compounds. After nuclear spin susceptibility revealed the existence of optical spin orientation of CB electrons in Si [2], the interest quickly moved toward semiconductors like GaSb [22], whose direct gap enabled the use of optical spectroscopy facilitating all-optical spin investigations schemes. There had been a substantial lack of progress on group IV materials for almost 40 years until Jonker et al. observed circular polarization of the light emitted in Si by the recombination of spin-polarized carriers injected through a tunnel barrier from a ferromagnetic contact [23]. This work unfolded the true potential of group IV semiconductors to lead a practical integration of spintronics onto the sought-after semiconductor technology [24, 25]. Furthermore, the effectiveness of this blueprint stimulated theoretical investigations aimed at refining the physical picture of spin-dependent selection rules, as introduced in Sect. 5.1, to correctly describe indirect gap semiconductors. At first, this was accomplished by implementing methods to model vertical transitions over the whole Brillouin zone [26] and later by including also the relevant phonon-mediated processes [27]. All these notable efforts revitalized the practical application of optical detection methods to group IV materials.

We are unaware of investigations of spin-dependent properties of bulk Ge prior to the observation in 1983 of optically oriented photoelectrons by Allenspach and collaborators [28]. In this work, a Cs-coated Ge(001) crystal was utilized as a proof of principle of an experimental approach capable of determining the symmetry-related orbital wave function mixing caused by spin-orbit interaction. Surprisingly, this work remained mostly unnoticed and optical spin orientation was no longer applied to Ge until 2009, when differential transmission measurements unveiled optically driven spin currents in (111)-oriented Ge wafers [29]. Ultrafast polarization-resolved pump-probe data later demonstrated that the spin relaxation of holes in bulk Ge occurs in about 700 fs at room temperature [30]. Since then, the field has blossomed and, contrary to Si, PL became an essential component of the experimental toolkit that was fruitfully utilized in shining light on the spin dynamics [31, 32].

After Si [26], Rioux and Sipe extended the 30-band $k{\cdot}p$ method also to Ge, providing the first detailed theoretical investigations of the optical spin orientation

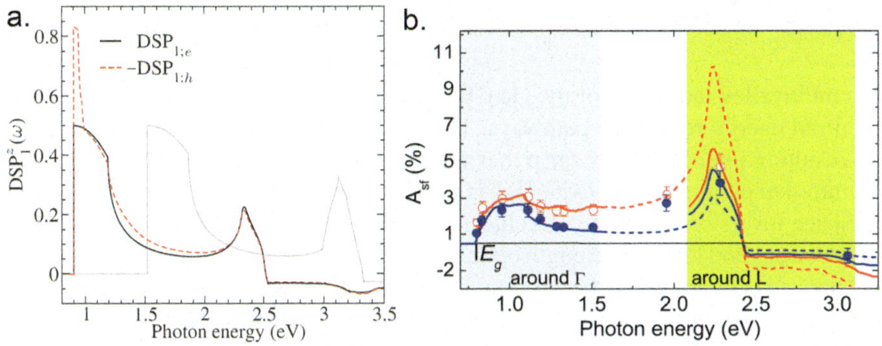


**Fig. 5.2** **a** Calculated degree of spin polarization (DSP) for bulk Ge as a function of the excitation energy for electrons and holes under left-handed $\sigma^-$ excitation [33]. **b** Spin filtering asymmetry signal of a spin photodiode versus photon energy [34]. Reprinted figure with permission from [33]. Copyright (2010) by the American Physical Society. Reprinted figure with permission from [34]. Copyright (2016) by the American Physical Society

process related to vertical transitions in bulk materials [33]. The authors demonstrated that in Ge, the degree of spin polarization, DSP, has a spectral behavior that is qualitatively similar to the one of GaAs [26]. As shown in Fig. 5.2, the DSP of holes was found to be well above 80% and sharply reduced when the excitation energy exceed by 80 meV the direct absorption edge, which is located at 0.89 eV. At such photon energies, the coherence between the HH and LH hole states is lost because their splitting reaches a cut-off value of 30 meV [33].

Specifically, $k{\cdot}p$ perturbation theory demonstrated that the electron DSP, that is $S_o$ (see Eq. 5.2), decreases from a maximum of 50% down to zero as the photon energy approaches the onset of the absorption from the SO band (see Fig. 5.2) [33]. The calculations clarified that such depolarization is mainly caused by the contribution of LH states with increasing nonzero momentum rather than to the activation of the SO band, as commonly argued by the zone-center model introduced in Sect. 5.1 [33]. Surprisingly, Rioux and Sipe predicted at 2.3 eV a nonvanishing electron spin population, i.e., >20%. It was suggested that this local $S_o$ maximum stems from the joint density of states, which results in net electron spin polarization at the $E_1$ critical point [33].

To date, very little experimental work has been done to explore excitation energies that optically couple energy levels far from the Γ point. Rinaldi and coworkers interestingly conducted polarization-sensitive photocurrent measurements at room temperature in Ge photodiodes fabricated utilizing ferromagnetic contacts [34]. This work derived a spectrally resolved degree of spin polarization over a wide energy window, namely 0.8–3.1 eV (see Fig. 5.2b), confirming a sizeable photon helicity and demonstrating maximum sensitivity at about 2.3 eV, i.e., far away from the $E_\Gamma$ threshold. This effect was rationalized following Ref. [35] and ascribed to the efficient direct optical pumping at the $L$ valley of the Ge CB [34].

### 5.2.1 Energy Relaxation and Spin Dynamics

The multivalley landscape of the Ge CB allows photogenerated electrons to sample manifold energy relaxation pathways. The electron kinetics is indeed puzzling and the resulting phenomena by far richer than in ordinary direct gap semiconductors. An intuitive picture for the dynamics is provided in Fig. 5.3.

Space inversion symmetry precludes deformation potential interaction between CB electrons and long-wavelength optical phonons [35]. Concomitantly, the Fröhlich mechanism is inhibited by the covalent nature of the Ge bonds. Optically filled states at the zone center are thus occupied by electrons efficiently coupled to acoustic phonons. Such levels are thereby depopulated on an ultrafast time scale, that is, within hundreds of fs [29, 36]. Thermalization towards the bottom of the $\Gamma$ valley turns out to be significantly less efficient than in direct gap semiconductors, although radiative recombination within the optically coupled region at $\Gamma$ can still occur. The majority of photogenerated electrons will eventually reach the absolute minimum of the CB, located at the $L$ valley (dashed arrows in Fig. 5.3). The transfer likely follows a detour through dark $X$ states (dotted arrows in Fig. 5.3) due to the favorable large density of states and the energy proximity to the $\Gamma$ valley (the energy separation $\Delta_{\Gamma-X}$ is a few tens of meV) [32].

Optical spin orientation and the activation of specific energy relaxation channels by extrinsic means, for example, temperature and doping, can drastically affect the spin kinetics of CB electrons. This spontaneously manifests itself in changes in the spin-dependent radiative recombination events, which can then be captured by PL spectroscopy. Such a method is also advantageous because of the simultaneous access to both direct and indirect emissions, whose spectral resolution is guaranteed by a relatively large gap ($\Delta_{\Gamma-L}$ ~140 meV). Above all, carrier lifetimes at the $\Gamma$- and $L$-points differ by more than 6 orders of magnitude ($\tau_\Gamma < 1$ ps, $\tau_L < 1$ μs) [37, 38]. This notable property endows spectrally resolved PL with the additional possibility to

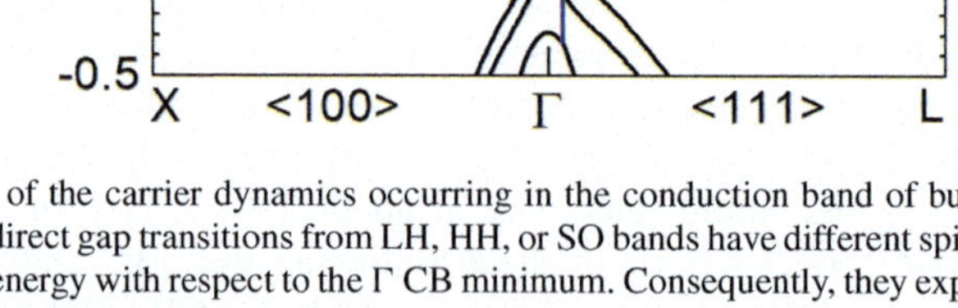


**Fig. 5.3** Sketch of the carrier dynamics occurring in the conduction band of bulk Ge. Electrons excited through direct gap transitions from LH, HH, or SO bands have different spin orientation and different excess energy with respect to the $\Gamma$ CB minimum. Consequently, they experience different energy relaxation pathways. Electrons can recombine both from the $\Gamma$ valley or scatter to the side valleys and reach the $L$-valley CB minimum

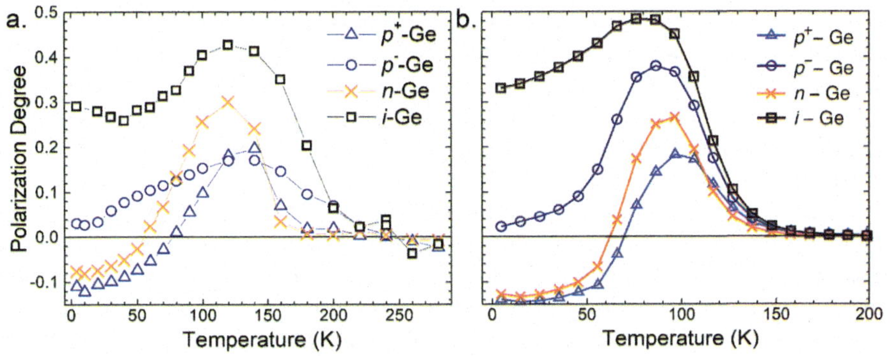


**Fig. 5.4** **a** Experimental and **b** calculated circular polarization degree of direct recombination as a function of doping and temperature in bulk Ge. Specifically, $p^+$-Ge is a Ge:Ga wafer, with an acceptor concentration of $3.6 \times 10^{18}$ $cm^{-3}$, n-Ge is a *n*-type Ge:As, with a donor concentration of $8.3 \times 10^{16}$ $cm^{-3}$, $p^-$-Ge is a *p*-type wafer with a doping concentration of $1.4 \times 10^{15}$ $cm^{-3}$, and i-Ge is a sample with a resistivity of 47 Ω cm [32]

capture time-tagged snapshots of the spin dynamics even without the application of time-resolved techniques [32, 39, 40]. The measurement of the circular polarization associated with optical transitions can thus be very informative about the relaxation process. It should be noted, however, that the very long carrier lifetime at the *L*-valley sets a very demanding constraint on the electron spin lifetime required to observe a nonvanishing circularly polarized emission from transitions across the indirect gap (see Eq. 5.7). Reports on the observation of *L*-point emission under optical spin orientation will be described in detail in the next section, while here, we will turn our attention to the ultrafast Γ-point transitions.

The circular polarization degree of the direct gap PL measured over a wide temperature range is shown in Fig. 5.4a for bulk (001)Ge with either *p*- or *n*-type doping [32]. The excitation energy was chosen to selectively promote electrons at the Γ-valley, i.e., 1.165 eV, and the pump was set to be left-handed circularly polarized. The sign of $\rho$ was defined consistently with the sign of the Stokes parameters, which in a polarimetric analysis determines the light helicity (see [31, 41, 42] for specific details). At cryogenic temperatures, PL reveals a net circular polarization for all the doping content of the Ge wafers, providing a compelling demonstration of optical spin orientation. In highly doped samples (>$10^{15}$ $cm^{-3}$) and below 90 K, the direct gap PL is copolarized, i.e., of negative sign, with respect to the excitation, while the luminescence of intrinsic and mildly doped wafers (<$10^{15}$ $cm^{-3}$) turns out to be counterpolarized. This implies that, despite the very same excitation conditions, impurities alone can strikingly reverse the population of the spin states that contribute to the $E_\Gamma$ radiative recombination.

The circular polarization degree of the direct gap transition persists at higher temperatures. Specifically, between 90 and 170 K, the photon helicity flips its sign in highly doped samples, whereas it remains positive at low impurity levels. Consequently, in this intermediate temperature regime the direct gap PL is always counterpolarized with respect to the excitation and $\rho$ approaches the largest measurable values in all the wafers [32]. Such data demonstrate that the circular polarization and the concomitant electron spin population is washed out as the temperature is increased above 170 K.

In Ref. [32], such puzzling experimental findings were rationalized by a direct comparison with Monte Carlo calculations. The results of the latter are summarized in Fig. 5.4b. This combined analysis offered decisive insights into the spin and energy relaxation mechanisms of the out-of-equilibrium spin ensemble in bulk Ge. It was shown that two electron populations with spin pointing in opposite directions can be simultaneously photogenerated nearby the zone center by infrared excitation. Notably, these two spin populations, being optically coupled either to SO or HH-like states, possess also different excess energies with respect to the $\Gamma$-edge of the CB, see Fig. 5.3. The relative contributions of these low- and high-energy electron populations to the direct gap luminescence strongly depend on excitation, temperature, and doping conditions [32], whose influence is ultimately crucial in modifying the carrier kinetics. Since it is highly likely for high-energy electrons to be transferred out of the $\Gamma$ valley during energy relaxation, the direct gap luminescence is eventually governed by low-energy electrons originating from the SO band. This physical picture explains the positive and large $\rho$ value observed in the intrinsic sample at low temperatures (despite the relatively small fraction of low-energy electrons at cryogenic temperatures). Owing to temperature-induced bandgap narrowing, the weight of these spin states becomes increasingly important at higher temperatures and explains the positive sign of $\rho$ between 80 and 110 K. The role of energy relaxation through the satellite $X$-valleys was, however, largely underestimated in previous literature works but it is pivotal to rationalize the overall kinetics when Ge is doped. In this case, collisions between the photogenerated electrons and the background charges introduced either via acceptors or donors is turned on. Coulomb-mediated interaction promotes thermalization of hot electrons within the $X$ valleys. While dwelling at the $X$ bottom, intervalley scattering can occasionally transfer some carriers back to the $\Gamma$ valley (see dotted arrow in Fig. 5.3). After such a round trip these electrons can finally contribute to the $\Gamma$-point radiative recombination, reducing the absolute value of $\rho$ and even changing its sign when they outsize the pristine low-energy electrons population.

Above 120 K, an electron backflow establishes from the $L$-valley due to thermal activation. In this temperature regime, electrons at the zone edge are unpolarized because their spin relaxation time (ns range) is much shorter than the carrier lifetime (~$\mu$s). Electrons from the $L$-valley can visit the $\Gamma$-valley with a Boltzmann-like probability, dependent on the energy separation between the $L$ and $\Gamma$ valley, and thereby experience radiative recombination from the latter valley. These events already play a role at 120 K and become progressively important as the temperature increases, leading to a complete depolarization of the direct emission [32].

Very recently, the $X$-to-$\Gamma$ backward scattering process and its effect on the polarization of the radiative emission have been further addressed in bulk Ge. In particular, Ref. [39] demonstrated that optical pumping suffices to govern the kinetics of spin-polarized carriers and, thus, the chirality of the radiative recombination. In this context, the $X$-to-$\Gamma$ backward scattering is dynamically governed by background charges introduced and finely controlled by the pump power. In lightly doped materials, the change of the intervalley scattering rate via the density of photogenerated carriers yields a straightforward modulation of the PL polarization without resorting to any external factor, such as magnetic fields or optical retarders. It was shown that the polarized component of the emission can be conveniently swept through right and left circular polarization, eventually giving linearly polarized light at an intermediate pump power regime. An energy- and polarization-resolved analysis of the direct gap PL under the latter excitation condition unveils, however, a subtle but intriguing scenario. Indeed, a modest energy detuning of few tens of meV from the peak maximum remarkably demonstrates circularly polarized emission (see Fig. 5.5). Chiefly, the high-energy tail of the PL peak yields a right-handed circular polarization eigenstate, whilst the low-energy side unambiguously discloses the pattern of a counterpolarized left-handed eigenstate. This evidence corroborates the modeling of the kinetics of the spin-polarized carriers anticipated in Fig. 5.3. Above all, it clarifies that the linearly polarized photons emitted at the PL peak are due to a genuine superposition of the circular basis, stemming from the spin-resolved energy spectrum of the $\Gamma$-valley electrons. Besides further strengthening the physical picture of the

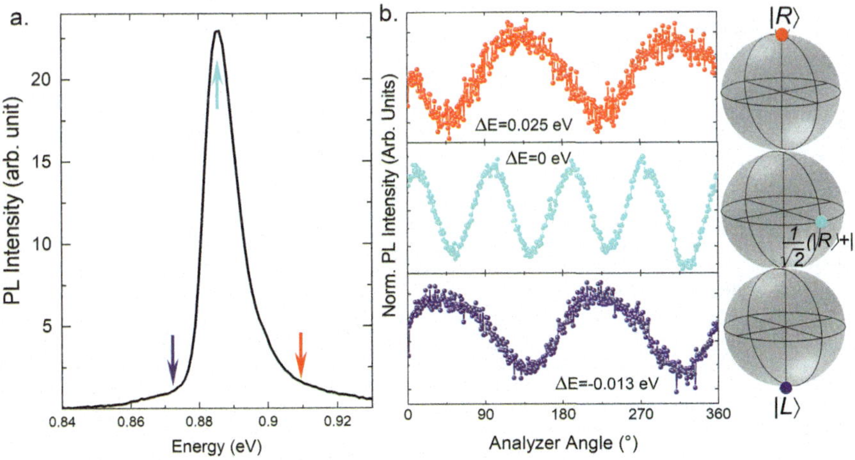


**Fig. 5.5** **a** Low temperature direct gap PL of bulk Ge under intermediate excitation power density (1.2 kW/cm$^2$). **b** Polarimetry measurements showing modulation of the PL peak intensity as derived in a Stokes analysis of the light polarization [39]. The data are spectrally resolved for photons emitted at the PL peak (cyan curve), at a positive (red curve) and at negative (violet curve) energy detuning $\Delta E$ with respect to the energy of the maximum. The intensity modulation patterns allows the determination of the state of light polarization as summarized by the corresponding representation on the Poincaré sphere [39]

optically reconfigurable polarization of the direct gap PL, the linear polarization is a direct manifestation of the fact that the fast spin-dependent relaxation pathways can yield a coherent superposition of two cross-polarized spin ensembles dwelling at the bottom of the Γ-valley. This unique spin dynamics pertaining to the multivalley CB of Ge introduces novel solutions that can accelerate the unification of spintronics and optoelectronics concepts leading to a new research frontier at the intersection between traditionally distant fields, such as magnetism, electronics, and photonics [43].

In another work, the dynamics of spin-polarized carriers in Ge was further addressed via photoluminescence excitation spectroscopy [40]. By using a tunable laser source, $\rho$ was mapped out as a function of the excitation energy. These results are summarized in Fig. 5.6a. It should be noted that here the excitation was right-handed ($\sigma^+$), hence a positive polarization corresponds to PL that is copolarized with respect to the laser.

From Fig. 5.6a it possible to observe that a polarization degree as high as ~22% can be achieved for excitation photon energies almost resonant with the direct gap threshold. Given the ultrafast lifetime of Γ-valley electrons this result implies via Eq. 5.4 a $S_0$ value in good agreement with the spin polarization theoretically predicted at such photon energies in bulk Ge (See Ref [33] and Fig. 5.2). It was observed that in the low doping regime, the fraction of electrons suffering Γ-$X$-Γ scattering is expected to be reduced, thus decreasing the average lifetime of electrons recombining

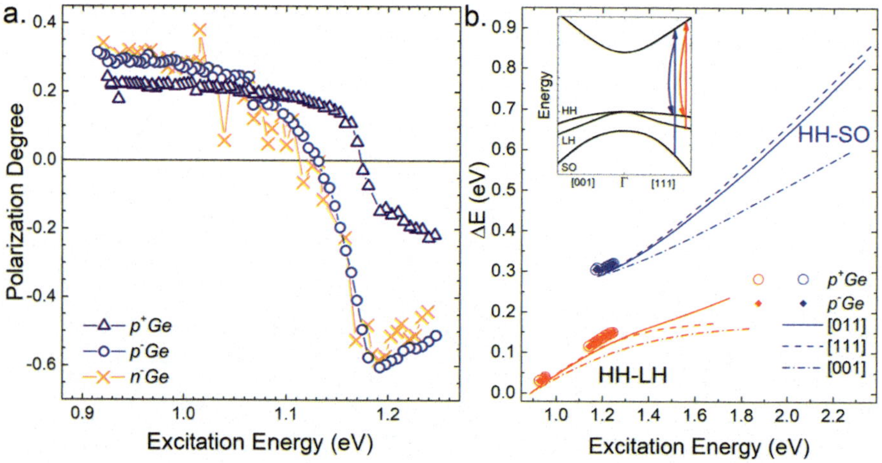


**Fig. 5.6** **a** Polarization degree of the PL emission as a function of the excitation energy. Data are taken at 6 K in bulk Ge. Specifically, $p^+$Ge is a $p$-type Ge:Ga wafer with an acceptor concentration of $3.6 \times 10^{18}$ cm$^{-3}$, n-Ge is a $n$-type with a donor concentration of $2.2 \times 10^{16}$ cm$^{-3}$, and p-Ge is a $p$-type wafer with a doping concentration of $2.2 \times 10^{16}$ cm$^{-3}$. **b** Energy difference between the electronic Raman scattering peaks and the laser energy for $p^+$Ge and $p^-$Ge samples as a function of the incident photon energy. The lines report the calculated Raman peak positions due to HH-to-LH and HH-to-SO scattering along [001] (dashed-dotted), [011] (solid) and [111] (dashed) directions. The inset shows a $k \cdot p$ band structure of Ge bulk and the arrows describe the Raman processes [40]

radiatively via the direct gap transitions. This mechanism possibly leads to a revision of the traditional assumption of a negligible hole spin polarization and results in a circular polarization degree of the PL as high as 35% [40]. When the energy of the optical pump increases, the circular polarization degree of the direct gap PL varies from 35 to −60% at about 1.2 eV. Helicity inversion shown in Fig. 5.5 heralds the dominance at the Γ-valley either of the electron spin population optically coupled to the SO or coupled to the HH states. In line with the model of the energy relaxation, we can find reassurance that the impurity content can increase the weight of the $\sigma^+$ copolarized contribution to the direct transition by noting that doping blue shifts the helicity crossover and yields a smaller magnitude of the polarization degrees in the negative $\sigma^-$ regime.

Optical spin orientation was also profitably utilized in Ref. [40] to disclose the emergence of resonant electronic Raman scattering (ERS) in bulk Ge. Giving that the ERS spectral features cannot be observed in *n*-type wafers, they were ascribed to light-scattering phenomena due to electronic excitations occurring in the VB rather than to exciton-like band-to-band or band-to-impurity recombination. These findings point out how polarization-resolved PL can be elegantly utilized to reconstruct the VB dispersion [40]. Figure 5.6b shows that the measurement of the energy dependence of ERS in the vicinity of the direct gap yields an accurate spectroscopic tool for a complete mapping of the VB dispersion and can be applied also to other semiconducting materials.

## 5.2.2 *Spin Lifetime*

### 5.2.2.1 Magneto-Optics

A viable radio-frequency (RF) technique relying on optical spin orientation was initially introduced by Guite and Venkataraman to detect spins and to measure their lifetime in bulk Ge [44]. In this method, the spin polarization is injected by photoexcitation across the direct gap using circularly polarized light, whose energy is sufficient to excite electrons from the HH and LH bands, but not from the spin-orbit SO band. Due to the selection rules described in the previous sections, the photogenerated electron population is spin polarized. As in the pioneering experiment on Si [2], spin orientation is not directly probed by an optical means. Instead, if the polarization of the incident light is modulated by means of an electro-optic modulator (EOM), the magnetic moment of the spin ensemble is also modulated, and it can be detected by a sensitive pickup coil coupled to a lock-in amplifier. The corresponding RF signal is then measured as a function of transverse external magnetic field (Voigt geometry) to determine the spin lifetime via the Hanle effect [44]. The setup is schematically shown in Fig. 5.7a.

As shown in Fig. 5.7b, a magnetic field of a few mT was sufficient to observe Hanle depolarization in (100)-oriented *n*-type bulk Ge [45]. Such fields are weaker than the one needed for GaAs, stemming from a longer spin lifetime. A maximum

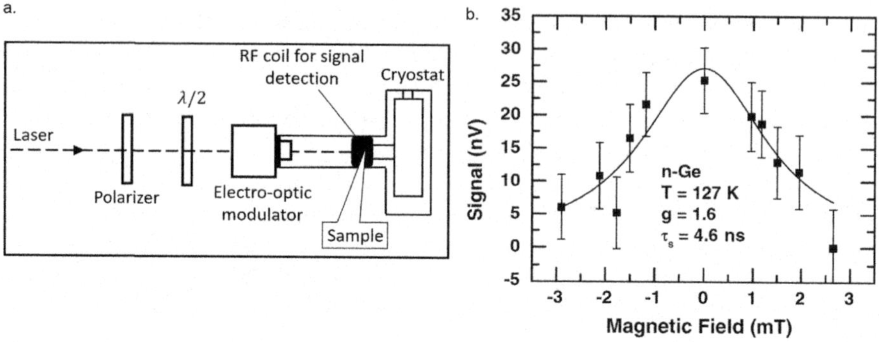


**Fig. 5.7** **a** Sketch of the RF technique presented in Ref. [45] and **b** Hanle measurement on Ge at low temperature. Reprinted figure with permission from [45, 33]. Copyright (2011) by the American Physical Society

spin orientation efficiency of 4.8% was shown and a spin lifetime of $(4.6 \pm 1.0)$ ns was derived at 127 K by assuming an effective $g$-factor of 1.6 [46].

The same experimental technique was lately employed to address the temperature dependence of the spin lifetime [46], showing a monotonous decrease in the 100–200 K range. The experimental data compared satisfactorily with the theory [10] although a shorter spin lifetime was retrieved. The difference was mainly attributed to an additional relaxation channel opened by impurity scattering.

Time-resolved Faraday rotation (TRFR) experiments in the Voigt configuration also proved useful to study the ensemble spin coherence and relaxation times in bulk Ge [13, 47, 48]. The precession of the photoexcited spins about the axis of the external field can be measured as an oscillatory-induced Faraday signal, i.e., the rotation of the polarization angle of a linearly polarized light beam upon passage through the medium.

In the work of Hautmann et al. [47], the excitation was a 60 fs pulsed laser source tunable between 1200 and 1600 nm, which was spatially separated in a circularly polarized beam, used as pump pulses, and a linearly polarized probe. Different signal strengths, effective $g$-factors, as well as coherence times permit to differentiate electron and hole dynamics.

TRFR experiments were carried out at 8 K on a (111)-oriented and undoped Ge wafer [47]. The authors subsequently analyze hole and electron spin coherence after excitation via the indirect bandgap by optimizing the pump conditions to selectively observe one of the two charged species. First, coherent spin precessions of holes resulted in damped oscillations with a decay time of tens of ps. The hole spin coherence was found to markedly depend on the external magnetic field and excitation power density, approaching 150 ps in the absence of a perturbing field [47]. This eventually allowed the estimation of 5.5 for the effective $g$-factor of holes [47].

The analysis of FR at high excitation intensities, where hole spins decay in $<10$ ps, led to the emergence of a remarkably robust electron spin signal. The FR transients shown in Fig. 5.8a are extracted from the difference between the FR signals for $\sigma^+$ and

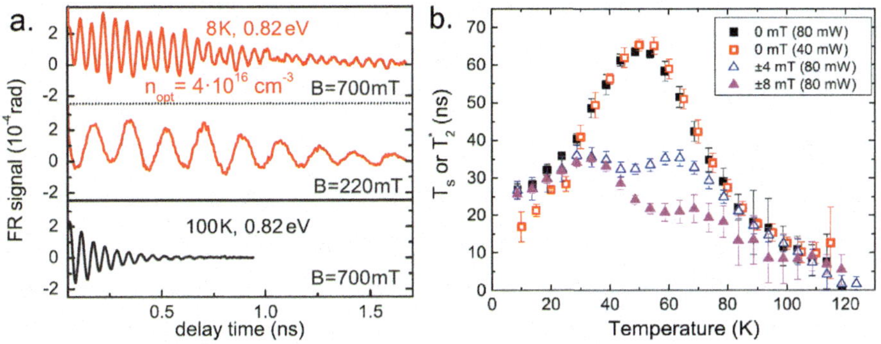


**Fig. 5.8** **a** Faraday rotation (FR) signal observed in Ge [47] at different magnetic fields (upper and middle panels) and different temperatures (lower panel). **b** Temperature dependence of the electron spin lifetime $T_S$ at zero magnetic field (squares) and coherence time $T_2^*$ for $B \neq 0$ (triangles) [48]. Reprinted figure with permission from [47]. Copyright (2011) by the American Physical Society. Reprinted figure with permission from [48]. Copyright (2014) by the American Physical Society

$\sigma^-$ polarized excitation to remove a slowly varying background from the data and are shown for different magnetic field magnitudes and temperatures. The Fourier analysis of temporal beating in the transient reveals a dominant frequency corresponding to an effective $g$-factor of about 1.83 [47]. Two additional frequency components were also observed because the surfaces of constant energy for $L$-valley electrons in Ge are ellipsoids and result in different cyclotron masses for different angles of an external magnetic field with respect to their principal axes. The obtained electron $g$-factors are in reasonable agreement with the available electron spin resonance data and theoretical predictions [49, 50].

In a follow-up paper [13], TRFR was performed along different crystallographic orientation, observing that the electron spin decoherence occurs within ~100 ps to ~10 ns depending on temperature, doping concentration, and orientation of the external magnetic field with respect to the $L$-valley ellipsoids of constant energy.

The temperature dependence of the electron spin lifetime (at zero field) and decoherence time was presented in Ref. [48] and is reported here in Fig. 5.8b. The maximum spin lifetime of about 65 ns was obtained at an intermediate temperature of ~50 K. At about 120 K the spin lifetime was found to be as long as ~5 ns and consistent with the theoretical predictions [10]. The slight decrease of the spin lifetime below 50 K and the deviation from the theoretical calculations were related by the authors to an efficient spin relaxation mechanism due to impurity scattering [48].

#### 5.2.2.2 Luminescence Data

In indirect gap semiconductors, optical transitions are mediated by electron–phonon interactions. This coupling mechanism has to be considered to properly describe the

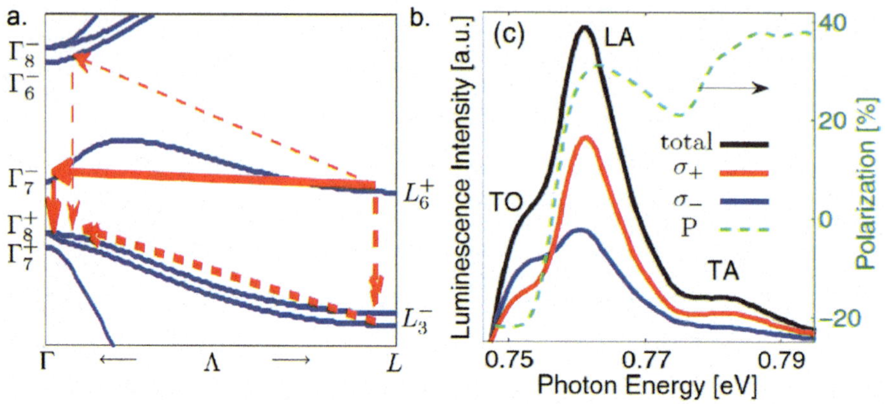


**Fig. 5.9** **a** Calculated band structure of bulk Ge. Arrows represent virtual paths for phonon-assisted optical transitions via intermediate states. **b** Calculated polarized luminescence and resulting polarization degree due to radiative recombination of spin-up electrons in unstrained Ge at 77 K. The light propagation is parallel to the spin orientation. Reprinted figure with permission from [51]. Copyright (2010) by the American Physical Society

relationship between the spin polarization of electrons and the circular polarization degree of the luminescence of the indirect transitions. The spin-dependent selection rules were theoretically investigated both in Si and Ge [27, 51].

As shown in Fig. 5.9a, in bulk Ge phonon-assisted optical transitions from an initial to a final state can proceed via several virtual paths. Longitudinal acoustic (LA) modes can be involved in the $L_6^+ \rightarrow \Gamma_7^- \rightarrow \Gamma_8^+$ optical transition [51]. Even though edge-to-edge optical transitions with the transverse acoustic (TA) mode are forbidden, high order transitions between states near $L_6^+ \rightarrow \Gamma_7^-$ do not vanish completely, thus LA and TA modes follow the same selection rules and demonstrate the same polarization sign of the emission (see Fig. 5.9) [51].

On the contrary, the transverse optical (TO) mode is symmetry allowed only via other specific intermediate states. The use of group theory in Ref. [51] enabled the demonstration that the $\rho$ sign of the LA- and TA-assisted transitions is opposite with respect to the one corresponding to the TO mode. Such a finding is summarized in Fig. 5.9b. Moreover, the intensity ratio between HH and LH for both the LA and TA peaks was predicted to be 3:1 as in direct gap transitions (see Fig. 5.1). This result derives from a notable property akin to the optical spin orientation process at the zone center (see Sect. 5.1), namely LA- and TA-assisted transitions involve electron-phonon interaction between almost pure spin eigenstates pertaining, in this specific case, to the lowermost conduction band [51].

The spin-dependent selection rules for phonon mediated transitions were experimentally proved in Si and Ge by means of PL spectroscopy [38, 52]. In bulk Ge, the PL spectrum measured at 4 K in a (001)-oriented, nominally intrinsic wafer is shown in Fig. 5.10a. Three bright emissions are well-resolved. Their energy ordering corresponds to exciton recombination involving the emission of TA, LA and TO phonons

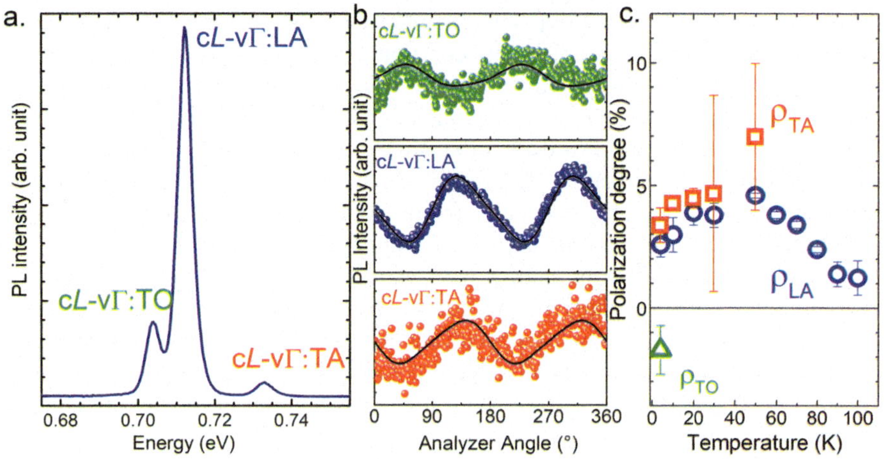


**Fig. 5.10** **a** Low temperature PL spectrum of the indirect gap recombination of intrinsic bulk Ge. **b** Intensity modulation of each phonon replica as a function of the polarization analyzer angle and **c** polarization degree of each phonon replica as a function of temperature [38]. Adapted from [38], with the permission of AIP Publishing

whose energy is 7, 27 and 36 meV, respectively. The polarization state of each of these three spectral features was probed in a single shot via polarimetry measurements. The emission was found to be circularly polarized demonstrating that the relationship between spin relaxation and electron lifetime is not unfavorable in group IV materials as previously thought. Besides providing a compelling proof of optically oriented electron spins, the PL analysis unveils that at cryogenic temperatures the LA and TA peaks are both copolarized with respect to the excitation, whilst the TO is counterpolarized (see Fig. 5.10b), which is in full agreement with the theoretical prediction [51].

Another significant result was the unprecedented observation of a concomitant circular polarization at both zone-edge and zone-center transitions. This finding further corroborates the physical picture of the energy relaxation process as described in the previous sections. Indeed, high-energy electrons originating from the top of the VB scatter preferentially out of the optically coupled Γ region. In weakly doped bulk Ge, collective and binary collision events with the Fermi-Dirac distributed background charges are negligible, thereby quenching the Coulomb-induced pipeline that yields carrier backflow towards the zone center. The majority of negative charges residing at zone center are thus low-energy electrons that were directly promoted at the edge of the Γ-valley from pristine SO states (see Fig. 5.3). Energy relaxation mechanisms will eventually yield a valley-dependent spin polarization: the optically oriented spins at the zone center are indeed cross-polarized with respect to the spin-polarized electron ensemble collected at energy-degenerate *L*-point global minima. Such puzzling phenomena, which stem from the rich spin dynamics sustained by Ge, should reflect itself in a counterpolarized helicity for the emissions occurring at the two band-edges and was astonishingly confirmed by experimental PL data [38].

The comparison between the expected degree of spin polarization and the measured circular polarizations for the phonon-assisted optical transitions provides additional means to gather information about spin-loss mechanisms. Since the measured circular polarization degrees, albeit small, are nonzero, $\tau_S$ is expected to be comparable with $\tau$, i.e., a sizeable fraction of the electron spins are still aligned prior to radiative recombination across the indirect bandgap (Eq. 5.7).

In their work, Giorgioni and coworkers combined optical spin orientation measurements with the study of the decay curve of the indirect PL emission to estimate spin and carrier dynamics occurring in bulk Ge [38]. A lower limit estimate for the spin relaxation time was set in hundreds of ns by leveraging the measurement of the circular polarization degrees with the determination of the effective carrier lifetime. The latter was found to be of about 757 ns at 14 K.

Figure 5.11 summarizes all the available theoretical and experimental values of the spin-related lifetime reported to date at different lattice temperatures in Ge [14, 38, 46, 48, 53–56]. Empty (full) symbols are experimental data obtained by electrical (optical) methods, whereas lines represents calculations [10, 11]. It is worth noticing that electrical techniques relying on three terminal devices and nonlocal detection schemes provide sub ns values over the whole temperature range. This can be possibly rationalized by considering that the achievement of electrical spin injection and detection requires the use of heavily doping layers, the presence of several interfaces related to the fabrication of ferromagnetic gate stacks and the possible percolation

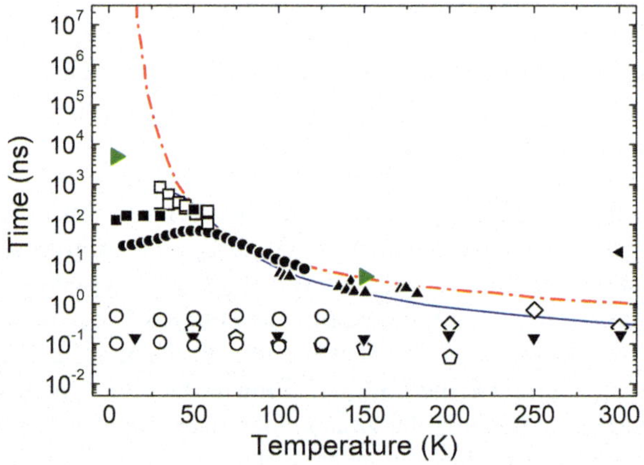


**Fig. 5.11** Temperature dependence of the spin-related lifetime. The red (dashed) and blue (solid) lines represent the results of theoretical calculations reported in [10, 11], respectively. Open symbols are experimental data obtained by electrical measurement in [14] (□), [53] (◊), [54] (⬠), [55] (○) and [57] (△). Full symbols are data points obtained by optical methods and reported in [38] (■), [46] (▲), [49] (●), [56] (▼) and [58] (◄). Green full symbols ► report spin relaxation times obtained in Ge quantum wells by optical methods in [31, 59]. Adapted from [38], with the permission of AIP Publishing

of stray fields [53–55]. All these factors can contribute with extrinsic relaxation channels that can be the limiting factors, particularly in the low temperature regime. Similar issues might affect also electrical detection in spin photodiodes [56] (for further details about spin-photodiode, see Sect. 5.5.1).

Unwanted interfacial effects can be strongly mitigated in metal-semiconductor devices relying on spin injection of hot electrons [60]. In contrast to electrical methods, optical-based techniques [38, 46, 48] are also inherently less prone to experimental artifact and spurious signals because they are contactless and do not need impactful preparation or sample processing.

Figure 5.11 shows that, above 50 K magneto-optical measurements (full symbols) [46, 48] and transport data acquired under the ballistic regime [14] (open squares) retrieve the tens-of-ns long spin lifetime predicted by the theory [10]. At lower temperatures, specifically below 30 K, PL data [38] (full squares) show electron spin relaxation times, which are hundreds of ns long and demonstrate a lengthening of almost one order of magnitude over the longest spin lifetime previously reported by TRFR experiments [48]. It should be noted, however, that, in the low temperature regime, the experimental findings are markedly shorter than the theoretical predictions [10, 11], where the peculiar relaxation mechanism due to intrinsic electron–phonon interaction was shown to boast an extremely long spin relaxation time for CB electrons (about 10 ms at $T = 10$ K).

There remains a large gap between theory and experiment. The reason for such a discrepancy is still an open question and may be due to the emergence of novel spin relaxation channels at cryogenic temperatures that limit the observable spin relaxation times in bulk material and outweigh the intrinsic EY mechanism. The possible origin of these extrinsic processes might be traced back to the subtle role played by impurities. Song et al. [61] put forward a donor-driven spin relaxation, in which the spin-flip of CB electrons is governed by scattering off the central-cell potential of the impurity. These phenomena can be effective particularly at low temperatures and in Ge samples having a relatively high concentration of donor impurities.

## 5.3 Ge Heterostructures

### *5.3.1 Strained Ge Epilayer*

Besides enhancing the direct gap recombination in photonic applications [62], strain can be introduced and utilized as an effective degree of freedom for tailoring spin-dependent phenomena in group IV semiconductors. It is expected that the removal of the Γ-point HH/LH degeneracy can substantially increase the initial spin population injected through optical spin orientation up to the limit of complete polarization [3]. In addition, the application of strain along specific crystallographic directions can lift the degeneracy of the CB minima in both Si and Ge. Intriguingly, as discussed

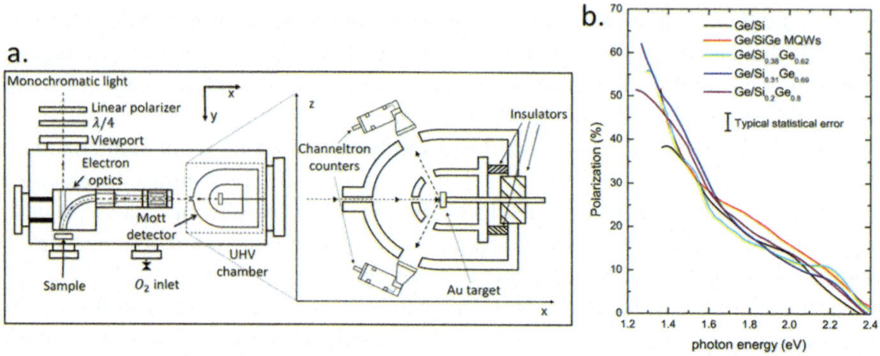


**Fig. 5.12** **a** Sketch of the experimental apparatus for spin-polarized photoemission measurements with the Mott detector rotated by 90° with respect to the apparatus. **b** Electron spin polarization as a function of the exciting photon energy from activated epitaxial Ge thin films. Reprinted from [65], with the permission of AIP Publishing

in Sect. 5.1, theoretical works pointed out that strain engineering can be leveraged to suppress the leading spin relaxation channel given by intervalley scattering of CB electrons and decisively lengthen their spin lifetime and diffusion length [10, 11].

The optical spin orientation process in compressively-strained Ge epilayers was initially explored by means of photoemission spectroscopy [63–66]. The spin polarization of electrons photoemitted from epitaxial thin films of Ge deposited on SiGe-buffered Si wafers can be obtained by means of Mott polarimetry. A schematic representation of the experimental apparatus used for such measurements is shown in Fig. 5.12a. The photoelectron spin polarization is along the $y$-axis, i.e., the quantization axis given by the light propagation direction. The initially longitudinally polarized beam is transformed into a transversally polarized one by the 90° rotator before being accelerated into the Mott detector. The latter is schematically represented in Fig. 5.12a. The spherical retarding field Mott polarimeter is rotated by 90° with respect to the apparatus sketched in the figure. The polarization values observed in the epitaxial Ge strained layers are consistently larger than those from the (unstrained) bulk-like sample and notably above the 50% limit attainable in bulk semiconductors (see Fig. 5.12b) [63]. Such findings can be explained by considering that biaxial compressive strain lifts the degeneracy between HH and LH states (see Fig. 5.12b). HH-CB transitions, populating only a given spin channel, can be selectively excited without contribution from the opposite spin channel due to LH-CB transitions. In principle, this should lead to a complete polarization of CB electrons, i.e., $S = 100\%$. The photoemission process, however, takes place away from the $\Gamma$ point, where the band orbital mixing between the LH and SO band is strong. The compressive strain increases this interaction, even in proximity of the $\Gamma$ point. This unavoidably increases the SO character of the LH states. This enhancement of the band orbital mixing is solely driven by the in-plane compressive strain acting on the Ge epilayer and not by any symmetry reduction along the [001] photoemission direction. As a result, the degree of spin polarization approaches 62%, which is nevertheless larger than the 50% bulk limit attainable in a fully relaxed layer. A further optimization of compressive strain in $Si_{0.18}Ge_{0.82}/Si_{0.4}Ge_{0.6}$ heterojunctions led subsequent works to achieve a spin polarization as high as 72% [65, 66]. In this case, the vacuum energy level was notably lowered below the CB edge, thus allowing direct access to photoemitted electrons resulting from lowest energy HH band transitions, for which compressive strain advantagously reduces the band mixing with LH and SO states.

The highest PL polarization value measured so far in strained Ge was reported by Vitiello et al. on Ge layers grown on Si(001) [67]. The mismatch in the thermal expansion coefficients between the Si substrate and the epitaxial Ge film spontaneously induces a biaxial tensile strain upon cooling the heterostructures from the deposition temperature, typically above 500 °C, down to room or cryogenic temperatures. Ref. [67], the impact of biaxial tensile strain on the spin-dependent direct gap radiative recombination was assessed by means of optical spin orientation, demonstrating a PL polarization degree as high as of 85%. Such a result was achieved at liquid-He temperatures in a not-intentionally doped sample and despite the laser energy (~1.1 eV) being tuned far off the resonance excitation of the fundamental direct gap transition (~0.86 eV, see Fig. 5.13a).

Polarization-resolved PL spectra, shown in Fig. 5.13a, unexpectedly reveal two peaks, that demonstrate opposite circular polarization. Specifically, the lowest (highest) energy component of the doublet is copolarized (counterpolarized) with the laser excitation. The doublet has been ascribed to direct gap recombination of electrons with the strain-split LH and HH. Owing to the tensile strain in the Ge-on-Si

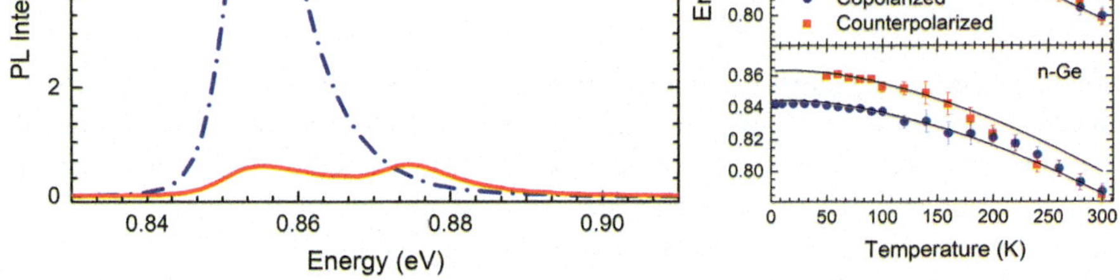


**Fig. 5.13** **a** PL spectra in copolarized and counterpolarized configuration with respect to the excitation of a not intentionally doped Ge-on-Si film (i-Ge) at 4 K. **b** Experimental PL peaks position as a function of the temperature and predicted Varshni-like behavior (solid lines). n-Ge and p-Ge are Ge-on-Si heterostructures that were *in situ* doped either with $P$ or $B$ impurities, respectively. The doping content was the same for the two samples, namely $4.5 \times 10^{17}$ $cm^{-3}$ [67]

heterosystem, the fundamental transition involves the LH band. Tight-binding calculations clarified that the spin polarization of CB electrons optically coupled to SO states is 90% [67]. The striking agreement of this value with the circular polarization degree associated with the low-energy PL peak corroborates the physical picture according to which the direct gap recombination is dominated by the small fraction (about 5%) of highly polarized electrons photogenerated from the deep SO states. Electrons excited from HH and LH VB states possess a larger kinetic energy. These are chiefly scattered out of the zone center and thus do not sizeably contribute to the direct gap luminescence.

A set of Ge epitaxial layers with the same tensile strain was also studied as a function of the doping and lattice temperature [67]. Polarization-resolved PL enabled a clear-cut spectral resolution of the LH and HH components of the direct gap emission and was crucial to unveil the occupation of the available hole states. The measured PL peak positions for the polarization-resolved LH and HH transition (blue and red data points) and the predicted (solid line) emission energies are compared in Fig. 5.13b. The latter have been obtained by considering the bandgap-induced changes due to temperature, doping and the concomitant strain-induced redshift and VB splitting at a given lattice temperature [67]. The upper panel of Fig. 5.13b shows the peak positions for a 3 $\mu$m thick Ge:B epitaxial layer. Below 200 K, the peak positions agree well with the predicted ones, while at higher temperatures, the experimentally observed peaks can no longer be resolved and merge onto the higher energy component of the doublet, thus suggesting that in $p$-type doped Ge, recombination through HH states dominates in the high temperature regime, despite the unfavorable tensile strain condition. Intermediate and upper panel of Figure 5.13b further demonstrate that in not-intentionally doped and $n$-type doped layers, the peak positions remain fully consistent with the lower energy component of the doublet, namely the LH transition.

### 5.3.2 *Quantum Confined Heterostructures: Ge/SiGe Quantum Wells*

Epitaxial Ge-based heterostructures open interesting perspectives toward the implementation of photonic architectures with a tunable emission/absorption wavelength. In this context, optical and structural properties of Ge-based multiple quantum well (MQW) heterostructures have been widely studied throughout the last decade, particularly because of the notable type I band alignment [68–74]. Nevertheless, it was only during the last few years that their spin-dependent properties were actually addressed. This was chiefly enabled thanks to optical spin orientation.

In Sect. 5.1, the description of valence states in terms of pure spin states revealed that besides strain, HH-LH degeneracy can be suitably removed and tailored by quantum confinement, so that electron spin polarization can in principle reach 100% when the excitation energy is precisely tuned at the $E_\Gamma$ absorption edge. However, as previously pointed out, this minimal model strictly holds at the zone center. The mixing of VB states has to be taken into account to describe optical excitations related to states with nonvanishing momentum. In addition, if a strain field is present, as usually happens in SiGe heterostructures, it can possibly contribute to reduce the symmetry and to enhance the coupling of some of the VB states.

Virgilio and Grosso [75] were the first to apply tight-binding calculations to explore the possibility of obtaining optical spin orientation in strained-balanced Ge/SiGe MQWs. The calculated spectrally-resolved spin polarization $S$ is shown in Fig. 5.14a. An almost full polarization, i.e., 96%, is predicted in correspondence with the fundamental transition.

In Fig. 5.14a, the absorption coefficient of circularly polarized light is directly compared to the calculated spin polarization of CB electrons. By doing so, the main features of the polarization spectrum can be associated with the transitions involving the specific VB and CB states at the Γ-point. Peaks in the spectrally-resolved spin polarization are found at the HH1-cΓ1 and HH1-cΓ2 resonances, while polarization dips are ascribed to the LH1-cΓ1, LH3-cΓ1, and SO-cΓ1 transitions [75]. It should be noted that, as opposed to III–V semiconductor QWs, no sign inversion of the electron spin polarization was pointed out at the energy threshold involving LH

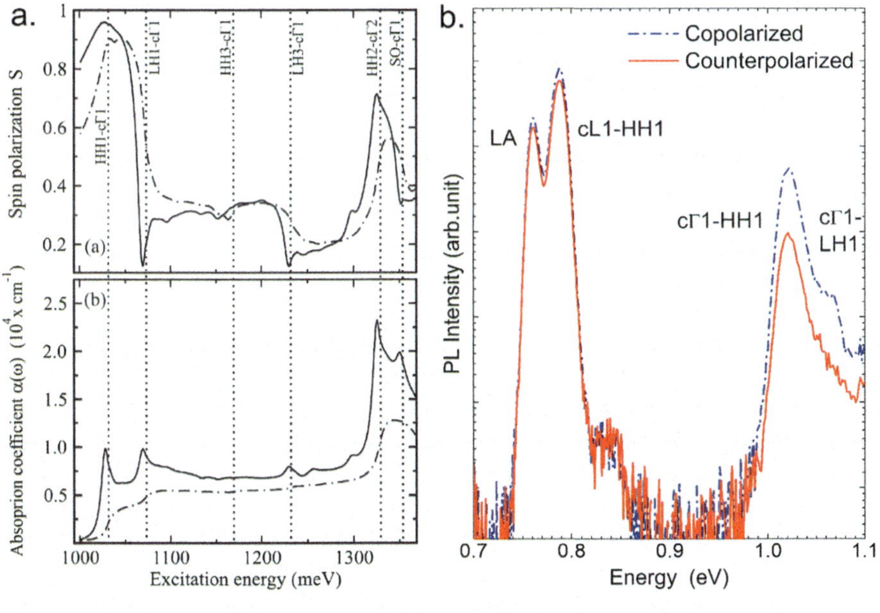


**Fig. 5.14** **a** Upper panel: spin polarization as a function of the energy for the CB electrons optically excited by circularly polarized light, incident normally to the QW plane. Lower panel: absorption spectrum. Vertical dotted lines report the relevant interband transitions. Adapted figure with permission from [75]. Copyright (2009) by the American Physical Society. **b** Spectra of copolarized (dash-dotted blue line) and counterpolarized (solid red line) PL emission with respect to circularly polarized excitation at 1.165 eV for a Ge/SiGe MQW sample at 4 K [31]

states. The tight-binding model presented in Ref. [75] provides precious support to rationalize the experiments on optical injection of spins in group IV heterostructures, as discussed in the following.

The first experimental proof of optical spin orientation in Ge QWs was successfully carried out in 2012 through polarization-resolved PL spectroscopy [31]. PL measurements revealed a circularly polarized emission with $\rho = 8\%$ at the indirect no-phonon line. Similarly, $\rho = 6\%$ was reported for the LA phonon replica [31]. Such a compelling helicity observation suggested that the electron spin relaxation time is longer than the carrier lifetime (see Sect. 5.1). By comparing the recombination lifetime [72] with the sizeable polarization degree measured in a Ge/SiGe MQWs, it was inferred that electrons feature a spin lifetime that can exceed 5 ns below 150 K [31]. Such an estimate was nicely confirmed by $k{\cdot}p$ simulations of the intrinsic relaxation mechanisms as summarized in Fig. 5.11 [10, 31]. Theory predicts that the electron spin lifetime at the bottom of the conduction band is indeed exceedingly long below 30 K, being governed by intravalley spin relaxation processes, whereas a shortening is expected at higher temperatures when intervalley scattering sets in [10].

PL offers simultaneous access to direct and indirect recombinations (Fig. 5.14b). This notable property together with the ultrafast electron transfer out of the center of the Brillouin zone allowed steady-state optical spectroscopy to resolve the concomitant spin dynamics of the two charged species. The analysis of the direct gap emission was pivotal to investigate the spin lifetime of holes. In particular, the circular polarization degree of the direct gap PL was shown to exceed the theoretical bulk limit, yielding about 37% and 85% for HH and LH transitions, respectively. The first value is in full agreement with the electron spin polarization predicted at the excitation energy used in the experiment [75]. This implies that a complete spin relaxation of the HH states has occurred on a time scale faster than the electron depopulation of the Γ-valley, namely 0.5 ps [76]. The opposite holds for the LHs, as suggested by the associated large $\rho$ value. Parity-conserving scattering events indeed guide the LH energy relaxation, contributing to the HH rather than LH depolarization.

Later on, the ultrafast hole spin dynamics was confirmed and addressed in detail in Ge/SiGe QWs by means of a time-resolved pump-probe method [77]. The unique possibility of lifting HH-LH degeneracy was pointed out to be useful in avoiding mixing of the VB states. A robust hole DSP of 80% was thus attained and the spin relaxation time for HH was shown to strongly depend upon the pump fluency and to approach 2.1 ps at 10 K [77].

The polarization of the direct gap emission from Ge/SiGe multiple quantum wells has been also studied in the 4–300 K temperature range as a function of the well thicknesses [78]. According to the spin-dependent optical selection rules, the expected polarization of the HH1-cΓ1 emission is given by two opposite contributions: a copolarized circular one deriving from the pristine absorption from HH states and a counterpolarized circular one deriving from the initial excitation of LH states.

In a 7 nm-thick QW sample, the HH1-cΓ1 excitonic peak exhibits copolarised emission with respect to the excitation (Fig. 5.15b, upper panel). The reason is that, with this well width, only the HH1-cΓ1 excitonic transition is activated by the optical pump, thus providing copolarized emission. The almost resonant condition is

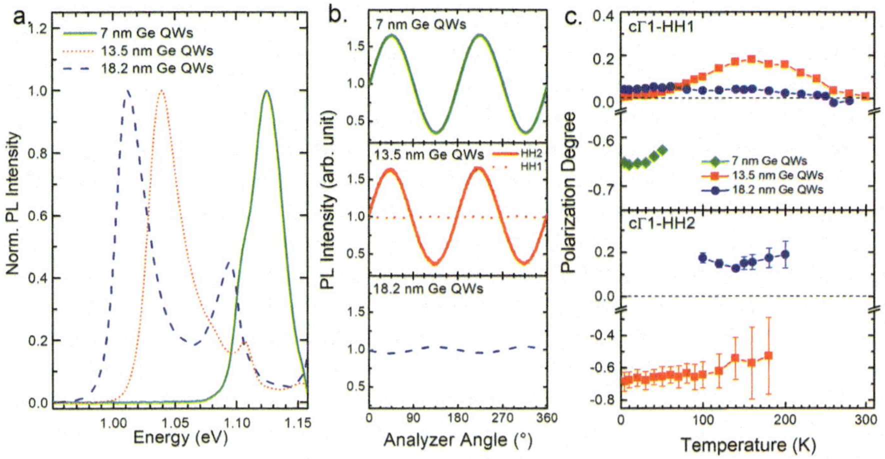


**Fig. 5.15** **a** PL spectra of Ge/SiGe MQWs with well width of 7 nm (green solid line), 13 nm (red dotted line), and 18 nm (blue dashed line). **b** Polarimetric analysis of the samples. **c** Temperature dependence of polarization degree for HH1-cΓ1 (HH2-cΓ1) transitions are reported in the upper (lower) panel. Positive (negative) values refer to counterpolarized (copolarized) with respect to the excitation. Adapted from Ref. [78], with the permission of AIP Publishing

reflected by a high polarization of about of 65%. In a sample with larger QW thickness, i.e., about 13 nm, the excitation activates not only the HH1-cΓ1 but also the LH1-cΓ1 and HH2-cΓ1 transitions (middle panel of Fig. 5.15b). In the latter transition case, the excess energy of the excitation is comparable with the excess energy of the HH1-cΓ1 transition in the 7-nm-thick QW sample. Indeed, this emission is copolarized with a high polarization of 68%. In the case of HH1-cΓ1 emission, on the contrary, electrons are excited from both the HH1 and the LH1 levels, providing two components to the HH1-cΓ1 PL line with opposite polarization yielding almost null polarization. This result evidences that HH1 spin sublevels are at least equally occupied, thus suggesting a depolarization of the heavy holes faster than the characteristic depopulation time, expected to be in the order of hundreds of fs [78]. In the last sample, the confinement energy is further reduced due to the larger thickness of the QWs, namely 18 nm, and the effect of electrons originating from the LH1 subband on the HH1-cΓ1 emission polarization increases, resulting in a counterpolarized PL peak (Fig. 5.15b lower panel), with a low $\rho$ value (8%). This work notably demonstrated a complete reversal of the PL helicity by sweeping the excitation energy throughout the LH absorption [78]. Such a finding differs, however, from the theoretical work described above [75]. Although these samples present slightly different strain level and composition of the barriers with respect to those considered in Ref. [75], the discrepancy between the experiment and theory were ascribed to the accuracy of the theoretical description of the valence-band warping, that would require a second-nearest-neighbors parametrization [78]. The experimental results demonstrate that the polarization type and degree strongly depend on the excitation of electrons from the HH and LH bands, thus providing an effective degree of freedom to control the polarization of the direct interband emission.

In addition, the analysis of the temperature dependence of the polarization degree, summarized in Fig. 5.15c, highlights spin depolarization mechanisms. In the thinnest sample, the HH1-c$\Gamma$1 emission is copolarized, and the $\rho$ value is nearly constant (65%) in the low temperature range. The same applies to the HH2-c$\Gamma$1 emission in the second sample, which is characterized by a similar $\rho$ value and by an analogous excess energy. $\rho$ values of this emission are measured up to 180 K, because above this temperature, the transition is thermally quenched. The HH1-c$\Gamma$1 transition of this sample is countercircularly polarized up to RT, and it displays a maximum of 20% at about 160 K. As mentioned before, the countercircular polarization is due to the dominance of the recombination channel activated through absorption from the LH states. When the temperature is increased, this mechanism is boosted due to bandgap shrinkage, yielding an enhancement of the countercircular polarization. Finally, in the sample with the thickest QWs, both the analyzed emissions are almost temperature independent, although above 160 K a weak luminescence depolarization is observed. Approaching RT, spin relaxation is expected due to phenomena such as electron–phonon backscattering from the $L$-valley.

Very recently, spin relaxation mechanisms in Ge/SiGe QWs have been studied in detail at cryogenic temperatures. Specifically, Ref. [59] provided a direct measurement by leveraging the time decay of the polarized PL emission.

After optical spin orientation, electron spin polarization is not extinguished during the transfer out of the $\Gamma$-valley towards the bottom of the $L$-valley. Eventually, it governs the radiative recombination with the unpolarized hole-pocket at $\Gamma$, yielding circularly polarized emission. Figure 5.16a reports the intensity decay versus time of the no-phonon PL peak, with the copolarized (blue open dots) and counterpolarized (red full dots) emissions with respect to different excitation pump powers. The different intensities of the two helicity-resolved PL components demonstrate a net circular polarization of the emission. Giving that the time decay of the optically oriented nonequilibrium population of the electron spins reflects itself by the time-dependent depolarization of the PL, these measurements provide a direct access to the spin relaxation time. Figure 5.16a demonstrates that by increasing the pump power, the PL polarization decreases at a faster pace, heralding shorter $\tau_S$ values. The density-dependent characteristics of the spin relaxation time shown in Fig. 5.16b suggests the emergence of depolarization induced by the electron–hole exchange interaction [59]. Possibly, this spin-relaxation channel is strengthened in the QWs by the spatial confinement of the carriers. Given that the PL intensity precludes a measurement in the lower carrier density regime, Giorgioni et al. confirmed that the spin relaxation time of CB electrons is in the $\mu$s regime by conducting electron spin resonance (ESR) measurements [59].

Concerning ESR, it has to be noted that the modeling of the resonance lines additionally provided the authors with $T_2^*$. The ensemble spin dephasing turned out to be ~20 ns, which is about two times longer than the hyperfine-limited dephasing times of electrons bound to shallow donors [17]. Moreover, $T_2^*$ decreases in thinner

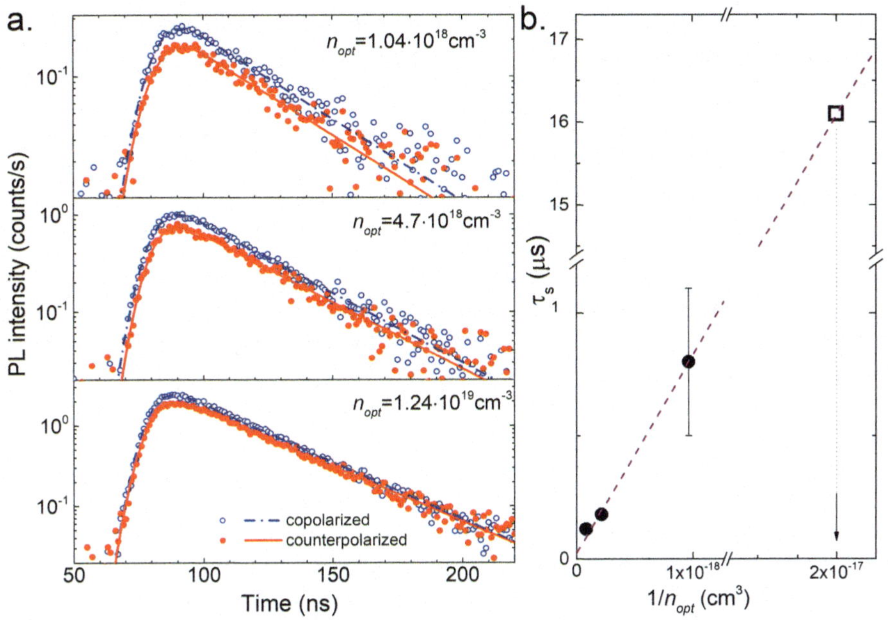


**Fig. 5.16** **a** Time-resolved PL curves showing the intensity decay from a 20-nm-thick QW at 4 K. Co-circular and counter-circular emissions with respect to the excitation have been obtained under 1.165 eV circularly polarized laser excitation. Data (dots) and fits (lines) of the PL for the various densities of the photogenerated carriers ($n_{opt}$) are reported in the three panels. **b** $\tau_S$ values (dots) obtained by using the data in **a** and a linear fit (dashed line) are reported as a function of the inverse of the photogenerated carriers [59]

QWs, due to an enhancement in the electron localization, with a correspondingly increasing efficiency in the spin dephasing.

It is worth noticing that the spin relaxation times derived in Ge/SiGe MQWs are substantially longer than the one reported in the previous sections for CB electrons in bulk Ge at the same temperatures (see Fig. 5.11). This demonstrates that quantum confinement possibly provides an additional mechanism concurring with the lengthening of the spin lifetime that arises when the CB electrons reside in the QWs rather than in bulk material. Indeed, by spatially separating mobile electrons from their parent donor atoms via remote doping in the SiGe barriers, it is possible to avoid impurity-induced spin relaxation, which is otherwise dominant, putting forward the beneficial role of confinement in enriching further the intriguing spin dynamics in the Ge/SiGe QWs.

ESR led to the additional important observation of a highly anisotropic electron Landé *g*-factor and provided the first experimental demonstration of the confinement-induced control over the *g*-factor in Ge, as theoretically predicted in 2003 [79]. Owning to spin-orbit coupling, Ge offers indeed a larger *g*-factor tuning range than Si. This opens unexplored pathways for future studies of confinement-induced tailoring of the spin physics in technologically relevant materials, such as group IV semiconductors.

## 5.4 Alloying Ge with Sn

The hybridization of the spin properties of Ge with the peculiar electronic states offered by the topological insulator $\alpha$-Sn, a heavier group IV element [80, 81], can further enrich the spin-dependent phenomena hitherto observed in Ge. Alloying Ge with Sn can, in principle, introduce a novel degree of freedom to tailor the spin susceptibility to external fields, besides offering the key manipulation of spin electronic effects comprising novel quantum phases [82].

Despite such exciting prospects, the genuine potential of $Ge_{1-x}Sn_x$ solid solutions remain presently unexplored. The main reason is that the large lattice mismatch and the low equilibrium solubility of Sn in Ge have impeded the direct large-scale deposition on Si wafers of high quality epitaxial heterostructures, especially at Sn-rich molar fractions [83]. Only very recent advances in out-of-equilibrium crystal growth techniques [83–86] have yielded reliable epitaxial films. Recently, lasing action in $Ge_{1-x}Sn_x$ has been discovered, identifying group IV candidates boasting a direct bandgap [87–89]. The contrasting theoretical [90–92] and experimental [87, 93–95] data accumulated in the literature proved the crucial resolution of the indirect-to-direct crossover to be nontrivial and highly debated in semiconductors based on group IV materials. The PL intensity is, to a good approximation, proportional to the quantum efficiency, and thus prone to the intertwined radiative and nonradiative processes.

Ubiquitous defects, such as dislocations, have been shown to have a crucial impact on the recombination dynamics, suggesting that the direct measurements of the carrier lifetime is needed to precisely resolve the picture of the carrier kinetics [87].

Figure 5.17a reports low-temperature PL of a Ge-rich $Ge_{0.95}Sn_{0.05}$ epitaxial layer grown under coherent compressive strain on a Ge-buffered Si substrate. The spectrum reveals an intense peak at about 0.64 eV, which is attributed to band-to-band transitions through the indirect $Ge_{0.95}Sn_{0.05}$ film [16, 96]. Figure 5.17b shows the PL transient measured at 7 K and at various pump fluencies by De Cesari et al. [17]. A nonmonoexponential decay time is observed, and particularly under low power excitation conditions, the early phase of the recombination process exhibits a rather slow decay, which becomes steeper after 4 ns. This puzzling PL transient can be accounted for by considering two competing recombination mechanisms, whose relative importance in dictating the carrier dynamics is determined by the instantaneous density of the out-of-equilibrium photogenerated carriers. Such a scenario was supported by a detailed analysis of the PL transient as a function of the pump power density (Fig. 5.17b) and lattice temperature (Fig. 5.17c) [17]. Specifically, the

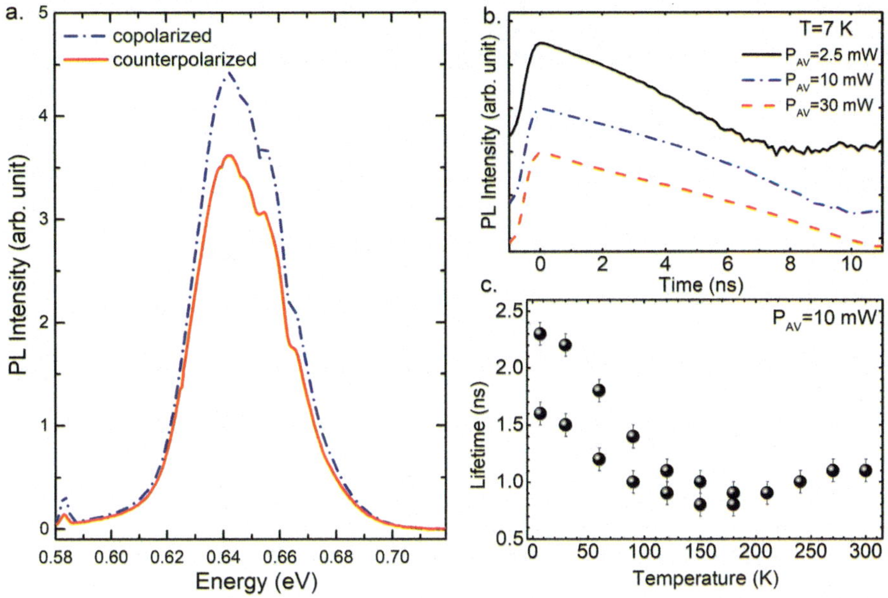


**Fig. 5.17** **a** Low temperature copolarized and counterpolarized PL emission, with respect to the excitation, of $Ge_{0.95}Sn_{0.05}$ epilayer. **b** Time decay of PL at different pump powers ($P_{AV}$). **c** Temperature dependence of the carrier lifetime [17]

temperature-induced changes of the carrier lifetime, which remains in the ns range up to room temperature, were satisfactorily modeled within the Shockley-Read-Hall framework, disclosing the presence of shallow traps localized at about 13 and 17 meV above the Fermi level [17].

### *5.4.1 Spin Relaxation and Spin Dephasing Time in GeSn*

Optical spin orientation was subsequently applied to generate an out-of-equilibrium ensemble of spins [16]. Figure 5.17a shows helicity-resolved PL of the coherent $Ge_{0.95}Sn_{0.05}$ epilayer at low temperature, indicating a circular polarization degree of about 12% [16]. It was shown that the steady-state circular PL polarization degree is robust over a wide temperature range and remarkably large even at room temperature. The helicity-resolved PL dynamics, following a circularly polarized excitation with a pulsed laser, further demonstrates an amplitude imbalance and provides an additional evidence for the successful optical spin injection in the $Ge_{0.95}Sn_{0.05}$ epilayer (Fig. 5.18a). Above all, these PL data openly manifest a rather long spin relaxation time. The decay rate of the polarization degree was shown to strikingly increase with the laser power density, shortening the spin relaxation time from about 60 to 10 ns [16]. This power dependence of $\tau_S$ points toward the occurrence of optically

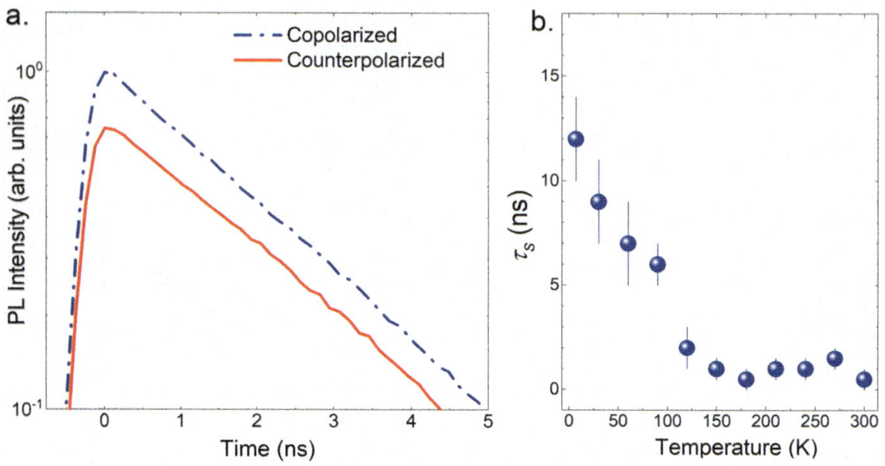


**Fig. 5.18** **a** Temporal evolution of the polarization-resolved PL of $Ge_{0.95}$ $Sn_{0.05}$ under right-handed circularly polarized excitation. The copolarized component is shown as the blue dashed line, whereas the counterpolarized component is reported as a red solid line. **b** Temperature dependence of the spin relaxation time $\tau_S$ [72]

induced spin-flip processes and suggests the opening also in $Ge_{1-x}Sn_x$ of a low temperature relaxation channel caused by exchange interaction among electrons and photogenerated holes.

The systematic investigation of the spin relaxation time as a function of the temperature is summarized in Fig. 5.18b, demonstrating that $\tau_S$ in the $Ge_{0.95}Sn_{0.05}$ epilayer remains in the nanosecond regime even in the high temperature range, thus comparing favorably with the values reported in III-V compounds and bulk Ge [10, 16] (see Fig. 5.11). Such a result provides a strong indication that Dyakonov-Perel relaxation does not play a major role in the spin dynamics in the diluted tin content regime.

If the electron spin dynamics is measured under the presence of an external magnetic field ($B$) in the Voigt geometry (i.e., $B$ along the [110] direction), intensity oscillations of the PL circular polarization components can be clearly observed (see Fig. 5.19a). As discussed in Sect. 5.2.2, these oscillations stem from the Larmor precession of the electron spins in the transverse magnetic field. De Cesari et al. also applied for the first time spin quantum beat spectroscopy to group IV materials, specifically to the $Ge_{0.95}Sn_{0.05}$/Ge/Si heterostructure.

The time dependence of the circular polarization degree was obtained by varying the strength of the applied field, as shown in the inset of Fig. 5.19a. It is worth noticing that the magnetic field amplitude affects both the period of the quantum beats oscillations and their damping. From the former, the Larmor precession frequency $\Omega$ was extracted at each field. Figure 5.19b demonstrates a linear relationship between $\Omega$ and $B$, allowing the determination of the effective Landé $g$-factor of CB electrons. It was found that $g_e = 1.48 \pm 0.01$ [16], which is in line with the expected Ge-like nature of the $L$-valley minima.

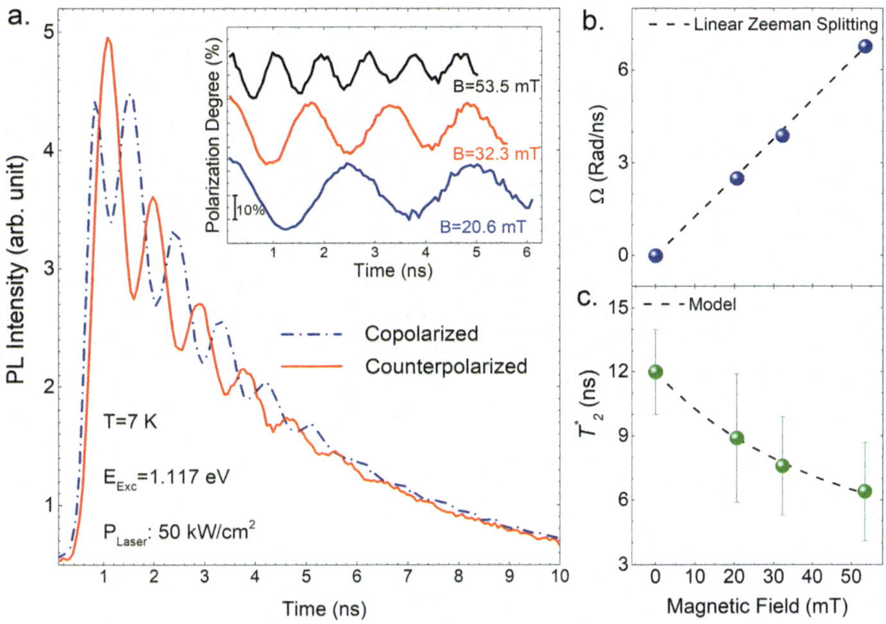


**Fig. 5.19** **a** Time evolution of the PL intensity of a $Ge_{0.95}Sn_{0.05}$ sample measured at 7 K for a right-handed circularly polarized laser excitation energy of 1.165 eV. In the inset: magnetic field dependence of the dynamics of the $\rho_{circ}$. The lines have been shifted for clarity. **b** Dependence of the Larmor frequency $\Omega$ on the magnetic field strength. The dashed line corresponds to the linear dependence determined by the Zeeman splitting. **c** Ensemble spin coherence time $T_2^*$ as a function of the external magnetic field

The damping of the oscillations, on the other hand, discloses the presence of the dephasing time of the spin ensemble, $T_2^*$. Figure 5.19c reports the $T_2^*$ values derived from the PL data. The dashed line in Fig. 5.19c is a modeling of the dephasing mechanisms. The cause of decoherence was ascribed to the microscopic $g$-factor spread $\Delta g \cong 2.1 \times 10^{-2}$ in the alloy layer associated with the random effective magnetic fields given by local changes in composition and strain [16, 97].

## 5.5 Future Perspective

The intriguing aspect of controlling light polarization and spin relaxation mechanisms by precisely tuning structural and electronic properties of Ge-based architectures, harnesses creative, and unprecedented strategies to overcome research roadblocks toward the implementation of a new research frontier, namely spin-optronics or spin-optoelectronics (SOE): a field at the crossroads of photonics, spintronics, quantum, and nano-technologies.

Information and communication technologies could take radical advantage of the interplay between light polarization and spin physics. In particular, the possibility of controlling and reading the helicity of photons has practical prospects for the implementation of new communication protocols, thereby boosting a wealth of applications ranging from the transmission of encrypted data, reconfigurable optical interconnection systems, and the realization of novel optical switches and modulators [25, 98, 99].

In the following, we will briefly review the two basic building blocks that are needed to attain a real-life deployment of SOE, namely emitters of circularly polarized light and detectors of the state of light polarization. Specifically, we will focus on achievements obtained using group IV semiconductors in the quest for the monolithic integration of SOE functionalities on state-of-the-art electronic circuitry.

### *5.5.1 Spin Photodiode and Spin-LEDs*

Along the path of realizing a fully scalable technological platform for spin-optoelectronics, different elements are needed [100], such as spin-based light-emitting-diodes (spin-LEDs) [101, 102] and spin-photodetectors (spin-PDs), [103, 104] converting electrical signals into photon helicity and vice versa. Conventionally, both kinds of devices are realized by means of a ferromagnetic (FM) contact and an oxide tunnel barrier, to provide injection or detection (filtering) of spin-polarized carriers, respectively.

In spin-PD devices, after spin-polarized generation in the semiconductor by means of optical spin orientation, the photogenerated electrons and holes separate each other and move in opposite directions under the action of an electric field produced by an external bias voltage. In particular, in forward (reverse) bias, electrons (holes) tend to move toward the oxide barrier and undergo tunneling toward the metal. The transmission of the barrier depends on the relative orientation between the carrier spin polarization and the FM magnetization, i.e., photocarriers suffer a different resistance for a spin parallel or antiparallel to the FM magnetization. This finally leads to a modulation of the electrical response of the device [105].

Rinaldi and coworkers demonstrated a spin-photodiode device based on epitaxial Fe/MgO/Ge(001) heterojunctions [106] capable of working at room temperature. They showed an electrical detection of the spin polarization of photogenerated carriers by circularly polarized photons at a wavelength of 1300 nm. A variation of the photocurrent of about 6% upon reversal of the incident light helicity was measured both in reverse and forward bias for light intensity down to 200 $\mu$W.

It has to be noted that the spin-PD operation principle requires that the magnetization is out-of-plane to be selective with respect to the direction of the quantization axis, which is defined by the light propagation. To this purpose, Rinaldi et al. [106] applied an external magnetic field of up to 1.2 T to drive the magnetization of the 10-nm-thick Fe layer in the out-of-plane configuration. However, this appears to be a strong limitation for future practical device applications. In their work, Djeffal et al. [107] investigated Co–Fe–B/MgO/Ge(001) spin photodiodes operating at zero magnetic field: a photocurrent helicity asymmetry of about 0.9% was measured at 9 K and remains detectable (about 0.1%) at room temperature. Since the FM demonstrates a spontaneous out-of-plane magnetization, these results constitute an important step toward the realization of Ge-based spin-PDs compatible with the conventional optoelectronic platforms.

A spin-LED should ideally consist of a semiconducting core where spin-oriented carriers are introduced by means of the FM polarizing contacts and recombine radiatively generating photons with a well-defined angular momentum, for example, circularly polarized [100]. The more stringent requirement for any practical implementation of such emitters is the efficient electrical spin injection at room temperature without high external magnetic fields. Spin-LEDs have been successfully fabricated in III-V-based semiconductors such as GaAs or GaN [101, 102, 108, 109] and spin-based lasers were very recently shown to outperform conventional vertical-cavity surface-emitting laser (VCSEL) structure in terms of ultrafast operations and low-energy consumption [99]. A spin-laser with a modulation frequency above 200 GHz was obtained, exceeding by nearly an order of magnitude the best conventional semiconductor lasers [99]. At the same time, an order of magnitude better energy efficiency is observed. The realization of ultrafast spin-lasers relies on a short carrier spin relaxation time, previously considered detrimental for spintronic applications [5].

To date, the efforts in the field of electrical spin injection in Ge led to the investigation of a room temperature spin-LED [110] relying on a Fe/$GeO_2$ contact. Such a Ge-based LED presents a circular polarization dependence of the direct gap electroluminescence although the use of a 4 T magnetic field is needed. Moreover, similar results were reported for a control device with a nonmagnetic Au contact, suggesting that the contribution of the electrically injected spins, if present, is relatively weak. This field is still in its infancy and further work needs to be done. However, given the striking results demonstrated in III–V compounds and the rapid progress in GeSn lasers, we can expect that the implementation of a spin-laser will be soon accomplished also in group IV materials, opening a viable route in the bourgeoning field of SOE.

### 5.5.2 Spin-Charge Interconversion Phenomena

The effects that allow the mutual transformation of a charge current into a flow of spins are generically named spin-charge interconversion (SCI) phenomena. The mechanisms that permit SCI have recently attracted attention because they provide a viable tool to probe spin injection via electric means and because they can be exploited in novel dissipationless devices based on pure spin current [111]. As we will discuss here, such a property can be fruitfully utilized also in the SOE arena for the implementation of efficient spin-sensitive photodetectors.

In solids, the spin-orbit interaction (SOI) affects the transport properties depending on the spin state of the charge carriers. The possibility of generating a spin accumulation from a charge current was first revealed by Dyakonov and Perel [112] and then by Hirsch [113]. The phenomenon was named the spin-Hall effect (SHE), in analogy with the well-known Hall effect. The spin-charge conversion, performed via SHE, usually takes place in the bulk of materials; therefore, we can consider it as a three-dimensional SCI phenomenon.

When a charge current density $\boldsymbol{j}$ is flowing in a material with SOI, the SHE generates a pure spin current density $\boldsymbol{j}_s$. The latter is perpendicular to both $\boldsymbol{j}$ and the direction of the spin polarization of the carriers $\boldsymbol{S}$. The phenomenological relation describing SHE is [4]:

$$\boldsymbol{j}_s = \gamma \boldsymbol{j} \times \boldsymbol{S} \tag{5.13}$$

$\gamma$ being the *spin-Hall angle*, which eventually determines the efficiency of SCI. Frequently, the $\gamma$ parameter is expressed as the ratio between a *spin-Hall conductivity* $\sigma_{\mathrm{SH}} = \gamma\sigma_{\mathrm{C}}$ and the electrical conductivity $\sigma_{\mathrm{C}}$. The time reversal of the SHE is the *inverse spin-Hall effect* (ISHE). In this case, the flow of a spin current density generates a charge current density:

$$\boldsymbol{j} = \gamma \boldsymbol{j}_s \times \boldsymbol{S} \tag{5.14}$$

It is worth mentioning that ISHE can be exploited to detect spin currents. In materials where SOI, and hence the spin-Hall angle $\gamma$, is large, ISHE can efficiently convert the spin current into a measurable charge current. When not dealing with materials where $\gamma$ is large enough for an efficient detection, it is possible to transfer the spin population to an adjacent material with a relevant SOI, where efficient SCI might occur. Microscopically, three different mechanisms can give rise to SCI. The first one is named *skew scattering* and is associated with scattering centers, represented for instance by doping impurities. In this case, the propagation direction of a carrier after the scattering event is sensitive to its spin-polarization state, especially if the scattering center possesses a large atomic number and is therefore endowed by a large SOI [4].

The *side jump* mechanism is a lateral displacement of the carrier, depending on its spin state, during the scattering event [4]. Its origin is related to the local distortion of the incoming carrier wavefunction given by the impurity. Due to their link with collisions, *skew scattering* and *side jump* are defined as *extrinsic* mechanisms.

Spin-polarized carriers can also undergo *intrinsic* SCI. The latter originates from the geometrical phase, referred to as *Berry phase*, acquired by the wavefunction due to the introduction of relativistic effects in the band structure, which become relevant in systems with large SOI [114].

In spin-to-charge conversion measurements, the electric signal generated by the conversion of a spin current via the ISHE is detected after spin current injection via optical spin orientation, as employed by Bottegoni et al. in Pt/Ge junctions [115]. In a typical experimental apparatus for photoinduced spin-charge conversion measurements, a monochromatic light beam passes through a polarizer and a photoelastic modulator (PEM), which modulates the circular polarization of the light. The light is then focused on the sample by a lens or an objective and the detected electric signal is demodulated by a lock-in amplifier at the PEM frequency. In case of Ref. [115], a high-SOI material (Pt) continuous film was put in direct contact with Ge, which is exploited as a spin generator by means of the optical spin orientation technique, while the Pt works as a spin detector. In order to detect the charge current in the Pt, generated by the SCI, two ohmic contacts (typically Au/Ti) are deposited on the top of the Pt layer. Giving that spin population is injected into the semiconductor and then transferred to the Pt, the spin current flows from the semiconductor to the metal and is perpendicular to the ohmic contact plane. According to Eq. 5.13, the injection of an in-plane spin component is mandatory to detect a spin-related charge current in the metal film. A reversal of the light helicity is reflected in a reverse ISHE signal. The spin polarization of the photoexcited electrons is parallel to the direction of the light wavevector inside Ge. In order to obtain a significant in-plane projection of the spin, it is convenient to shine light onto the semiconductor at a grazing incidence. Notably, the selectivity to polarized electrons generates an angular dependence of the spin-to-charge conversion signal, which is a viable tool to confirm the spin-related nature of the measurable electric signal without the demanding fabrication of ferromagnetic contacts [115].

The photon energy dependence of the ISHE signal was also addressed in Pt/Ge [116]. As expected, the ISHE photovoltage signal, $\Delta V_{\text{ISHE}}$, mimics the electron spin polarization: it reaches a maximum when the photon energy is tuned around the direct energy gap of Ge and decreases as transitions toward the SO band are allowed. Such results point toward the exploitation of such mechanisms for the implementation of polarization-sensitive detectors over a wide spectral window.

The illumination of the sample at grazing incidence, as described above, degrades the spatial resolution of the experimental setup and is not compatible with conventional working conditions of photodetectors that typically require light propagation parallel to the surface normal of the device. A viable route to overcome these issues consists in a metal grating deposited on the top of the illuminated facet [117].

When the circular polarized laser beam impinges along the $z$ direction (see Fig. 5.20a), hence orthogonally onto one edge of a metallic ridge, the component of the field which is perpendicular to the edge ($E_x$) induces an electric dipole in the metal, generating a near field in the semiconductor with a component directed along $z$ [58]. The latter is coupled to the $E_y$ component of the propagating wave. Since $E_z$ and $E_y$ have a phase shift of $\pi/2$, an elliptically polarized electric field is produced in the $yz$ plane. Upon absorption of this field, a spin-oriented electron population is generated in the CB of the underlying semiconductor with a spin polarization along the $x$-axis. Moreover, the direction of $E_y$ is opposite at opposite edges of the strip line, thus flipping the spin of the photogenerated carriers. Opposite spin polarizations are therefore obtained at opposite edges of the metal grating.

The reflectivity map and the spin-related electrical signal obtained with such a device are shown in Fig. 5.20b, c, and e. Since it is possible to correlate the spin

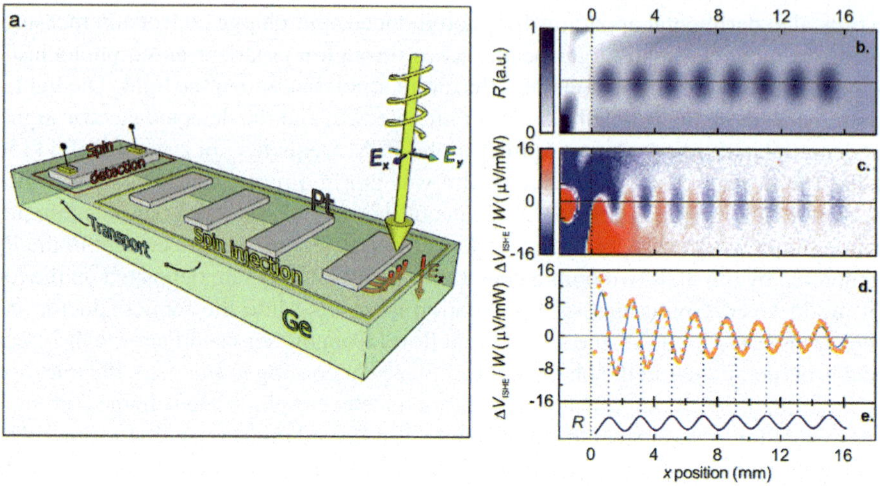


**Fig. 5.20** **a** Sketch of a lateral device for optical spin generation and nonlocal electrical spin detection [58]. To obtain optical generation of in-plane spin-polarized electrons, the light beam is focused at the edges of a Pt pad, and the y component (green arrow) of the circularly polarized illuminating field induces antiphase oscillating charges that, in turn, generate a static-like near field (red field lines) with a strong $z$ component that, combining with the $\pi/2$ dephased $x$ component of the illuminating field (blue arrows) produces a circularly polarized electric field. A complementary spin polarization is generated at the opposite edge. **b** Reflectivity (R) optical images and **c** corresponding simultaneously recorded ISHE voltage signals normalized to the laser power W. **d** Voltage signal and **e** reflectivity profiles along x across the centers of the Pt pads. In **d** solid lines correspond to fits and the dots are the experimental data reported in Ref. [58]. Reprinted figure with permission from [58]. Copyright (2017) by the American Physical Society

generation point to the corresponding signal obtained at the far-away ISE detector, this spin injection/detection schemes allow for the additional direct measurement of the diffusive spin paths in Ge. Notably, the amplitude of the signal oscillates around zero as the laser beam illuminates neighboring edges of the metal grating and decreases as the light beam moves away from the ISHE detector. These findings were interpreted within the frame of a one-dimensional diffusion model, yielding at room temperature a spin lifetime $\tau_S \approx 20$ ns and spin diffusion length of about 10 μm.

## 5.6 Conclusion

We discussed progress in the injection of spins in Ge-based heterostructures by optical means and emphasized the importance of this approach in gathering a central understanding of the spin physics as well as light-matter interaction. Investigations based on the optical spin orientation process pointed out a rich spin dynamic and the compelling role of the multivalley nature of the CB of Ge in governing it. These findings can be instrumental in the quest for the joint implementation of advanced spintronic and photonic functionalities onto the existing Si-based technology.

## References


1. P. Rouard, P. Bousquet, Quelques Résultats Récents Concernant Les Techniques de Préparation et L'étude Optique Des Couches Minces Solides. Opt. Acta (Lond). **16**, 675 (1969)
2. G. Lampel, Nuclear dynamic polarization by optical electronic saturation and optical pumping in semiconductors. Phys. Rev. Lett. **20**, 491 (1968)
3. M.I. D'yakonov, V.I. Perel, Theory of optical spin orientation of electrons and nuclei in semiconductors, in *Optical Orientation*, ed. by F. Meier, B.P. Zakharchenya (North Holland, Amsterdam, 1984)
4. M.I. D'yakonov, *Spin Physics in Semiconductors* (Spinger-Verlag, Berlin, 2004)
5. I. Žutić, J. Fabian, S. Das Sarma, Spintronics: fundamentals and applications. Rev. Mod. Phys. **76**, 323 (2004)
6. A. Corney, *Atomic and Laser Spectroscopy* (1977)
7. Z. Wilamowski, H. Malissa, F. Schäffler, W. Jantsch, G-Factor tuning and manipulation of spins by an electric current. Phys. Rev. Lett. **98**, 1 (2007)
8. Z. Wilamowski, N. Sandersfeld, W. Jantsch, D. Többen, F. Schäffler, Screening breakdown on the route toward the metal-insulator transition in modulation doped Si/SiGe quantum wells. Phys. Rev. Lett. **87**, 19 (2001)
9. J. Fabian, A. Matos-Abiague, C. Ertler, P. Stano, I. Žutić, Semiconductor spintronics. Acta Phys. Slovaca **57**, 565 (2007)
10. P. Li, Y. Song, H. Dery, Intrinsic spin lifetime of conduction electrons in germanium. Phys. Rev. B **86**, 85202 (2012)
11. J.-M. Tang, B.T. Collins, M.E. Flatté, Electron spin-phonon interaction symmetries and tunable spin relaxation in silicon and germanium. Phys. Rev. B **85**, 45202 (2012)
12. P. Li, H. Dery, Spin-orbit symmetries of conduction electrons in silicon. Phys. Rev. Lett. **107**, 1 (2011)
13. C. Hautmann, M. Betz, Magneto-optical analysis of the effective *g* tensor and electron spin decoherence in the multivalley conduction band of bulk germanium. Phys. Rev. B **85**, 121203(R) (2012)
14. P. Li, J. Li, L. Qing, H. Dery, I. Appelbaum, Anisotropy-driven spin relaxation in germanium. Phys. Rev. Lett. **111**, 257204 (2013)
15. G.E. Pikus, A.N. Tikov, Spin relaxation, in *Optical Orientation*, ed. by F. Meier, B.P. Zakharchenya (North Holland, Amsterdam, 1984)
16. S. De Cesari, A. Balocchi, E. Vitiello, P. Jahandar, E. Grilli, T. Amand, X. Marie, M. Myronov, F. Pezzoli, Spin-coherent dynamics and carrier lifetime in strained Ge(1-x)Sn(x) semiconductors on silicon. Phys. Rev. B **99** (2019)
17. A.J. Sigillito, R.M. Jock, A.M. Tyryshkin, J.W. Beeman, E.E. Haller, K.M. Itoh, S.A. Lyon, Electron spin coherence of shallow donors in natural and isotopically enriched germanium. Phys. Rev. Lett. **115** (2015)
18. A.J. Sigillito, J.C. Loy, D.M. Zajac, M.J. Gullans, L.F. Edge, J.R. Petta, Site-selective quantum control in an isotopically enriched $^{28}Si/Si_{0.7}Ge_{0.3}$ quadruple quantum dot. Phys. Rev. Appl. **11**, 1 (2019)
19. C.P.J. Poole, *Electron Spin Resonance: A Comprehensive Treatise on Experimental Technique* (Dover Pubblications, 1996)
20. W. Hanle, Über magnetische beeinflussung der polarisation der resonanzfluoreszenz. Zeitschrift Für Phys. **30**, 93 (1924)

21. A.P. Heberle, W.W. Rühle, K. Ploog, Quantum beats of electron larmor precession in GaAs wells. Phys. Rev. Lett. **72**, 3887 (1994)
22. R.R. Parsons, Band-to-band optical pumping in solids and polarized photoluminescence. Phys. Rev. Lett. **23**, 1152 (1969)
23. B.T. Jonker, G. Kioseoglou, A.T. Hanbicki, C.H. Li, P.E. Thompson, Electrical spin-injection into silicon from a ferromagnetic metal/tunnel barrier contact. Nat. Phys. **3**, 542 (2007)
24. S.P. Dash, S. Sharma, R.S. Patel, M.P. De Jong, R. Jansen, Electrical creation of spin polarization in silicon at room temperature. Nature **462**, 491 (2009)
25. R. Jansen, Silicon spintronics. Nat. Mater. **11**, 400 (2012)
26. F. Nastos, J. Rioux, M. Strimas-Mackey, B.S. Mendoza, J.E. Sipe, Full band structure LDA and Kp calculations of optical spin-injection. Phys. Rev. B—Condens. Matter Mater. Phys. **76**, 205113 (2007)
27. P. Li, H. Dery, Theory of spin-dependent phonon-assisted optical transitions in silicon. Phys. Rev. Lett. **105**, 37204 (2010)
28. H. Allenspach, F. Meier, D. Pescia, Experimental symmetry analysis of electronic states by spin-dependent photoemission. Phys. Rev. Lett. **51**, 2148 (1983)
29. E.J. Loren, B.A. Ruzicka, L.K. Werake, H. Zhao, H.M. Van Driel, A.L. Smirl, Optical injection and detection of ballistic pure spin currents in Ge. Appl. Phys. Lett. **95** (2009)
30. E.J. Loren, J. Rioux, C. Lange, J.E. Sipe, H.M. van Driel, A.L. Smirl, Hole spin relaxation and intervalley electron scattering in germanium. Phys. Rev. B **84**, 214307 (2011)
31. F. Pezzoli, F. Bottegoni, D. Trivedi, F. Ciccacci, A. Giorgioni, P. Li, S. Cecchi, E. Grilli, Y. Song, M. Guzzi, H. Dery, G. Isella, Optical spin injection and spin lifetime in Ge heterostructures. Phys. Rev. Lett. **108**, 156603 (2012)
32. F. Pezzoli, L. Qing, A. Giorgioni, G. Isella, E. Grilli, M. Guzzi, H. Dery, Spin and energy relaxation in germanium studied by spin-polarized direct-gap photoluminescence. Phys. Rev. B **88**, 45204 (2013)
33. J. Rioux, J.E. Sipe, Optical injection and control in germanium: thirty-band Kp theory. Phys. Rev. B **81**, 155215 (2010)
34. C. Rinaldi, M. Cantoni, M. Marangoni, C. Manzoni, G. Cerullo, R. Bertacco, Wide-range optical spin orientation in Ge from near-infrared to visible light. Phys. Rev. B—Condens. Matter Mater. Phys. **90**, 161304 (2014)
35. P.Y. Yu, M. Cardona, *Fundamentals of Semiconductors: Physics and Materials Properties*, 4th ed. (2010)
36. G. Mak, H.M. Van Driel, Femtosecond transmission spectroscopy at the direct band-edge of germanium. Phys. Rev. B **49**, 16817 (1994)
37. K. Tanaka, H. Ohtake, T. Suemoto, Determination of intervalley scattering time in germanium by. Phys. Rev. Lett. **71**, 1935 (1993)
38. A. Giorgioni, E. Vitiello, E. Grilli, M. Guzzi, F. Pezzoli, Valley-dependent spin polarization and long-lived electron spins in germanium. Appl. Phys. Lett. **105** (2014)
39. S. De Cesari, R. Bergamaschini, E. Vitiello, A. Giorgioni, F. Pezzoli, Optically reconfigurable polarized emission in germanium. Sci. Rep. **8**, 1 (2018)
40. F. Pezzoli, A. Balocchi, E. Vitiello, T. Amand, X. Marie, Optical orientation of electron spins and valence-band spectroscopy in germanium. Phys. Rev. B **91**, 201201 (2015)
41. D.H. Goldstein, *Polarized Light*, 3rd edn. (CRC Press, 2010)
42. E. Collett, *Field Guide to Polarization* (SPIE Press, 2005)
43. M. Oestreich, J. Hubner, D. Hagele, M. Bender, N. Gerhardt, M. Hofmann, W.W. Ruhle, H. Kalt, T. Hartmann, P. Klar, Spintronics: spin electronics and optoelectronics in semiconductors. Adv. Solid State Phys **41**, 173 (2001)
44. C. Guite, V. Venkataraman, High sensitivity detection of radio-frequency modulated magnetic moment in semiconductors. Rev. Sci. Instrum. **82**, 103905 (2011)
45. C. Guite, V. Venkataraman, Measurement of electron spin lifetime and optical orientation efficiency in germanium using electrical detection of radio frequency modulated spin polarization. Phys. Rev. Lett. **107**, 166603 (2011)

46. C. Guite, V. Venkataraman, Temperature dependence of spin lifetime of conduction electrons in bulk germanium. Appl. Phys. Lett. **101**, 252404 (2012)
47. C. Hautmann, B. Surrer, M. Betz, Ultrafast optical orientation and coherent larmor precession of electron and hole spins in bulk germanium. Phys. Rev. B—Condens. Matter Mater. Phys. **83**, 1 (2011)
48. J. Lohrenz, T. Paschen, M. Betz, Resonant spin amplification in intrinsic bulk germanium: evidence for electron spin lifetimes exceeding 50 ns. Phys. Rev. B **89**, 121201(R) (2014)
49. L.M. Roth, G factor and donor spin-lattice relaxation for electrons in germanium and silicon. Phys. Rev. **118**, 1534 (1960)
50. J.C. Hensel, G. Feher, Cyclotron resonance experiments in uniaxially stressed silicon: valence band inverse mass parameters and deformation potentials. Phys. Rev. **129**, 1041 (1963)
51. P. Li, D. Trivedi, H. Dery, Spin-dependent optical properties in strained silicon and germanium. Phys. Rev. B **87**, 115203 (2013)
52. N. Sircar, D. Bougeard, Experimental investigation of the optical spin-selection rules in bulk Si and Ge/Si quantum dots. Phys. Rev. B—Condens. Matter Mater. Phys. **89**, 1 (2014)
53. A. Jain, J.-C. Rojas-Sanchez, M. Cubukcu, J. Peiro, J.C. Le Breton, E. Prestat, C. Vergnaud, L. Louahadj, C. Portemont, C. Ducruet, V. Baltz, A. Barski, P. Bayle-Guillemaud, L. Vila, J.-P. Attané, E. Augendre, G. Desfonds, S. Gambarelli, H. Jaffrès, J.-M. George, M. Jamet, Crossover from spin accumulation into interface states to spin injection in the germanium conduction band. Phys. Rev. Lett. **109**, 106603 (2012)
54. K. Hamaya, Y. Baba, G. Takemoto, K. Kasahara, S. Yamada, K. Sawano, M. Miyao, Qualitative study of temperature-dependent spin signals in n-Ge-based lateral devices with Fe3Si/N+ -Ge schottky-tunnel contacts. J. Appl. Phys. **113**, 183713 (2013)
55. Y. Fujita, M. Yamada, S. Yamada, T. Kanashima, K. Sawano, K. Hamaya, Temperature-independent spin relaxation in heavily doped *n*-type germanium. Phys. Rev. B **94**, 1 (2016)
56. C. Rinaldi, S. Bertoli, M. Asa, L. Baldrati, C. Manzoni, M. Marangoni, G. Cerullo, M. Bianchi, R. Sordan, R. Bertacco, M. Cantoni, Determination of the spin diffusion length in germanium by spin optical orientation and electrical spin injection. J. Phys. D. Appl. Phys. **49** (2016)
57. Y. Zhou, W. Han, L.-T. Chang, F. Xiu, M. Wang, M. Oehme, I.A. Fischer, J. Schulze, R.K. Kawakami, K.L. Wang, Electrical spin injection and transport in germanium. Phys. Rev. B **84**, 125323 (2011)
58. C. Zucchetti, F. Bottegoni, C. Vergnaud, F. Ciccacci, G. Isella, L. Ghirardini, M. Celebrano, F. Rortais, A. Ferrari, A. Marty, M. Finazzi, M. Jamet, Imaging spin diffusion in germanium at room temperature. Phys. Rev. B **96**, 014403 (2017)
59. A. Giorgioni, S. Paleari, S. Cecchi, E. Vitiello, E. Grilli, G. Isella, W. Jantsch, M. Fanciulli, F. Pezzoli, Strong confinement-induced engineering of the *g* factor and lifetime of conduction electron spins in Ge quantum wells. Nat. Commun. **7**, 1 (2016)
60. I. Appelbaum, B. Huang, D.J. Monsma, Electronic measurement and control of spin transport in silicon. Nature **447**, 295 (2007)
61. Y. Song, O. Chalaev, H. Dery, Donor-driven spin relaxation in multivalley semiconductors. Phys. Rev. Lett. **113**, 167201 (2014)
62. X. Sun, J. Liu, L.C. Kimerling, J. Michel, Direct gap photoluminescence of *n*-type tensile-strained Ge-on-Si. Appl. Phys. Lett. **95** (2009)
63. F. Bottegoni, A. Ferrari, G. Isella, S. Cecchi, M. Marcon, D. Chrastina, G. Trezzi, F. Ciccacci, Ge/SiGe heterostructures as emitters of polarized electrons. J. Appl. Phys. **111**, 063916 (2012)
64. F. Bottegoni, A. Ferrari, G. Isella, M. Finazzi, F. Ciccacci, Enhanced orbital mixing in the valence band of strained germanium. Phys. Rev. B **85**, 245312 (2012)
65. A. Ferrari, F. Bottegoni, G. Isella, S. Cecchi, F. Ciccacci, Epitaxial $Si_{1-x}Ge_x$ alloys studied by spin-polarized photoemission. Phys. Rev. B **88**, 115209 (2013)
66. A. Ferrari, F. Bottegoni, S. Cecchi, G. Isella, F. Ciccacci, Optical spin orientation in group-IV heterostructures. J. Appl. Phys. **113**, 17 (2013)
67. E. Vitiello, M. Virgilio, A. Giorgioni, J. Frigerio, E. Gatti, S. De Cesari, E. Bonera, E. Grilli, G. Isella, F. Pezzoli, Spin-dependent direct gap emission in tensile-strained Ge films on Si substrates. Phys. Rev. B—Condens. Matter Mater. Phys. **92**, 201203(R) (2015)

68. Y.-H. Kuo, Y.K. Lee, Y. Ge, S. Ren, J.E. Roth, T.I. Kamins, D.A.B. Miller, J.S. Harris, Strong quantum-confined stark effect in germanium quantum-well structures on silicon. Nature **437**, 1334 (2005)
69. M. Bonfanti, E. Grilli, M. Guzzi, M. Virgilio, G. Grosso, D. Chrastina, G. Isella, H. Von Känel, A. Neels, Optical transitions in Ge/SiGe multiple quantum wells with Ge-rich barriers. Phys. Rev. B—Condens. Matter Mater. Phys. **78**, 041407(R) (2008)
70. D.J. Paul, 8-Band k.p modeling of the quantum confined stark effect in Ge quantum wells on Si substrates. Phys. Rev. B **77**, 155323 (2008)
71. E. Gatti, E. Grilli, M. Guzzi, D. Chrastina, G. Isella, H. Von Känel, Room temperature photoluminescence of Ge multiple quantum wells with Ge-rich barriers. Appl. Phys. Lett. **98**, 031106 (2011)
72. A. Giorgioni, E. Gatti, E. Grilli, A. Chernikov, S. Chatterjee, D. Chrastina, G. Isella, M. Guzzi, Photoluminescence decay of direct and indirect transitions in Ge/SiGe multiple quantum wells. J. Appl. Phys. **111**, 13501 (2012)
73. E. Gatti, F. Isa, D. Chrastina, E.M. Gubler, F. Pezzoli, E. Grilli, G. Isella, Ge/SiGe quantum wells on Si(111): growth, structural, and optical properties. J. Appl. Phys. **116**, 43518 (2014)
74. P. Chaisakul, D. Marris-Morini, J. Frigerio, D. Chrastina, M.-S. Rouifed, S. Cecchi, P. Crozat, G. Isella, L. Vivien, Integrated germanium optical interconnects on silicon substrates. Nat. Photonics **8**, 482 (2014)
75. M. Virgilio, G. Grosso, Optical spin orientation in strained Ge/SiGe quantum wells: a tight-binding approach. Phys. Rev. B **80**, 205309 (2009)
76. C. Lange, N.S. Köster, S. Chatterjee, H. Sigg, D. Chrastina, G. Isella, H. Von Känel, M. Schäfer, M. Kira, S.W. Koch, Ultrafast nonlinear optical response of photoexcited Ge/SiGe quantum wells: evidence for a femtosecond transient population inversion. Phys. Rev. B—Condens. Matter Mater. Phys. **79**, 201306(R) (2009)
77. C. Lange, G. Isella, D. Chrastina, F. Pezzoli, N.S. Koester, R. Woscholski, S. Chatterjee, Spin band-gap renormalization and hole spin dynamics in Ge/SiGe quantum wells. Phys. Rev. B **85**, 241303(R) (2012)
78. A. Giorgioni, F. Pezzoli, E. Gatti, S. Cecchi, C. Kazuo Inoki, C. Deneke, E. Grilli, G. Isella, M. Guzzi, Optical tailoring of carrier spin polarization in Ge/SiGe multiple quantum wells. Appl. Phys. Lett. **102**, 12408 (2013)
79. F.A. Baron, A.A. Kiselev, H.D. Robinson, K.W. Kim, K.L. Wang, E. Yablonovitch, Manipulating the $L$-valley electron $g$ factor in SiGe heterostructures. Phys. Rev. B—Condens. Matter Mater. Phys. **68**, 1 (2003)
80. A. Barfuss, L. Dudy, M.R. Scholz, H. Roth, P. Höpfner, C. Blumenstein, G. Landolt, J.H. Dil, N.C. Plumb, M. Radovic, A. Bostwick, E. Rotenberg, A. Fleszar, G. Bihlmayer, D. Wortmann, G. Li, W. Hanke, R. Claessen, J. Schäfer, Elemental topological insulator with tunable Fermi level: strained $\alpha$-Sn on Insb(001). Phys. Rev. Lett. **111**, 157205 (2013)
81. Y. Ohtsubo, P. Le Fèvre, F. Bertran, A. Taleb-Ibrahimi, Dirac cone with helical spin polarization in ultrathin $\alpha$-Sn(001) films. Phys. Rev. Lett. **111**, 216401 (2013)
82. H.S. Lan, S.T. Chang, C.W. Liu, Semiconductor, topological semimetal, indirect semimetal, and topological dirac semimetal phases of $Ge_{1-x}Sn_x$ alloys. Phys. Rev. B **95**, 201201 (2017)
83. S. Wirths, D. Buca, S. Mantl, Si-Ge-Sn alloys: from growth to applications. Prog. Cryst. Growth Charact. Mater. **62**, 1 (2016)
84. V.R. D'Costa, Y.Y. Fang, J. Tolle, J. Kouvetakis, J. Menéndez, Tunable optical gap at a fixed lattice constant in group-IV semiconductor alloys. Phys. Rev. Lett. **102**, 1 (2009)
85. B. Vincent, F. Gencarelli, H. Bender, C. Merckling, B. Douhard, D.H. Petersen, O. Hansen, H.H. Henrichsen, J. Meersschaut, W. Vandervorst, M. Heyns, R. Loo, M. Caymax, Undoped and *in-situ* B doped GeSn epitaxial growth on Ge by atmospheric pressure-chemical vapor deposition. Appl. Phys. Lett. **99**, 6 (2011)
86. M.R. Bauer, C.S. Cook, P. Aella, J. Tolle, J. Kouvetakis, P.A. Crozier, A.V.G. Chizmeshya, D.J. Smith, S. Zollner, SnGe superstructure materials for Si-based infrared optoelectronics. Appl. Phys. Lett. **83**, 3489 (2003)

87. S. Wirths, R. Geiger, N. Von Den Driesch, G. Mussler, T. Stoica, S. Mantl, Z. Ikonic, M. Luysberg, S. Chiussi, J.M. Hartmann, H. Sigg, J. Faist, D. Buca, D. Grützmacher, Lasing in direct-bandgap GeSn alloy grown on Si. Nat. Photonics **9**, 88 (2015)
88. S. Al-Kabi, S.A. Ghetmiri, J. Margetis, T. Pham, Y. Zhou, W. Dou, B. Collier, R. Quinde, W. Du, A. Mosleh, J. Liu, G. Sun, R.A. Soref, J. Tolle, B. Li, M. Mortazavi, H.A. Naseem, S.Q. Yu, An optically pumped 2.5 $\mu$m GeSn laser on Si operating at 110 K. Appl. Phys. Lett. **109** (2016)
89. J. Margetis, S. Al-Kabi, W. Du, W. Dou, Y. Zhou, T. Pham, P. Grant, S. Ghetmiri, A. Mosleh, B. Li, J. Liu, G. Sun, R. Soref, J. Tolle, M. Mortazavi, S.Q. Yu, Si-based GeSn lasers with wavelength coverage of 2–3 $\mu$m and operating temperatures up to 180 K. ACS Photonics **5**, 827 (2018)
90. E.A. Fitzgerald, P.E. Freeland, M.T. Asom, W.P. Lowe, R.A. Macharrie, B.E. Weir, A.R. Kortan, F.A. Thiel, Y.-H.H. Xie, A.M. Sergent, S.L. Cooper, G.A. Thomas, L.C. Kimerling, Epitaxially stabilized $Ge_xSn_{1-x}$ diamond cubic alloys. J. Electron. Mater. **20**, 489 (1991)
91. P. Moontragoon, R.A. Soref, Z. Ikonic, The direct and indirect bandgaps of unstrained Si $_x$ Ge $_{1-x-y}$ Sn $_y$ and their photonic device applications. J. Appl. Phys. **112**, 073106 (2012)
92. G. He, H.A. Atwater, Interband transitions in $Sn_xGe_{1-x}$ alloys (1997)
93. M.-Y. Ryu, T.R. Harris, Y.K. Yeo, R.T. Beeler, J. Kouvetakis, Temperature-dependent photoluminescence of Ge/Si and $Ge_{1-y}Sn_y$/Si, indicating possible indirect-to-direct bandgap transition at lower Sn content. Appl. Phys. Lett. **102**, 171908 (2013)
94. S.A. Ghetmiri, W. Du, J. Margetis, A. Mosleh, L. Cousar, B.R. Conley, L. Domulevicz, A. Nazzal, G. Sun, R.A. Soref, J. Tolle, B. Li, H.A. Naseem, S.Q. Yu, Direct-bandgap GeSn grown on silicon with 2230 nm photoluminescence. Appl. Phys. Lett. **105** (2014)
95. J.D. Gallagher, C.L. Senaratne, J. Kouvetakis, J. Menéndez, Compositional dependence of the bowing parameter for the direct and indirect band gaps in $Ge_{1-Y}Sn_y$ alloys. Appl. Phys. Lett. **105**, 142102 (2014)
96. F. Pezzoli, A. Giorgioni, D. Patchett, M. Myronov, Temperature-dependent photoluminescence characteristics of GeSn epitaxial layers (2016)
97. A.D. Margulis, V.A. Margulis, Spin relaxation of conduction electrons in semiconductors in a strong magnetic field. The precession mechanism. Sov. Phys. Solid State **25**, 918 (1983)
98. S. Manipatruni, D.E. Nikonov, I.A. Young, Beyond CMOS computing with spin and polarization. Nat. Phys. **14**, 338 (2018)
99. M. Lindemann, G. Xu, T. Pusch, R. Michalzik, M.R. Hofmann, I. Žutić, N.C. Gerhardt, Ultrafast spin-lasers. Nature **568**, 212 (2019)
100. M. Holub, P. Bhattacharya, Spin-polarized light-emitting diodes and lasers. J. Phys. D Appl. Phys. **40**, R179 (2007)
101. R. Flederling, M. Kelm, G. Reuscher, W. Ossau, G. Schmidt, A. Waag, L.W. Molenkamp, Injection and detection of a spin-polarized current in a light-emitting diode. Nature **402**, 787 (1999)
102. Y. Ohno, D.K. Young, B. Beschoten, F. Matsukura, H. Ohno, D.D. Awschalom, Electrical spin injection in a ferromagnetic semiconductor heterostructure. Nature **402**, 790 (1999)
103. A. Hirohata, Y.B. Xu, C.M. Guertler, J.A.C. Bland, Spin-dependent electron transport at the ferromagnet/semiconductor interface. J. Appl. Phys. **85**, 5804 (1999)
104. R. Farshchi, M. Ramsteiner, J. Herfort, A. Tahraoui, H.T. Grahn, Optical communication of spin information between light emitting diodes. Appl. Phys. Lett. **98**, 162508 (2011)
105. M. Cantoni, C. Rinaldi, Light helicity detection in MOS-based spin-photodiodes: an analytical model. J. Appl. Phys. **120**, 104505 (2016)
106. C. Rinaldi, M. Cantoni, D. Petti, A. Sottocorno, M. Leone, N.M. Caffrey, S. Sanvito, R. Bertacco, Ge-based spin-photodiodes for room-temperature integrated detection of photon helicity. Adv. Mater. **24**, 3037 (2012)
107. A. Djeffal, F. Cadiz, M. Stoffel, D. Lagarde, X. Gao, H. Jaffrès, X. Devaux, S. Migot, X. Marie, H. Rinnert, S. Mangin, J.-M. George, P. Renucci, Y. Lu, Co-Fe-B spin photodiode operating at telecommunication wavelength with zero applied magnetic field. Phys. Rev. Appl. **10**, 10 (2018)

108. S.H. Liang, T.T. Zhang, P. Barate, J. Frougier, M. Vidal, P. Renucci, B. Xu, H. Jaffrès, J.M. George, X. Devaux, M. Hehn, X. Marie, S. Mangin, H.X. Yang, A. Hallal, M. Chshiev, T. Amand, H.F. Liu, D.P. Liu, X.F. Han, Z.G. Wang, Y. Lu, Large and robust electrical spin injection into GaAs at zero magnetic field using an ultrathin CoFeB/MgO injector. Phys. Rev. B—Condens. Matter Mater. Phys. **90**, 085310 (2014)
109. J.Y. Chen, C.Y. Ho, M.L. Lu, L.J. Chu, K.C. Chen, S.W. Chu, W. Chen, C.Y. Mou, Y.F. Chen, Efficient spin-light emitting diodes based on InGaN/GaN quantum disks at room temperature: a new self-polarized paradigm. Nano Lett. **14**, 3130 (2014)
110. S. Iba, H. Saito, S. Yuasa, Y. Yasutake, S. Fukatsu, Fabrication of Ge-based light-emitting diodes with a ferromagnetic metal/insulator tunnel contact. Japan. J. Appl. Phys. **54**, 04DM02 (Japan Society of Applied Physics, 2015)
111. F. Bottegoni, C. Zucchetti, G. Isella, M. Bollani, M. Finazzi, F. Ciccacci, Spin-charge interconversion in heterostructures based on group-IV semiconductors. Riv. Del Nuovo Cim (2020)
112. M.I. Dyakonov, V.I. Perel, Current-induced spin orientation of electrons in semiconductors. Phys. Lett. A **35**, 459 (1971)
113. J.E. Hirsch, *Spin Hall Effect*, vol. 83 (American Physical Society, 1999)
114. E.I. Rashba, Side jump contribution to spin-orbit mediated hall effects and berry curvature. Semiconductors **42**, 905 (2008)
115. F. Bottegoni, A. Ferrari, F. Rortais, C. Vergnaud, A. Marty, G. Isella, M. Finazzi, M. Jamet, F. Ciccacci, Spin diffusion in Pt as probed by optically generated spin currents. Phys. Rev. B—Condens. Matter Mater. Phys. **92**, 214403 (2015)
116. F. Bottegoni, C. Zucchetti, G. Isella, E. Pinotti, M. Finazzi, F. Ciccacci, Modeling the photo-induced inverse spin-hall effect in Pt/semiconductor junctions. J. Appl. Phys. **124**, 033902 (2018)
117. F. Bottegoni, M. Celebrano, M. Bollani, P. Biagioni, G. Isella, F. Ciccacci, M. Finazzi, Spin voltage generation through optical excitation of complementary spin populations. Nat. Mater. **13**, 790 (2014)

# Part II
# Advances in Integration Architectures

# Chapter 6
# Subwavelength Silicon Photonics

**Hon Ki Tsang, Xia Chen, Zhenzhou Cheng, Wen Zhou, and Yeyu Tong**

**Abstract** Subwavelength gratings refer to periodic structures that have a period less than half the wavelength of light in the material so that no Bragg diffraction mode is supported. Instead, the light will propagate as if it was in a homogeneous material with anisotropic refractive indices. Subwavelength gratings have attracted great interest recently, as they provide a useful degree of freedom for the crafting of the effective refractive index of the material in photonic devices. In this chapter, we will introduce some of the applications of subwavelength structures for silicon photonics devices. We start by introducing the background theory of subwavelength gratings and then discuss their applications for the engineering of waveguide grating couplers, suspended membrane devices for mid-infrared (mid-IR) wavelengths, and their use with numerical optimization techniques for optimizing photonic devices. We shall discuss the classic effective medium theory (EMT) for subwavelength gratings and show how EMT can reduce time-consuming three-dimensional (3D) numerical optimizations to an effective two-dimensional (2D) optimization problem.

H. K. Tsang (✉) · Y. Tong
Department of Electronic Engineering, The Chinese University of Hong Kong, Shatin NT, Hong Kong SAR, People's Republic of China
e-mail: hktsang@ee.cuhk.edu.hk

Y. Tong
e-mail: yytong_ee@link.cuhk.edu.hk

X. Chen
Optoelectronics Research Centre, University of Southampton, Southampton SO17 1BJ, UK
e-mail: xia.chen@soton.ac.uk

Z. Cheng
School of Precision Instruments and Optoelectronics Engineering, Tianjin University, Tianjin 300072, China
e-mail: zhenzhoucheng@tju.edu.cn

W. Zhou
Department of Materials, University of Oxford, Parks Road, Oxford OX1 3PH, UK
e-mail: wen.zhou@materials.ox.ac.uk



D. J. Lockwood and L. Pavesi (eds.), *Silicon Photonics IV*,
Topics in Applied Physics 139, https://doi.org/10.1007/978-3-030-68222-4_6

## 6.1 Introduction

The advanced semiconductor fabrication technologies which are offered by commercial foundries can routinely fabricate subwavelength structures in silicon photonic circuits. Light incident on structures which have dimensions similar to or larger than the optical wavelength is scattered strongly at each abrupt boundary between two regions of different refractive index. This is useful for reducing the size of photonic devices and is used in high-refractive-index contrast silicon photonics for optical waveguides, gratings, ring resonators, photonic crystals, and other passive devices. On the other hand, if the dimensions of the photonic structures are much smaller than the optical wavelength, they will behave like a homogeneous material that can have anisotropic optical properties that may be described with a permittivity tensor [1]. Subwavelength gratings refer to periodic structures that have a period less than half the wavelength of light in the material so that no Bragg diffraction mode is supported. Instead, the light will propagate as if it was in a homogeneous material with anisotropic refractive indices.

Subwavelength gratings have attracted great interest recently, as they provide a useful degree of freedom for the crafting of the effective refractive index of the material in photonic devices. Within the top silicon layer of the silicon-on-insulator (SOI) platform, we can make subwavelength structures that can have an effective refractive index at any desired value between the index of silicon and the cladding material (typically silicon dioxide with an index of 1.44). This flexible control of the effective index can be realized with a single etch step to make the appropriately designed subwavelength grating.

Subwavelength grating waveguides were first demonstrated in 2010 [2]. Unlike other periodic waveguides such as line-defects in a 2D photonic crystal lattice, a subwavelength grating waveguide confines the light as a conventional index-guided structure and does not exhibit optically resonant behavior. The experimentally measured propagation losses were as low as 2.1 dB/cm and compare favorably with conventional silicon nanowires. In 2009, subwavelength gratings were also proposed for the engineering of waveguide grating couplers [3] and experimentally demonstrated [4]. Most subwavelength gratings couplers can be fabricated in one single etch step. The use of subwavelength gratings enables many possible design optimizations of the waveguide grating coupler, including improving their coupling efficiency with grating apodization, compensating waveguide birefringence with the birefringence of the subwavelength grating to enable polarization-independence and increasing the optical bandwidth for wideband grating couplers. Subwavelength-structured inverse tapers were also demonstrated [5]. Later, it was realized that subwavelength structures that were etched through to the buried oxide (BOX) could also be used to realize suspended photonic waveguide devices, in which the BOX is locally etched away by buffered hydrofluoric acid which could attack the BOX via the deep-etched holes forming the subwavelength grating [6]. The subwavelength structures can thus be used to facilitate the fabrication of mid-IR photonic sensors that are highly sensitive to changes in the refractive index produced by gases or solutions.

## 6.2 Effective Medium Theory

In 1954, the Soviet physicist Sergei M. Rytov published his theoretical paper on the propagation of light in periodic structures which have periodicity along the $z$-axis of $d \ll \frac{\lambda}{2\pi n}$ where $\lambda$ is the wavelength of light in vacuum [1] and $n$ is the refractive index of the material and may be anisotropic. Rytov considered the cases of light propagation perpendicular to the axis of periodicity and along the axis of periodicity and showed for sufficiently long wavelengths, and an inhomogeneous (stratified) medium could behave as a homogenous anisotropic medium which has effective permeabilities and permittivities dependent only on the polarization of light and the original materials and their relative mix.

### *6.2.1 Rytovs's Equations and Application in Periodic Dielectric Subwavelength Gratings*

For a general one-dimensional (1D) periodic photonic lattice as shown in Fig. 6.1, when the wavelength of light is much longer than the period of the lattice, Rytov's EMT can be used to calculate the effective permittivity and permeability [1] and can be expressed as follows:

For the transverse magnetic (TM) polarized light:

$$\bar{\varepsilon} = \frac{a\varepsilon_1 + b\varepsilon_2}{a + b}. \tag{6.1}$$

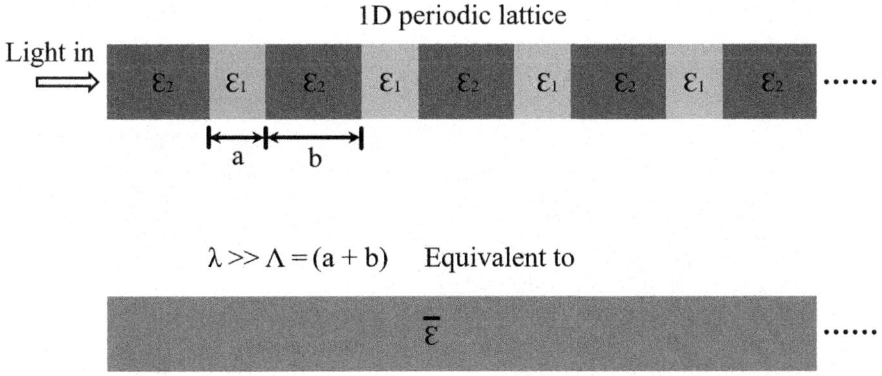


**Fig. 6.1** Schematic of a 1D periodic photonic lattice with periodically alternating two mediums with relative permittivity of $\varepsilon_1$ and $\varepsilon_2$ (and permeability of $\mu_1$ and $\mu_2$) in the light propagation direction. A 1D periodic lattice can be an equivalent slab with permittivity of $\bar{\varepsilon}$ (and permeability of $\bar{\mu}$) when the wavelength of light is much longer than the period of the lattice

$$\frac{1}{\bar{\mu}} = \frac{a/\mu_1 + b/\mu_2}{a+b}. \tag{6.2}$$

For the transverse electric (TE) polarized light:

$$\frac{1}{\bar{\varepsilon}} = \frac{a/\varepsilon_1 + b/\varepsilon_2}{a+b}. \tag{6.3}$$

$$\bar{\mu} = \frac{a\mu_1 + b\mu_2}{a+b}. \tag{6.4}$$

For dielectric mediums,

$$\bar{\mu} = \mu_1 = \mu_2 = 1 \quad \text{and} \quad n = \sqrt{\bar{\varepsilon}}.$$

As a result, a 1D periodic photonic lattice can be treated as an equivalent slab when light propagates in the subwavelength regime. Specifically, the above Rytov's EMT theory can be applied not only in the subwavelength grating waveguide [2] when the light propagation direction is parallel to the orientation of the subwavelength lattice (longitudinal application) but also applicable in designing 2D subwavelength gratings for fiber-chip grating couplers (GC) when the light propagation direction is perpendicular to that of the subwavelength lattice (lateral application). A detailed introduction of subwavelenth fiber-chip GCs will be presented in the following section. Here, we will illustrate the theoretical treatment to estimate the effective refractive index of the grating lines made of subwavelength structures.

A uniform 2D subwavelength grating is shown in Fig. 6.2. For simplicity, we use periodically arranged square-shaped holes to form the subwavelength gratings. The light is propagating in the $x$-direction. The grating period $\Lambda_x$ along the light propagation direction is normally similar to the effective optical wavelength in the medium, governed by the phase-matching condition for the grating coupler (see Sect. 6.3). On the other hand, the period of the grating holes in the $y$-axis $\Lambda_y$ must be

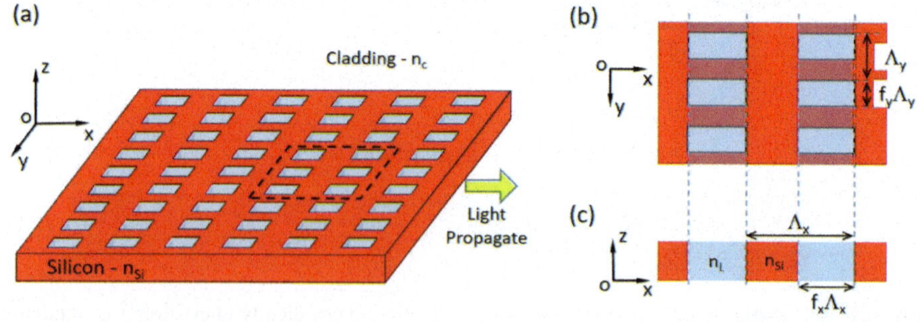


**Fig. 6.2** **a** 3D schematic illustration of a 2D subwavelength grating with lateral subwavelength structures on a silicon-on-insulator platform. **b** Schematic top view of part of the grating. **c** Schematic side-view of the part of the grating

much smaller than the effective wavelength of the propagating light in the medium. Thus, the shaded region in the *y*-direction as shown in Fig. 6.2b can be treated as an equivalent homogenous material with a lower effective index than that of silicon. This allows the EMT at zeroth-order to be employed to approximate the structure by a groove with a lower refractive index of $n_L$ as shown in Fig. 6.2c [5]:

$$\frac{1}{n_{L\_TE}^{(0)}} = \left[\frac{f_y}{n_c^2} + \frac{(1-f_y)}{n_{Si}^2}\right]^{1/2}, \tag{6.5}$$

$$n_{L\_TM}^{(0)} = [n_c^2 f_y + n_{Si}^2(1-f_y)]^{1/2}, \tag{6.6}$$

where $n_{L\_TE}^{(0)}$ and $n_{L\_TM}^{(0)}$ are the refractive indices of the equivalent groove derived by zeroth-order EMT for the TE- and TM-polarized optical waves, respectively. $n_c$ equals to index of the etched holes, which is typically filled with the cladding material (silicon dioxide with $n_c = 1.45$); $n_{Si}$ equals to index of silicon (3.48). The lateral fill factor of the grating holes is denoted by $f_y$.

However, this zeroth-order approximation is only accurate when the period-to-wavelength ratio, defined as $R = n_{eff}\Lambda_y/\lambda$, is much smaller than 1, where $n_{eff}$ is the mode effective index of the light guided in the slab waveguide (we also use $n_{eff\_TE}$ and $n_{eff\_TM}$ to denote the same value for the TE and TM mode, respectively). More accurate approximations employing the second-order EMT can be used:

$$n_{L\_TE}^{(2)} = n_{L\_TE}^{(0)}\left[1 + \frac{\pi^2}{3}R^2 f_y^2(1-f_y)^2(n_c^2 - n_{Si}^2)^2\left(\frac{n_{L\_TM}^{(0)}}{n_{eff\_TE}}\right)^2\left(\frac{n_{L\_TE}^{(0)}}{n_c n_{Si}}\right)^4\right]^{1/2}, \tag{6.7}$$

$$n_{L\_TM}^{(2)} = n_{L\_TM}^{(0)}\left[1 + \frac{\pi^2}{3}R^2 f_y^2(1-f_y)^2\left(\frac{n_c^2 - n_{Si}^2}{n_{eff\_TM}n_{L\_TM}^{(0)}}\right)^2\right]^{1/2}. \tag{6.8}$$

where $n_{L\_TE}^{(2)}$ and $n_{L\_TM}^{(2)}$ are the refractive indices of the approximated groove derived by EMT with second-order approximation for the TE- and TM-polarized optical waves, respectively. These equations give more accurate results for the subwavelength grating couplers when the lateral feature size (which may be limited by the fabrication technology) is slightly smaller than the wavelength. One example set of calculated results is shown in Fig. 6.3 with subwavelength gratings fabricated on SOI with a 340 nm top silicon thickness and $\Lambda_y = 400$ nm [5].

As shown in Fig. 6.3, refractive indices of the fundamental TE and TM modes in the subwavelength grating can be widely and continuously tunable between $n_{Si}$ and $n_c$. And refractive indices of the fundamental TE mode are smaller than that of TM mode with a same lateral fill factor due to the slot waveguide effect, i.e., light is more confined in the medium with a low refractive index in the lateral subwavelength grating for the TE mode. The above properties offer a new approach for the design of

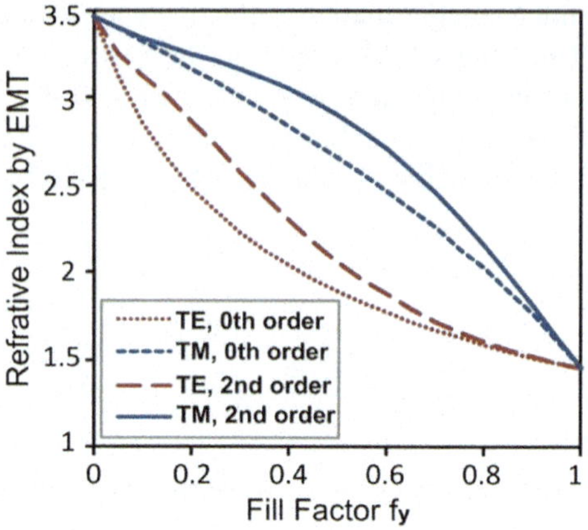


**Fig. 6.3** Theoretical calculation of refractive indices of the fundamental TE and TM modes versus lateral fill factor based on the zeroth- and second-order EMT, for subwavelength gratings fabricated on SOI with a 340 nm top silicon thickness and $\Lambda_y = 400$ nm. Reproduced with permission from [5]. Copyright © 2011 by the Optical Society

polarization-independent, wideband, apodized subwavelength grating couplers with low back reflection as introduced in the later sections of this chapter.

## *6.2.2 Spectral Range of Validity of EMT*

A subwavelength grating is a basic periodic structure in which light propagation satisfies the Floquet–Bloch theory, and the periodic potential supports solutions that have the same periodicity as the lattice, such as a periodically modulated plane wave. [7–9]. Generally, such solutions can be expressed as: $E(x, y) = E(x + \Lambda, y)$, in which $E(x, y) = f(y) \cdot \exp(ikx)$ and $E(x + \Lambda, y) = f(y) \cdot \exp(ikx) \cdot \exp(ik\Lambda)$. Thus, $E(x + \Lambda, y) = E(x, y) \cdot \exp(ik\Lambda)$. Considering a Bloch mode propagating in a unit cell of a subwavelength grating, it can be expressed as $E(x + \Lambda, y) = E(x, y) \cdot \exp(j\phi)$, where the phase shift $\phi = 2\pi n_{\text{eff}}\Lambda/\lambda_0$. It introduces a Floquet–Bloch periodic boundary condition in solving Maxwell's equations: $[\nabla \times \nabla \times -k_0^2 \varepsilon_r]\, E(x, y) = -i\omega\mu_0 j$, where $\varepsilon_r$ is the spatial distribution of the relative permittivity, $k_0 = \omega/c$ is the free-space wavenumber, $E(x, y)$ is the electric field distribution, and $j$ is the electric current density. Numerical calculation of eigenmodes of the subwavelength grating can be performed using the 3D finite-element method (FEM) [10]. By setting the phase-shifting term $\phi$ at Floquet–Bloch periodic boundaries, one can calculate $n_{\text{eff}}$ for the solved eigenfrequency $\omega$ from the definition $\phi = 2\pi n_{\text{eff}}\Lambda/\lambda_0$. Thus, the dispersion relationship in the frequency-wavevector $(\omega, k_0 n_{\text{eff}})$ plane can be calculated by sweeping $\phi$, as shown schematically in Fig. 6.4. Alternatively, without applying the Floquet–Bloch theory, one can numerically calculate the $n_{\text{eff}}$ of its equivalent homogeneous waveguide with a refractive index $n_{\text{L\_TE(TM)}}^{(0/2)}$ calculated by the zeroth- or second-order EMT under the assumption of light operating in the subwavelength regime ($\Lambda \ll \lambda$).

If the frequency is close to the bandgap of the photonic periodic structure, the band tends to be flattened with a reduced slope or an enlarged group index due

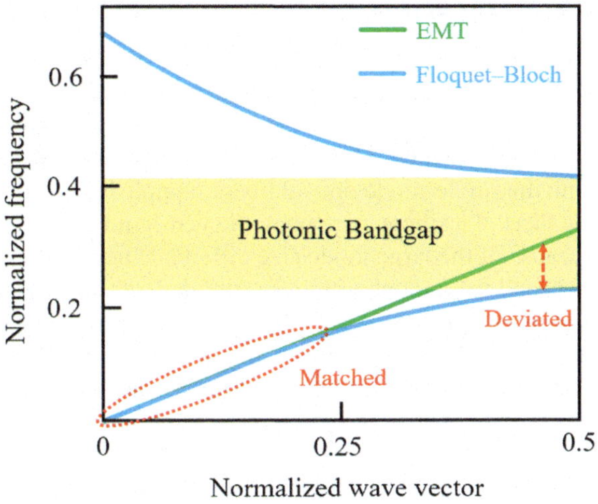


**Fig. 6.4** Schematic dispersion diagram of eigenmodes calculated based on Floquet–Bloch theory and EMT for a periodic subwavelength grating

to the slow light effect, which however, is absent in the homogeneous waveguide under EMT approximation As a result, there is a large deviation between effective indices of the subwavelength grating and its equivalent homogeneous waveguide. One rigorous way to find the applicable frequency region of EMT for a subwavelength grating with a certain period and duty cycle is by plotting the dispersion diagram of the subwavelength grating and its equivalent homogeneous waveguide [11]. As described above, the effective indices of the subwavelength grating and its equivalent waveguide can be calculated numerically based on the Floquet–Bloch theory and EMT, respectively. In the next step, the applicable frequency range for EMT to be accurate is in "matched" region indicated by the dotted line in the dispersion diagram of Fig. 6.4. Alternatively, a simple rule for checking the applicability of EMT in the subwavelength grating is by comparing effective indices calculated based on Floquet–Bloch theory and EMT. If there is a deviation, the period of the subwavelength grating can be decreased to rigorously satisfy the subwavelength condition at a certain wavelength.

## 6.3 Subwavelength Waveguide Grating Couplers

In this section, we introduce the basic techniques for designing subwavelength grating couplers. We will first give a basic introduction of a diffractive grating coupler, followed by a brief review of the work on subwavelength grating couplers in the literature. Then, we shall pick a few typical subwavelength grating coupler designs and introduce them in more detail. More review papers [11, 12] can also be found in the literature in this research field.

### 6.3.1 Introduction

Improving the coupling efficiency between fiber and silicon photonic circuits is a key challenge to increase the power efficiency and overall performance of silicon photonics devices. Most silicon photonic circuits require an interface to connect with the single-mode optical fibers. Simple butt-coupling of the single-mode fiber to the facet of a silicon nanophotonic waveguide will give a high optical loss (>20 dB), because of the large mode size difference between the optical fibers (~10 μm diameter) and the nanophotonic waveguides (~0.4 μm diameter). Diffractive waveguide grating couplers [13–15] were proposed and are getting very popular among the research and industry community in recent years [16], because of their many advantages, such as simpler back-end processing, since they do not require polishing of facets, the flexibility to place the optical input/output anywhere on the chip, and the compatibility with wafer-scale testing before dicing [17].

A typical grating coupler [18] is shown in Fig. 6.5. The optical signal from the nanophotonic waveguide is first expanded laterally by an adiabatic taper to a waveguide of about 10 μm width, which matches the mode size of optical fiber in the y-axis. The light is then coupled out by diffraction of shallow-etched gratings into the optical fiber. The etch depth of the grating was calculated to achieve an optimized coupling strength and diffraction directionality. The period and the diffraction angle $\theta$ are governed by the phase-matching condition:

$$k_o n_{\text{eff\_avg}} = k_o n_c \sin\theta + q\frac{2\pi}{\Lambda}. \tag{6.9}$$

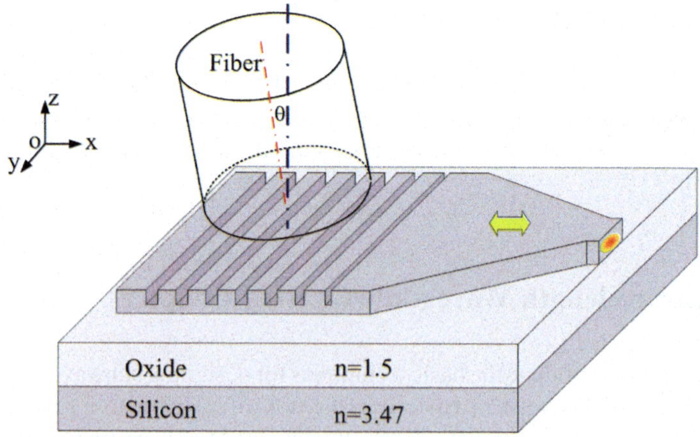


**Fig. 6.5** Schematic illustration of an apodized waveguide grating coupler between an optical fiber and nanophotonic waveguide. $\theta$ is the angle of the fiber axis to the surface normal of the SOI wafer. The yellow arrow indicates the direction of the propagation of light

where $k_o = 2\pi/\lambda$, $n_c$ is the refractive index of the cladding (this is equal to 1.45 for oxide), $\theta$ is the angle of the output light to the surface normal of the SOI wafer, $n_{\text{eff\_avg}}$ is the average mode effective index in the grating area, $q$ is an integer (equals 1 for grating couplers), and $\Lambda$ is the waveguide grating period. For most of the grating couplers demonstrated so far, the optical fiber is orientated with an angle $\theta$ at about 10°. This is to avoid the large second-order Bragg back reflection, which would otherwise reflect about half of the optical power back into the waveguide. Apart from the typical shallow-etched gratings introduced above, alternative designs of fiber-chip grating couplers that have been explored include slanted gratings [19] and metallic gratings [20].

For the conventional grating couplers mentioned previously, the lithography and etching processes needed to fabricate the grating structures must be distinct from the corresponding waveguide fabrication steps because of the different etch depths required for the grating couplers. The total number of fabrication steps and the cost can be reduced if we make the grating couplers with the same deep-etch step as used for the waveguides. However, for those conventional gratings formed by continuous lines, the refractive index step introduced would normally be too large if they are etched to the depths typically needed for waveguide fabrication. The grating strength in this case would be too strong to achieve an efficient coupling, leading to a low mode matching efficiency and a large back reflection loss. Subwavelength structures therefore can be used to adjust the effective index of the deep-etched grating lines and to reduce the index step.

Such subwavelength devices can be fabricated in the same lithography step as conventional waveguides, while we are still able to adjust the effective index of the grating lines. Highly efficient subwavelength grating couples were developed at both near-infrared (telecommunication) and mid-infrared wavelengths. Some of them are among the grating couplers achieving the best coupling efficiencies up to date. As an example, a fully etched apodized subwavelength grating coupler with an aluminum mirror demonstrated a coupling efficiency of −0.58 dB in 2014 [21].

### 6.3.2 A Brief Review

Subwavelength grating couplers were first proposed in 2009 [3]. The first experimental demonstration of the subwavelength grating coupler was carried out independently in the same year and reported a coupling efficiency of −4.69 dB (34%) with uniform round-shaped subwavelength structures [4]. Grating couplers based on a fully etched photonic crystal structure were reported in 2010 with −3.77 dB (42%) coupling efficiency [22], which can also be categorized as uniform subwavelength grating couplers based on its working principle. By further optimizing the buffer thickness of uniform subwavelength grating couplers, a coupling efficiency of −2.29 dB (59%) was reported in [23]. Uniform subwavelength grating couplers in a silicon membrane on a glass substrate with a coupling efficiency of −4.07 dB were described in [24]. An implementation of a blazed grating coupler using a subwavelength grating (SWG) structure was also proposed in [25], with a simulated coupling efficiency of −1.61 dB (69%).

In order to increase the matching efficiency with the optical fiber mode, subwavelength gratings can be apodized by varying the coupling strength along the grating. An apodized subwavelength grating was demonstrated experimentally with a coupling efficiency of −3.7 dB [26]. In 2012, an apodized design with focusing and a coupling efficiency of −3 dB was presented [27]. Using a triangular array of circular holes, apodization, and an optimized buffer thickness, an efficiency of −1.74 dB (67%) was achieved in [28]. It is worth noting that optimal grating apodization also reduces back-reflections from the grating coupler since the refractive index discontinuities are smoothed.

The coupling efficiency of apodized subwavelength grating couplers can be further improved by fabricating a metal reflector in the thinned substrate. The metal can reflect the light diffracted to the substrate back into the optical fiber. Coupling efficiencies of −0.58 dB [21] and −0.69 dB (85.3%) [29] were experimentally demonstrated with slightly different designs, respectively. Both are better than 1 dB, which is usually considered the threshold for many commercial applications. Such processes can also be used in combination with subwavelength structures to create perfectly vertical grating couplers (PVGC) [30] that allow more straightforward packaging.

Subwavelength grating structures can even yield polarization-insensitive couplers [5]. An experimental demonstration with the additional ability to focus the light in the chip plane was reported in [31]. We will introduce the details of polarization-insensitive couplers later in this section.

A considerable practical constraint of grating couplers is their limited spectral bandwidth of approximately 35 nm (1 dB bandwidth) near a wavelength of 1.55 μm, because the momentum-matching requirement in the grating equation imposes a variation of the diffraction angle with wavelength. This variation is proportional to the grating refractive index. By using subwavelength structures to decrease the effective refractive index, a 73 nm 1 dB bandwidth was demonstrated experimentally with −5.6 dB coupling efficiency in 2012 [32]. A focusing grating coupler with a similar structure was reported in [33] for which the bandwidth is ~90 nm (3 dB bandwidth) with a peak coupling efficiency of −3.5 dB. By reducing the dispersion of the subwavelength gratings, a further bandwidth (3 dB) improvement of up to 117 nm can be achieved [34]. In 2015, a 1-dB bandwidth of 90 nm has been demonstrated, albeit at the expense of coupling efficiency [35]. Thus, achieving simultaneous broadband and high-efficiency operation is a challenge. Prism-assisted subwavelength grating couplers could potentially provide such a solution [36].

### 6.3.3 Uniform Subwavelength Grating Couplers

The most basic uniform subwavelength grating coupler is shown in Fig. 6.6. Along the $x$-direction, the device acts as a conventional (diffractive) grating. However, the lower

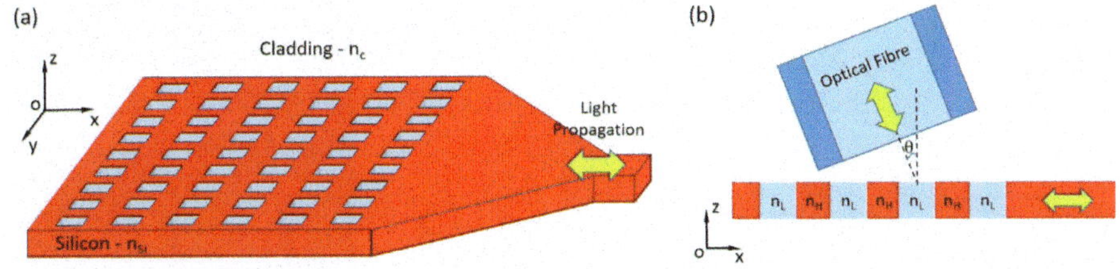


**Fig. 6.6** Illustration of uniform subwavelength grating couplers. **a** Subwavelength (non-diffractive) structures along the *y*-direction are used to control the equivalent refractive index in the grating trenches, thereby controlling the strength of the grating and enhancing the coupling efficiency. **b** Two-dimensional model of a fiber-chip grating coupler. Light traveling along the *x*-axis in the waveguide mode is scattered upwards and downwards by a diffractive grating. A fiber situated above the grating captures a fraction of the diffracted light

index part of the grating lines (or grooves) along the lateral direction (*y*-axis) has a subwavelength period and thus acts as an artificial homogeneous media, synthesizing the equivalent indexes of the grating grooves. The refractive indexes can be approximately estimated by EMT formulas as introduced in the previous section. Then, the effective index for the optical mode can be obtained by treating each subwavelength grating structure as a two-dimensional multilayer slab and computing the effective index of its fundamental mode [3]. By changing the size of the silicon segments of the subwavelength grating, the equivalent index can be controlled, yielding an optimized effective index difference for the diffractive grating.

Designing a uniform coupler is then straightforward: (i) through simulation of the grating structure in Fig. 6.6b, the value of $n_L$ that yields the highest coupling efficiency is determined; (ii) using the data calculated by the EMT as a look-up table, the physical dimensions of the SWG structure that yields the desired value of $n_L$ are obtained. The first implementation of a uniform subwavelength coupler [4] was designed in this way, although the square-shaped holes were replaced by circular-shaped holes with a similar size to satisfy the foundry design rules.

The fabricated subwavelength nanoholes grating couple is shown in Fig. 6.7. It was fabricated on SOI with a 220 nm top silicon thickness and 2 μm BOX thickness. The nanoholes grating was fabricated on a 10 μm wide waveguide. A 500 μm long adiabatic taper was used to connect the grating coupler with the 500 nm wide nanophotonic waveguide. The fiber was tilted with an 8-degree angle ($\theta$). A 34% coupling efficiency was experimentally measured.

### 6.3.4 Apodized Subwavelength Grating Couplers

An apodized subwavelength grating was proposed [3], and subsequently demonstrated, with better mode matching for an improved coupling efficiency [26]. Unlike in the uniform grating, in the continuously apodized coupler, as shown in Fig. 6.8, the

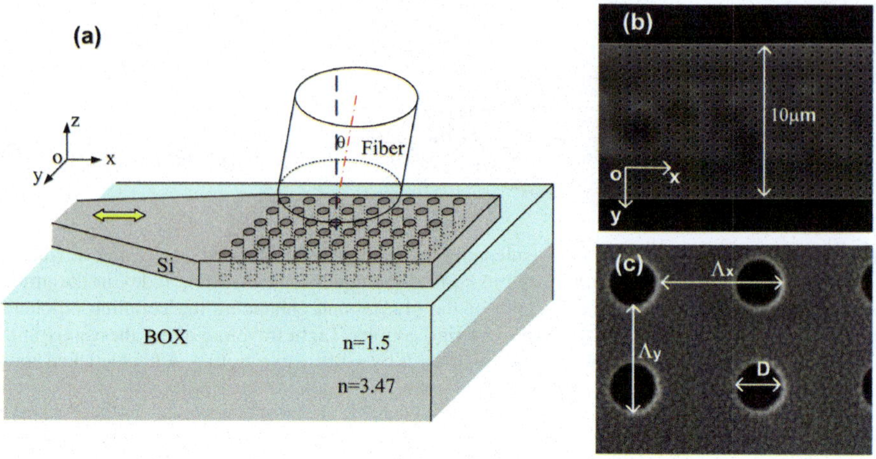


**Fig. 6.7** **a** Schematic picture of the grating coupler for coupling between fibers and nanophotonic wire waveguide on SOI, with $\theta = 8°$. **b** Scanning electron microscope (SEM) image of the fabricated array of nanoholes on the 10 μm-width waveguide. The silicon dioxide on top was removed. **c** Nanoholes array with the holes diameter $D$, grating period $\Lambda_x$, and transversal holes period $\Lambda_y$ defined ($D = 200$ nm,$\Lambda_x = 610$ nm, and $\Lambda_y = 500$ nm) [4]

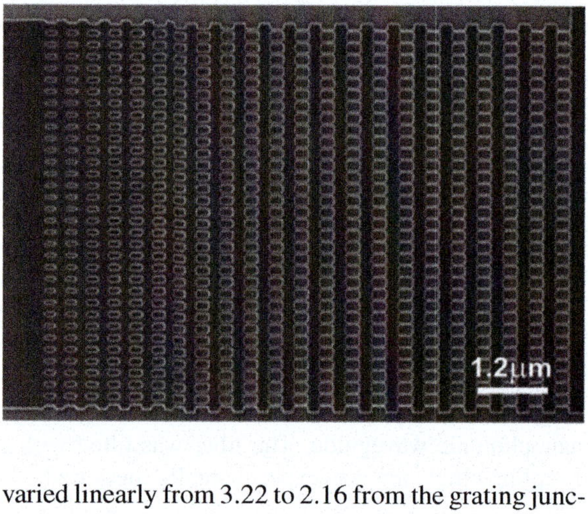


**Fig. 6.8** SEM image of an apodized fiber-chip grating coupler using fully etched subwavelength grating trenches. The apodization gradually decreases the equivalent index in the trenches, producing a Gaussian-like radiated beam that matches the fiber mode [12]

equivalent medium index was varied linearly from 3.22 to 2.16 from the grating junction with the lateral taper to the end of the grating. These effective indices apodize the strength of the grating to produce a near-Gaussian radiated field. The pitch in the longitudinal direction ($x$-axis) was varied along the grating according to the phase-matching condition so that all grating periods contribute light to the radiated beam with the correct phase. The total length of the coupler is 20 μm, and a 50% duty cycle is used in the x-direction. A coupling efficiency of −3.7 dB was experimentally demonstrated, with a 1 dB bandwidth of 40 nm. Since the change in the refractive index between the waveguide and the grating coupler is reduced, back reflections are also reduced to ~1.5%. It is worth to note that the minimum dimension was limited to 100 nm because of design rules in the 193 nm deep-ultraviolet (DUV) lithography used for the fabrication. This sacrifices the coupling efficiency slightly. There is also a significant portion of power diffracted to the silicon substrate. Therefore, the coupling efficiency can be greatly improved by adding a back reflector, as demonstrated years later [21, 29].

### 6.3.5 Polarization-Independent Grating Couplers

A polarization-independent fiber-chip grating coupler has been demonstrated, utilizing subwavelength structures for refractive index engineering [5]. Typically, in a conventional 1D grating, the TE mode effective index is larger than the TM mode index, because of the larger confinement factor, for both the grooves and teeth. Therefore, the grating period for coupling to the TE mode of the waveguide is thus always smaller than the period required for the TM mode according to the phase-matching condition (**6.9**). In order to realize a polarization-independent operation, a novel structure would be required to reverse the normal case and make an effective index for the TM mode equal to or larger than the one for the TE mode in the grating region. This can be achieved with subwavelength structures, as shown in the calculated result in Fig. 6.3. The effective index of the subwavelength structure for the TE mode is lower than that for the TM mode, which is opposite to the case in a conventional slab waveguide. Therefore, by balancing the ratio of the subwavelength structures (with a larger effective index for the TM mode) and conventional slab waveguide sections (with a larger effective index for the TE mode), we can achieve the same average effective index for both the TE and TM mode.

The detailed design and optimization can be found in our original paper [5]. The coupling efficiency for the optimized grating designs was consistently calculated by both 2D and 3D finite-difference time-domain (FDTD) simulations. Peak coupling efficiency of over 64% is predicted for both the TE and TM mode, including a bottom reflector. The 3 dB optical bandwidth is about 65 nm.

### 6.3.6 Wideband Grating Couplers

A wideband grating coupler using the subwavelength structure is shown in Fig. 6.9 [32]. The wavelength-dependent change of coupling efficiency of a grating coupler is mainly dependent on the diffraction angle, as fibers only receive optical power within its numerical aperture (NA). The following equation can be derived from the phase-matching equation to examine the bandwidth of the coupling efficiency [32]:

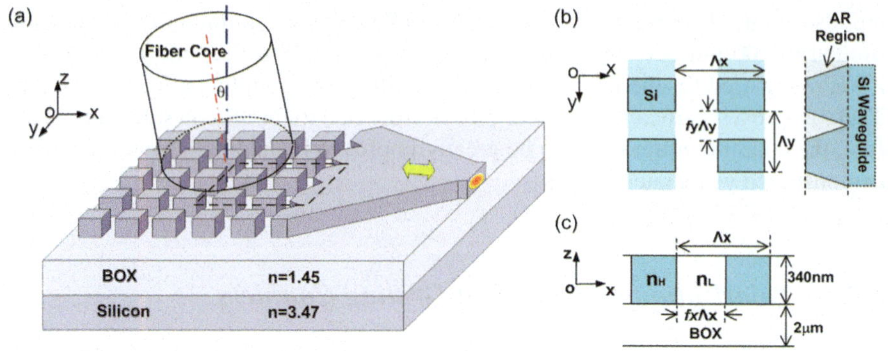


**Fig. 6.9** **a** Three-dimensional schematic illustration of the wideband subwavelength grating coupler. **b** Schematic top view of the cycled region of the grating. An anti-reflective structure is adopted to reduce the back reflection. **c** Schematic side-view of the grating [32]. Reprinted with permission from [32] © The Optical SocietyReprinted with permission from [32] © The Optical Society

$$\Delta\lambda_{1\mathrm{dB}} = \eta_{1\mathrm{dB}}\left|\frac{d\lambda}{d\theta}\right| = \eta_{1\mathrm{dB}}\left|\frac{-\Lambda n_c \cos\theta}{1-\Lambda\frac{dn_{\mathrm{eff}}(\lambda)}{d\lambda}}\right| \tag{6.10}$$

As the effective index would decrease with the increase of wavelength for normal waveguides, the dispersion $dn_{\mathrm{eff}}(\lambda)/d\lambda$ is negative in. For silicon waveguides with similar dimensions, the dispersion at a fixed wavelength would normally increase (approaching zero) while reducing the effective index of the grating region, as the light is loosely guided. The period of the grating would also increase while reducing the effective index. Both the increased dispersion and period will lead to a larger bandwidth. Therefore, under the constraints of the same material system and similar coupling scheme (same coupling angle) for 1550 nm wavelength, an obvious way to increase the optical bandwidth of the grating coupler is to reduce the effective index of the grating coupler.

We found that the grating effective index can be dramatically reduced using the subwavelength pillar structure as shown in Fig. 6.10, while keeping an optimized coupling strength. Compared to the previous designs which set the teeth index $n_H$ to 3.12 (for a 340 nm thick top silicon layer) and apply subwavelength structures for the grooves for coupling strength optimization, this design was created by first setting the grooves index $n_L$ to 1.45, which is the lowest value possible as it is simply filled with cladding material. The effective index of the grating teeth $n_H$ was then reduced by subwavelength structure to obtain an optimized coupling strength. This alternative design approach using subwavelength pillars can give a much lower average effective index for the grating coupler and thus increase its bandwidth. The measured 1 dB bandwidth was dramatically increased to about 73 nm, which is twice the bandwidth demonstrated on a similar platform, with a coupling efficiency of −5.6 dB (27.5%).

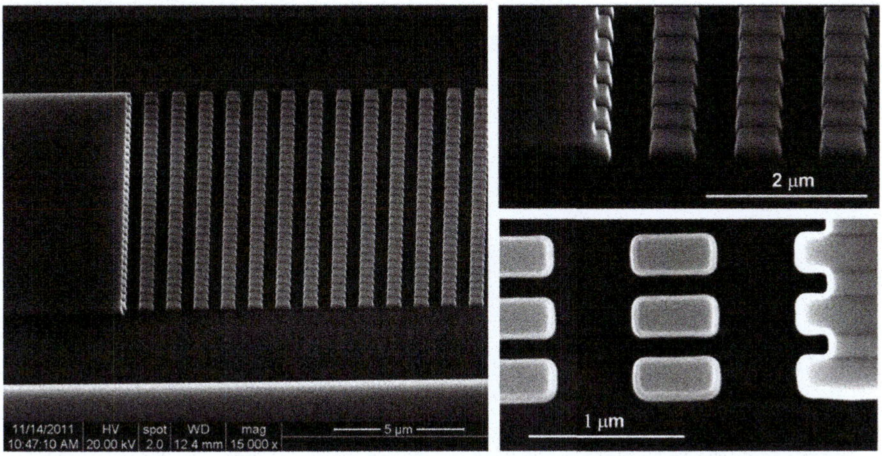


**Fig. 6.10** SEM images of the fabricated subwavelength wideband grating coupler. The cladding oxide was removed before taking these images [32]. Reprinted with permission from [32] © The Optical Society

### *6.3.7 Focusing Apodized Subwavelength Grating Coupler*

In 2012, we experimentally demonstrated a novel TM-mode apodized focusing subwavelength grating coupler [27]. The design of the apodized subwavelength grating coupler is shown in Fig. 6.11a, b. In our work, we used an SOI wafer with a top-silicon-layer thickness of 340 nm and a BOX thickness of 2 µm. The angle between the fiber axis and the waveguide planar was 10°. We fixed the lateral subwavelength period ($\Lambda x$) as 400 nm and the longitudinal grating fill factor ($f_Y$) as 0.6. Then, we used 2D FDTD software to calculate the coupling strength of the subwavelength grating couplers with different lateral subwavelength fill factors ($f_X$) and longitudinal grating periods ($\Lambda_Y$). After that, we designed the apodized subwavelength grating coupler to match the profile of diffracted light with the Gaussian-shaped profile $G(y)$ of a single-mode fiber with a 10.4 µm mode diameter, as shown in Fig. 6.11a. The $\Lambda_Y$ and $fx$ we employed are summarized in Table 6.1. With this design, the grating diffracted mode profile can match well with the Gaussian-shaped profile, as shown in Fig. 6.11b. The field overlap achieved was 98% between the diffracted light and the fundamental mode of the single-mode fiber. The 2D FDTD simulations predicted a maximum efficiency of −1.7 dB with a ~50 nm 3 dB bandwidth and a −14 dB back reflection for TM-mode polarized light coupling.

The experimental results of the apodized focusing subwavelength grating coupler are shown in Fig. 6.11c, d. Based on the above design, we used the focusing grating structure to future reduce the device footprint. The SEM image of the fabricated device is shown in Fig. 6.11c. The minimum feature size was more than 100 nm in the grating coupler; therefore, it is possible to use deep-ultraviolet photolithography to fabricate such the device in large volume and at low cost. As shown in Fig. 6.11d,

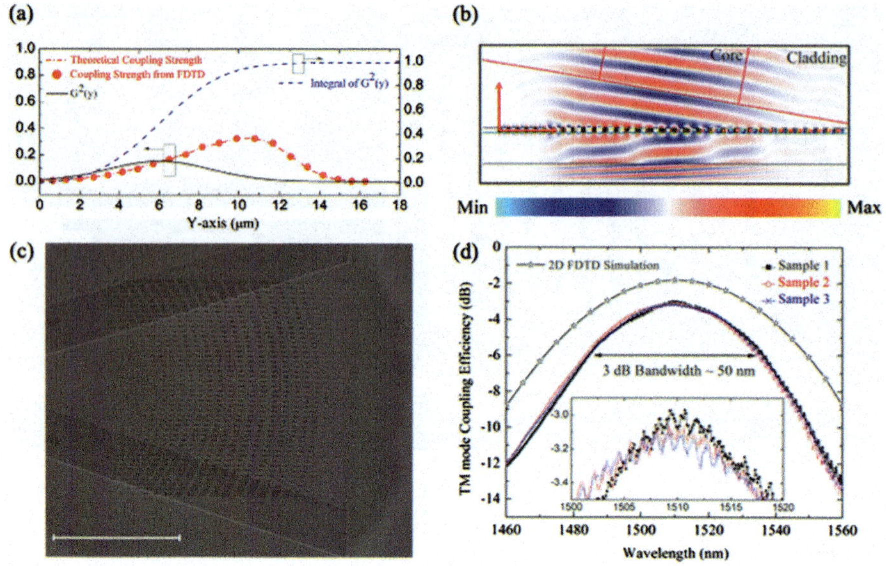


**Fig. 6.11** Design and demonstration of the apodized focusing subwavelength grating coupler. **a** Design of the device. **b** Simulation results. **c** SEM image. **d** Experimental measurements. Reproduced from [27] with the permission of AIP Publishing

**Table 6.1** Parameters of the apodized subwavelength grating coupler

| No. | 1 | 2 | 3 | 4 | 5 | 6 | 7 | 8 | 9 | 10 | 11 |
|---|---|---|---|---|---|---|---|---|---|---|---|
| $\Lambda_Y$ (nm) | 663 | 668 | 687 | 702 | 719 | 736 | 760 | 783 | 809 | 825 | 843 |
| *fx* | 0.26 | 0.28 | 0.36 | 0.41 | 0.46 | 0.5 | 0.54 | 0.57 | 0.6 | 0.62 | 0.64 |
| No. | 12 | 13 | 14 | 15 | 16 | 17 | 18 | 19 | 20 | 21 | |
| $\Lambda_Y$ (nm) | 863 | 873 | 883 | 883 | 873 | 833 | 791 | 736 | 696 | 663 | |
| *fx* | 0.66 | 0.67 | 0.68 | 0.68 | 0.67 | 0.63 | 0.58 | 0.5 | 0.39 | 0.26 | |

the maximum device exhibits more than a −3.0 dB peak coupling efficiency with a ~50 nm 3-dB bandwidth, which agrees with the 2D FDTD simulation. Also, we have fabricated numerous other couplers with the apodized focusing subwavelength grating coupler design, which gave consistently good reproducibility. Moreover, the apodized focusing subwavelength grating coupling profile shows weak Fabry–Perot (F-P) ripples, indicating a weak back reflection in the grating.

### *6.3.8 Broadband Focusing Subwavelength Grating Coupler*

The spectral bandwidth is another important figure-of-merit (FOM) of grating couplers. For example, in the telecommunication band, the 3-dB bandwidth of subwavelength grating couplers based on nanohole structures is usually about 50 nm. To overcome this limitation, in 2012, we demonstrated a novel fishbone grating coupler to achieve broadband light coupling between optical fibers and silicon waveguides [33]. The design principle of the broadband focusing subwavelength grating coupler is shown in Fig. 6.12a. Generally, subwavelength grating couplers based on nanohole structures are composed of silicon (high-refractive-index (RI) region) and effective medium 1 (EM1) (low RI region) created by the subwavelength structure. In the fishbone grating, we used another low RI EM2 to replace the high RI silicon region; therefore, the EM1 and EM2 were designed as the high RI and the low RI regions, respectively. Similarly, we designed the apodized grating structure to improve the coupling efficiency of the fishbone grating. Finally, we used the phase-matching formula to construct the focusing grating structure to provide a compact grating footprint. The fishbone grating was designed on an SOI wafer which has a 340-nm thick top silicon layer and 2.0-μm BOX. The angle between the fiber axis and the waveguide planar was 10 degrees. We set the lateral subwavelength periods ($\Lambda_{EM1x}$) and ($\Lambda_{EM2x}$) as 400 nm, lateral subwavelength fill factor ($f_{EM1x}$) as 0.4, and

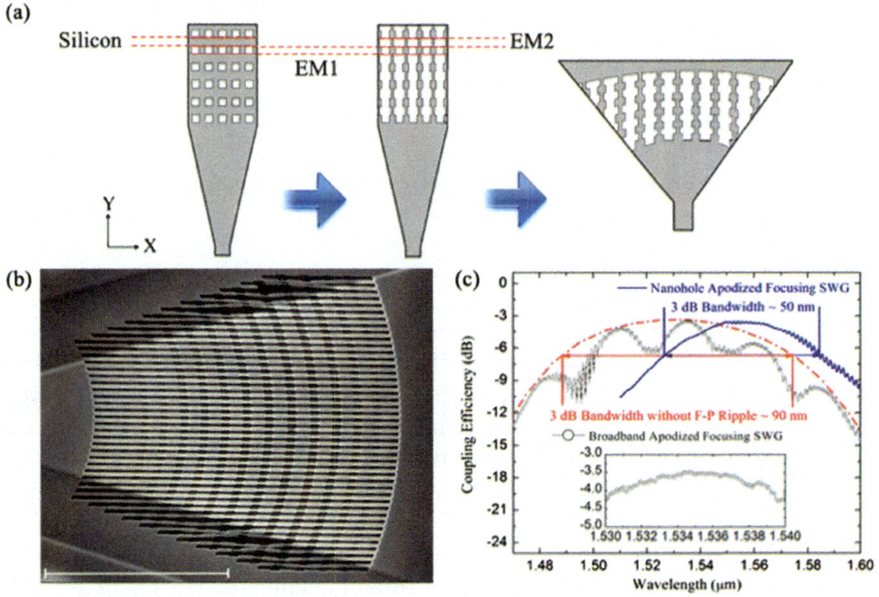


**Fig. 6.12** Design and demonstration of the broadband focusing subwavelength grating coupler. **a** Schematic of the device. **b** SEM image. **c** Experimental measurements. Reprinted with permission from [33] © The Optical Society

**Table 6.2** Parameters of the broadband apodized subwavelength grating coupler

| No. | 1 | 2 | 3 | 4 | 5 | 6 | 7 | 8 |
|---|---|---|---|---|---|---|---|---|
| Λy (nm) | 950 | 960 | 988 | 1021 | 1043 | 1060 | 1102 | 1126 |
| $f_{EM2x}$ | 0.63 | 0.64 | 0.67 | 0.71 | 0.74 | 0.76 | 0.80 | 0.82 |
| No. | 9 | 10 | 11 | 12 | 13 | 14 | 15 | |
| Λy (nm) | 1150 | 1186 | 1198 | 1162 | 1102 | 1036 | 979 | |
| $f_{EM2x}$ | 0.84 | 0.87 | 0.88 | 0.85 | 0.80 | 0.73 | 0.66 | |

longitudinal grating fill factor ($f_Y$) as 0.6. The longitudinal grating periods ($\Lambda_Y$) and lateral subwavelength fill factors ($f_{EM1x}$) we used in the design are listed in Table 6.2.

An experimental demonstration of the device is shown in Fig. 6.12b, c. Figure 6.12b presents the SEM image of the fabricated broadband apodized focusing subwavelength grating coupler. Two gratings were connected to a 150-μm long waveguide. The device transmission measurement is shown in Fig. 6.12c There is a small reflection between the two gratings, as evident from the small F-P ripple with a period of ~1.5 nm. While a large fluctuation with a period of ~30 nm exists in the measurement, which should be caused by the F-P resonance within the grating between the front and the rear ends. This fluctuation may be reduced by adding an anti-reflection section at the end of the grating. The maximum efficiency reaches −3.5 dB with a possible 3-dB bandwidth of ~90 nm for TM-mode light coupling as indicated by the plotted coupling profile in Fig. 6.12c. Compared with the subwavelength grating couplers based on nanohole structures, the bandwidth of the fishbone grating has a more than 80% improvement.

### *6.3.9 Polarization-Insensitive Focusing Subwavelength Grating Coupler*

In 2014, we experimentally demonstrated a novel polarization-insensitive grating coupler based on subwavelength structures [31]. We first optimized the apodized grating coupler based on an SOI wafer with a top-silicon-layer thickness of 340 nm and a BOX thickness of 2 μm. The angle between the fiber axis and the waveguide planar was chosen as 10 degrees. The lateral subwavelength period ($\Lambda x$) and longitudinal grating fill factor ($f_Y$) were fixed as 400 nm and 0.6, respectively. The longitudinal grating periods ($\Lambda_Y$) and lateral subwavelength fill factors ($f_X$) we used in the design are listed in Table 6.3.

We fabricated and measured grating couplers with gradually reducing fx, as shown in Fig. 6.13a–d. For the TM mode, the maximum coupling efficiency of −3.2 dB with a 1-dB bandwidth of ~28 nm was experimentally demonstrated. With the sane grating, the measured maximum coupling efficiency reached −4.3 dB with a 1-dB bandwidth of ~58 nm for the TE mode. We theoretically verified the grating coupling profiles in Fig. 6.13b by using 3D FDTD simulation software. The simulation results

**Table 6.3** Parameters of the polarization-insensitive subwavelength grating coupler

| No. | 1 | 2 | 3 | 4 | 5 | 6 | 7 | 8 | 9 | 10 |
|---|---|---|---|---|---|---|---|---|---|---|
| $\Lambda_Y$ (nm) | 655 | 668 | 687 | 702 | 719 | 736 | 760 | 783 | 809 | 843 |
| *fx* | 0.22 | 0.28 | 0.36 | 0.41 | 0.46 | 0.5 | 0.54 | 0.57 | 0.6 | 0.62 |
| No. | 11 | 12 | 13 | 14 | 15 | 16 | 17 | 18 | 19 | 20 |
| $\Lambda_Y$ (nm) | 863 | 873 | 883 | 883 | 873 | 833 | 791 | 736 | 696 | 663 |
| *fx* | 0.64 | 0.66 | 0.67 | 0.68 | 0.67 | 0.63 | 0.58 | 0.5 | 0.39 | 0.26 |

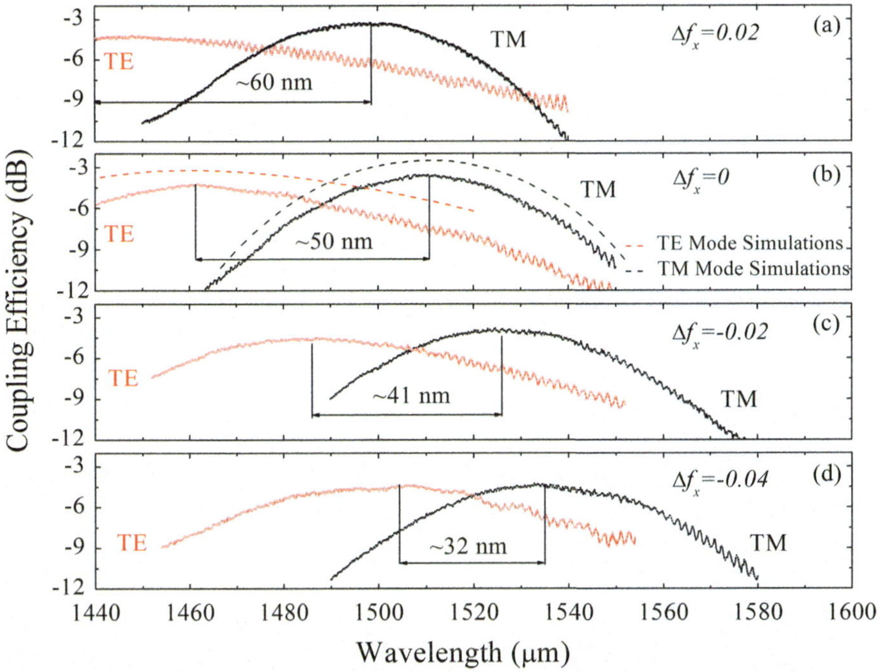


**Fig. 6.13** Experimental demonstrations of the polarization insensitive focusing subwavelength grating coupler. **a–d** represent grating coupling profiles with different subwavelength structure sizes. Reprinted with permission from [31] © The Optical Society

show that both the TE and TM mode coupling profiles have redshifts when gradually reducing fx. Moreover, the experimental measurements agree well with the 3D FDTD simulations, as shown in Fig. 6.13b. Moreover, the F-P ripples are more obvious for the TE mode than the TM mode, indicating that the back reflection of the TE mode-polarized light is stronger than that of the TM mode, which also agrees well with the 3D FDTD simulations.

## 6.4 Use of SWGs for Waveguide Devices and Mid-IR Photonics

In this section, we introduce the progress of using subwavelength structures to develop waveguide devices, namely, waveguides, anti-reflective (AR) layers, directional couplers, and suspended membrane devices for mid-IR photonics. We first briefly discuss the history and state-of-the-art techniques of SWG-based waveguide devices and then mainly focus on our previous studies of mid-IR SWG devices.

As we have discussed before, compared to photonic crystal structures, light propagating behavior in a subwavelength structure could be conveniently described by using a refractive index of an equivalent homogeneous medium. Generally, if the light frequency is much lower than the photonic bandgap of the periodic structure, the propagation constant monotonically increases with the light frequency. This property indicates that the subwavelength structure could be used for guiding light as an equivalent homogeneous medium. The first experimental results of silicon SWG waveguides were published in 2010 [2], as shown in Fig. 6.14a. In this work, the authors demonstrated that the propagating optical loss of the silicon SWG waveguide could be as low as 2.1 dB/cm with negligible polarization and wavelength-dependent optical loss. Based on the same principle, researchers also demonstrated

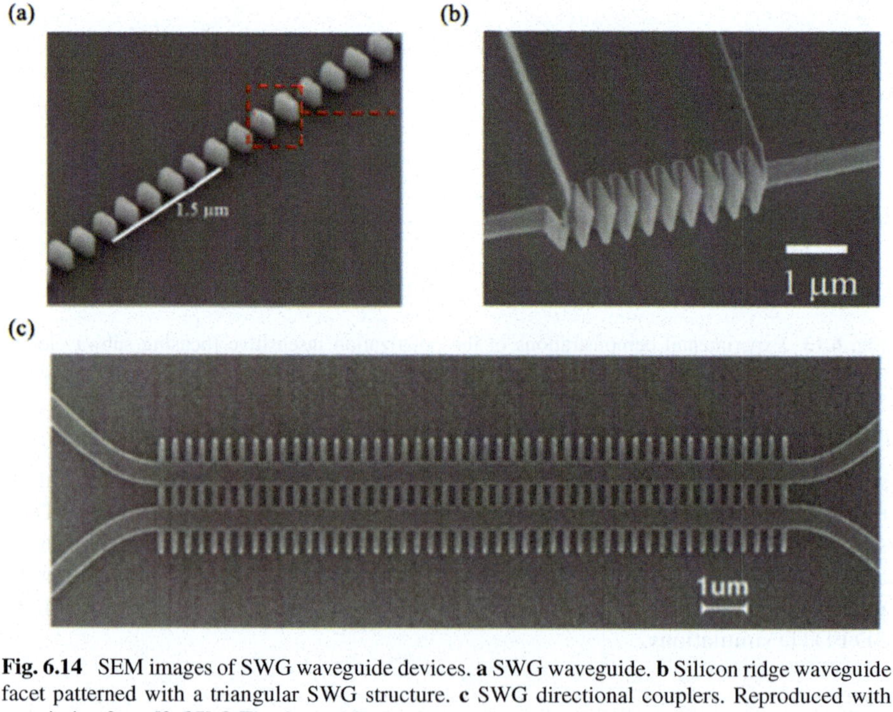


**Fig. 6.14** SEM images of SWG waveguide devices. **a** SWG waveguide. **b** Silicon ridge waveguide facet patterned with a triangular SWG structure. **c** SWG directional couplers. Reproduced with permission from [2, 37] © The Optical Society. Reproduced from [38] with the permission of IEEE

various waveguide-integrated passive devices, namely, AR layers [37], and direction couplers [38], as shown in Fig. 6.14b, c. Due to the merits of high RI sensitivity, broad operating bandwidth, and strong light-matter interaction, SWG devices are expected to be a promising candidate for developing on-chip sensing circuits and wide spectral bandwidth optoelectronic devices.

On the other hand, mid-IR silicon photonics has tremendous applications in biochemical sensing, free-space communication, and thermal imaging. However, mid-IR devices cannot be directly designed and fabricated on an SOI wafer because of the strong absorption of the BOX in the mid-IR spectral region. To overcome this limitation, several novel platforms have been demonstrated, among which SWG devices play a significant role.

SWGs have been employed for light coupling from mid-IR fibers to silicon waveguides. In 2012, an SWG coupler was fabricated on the silicon on sapphire (SOS) platform [39], as shown in Fig. 6.15a and a focusing SWG coupler on the suspended

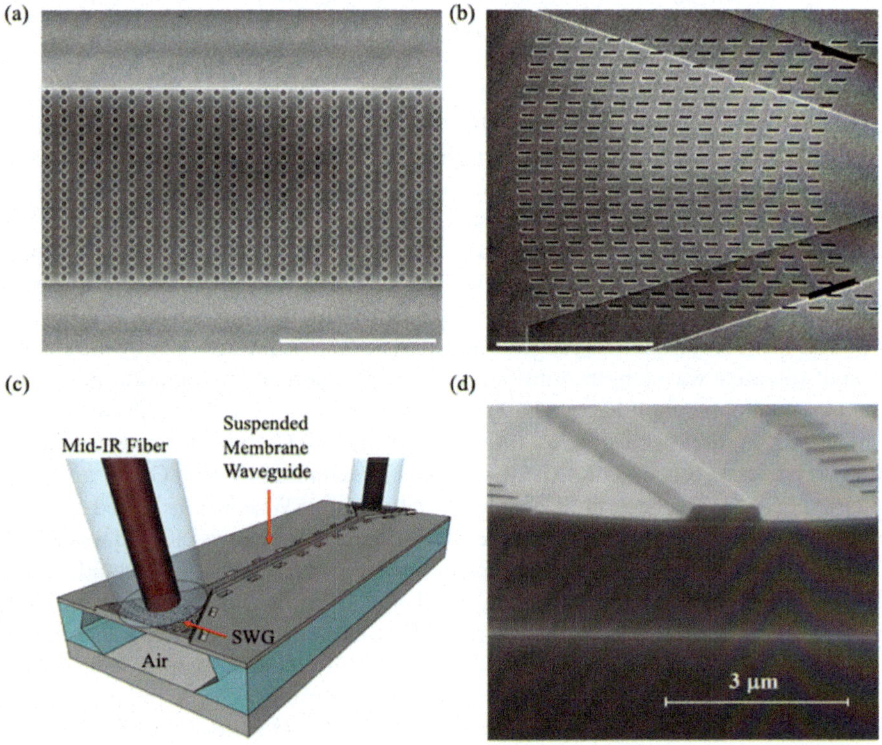


**Fig. 6.15** Mid-IR SWG couplers on the silicon-on-sapphire (SOS) and suspended membrane platforms. **a** SEM image of the mid-IR SWG coupler on the SOS platform. **b** SEM image of the mid-IR focusing SWG coupler on the suspended membrane platform. **c** Schematic of the suspended membrane waveguide integrated with the focusing SWG couplers. **d** SEM image of the suspended membrane waveguide cross-section. Reproduced from [39, 40] with the permission of IEEE

membrane platform [40], as shown in Fig. 6.15b, respectively. For the SWG coupler on the SOS wafer, the design was fabricated on the 600-nm thick top silicon layer. For operation at a wavelength of 2.7 μm, the TM-mode nanohole SWG coupler was formed by etching holes with a 253 nm radius and 1270 nm period. The maximum coupling efficiency was demonstrated to be 11.6% with an incident angle of 12.5°. The experimental measurements of the coupling efficiency are ~20% lower than the simulation results, which was due to uncontrolled polarization states in the mid-IR fiber laser. For the focusing SWG coupler on the suspended membrane platform, the grating coupler was fabricated on the top-silicon-layer thickness of 340 nm and BOX thickness of 2 μm. With the lateral subwavelength period of 800 nm, the TE mode focusing SWG couplers with lateral subwavelength fill factors of 0.25, 0.33, and 0.4 were fabricated and measured at a wavelength of 2.7 μm. Under the 10-degree incident angle, a maximum coupling efficiency of 24.7% was measured. Similarly, there is about a 20% coupling efficiency difference between theoretical simulations and experimental results due to the unoptimized polarization of the incident light. In the suspended membrane waveguide coupled with the SWG, shown in Fig. 6.15c, the BOX below the grating coupler and waveguide region has been entirely etched away, and the waveguide is supported by two-side cantilevers, as shown in Fig. 6.15d. This suspended structure overcomes the problem of the BOX absorption to mid-IR light and significantly expand the spectral bandwidth to wavelengths as long as 8 μm.

An alternative approach to the suspended membrane waveguide is to use a subwavelength-structured fully suspended slot waveguide (FSSWG) for mid-IR applications [10]. 1D SWGs were used to support slotted waveguide cores. The SWGs in the FSSWG can be treated as a lateral waveguide cladding with an equivalent and homogeneous RI [41]. Fabricated symmetric FSSWGs had a propagation loss of 7.9 dB/cm at the wavelength of 2.25 μm. With the removal of the BOX, the FSSWG has a broad spectral range of transparency that is limited only by the absorption of silicon at wavelengths longer than 8 μm. Numerical investigation shows that the maximal absorption loss is 1.89 dB/cm at 6.9 μm in a wavelength range of 2–8 μm [42]. Figure 6.16a shows a tilted view of the SWG coupler connecting to the FSSWG. Figure 6.16b shows the top view of a symmetric FSSWG and an asymmetric FSSWG. To check the mechanical stability, FSSWGs were cleaved from the middle and released by wet-etching the BOX. Figure 6.16c shows cleaved waveguide ends without vertical misalignment for the two separated silicon waveguide cores. Figure 6.16d–h presents the top-view SEM images of the FSSWG bend, FSSWG racetrack resonator, and strip-to-slot mode converter, respectively. Note that light is coupled into the resonator based on a symmetric directional coupler as shown in the green solid box in Fig. 6.16f. By using asymmetric FSSWGs, the propagation loss, bending loss, and intrinsic optical quality factor ($Q$ factor) were improved to 2.8 dB/cm, 0.15 dB/90°, and 12,600, respectively, compared with those based on the symmetric FSSWGs. The average conversion efficiency of a strip-to-slot mode converter is 95.4% in a 170-nm bandwidth. In addition, fully suspended nanophotonic waveguide resonators with a high-$Q$ factor of 40,600 and tailorable operational bandwidth were also demonstrated.

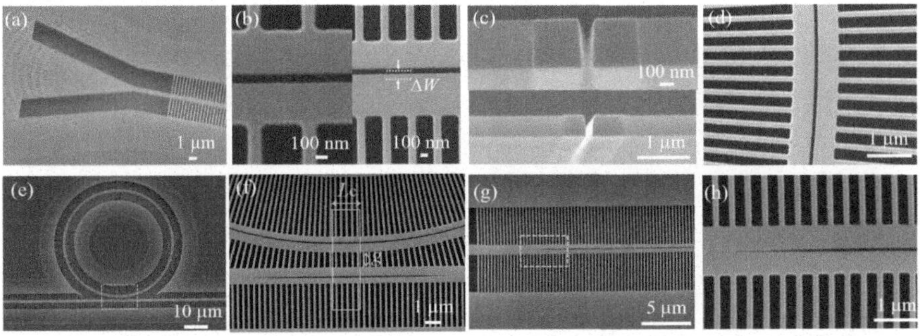


**Fig. 6.16** SEM images of fabricated fully suspended slot waveguide (FSSWG) devices. **a** Tilted view of the SWG coupler connecting to the FSSWG. **b** Top view of the symmetric FSSWG and an asymmetric FSSWG. **c** Cross-sectional view of cleaved waveguide ends of the symmetric FSSWG and an asymmetric FSSWG. d FSSWG bend. **e** FSSWG racetrack resonator side-coupled to a bus waveguide. **f** Zoom-in of the coupling region indicated by the dashed rectangle in e. **g** A strip-to-slot mode converter. **h** Zoom-in of the inverse slot taper indicated by the dashed rectangle in g. Adapted with permission from [10] © The Optical Society and [42] with the permission of AIP Publishing

Besides silicon devices, SWGs have been also studied in mid-IR germanium photonics. The mid-IR fingerprint spectral region, spanning approximately 8–14 μm, has numerous applications in biochemical sensing and spectroscopy. However, silicon PICs are applicable only at wavelengths below 8 μm, which is moderated by the transmittance of waveguide or substrate materials. To overcome this problem, germanium photonic integrated circuits (PICs) have been studied in the past few years. Since 2017, we have developed novel mid-IR germanium PICs including focusing SWG couplers [43], photonic crystal cavities [44, 45], and microring resonators [46]. The germanium SWG couplers were designed on a germanium-on-insulator (GOI) wafer with a top germanium layer thickness of 600 nm. We chose the lateral subwavelength period ($\Lambda x$) as 600 nm to satisfy the subwavelength condition at a wavelength of 2.4 μm. After the optimization, with the longitudinal grating period ($\Lambda_Y$) of 1196 nm, longitudinal grating fill factor ($f_Y$) of 0.6, and lateral subwavelength fill factor ($fx$) of 0.83, the 2D simulation shows that the TM-mode SWG couplers have a maximum coupling efficiency of −4.1 dB with the 1-dB bandwidth of ~66 nm. We then fabricated the focusing SWG couplers as shown in Fig. 6.17a. Experimental measurements show that the focusing SWG coupler has a maximum efficiency of −11 dB and a 1-dB bandwidth of ~58 nm and coupled the TM-polarized light at a center wavelength of 2.37 μm, as shown in Fig. 6.17b. The measured bandwidth of the focusing SWG coupler agrees with the 3D FDTD simulation, but the measured coupling efficiency is much lower than the simulation. The large discrepancy between the experimental measurement and simulation results is because the light emitted from the input fiber was elliptically polarized. Moreover, we demonstrated the germanium SWG couplers integrated with suspended membrane waveguides (Fig. 6.17c), photonic crystal cavity with Fano resonance (Fig. 6.17d), photonic crystal nanobeam (Fig. 6.17e), and microring resonators, respectively. At

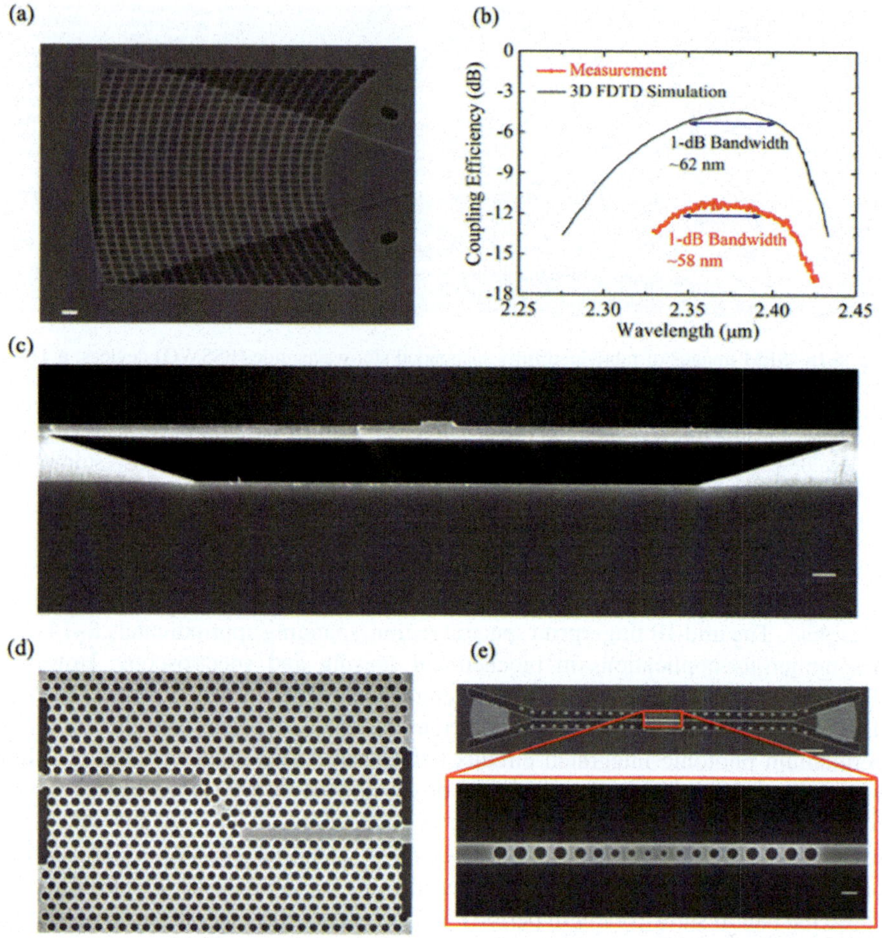


**Fig. 6.17** Demonstrations of the mid-IR germanium suspended membrane devices. **a** SEM image of the focusing SWG coupler. **b** 3D FDTD simulation and measurement of the focusing SWG coupler. **c** SEM image of the suspended membrane waveguide cross-section. **d** SEM image of photonic crystal waveguide with Fano resonance. **e** SEM image of the photonic crystal nanobeam cavity. Reproduced with permission from [43–45] © The Optical Society

a wavelength of 2.4 μm, we demonstrated the maximum Q factors of ~18,000 and ~57,000 for the photonic crystal nanobeam and microring resonator, respectively. The suspended membrane germanium PICs open new doors for the production of mid-IR devices in long-wavelength regions.

## 6.5 Numerical Optimization for the Design of Subwavelength Structures

Numerical optimization of a particular design should start with a clear definition of what combination of performance parameters are to be optimized. There exist many different general FOMs of an integrated photonic component: Examples of performance parameters that may be used in the FOMs include insertion loss, back reflection, optical bandwidth, polarization dependence, optical dispersion, effective area, and nonlinearity. Many of these FOMs are dependent on different parameters of the subwavelength structures on the planar waveguide circuit. Moreover, the minimum dimension and critical feature size of a particular design determine the fabrication tolerance and large-volume manufacturability of photonic components. Beyond using our understanding of the underlying physics of a particular functional element, as elaborated in many examples above, numerical optimization can be an integral part of the design flow that can further improve device performance by numerical optimization tools. Within the large design parameter space that is typically associated with the design of photonic components, numerical optimization offers the possibility to search for a solution with the best combination of FOMs. Precise balancing of the various FOMs is also allowed in the iteration process, which can be fully automated. The numerical optimization can improve the device performance, fabrication robustness and may also lead to more compact device footprints.

Integrated photonic components such as the subwavelength grating coupler, beam splitter, waveguide crossing, and wavelength-, polarization-, and mode-handling devices have been reported using the various optimization tools. The reported optimization toolbox includes the particle-swarm algorithm [30, 47, 48], genetic optimization [49–53], and machine-learning methods [54–56]. Intricate geometric waveguide patterns can be generated on the SOI platform by the inverse design and topology optimization [57–62]. Benefitting from the growing computational resources, the optimization approaches are being developed toward the ability to seek a multitude of optimum parameters to enable more complex FOMs criteria, while still considering the fabrication robustness in volume production.

This section elaborates on four applications of numerical optimization for the design of silicon photonic components. We describe the application of genetic optimization in the design of a diffractive waveguide grating coupler with an objective function that enables perfectly vertical coupling, high coupling efficiency, wide optical bandwidth, low back reflection, and fabrication robustness [51]. Taking advantage of the EMT discussed previously, the numerical optimization can be practically applied for the design of a subwavelength dual-polarization grating coupler for use with spatial-division multiplexing (SDM) in a few-mode fiber (FMF) [52, 53]. Next, design and numerical optimization of dual-wavelength-band focusing subwavelength grating couplers are discussed [69, 70]. Finally, a numerically designed subwavelength hyperuniform disordered silicon photonic (HUDSiP) polarizer is discussed [63].

### 6.5.1 Perfectly Vertical Grating Coupler for Multi-core Fiber

Typical grating couplers rely on the tilted configuration, which needs a small coupling angle relative to the chip-surface normal to reduce reflection into the optical waveguide due to the second-order Bragg diffraction. Nevertheless, perfectly vertical coupling takes the advantages of an easier optical alignment in the packing process. It has no spatial-channel-dependent loss theoretically when directly coupled to a multi-core fiber (MCF) in the spatial-division multiplexing (SDM) system. The number of spatial channels can be scaled by a grating array depending on the design of the MCF, as shown in Fig. 6.18a for a 7-channel MCF. Therefore, the perfectly vertical coupling may be more suitable as a spatial multiplexer for MCF compared with the off-normal configuration.

The PVGC is designed for coupling to the TE mode in a silicon waveguide from a standard single-mode fiber (SMF) with a mode field diameter (MFD) of 10.4 μm at 1550 nm. A polysilicon layer with a thickness of 160 nm is deposited on the SOI wafer to increase the grating directionality of over 80%, as implemented in [64, 65]. The SOI wafer has a 220-nm thick top silicon layer and a 2-μm thick BOX layer. Dry etching of 70 nm into the bottom crystalline silicon is executed for the grating grooves. The top cladding is silicon dioxide.

Ideally, a PVGC is expected to possess a high coupling efficiency, wide optical bandwidth, low back reflection, and fabrication robustness. The structural parameters of a diffractive grating include the length of the grating teeth and grooves. The genetic

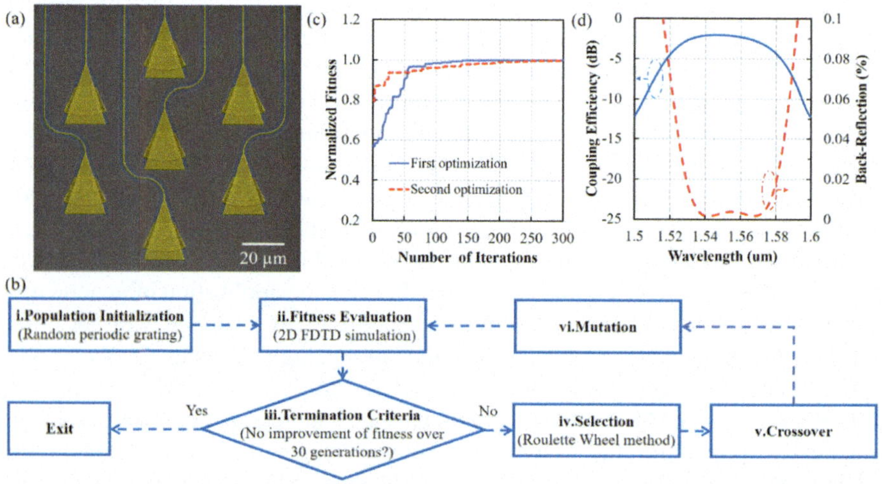


**Fig. 6.18** **a** SEM image of the PVGC array for a 7-channel MCF. **b** Optimization flowchart of the genetic algorithm for the PVGC. **c** Evolutional diagram of the PVGC showing the normalized fitness versus the number of generations. **d** Simulated coupling efficiency and back reflection of the PVGC by a 2D FDTD. Reproduced with permission from [51] © The Optical Society

algorithm is used for the optimization of structural parameters oriented by a custom-defined fitness [66]. The flowchart of the genetic algorithm for PVGC is illustrated in Fig. 6.18b. The optimization loop is fed by a random periodic grating in the step of (i) population initialization. Each population is a vector of a point in the search space. To take into account of all the objectives, the fitness $F$ is defined by (**6.11**) for (ii) evaluation,

$$F = \sum_{i=1}^{N} CE(\lambda_i) - \eta \times BR(\lambda_i) \tag{6.11}$$

where $\lambda$ is the optical wavelength, *CE* is the coupling efficiency evaluated by 2D-FDTD simulation, *BR* denotes the back reflection into the silicon waveguide, and $\eta$ is a weighting factor to balance the improvement in numerical iterations. The fitness $F$ includes simulation performance covering the C-band from 1530 nm to 1565 nm. (iii) Determine the termination of iterations according to the criteria that no improvement has been obtained in the most-recent 30 generations. In the step (iv) selection, populations are kept using the Roulette-Wheel selection method [66]. The (iv) crossover function is implemented at an 80% probability to intermix the populations with each other. The reproduced populations would experience (vi) mutation at a probability of 10%, where the structural parameters experience random variations. In order to possess a robust fabrication performance, the minimum feature size is restricted to be above the requirement using the 193 nm DUV lithography for large-volume manufacturing by the commercial silicon photonics foundries.

The evolutional diagram showing the normalized fitness in different optimization generations is presented in Fig. 6.18c, confirming the progress of convergence. $\eta$ is set as zero in the first round of optimization and results in a large back reflection above −10 dB. By setting $\eta$ at a tentative value of 4 in the second round of optimization, the optical power reflected into the SOI waveguide can be below −20 dB from 1532 to 1576 nm as shown in Fig. 6.18d. The 2D-FDTD simulation predicts the optimized diffractive grating has a peak coupling efficiency of −2.0 dB (63.0%) at 1546 nm. The 1-dB optical bandwidth is 43 nm, and the 3-dB bandwidth is 64 nm, as presented in Fig. 6.18d. The structural parameters after two rounds of genetic optimization of the PVGC are summarized in Table 6.4.

**Table 6.4** Parameters of the perfectly vertical grating coupler

| No. | 1 | 2 | 3 | 4 | 5 | 6 | 7 | 8 | 9 | 10 | 11 | 12 |
|---|---|---|---|---|---|---|---|---|---|---|---|---|
| Teeth (nm) | 270 | 252 | 209 | 221 | 225 | 212 | 217 | 228 | 276 | 296 | 330 | 325 |
| Grooves (nm) | 236 | 301 | 351 | 353 | 365 | 373 | 376 | 363 | 296 | 254 | 237 | 247 |
| No. | 13 | 14 | 15 | 16 | 17 | 18 | 19 | 20 | 21 | 22 | 23 | 24 |
| Teeth (nm) | 294 | 292 | 316 | 332 | 308 | 310 | 324 | 293 | 215 | 316 | 321 | 269 |
| Grooves (nm) | 266 | 276 | 206 | 232 | 256 | 255 | 219 | 258 | 340 | 261 | 264 | 294 |

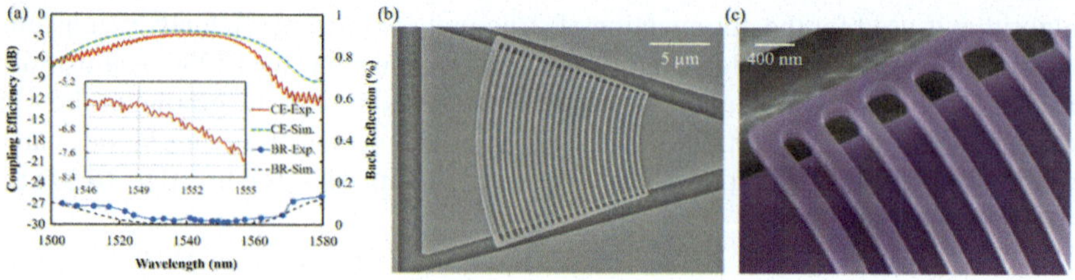


**Fig. 6.19** **a** Spectrum of the coupling efficiency (CE) and back reflection (BR) of the PVGC by simulation (Sim.) and experiment (Exp.). A zoom-in view of the FP ripples is shown in the inset. **b** SEM image of the focusing PVGC. **c** SEM image of the PVGC with a polysilicon layer. Reproduced with permission from [51] © The Optical Society

The experimental coupling efficiency is measured with two focusing PVGCs connected by a strip waveguide. The performance of a single grating coupler is depicted by the solid red line in Fig. 6.19a. The grating coupler has a peak coupling efficiency of −2.7 dB (53.7%) at 1539.2 nm. The measured 1-dB bandwidth is 33 nm, and the 3-dB bandwidth is 56 nm. Effective suppression of the back reflection is achieved around the central spectral region with an obviously reduced F-P ripple size of the two PVGCs, presented in the inset in Fig. 6.19a. The back reflection spectrum is shown by the solid blue line, which is extracted from the F-P ripples in the measured coupling efficiency spectrum. The optimum back reflection is less than −13 dB in the C-band. The minimum measured ripple size is about 0.26 dB, which represents a reflection of about −18 dB. SEM images of the grating coupler array and the top view are shown in Fig. 6.19b, c after removing the top oxide layer.

### *6.5.2 Subwavelength Dual-Polarization Grating Coupler for Few-Mode Fiber*

An alternative and popular scheme of the SDM communication system is based on the multimode fiber or FMF employing the different fiber modes. Diffractive waveguide grating couplers can be used as a mode multiplexer as conceptually illustrated in Fig. 6.20a–d. To increase the number of multiplexing channels via a single component, a 2D grating coupler is adopted by combining the TE modes in two orthogonally oriented waveguides. The waveguide grating is symmetric for the two orthogonal polarizations so that a total of four optical channels including $LP_{01-x}$, $LP_{11x}$, $LP_{01-y}$, and $LP_{11-y}$ can be coupled into the FMF. The simultaneous diffraction of two TE modes via a single grating is allowed because their effective refractive index in the wide waveguide, which has an expanded width of more than 10 μm, is about the same. Similar to the case in the previous example, the subwavelength 2D grating coupler is expected to have a high coupling efficiency, wide optical bandwidth, low back reflection, and fabrication robustness.

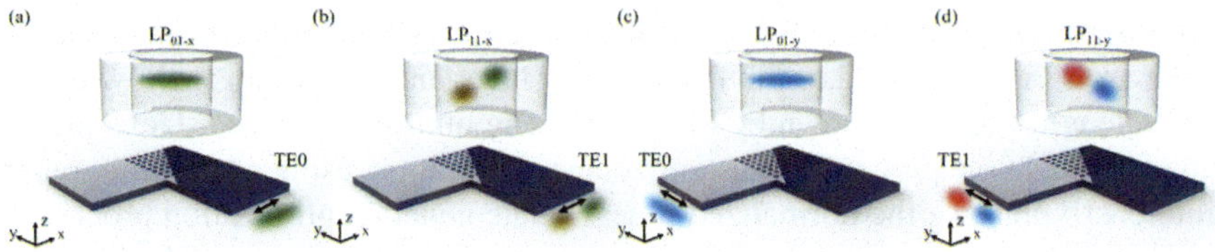


**Fig. 6.20** Schematic diagrams of the mode excitation in a two-mode few-mode fiber (FMF) via a dual-polarization subwavelength silicon waveguide grating. **a** $LP_{01-x}$ mode. **b** $LP_{11-x}$ mode. **c** $LP_{01-y}$ mode, and **d** $LP_{11-y}$ mode. Reproduced from [52] with the permission of IEEEReproduced from [52] with the permission of IEEE

However, the logic of numerical optimization is hindered by the complexity of the 3D FDTD simulation, which is required for a dual-polarization grating coupler. Each 3D FDTD simulation of the subwavelength grating with a footprint larger than dozens of microns requires several hours to finish, resulting in an extremely long time for numerical iterations, let alone the convergence of optimization. This obstacle can be reduced by employing the EMT and 2D FDTD simulations.

The adoption of the 2D FDTD simulation for a dual-polarization subwavelength grating is under the premise of two conditions. The first condition is that EMT denoted by (6.7)–(6.8) can be used to estimate the refractive index of the subwavelength region in 2D FDTD simulation as, shown by Fig. 6.21a. Secondly, the subwavelength grating is symmetric for the two orthogonal polarizations, so that performance of only a single polarization is needed. As the modes $TE_0$ and $TE_1$ in the wide grating region have similar effective refractive indexes, the FOMs of the dual-polarization subwavelength grating in the optimization process can be simplified to the performance of the fundamental TE mode in a single polarization. In other words, 2D FDTD simulation is feasible for subwavelength grating optimization instead of relying on the time-consuming 3D FDTD.

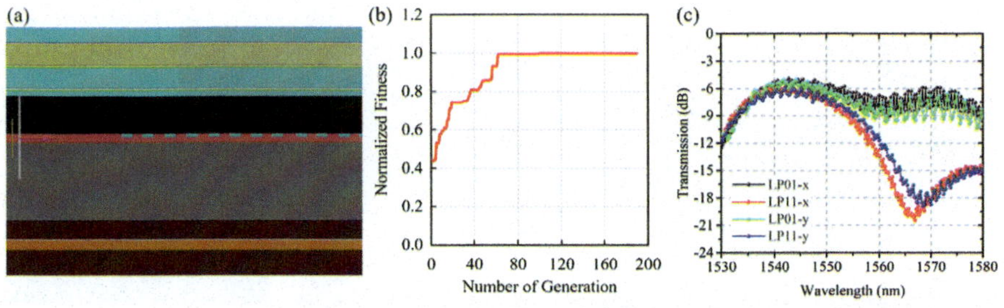


**Fig. 6.21** **a** Exported XY view for 2D FDTD simulation of the subwavelength grating coupler by *Lumerical*. **b** Evolutional diagram of the subwavelength grating coupler showing the normalized fitness versus the number of generations. **c** Experimental coupling efficiency of the subwavelength grating coupler for mode $LP_{01-x}$, $LP_{11-x}$, $LP_{01-y}$, and $LP_{11-y}$. Reproduced from [52] with the permission of IEEE

The high-refractive-index region and the low-refractive-index region of the grating coupler are formed by the 220-nm thick silicon layer and the 70-nm shallow-etched subwavelength holes, as shown by Fig. 6.21a. The minimum feature size is limited in the optimization process to sustain robust fabrication using photolithography. Figure 6.21b presents the normalized fitness in the genetic optimization process. In 2D FDTD simulation, a square-shaped subwavelength hole is used. However, the square-shaped hole is converted to the round-shaped hole in the mask layout, and fabrication with the same area as a round-shaped hole allows fabrication using the standard DUV photolithography with less stringent requirements, as discussed in the design of a nanohole grating coupler.

The experimental coupling efficiency of the dual-polarization waveguide grating is characterized by measuring the fiber-waveguide-fiber transmission with an SMF as the input and a two-mode FMF as the output. The coupling spectrums of the $LP_{01-x}$, $LP_{11x}$, $LP_{01-y}$, and $LP_{11-y}$ using a subwavelength dual-polarization waveguide grating are shown in Fig. 6.21c. The $LP_{01}$ mode has a peak coupling efficiency of −4.9 dB. The $LP_{11}$ mode has a peak coupling efficiency of −5.8 dB. The 3-dB bandwidths are 23.2 nm and 21.6 nm for the two modes, respectively.

On-chip mode multiplexing of the $TE_0$ mode and $TE_1$ mode is realized by the asymmetrical directional coupler based on the phase matching between the parallelly spaced waveguides, as shown in Fig. 6.22a. The effective refractive index of the $TE_0$ mode in the narrower waveguide should be close to that of the $TE_1$ mode in the wider waveguide [67, 68]. To verify the selective excitation of different modes and polarizations in FMF via the subwavelength waveguide grating coupler, an infrared camera with a 40 × lens is used at the end of the FMF to capture the output field profiles, as shown in Fig. 6.22c.

Thanks to the EMT and 2D FDTD, the simulation time of the subwavelength 2D grating coupler can be significantly reduced to about 10 s rather than more than 5 h using 3D FDTD under the same mesh order. Numerical iterations in the multiple parameter space are thus feasible for device optimization.

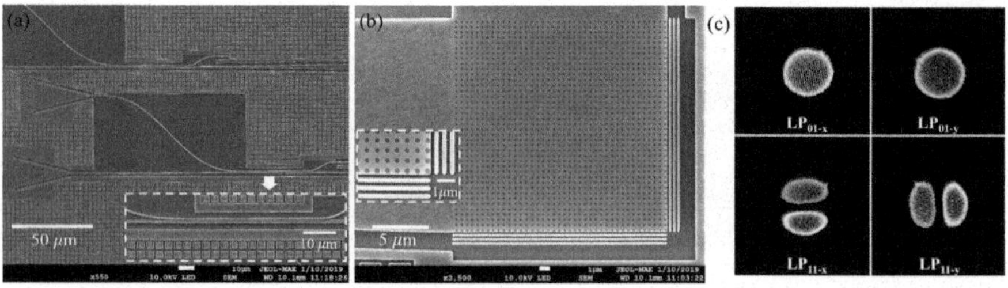


**Fig. 6.22** **a** SEM image of the input single-mode gratings and asymmetric directional couplers for on-chip mode multiplexing with a zoom-in view shown in the inset. **b** SEM image of the fabricated subwavelength waveguide grating coupler with a zoom-in view shown in the inset. **c** Output field profiles of the FMF when $LP_{01-x}$, $LP_{11x}$, $LP_{01-y}$, and $LP_{11-y}$ are selectively coupled by the subwavelength waveguide grating coupler. Reproduced from [52] with the permission of IEEE

### 6.5.3 Dual-Wavelength-Band Focusing Subwavelength Grating Couplers (DWB FSWGCs)

To overcome inherent bandwidth limitation, a new type of surface grating coupler (GC) was proposed, which can simultaneously couple TE-polarized light that is spectrally separated in two wavelength bands with peaks located at ($\lambda_1$, $\lambda_2$) and a tailorable spectral separation ($\Delta\lambda = \lambda_2 - \lambda_1$) from a single-mode fiber into an SOI waveguide under the same diffraction angle $\theta$ [69]. The focusing dual-wavelength-band (DWB) subwavelength grating coupler was designed on an SOI suspended membrane waveguide platform [40]. Figure 6.23a shows the schematic. Figure 6.23b shows an SEM image of a fabricated $GC_4$. The focal length is 12 μm in the –*x*-direction. To avoid higher-order mode excitation, an inverse taper was used [71].

Figure 6.19c shows the design steps of DWB subwavelength grating couplers with flexibly tailored peaks at ($\lambda_1$, $\lambda_2$). Black, white, and gray regions represent silicon, air, and SWG, respectively. $GC_{1-2}$ are uniform single-wavelength-band GCs, which are designed based on the phase-matching condition [72] to have trial peaks at ($\lambda_{u1}$, $\lambda_{u2}$) and the same $\theta$. Note that $\lambda_{u1}$ ($\lambda_{u2}$) was initially designed longer than the targeted $\lambda_1$ ($\lambda_2$). $GC_3$ is a DWB GC generated by merging SWG regions in $GC_1$ and $GC_2$, whose period and fill factor in the *x*-direction were tuned to create $GC_3$ diffracting light at the targeted ($\lambda_1$, $\lambda_2$) under $\theta$ and with balanced coupling efficiencies (CEs), simultaneously. A flowchart is shown in Fig. 6.23d, in which a modified direct search algorithm [73] is used for optimizing the figure-of-merit (FOM) of $GC_3$, which is defined as the multiplication of coupling efficiencies at the targeted two peaks

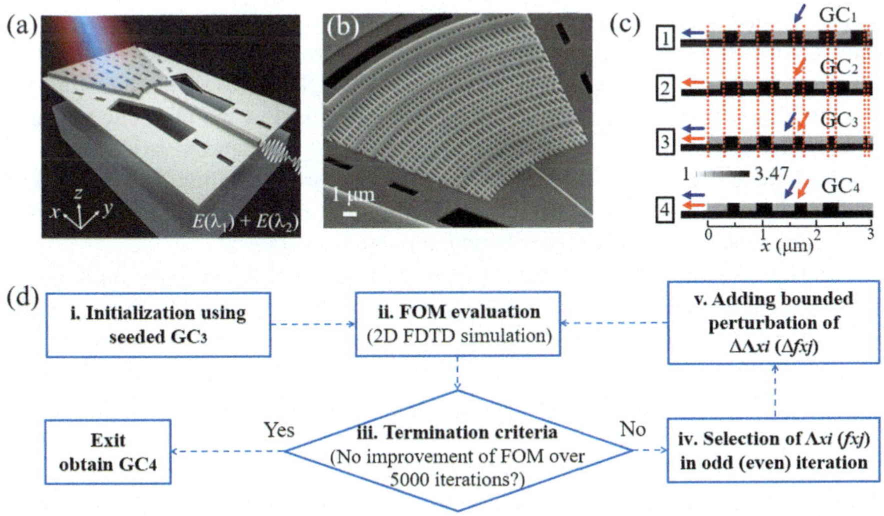


**Fig. 6.23** **a** 3D illustration of a DWB FSWGC. **b** SEM image of a fabricated DWB FSWGC. **c** Designing steps 1–4 for DWB FSWGCs. **d** Optimization flowchart of a modified direct search algorithm used in step 4. Reproduced with permission from [69, 70] © The Optical Society

CE($\lambda_1$)·CE($\lambda_2$), and lastly to obtain an optimized $GC_4$. In the flowchart, $\Lambda_{xi}$ ($f_{xj}$) is a period (fill factor) of the $i$($j$)-th diffractive unit in the $x$-direction, and $\Delta\Lambda_{xi}$ ($\Delta f_{xj}$) is a bounded perturbation in each looped optimization step. After 2D FDTD simulation in each iteration, FOM is calculated, if there is (no) improvement, the perturbation is (discarded) stored. If there is no improvement in FOM over 5000 iterations, the algorithm will be automatically stopped and output the final structure of $GC_4$.

Designed on 340-nm and 500-nm SOI wafers, the numerical simulation predicts a wavelength separation $\Delta\lambda$ of $GC_4$ covering a range of 50–350 nm and 250–1250 nm when CE > 30%, respectively [69, 70]. As proof of concept, using a 340-nm SOI, measured peaks of two designed $GC_4$ are at (1486.0, 1594.5) nm with CEs of (18.3%, 20.14%) and (1481.5, 1661.5) nm with CEs of (14.5%, 17.5%) [69]. In another example, using a 500-nm SOI, measured peaks of two $GC_4$ are at (1.56, 2.255) μm with CEs of (20.2%, 25.8%) and (1.48, 2.33) μm with CEs of (20.6%, 26.9%) [70]. With high CEs and a tailorable $\Delta\lambda$, DWB FSWGCs are promising candidates for light coupling and wavelength multiplexing in the conventional telecommunication bands and the emerging 2-μm wavelength band.

### *6.5.4 Hyperuniform Disordered Silicon Photonic (HUDSiP) Polarizers*

Another numerically designed device covered in this chapter is a HUDSiP polarizer. A collective coordinate numerical optimization method can be used for the generation of hyperuniform disordered solid (HUDS) patterns [74–76]. Based on the exotic properties of HUDS [77, 78], a state-of-the-art TM-pass HUDSiP polarizer was demonstrated [63]. Operating in the subwavelength region, the TM-polarized light has a low insertion loss in a HUDSiP polarizer [11, 72]. Due to strong birefringence in a thin (220-nm) SOI, the TE-polarized light is blocked by the photonic bandgap-based reflection, scattering, and radiative band-edge resonances validated by the dipole cloud approach [79]. These combined effects produce a much wider working bandwidth than those of a hyperuniform disordered network polarizer [80] and a hyperuniform disordered photonic bandgap polarizer [81].

Figure 6.24a shows optical micrographs of fabricated devices. A single device consists of TE- and TM-mode subwavelength grating couplers, a pair of directional couplers for (de)multiplexing of orthogonally polarized modes into (out of) a bus waveguide, and a HUDSiP polarizer is at the middle of a bus waveguide. Figure 6.24b, c shows the top-view SEM images of a 12.9-μm-long HUDSiP polarizer. The measured average transmission of the TE mode and the TM mode, average extinction ratio, insertion loss of the TM mode at 1550 nm, and bandwidth are, respectively (−39.5 dB, −0.7 dB, 38.8 dB, 0.5 dB, >210 nm) and (−42.2 dB, −0.8 dB, 41.4 dB, 0.4 dB, >40 nm) for the 8.6-μm-long and 12.9-μm-long HUDSiP polarizers. HUDSiP polarizers have the state-of-the-art figure-of-merits (including

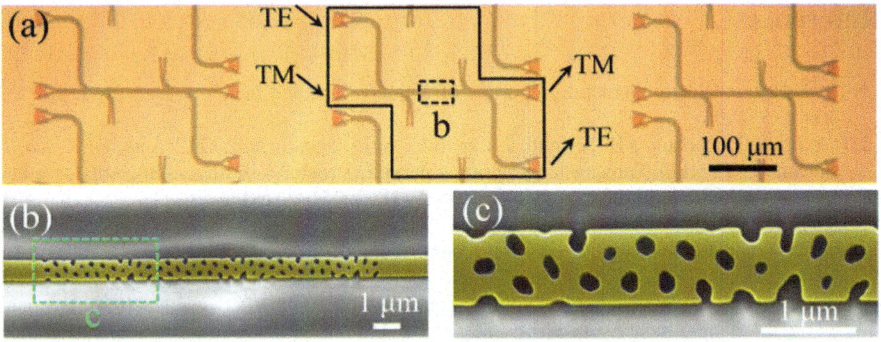


**Fig. 6.24** Micrograph images of fabricated devices. **a** Device array. **b** Top-view SEM image of a HUDSiP polarizer. **c** Zoom-in at the part indicated by a dashed rectangle in b. Reproduced from [63] with the permission of IEEE

working bandwidth, extinction ratio, and insertion loss) compared with those of previously reported polarizers [63].

## 6.6 Conclusion

In this chapter, we have described the use of effective medium theory and subwavelength gratings in the design of photonic devices. We showed how SWG are useful for increasing the bandwidth of grating couplers, apodizing the grating strength for mode matching, compensating waveguide birefringence and providing the holes for the local etching of the BOX in a suspended membrane waveguide and fully suspended slot waveguides and enabling the fabrication of devices operating in the mid-IR spectral region, and avoid limitations from the optical absorption of the buried oxide layer. In the optimization of the design of passive optical components, numerical techniques are invariably used, and the EMT is a particularly useful approximation, which can enable 3D structures to be modeled by equivalent 2D simulations in numerical optimization algorithms. We discussed their use in the engineering of perfectly vertical grating couplers and dual-wavelength-band grating couplers. In the context of subwavelength devices, we also briefly discussed the hyperuniform disordered structures formed by etching subwavelength holes in a hyperuniform disordered structure to form broadband integrated polarizers. Subwavelength engineering offers a highly useful additional degree of freedom in the design and optimization of photonic devices and will doubtless remain useful for the implementation of new passive devices based on a combination of physical principles (e.g., metasurfaces, cavities, dispersion engineering) and numerical optimizations in the future.

## References


1. S.M. Rytov, Electromagnetic properties of a finely stratified medium. Sov. Phys. JEPT **2**, 466–475 (1956)
2. P.J. Bock, P. Cheben, J.H. Schmid, J. Lapointe, A. Delâge, S. Janz, G.C. Aers, D.-X. Xu, A. Densmore, T.J. Hall, Subwavelength grating periodic structures in silicon-on-insulator: a new type of microphotonic waveguide. Opt. Express **18**, 20251–20262 (2010)
3. R. Halir, P. Cheben, S. Janz, D.-X. Xu, Í. Molina-Fernández, J.G. Wangüemert-Pérez, Waveguide grating coupler with subwavelength microstructures. Opt. Lett. **34**, 1408–1410 (2009)
4. X. Chen, H.K. Tsang, Nanoholes grating couplers for coupling between silicon-on-insulator waveguides and optical fibers. IEEE Photonics J. **1**, 184–190 (2009)
5. X. Chen, H.K. Tsang, Polarization-independent grating couplers for silicon-on-insulator nanophotonic waveguides. Opt. Lett. **36**, 796–798 (2011)
6. Z. Cheng, X. Chen, C.Y. Wong, K. Xu, C.K.Y. Fung, Y.M. Chen, H.K. Tsang, Focusing subwavelength grating coupler for mid-infrared suspended membrane waveguide. Opt. Lett. **37**, 1217–1219 (2012)
7. G. Floquet, Sur les équations différentielles linéaires à coefficients périodiques. Ann. Sci. LÉcole Norm. Supér. **12**, 47–88 (1883)
8. F. Bloch, Über die Quantenmechanik der Elektronen in Kristallgittern. Z. Für Phys. **52**, 555–600 (1929)
9. L. Brillouin, "Wave propagation in periodic structures: electric filters and crystal lattices," Dover Publications (2003) https://cds.cern.ch/record/106186
10. W. Zhou, Z. Cheng, X. Wu, B. Zhu, X. Sun, H.K. Tsang, Fully suspended slot waveguides for high refractive index sensitivity. Opt. Lett. **42**, 1245–1248 (2017)
11. P. Cheben, R. Halir, J.H. Schmid, H.A. Atwater, D.R. Smith, Subwavelength integrated photonics. Nature **560**, 565–572 (2018)
12. R. Halir, P.J. Bock, P. Cheben, A. Ortega-Moñux, C. Alonso-Ramos, J.H. Schmid, J. Lapointe, D.-X. Xu, J.G. Wangüemert-Pérez, Í. Molina-Fernández, S. Janz, Waveguide sub-wavelength structures: a review of principles and applications. Laser Photonics Rev. **9**, 25–49 (2015)
13. T.W. Ang, G.T. Reed, A. Vonsovici, A.G.R. Evans, P.R. Routley, M.R. Josey, Effects of grating heights on highly efficient unibond SOI waveguide grating couplers. IEEE Photonics Technol. Lett. **12**, 59–61 (2000)
14. D. Taillaert, W. Bogaerts, P. Bienstman, T.F. Krauss, P. Van Daele, I. Moerman, S. Verstuyft, K. De Mesel, R. Baets, An out-of-plane grating coupler for efficient butt-coupling between compact planar waveguides and single-mode fibers. IEEE J. Quantum Electron. **38**, 949–955 (2002)
15. D. Taillaert, F.V. Laere, M. Ayre, W. Bogaerts, D.V. Thourhout, P. Bienstman, R. Baets, Grating couplers for coupling between optical fibers and nanophotonic waveguides. Jpn. J. Appl. Phys. **45**, 6071 (2006)
16. X. Chen, M.M. Milosevic, S. Stanković, S. Reynolds, T.D. Bucio, K. Li, D.J. Thomson, F. Gardes, G.T. Reed, The emergence of silicon photonics as a flexible technology platform. Proc. IEEE **106**, 2101–2116 (2018)
17. X. Chen, C. Li, H.K. Tsang, Device engineering for silicon photonics. NPG Asia Mater. **3**, 34–40 (2011)
18. X. Chen, C. Li, C.K.Y. Fung, S.M.G. Lo, H.K. Tsang, Apodized waveguide grating couplers for efficient coupling to optical fibers. IEEE Photonics Technol. Lett. **22**, 1156–1158 (2010)
19. J. Schrauwen, F. Van Laere, D. Van Thourhout, R. Baets, Focused-Ion-Beam fabrication of slanted grating couplers in silicon-on-insulator waveguides. IEEE Photonics Technol. Lett. **19**, 816–818 (2007)
20. S. Scheerlinck, J. Schrauwen, F.V. Laere, D. Taillaert, D.V. Thourhout, R. Baets, Efficient, broadband and compact metal grating couplers for silicon-on-insulator waveguides. Opt. Express **15**, 9625–9630 (2007)

21. Y. Ding, C. Peucheret, H. Ou, K. Yvind, Fully etched apodized grating coupler on the SOI platform with −0.58 dB coupling efficiency. Opt. Lett. **39**, 5348–5350 (2014)
22. L. Liu, M. Pu, K. Yvind, J.M. Hvam, High-efficiency, large-bandwidth silicon-on-insulator grating coupler based on a fully-etched photonic crystal structure. Appl. Phys. Lett. **96**, 051126 (2010)
23. X. Xu, H. Subbaraman, J. Covey, D. Kwong, A. Hosseini, R.T. Chen, Complementary metal–oxide–semiconductor compatible high efficiency subwavelength grating couplers for silicon integrated photonics. Appl. Phys. Lett. **101**, 031109 (2012)
24. H. Subbaraman, X. Xu, J. Covey, R.T. Chen, Efficient light coupling into in-plane semiconductor nanomembrane photonic devices utilizing a sub-wavelength grating coupler. Opt. Express **20**, 20659–20665 (2012)
25. J. Yang, Z. Zhou, H. Jia, X. Zhang, S. Qin, High-performance and compact binary blazed grating coupler based on an asymmetric subgrating structure and vertical coupling. Opt. Lett. **36**, 2614–2617 (2011)
26. R. Halir, P. Cheben, J.H. Schmid, R. Ma, D. Bedard, S. Janz, D.-X. Xu, A. Densmore, J. Lapointe, Í. Molina-Fernández, Continuously apodized fiber-to-chip surface grating coupler with refractive index engineered subwavelength structure. Opt. Lett. **35**, 3243–3245 (2010)
27. Z. Cheng, X. Chen, C.Y. Wong, K. Xu, H.K. Tsang, Apodized focusing subwavelength grating couplers for suspended membrane waveguides. Appl. Phys. Lett. **101**, 101104 (2012)
28. Y. Ding, H. Ou, C. Peucheret, Ultrahigh-efficiency apodized grating coupler using fully etched photonic crystals. Opt. Lett. **38**, 2732–2734 (2013)
29. D. Benedikovic, P. Cheben, J.H. Schmid, D.-X. Xu, B. Lamontagne, S. Wang, J. Lapointe, R. Halir, A. Ortega-Moñux, S. Janz, M. Dado, Subwavelength index engineered surface grating coupler with sub-decibel efficiency for 220-nm silicon-on-insulator waveguides. Opt. Express **23**, 22628–22635 (2015)
30. T. Watanabe, M. Ayata, U. Koch, Y. Fedoryshyn, J. Leuthold, Perpendicular grating coupler based on a blazed antiback-reflection structure. J. Light. Technol. **35**, 4663–4669 (2017)
31. Z. Cheng, H.K. Tsang, Experimental demonstration of polarization-insensitive air-cladding grating couplers for silicon-on-insulator waveguides. Opt. Lett. **39**, 2206–2209 (2014)
32. X. Chen, K. Xu, Z. Cheng, C.K.Y. Fung, H.K. Tsang, Wideband subwavelength gratings for coupling between silicon-on-insulator waveguides and optical fibers. Opt. Lett. **37**, 3483–3485 (2012)
33. Z. Cheng, X. Chen, C.Y. Wong, K. Xu, H.K. Tsang, Broadband focusing grating couplers for suspended-membrane waveguides. Opt. Lett. **37**, 5181–5183 (2012)
34. X. Xu, H. Subbaraman, J. Covey, D. Kwong, A. Hosseini, R.T. Chen, Colorless grating couplers realized by interleaving dispersion engineered subwavelength structures. Opt. Lett. **38**, 3588–3591 (2013)
35. Y. Wang, W. Shi, X. Wang, Z. Lu, M. Caverley, R. Bojko, L. Chrostowski, N.A.F. Jaeger, Design of broadband subwavelength grating couplers with low back reflection. Opt. Lett. **40**, 4647–4650 (2015)
36. A. Sánchez-Postigo, J.G. Wangüemert-Pérez, J.M. Luque-González, Í. Molina-Fernández, P. Cheben, C.A. Alonso-Ramos, R. Halir, J.H. Schmid, A. Ortega-Moñux, Broadband fiber-chip zero-order surface grating coupler with 0.4 dB efficiency. Opt. Lett. **41**, 3013–3016 (2016)
37. J.H. Schmid, P. Cheben, S. Janz, J. Lapointe, E. Post, D.-X. Xu, Gradient-index antireflective subwavelength structures for planar waveguide facets. Opt. Lett. **32**, 1794–1796 (2007)
38. Y. Wang, Z. Lu, M. Ma, H. Yun, F. Zhang, N.A.F. Jaeger, L. Chrostowski, Compact broadband directional couplers using subwavelength gratings. IEEE Photonics J. **8**, 1–8 (2016)
39. Z. Cheng, X. Chen, C.Y. Wong, K. Xu, C.K.Y. Fung, Y.M. Chen, H.K. Tsang, Mid-Infrared grating couplers for silicon-on-sapphire waveguides. IEEE Photonics J. **4**, 104–113 (2012)
40. Z. Cheng, X. Chen, C.Y. Wong, K. Xu, H.K. Tsang, Mid-infrared suspended membrane waveguide and ring resonator on silicon-on-insulator. IEEE Photonics J. **4**, 1510–1519 (2012)
41. J.S. Penadés, C. Alonso-Ramos, A.Z. Khokhar, M. Nedeljkovic, L.A. Boodhoo, A. Ortega-Moñux, I. Molina-Fernández, P. Cheben, G.Z. Mashanovich, Suspended SOI waveguide with sub-wavelength grating cladding for mid-infrared. Opt. Lett. **39**, 5661–5664 (2014)

42. W. Zhou, Z. Cheng, X. Wu, X. Sun, H.K. Tsang, Fully suspended slot waveguide platform. J. Appl. Phys. **123**, 063103 (2018)
43. J. Kang, Z. Cheng, W. Zhou, T.-H. Xiao, K.-L. Gopalakrisna, M. Takenaka, H.K. Tsang, K. Goda, Focusing subwavelength grating coupler for mid-infrared suspended membrane germanium waveguides. Opt. Lett. **42**, 2094–2097 (2017)
44. T.-H. Xiao, Z. Zhao, W. Zhou, M. Takenaka, H.K. Tsang, Z. Cheng, K. Goda, Mid-infrared germanium photonic crystal cavity. Opt. Lett. **42**, 2882–2885 (2017)
45. T.-H. Xiao, Z. Zhao, W. Zhou, M. Takenaka, H.K. Tsang, Z. Cheng, K. Goda, High-Q germanium optical nanocavity. Photonics Res. **6**, 925–928 (2018)
46. T.-H. Xiao, Z. Zhao, W. Zhou, C.-Y. Chang, S.Y. Set, M. Takenaka, H.K. Tsang, Z. Cheng, K. Goda, Mid-infrared high-Q germanium microring resonator. Opt. Lett. **43**, 2885–2888 (2018)
47. Y. Ma, Y. Zhang, S. Yang, A. Novack, R. Ding, A.E.-J. Lim, G.-Q. Lo, T. Baehr-Jones, M. Hochberg, Ultralow loss single layer submicron silicon waveguide crossing for SOI optical interconnect. Opt. Express **21**, 29374–29382 (2013)
48. Q. Zhong, V. Veerasubramanian, Y. Wang, W. Shi, D. Patel, S. Ghosh, A. Samani, L. Chrostowski, R. Bojko, D.V. Plant, Focusing-curved subwavelength grating couplers for ultra-broadband silicon photonics optical interfaces. Opt. Express **22**, 18224–18231 (2014)
49. J. Covey, R.T. Chen, Efficient perfectly vertical fiber-to-chip grating coupler for silicon horizontal multiple slot waveguides. Opt. Express **21**, 10886–10896 (2013)
50. Z. Yu, H. Cui, X. Sun, Genetic-algorithm-optimized wideband on-chip polarization rotator with an ultrasmall footprint. Opt. Lett. **42**, 3093–3096 (2017)
51. Y. Tong, W. Zhou, H.K. Tsang, Efficient perfectly vertical grating coupler for multi-core fibers fabricated with 193-nm DUV lithography. Opt. Lett. **43**, 5709–5712 (2018)
52. Y. Tong, W. Zhou, X. Wu, H.K. Tsang, Efficient mode multiplexer for few-mode fibers using integrated silicon-on-insulator waveguide grating coupler. IEEE J. Quantum Electron. **56**, 1–7 (2020)
53. Y. Tong, X. Zhou, Y. Wang, C.-W. Chow, H. K. Tsang, Bridging the graded-index few-mode fibre with photonic integrated circuits via efficient diffraction gratings in European Conference on Integrated Optics, 22nd edition, Paris (2020)
54. M. Turduev, E. Bor, C. Latifoglu, I.H. Giden, Y.S. Hanay, H. Kurt, Ultracompact photonic structure design for strong light confinement and coupling into nanowaveguide. J. Light. Technol. **36**, 2812–2819 (2018)
55. J. Peurifoy, Y. Shen, L. Jing, Y. Yang, F. Cano-Renteria, B.G. DeLacy, J.D. Joannopoulos, M. Tegmark, M. Soljačić, Nanophotonic particle simulation and inverse design using artificial neural networks. Sci. Adv. **4**, eaar4206 (2018)
56. D. Melati, Y. Grinberg, M. Kamandar Dezfouli, S. Janz, P. Cheben, J. H. Schmid, A. Sánchez-Postigo, D.-X. Xu, Mapping the global design space of nanophotonic components using machine learning pattern recognition. Nat. Commun. **10**, 4775 (2019)
57. J.S. Jensen, O. Sigmund, Topology optimization for nano-photonics. Laser Photonics Rev. **5**, 308–321 (2011)
58. A.Y. Piggott, J. Lu, K.G. Lagoudakis, J. Petykiewicz, T.M. Babinec, J. Vučković, Inverse design and demonstration of a compact and broadband on-chip wavelength demultiplexer. Nat. Photonics **9**, 374–377 (2015)
59. B. Shen, P. Wang, R. Polson, R. Menon, An integrated-nanophotonics polarization beamsplitter with 2.4 $\times$ 2.4 $\mu$m 2 footprint. Nat. Photonics **9**, 378–382 (2015)
60. L.F. Frellsen, Y. Ding, O. Sigmund, L.H. Frandsen, Topology optimized mode multiplexing in silicon-on-insulator photonic wire waveguides. Opt. Express **24**, 16866–16873 (2016)
61. S. Molesky, Z. Lin, A.Y. Piggott, W. Jin, J. Vucković, A.W. Rodriguez, Inverse design in nanophotonics. Nat. Photonics **12**, 659–670 (2018)
62. A.Y. Piggott, E.Y. Ma, L. Su, G.H. Ahn, N.V. Sapra, D. Vercruysse, A.M. Netherton, A.S.P. Khope, J.E. Bowers, J. Vučković, Inverse-designed photonics for semiconductor foundries. ACS Photonics **7**, 569–575 (2020)
63. W. Zhou, Y. Tong, X. Sun, H.K. Tsang, Ultra-broadband hyperuniform disordered silicon photonic polarizers. IEEE J. Sel. Top. Quantum Electron. **26**, 1–9 (2020)

64. G. Roelkens, D. Vermeulen, D. Van Thourhout, R. Baets, S. Brision, P. Lyan, P. Gautier, J.-M. Fédéli, High efficiency diffractive grating couplers for interfacing a single mode optical fiber with a nanophotonic silicon-on-insulator waveguide circuit. Appl. Phys. Lett. **92**, 131101 (2008)
65. D. Vermeulen, S. Selvaraja, P. Verheyen, G. Lepage, W. Bogaerts, P. Absil, D.V. Thourhout, G. Roelkens, High-efficiency fiber-to-chip grating couplers realized using an advanced CMOS-compatible silicon-on-insulator platform. Opt. Express **18**, 18278–18283 (2010)
66. S. Amaran, N.V. Sahinidis, B. Sharda, S.J. Bury, Simulation optimization: a review of algorithms and applications. Ann. Oper. Res. **240**, 351–380 (2016)
67. Y. Ding, J. Xu, F.D. Ros, B. Huang, H. Ou, C. Peucheret, On-chip two-mode division multiplexing using tapered directional coupler-based mode multiplexer and demultiplexer. Opt. Express **21**, 10376–10382 (2013)
68. D. Dai, J. Wang, Y. Shi, Silicon mode (de)multiplexer enabling high capacity photonic networks-on-chip with a single-wavelength-carrier light. Opt. Lett. **38**, 1422 (2013)
69. W. Zhou, Z. Cheng, X. Sun, H.K. Tsang, Tailorable dual-wavelength-band coupling in a transverse-electric-mode focusing subwavelength grating coupler. Opt. Lett. **43**, 2985–2988 (2018)
70. W. Zhou, H.K. Tsang, Dual-wavelength-band subwavelength grating coupler operating in the near infrared and extended shortwave infrared. Opt. Lett. **44**, 3621–3624 (2019)
71. C. Alonso-Ramos, A. Ortega-Moñux, L. Zavargo-Peche, R. Halir, J. de Oliva-Rubio, I. Molina-Fernández, P. Cheben, D.-X. Xu, S. Janz, N. Kim, B. Lamontagne, Single-etch grating coupler for micrometric silicon rib waveguides. Opt. Lett. **36**, 2647–2649 (2011)
72. W. Zhou, Z. Cheng, X. Chen, K. Xu, X. Sun, H. Tsang, Subwavelength engineering in silicon photonic devices. IEEE J. Sel. Top. Quantum Electron. **25**, 1–13 (2019)
73. X. Wen, K. Xu, Q. Song, Design of a barcode-like waveguide nanostructure for efficient chip-fiber coupling. Photonics Res. **4**, 209–213 (2016)
74. S. Torquato, F.H. Stillinger, Local density fluctuations, hyperuniformity, and order metrics. Phys. Rev. E **68**, 041113 (2003)
75. R.D. Batten, F.H. Stillinger, S. Torquato, Classical disordered ground states: Super-ideal gases and stealth and equi-luminous materials. J. Appl. Phys. **104**, 033504 (2008)
76. O.U. Uche, F.H. Stillinger, S. Torquato, Constraints on collective density variables: two dimensions. Phys. Rev. E Stat. Nonlin. Soft Matter Phys. **70**, 046122 (2004)
77. M. Florescu, S. Torquato, P.J. Steinhardt, Designer disordered materials with large, complete photonic band gaps. Proc. Natl. Acad. Sci. **106**, 20658–20663 (2009)
78. W. Man, M. Florescu, E.P. Williamson, Y. He, S.R. Hashemizad, B.Y.C. Leung, D.R. Liner, S. Torquato, P.M. Chaikin, P.J. Steinhardt, Isotropic band gaps and freeform waveguides observed in hyperuniform disordered photonic solids. Proc. Natl. Acad. Sci. **110**, 15886–15891 (2013)
79. M. Lee, J. Lee, S. Kim, S. Callard, C. Seassal, H. Jeon, Anderson localizations and photonic band-tail states observed in compositionally disordered platform. Sci. Adv. **4**, e1602796 (2018)
80. W. Zhou, Z. Cheng, B. Zhu, X. Sun, H.K. Tsang, Hyperuniform disordered network polarizers. IEEE J. Sel. Top. Quantum Electron. **22**, 288–294 (2016)
81. W. Zhou, Y. Tong, X. Sun, H.K. Tsang, Hyperuniform disordered photonic bandgap polarizers. J. Appl. Phys. **126**, 113106 (2019)

# Chapter 7
# Non-Hermitian Physics and Engineering in Silicon Photonics

**Changqing Wang, Zhoutian Fu, and Lan Yang**

**Abstract** Driven by the great needs for low-cost and scalable advanced optoelectronic systems that could leverage the existing infrastructure already developed for the semiconductor industry, silicon photonics has been extensively explored as a platform with system-level integration to host numerous devices and systems with various functionalities, including lasers, modulators, filters, isolators, wavelength division multiplexing (WDM) transceivers, etc. Recently, non-Hermitian physics, which breaks the conventional scope of quantum mechanics based on Hermitian Hamiltonians, has been widely explored in the platform of silicon photonics. With judicious designs of refractive index, modal coupling and gain–loss distribution, unconventional control and manipulation of light flow and nonlinear effects could be achieved. As we will discuss in this chapter, the unconventional properties of exceptional points and parity-time symmetry realized in silicon photonics have created new opportunities for ultrasensitive sensors, laser engineering, control of light propagation, topological mode conversion, etc. The marriage between the quantum non-Hermiticity and classical silicon platforms not only inspires numerous studies on the fundamental physics but also enriches the potential functionalities of the integrated photonic systems.

Driven by the great needs for low-cost and scalable advanced optoelectronic systems that could leverage the existing infrastructure already developed for the semiconductor industry, silicon photonics has been extensively explored as a platform with system-level integration to host numerous devices and systems with various functionalities, including lasers, modulators, filters, isolators, wavelength division multiplexing (WDM) transceivers, etc. Recently, non-Hermitian physics, which breaks the conventional scope of quantum mechanics based on Hermitian Hamiltonian, has been widely explored in the platform of silicon photonics. With judicious designs of refractive index, modal coupling and gain–loss distribution, unconventional control

C. Wang · Z. Fu · L. Yang (✉)
Department of Electrical and Systems Engineering, Washington University, St. Louis, MO 63130, USA
e-mail: yang@seas.wustl.edu

 
D. J. Lockwood and L. Pavesi (eds.), *Silicon Photonics IV*,
Topics in Applied Physics 139, https://doi.org/10.1007/978-3-030-68222-4_7

and manipulation of light flow and nonlinear effects could be achieved. As we will discuss in this chapter, the unconventional properties of exceptional points and parity-time symmetry realized in silicon photonics have created new opportunities for ultra-sensitive sensors, laser engineering, control of light propagation, topological mode conversion, etc. The marriage between the quantum non-Hermiticity and classical silicon platforms not only inspires numerous studies on the fundamental physics but also enriches the potential functionalities of the integrated photonic systems.

## 7.1 Introduction

Non-Hermitian physics describes open quantum systems that interact with the environment in the form of matter or energy exchange. In quantum mechanics, the Hamiltonian ($H$) describing a closed quantum system is a Hermitian operator satisfying $H^\dagger = H$[1]. Its eigenvalues are real, and the corresponding eigenstates are orthogonal to each other, which provide a complete orthonormal basis to describe any state of the system in the Hilbert space. However, such argument conveys no information of the spectral behavior and state description when the system interacts—often unavoidably—with the environment. The most common form of interaction is the energy dissipation, i.e., the energy transfers from the system to the surrounding environment, such as decoherence in atomic states, or optical absorption/radiation/scattering loss in confined optical structures. Even though the density matrix description and the master equation approach [2–4] offer standard tools for describing behavior of open systems in different situations, in many cases, concise descriptions directly offered by Hamiltonians are more favorable.

Toward the end of the twentieth century, the traditional scenario of quantum theory was expanded by introducing a non-Hermitian Hamiltonian, which satisfies $H^\dagger \neq H$, to describe an open system [5–7]. The eigenvalues are in general complex, and the associated eigenstates can be non-orthogonal. In particular, non-Hermitian degeneracies happen at an exceptional point (EP) where two or more eigenvalues and corresponding eigenstates coalesce simultaneously. Besides, much interest was aroused to a special family of open systems preserving parity-time ($PT$) symmetry [8–14], whose Hamiltonians commute with the joint operation of parity operator ($P$) and time-reversal operator ($T$), i.e., $[H, PT] = 0$. $PT$-symmetric systems are found to exhibit real eigenspectra out of complex potential under a certain range of parameters, while the eigenstates are non-orthogonal to each other. In short, the introduction of the non-Hermitian approach has opened up new routes to the exploration of physical properties of open quantum systems and established a nascent research field that facilitates the understanding of dissipation/amplification in a variety of physical systems.

About a decade later, non-Hermitian physics initiated a new wave of ripples in classical systems [13], including electronics [15, 16], metamaterial [17], acoustics [18], and especially optics [19–23]. Helmholtz equations under paraxial approximation can be viewed in analogy to the Schrödinger equation, with the quantum potential replaced by the optical index profiles [24]. The real and imaginary parts of the potential can be realized by optical refractive index and the optical gain/loss, respectively, which are quite feasible with the modern nanophotonic technique. As a result, a series of studies have been carried out for optical systems operating around EPs, where unconventional physical phenomena have been unveiled, including spectral singularity, chiral modes, unidirectional reflection, and so on, creating new opportunities to sensing, mode engineering, and optical signal processing. Interests in $PT$-symmetry also boomed due to its potential applications in engineering on-chip lasers and light transport, such as single-mode lasing and nonreciprocal light propagation.

Silicon photonics has played an indispensable role in the progress of non-Hermitian physics in optical systems. They offer unique advantageous features for optical design, including integratability, low cost, scalability, room-temperature operation, flexible control of optical parameters (field intensity, phase, polarization, refractive indices, mode coupling strengths, and optical damping rates), compatibility with other degrees of freedom, etc. [25]. Especially, many ideas were implemented into various optical structures such as waveguides and resonators, as well as photonic crystals. In such structures, optical resonant frequencies can be controlled via the thermo-optic effect [26, 27], while the coupling between different units can be introduced and adjusted by spatial positioning [28], backscattering engineering [28, 29], or geometric deformation [30–33]. The optical gain can be introduced by doping rare-earth ions in silica material [34], while the loss can be engineered by external absorbers [35] or the coupling loss induced by incoming/outcoming channels [36]. In both waveguides and resonators, the spatial deposition of absorbing materials, such as germanium (Ge) and chromium (Cr) onto silicon-based structures, enable local engineering of refractive indices. In addition, by coating silica optical devices with other transparent optical materials such as polymer [37–41] or silk fibroin [42, 43], one can modify optical, thermal, and mechanical properties to achieve novel nonlinear behavior. As a result, the silicon photonics platforms provide a fertile ground to the studies of non-Hermitian physics. In return, non-Hermitian physics unleashes the full potential of silicon photonics for novel functionalities and applications spanning from sensing, lasing and optical information processing to the control of nonlinear optical behavior.

In this chapter, we review the studies of non-Hermitian physics and the associated engineering progress in silicon photonics. The chapter is organized as follows. In Sect. 7.2, we introduce the theory on how non-Hermitian physics can be realized in optical systems in an analogous manner to quantum scenarios. In Sect. 7.3, we discuss the eigenspectra of non-Hermitian systems, where we explain the spectral singularity associated with the exceptional points, and how they lead to ultrasensitive sensors. In Sect. 7.4, we review the studies relevant to the mode interaction and eigenstate features in non-Hermitian optical systems. In particular, the supermodes in $PT$-symmetric systems and the chiral eigenmodes at EPs enable a series of unconventional lasing behavior. In Sect. 7.5, we turn to the discussion of optical scattering properties enabled by EPs and PT symmetry, i.e., the system's response to the incident optical waves. Such properties lay the foundation for various features of light propagation including directionality, nonreciprocity, and group delay. The topological properties associated with the eigenspectra near EPs are discussed in Sect. 7.6, which pave a new way to control and switch optical modes by topological operations. Finally, we conclude the chapter by offering an outlook in Sect. 7.7.

## 7.2 Non-Hermitian Physics: From Quantum Mechanics to Optics

### *7.2.1 Non-Hermitian Physics in Quantum Mechanics*

Non-Hermitian quantum mechanics is an alternative formalism to standard Hermitian quantum mechanics and presents a unique approach to describe open quantum systems. The non-Hermitian theory is introduced to quantum mechanics mainly for two reasons [5]. First, in quantum mechanics, many problems can be treated in a simplified way, either analytically or numerically, by non-Hermitian quantum mechanics than using the standard quantum theory. Second, many problems, such as classical mechanical statistics and propagation of electromagnetic waves, can be described by a Hamiltonian in analogy to the quantum description. The analogous Hamiltonian, however, may not respect Hermiticity, and thereby in many cases, the analogous quantum problem can only be described by non-Hermitian theory and have no counterpart in the standard Hermitian quantum mechanics. In this chapter, we will mainly focus on the second motivation with phenomena found in silicon photonics being the main topic of interest. More interesting phenomena unveiled by non-Hermitian physics can be found in related books and reviews [5, 13].

One of the most important ways of introducing non-Hermitian formalism to quantum systems is to consider complex potentials in Schrödinger equations. It is known that in standard quantum mechanics, the quantum potential can only be a real function, conserving the total particle number or stored energy, while the non-conservative open features can only be depicted by more complicated formalism using a density matrix approach. Non-Hermitian quantum theory makes sense of complex potentials that break Hermiticity of Hamiltonians which represent the conservative property of quantum systems. To be more specific, the Schrödinger equation describing the evolution of a quantum wavefunction $\Psi(\boldsymbol{r}, t)$ in a time-invariant complex potential is given by

$$i\hbar\frac{\partial}{\partial t}\Psi(\boldsymbol{r}, t) = \left[-\frac{\hbar^2}{2m}\nabla^2 + V(\boldsymbol{r})\right]\Psi(\boldsymbol{r}, t), \tag{7.1}$$

where the complex potential $V(\boldsymbol{r})$ implies non-conserved probability $\int|\Psi(\boldsymbol{r}, t)|^2 d\boldsymbol{r}$.

Non-Hermitian quantum theory not only offers an alternative approach for open quantum problems but also introduces unconventional physical concepts and phenomena, such as self-orthogonal eigenstates and parity-time symmetry. One can find a considerable disparity between Hermitian and non-Hermitian systems.

Closed systems with Hermitian Hamiltonians always have real eigenvalues and orthogonal eigenstates under which the Hamiltonians are represented by diagonalized matrices. Their eigenstates $|\psi_i\rangle$ are orthonormal under the inner product, i.e. $\langle\psi_i|\psi_j\rangle = \delta_{i,j}$. Non-Hermitian systems, however, can possess complex eigenvalues and non-orthogonal eigenstates. Let us consider a non-Hermitian operator $A$ with eigenvalues $\lambda_i$ and the corresponding eigenvectors $|\phi_i)$, i.e., $A|\phi_i) = \lambda_i|\phi_i)$. In order to find an orthonormal eigenbasis, we define a bi-orthogonal product (c-product) in replacement of the inner product, such that $\left(\phi_i|\phi_j\right) = \delta_{i,j}$ [5].

A peculiar situation occurs at non-Hermitian degeneracies where two or more non-orthogonal eigenvectors coalesce, which is also named as a branch point, or an EP. The bi-orthogonality yields $(\phi_{\mathrm{EP}}|\phi_{\mathrm{EP}}) = 0$, i.e., the eigenstate at an EP is self-orthogonal. Such self-orthogonality is responsible for a great many singular phenomena, including the divergence of expectation values of non-Hermitian operators, creation of probabilities, Berry phase accumulation by adiabatically encircling EPs, and so on [5].

On the other hand, the complex potential can generate purely real eigenvalues, if preserving symmetry under a joint operation of parity ($P$) and time-reversal ($T$) operators, known as parity-time ($PT$) symmetry [8]. The condition for a parity-time symmetric potential is given as

$$PTV(\boldsymbol{r}) = V(\boldsymbol{r}), \tag{7.2}$$

which yields a requirement for the complex potential

$$V(\boldsymbol{r}) = V^*(-\boldsymbol{r}). \tag{7.3}$$

So far, we have provided the basic concepts of non-Hermitian quantum mechanics, including the definitions of EPs and $PT$ symmetry, in the context of open quantum systems. Below we show how these quantum concepts and the associated phenomena can be readily mimicked and applied in classical optical systems. We classify the optical realization of non-Hermitian physics, including $PT$ symmetry and EPs into three categories: wave propagation under paraxial approximation in a transverse complex potential, wave scattering in a longitudinal complex potential, and discrete optics with a finite-dimension matrix form of Hamiltonians.

### 7.2.2 *Paraxial Propagation of Electromagnetic Fields in a Transverse Complex Potential*

The classical realization of non-Hermitian physics can be first seen by investigating the propagation of electromagnetic waves in an optical medium with planar structures

[24, 44, 45]. We start our analysis from a one-dimensional case shown in Fig. 7.1a, where the field propagation occurs in a 1D potential along $z$ axis and uniformity is assumed for the potential in the $y$ direction. Without loss of generality, we consider a TE wave with an electric field $\boldsymbol{E}(x, z)e^{-i\omega t}$, for which the Helmholtz equation is given by

$$\left[\nabla^2 + k^2\epsilon(x, z)\right]\boldsymbol{E}(x, z) = 0. \tag{7.4}$$

If the medium is uniform along the $z$ direction, we can rewrite the Helmholtz equation under the paraxial approximation $\boldsymbol{E}(x, z) \sim E(x)e^{ikz}\hat{z}$, where $\hat{z}$ is the direction of the electric field,

$$ik\frac{\partial}{\partial z}E(x) = -\frac{\partial^2}{\partial x^2}E(x) - \frac{\omega^2}{c^2}\epsilon(x)E(x), \tag{7.5}$$

which takes a similar form to that of the Schrödinger equation. The time evolution is replaced by propagation along the $z$-direction. The potential $V(x)$ in Schrödinger equations is represented by a term consisting of the complex permittivity $\epsilon(x)$, or equivalently the squared refractive index $n(x)^2$. Thereby, the propagation of the electromagnetic field under paraxial approximation is in analogy to the quantum evolution of a wavefunction within a 1D complex potential. It is now apparent that the concept of PT symmetry can be introduced to the optical context with the spatial distribution of complex permittivity satisfying

$$\epsilon(x) = \epsilon^*(-x), \tag{7.6}$$

which can be realized in linear media by designing the refractive index distribution

$$n(x) = n^*(-x), \tag{7.7}$$

indicating that the complex potential has symmetric real parts and antisymmetric imaginary parts (Fig. 7.1b), i.e.,

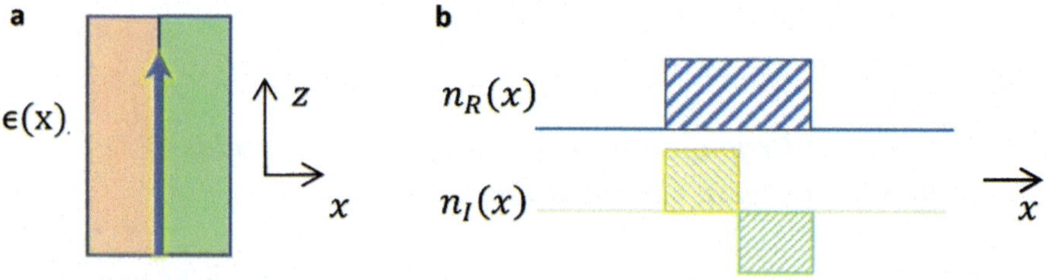


**Fig. 7.1** Realization of parity-time ($PT$) symmetry in a transverse optical potential. **a** Wave propagation in a planar waveguide which has a distribution of electric permittivity $\epsilon(x)$ in the transverse ($x$) direction. **b** An example distribution of the real and imaginary part of the refractive index as a function of $x$ to mimic a $PT$-symmetric potential

$$n_R(x) = n_R(-x), n_I(x) = -n_I(-x). \tag{7.8}$$

Similarly, in the two-dimensional potential, the condition for $PT$-symmetry is modified to

$$n(x, y) = n^*(-x, -y). \tag{7.9}$$

The antisymmetric imaginary parts indicate that the spatial balance between gain and loss is required for $PT$-symmetric optic structures.

Equations (7.7)–(7.9) provide the basic guidance for engineering non-Hermitian behavior and $PT$-symmetry in optical structures, with the real and imaginary parts of the refractive index governing the oscillating behavior and amplification/dissipation features of the electromagnetic wave. However, as we can see, the above discussion focuses on the situation that the optical potential remains uniform along the direction of wave propagation, which intends to mimic a time-invariant quantum complex potential. A natural question to ask is whether and how a $PT$-symmetric system can be found if the wave encounters variance of refractive index during propagation.

### *7.2.3 Wave Scattering in a Longitudinal Complex Potential*

In this section, we turn to the discussion of the physical realization of wave scattering in a longitudinal complex optical potential [46]. For a planar waveguide shown in Fig. 7.2, the optical potential is modulated along the $z$ axis which is the direction of propagation in contrast to the situation in the last section where the refractive index is only modified transversely. We assume that the waveguide has a resonant frequency $\omega_0$, and the linear uniform optical medium has a plasma frequency $\omega_p$ and damping constant $\delta$. The relative permittivity is then given by

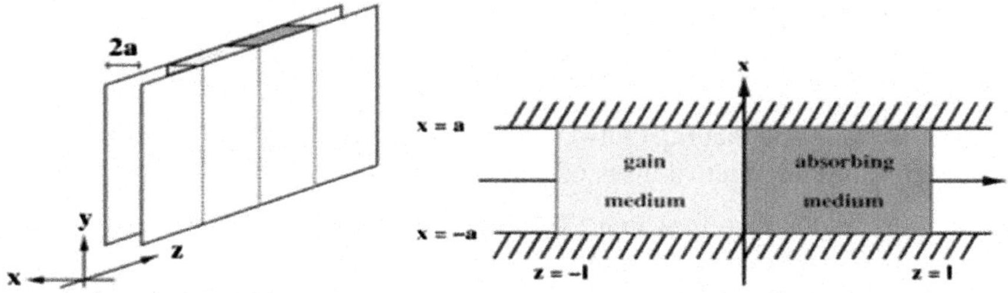


**Fig. 7.2** Realization of $PT$-symmetry in an optical scattering potential. Schematic diagram for the planar slab waveguide in the three-dimensional view and the top view which are shown in the left and right panels, respectively. Figures from Ref. [46]

$$\epsilon(z,\omega) = 1 - \xi(z)\frac{\omega_p^2}{\omega^2 - \omega_0^2 + 2i\delta\omega}, \tag{7.10}$$

where $\xi(z)$ regulates the gain/loss distribution along the $z$ direction. For example, the gain medium has $\xi(z) = -1$ and the absorbing medium has $\xi(z) = 1$. In the vacuum, $\xi(z) = 0$.

Considering a single-frequency TM mode with the electric field $E_y(x, y, z, \omega) = -i\omega\hat{B}_x(x)\phi(z,\omega)$ and the magnetic field $B_x(x, y, z, \omega) = \hat{B}_x(x)\frac{\partial\phi}{\partial x}(z,\omega)$, $B_z(x, y, z, \omega) = -\frac{\partial\hat{B}_x}{\partial x}(x)\phi(z,\omega)$, we can derive the following relations based on the Maxwell's equations

$$\left(c^2\frac{\partial^2}{\partial x^2} + \omega_c^2\right)\hat{B}_x(x) = 0, \tag{7.11}$$

$$\left[\frac{\omega^2\epsilon(z,\omega) - \omega_c^2}{c^2} + \frac{\partial^2}{\partial z^2}\right]\phi(z,\omega) = 0, \tag{7.12}$$

with $\omega_c$ being the cutoff frequency.

Now we consider the frequency $\omega$ to be slightly larger than the cutoff frequency $\omega_c$, i.e., $\omega = \omega_c + \Delta\omega$, as well as the assumptions $\frac{\omega_p^2}{\delta} \ll \delta, \omega_c$ and $\Delta\omega \ll \delta, \omega_c$. One can find that

$$\frac{\partial^2}{\partial x^2}\phi(z,\omega_c+\Delta\omega) = -\left(i\xi(z)\frac{\omega_c\omega_p^2}{2c^2\delta} + 2\Delta\omega\frac{\omega_c}{c^2}\right)\phi(z,\omega_c+\Delta\omega). \tag{7.13}$$

The approximation made so far has reduced all higher-order terms of $\Delta\omega/\omega_c$ so that the time evolution term is linearized to a simple form. Reconverting Eq. (7.13) to the time domain with the definition $\psi(z,t) = \int d\Delta\omega\phi(z,\omega_c+\Delta\omega)e^{-i\Delta\omega t}$, we get

$$i\hbar\frac{\partial\psi}{\partial x}(z,t) = -\frac{c^2}{2\omega_c}\frac{\partial^2}{\partial z^2}\psi(z,t) - i\xi(z)\frac{\omega_p^2}{4\delta}\psi(z,t). \tag{7.14}$$

It is not hard to find that Eq. (7.14) takes the form of a Schrödinger equation with the definition of the effective mass of photon $m = \frac{\hbar\omega_c}{c^2}$ and the effective potential $V(z) = -i\xi(z)\frac{\omega_p^2\hbar}{4\delta}$. Therefore, non-Hermitian phenomena can be generated by judiciously designing the optical structures. For example, the condition of $PT$ symmetric scattering potential for the mode with near-cutoff frequency is $V(z) = V^*(-z)$. This can be achieved, for instance, by engineering the gain/loss distribution in the waveguide to realize a simple form of $\xi(z)$ as

$$\xi(z) = \begin{cases} -1 & \text{gain medium} & -l < z < 0 \\ 1 & \text{absorbing medium} & 0 < z < l \\ 0 & \text{vaccum} & |z| > l \end{cases}. \tag{7.15}$$

### 7.2.4 Non-Hermitian Optical Waveguides and Resonators

Being the most indispensable elements in photonic devices and systems, guided wave optical structures such as waveguides and resonators are common platforms for the realization of non-Hermitian behavior and functionalities. Moreover, the aforementioned wave propagation under paraxial approximation can be readily applied to optical transport in guided wave structures. Therefore, in this section, we take a further step to introduce the coupled-mode theory describing the non-Hermitian guided wave optical structures.

Without loss of generality, let us consider a 1D coupled waveguide system consisting of two waveguides with a difference in their uniformly distributed loss (or gain). By finding the extremum of Lagrangian with respect to the mode amplitudes $a_1(z)$ and $a_2(z)$, one can derive the coupled-mode equations [47–49]

$$i\frac{da_1}{dz} = -i\gamma_1 a_1 + \kappa_{12} a_2, \tag{7.16}$$

$$i\frac{da_2}{dz} = -i\gamma_2 a_2 + \kappa_{21} a_1, \tag{7.17}$$

with $\gamma_1$ and $\gamma_2$ representing the loss rates for the first and second waveguide modes, respectively, and $\kappa$ being the coupling strength between the two modes. It follows that by writing $i\frac{da}{dz} = Ha$, where $a = \begin{pmatrix} a_1 \\ a_2 \end{pmatrix}$, one can derive a Hamiltonian matrix out of the coupled-mode equations [13]

$$H = \begin{pmatrix} -i\gamma_1 & \kappa_{12} \\ \kappa_{21} & -i\gamma_2 \end{pmatrix}. \tag{7.18}$$

Note that $H \neq H^\dagger$ implies the non-conservation of energy induced by, for instance, the existence of loss, even though we usually have $\kappa_{12} = \kappa_{21}^*$ for energy-conserved coupling. If one waveguide is engineered to have a loss and the other has an equal amount of gain, i.e., $\gamma_1 = -\gamma_2 > 0$, then the system respects $PT$-symmetry.

On the other hand, optical resonators are another set of ideal candidates for implementing non-Hermitian potential. As a counterpart to the spatial propagation dynamics in coupled waveguides, the coupled resonators have dynamic temporal evolution described by temporal-coupled-mode theory [50]

$$i\frac{da_1}{dt} = \left(\omega_1 - i\frac{\gamma_1}{2}\right)a_1 + \kappa_{12}a_2\ ,$$
$$i\frac{da_2}{dt} = \left(\omega_2 - i\frac{\gamma_2}{2}\right)a_2 + \kappa_{21}a_1\ , \qquad (7.19)$$

where $\omega_{1,2}$ are resonant frequencies of the mode 1 (with field amplitude $a_1$) and mode 2 (with field amplitude $a_2$), $\gamma_{1,2}$ are the intrinsic loss rates of the resonators, and $\kappa_{12}$ ($\kappa_{21}$) is the coupling strength from mode 2 to mode 1 (mode 1 to mode 2). For two directly coupled resonators, we have $\kappa_{12} = \kappa_{21}^* = \kappa$ due to conservation of energy. The phase factor of a coupling constant usually does not exhibit physical meaning unless the relative phase between different coefficients is considered [51–53]. When the resonant frequencies of the two resonators are aligned such that $\omega_1 = \omega_2$, it is also possible to manage the two resonators to realize a $PT$-symmetric system by setting a gain–loss balance, or a passive $PT$-symmetric system if gain and loss are not balanced. The eigenvalues are given by

$$\lambda_{\pm} = \frac{\omega_1 + \omega_2}{2} - i\frac{\gamma_1 + \gamma_2}{4} \pm \sqrt{|\kappa|^2 - \left(\frac{\gamma_1 - \gamma_2}{4}\right)^2}\ . \qquad (7.20)$$

If the coupling strength $\kappa$ is larger than a critical value $\kappa_{th} = \frac{\gamma_1 - \gamma_2}{4}$, then the two eigenvalues have two real parts and identical imaginary parts, corresponding to two supermodes with the same gain/loss feature but split resonant frequencies. If $\kappa < \kappa_{th}$, however, the two eigenvalues overlap in their real parts, but have different imaginary parts, leading to two optical supermodes with identical resonant frequency but disparate gain/loss. At the phase transition point where $\kappa = \frac{\gamma_1 - \gamma_2}{4}$, the eigenvalues are degenerate, and the associated eigenstates also coalesce, leading to the emergence of an EP. The aforementioned phase transition under varied coupling strength or gain/loss rates defines different regimes for $PT$-symmetric systems and has been observed in both optical waveguides [47] and resonators [54].

## 7.3 Spectral Singularity and Enhanced Sensing

The EP of a non-Hermitian system is characterized with the coalescence of both eigenvalues and eigenvectors, and the behaviors of eigenvalues and eigenvectors are of great interest when a small perturbation is introduced to the system. Generalized perturbation theory on the eigenvalue bifurcation at an $N$th-order EP yields a unique Nth-root dependence on the perturbation strength [55]. This mathematical result, if implemented in the application of sensors, could benefit the sensing performance by amplifying the sensor response to tiny perturbations according to the $N$th-root law. However, the realization of an EP in sensors may pose a problem, because it typically requires fine control of parameters and its status needs to be sustained for long enough. Fortunately, microcavity sensors based on silicon photonics provide unprecedented tunability in terms of the system parameters and therefore become an ideal candidate for the demonstration of EP-enhanced sensors. In this section, we will discuss the theory on the enhancement of response in EP sensors, as well as how EPs are experimentally realized and maintained in these silicon-photonic systems.

### *7.3.1 Spectral Singularity at Exceptional Points*

In the systems with non-Hermitian Hamiltonians, the spectral singularities associated with EPs manifest themselves as the simultaneous coalescence of several eigenvalues and the corresponding eigenvectors. The Hamiltonian becomes non-diagonalizable at these singularities and can only be transformed into Jordan canonical form by a similarity matrix, where the order of EPs determines the length of Jordan block within the Jordan canonical form. For example, an $N$th-EP corresponds to a Jordan block of length $N$ that takes the form below

$$J_\lambda(N) = \begin{bmatrix} \lambda & 1 & \cdots & 0 \\ 0 & \lambda & \cdots & 0 \\ \vdots & & \ddots & \vdots \\ 0 & 0 & \cdots & 1 \\ 0 & 0 & \cdots & \lambda \end{bmatrix}. \tag{7.21}$$

It is of a natural instinct to investigate how perturbation to the system will cause a bifurcation of eigenvalues and eigenvectors. Square-root and third-root dependence in the bifurcation with respect to the perturbation strength have been found in the second- and third-order EPs [56, 57], respectively. In a more general case, it has also been shown that the perturbative behavior of a Jordan block of length $N$ follows an $N$th-root relation [55]. Under a small perturbation $\epsilon$ to the elements of Jordan block, the eigenvalue can be expressed in the form of Puiseux series

$$\lambda(\epsilon) = \lambda(0) + \lambda_1 \sqrt[N]{\epsilon} + \mathcal{O}\left(\sqrt[N]{\epsilon}\right). \tag{7.22}$$

The above relation holds as long as the first term $\lambda_1$ is not vanishing, which is usually the case for most physical perturbations that could affect the systems in reality. Since the bifurcation response proportional to the $N$th-root of a sufficiently small perturbation is much larger than the linear case ($\sqrt[N]{\epsilon} \gg \epsilon$ when $\epsilon \to 0$), steering the sensor system to an EP was proposed as a novel method to enhance the response of sensors to small signals [58–64]. In practice, EPs of lower order are easier to implement, and their enhancing effects have been demonstrated in several different types of photonic sensors, including nanoparticle sensors [65], gyroscopes [66], thermal sensors [67], and even an immuno-assay nanosensor in a plasmonic system [68]. The great flexibility of parameters in the silicon photonic systems facilitates the realization of EPs and paves a new way to improve sensor performance, which will be discussed in the following sections.

### 7.3.2 EP-Enhanced Nanoparticle Sensor

The unique $N$th-root topology in the vicinity of an EP can be utilized to enhance the sensor performance, for example, EP-enhanced nanoparticle sensing based on an optical microcavity [65]. To understand the amplification of sensor response at EPs, let us consider the conventional case (a microcavity without EP) first. When a conventional microcavity is subject to a perturbation induced by some scatterers, like nanoparticles, the Hermitian degeneracy (also known as a diabolic point, DP) between clockwise (CW) and counterclockwise (CCW) traveling modes is lifted, leading to a mode-splitting in the microcavity [28, 69–72]. The strength of mode-splitting at DPs is proportional to the strength of perturbation, which can be derived from the Hamiltonian: Starting from a general scenario—an unperturbed system consisting of a microcavity and $M$ scatterers. The $(M + 1)$th scatterer is introduced as a perturbation, then the total effective Hamiltonian is [58, 73]

$$H = \begin{bmatrix} \Omega^{(M)} & A^{(M)} \\ B^{(M)} & \Omega^{(M)} \end{bmatrix} + \begin{bmatrix} V + U & (V - U)e^{-i2m\beta} \\ (V - U)e^{i2m\beta} & V + U \end{bmatrix}. \tag{7.23}$$

The first term describes the unperturbed system, with the off-diagonal elements $A^{(M)}$ and $B^{(M)}$ representing the intrinsic backscattering. The second term is the Hamiltonian of perturbation. $2V$ and $2U$ are the complex frequency shifts for positive- and negative-parity modes due to the perturbation; $m$ is the azimuthal mode order and $\beta$ is the relative angle of the $(M + 1)$th scatterer. Conventional microcavity sensors working at DPs have no intrinsic backscattering between CW and CCW modes, i.e., $A^{(M)} = B^{(M)} = 0$. Therefore, the complex frequency splitting caused by a perturbation can be computed as $\Delta\Omega_{\text{DP}} = 2(V - U) = 2\epsilon$. Here we define $(V - U)$ as the complex perturbation strength $\epsilon$, so we show that the surface of eigenfrequencies for conventional sensors has a linear (cone-shaped) topology around DPs (Fig. 7.3a).

However, sensors operating at EPs have a completely different topology, which can help enhance sensitivity for small perturbations. Considering the same $2 \times 2$ Hamiltonian matrix above, if the system is at an EP, its backscattering will be fully asymmetric. There is either no backscattering from CCW to CW modes ($A^{(M)} = 0$), or from CW to CCW modes ($B^{(M)} = 0$). In the case of $B^{(M)} = 0$, for example, the complex frequency splitting at a second-order EP will be

$$\Delta\Omega_{\text{EP}} = 2\epsilon\sqrt{1 + \frac{A^{(M)}e^{i2m\beta}}{\epsilon}} \approx 2e^{im\beta}\sqrt{A^{(M)}\epsilon}, \tag{7.24}$$

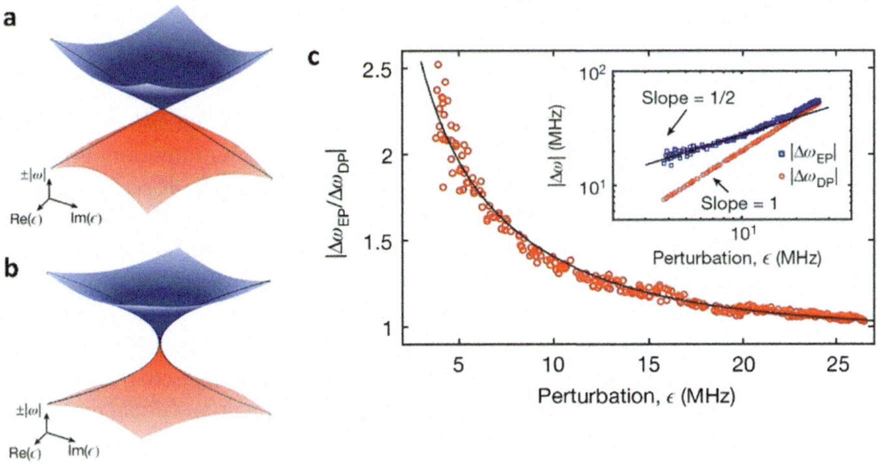


**Fig. 7.3** a-b, Surfaces of the eigenfrequencies of **a** DP sensors and **b** EP sensors. **c** Measured EP enhancement factor upon the variation of perturbation strength. The inset shows the log-log plots of the mode-splitting in the EP and DP sensors. Figures from Ref. [65]

which gives rise to a square-root topology (Fig. 7.3b). As long as the intrinsic backscattering is much larger than the perturbation ($\left|A^{(M)}\right| \gg |\epsilon|$), EP sensors will have a larger response compared to conventional DP sensors

$$|\Delta\Omega_{\mathrm{EP}}| = 2\sqrt{A^{(M)}\epsilon} > |\Delta\Omega_{\mathrm{DP}}| = 2\epsilon. \tag{7.25}$$

This implies that the sensitivity of sensor to sufficiently small perturbation can be enhanced by operating the sensor at EPs.

EP-enhanced nanoparticle sensing has been demonstrated in a silica microcavity-scatterer system [65]. To steer the microcavity to an EP, two silica fiber tips (nanotips) were used as scatterers. The nanotips were mounted on two translation stages, respectively, which can adjust their positions finely and thus tune the backscattering. The EP sensor is realized when the mode-splitting by the first nanotip disappears in the presence of the second nanotip, judged by observing unidirectional suppression of reflection signal [29, 74]. Then, a third nanotip was introduced as the target perturbation to characterize the sensor, and the perturbation strength is varied by tuning the position of the third nanotip within the optical mode volume. The observed mode-splitting in the EP sensor is larger than that in the DP sensor for any perturbation that is weaker than 25 MHz, and the enhancement factor can go up to 2.5 for perturbation strength less than 5 MHz (Fig. 7.3c). The inset shows the log-log plots of the amount of mode-splitting as the perturbation strength varies for the EP (blue dots) and DP sensor (red dots). The log-log plot for EP sensor has a slope of 1/2, which validates the existence of a second-order EP and its square-root topology.

### 7.3.3 EP-Enhanced Gyroscope

The sensitivity enhancing effect at EPs has also been exploited in silicon-photonic gyroscope based on a Brillouin laser [66], which consists of a silica disk resonator on a silicon chip coupled to a fiber taper. Brillouin scattering is an important nonlinear process that describes the scattering of photons from phonons. It occurs spontaneously at a low-power level, and its stimulated emission excited by a high-power pump can even lead to stimulated Brillouin lasers (SBLs). In silica disk resonators, first- and higher-order SBLs can be excited at a milliwatt pump power, thanks to the ultrahigh optical quality ($Q$)-factor and the fine control of the resonator size [75]. Conventionally (without EPs), when the whole system rotates, the Sagnac-induced frequency shifts for counter-propagating SBLs are opposite, leading to the formation of a beat note in the emission field, with the beating frequency proportional to the rotation rate [76, 77]. In the EP gyroscope, CW and CCW SBLs in the same cavity mode are excited by sending two pump waves with opposite directions into the resonator (Fig. 7.4a). To achieve a second-order EP, it is critical to induce dissipative coupling (with purely imaginary coupling strength) between these two counter-propagating lasing modes, which is achieved by the scattering at the fiber-taper waveguide.

To analyze the EP-enhanced Sagnac effect predicted by previous theoretical work [61, 62], we consider the Hamiltonian that governs the time evolution of CW and CCW SBL modes subject to an angular rotation rate of $\Omega$:

$$H = \begin{bmatrix} \omega_0 + \frac{\gamma}{\Gamma}\Delta\Omega_1 & i\kappa \\ i\kappa & \omega_0 + \frac{\gamma}{\Gamma}\Delta\Omega_2 \end{bmatrix} + \begin{bmatrix} -\frac{1}{2}\Delta\omega_{\text{sagnac}} & 0 \\ 0 & \frac{1}{2}\Delta\omega_{\text{sagnac}} \end{bmatrix}, \tag{7.26}$$

where $\omega_0$ is the Stokes cavity mode without pump, $\gamma$ is the cavity damping rate, $\Gamma$ is the bandwidth of Brillouin gain, $\kappa$ is the dissipative coupling rate, $\Delta\Omega_j = \omega_{pj} - \omega_s - \Omega_{\text{phonon}}$, for $j = 1, 2$, is the frequency mismatch of Brillouin scattering

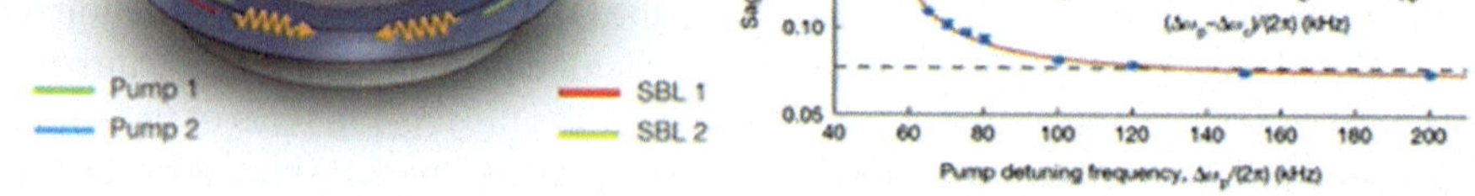


**Fig. 7.4** **a** Schematic of an EP gyroscope. Two SBL modes (*red and yellow arrows*) excited by two pumps (*green and blue arrows*) have dissipative coupling with each other. **b** Experimentally measured Sagnac scale factor as a function of pump detuning (*blue dots*), compared with the theoretical results (*red curve*). The inset is a log-log plot near the EP. Figures from Ref. [66]

corresponding to CW and CCW lasing modes [78], where $\omega_{pj}$ is the pump frequency, $\omega_s$ is the Stokes lasing frequency and $\Omega_{\text{phonon}}$ is the Brillouin shift frequency. The second term describes the Sagnac effect that induces a frequency difference between CW and CCW modes $\Delta\omega_{\text{sagnac}} = 2\pi D\Omega/(n_g\lambda)$, where $D$ is the resonator diameter, $n_g$ is the group index of the unpumped cavity mode and $\lambda$ is the laser wavelength. The beating frequency is evaluated from the difference of Hamiltonian's eigenfrequencies

$$\Delta\omega_s = \frac{\gamma/\Gamma}{1+\gamma/\Gamma}\sqrt{\left(\Delta\omega_p + \Gamma\Delta\omega_{\text{sagnac}}/\gamma\right)^2 - \Delta\omega_c^2}, \tag{7.27}$$

where $\Delta\omega_p = \omega_{p2} - \omega_{p1}$ is the pump detuning and $\Delta\omega_c = 2\Gamma\kappa/\gamma$ is the critical pump detuning to reach an EP. To quantify the EP enhancement, the Sagnac scale factor for a low rotation rate is defined as

$$S = \left.\frac{\partial\Delta\omega_s}{\partial\Omega}\right|_{\Omega\to 0} = \frac{1}{1+\gamma/\Gamma}\frac{\Delta\omega_p}{\sqrt{\Delta\omega_p^2 - \Delta\omega_c^2}}\frac{2\pi D}{n_g\lambda}. \tag{7.28}$$

Therefore, if the gyroscope operates near an EP, i.e., $\Delta\omega_p \to \Delta\omega_c$ in the regime of $|\Delta\omega_p| > \Delta\omega_c$, its small-signal Sagnac scale factor will be much larger than the conventional value $2\pi D/(n_g\lambda)$.

In the experiment, the disk resonator is packaged in a box and experiences a low rotation rate of 410°/h produced by a piezoelectric stage. The measured Sagnac scale factor upon the variation of pump detuning is shown in Fig. 7.4b. The experiment result (blue dots) matches well with the theoretical prediction (red curve), and the Sagnac factor can be enhanced by a factor of 4 near the EP compared to the conventional value (black dash line). Additionally, a log-log plot of five data points near the EP (inset of Fig. 7.4b) exhibits a slope of −1/2, which confirms the existence of $1/\sqrt{\Delta\omega_p^2 - \Delta\omega_c^2}$ factor in the Sagnac scale factor predicted by the theory.

It is also worth noting that EP gyroscope demonstrated here is only exceptional in its large response compared to conventional ones. But the achievement of an exceptional signal-to-noise ratio (SNR) in EP sensors is not straightforward [79, 80]. Recently, it has been experimentally demonstrated that the SNR of EP gyroscope is not improved [81], as the enhanced Sagnac scale factor and the Petermann linewidth broadening cancel each other. This is essentially because the non-orthogonality of mode at an EP will lead to a broadening of SBL linewidth. The enhancement of laser linewidth near an EP is assessed with the Petermann factor [82–85], which poses a fundamental limit to the improvement of EP gyroscope based on SBLs. However, by adopting quantum quadrature measurement scheme or nonreciprocal approach, it is expected that EP amplifying sensors can exhibit enhanced SNR that breaks the quantum noise limit [86, 87].

## 7.4 Mode Interactions and Lasing Effects

It is not hard to see that the interactions between optical modes have played critical roles in non-Hermitian photonic devices and systems. Conversely, the non-Hermitian features can exert significant influence on the optical modes and their interactions. As discussed above, by coupling optical modes with gain/loss features in confined optical structures, non-Hermitian optical systems render not only complex eigenspectra but also non-orthogonal eigenmodes, i.e., the eigenstates of the systems are no longer orthogonal. This is true even for $PT$-symmetric systems in the unbroken regime where real eigenfrequencies are obtained. What is most interesting is the eigenstate behavior at an EP, where two eigenstates not only coalesce but also possess a special chirality, as will be discussed in this section.

One of the most intriguing results of mode interactions can be found in lasing and emission behavior of non-Hermitian optical systems. Silicon photonic platforms have provided versatile designs of microlasers with high efficiency and low threshold [88]. However, the design of lasers with single-mode operation, controllable emission direction, and stability remain challenging and highly rely on the development of novel physics and optical structures. We will examine several counterintuitive phenomena in lasing and emission, which leverage unconventional physical properties of optical modes such as chirality and non-orthogonality, as well as singular behavior in eigenspectra evolution around the phase transition points.

### *7.4.1 Chiral Modes at Exceptional Points*

The most exotic phenomena arising from the interaction of optical modes in non-Hermitian systems can be found at EPs. When two eigenstates of a non-Hermitian system coalesce at an EP, one can obtain a self-orthogonal eigenstate which is a superposition of the original eigenvector. For the system comprising of two modes, the eigenstate is typically in the form of $\begin{pmatrix} 1 \\ \pm i \end{pmatrix}$, with a typical $\pm\frac{\pi}{2}$ phase shift between the two components (either the positive or negative sign is taken depending on the condition of the system). Such form of eigenstate possesses a particular handedness, which can be defined as chirality [89]. One can visualize the concept of chirality by postulating that the wavefunctions in the original basis consist of two perpendicular polarization directions, and then at an EP, the eigenstate takes the form of a circularly polarized wave and rotates in either CW or CCW direction.

In optical resonators, EPs with specific chiral eigenmodes have been engineered and verified experimentally [90]. Intrinsically, a whispering gallery mode (WGM) microresonator made in silica supports degenerate optical modes propagating along the circular boundary of the device in both CW and CCW directions. A nanotip made by etching a fiber-taper end into a cone shape is positioned close to the rim of the resonator and perturbs the evanescent field, inducing backscattering (with strength $\epsilon_1$)

of the optical field and thereby coupling the CW and CCW optical modes. The system under perturbation supports symmetric and antisymmetric standing wave supermodes, each constituting CW and CCW components. A second nanotip separated from the first by an azimuthal distance $\beta$ can be exploited to break the chiral symmetry of the optical structure by introducing an additional perturbation $\epsilon_2$ to the evanescent field. The non-Hermitian Hamiltonian of the system under the perturbation of the two nanoscatters can be derived [58, 73]

$$H = \begin{pmatrix} \omega_0 - i\frac{\gamma_0}{2} + \epsilon_1 + \epsilon_2 & \epsilon_1 + \epsilon_2 e^{-2im\beta} \\ \epsilon_1 + \epsilon_2 e^{2im\beta} & \omega_0 - i\frac{\gamma_0}{2} + \epsilon_1 + \epsilon_2 \end{pmatrix}, \tag{7.29}$$

where the CW and CCW modes with azimuthal mode number $m$ have identical resonant frequency $\omega_0$ and intrinsic loss rate $\gamma_0$. The supermodes are given by $\psi_\pm = \sqrt{A}\psi_{ccw} \pm \sqrt{B}\psi_{cw}$, where $A = \epsilon_1 + \epsilon_2 e^{-2im\beta}$ and $B = \epsilon_1 + \epsilon_2 e^{2im\beta}$. By manipulating the position of the second nanotip, one can engineer one of the non-diagonal element ($A$ or $B$) to become zero, leading to an EP with a singular form of $H$ (one non-diagonal element vanishes) and only one eigenstate $\psi_{EP} = \sqrt{A}\psi_{ccw}$ or $\psi_{\text{EP}} = \sqrt{B}\psi_{cw}$. Such eigenstate is purely chiral with one propagation direction, by which we can define chirality $-1$ for a CCW eigenmode and chirality 1 for a CW eigenmode [90]. By tuning the relative phase angle $\beta$, one can change the value of chirality continuously and periodically between $-1$ and 1, due to the periodic variation in the phase of the non-diagonal coupling element in the Hamiltonian (Fig. 7.5b).

### *7.4.2 Unidirectional Lasing*

The chirality at an EP can manifest itself with the assistance of optical modes operating beyond the lasing threshold which exhibits directional laser emission. The chiral mode at an EP which rotates in one direction has been demonstrated in a silica microresonator with gain, and unidirectional lasing behavior has been observed [90, 91]. The gain medium is introduced by spin coating sol-gel solutions containing $Er^{3+}$ ions to a silicon wafer and forming a layer of silica material with about 2 $\mu$m thickness by thermal annealing [34, 92–94]. The silica microtoroid is then made in the standard photolithography, wet and dry etching, and reflow procedures. In experiments, the device is characterized by coupling to double fiber-taper waveguides with the configuration shown in Fig. 7.5a. By injecting pump light in the 1450 nm band from port 1, the device can lase around the wavelength of 1550 nm with detectable lasing emission from port 3 (Fig. 7.5c). Once the microlaser is perturbed by a nanotip, bidirectional laser emission can be collected from port 3 and port 4, because the scatterer induces coupling between the CW and CCW optical modes and forming standing wave lasing fields (Fig. 7.5d). However, if a second nanotip is applied and tunes the resonator to an EP where backscattering from CW to CCW directions vanishes, then the only surviving eigenstate is in the CW direction, and thereby the laser emission

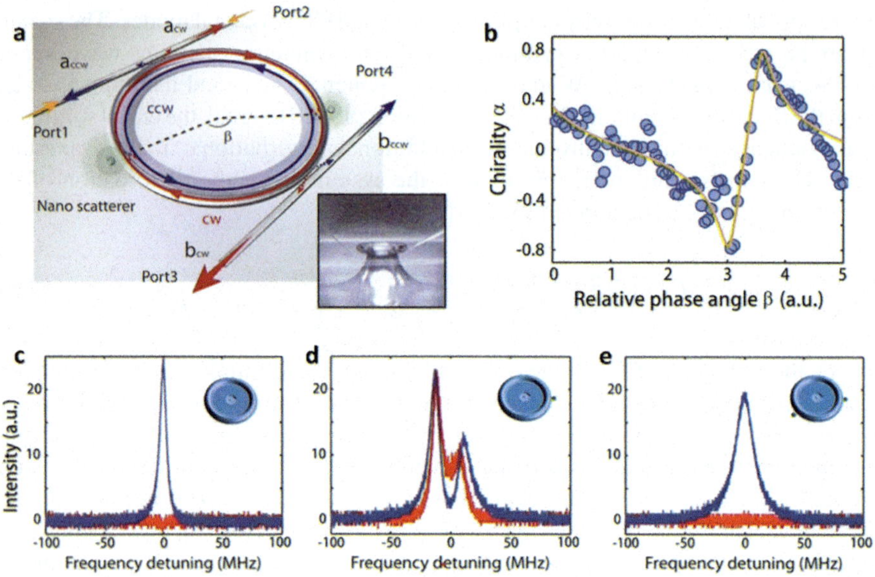


**Fig. 7.5** Chiral modes and unidirectional lasing at an EP. **a** Schematic diagram for realizing chiral modes at EPs in a silica microresonator. The resonator is coupled to two fiber-taper waveguides and perturbed by two silica nanotips which can adjust the backscattering between CW and CCW modes. **b,** Measured chirality as a function of the relative phase angle between the two nanotips. **c-e** Measured light intensity from port 3 (red) and port 4 (blue), for (**c**) no nanotip perturbation, (**d**) one-nanotip perturbation, and (**e**) an EP achieved by two-nanotip perturbation. Figures from Ref. [90]

is only detected from port 3 (Fig. 7.5e). Further adjusting the position of the second scatterer to engineer another EP with CCW eigenmode and reversing the chirality will lead to unidirectional lasing in only the CCW direction.

The chiral mode can also be exploited to generate vortex beam and orbital angular momentum lasers if the amplifying mode emits vertically out of the plane of the circular optical path in the cavity [95]. In addition, some behavior opposite to what is described in this part has also been reported. For instance, in a system where the backscattering ratio between CW and CCW modes can be designed precisely, spontaneous emission of emitters can be coupled to a Jordan vector of the EP microcavity instead of the channel aligned with the eigenstate [96].

### *7.4.3 Single-Mode Lasers*

Cavities in lasers, which provide wavelength selection and coherent amplification of light, usually support multiple resonances and optical modes [97]. Due to the homogeneous broadening of gain spectrum of the active medium [98], adding gain to the desired mode is often accompanied by generating lasing in the adjacent modes, giving rise to mode competition and laser instability. Therefore, to achieve single-mode operation for lasers is one of the critical tasks in the design and management of lasing systems.

By leveraging the threshold of $PT$-symmetry breaking, single-mode lasing operations can be established in microcavities which support multiple modes in the gain spectrum [99, 100]. The system is composed of coupled active and passive microresonators with identical geometry, thus with the same resonant frequencies in the spectra. Due to the inhomogeneously broadened gain profile, the different modes in the active resonator will experience a different amount of gain. The strong coupling between the optical modes in the two resonators will cause frequency splitting, generating pairs of modes in the eigenspectra of the whole system. The maximum gain is first tuned to be equal to the loss in the other resonator, so that the system stays in the $PT$ unbroken phase, where each pair of mode will remain non-amplifying, due to the fact that the two supermodes have the same imaginary part of the eigenfrequency and remain below the lasing threshold. By increasing the pump power and the gain in the active resonator, one pair of mode with the largest gain will pass the $PT$-transition threshold and enter the $PT$-broken regime, leaving one supermode with amplification and the other with dissipation. Therefore, only the supermode with net gain will lase, whereas the supermodes in other mode pairs still remain in the $PT$-unbroken regime and stay below the lasing threshold. Such lasers have been realized in various semiconductor platforms [99, 100] and provide novel routes to laser mode management and control in silicon photonics.

### *7.4.4 Revival of Lasing by Loss*

More unconventional lasing behaviors can be engineered by leveraging non-Hermiticity of open lasing systems, where the phase transition at EPs is the root cause of bizarre lasing effects. One example can be found in a photonic molecule with two coupled microresonators, where increasing the loss rate of one microresonator can induce the suppression and even the revival of laser emission in another, contradictory to the common sense that loss always acts as a negative and detrimental factor to laser operations [35].

In the experimental demonstration of this effect, two silica microtoroid resonators are coupled via evanescent field and probed by fiber-taper waveguides (Fig. 7.6a). One resonator is active with the assistance of Raman gain. The Raman nonlinear effect which is commonly found in silica material induces frequency shift of the pump light, and, with the help of a cavity, can induce coherent amplification of the Raman signal light, leading to stimulated Raman emission, i.e., the Raman lasing [88]. The other resonator is purely lossy and subject to external loss exerted by a chromium (Cr) absorber. The Raman laser in this scenario is thus embedded in a non-Hermitian context with the interaction between optical gain and loss units. It will undergo unconventional enhancement and reduction with the change of the external

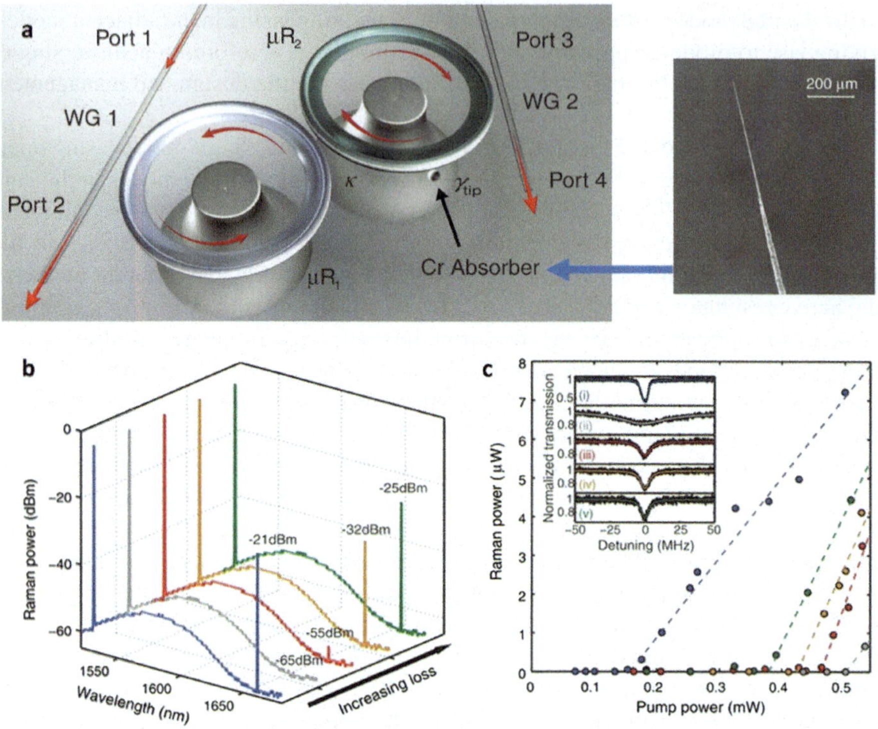


**Fig. 7.6** Revival of Raman laser by external loss. **a** Schematic diagram of the coupled microtoroid resonators $\mu R_1$ and $\mu R_2$ coupled to two waveguides. A Chromium (Cr) absorber (inset on the right) is used to tune the loss of $\mu R_2$. **b** Raman lasing spectra when the loss of $\mu R_2$ is increased by adjusting the position of the Cr absorber. **c** Variation of Raman laser power with the pump power for different loss of $\mu R_2$. The inset shows the transmission spectra near the pump frequency with loss increase from top to bottom. Each color in **b** and **c** corresponds to a value of an induced loss on $\mu R_2$. Figures from Ref. [35]

loss on one of the resonators. When the loss rate of the second resonator is increased, the Raman lasing threshold first increases corresponding to the suppression of lasing, and then decreases, leading to the revival of lasing (Fig. 7.6b, c). The change from laser suppression to laser revival occurs around the phase transition point, i.e., the EP. One can understand the phenomenon based on the field distribution in the two resonators. When the loss in the second resonator is relatively small, the system supports two supermodes with identical loss rates but different resonances, and the energy distribution in the two resonators is symmetric. With more loss introduced to the lossy resonator by the Cr absorber, the system will undergo a phase transition passing the EP and enter the broken regime where the supermodes with amplification and dissipation are localized in the active and the lossy resonator, respectively. The higher contrast of gain and loss between the two units intensifies the mode localization. Therefore, when the external loss increases, the field intensity in the lossy resonator becomes even smaller while that in the active resonator is enlarged. Such redistribution of the optical energy among the system vividly presents the process where the field in a lossy unit sacrifices for the survival of the active counterpart as the imbalance of dissipation among the system is amplified.

Such unconventional lasing behavior managed by the loss is a direct result of the interaction between optical modes with gain and loss, as well as the redistribution of optical field among the whole system combined with the phase transition. A similar observation has also been made in other optical structures and platforms [49, 101].

### *7.4.5 Petermann Factor and Laser Linewidth*

The linewidths of lasers determined by quantum noise can be significantly influenced by modal coupling in laser systems. With the interaction between the optical modes with different gain/loss features, non-Hermitian systems render non-orthogonal eigenmodes and consequently have nontrivial effects on the laser linewidths. It has been predicted and demonstrated that the non-orthogonality of the eigenmodes could lead to amplified quantum noise in lasers, as well as a significant enhancement of laser linewidth quantified by the Petermann factor beyond the Schawlow–Townes quantum limit [82–85, 102]. Moreover, at an EP where eigenmodes are self-orthogonal, the laser linewidth is predicted to be extremely broadened [103].

Such effect has been investigated for a phonon laser operating at an EP realized in a silica microresonator [104]. Microtoroid structures with silica microdisks supported by a silicon pillar can support mechanical oscillations, which can be excited by radiation pressure induced by intracavity optical fields [105]. A phonon laser is based on the coherent amplification of a mechanical mode by the mechanical gain, can be provided by the two optical supermodes with split frequencies that form a two-level structure [106]. If the split eigenspectrum has certain linewidths covering a range that can trigger the mechanical oscillation supported by the microtoroid structure, then the two-level system acts as a gain medium and interacts coherently with the mechanical mode, thus building a phonon laser scheme (Fig. 7.7a). In such a phonon laser system, an EP for the optical modes can be engineered by tuning the optical loss rate of one resonator, and subsequently it will lead to intriguing phonon lasing behavior. When increasing the loss of a resonator externally by a Cr nanotip, the phonon lasing threshold is first lifted and then falls abruptly around the EP (Fig. 7.7c), before finally reviving slowly. Besides, the linewidth of the phonon laser is significantly broadened near the EP (Fig. 7.7d). The broadening of the linewidth around the EP originates from the excessive noise in the optical gain medium resulting from the strongly non-orthogonal eigenstates.

The enhanced noise in EP laser systems has a significant influence on the performance of EP sensors which rely on enhanced lasing mode-splitting. It has been shown in a high-$Q$ silica microresonator that the ring-laser serving as a gyroscope does not offer enhanced signal-to-noise ratio compared to conventional ones, as a result of the mode non-orthogonality and Petermann-factored noise [81]. However, a novel

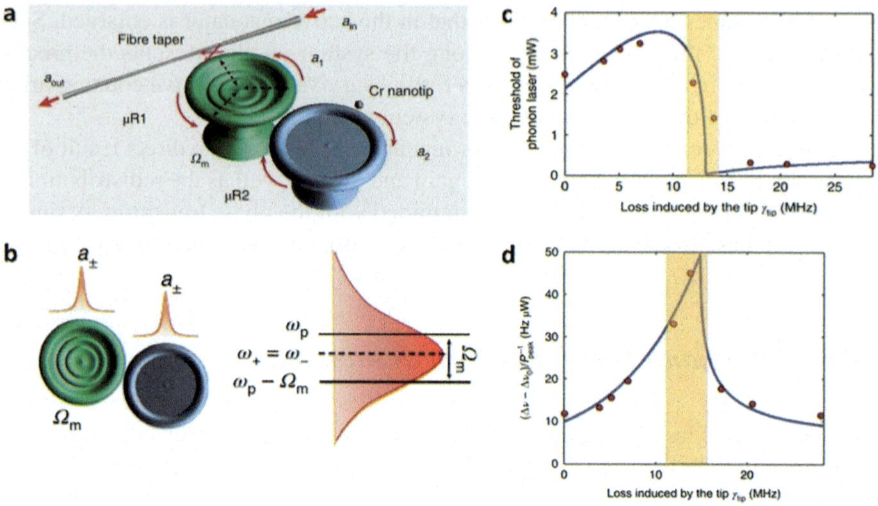


**Fig. 7.7** A phonon laser at an EP. **a** Schematic diagram for coupled microtoroid resonators supporting WGMs with field amplitudes $a_1$ and $a_2$ coupled to a fiber-taper waveguide. The green resonator $\mu R_1$ supports mechanical oscillation with a resonant frequency $\Omega_m$. The blue resonator $\mu R_2$ is perturbed by a Cr nanotip so that the optical loss rate can be tuned externally. **b** Optical supermode and spectrum when the system is at the EP. **c** The measured threshold of the phonon laser varying with the external loss induced by the nanotip. **d** The measured phonon laser linewidth as a function of the external loss induced by the nanotip. Figures from Ref. [104]

detection scheme that acquires full information of quadratures presents opportunities for overcoming the limitations of signal-to-noise ratio imposed by the excessive noise [87].

### *7.4.6 Other Non-Hermitian Lasing Behavior*

The inverse operation of a laser, which is referred to a coherent perfect absorber (CPA), can also exhibit intriguing features in the non-Hermitian setting. It is proposed and demonstrated that a $PT$-symmetric waveguide with gain/loss modulation operating in the broken phase could simultaneously function as a laser and a CPA [107, 108]. Furthermore, the degenerate CPA solutions coalescing to an EP can exhibit anomalous lineshape in the absorption spectrum and chiral absorbing feature [109].

Phonon laser operation is also investigated theoretically for an $PT$-symmetric optical medium [110], where EPs are found to enhance the intracavity photon number and thus the optical pressure, leading to a lower threshold in phonon lasing. Moreover, it is proposed that optical amplifiers operating at EPs can show a better gain-bandwidth scaling [111]. In addition, EPs are also predicted to be able to control and turn off coherent emission above the lasing threshold in laser systems [112].

We expect a near-future demonstration of these predictions in the silicon photonic platforms.

## 7.5 Scattering Properties and Light Propagation

Scattering properties of light describe the relation betweenship output and input optical signals in optical devices and systems, laying the foundation of a great many applications in optical communication and information processing. Optical devices in silicon and silica material such as optical fibers, on-chip waveguides, microresonators, and photonic lattices have played indispensable roles in guiding light transport due to their transparency and low material loss. Furthermore, silicon/silica photonic structures allow for strong nonlinear optical effects [113–115] such as Kerr effect, optothermal effect, Raman scattering and optomechanics [105], which significantly influence optical propagation and dynamics. By leveraging EPs and $PT$-symmetry in optical structures, several unconventional scattering properties emerge such as unidirectional reflection and modulated electromagnetically induced transparency (EIT). In addition, the integration of $PT$-symmetry with nonlinear optical effect [13] enables novel methods of designing nonreciprocal light transport with high performance.

### *7.5.1 Unidirectional Zero Reflection at Exceptional Points*

We consider a one-dimensional $PT$-symmetric photonic heterostructure, where complex index modulation yields spatial separation of gain and loss regions. The relation between the output and input electromagnetic waves can be described by a scattering ($S$) matrix as

$$\begin{pmatrix} a_{\text{out},L} \\ a_{\text{out},R} \end{pmatrix} = S \begin{pmatrix} a_{\text{in},L} \\ a_{\text{in},R} \end{pmatrix} = \begin{pmatrix} r_L & t \\ t & r_R \end{pmatrix} \begin{pmatrix} a_{\text{in},L} \\ a_{\text{in},R} \end{pmatrix}, \tag{7.30}$$

where $a_{in,L(R)}$ represents the input optical wave from the left (right), and $a_{out,L(R)}$ is the output optical wave from the left (right). The identical off-diagonal elements represent reciprocity of light transport. Generalized unitary relation leads to the conservation relation [116]

$$|T - 1| = \sqrt{R_L R_R}, \tag{7.31}$$

where $T = |t|^2$, $R_L = |r_L|^2$, and $R_R = |r_R|^2$. Specifically, the transmission rate $T$ is smaller than 1 when the system is in the $PT$-symmetric unbroken regime, and larger than 1 when the $PT$-symmetric phase is broken. Peculiar scattering phenomena can happen at the phase transition point, i.e., the EP, in the $PT$-symmetric 1D scattering potential, where the eigenspectra turned from real to complex and the transmission $T$ is equal to 1. As a result, the vanishing left-hand side of the conservation relation yields a zero reflection rate for either $R_L$ or $R_R$, leading to unidirectional reflectionless light transport [117]. The theory can also be extended to passive $PT$-symmetric systems.

Such a phenomenon can be observed in synthetic silicon photonic structures with modulation of complex refractive index along the light propagation direction. For instance, Feng et al. demonstrated a unidirectional reflectionless Bragg grating structure based on a Si waveguide that is embedded inside $SiO_2$ (Fig. 7.8a) [118]. The realization of a passive $PT$-symmetric potential is achieved by introducing a periodic modulation of dielectric permittivity in the $z$ direction $\Delta\epsilon = \cos(qz) - i\delta \sin(qz)$, where $q = 2k_1$ and $4n\pi/q + \pi/q \leq z \leq 4n\pi/q + 2\pi/q$, with $k_1$ being the wavevector of the fundamental mode. An EP occurs when the parameter $\delta$ is equal

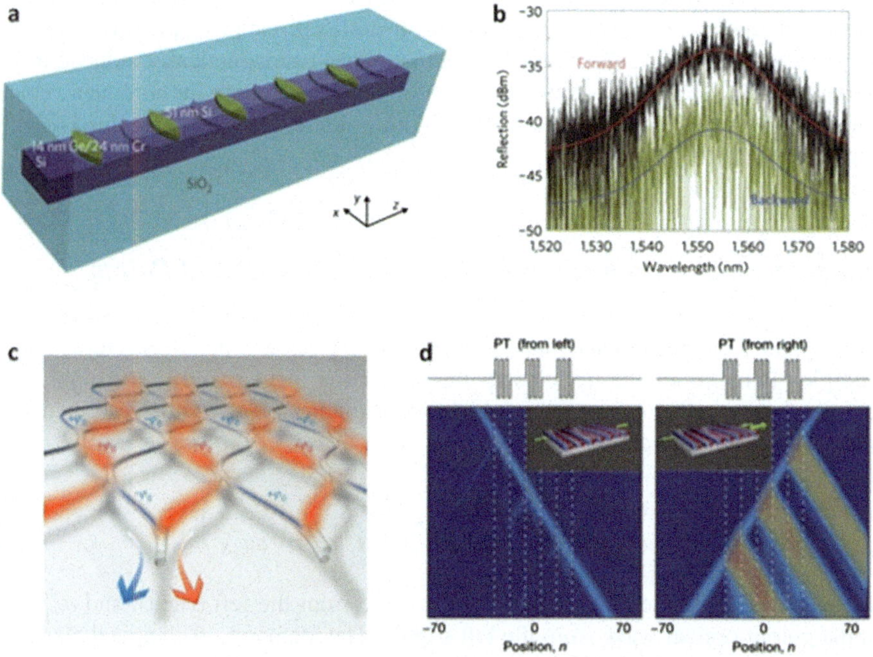


**Fig. 7.8** Unidirectional reflectionless light propagation at exceptional points. **a** A passive $PT$-symmetric Brag structure made of a 800-nm-wide Si waveguide embedded in $SiO_2$ with periodic modulation of the complex refractive index engineered by periodically structured Ge/Cr and Si layers. **b** Measured reflection spectrum for the device in (**a**) from both forward and backward directions. **c** $PT$-symmetric photonic lattice made of a periodic layout of fiber loops with gain (red) and loss (blue). The phase shift $\pm\varphi_0$ imposes a symmetric real part of the potential. **d** Probing the grating at an EP by left and right incoming light beams. The structure is invisible from the left, but visible from the right. Color scale: logarithm, red for high intensity and blue for low intensity. Figures **a** and **b** from Ref. [118], and figures **c** and **d** from Ref. [119]

to 1. The changes of the real and imaginary parts of the refractive index are, respectively, introduced by depositing Si and germanium (Ge)/chrome (Cr) bilayer combo structures on the Si waveguide. An on-chip waveguide directional coupler made of Si is fabricated along with the $PT$-symmetric waveguide to measure the reflection rates from both forward and backward directions. The measured forward reflection is much larger than the backward reflection over a broad telecom band (Fig. 7.8b), verifying the unidirectional reflectionless light transport at an EP.

The suppression of reflection from a direction indicates that the object is invisible from that particular direction. Conventionally, invisibility can be engineered by encompassing an object with a cloak medium. Here the suppression of reflection from one side at EPs can be exploited to create optical structures with complete unidirectional invisibility over a broad frequency range [120, 121]. It has been experimentally demonstrated that a temporal $PT$-symmetric photonic lattice consisting of optical fiber loops can be made invisible from one side (Fig. 7.8c, d) [119]. Furthermore, the phenomenon is found to be robust against Kerr nonlinearity and perturbations [120] and has been proposed in two-layer slab structures [122], one-dimensional photonic crystals [123, 124], and various other platforms [125].

### *7.5.2 Nonreciprocal Light Transport in Nonlinear Parity-Time Symmetric Systems*

Nonreciprocal light propagation breaks the symmetry of the scattering matrices, creating asymmetric transmission of light propagating in opposite directions, allowing light to propagate only in one direction and completely blocking the opposite transmission [126]. Nonreciprocal phenomena have found widespread adoption in photonic applications such as building isolators protecting lasers from the damaging effect of reflection signals, as well as designing circulators that route directional light propagation among different ports. To realize nonreciprocal light transport, the Lorentz reciprocity must be broken [127], typically with the assistance from magneto-optic effect, optical nonlinearity, or temporal modulation of materials. As the magneto-optical effect is weak in many materials, bulky structures are needed, which add difficulty to the integration of nonreciprocal devices. As a result, much attention has been paid to nonlinear [128–132] and time-dependent [133–135] effects in the attempt to break Lorentz reciprocity.

The nonreciprocal light transport is observed in a $PT$-symmetric nonlinear system as a result of the strongly enhanced nonlinearity in the $PT$-broken regime [54]. The system is composed of coupled WGM microtoroid resonators made of silica with different gain/loss features and two fiber-taper waveguides as input and output channels. Resonators are fabricated on the edges of two substrates so that the coupling strength between them can be adjusted by manipulating their spatial gap via nanopositioners (Fig. 7.9a, b). The gain is introduced to the first resonator by $Er^{3+}$ ion dopants and tuned by the pump light so that the net gain balances the intrinsic

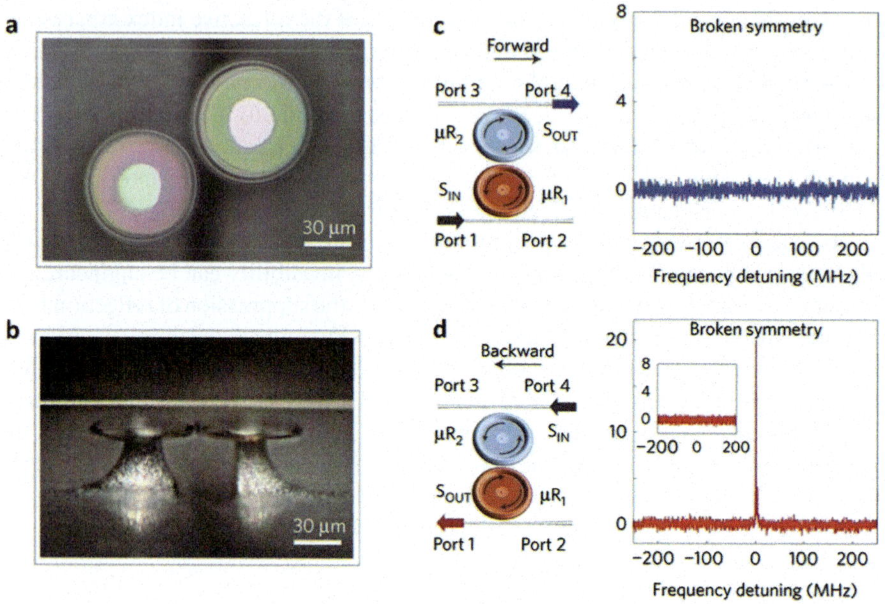


**Fig. 7.9** Nonreciprocal light propagation in nonlinear $PT$-symmetric photonic molecules. **a-b** Top view (**a**) and side view (**b**) of the coupled silica microtoroid resonators with gain and loss. **c** Schematic for the forward light propagation and the measured forward transmission spectrum in the $PT$-broken regime. **d** Schematic for the backward light propagation and the measured backward transmission spectrum in the PT-broken regime. Figure from Ref. [54]

loss in the second resonator. The $PT$ phase transition can be observed by varying the coupling strength via adjusting the gap between the two resonators. Nonlinear effects such as gain saturation and Kerr effects can happen in silica microresonators and be triggered by large intracavity optical intensity. In the $PT$ unbroken regime, the system has two supermodes without amplification or dissipation, and thus exhibits linear response and reciprocity in the transmission when probed by small power signals (Fig. 7.9c). However, in the $PT$-broken regime, the system supports two supermodes with amplification and dissipation which are localized in the active and passive resonators, respectively, regardless of the input direction. As a result, the nonlinearity in the active resonator is significantly enhanced by the localized optical field, giving rise to strong nonreciprocity. A strong contrast occurs in the forward and backward transmission signals where the output signals are collected from the lossy and active resonator sides, respectively (Fig. 7.9d). A complete suppression of the forward transmission is observed, with the threshold of the nonreciprocal light transport as low as $1\,\mu W$.

Due to the giant nonlinearity enhanced by gain medium and $PT$-phase breaking [136], the $PT$-symmetric resonator systems which operate near the boundary of stability is an ideal platform to study versatile nonlinear static or dynamic behavior. The rich tuning degrees of freedom such as waveguide–resonator coupling, probe power and frequency detuning, make it possible to engineer actively controlled optical isolators [137, 138]. The nonreciprocal wave transport enabled by $PT$-symmetry has been explored in mechanical [136], acoustic [139], and electronic [15] systems as well.

### *7.5.3 Electromagnetically Induced Transparency in Non-Hermitian Systems*

Electromagnetically induced transparency (EIT) describes the phenomenon that light can pass through an opaque dielectric medium due to the destructive interference established by strong coherent light-matter interaction. The strong cancellation of absorption enabled by EIT leads to a vast change of material dispersion which gives rise to slowing down of the group velocity of light [140–142]. Such slow light behavior plays an indispensable role in optical memory and storage [143]. EIT was first proposed and demonstrated in atomic systems that support $\Lambda$-type energy levels [144, 145], where the absorption of probe light resonant with atomic electron transition between two energy levels can be eliminated due to the presence of the strong coupling light which couples the excited level to another metastable level. Optical analogues of EIT has been studied in various linear optical systems [146–150], with the advantage of on-chip integration and room temperature operation. Typically, coupled optical elements are involved in order to mediate photon absorption via destructive interference of optical fields. For example, coupled microresonators with contrasted $Q$ factors can be utilized to create a photon energy level structure that mimics $\Lambda$-type atomic energy levels, which allow for EIT operation without the need of strong coherent coupling light [151–153]. Furthermore, optomechanically induce transparency (OMIT) can be found in optomechanical cavities thanks to the interaction between red-detuned pump light, probe light on resonance with the cavity, and the mechanical oscillation [154–156]. As we will see, the non-Hermitian properties of the system can offer large degrees of freedom for technical implementation and operational capability in EIT studies.

EPs have been found to offer a novel approach to control EIT in non-Hermitian silica optical structure. The eigenmode at EPs with certain chirality has been exploited to realize switching on/off EIT in an indirectly coupled WGM microresonator system [157]. The two microtoroid resonators with disparate $Q$ factors are coupled to the same fiber-taper waveguide and probed by injecting single-mode laser with continuous wavelength scanning (Fig. 7.10a). In an ideal situation where no scatterer exists in both resonators, they are not coupled to each other, due to the fact that light can propagate only in one direction so that any light coming out of the first resonator cannot return back into its cavity mode. Nevertheless, backscattering in resonators is ubiquitous due to particle accumulation and surface roughness. Therefore, the optical field in each resonator can be reflected from CW into CCW direction (or inversely) so that an optical path loop is formed in the system. With an optimized

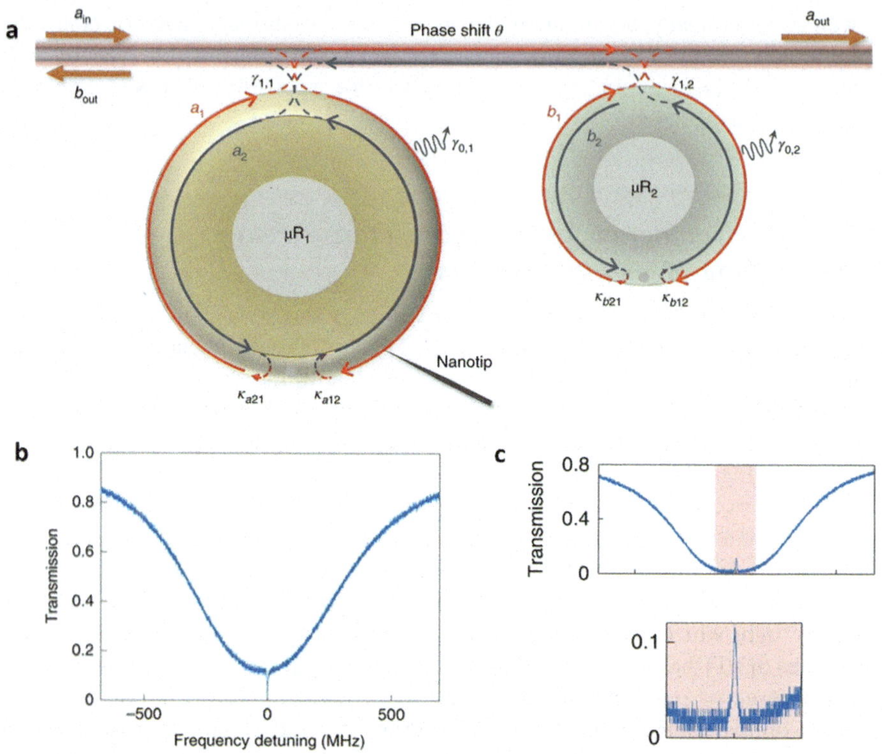


**Fig. 7.10** Electromagnetically induced transparency (EIT) controlled by chiral modes at exceptional points. **a** Indirectly coupled microresonators for realizing the optical analogue of EIT. By tuning the phase $\theta$, destructive interference can happen in the optical path loop established by backscattering on the two resonators. $\mu R_1$ has a smaller optical loss than $\mu R_2$, and is perturbed by a nanotip so that its chirality can be adjusted and exceptional points can be realized. **b** Transmission spectrum under the condition that $\mu R_1$ is at an exceptional point with chirality $-1$. **c,** Transmission spectrum under the condition that $\mu R_1$ is at an exceptional point with chirality 1. The inset on the lower panel shows a close up of the central transparency window. Figure from Ref. [157]

phase accumulation on the loop, which can be tuned by the distance between the two resonators, destructive interference can occur, leading to suppression of cavity field and transparency in the forward transmission spectrum. The transparency window exhibits splitting due to the lift of degeneracy in the resonator with the higher $Q$ factor. By an additional nanotip, one can tune $\mu R_1$ to an EP with a chirality of $-1$ and a CCW eigenmode, or another type of EP with a chirality of 1 and a CW eigenmode, which affects EIT very differently.

For a chirality of $-1$, the reflection from CCW to CW vanishes, and the optical path loop is broken. Thus, EIT is switched off, and a narrow absorption dip appears in the transmission spectrum (Fig. 7.10b). Such an effect is named as EP-assisted absorption

(EPAA). By adjusting the position of the nanotip and reversing the chirality to 1, the reflection from CCW to CW becomes present and destructive interference can occur with precise tuning of the distance between the two resonators. As a result of the degenerate eigenmode at the EP, a narrow transparency window appears in the transmission spectrum, leading to a standard EIT phenomenon (Fig. 7.10c), which is named as EP assisted transparency (EPAT). Therefore, this scheme provides a method of leveraging non-Hermitian eigenstates for the control of the all-optical analogue of EIT in optical resonators.

Non-Hermitian features have also been proposed to modify OMIT processes in cavity optomechanical systems. For example, OMIT can be engineered in a WGM resonator with its chirality being adjusted by two scatterers [158]. By adjusting the position of one scatterer, one can change the chirality of the optical field which modifies the group velocity of light and enable switching between slow and fast light. In a system of coupled active and passive resonators, the requirement of the driving amplitude and the optomechanical coupling strength for OMIT can be greatly relaxed due to the fact that the intracavity optical field is enhanced by the presence of optical gain [159]. The non-Hermitian properties thus offer a wealth of new avenues in the realization of dispersion modification, slow light manipulation, and optical storage [160, 161].

## 7.6 Topological Features and Mode Switching

The special topological features of EPs not only attract attention in the applications of sensors but also unveil profound physics in the dynamical behavior of eigenstates under variation of parameters. For example, the neighborhood of a second-order EP in the parameter space was found to resemble the Riemann surface of $\sqrt{z}$. Any cyclic parametric loop around the EP has a $4\pi$ periodicity, and adiabatic evolution along such loop is predicted to have many interesting effects, such as state transfer [162–164] and geometric phase [165]. More significantly, it was also shown with both numerical and theoretical methods, that anti-adiabatic jump is bound to happen in a bidirectional encircling of an EP. No matter how smooth the evolution is, there will always be a breakdown of adiabaticity at least for one mode, leading to a chiral exchange of modes. Hence, the final state is only determined by the encircling direction in parameter space, regardless of the initial state. Experimental demonstration of this phenomenon has been done in several systems, including a silicon-waveguide system where the time evolution of parameters can be conveniently implemented. Asymmetric mode switching by encircling EPs paves a new way to manipulate the evolution of modes in non-Hermitian systems.

### 7.6.1 Dynamics of Encircling EPs

An adiabatic process, in the context of quantum mechanism, is a process that allows the system to stay in the instantaneous eigenstates when the variation of time-dependent Hamiltonian $H(t)$ is sufficiently slow. As the quantum adiabatic theorem states [166, 167], if the system is initially at the $n$th eigenstate of the initial Hamiltonian $H(0)$, it will always be at the $n$th eigenstate of $H(t)$ at any given time $t$ in the adiabatic process, only picking up some phase factors. In a more general case, the Hamiltonian is defined by some parameters $\boldsymbol{\lambda}(t)$ that evolve with a time period of $T$, that is, $\boldsymbol{\lambda}(0) = \boldsymbol{\lambda}(T)$. Adiabatic evolution of $\boldsymbol{\lambda}(t)$ forms a closed loop in the parameter space, and the state of the system will acquire an additional phase (Berry phase) that is connected to the geometric property of the loop [168–171].

In the context of non-Hermitian Hamiltonian, the study of adiabatic processes predicts new phenomena. In the vicinity of a second-order EP associated with a non-Hermitian Hamiltonian, the parameter space has a self-intersecting topology with a $4\pi$ periodicity [172]. This can be shown using Puiseux series to express the two orthogonal states near an EP [5],

$$\psi_{\pm}(\lambda) \approx A_{\pm}\left(\psi_{\mathrm{EP}} \pm \sqrt{\lambda}\psi_1\right), \tag{7.32}$$

where $\lambda$ is the parameter and EP is defined at the origin ($\lambda = 0$). Here we only investigate a second-order EP, so the system parameter $\lambda$ is assumed to be a complex number instead of a vector with more elements. To satisfy the orthogonality and unitary normalization conditions, the amplitudes are evaluated as $A_+ = -iA_- = \left[2\sqrt{\lambda}\langle\psi_1|\psi_{EP}\rangle\right]^{-1/2}$. It seems straightforward that under stationary condition, encircling an EP (by choosing a circle of radius $R$ around the EP, that is, $\lambda = Re^{i\varphi}$) will result in an exchange of eigenstates and acquisition of Berry phase [165],

$$\begin{aligned} \psi_{\pm}(\lambda(\varphi = 2\pi)) &= \mp\psi_{\mp}(\lambda(\varphi = 0)), \\ \psi_{\pm}(\lambda(\varphi = 4\pi)) &= e^{i\pi}\psi_{\pm}(\lambda(\varphi = 0)), \end{aligned} \tag{7.33}$$

which has been verified experimentally in various systems [173, 174]. Further work, however, found that the small non-adiabatic coupling would inevitably lead to an abrupt transition between the states, preventing adiabatic mode switching to be accomplished in both directions of the closed loop [175–177].

Taking a non-Hermitian two-mode system for example, its dynamics is described by a Schrödinger-type equation $i\mathrm{d}\psi/\mathrm{d}t = H\psi$, where $\psi$ is the amplitude of the modes. The system has a Hamiltonian in the form

$$H = \begin{bmatrix} \omega_1 - i\frac{\gamma_1}{2} & \kappa \\ \kappa & \omega_2 - i\frac{\gamma_2}{2} \end{bmatrix}, \tag{7.34}$$

where $\omega_j$ and $\gamma_j$, for $j = 1, 2$, are the resonant frequencies and loss rates of the modes, respectively, and $\kappa$ is the coupling rate between two modes. Then the parameter conditions for an EP are $\Delta\omega \equiv \omega_2 - \omega_1 = 0$ and $\kappa = |\gamma_2 - \gamma_1|/4$. The surface of eigenfrequencies with respect to $\Delta\omega$ and $\kappa$ has two Riemann sheets, on which dynamic encircling the EP with different initial states and directions is considered. If the initial state is on the red sheet (Fig. 7.11a), counterclockwise encircling the EP can be overall adiabatic and yields a state exchange, while the clockwise one experiences a breakdown of adiabaticity and goes back to the original state; similar phenomenon also happens when the initial state is on the blue sheet (Fig. 7.11b). Therefore, encircling an EP gives rise to a unique chirality that the final state is determined by the encircling direction, regardless of the initial state.

The chiral exchange of states by encircling EPs was first demonstrated in two kinds of platforms. The first one is based on the metallic microwave waveguide [178], in which spatial deformation of the waveguide and a special absorption region are designed to steer the system around its EP. The second experiment is conducted in an optomechanical system [179], where the mechanical EP of a membrane inside a cryogenic optical cavity is enclosed by changing laser parameters. These experiments not only inspire more theoretical work and application proposals [180–184], but also provide practical ideas to precisely vary system parameters while maintaining the EPs. On the other hand, silicon photonics offers huge flexibility in the tuning of parameters and thus is also a promising candidate in the realization of topological operation around EPs. As we will discuss in the next section, demonstration of chiral mode switching can be done in a silicon platform, without the need of any cryogenic condition or complicated absorption region.

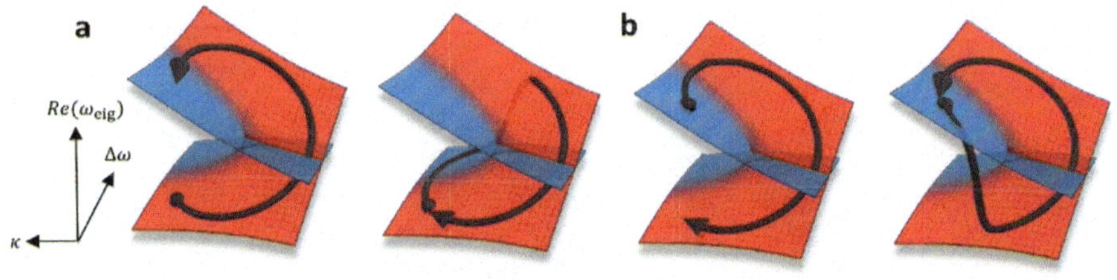


**Fig. 7.11** Encircling a second-order EP starting from the state in **a** the lower or **b** the upper sheets. Due to the breakdown of adiabaticity, encircling the EP in both directions yields a chiral switching of modes. Figures from Ref. [178]

### 7.6.2 Asymmetric Mode Switching

The demonstration of chiral exchange of states by encircling EPs relies on precise control of system parameters, which was also achieved in the silicon-based waveguides [185]. The system consists of two coupled channel waveguides and a slab-waveguide patch on a silica-encapsulated silicon platform (Fig. 7.12a). Each of the channel waveguides supports a fundamental guided mode, and the slab-waveguide patch is coupled to second channel waveguide to control the radiation loss. The time-dependent non-Hermitian Hamiltonian that governs the mode evolution $d\psi/dt = iH(t)\psi$ is taken to be traceless by a gauge transformation, that is,

$$H(t) = \begin{bmatrix} p(t) + iq(t) & 1 \\ 1 & -p(t) - iq(t) \end{bmatrix}. \quad (7.35)$$

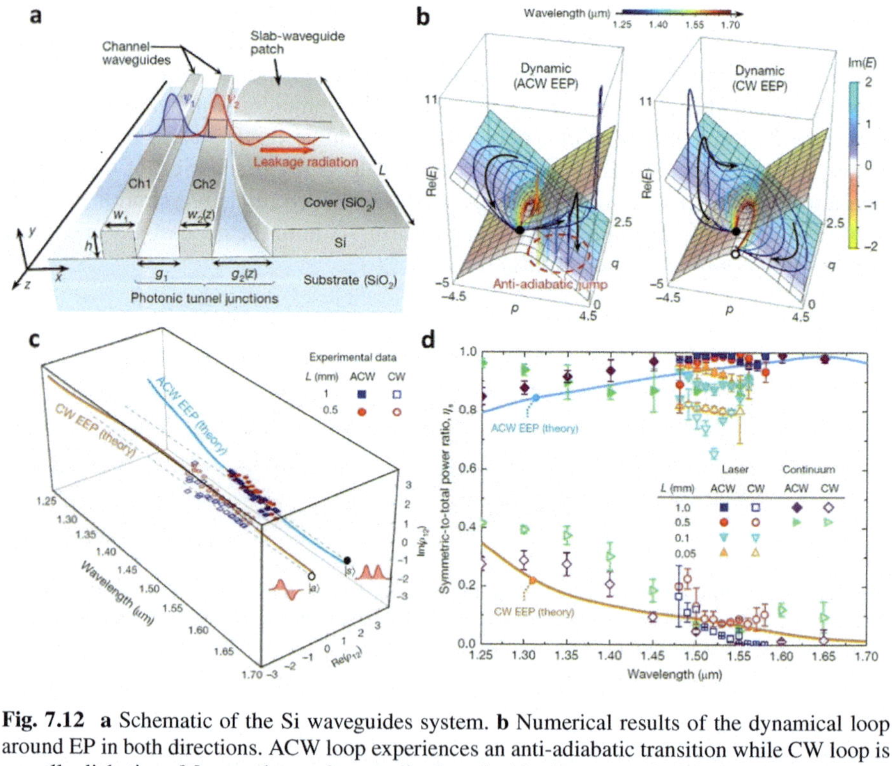


**Fig. 7.12** **a** Schematic of the Si waveguides system. **b** Numerical results of the dynamical loop around EP in both directions. ACW loop experiences an anti-adiabatic transition while CW loop is overall adiabatic. **c** Measured complex amplitude ratio (*dots*) compared with the theory (*curves*). The output mode of ACW loop is almost the same as the input, while the output of CW loop is an asymmetric mode. **d** Measured power ratio of symmetric mode with different device lengths and different light sources (laser and supercontinuum source). Figures from Ref. [185]

Since the temporal behavior is now transferred to the propagation behavior along the waveguide system, the reduced parameters $p(t)$ and $q(t)$ denote the real and imaginary wavevectors, representing the difference in propagation constant and loss, respectively. The unitary off-diagonal terms correspond to the constant coupling rate. The parameters are controlled by the spatial profile of waveguides to form a loop that smoothly encloses the EP at $p = 0, q = 1$. The dynamics of encircling the EP is numerically computed in both anticlockwise (ACW) and CW orientations, over a broad wavelength range from 1.25 μm to 1.70 μm. Shown in Fig. 7.12b, the ACW loops for different wavelengths are all characterized by an anti-adiabatic jump and return to the initial state. On the other hand, the CW loops result in an exchange of states and overall have an adiabatic result, despite that they do not strictly follow the ideally stationary route in the parameter space.

In the experiment, standard nanolithography techniques are used to fabricate the devices for ACW and CW operations. The initial state is always chosen to be a symmetric state $|s\rangle = |\psi_1\rangle + |\psi_2\rangle$, and the final state is measured by fitting the diffraction patterns of the output light. The final state $|\psi_{\text{final}}\rangle = a_1|\psi_1\rangle + a_2|\psi_2\rangle$ is assessed by two quantities: the complex amplitude ratio $\rho_{12} = a_1/a_2$, and the power ratio of symmetric mode that is defined as

$$\eta_s = \left| \frac{\langle s|\psi_{\text{final}}\rangle}{\langle \psi_{\text{final}}|\psi_{\text{final}}\rangle} \right|^2 = \frac{|1+\rho_{12}|^2}{1+|\rho_{12}|^2}. \tag{7.36}$$

The measurements are repeated with devices in different lengths $L = 0.5, 1.0$ mm and under various laser wavelengths ranging from 1.48 to 1.58 μm. The results show that the complex amplitude ratio matches well with the theory (Fig. 7.12c). Over this 100 nm band, the final state for ACW loop is mostly symmetric, but for CW it is switched to be asymmetric, and the spectrally averaged extinction ratio of both directions $[\eta_s]_{ACW}/[\eta_s]_{CW}$ is computed to be as large as 20.1 dB for $L = 1.0$ mm. If the device length is reduced such that adiabaticity is always broken, the mode transfer should not happen and the experimental power ratio deviates a lot from the theoretical curve (*cyan and orange dots* in Fig. 7.12d, corresponding to $L = 0.1, 0.05$ mm). Besides using the laser as excitation, a supercontinuum source is also used in the experiment. The measured power ratio under the continuum excitation exhibits the same trend of chiral mode transfer, but now in a broader spectrum that covers the entire optical communication bands (*purple and green dots* in Fig. 7.12d). The broadband asymmetric mode switching demonstrated in the Si waveguides systems shows the potential of more applications utilizing the dynamics around EPs in the optical telecommunication.

## 7.7 Conclusion and Outlook

In this chapter, we have summarized and reviewed the initial studies of non-Hermitian optics based on silicon photonic platforms, emphasizing throughout the fundamental physical phenomena and the associated novel applications. Transforming from quantum mechanics to classical electromagnetic waves, the broken Hermiticity of a Hamiltonian has brought forth rich opportunities for engineering photonic systems with unconventional optical behavior by judicious design of optical amplification/dissipation and coupling between photonic structures. The silicon photonic platforms fabricated with standard techniques, full integrability, and strong scalability breed the optical structures to meet those requirement, including coupled guided wave structures with gain/loss distribution, scatterer perturbed microresonators, Bragg gratings with periodic index modulation, synthetic photonic lattices, etc. Grounded on the rich light-matter interactions and versatile approaches of manipulating material properties in these devices and systems, numerous counterintuitive non-Hermitian properties have been revealed and demonstrated ranging from spectral singularities, chiral modes, non-orthogonal eigenstates, field localization, to unidirectional reflection/invisibility, enhanced nonlinearity and exotic topology. Consequently, significant progress has taken place in photonic applications such as ultrasensitive optical sensors, unconventional lasing behavior, unidirectional/nonreciprocal light propagation, modified EIT, asymmetric mode switches, and so on.

Though vast achievement has been witnessed, we envision that great potential opportunities of future development of non-Hermitian silicon photonics lie in its incorporation with quantum optics domain and topological photonics.

Integrated photonic circuits fabricated on silicon chips have become an important platform that enables the implementation of quantum optical devices and networks for quantum sensing, computation and communication [186]. The introduction of non-Hermitian physics into quantum silicon photonics will not only shed interesting light on the true quantum nature of non-Hermitian notions but also pave the way for the development of important cutting-edge quantum technology.

$PT$-symmetry has been implemented in the quantum optics domain based on an on-chip waveguide directional coupler fabricated by the cutting-edge femtosecond direct laser writing on a fused silica wafer [187]. The directional coupler has been conventionally leveraged to generate the two-photon Hong–Ou–Mandel (HOM) interference, where the two photons become indistinguishable marked by a HOM dip in the coincidence when the propagation distance equates the coupling length [188]. Here, the system renders passive $PT$-symmetry with disparate dissipation in two waveguides, where additional bending loss is introduced to the left waveguide by a slight sinusoidal modulation of the geometry. The result shows a shifted position of HOM dip to a shorter propagation distance, which is attributed to the modified quantum dynamics by the possibility of photon loss.

Up to now, the non-Hermitian physics in quantum optics remains as a nontrivial issue. There have been numerous proposals and open questions along this route. First, it has been shown that it is impossible to introduce full $PT$-symmetry to quantum optics in a similar fashion to that in classical optics, because the noise introduced by quantum gain medium will break the $PT$-symmetry [189]. However, it is possible to exploit asymmetric dissipation to realize passive $PT$-symmetry [190, 191], or even leverage unitary physics of squeezing to implement non-Hermitian quantum setting in absence of dissipation [192]. Second, non-Hermitian singularity, i.e., EPs, may generate exotic effects in the quantum dynamics and fluctuation such as quantum decoherence [193, 194], entanglement [195], information retrieval [196], and quantum noise behavior [80, 86, 87, 197]. Third, it would be interesting to adapt the non-Hermitian setting to nonlinear properties of silicon/silica material, such as Kerr effect [198] and optomechanical interaction [199, 200], which may significantly affect photon dynamics and light-matter interaction within the quantum limit [201]. With the development in silicon photonic technique especially the capability of manipulating nonclassical light such as single-photon sources, squeezed states and entanglement, the in-depth exploration into these questions and experimental implementation will attract increasing attention and bring revenue to the fundamental study of non-Hermitian physics and quantum optics, as well as the engineering design of integrated quantum devices and networks.

Topological photonics [202, 203] is an emerging branch of photonic research that revolutionizes our understanding of the transport and control of light. The essential ideas of topological photonics are adopted from the studies of electronic topological insulators [204], the latter of which is a new phase of matter with insulating bulk and conductive, robust edge states. The earliest demonstrations verifying the feasibility of topological photonics were performed in the platforms of photonic waveguide lattices [205] and resonator arrays [206]. It is not hard to anticipate that the rich physics in non-Hermitian systems will bring more intriguing stories in topological photonics. Recent years have witnessed a rapid burgeoning of the research work on incorporating these two fields. $PT$-symmetric systems, as demonstrated before, can support their own topological interface modes [207], or selectively enhance the interface mode while suppressing the extended modes [208]. In a more recent experiment based on the silicon photonic platform, the reappearance of interface mode via $PT$-symmetric phase transition has also been observed in a lattice of silicon waveguides [209]. These non-Hermitian topological states can be classified in a more general framework [210], which is helpful for finding new topological phases in future experiments. In the non-Hermitian context, conventional bulk-boundary correspondence may be broken, and the topological charges associated with EPs are half-integers [211–213], and new kinds of topological transition could be realized without closing the gap or breaking usual symmetries.

On the other hand, the combination of topological photonics and active media provides innovative ideas in the design of optical components. Topological photonic lasers have been achieved in a greats variety of systems [214–220], which have advantages including robustness to scattering, high efficiency and single-mode operation. Follow-up theoretical proposals predict more novel effects and applications, such as robust extended modes [221], phase transition to multimode operation [222], and frequency comb generation [223]. Another recent experiment on the reconfigurable topological light steering demonstrated the ability to control the flow of light at will [224], which could further facilitate novel designs of optical circuitry with arbitrary light routing. Since silicon/silica materials offer convenient hybridization with gain/loss, we expect more complex optical networks to be built based on silicon photonic platforms in the future.

**Acknowledgement** This authors acknowledge the funding support from the National Science Foundation(NSF) grant No. EFMA1641109.

## References

1. L. D. Landau, E. M. Lifshitz, *Quantum Mechanics: Non-Relativistic Theory* (Pergamon Press, New York, 1991)
2. F. Breuer, Heinz-Peter, Petruccione, *The Theory of Open Quantum Systems* (Oxford University Press on Demand, 2002)
3. Á. Rivas, S.F. Huelga, *Open Quantum Systems. An Introduction* (2011)
4. G.S. Agarwal, Prog. Opt. **11**, 1 (1973)
5. N. Moiseyev, *Non-Hermitian Quantum Mechanics* (Cambridge University Press, Cambridge, England, 2011)
6. I. Rotter, J. Phys. A: Math. Theor. **42**, 153001 (2009)
7. C.M. Bender, Rep. Prog. Phys. **70**, 947 (2007)
8. C.M. Bender, S. Boettcher, Phys. Rev. Lett. **80**, 5243 (1998)
9. C.M. Bender, S. Boettcher, P.N. Meisinger, J. Math. Phys. **40**, 2201 (1999)
10. C.M. Bender, Contemp. Phys. **46**, 277 (2005)
11. C.M. Bender, Europhys. News **47**, 18 (2016)
12. R. El-Ganainy, K.G. Makris, M. Khajavikhan, Z.H. Musslimani, S. Rotter, D.N. Christodoulides, Nat. Phys. **14**, 11 (2018)
13. V.V. Konotop, J. Yang, D.A. Zezyulin, Rev. Mod. Phys. **88**, 1 (2016)
14. D. Jiankeeyang, *Parity-Time Symmetry and Its Applications* (Springer, 2018)
15. J. Schindler, A. Li, M.C. Zheng, F.M. Ellis, T. Kottos, Phys. Rev. A **84**, 040101(R) (2011)
16. S. Assawaworrarit, X. Yu, S. Fan, Nature **546**, 387 (2017)
17. N. Lazarides, G.P. Tsironis, Phys. Rev. Lett. **110**, 053901 (2013)
18. X. Zhu, H. Ramezani, C. Shi, J. Zhu, X. Zhang, Phys. Rev. X **4**, 031042 (2014)
19. R. El-Ganainy, M. Khajavikhan, D.N. Christodoulides, S.K. Özdemir, Commun. Phys. **2**, 37 (2019)
20. A.A. Zyablovsky, A.P. Vinogradov, A.A. Pukhov, A.V Dorofeenko, A.A. Lisyansky, Physics-Uspekhi **57**, 1063 (2014)
21. L. Feng, R. El-Ganainy, L. Ge, Nat. Photonics **11**, 752 (2017)
22. M.-A. Miri, A. Alù, Science **363**, eaar7709 (2019)
23. S.K. Özdemir, S. Rotter, F. Nori, L. Yang, Nat. Mater. **18**, 783 (2019)
24. R. El-Ganainy, K.G. Makris, D.N. Christodoulides, Z.H. Musslimani, Opt. Lett. **32**, 2632 (2007)
25. B. Jalali, S. Fathpour, J. Light. Technol. **24**, 4600 (2006)
26. X. Jiang, L. Yang, Light Sci. Appl. **9**, 2047 (2020)
27. T. Carmon, L. Yang, K.J. Vahala, Opt. Express **12**, 4742 (2004)
28. A. Mazzei, S. Götzinger, D.S. Menezes, G. Zumofen, O. Benson, V. Sandoghdar, Phys. Rev. Lett. **99**, 173603 (2007)
29. J. Wiersig, Phys. Rev. A **84**, 063828 (2011)
30. H. Cao, J. Wiersig, Rev. Mod. Phys. **87**, 61 (2015)
31. J. Wiersig, A. Eberspächer, J.-B. Shim, J.-W. Ryu, S. Shinohara, M. Hentschel, H. Schomerus, Phys. Rev. A **84**, 23845 (2011)

32. X. Jiang, L. Shao, S.X. Zhang, X. Yi, J. Wiersig, L. Wang, Q. Gong, M. Loncar, L. Yang, Y.F. Xiao, Science **358**, 344 (2017)
33. X.-F. Jiang, Y.-F. Xiao, C.L. Zou, L. He, C.H. Dong, B.B. Li, Y. Li, F.W. Sun, L. Yang, Q. Gong, Adv. Mater. **24**, 260 (2012)
34. L. Yang, K.J. Vahala, Opt. Lett. **28**, 592 (2003)
35. B. Peng, S.K. Özdemir, S. Rotter, H. Yilmaz, M. Liertzer, F. Monifi, C.M. Bender, F. Nori, L. Yang, Science **346**, 328 (2014)
36. Y.D. Chong, L. Ge, H. Cao, A.D. Stone, Phys. Rev. Lett. **105**, 053901 (2010)
37. L. He, Y.F. Xiao, C. Dong, J. Zhu, V. Gaddam, L. Yang, Appl. Phys. Lett. **93**, 201102 (2008)
38. Y.F. Xiao, L. He, J. Zhu, L. Yang, Appl. Phys. Lett. **94**, 231115 (2009)
39. J. Zhu, G. Zhao, I. Savukov, L. Yang, Sci. Rep. **7**, 8896 (2017)
40. G. Zhao, Ş.K. Özdemir, T. Wang, L. Xu, E. King, G.L. Long, L. Yang, Sci. Bull. **62**, 875 (2017)
41. L. He, Y.-F. Xiao, J. Zhu, S.K. Özdemir, L. Yang, Opt. Express **17**, 9571 (2009)
42. L. Xu, X. Jiang, G. Zhao, D. Ma, H. Tao, Z. Liu, F.G. Omenetto, L. Yang, Opt. Express **24**, 20825 (2016)
43. Y. Liu, X. Jiang, C. Wang, L. Yang, Appl. Phys. Lett. **116**, 201104 (2020)
44. K.G. Makris, R. El-Ganainy, D.N. Christodoulides, Z.H. Musslimani, Phys. Rev. Lett. **100**, 103904 (2008)
45. Z.H. Musslimani, K.G. Makris, R. El-Ganainy, D.N. Christodoulides, Phys. Rev. Lett. **100**, 030402 (2008)
46. A. Ruschhaupt, F. Delgado, J.G. Muga, J. Phys. A: Math. Gen. **38**, L171 (2005)
47. C.E. Rüter, K.G. Makris, R. El-Ganainy, D.N. Christodoulides, M. Segev, D. Kip, Nat. Phys. **6**, 192 (2010)
48. Y. Chen, A.W. Snyder, D.N. Payne, IEEE J. Quantum Electron. **28**, 239 (1992)
49. A. Guo, G.J. Salamo, D. Duchesne, R. Morandotti, M. Volatier-Ravat, V. Aimez, G.A. Siviloglou, D.N. Christodoulides, Phys. Rev. Lett. **103**, 093902 (2009)
50. H. A. Haus, *Waves and Fields in Optoelectronics* (Prentice-Hall, 1984)
51. K. Srinivasan, O. Painter, Phys. Rev. A **75**, 023814 (2007)
52. C. Wang, Y.L. Liu, R. Wu, Y.X. Liu, Phys. Rev. A **96**, 13818 (2017)
53. T. Aoki, B. Dayan, E. Wilcut, W.P. Bowen, A.S. Parkins, T.J. Kippenberg, K.J. Vahala, H.J. Kimble, Nature **443**, 671 (2006)
54. B. Peng, Ş.K. Özdemir, F. Lei, F. Monifi, M. Gianfreda, G.L. Long, S. Fan, F. Nori, C.M. Bender, L. Yang, Nat. Phys. **10**, 394 (2014)
55. A.P. Seyranian, A.A. Mailybaev, *Multiparameter Stability Theory with Mechanical Applications* (World Scientific, 2003)
56. A.P. Seyranian, O.N. Kirillov, A.A. Mailybaev, J. Phys. A: Math. Gen. **38**, 1723 (2005)
57. G. Demange, E.M. Graefe, J. Phys. A: Math. Theor. **45**, 025303 (2012)
58. J. Wiersig, Phys. Rev. A **93**, 33809 (2016)
59. N. Zhang, S. Liu, K. Wang, Z. Gu, M. Li, N. Yi, S. Xiao, Q. Song, Sci. Rep. **5**, 1 (2015)
60. Z.P. Liu, J. Zhang, Ş.K. Özdemir, B. Peng, H. Jing, X.Y. Lü, C.W. Li, L. Yang, F. Nori, Y.X. Liu, Phys. Rev. Lett. **117**, 110802 (2016)
61. J. Ren, H. Hodaei, G. Harari, A.U. Hassan, W. Chow, M. Soltani, D. Christodoulides, M. Khajavikhan, Opt. Lett. **42**, 1556 (2017)
62. S. Sunada, Phys. Rev. A **96**, 033842 (2017)
63. W. Chen, J. Zhang, B. Peng, Ş.K. Özdemir, X. Fan, L. Yang, Photonics Res. **6**, A23 (2018)
64. P. Djorwe, Y. Pennec, B. Djafari-Rouhani, Phys. Rev. Appl. **12**, 024002 (2019)
65. W. Chen, Ş.K. Özdemir, G. Zhao, J. Wiersig, L. Yang, Nature **548**, 192 (2017)
66. Y.H. Lai, Y.K. Lu, M.G. Suh, Z. Yuan, K. Vahala, Nature **576**, 65 (2019)
67. H. Hodaei, A.U. Hassan, S. Wittek, H. Garcia-Gracia, R. El-Ganainy, D.N. Christodoulides, M. Khajavikhan, Nature **548**, 187 (2017)
68. J.H. Park, A. Ndao, W. Cai, L. Hsu, A. Kodigala, T. Lepetit, Y.H. Lo, B. Kanté, Nat. Phys. **16**, 462 (2020)

69. D.S. Weiss, V. Sandoghdar, J. Hare, V. Lefèvre-Seguin, J.-M. Raimond, S. Haroche, Opt. Lett. **20**, 1835 (1995)
70. M.L. Gorodetsky, A.D. Pryamikov, V.S. Ilchenko, J. Opt. Soc. Am. **17**, 1051 (2000)
71. T.J. Kippenberg, S.M. Spillane, K.J. Vahala, Opt. Lett. **27**, 1669 (2002)
72. J. Zhu, S.K. Özdemir, Y.F. Xiao, L. Li, L. He, D.R. Chen, L. Yang, Nat. Photonics **4**, 46 (2010)
73. J. Wiersig, Phys. Rev. Lett. **112**, 203901 (2014)
74. J. Zhu, Ş.K. Özdemir, L. He, L. Yang, Opt. Express **18**, 23535 (2010)
75. H. Lee, T. Chen, J. Li, K.Y. Yang, S. Jeon, O. Painter, K.J. Vahala, Nat. Photonics **6**, 369 (2012)
76. F. Zarinetchi, S.P. Smith, S. Ezekiel, Opt. Lett. **16**, 229 (1991)
77. J. Li, M.G. Suh, K. Vahala, Optica **4**, 346 (2015)
78. J. Li, H. Lee, T. Chen, K.J. Vahala, Opt. Express **20**, 20170 (2012)
79. W. Langbein, Phys. Rev. A **98**, 23805 (2018)
80. N.A. Mortensen, P.A.D. Gonçalves, M. Khajavikhan, D.N. Christodoulides, C. Tserkezis, C. Wolff, Optica **5**, 1342 (2018)
81. H. Wang, Y.H. Lai, Z. Yuan, M.G. Suh, K. Vahala, Nat. Commun. **11**, 1610 (2020)
82. K. Petermann, IEEE J. Quantum Electron. **15**, 566 (1979)
83. A.E. Siegman, Phys. Rev. A **39**, 1264 (1989)
84. Y.J. Cheng, C.G. Fanning, A.E. Siegman, Phys. Rev. Lett. **77**, 627 (1996)
85. M.V. Berry, J. Mod. Opt. **50**, 63 (2003)
86. H.K. Lau, A.A. Clerk, Nat. Commun. **9**, 4320 (2018)
87. M. Zhang, W. Sweeney, C.W. Hsu, L. Yang, A.D. Stone, L. Jiang, Phys. Rev. Lett. **123**, 180501 (2019)
88. H. Rong, A. Liu, R. Jones, O. Cohen, D. Hak, R. Nicolaescu, A. Fang, M. Paniccla, Nature **433**, 292 (2005)
89. W.D. Heiss, H.L. Harney, Eur. Phys. J. D **17**, 149 (2001)
90. B. Peng, Ş.K. Özdemir, M. Liertzer, W. Chen, J. Kramer, H. Yılmaz, J. Wiersig, S. Rotter, L. Yang, Proc. Natl. Acad. Sci. **113**, 6845 (2016)
91. S. Longhi, L. Feng, Photonics Res. **5**, B1 (2017)
92. L. He, Ş.K. Özdemir, L. Yang, Laser Photon. Rev. **7**, 60 (2013)
93. L. Yang, T. Carmon, B. Min, S.M. Spillane, K.J. Vahala, Cit. Appl. Phys. Lett **86**, 91114 (2005)
94. L. Yang, D.K. Armani, K.J. Vahala, Appl. Phys. Lett. **83**, 825 (2003)
95. P. Miao, Z. Zhang, J. Sun, W. Walasik, S. Longhi, N.M. Litchinitser, L. Feng, Science **353**, 464 (2016)
96. H.Z. Chen, T. Liu, H.Y. Luan, R.J. Liu, X.Y. Wang, X.F. Zhu, Y.B. Li, Z.M. Gu, S.J. Liang, H. Gao, L. Lu, L. Ge, S. Zhang, J. Zhu, R.M. Ma, Nat. Phys. **16**, 571 (2020)
97. A.L. Schawlow, C.H. Townes, Phys. Rev. **112**, 1940 (1958)
98. M. Asada, A. Kameyama, Y. Suematse, IEEE J. Quantum Electron. **20**, 745 (1984)
99. H. Hodaei, M.A. Miri, M. Heinrich, D.N. Christodoulides, M. Khajavikhan, Science **346**, 975 (2014)
100. L. Feng, Z.J. Wong, R.M. Ma, Y. Wang, X. Zhang, Science **346**, 972 (2014)
101. M. Brandstetter, M. Liertzer, C. Deutsch, P. Klang, J. Schöberl, H.E. Türeci, G. Strasser, K. Unterrainer, S. Rotter, Nat. Commun. **5**, 4034 (2014)
102. A. Pick, B. Zhen, O.D. Miller, C.W. Hsu, F. Hernandez, A.W. Rodriguez, M. Soljacic, S.G. Johnson, Opt. Express **25**, 12325 (2017)
103. S.Y. Lee, J.W. Ryu, J.B. Shim, S.B. Lee, S.W. Kim, K. An, Phys. Rev. A **78**, 015805 (2008)
104. J. Zhang, B. Peng, Ş.K. Özdemir, K. Pichler, D.O. Krimer, G. Zhao, F. Nori, Y. Liu, S. Rotter, L. Yang, Nat. Photonics **12**, 479 (2018)
105. M. Aspelmeyer, T.J. Kippenberg, F. Marquardt, Rev. Mod. Phys. **86**, 1391 (2014)
106. I.S. Grudinin, H. Lee, O. Painter, K.J. Vahala, Phys. Rev. Lett. **104**, 083901 (2010)
107. Y.D. Chong, L. Ge, A.D. Stone, Phys. Rev. Lett. **106**, 093902 (2011)
108. Z.J. Wong, Y.L. Xu, J. Kim, K. O'Brien, Y. Wang, L. Feng, X. Zhang, Nat. Photonics **10**, 796 (2016)

109. W.R. Sweeney, C.W. Hsu, S. Rotter, A.D. Stone, Phys. Rev. Lett. **122**, 093901 (2019)
110. H. Jing, S.K. Özdemir, X.Y. Lü, J. Zhang, L. Yang, F. Nori, Phys. Rev. Lett. **113**, 053604 (2014)
111. Q. Zhong, S.K. Özdemir, A. Eisfeld, A. Metelmann, R. El-Ganainy, Phys. Rev. Appl. **13**, 14070 (2020)
112. M. Liertzer, L. Ge, A. Cerjan, A.D. Stone, H.E. Türeci, S. Rotter, Phys. Rev. Lett. **108**, 173901 (2012)
113. Y. Li, X. Jiang, G. Zhao, L. Yang, ArXiv Prepr. ArXiv1809.04878 (2018)
114. T. J. A. Kippenberg, Nonlinear Optics in Ultra-High-Q Whispering-Gallery Optical Microcavities, California Institute of Technology, 2004
115. G. Lin, A. Coillet, Y.K. Chembo, Adv. Opt. Photonics **9**, 828 (2017)
116. L. Ge, Y.D. Chong, A.D. Stone, Phys. Rev. A **85**, 23802 (2012)
117. C. Shi, M. Dubois, Y. Chen, L. Cheng, H. Ramezani, Y. Wang, X. Zhang, Nat. Commun. **7**, 11110 (2016)
118. L. Feng, Y.L. Xu, W.S. Fegadolli, M.H. Lu, J.E.B. Oliveira, V.R. Almeida, Y.F. Chen, A. Scherer, Nat. Mater. **12**, 108 (2013)
119. A. Regensburger, C. Bersch, M.A. Miri, G. Onishchukov, D.N. Christodoulides, U. Peschel, Nature **488**, 167 (2012)
120. Z. Lin, H. Ramezani, T. Eichelkraut, T. Kottos, H. Cao, D.N. Christodoulides, Phys. Rev. Lett. **106**, 213901 (2011)
121. M. Sarısaman, Phys. Rev. A **95**, 13806 (2017)
122. Y. Shen, X.H. Deng, L. Chen, Opt. Express **22**, 19440 (2014)
123. J. Gear, F. Liu, S.T. Chu, S. Rotter, J. Li, Phys. Rev. A **91**, 33825 (2015)
124. K. Ding, Z.Q. Zhang, C.T. Chan, Phys. Rev. B **92**, 235310 (2015)
125. Y. Huang, Y. Shen, C. Min, S. Fan, G. Veronis, Nanophotonics **6**, 977 (2017)
126. W. Chen, D. Leykam, Y.D. Chong, L. Yang, MRS Bull. **43**, 443 (2018)
127. D. Jalas, A. Petrov, M. Eich, W. Freude, S. Fan, Z. Yu, R. Baets, M. Popović, A. Melloni, J.D. Joannopoulos, M. Vanwolleghem, C.R. Doerr, H. Renner, Nat. Photonics **7**, 579 (2013)
128. L. Fan, J. Wang, L.T. Varghese, H. Shen, B. Niu, Y. Xuan, A.M. Weiner, M. Qi, Science **335**, 447 (2012)
129. L. Bi, J. Hu, P. Jiang, D.H. Kim, G.F. Dionne, L.C. Kimerling, C.A. Ross, Nat. Photonics **5**, 758 (2011)
130. C.H. Dong, Z. Shen, C.L. Zou, Y.L. Zhang, W. Fu, G.C. Guo, Nat. Commun. **6**, 6193 (2015)
131. L.D. Tzuang, K. Fang, P. Nussenzveig, S. Fan, M. Lipson, Nat. Photonics **8**, 701 (2014)
132. J. Kim, M.C. Kuzyk, K. Han, H. Wang, G. Bahl, Nat. Phys. **11**, 275 (2015)
133. M.S. Kang, A. Butsch, P.S.J. Russell, Nat. Photonics **5**, 549 (2011)
134. Z. Yu, S. Fan, Nat. Photonics **3**, 91 (2009)
135. H. Lira, Z. Yu, S. Fan, M. Lipson, Phys. Rev. Lett. **109**, 033901 (2012)
136. J. Zhang, B. Peng, Ş.K. Özdemir, Y.X. Liu, H. Jing, X.Y. Lü, Y.L. Liu, L. Yang, F. Nori, Phys. Rev. B **92**, 115407 (2015)
137. X. Zhou, Y.D. Chong, Opt. Express **24**, 6916 (2016)
138. L. Chang, X. Jiang, S. Hua, C. Yang, J. Wen, L. Jiang, G. Li, G. Wang, M. Xiao, Nat. Photonics **8**, 524 (2014)
139. L. Shao, W. Mao, S. Maity, N. Sinclair, Y. Hu, L. Yang, M. Lončar, Nat. Electron. **3**, 267 (2020)
140. L. Vestergaard Hau, S. E. Harris, Z. Dutton, C. H. Behroozi, Nature **397**, 594 (1999)
141. B. Wu, J.F. Hulbert, E.J. Lunt, K. Hurd, A.R. Hawkins, H. Schmidt, Nat. Photonics **4**, 776 (2010)
142. A. Marx, X. Zhou, H. Huebl, A. Schliesser, R. Gross, F. Hocke, T.J. Kippenberg, Nat. Phys. **9**, 179 (2013)
143. A.I. Lvovsky, B.C. Sanders, W. Tittel, Nat. Photonics **3**, 706 (2009)
144. P. Project, A. Geodesy, Q. City, P. Project, E. Science, Q. City, K. E. Y. Words, Phys. Rev. Lett. **64**, 1107 (1990)
145. K.-J. Boller, A. Imamoğlu, S.E. Harris, Phys. Rev. Lett. **66**, 2593 (1991)

146. Q. Xu, P. Dong, M. Lipson, Nat. Phys. **3**, 406 (2007)
147. N. Liu, L. Langguth, T. Weiss, J. Kästel, M. Fleischhauer, T. Pfau, H. Giessen, Nat. Mater. **8**, 758 (2009)
148. N. Papasimakis, V.A. Fedotov, S.L. Prosvirnin, N.I. Zheludev, Phys. Rev. Lett. **101**, 253903 (2008)
149. Y.-C. Liu, B.-B. Li, Y.-F. Xiao, Nanophotonics **6**, 789 (2017)
150. M.F. Limonov, M.V. Rybin, A.N. Poddubny, Y.S. Kivshar, Nat. Photonics **11**, 543 (2017)
151. B. Peng, Ş.K. Özdemir, W. Chen, F. Nori, L. Yang, Nat. Commun. **5**, 5082 (2014)
152. D. U. Smith, H. Chang, K. A. Fuller, A. T. Rosenberger, R. W. Boyd, Phys. Rev. A - At. Mol. Opt. Phys. **69**, 063804 (2004)
153. C. Wang, X. Jiang, W. R. Sweeney, C. W. Hsu, Y. Liu, G. Zhao, B. Peng, M. Zhang, L. Jiang, A. D. Stone, L. Yang, Proc. Natl. Acad. Sci. **118**, e2012982118 (2021)
154. A.H. Safavi-Naeini, T.P.M. Alegre, J. Chan, M. Eichenfield, M. Winger, Q. Lin, J.T. Hill, D.E. Chang, O. Painter, Nature **472**, 69 (2011)
155. S. Weis, R. Rivière, S. Deléglise, E. Gavartin, O. Arcizet, A. Schliesser, T.J. Kippenberg, Science **330**, 1520 (2010)
156. C. Dong, V. Fiore, M.C. Kuzyk, H. Wang, Science **338**, 1609 (2012)
157. C. Wang, X. Jiang, G. Zhao, M. Zhang, C.W. Hsu, B. Peng, A.D. Stone, L. Jiang, L. Yang, Nat. Phys. **16**, 334 (2020)
158. H. Lü, C. Wang, L. Yang, H. Jing, Phys. Rev. Appl. **10**, 14006 (2018)
159. W. Li, Y. Jiang, C. Li, H. Song, Sci. Rep. **6**, 31095 (2016)
160. X.Y. Zhang, Y.Q. Guo, P. Pei, X.X. Yi, Phys. Rev. A **95**, 63825 (2017)
161. B. Wang, Z.-X. Liu, C. Kong, H. Xiong, Y. Wu, Opt. Express **27**, 8069 (2019)
162. O. Latinne, N.J. Kylstra, M. Dörr, J. Purvis, M. Terao-Dunseath, C.J. Joachain, P.G. Burke, C.J. Noble, Phys. Rev. Lett. **74**, 46 (1995)
163. R. Lefebvre, O. Atabek, M. Šindelka, N. Moiseyev, Phys. Rev. Lett. **103**, 123003 (2009)
164. O. Atabek, R. Lefebvre, M. Lepers, A. Jaouadi, O. Dulieu, V. Kokoouline, Phys. Rev. Lett. **106**, 173002 (2011)
165. A.A. Mailybaev, O.N. Kirillov, A.P. Seyranian, Phys. Rev. A **72**, 014104 (2005)
166. M. Born, Zeitschrift Für Phys. **40**, 167 (1927)
167. T. Kato, J. Phys. Soc. Japan **5**, 435 (1950)
168. B. Simon, Phys. Rev. Lett. **51**, 2167 (1983)
169. M.V. Berry, Proc. R. Soc. London. A. **392**, 45 (1984)
170. M.V. Berry, J. Phys. A: Gen. Phys. **18**, 15 (1985)
171. J.H. Hannay, J. Phys. A: Math. Gen. **18**, 221 (1985)
172. W.D. Heiss, J. Phys. A: Math. Theor. **45**, 444016 (2012)
173. C. Dembowski, H.D. Gräf, H.L. Harney, A. Heine, W.D. Heiss, H. Rehfeld, A. Richter, Phys. Rev. Lett. **86**, 787 (2001)
174. T. Gao, E. Estrecho, K.Y. Bliokh, T.C.H. Liew, M.D. Fraser, S. Brodbeck, M. Kamp, C. Schneider, S. Höfling, Y. Yamamoto, F. Nori, Y.S. Kivshar, A.G. Truscott, R.G. Dall, E.A. Ostrovskaya, Nature **526**, 554 (2015)
175. R. Uzdin, A. Mailybaev, N. Moiseyev, J. Phys. A: Math. Theor. **44**, 435302 (2011)
176. M.V. Berry, R. Uzdin, J. Phys. A: Math. Theor. **44**, 435303 (2011)
177. T.J. Milburn, J. Doppler, C.A. Holmes, S. Portolan, S. Rotter, P. Rabl, Phys. Rev. A **92**, 52124 (2015)
178. J. Doppler, A.A. Mailybaev, J. Böhm, U. Kuhl, A. Girschik, F. Libisch, T.J. Milburn, P. Rabl, N. Moiseyev, S. Rotter, Nature **537**, 76 (2016)
179. H. Xu, D. Mason, L. Jiang, J.G.E. Harris, Nature **537**, 80 (2016)
180. A.U. Hassan, B. Zhen, M. Soljačić, M. Khajavikhan, D.N. Christodoulides, Phys. Rev. Lett. **118**, 093002 (2017)
181. Y. Choi, C. Hahn, J.W. Yoon, S.H. Song, P. Berini, Nat. Commun. **8**, 14154 (2017)
182. X.L. Zhang, C.T. Chan, Phys. Rev. A **98**, 033810 (2018)
183. H. Wang, S. Assawaworrarit, S. Fan, Opt. Lett. **44**, 638 (2019)

184. X. L. Zhang, C. T. Chan, Dynamically encircling exceptional points in a three-mode waveguide system. Commun. Phys. **2**, 1 (2019)
185. J.W. Yoon, Y. Choi, C. Hahn, G. Kim, S.H. Song, K.Y. Yang, J.Y. Lee, Y. Kim, C.S. Lee, J.K. Shin, H.S. Lee, P. Berini, Nature **562**, 86 (2018)
186. D. Bonneau, J. W. Silverstone, M. G. Thompson, in *Silicon Quantum Photonics III*, edited by L. Pavesi, D. J. Lockwood (Springer, Berlin, 2016), p. 41
187. F. Klauck, L. Teuber, M. Ornigotti, M. Heinrich, S. Scheel, A. Szameit, Nat. Photonics **13**, 883 (2019)
188. C.K. Hong, Z.Y. Ou, L. Mandel, Phys. Rev. Lett. **59**, 2044 (1987)
189. S. Scheel, A. Szameit, EPL **122**, 34001 (2018)
190. Y.N. Joglekar, A.K. Harter, Photonics Res. **6**, A51 (2018)
191. J. Li, A.K. Harter, J. Liu, L. de Melo, Y.N. Joglekar, L. Luo, Nat. Commun. **10**, 855 (2019)
192. Y.X. Wang, A.A. Clerk, Phys. Rev. A **99**, 063834 (2019)
193. S. Dey, A. Raj, S.K. Goyal, Phys. Lett. A **383**, 125931 (2019)
194. B. Gardas, S. Deffner, A. Saxena, Phys. Rev. A **94**, 040101 (2016)
195. S. Chakraborty, A.K. Sarma, Phys. Rev. A **100**, 063846 (2019)
196. K. Kawabata, Y. Ashida, M. Ueda, Phys. Rev. Lett. **119**, 190401 (2017)
197. C. Chen, L. Jin, R. B. Liu, New J. Phys. **21**, (2019)
198. J. Peřina, A. Lukš, J.K. Kalaga, W. Leoński, A. Miranowicz, Phys. Rev. A **100**, 53820 (2019)
199. B. Jaramillo Ávila, C. Ventura-Velázquez, R. de J. León-Montiel, Y. N. Joglekar, B. M. Rodríguez-Lara, Sci. Rep. **10**, 1 (2020)
200. Y.-L. Liu, R. Wu, J. Zhang, Ş.K. Özdemir, L. Yang, F. Nori, Y.X. Liu, Phys. Rev. A **95**, 013843 (2017)
201. S. Longhi, Phys. Rev. Lett. **105**, 013903 (2010)
202. L. Lu, J.D. Joannopoulos, M. Soljačić, Nat. Photonics **8**, 821 (2014)
203. T. Ozawa, H.M. Price, A. Amo, N. Goldman, M. Hafezi, L. Lu, M.C. Rechtsman, D. Schuster, J. Simon, O. Zilberberg, I. Carusotto, Rev. Mod. Phys. **91**, 15006 (2019)
204. M.Z. Hasan, C.L. Kane, Rev. Mod. Phys. **82**, 3045 (2010)
205. M.C. Rechtsman, J.M. Zeuner, Y. Plotnik, Y. Lumer, D. Podolsky, F. Dreisow, S. Nolte, M. Segev, A. Szameit, Nature **496**, 196 (2013)
206. M. Hafezi, S. Mittal, J. Fan, A. Migdall, J.M. Taylor, Nat. Photonics **7**, 1001 (2013)
207. S. Weimann, M. Kremer, Y. Plotnik, Y. Lumer, S. Nolte, K.G. Makris, M. Segev, M.C. Rechtsman, A. Szameit, Nat. Mater. **16**, 433 (2017)
208. C. Poli, M. Bellec, U. Kuhl, F. Mortessagne, H. Schomerus, Selective enhancement of topologically induced interface states in a dielectric resonator chain. Nat. Commun. **6**, 1 (2015)
209. M. Pan, H. Zhao, P. Miao, S. Longhi, L. Feng, Photonic zero mode in a non-Hermitian photonic lattice. Nat. Commun. **9**, 1 (2018)
210. Z. Gong, Y. Ashida, K. Kawabata, K. Takasan, S. Higashikawa, M. Ueda, Phys. Rev. X **8**, 31079 (2018)
211. T.E. Lee, Phys. Rev. Lett. **116**, 1 (2016)
212. D. Leykam, K.Y. Bliokh, C. Huang, Y.D. Chong, F. Nori, Phys. Rev. Lett. **118**, 040401 (2017)
213. H. Shen, B. Zhen, L. Fu, Phys. Rev. Lett. **120**, 146402 (2018)
214. P. St-Jean, V. Goblot, E. Galopin, A. Lemaître, T. Ozawa, L. Le Gratiet, I. Sagnes, J. Bloch, A. Amo, Nat. Photonics **11**, 651 (2017)
215. B. Bahari, A. Ndao, F. Valini, A. El Amili, Y. Fainman, B. Kante, Science **358**, 636 (2018)
216. M. Parto, S. Wittek, H. Hodaei, G. Harari, M.A. Bandres, J. Ren, M.C. Rechtsman, M. Segev, D.N. Christodoulides, M. Khajavikhan, Phys. Rev. Lett. **120**, 113901 (2018)
217. Y. Ota, R. Katsumi, K. Watanabe, S. Iwamoto, Y. Arakawa, Commun. Phys. **1**, 86 (2018)
218. M. A. Bandres, S. Wittek, G. Harari, M. Parto, J. Ren, M. Segev, D. N. Christodoulides, M. Khajavikhan, Science **359**, eaar4005 (2018)
219. C. Han, M. Lee, S. Callard, C. Seassal, H. Jeon, Light Sci. Appl. **8**, 1 (2019)
220. H. Zhao, P. Miao, M. H. Teimourpour, S. Malzard, R. El-Ganainy, H. Schomerus, L. Feng, Topological hybrid silicon microlasers. Nat. Commun. **9**, 1 (2018)

221. S. Longhi, Ann. Phys. **530**, 1800023 (2018)
222. S. Longhi, Opt. Lett. **44**, 1190 (2019)
223. L. Pilozzi, C. Conti, Opt. Lett. **42**, 5174 (2017)
224. H. Zhao, X. Qiao, T. Wu, B. Midya, S. Longhi, L. Feng, Science **365**, 1163 (2019)

# Chapter 8
# Topological Photonics with Microring Lattices

**Shirin Afzal, Tyler James Zimmerling, and Vien Van**

**Abstract** Topological photonic insulators have attracted considerable attention due to their unique ability to transport light via topologically-protected edge states that are immune to defect scattering. Among the various potential applications, this property can be exploited to engineer robust photonic devices that are insensitive to fabrication imperfections. This chapter reviews the key concepts of topological insulator systems in one and two dimensions and their realizations using coupled microring resonator lattices. Particular emphasis will be placed on the treatment of microring lattices as periodically-driven systems and their Floquet topological characteristics. Experimental efforts in realizing Floquet microring lattices and demonstrating anomalous Floquet insulator behaviors will also be reviewed.

## 8.1 Introduction

The discovery of the quantum Hall effect [1] revealed the existence of a new class of electrical insulators, called topological insulators, whose electrical properties depend on the topological properties of their energy bands. Kohmoto and Thouless were the first to relate the quantized Hall conductance to one of the topological invariants, the Chern number, which also directly corresponds to the number of edge states that can exist at the sample boundary [2, 3]. In 2008, Haldane and Raghu formally extended topological insulator concepts to photonic systems by proposing a photonic crystal structure which could theoretically support topological edge states at photonic frequencies [4, 5]. In 2009, Wang et al. reported the first experimental observation

S. Afzal · T. J. Zimmerling · V. Van (✉)
Department of Electrical and Computer Engineering, University of Alberta,
Edmonton, AB T6G 2V4, Canada
e-mail: vien@ualberta.ca

S. Afzal
e-mail: safzal1@ualberta.ca

T. J. Zimmerling
e-mail: tzimmerl@ualberta.ca

 
D. J. Lockwood and L. Pavesi (eds.), *Silicon Photonics IV*,
Topics in Applied Physics 139, https://doi.org/10.1007/978-3-030-68222-4_8

of topological behavior at microwave frequencies in a 2D periodic lattice of gyromagnetic ferrite rods [6]. Their work has since initiated many other theoretical and experimental studies exploring topological behaviors in various bosonic systems, spanning a broad frequency range from acoustics to photonics [7–17].

A distinguishing feature of topological insulators is the existence of topologically-protected edge states at the material boundaries that are immune to defect scattering. This has led to the proposal of photonic devices based on topological photonic insulators (TPIs) that can transport light via edge modes that are insensitive to fabrication imperfections [18, 19]. In addition, the unique properties of TPIs provide novel ways to manipulate and control electromagnetic waves over a wide range of frequencies. Some of the potential applications of topological photonics are robust optical delay lines, optical isolators, and topological lasers [13, 20–25].

Various types of photonic lattices have been used to achieve topological behaviors, including metamaterials, coupled waveguide arrays, microring resonators, quasicrystals, and photonic crystals [11, 15, 26–29]. Among these structures, microring lattices are particularly suitable for implementing 1D and 2D TPIs since it is relatively easy to vary both the coupling strengths between adjacent resonators and the resonant frequencies (to realize on-site potentials) of individual elements of the lattice. Additionally, as we will show, these structures can be used to emulate periodically-driven quantum systems. This allows us to realize Floquet insulators, which are bandgap structures governed by periodically-varying Hamiltonians [14, 30, 31]. Floquet TPIs can exhibit nontrivial topological behaviors, such as anomalous Floquet insulator behavior, not present in static systems.

In this chapter, we will first introduce the key concepts of a topological insulator by examining 1D topological systems based on coupled microrings in Sect. 8.2. We will then show how a microring array can be described as a topological system with a periodically-varying Hamiltonian. Section 8.3 will be devoted to the treatment of 2D microring Floquet TPIs, with a review of our experimental efforts in realizing these systems in silicon photonics. We conclude the chapter in Sect. 8.4 with an outlook on future research directions in topological photonics.

## 8.2 Topological Photonic Insulators in 1D Microring Lattices

### *8.2.1 1D Microring Lattice as an SSH Topological Insulator*

The simplest topological insulator is the Su-Schrieffer-Heeger (SSH) system, which describes electrons hopping along a 1D lattice with alternating hopping amplitudes [32]. We can emulate such a system by a 1D chain of microrings with identical resonant frequencies and alternating energy coupling rates $\mu_a$ and $\mu_b$, as shown in Fig. 8.1a. Each unit cell in the lattice consists of two microrings, labeled A and B, with lattice constant $\Lambda \sim 4R$, where $R$ is the microring radius. Let $\tilde{a}_n(t)$ and $\tilde{b}_{n+1}(t)$

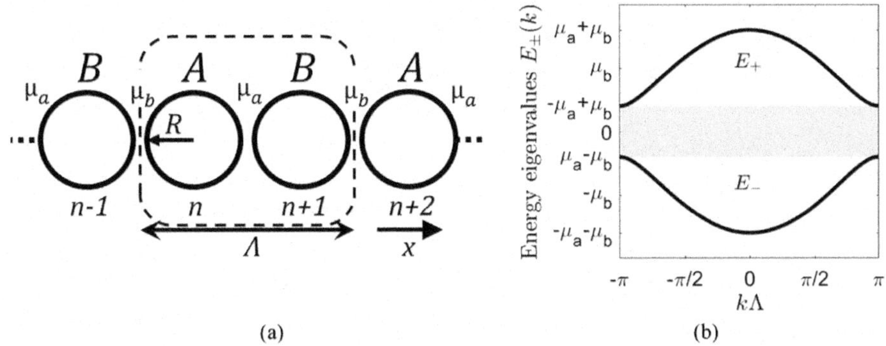


**Fig. 8.1** **a** Schematic of a 1D SSH microring lattice with alternating coupling strengths $\mu_a$ and $\mu_b$ and **b** its energy band diagram, labeled for $\mu_a < \mu_b$

denote the instantaneous energy amplitudes in resonators $n$ and $n+1$, respectively. The coupled mode equations for each unit cell are

$$\frac{d\tilde{a}_n}{dt} = -i\omega_0 \tilde{a}_n + i\mu_b \tilde{b}_{n-1} + i\mu_a \tilde{b}_{n+1} \quad (8.1)$$

$$\frac{d\tilde{b}_{n+1}}{dt} = -i\omega_0 \tilde{b}_{n+1} + i\mu_a \tilde{a}_n + i\mu_b \tilde{a}_{n+2} \quad (8.2)$$

where $\omega_0$ is the resonant frequency of the microrings. Letting $\tilde{a}_n(t) = a_n(t)e^{-i\omega_0 t}$ and $\tilde{b}_{n+1}(t) = b_{n+1}(t)e^{-i\omega_0 t}$, we can rewrite the above equations in the rotating frame of reference as

$$\frac{da_n}{dt} = i\mu_b b_{n-1} + i\mu_a b_{n+1} \quad (8.3)$$

$$\frac{db_{n+1}}{dt} = i\mu_a a_n + i\mu_b a_{n+2} \quad (8.4)$$

Since the lattice is spatially periodic (along the $x$-direction in Fig. 8.1a), we can apply Bloch theorem to write $a_{n+2} = a_n e^{ik\Lambda}$ and $b_{n-1} = b_{n+1}e^{-ik\Lambda}$. Substituting these expressions into (8.3) and (8.4), we get

$$i\frac{d}{dt}\begin{bmatrix} a_n \\ b_{n+1} \end{bmatrix} = -\begin{bmatrix} 0 & \mu_a + \mu_b e^{-ik\Lambda} \\ \mu_a + \mu_b e^{ik\Lambda} & 0 \end{bmatrix}\begin{bmatrix} a_n \\ b_{n+1} \end{bmatrix} \quad (8.5)$$

We recognize the above equation as having the form of the Schrodinger equation:

$$i\frac{d}{dt}|\psi(k)\rangle = H(k)|\psi(k)\rangle \quad (8.6)$$

where $H$ is the Bloch Hamiltonian of the periodic lattice of the form

$$H(k) = \begin{bmatrix} 0 & h(k) \\ h^*(k) & 0 \end{bmatrix} \tag{8.7}$$

with

$$h(k) = h_x(k) + ih_y(k) = -[\mu_a + \mu_b \cos(k\Lambda)] + i\mu_b \sin(k\Lambda) \tag{8.8}$$

We note that the SSH system has chiral symmetry, since its Hamiltonian satisfies the relation $\Gamma H \Gamma = -H$ for $\Gamma = \begin{bmatrix} 1 & 0 \\ 0 & -1 \end{bmatrix}$ (i.e., the Pauli matrix $\sigma_z$). Chiral symmetry restricts the Hamiltonian of the system to being anti-diagonal.

The energies of the Bloch states are given by the eigenvalues of the Hamiltonian, which for the 1D microring chain are

$$E_\pm(k) = \pm|h(k)| = \pm\sqrt{\mu_a^2 + \mu_b^2 + 2\mu_a\mu_b \cos(k\Lambda)} \tag{8.9}$$

The energy diagram of the lattice thus consists of two bands, separated by a bandgap $\Delta E = 2|\mu_a - \mu_b|$ (see Fig. 8.1b). Each energy band is periodic in $k$ with periodicity of $2\pi/\Lambda$. If we define $\mathbf{h}$ in (8.8) as a 2D vector in the $h_x - h_y$ plane, then as we sweep $k$ from $-\pi/\Lambda$ to $\pi/\Lambda$, the tip of the $\mathbf{h}(k)$ vector traces out a circle of radius $\mu_b$ centered at $(-\mu_a, 0)$ in the $h_x - h_y$ plane. Figure 8.2 depicts this trajectory of $\mathbf{h}$ for the three cases, $\mu_a > \mu_b$, $\mu_a = \mu_b$ and $\mu_a < \mu_b$. Since the energies of the two eigenstates are equal to $\pm|\mathbf{h}|$, it is clear that the bandgap closes only when $\mu_a = \mu_b$ since the corresponding circle passes through the origin in this case. In other words, the lattice behaves as an insulator for $\mu_a \neq \mu_b$. We also note that the circle excludes the origin for the case $\mu_a > \mu_b$ but includes the origin for the case $\mu_a < \mu_b$. We can specify this difference by defining a winding number $W$, which is an integer indicating the number of times the tip of the vector $\mathbf{h}$ winds around the origin as $k$ is swept over one Brillouin zone. We thus have

$$W = \begin{cases} 0, & \mu_a > \mu_b \\ 1, & \mu_a < \mu_b \end{cases} \tag{8.10}$$

Suppose we now subject an insulating microring lattice to small adiabatic changes in the coupling rates $\mu_a$ and $\mu_b$ such that the bandgap always remains open. The energy bands may become deformed but the energies of the two eigenstates never vanish. The path which the vector $\mathbf{h}$ traces out in the $h_x - h_y$ plane may no longer be a circle but it is still a closed loop. Furthermore, the condition that the circle does not cross the origin ($|\mathbf{h}| \neq 0$) implies that the winding number of the loop remains unchanged under these adiabatic variations. We call the winding number a topological invariant of the lattice.

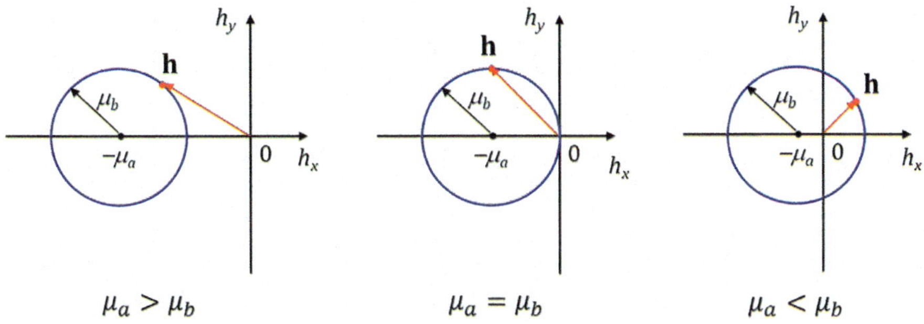


**Fig. 8.2** Trajectory of the tip of the vector $\mathbf{h}(k)$ in the $h_x - h_y$ plane as $k$ is swept over one Brillouin zone

In general, for a 1D chiral lattice with Hamiltonian of the form in (8.7), the winding number can be computed from [33]

$$w[h] = \frac{1}{2\pi i} \int_{\mathrm{BZ}} \frac{\mathrm{d}}{\mathrm{d}k} \left( \ln \frac{h}{|h|} \right) \mathrm{d}k \tag{8.11}$$

The lattice is a normal insulator if its winding number is zero and a topological insulator if its winding number is nonzero. As discussed above, the winding number remains unchanged under continuous adiabatic deformation of the lattice Hamiltonian. It also cannot change abruptly at an interface between two different insulators. A consequence of this is that at the edge of a truncated topological lattice, or at the interface between a topological insulator and a normal insulator, the bandgap must close, giving rise to an edge mode. It can be shown that the number of edge modes that can exist at a truncated end of a topological insulator is equal to its winding number. This relationship between the number of edge modes and the topological invariant of the bulk lattice is called the bulk-edge correspondence. This allows us to predict the behavior at the edge of an insulator based solely on the topological property of the bulk. Furthermore, the edge modes are said to be topologically protected since they depend only on the topological invariant of the bulk lattice and are unaffected by local variations at the interface. Thus, fabrication defects or other variations at the surface of the TPI do not affect the edge modes. This property of topological insulators, called topological protection, can be exploited to realize photonic devices that are robust to fabrication imperfections [13].

For the 1D microring lattice, we can determine the field distributions of the edge modes by considering an array with $\mu_a < \mu_b$ having a finite length of $N$ unit cells. Using (8.3) and (8.4), we obtain the matrix equation for the eigenstates of the array

$$-\begin{bmatrix} 0 & \mu_a & & & & & \\ \mu_a & 0 & \mu_b & & & & \\ & \mu_b & 0 & \mu_a & & & \\ & & \mu_a & 0 & \mu_b & & \\ & & & \ddots & \ddots & \ddots & \\ & & & & \mu_b & 0 & \mu_a \\ & & & & & \mu_a & 0 \end{bmatrix} \begin{bmatrix} a_1 \\ b_2 \\ a_3 \\ b_4 \\ \vdots \\ a_{2N-1} \\ b_{2N} \end{bmatrix} = E \begin{bmatrix} a_1 \\ b_2 \\ a_3 \\ b_4 \\ \vdots \\ a_{2N-1} \\ b_{2N} \end{bmatrix} \tag{8.12}$$

Since the edge modes exist in the bulk bandgap, they have energies close to zero. Setting $E = 0$ in the above equation, we solve for the amplitudes $a_n$ and $b_{n+1}$ to get [32]

$$\begin{aligned} a_{n+2} &= -\frac{\mu_a}{\mu_b} a_n, \qquad n = 1, 3, 5, \ldots, (2N-3) \\ a_{2N-1} &= -\frac{E}{\mu_a} b_{2N} \approx 0 \end{aligned} \tag{8.13}$$

$$\begin{aligned} b_{n+3} &= -\frac{\mu_b}{\mu_a} b_{n+1}, \qquad n = 1, 3, 5, \ldots, (2N-3) \\ b_2 &= -\frac{E}{\mu_a} a_1 \approx 0 \end{aligned} \tag{8.14}$$

The above expressions show that for a finite lattice of length $N$, there are no eigenmodes with zero energy. However, for small non-zero $E$, there are two sets of solutions: a decaying alternating geometric sequence $\{a_n\}$ with boundary condition $a_{2N-1} \approx 0$ representing the edge mode on the left boundary, and a growing alternating geometric sequence $\{b_{n+1}\}$ with boundary condition $b_2 \approx 0$, which corresponds to the edge mode on the right boundary. As the array length $N \to \infty$, the sequence of amplitudes $\{|a_n|\}$ can be approximated by the equation

$$\frac{\mathrm{d}|a(x)|}{\mathrm{d}x} = -\frac{1}{\Lambda}\left(1 - \frac{\mu_a}{\mu_b}\right)|a(x)| \tag{8.15}$$

where $|a(x)|$ is the mode amplitude at position $x = (n-1)\Lambda/2$ ($n$ = odd integer). The solution of the above equation gives the field distribution for the left edge mode:

$$a_n = (\pm) a_1 e^{-(n-1)/\zeta_L}, \qquad n = 1, 3, 5, \ldots, 2N-1 \tag{8.16}$$

where $\zeta_L = 2/(1 - \mu_a/\mu_b)$ is the decay length (in unit of $\Lambda/2$) of the left edge mode and the $\pm$ sign indicates an alternating sequence. Similarly we can obtain the field distribution for the right edge mode as

$$b_{n+1} = (\pm) b_{2N} e^{-(2N-n-1)/\zeta_R}, \quad n = 1, 3, 5, \ldots, 2N-1 \tag{8.17}$$

where $\zeta_R = 2/(\mu_b/\mu_a - 1)$ is the decay length of the right edge mode.

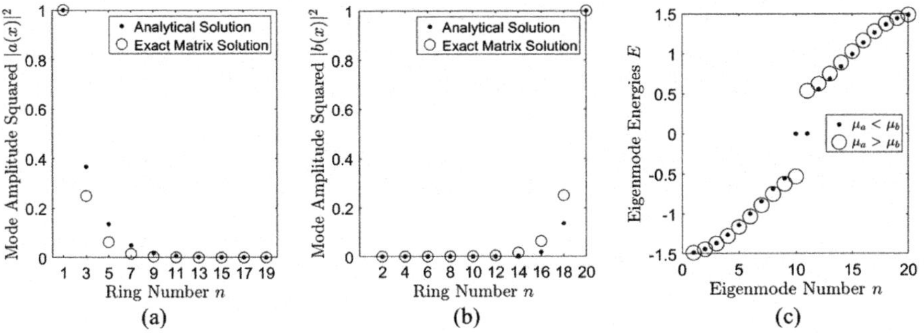


**Fig. 8.3** Light intensity distribution of the **a** left edge mode and **b** right edge mode of a finite 1D microring lattice ($N = 10$ unit cells) with $\mu_a/\mu_b = 0.5$. Dots are approximate solutions from (8.16) and (8.17) respectively; open circles are the exact solution from (8.12). **c** Energy spectra of a finite 1D microring lattice ($N = 10$ unit cells) with trivial (open circles) and nontrivial (dots) winding numbers

Figure 8.3a and b shows the plots of light intensity distributions $|a_n|^2$ and $|b_{n+1}|^2$, respectively, for a finite 1D microring lattice of length $N = 10$ unit cells with $\mu_a/\mu_b = 0.5$. The dots are the approximate solutions of (8.16) and (8.17) while the open circles are the exact solutions of the matrix equation in (8.12). It is clear that the solution for $|a_n|^2$ in Fig. 8.3a corresponds to the left edge mode while the solution for $|b_{n+1}|^2$ in Fig. 8.3b gives the right edge mode. Both modes have intensity decaying exponentially into the bulk lattice. Figure 8.3c shows the energy spectrum of the lattice for $\mu_a/\mu_b = 0.5$ and $\mu_a/\mu_b = 2$. For the case $\mu_a < \mu_b$, two gap modes can be seen appearing in the bulk bandgap near zero energy whereas for the case $\mu_a > \mu_b$, the bulk bandgap remains open with no gap modes. The two gap modes have equal energies but opposite sign, each corresponding to a hybrid combination of a left edge mode ($|a_n|^2$) and a right edge mode ($|b_{n+1}|^2$).

To summarize this section, we examined a simple 1D topological system that can be implemented using a chain of coupled microring resonators. We introduced several key concepts of topological insulators: topological invariants, the existence of edge modes in bulk bandgaps, and the bulk-edge correspondence linking the number of edge modes to the topological invariants of the bulk lattice. The 1D SSH microring lattice has been used to demonstrate topological lasers [21, 34, 35]. Other ways of realizing 1D microring TPIs include introducing periodically-varying coupling strengths along the microring chain and periodic resonant frequency detunings to emulate on-site potentials. A variant of the latter is a circular array of coupled microring resonators with periodic resonant frequency detunings [36], which can emulate Harper's equation describing electrons in a 2D lattice subjected to a perpendicular magnetic field, a problem intimately related to the quantum Hall effect and topological insulators.

The treatment of the 1D microring TPI in this section is based on the so-called energy coupling formalism [37], or tight binding approach in solid state physics. In this model the microring chain is described by a time-independent Bloch Hamiltonian so the system is considered to be static. In the next section, we treat the microring array using the more rigorous field coupling formalism, which tracks the field in each microring as it propagates around the resonator. This approach leads to the description of the microring array as a periodically-driven system, with a Hamiltonian which varies in the direction of light propagation in the microrings. This allows us to use coupled microrings to realize Floquet insulators, which can exhibit richer topological behaviors than static systems, as will be shown in Sect. 8.3.

### 8.2.2 1D Microring Lattice as a Floquet Topological Insulator

We consider again a 1D microring array with alternating field coupling coefficients $\kappa_a$ and $\kappa_b$, defined such that $\kappa_a^2$ and $\kappa_b^2$ give the fractions of power coupled between adjacent microrings. We assume that each microring supports only a clockwise or counterclockwise propagating wave, as shown in Fig. 8.4a. As in the SSH model, each unit cell in the lattice consists of two microrings, which are labeled A and B. As light circulates around each microring, it periodically couples with its neighbors via coupling coefficients $\kappa_a$ and $\kappa_b$. The lattice can thus be regarded as a periodically-varying system with a period equal to the microring circumference, $L = 2\pi R$. In order to derive its Hamiltonian, we transform the microring lattice into an equivalent array of coupled waveguides by cutting the microrings at the points indicated by the

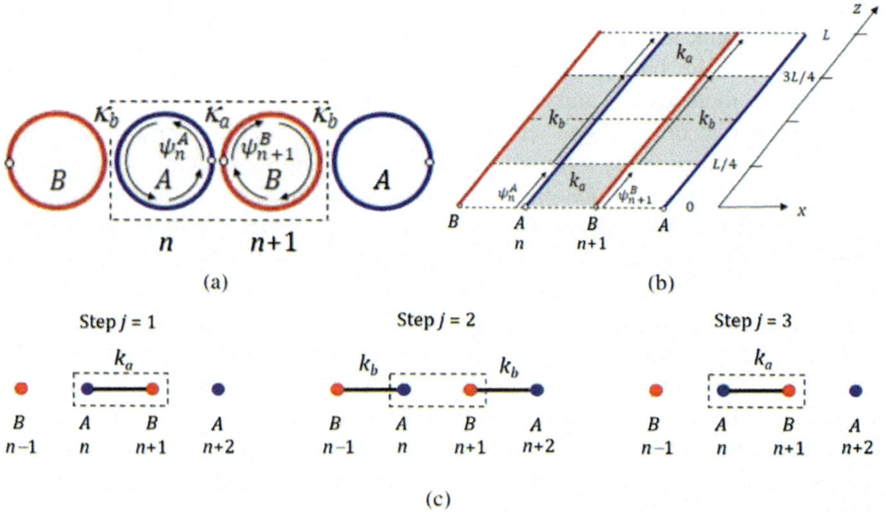


**Fig. 8.4** **a** Schematic of a 1D microring lattice as a periodically-driven system. **b** Equivalent coupled waveguide array of the microring lattice. **c** Hopping sequence showing 3 coupling steps in one period of evolution

small open circles in Fig. 8.4a and unrolling them into straight waveguides. In this manner, we obtain an array of periodically-coupled straight waveguides, as shown in Fig. 8.4b, with the direction of light propagation denoted by $z$ and each period equal to the microring circumference $L$. The shaded regions between waveguides indicate the regions of coupling between adjacent microrings. For simplicity, we have assumed that couplings between adjacent microrings are distributed over half a microring circumference with constant coupling strengths per unit length, $k_a$ and $k_b$, such that $\kappa_a = \sin(k_a L/2) = \sin\theta_a$ and $\kappa_b = \sin(k_b L/2) = \sin\theta_b$, where $\theta_a$ and $\theta_b$ are the coupling angles. This assumption simplifies the computation of the evolution operator of the system and represents only an adiabatic change, so it does not alter the topological characteristics of the lattice.

Over each period $L$, the fields $\psi_n^A$ and $\psi_{n+1}^B$ in resonators A and B of each unit cell execute three coupling steps, as shown in Fig. 8.4c. Using the Coupled Mode Theory, we can write the equations of motion for the fields $[\psi_n^A, \psi_{n+1}^B]$ along the direction of propagation $z$ over one period $L$ as:

$$\frac{d\psi_n^A}{dz} = i\beta\psi_n^A + ik_a(1)\psi_{n+1}^B + ik_b(2)\psi_{n-1}^B + ik_a(3)\psi_{n+1}^B \tag{8.18}$$

$$\frac{d\psi_{n+1}^B}{dz} = i\beta\psi_{n+1}^B + ik_a(1)\psi_n^A + ik_b(2)\psi_{n+2}^A + ik_a(3)\psi_n^A \tag{8.19}$$

where $\beta$ is the propagation constant and $k_{(a,b)}(j) = k_{(a,b)}$ in step $j$ but is 0 otherwise. Since the waveguide array is periodic in $x$ with period $\Lambda \sim 4R$, we apply the Bloch conditions $\psi_{n+2}^A = \psi_n^A e^{ik\Lambda}$ and $\psi_{n-1}^B = \psi_{n+1}^B e^{-ik\Lambda}$ to obtain

$$i\frac{d}{dz}|\psi(k,z)\rangle = [-\beta I + H_{\text{FB}}(k,z)]|\psi(k,z)\rangle \tag{8.20}$$

where $|\psi\rangle = [\psi^A, \psi^B]^T$, $I$ is the identity matrix and $H_{\text{FB}}$ is the Floquet-Bloch (FB) Hamiltonian given by

$$H_{\text{FB}}(k,z) = -\begin{bmatrix} 0 & k_a \\ k_a & 0 \end{bmatrix} \equiv H_1, \quad \text{for } 0 \le z < L/4 \text{ and } 3L/4 \le z < L \tag{8.21}$$

$$H_{\text{FB}}(k,z) = -\begin{bmatrix} 0 & k_b e^{-ik\Lambda} \\ k_b e^{ik\Lambda} & 0 \end{bmatrix} \equiv H_2, \quad \text{for } L/4 \le z < 3L/4 \tag{8.22}$$

We note that the FB Hamiltonian is periodic in $z$ with a period of $L$, i.e., $H_{\text{FB}}(k,z) = H_{\text{FB}}(k,z+L)$. In addition, it also satisfies the chiral symmetry condition,

$$\Gamma H_{\text{FB}}(k,z)\Gamma = -H_{\text{FB}}(k,-z) \tag{8.23}$$

with $\Gamma = \sigma_z$.

The solution of the Schrodinger equation in (8.20) can be expressed as

$$|\psi(k,z)\rangle = U(k,z)e^{i\beta z}|\psi(k,0)\rangle \tag{8.24}$$

where $U(k,z)$ is the evolution operator[1] given by

$$U(k,z) = \mathcal{T}e^{-i\int_0^z H_{\mathrm{FB}}(k,z')\mathrm{d}z'} \tag{8.25}$$

with $\mathcal{T}$ being the time-order operator.[2] Using the expressions for the FB Hamiltonian in (8.21) and (8.22), and the fact that the Hamiltonian is constant in each step, we obtain

$$U(k,z) = \begin{cases} U_1(k,z), & 0 \le z < L/4 \\ U_2(k,z-L/4)U_1(k,L/4), & L/4 \le z < 3L/4 \\ U_1(k,z-3L/4)U_2(k,L/2)U_1(k,L/4), & 3L/4 \le z < L \end{cases} \tag{8.26}$$

where

$$U_1(k,z) = e^{-iH_1 z} = \begin{bmatrix} \cos(k_a z) & i\sin(k_a z) \\ i\sin(k_a z) & \cos(k_a z) \end{bmatrix} \tag{8.27}$$

$$U_2(k,z) = e^{-iH_2 z} = \begin{bmatrix} \cos(k_b z) & i\sin(k_b z)e^{-ik\Lambda} \\ i\sin(k_b z)e^{ik\Lambda} & \cos(k_b z) \end{bmatrix} \tag{8.28}$$

Over one period $L$, the evolution operator is

$$U(k,L) = U_1(k,L/4)U_2(k,L/2)U_1(k,L/4) \equiv U_F(k) \tag{8.29}$$

which is called the Floquet operator. The Floquet operator provides a stroboscopic snapshot (at every interval $L$) of the state of the system. For the microring lattice, the Floquet operator can be expressed as

$$U_F(k) = \begin{bmatrix} u & iv \\ iv^* & u \end{bmatrix} \tag{8.30}$$

$$u = \tau_a\tau_b - \kappa_a\kappa_b\cos(k\Lambda)$$

$$v = \kappa_a\tau_b + \tau_a\kappa_b\cos(k\Lambda) - i\kappa_b\sin(k\Lambda)$$

where $\tau_a = \cos\theta_a$ and $\tau_b = \cos\theta_b$.

---

[1]For simplicity we have excluded the phase factor $e^{i\beta z}$ from the evolution operator since it does not affect the topological behavior of the system.

[2]The time-order operator is defined as:

$$\mathcal{T}e^{-i\int H(z')\mathrm{d}z'} = \lim_{\delta z\to 0} e^{-iH(z)\delta z}e^{-iH(z-\delta z)\delta z}e^{-iH(z-2\delta z)\delta z}\ldots e^{-iH(0)\delta z}$$

.

Since the evolution is periodic in $z$, we can apply the Floquet theorem to write

$$|\psi(k, z+L)\rangle = U_F(k)|\psi(k, z)\rangle = e^{-i\varepsilon L}|\psi(k, z)\rangle \tag{8.31}$$

where $\lambda = e^{-i\varepsilon L} = e^{-i\phi}$ is an eigenvalue of the Floquet operator and $\varepsilon$ is called a quasienergy of the lattice. The phase $\phi_n(k)$ of the $n$th eigenvalue $\lambda_n(k)$ thus gives the quasienergy band $\varepsilon_n(k)$, which is periodic in $k$ with periodicity of $2\pi/\Lambda$. Moreover, since we can add an integer multiple of $2\pi$ to the phase $\phi_n$ without changing the eigenvalue $\lambda_n$, we obtain an infinite number of quasienergy bands $\varepsilon_n(k) + 2m\pi/L$, $m \in \mathbb{Z}$, associated with each eigenvalue $\lambda_n$ of the Floquet operator.

For the Floquet operator in (8.30), the eigenvalues are

$$\lambda_\pm = u \pm i|v| = e^{\pm i\phi} \tag{8.32}$$

where $\phi = \cos^{-1}(u)$. We can also define an effective Hamiltonian, $H_{\text{eff}}$, in terms of the Floquet operator such that

$$U_F(k) = e^{-iH_{\text{eff}}(k)L} \tag{8.33}$$

The eigenvalues of $H_{\text{eff}}$ thus give the quasienergy bands $\varepsilon_n(k)$. For the microring lattice, the effective Hamiltonian is given by

$$H_{\text{eff}}(k) = -\begin{bmatrix} 0 & \varepsilon e^{-i\chi} \\ \varepsilon e^{i\chi} & 0 \end{bmatrix} = \begin{bmatrix} 0 & h_{\text{eff}}(k) \\ h^*_{\text{eff}}(k) & 0 \end{bmatrix} \tag{8.34}$$

where $\varepsilon = \phi/L$ and $e^{-i\chi} = v/|v|$. We note that the effective Hamiltonian has the same form as the Hamiltonian of the SSH system in (8.7) with its associated chiral symmetry. The eigenvalues of $H_{\text{eff}}$ are given by $\pm|h_{\text{eff}}| = \pm\varepsilon(k) = \pm\cos^{-1}(u)/L$, yielding two quasienergy bands in every $[(2m-1)\pi/L, (2m+1)\pi/L]$ interval, $m \in \mathbb{Z}$. Thus, unlike the SSH system, the 1D Floquet microring lattice has an infinite number of quasienergy bands, and within each period there are two bandgaps, centered at 0 and $\pi/L$, which we label as the 0 bandgap and $\pi$ bandgap, as depicted in Fig. 8.5. The 0 bandgap closes when $\kappa_a = \kappa_b$.

The winding numbers of the quasienergy bands can be computed using (8.11) with $h(k)$ replaced by $h_{\text{eff}}(k)$. Specifically, since $h_{\text{eff}} = -\varepsilon v/|v|$ and $\varepsilon$ is real, we get

$$W = w[h_{\text{eff}}] = w[v] = \begin{cases} 0, & \theta_a > \theta_b \\ \pm 1, & \theta_a < \theta_b \end{cases} \tag{8.35}$$

Thus, the lattice behaves as a normal insulator for $\kappa_a > \kappa_b$ and as a topological insulator for $\kappa_a < \kappa_b$, which is the same behavior as predicted by the SSH model.

**Dynamical Topological Invariant**: The winding numbers for the quasienergy bands in (8.35) are computed from the effective Hamiltonian (or equivalently, the Floquet operator), which does not depend on the propagation distance $z$. These winding numbers, which may be called static topological invariants, do not capture the exact details of the system evolution over one period and thus may not completely characterize the topological behaviors of the system. Another topological invariant, called the gap winding number, which depends on the complete evolution history of the system, is required to capture the full topological characteristics of the system [30, 38]. The gap winding number is computed for each bulk bandgap using the evolution operator and may thus be called the dynamical topological invariant of the system.

In order to determine the winding number associated with a bandgap, we first periodize the evolution operator by requiring that $U(k, L) = U(k, 0) = I$, i.e., the system must return to exactly the same state it was in after each period $L$. This is achieved by deforming the Hamiltonian of the system in an adiabatic manner without closing the bandgap. More specifically, if $\xi$ is some quasienergy value in the bandgap, the Hamiltonian should be deformed in such a way that there is always a gap around $\xi$ at every point $z$ in an evolution period. A common approach to periodize the evolution operator is to multiply it by an operator $V_\xi$ [30, 38],

$$U_\xi(k, z) = U(k, z)V_\xi(k, z) \tag{8.36}$$

where

$$V_\xi(k, z) = e^{iH_{\mathrm{eff},\xi}(k)z} \tag{8.37}$$

with the eigenvalues of $H_{\mathrm{eff},\xi}$ chosen between $\xi$ and $\xi + 2\pi/L$. It can be verified that $U_\xi(k, L) = I$ and $U_\xi(k, z)$ maintains a bandgap around $\xi$ for all values of $z$. Due to the chiral symmetry of the system, the periodized evolution operators $U_\xi$ for $\xi L = 0$ and $\xi L = \pi$ satisfy the following relations at the midpoint of the evolution period ($z = L/2$) [38]

$$\Gamma U_0(k, L/2)\Gamma = -U_0(k, L/2) \tag{8.38}$$

$$\Gamma U_\pi(k, L/2)\Gamma = U_\pi(k, L/2) \tag{8.39}$$

The above relations imply that $U_0(k, L/2)$ is anti-diagonal and $U_\pi(k, L/2)$ is diagonal:

$$U_0(k, L/2) = \begin{bmatrix} 0 & U_0^+(k) \\ U_0^-(k) & 0 \end{bmatrix}; \quad U_\pi(k, L/2) = \begin{bmatrix} U_\pi^+(k) & 0 \\ 0 & U_\pi^-(k) \end{bmatrix} \tag{8.40}$$

The winding numbers associated with the 0 and $\pi$ bandgaps are then given by the winding numbers of $U_0^+$ and $U_\pi^+$, respectively [38].

For the 1D microring lattice, the operator $V_0$ for $\xi L = 0$ (with quasienergies chosen as $\varepsilon(k)$ and $2\pi/L - \varepsilon(k)$) is

$$V_0(k, z) = \begin{bmatrix} \cos(\varepsilon z - \pi z/L) & -i\sin(\varepsilon z - \pi z/L)e^{-i\chi} \\ -i\sin(\varepsilon z - \pi z/L)e^{i\chi} & \cos(\varepsilon z - \pi z/L) \end{bmatrix} e^{i\pi z/L} \tag{8.41}$$

The operator $V_\pi$ for $\xi L = \pi$ (with quasienergies $-\varepsilon(k) + 2\pi/L$ and $\varepsilon(k) + 2\pi/L$) is

$$V_\pi(k, z) = \begin{bmatrix} \cos(\varepsilon z) & -i\sin(\varepsilon z)e^{-i\chi} \\ -i\sin(\varepsilon z)e^{i\chi} & \cos(\varepsilon z) \end{bmatrix} e^{i2\pi z/L} \tag{8.42}$$

At the midpoint of the evolution period, the periodized evolution operator is

$$U_\xi(k, L/2) = U(k, L/2)V_\xi(k, L/2) \tag{8.43}$$

where

$$U(k, L/2) = \begin{bmatrix} a & ib \\ ib^* & a^* \end{bmatrix} \tag{8.44}$$

$$a = \cos(\theta_a/2)\cos(\theta_b/2) - \sin(\theta_a/2)\sin(\theta_b/2)e^{-ik\Lambda}$$
$$b = \sin(\theta_a/2)\cos(\theta_b/2) + \cos(\theta_a/2)\sin(\theta_b/2)e^{-ik\Lambda}$$

Substituting the expressions for $V_0$ and $V_\pi$ from (8.41) and (8.42) into (8.43), we obtain the periodized evolution operators at $L/2$

$$U_0(k, L/2) = \begin{bmatrix} 0 & f_0 \\ f_0^* & 0 \end{bmatrix}; \quad U_\pi(k, L/2) = \begin{bmatrix} f_\pi & 0 \\ 0 & f_\pi^* \end{bmatrix} \tag{8.45}$$

where

$$f_0 = -a\cos(\phi/2)e^{-i\chi} - b\sin(\phi/2)$$
$$f_\pi = -a\cos(\phi/2) - b\sin(\phi/2)e^{i\chi} = f_0 e^{i\chi}$$

The winding numbers associated with the bandgaps at quasienergies 0 and $\pi/L$ can be computed using (8.11) to give[3]

$$w_0 = w[f_0] = \begin{cases} 0, & \theta_a > \theta_b \\ 1, & \theta_a < \theta_b \end{cases} \tag{8.46}$$

[3] Alternatively, using the relations $e^{-i\chi} = v/|v|$, $v = 2a^*b$ and $|v| = \sin\phi$, we can write $f_0$ as

$$f_0 = -\left[\frac{|a|^2}{\sin(\phi/2)} + \sin(\phi/2)\right]b$$

Since the terms in the square brackets are real, the winding number $w[f_0]$ is the same as $w[b]$, which can be obtained from the expression for $b$ in (8.44) to get the result in (8.46). For the winding number $w[f_\pi]$, using the relation $w[y_1 y_2^*] = w[y_1] - w[y_2]$ for any two functions $y_1$ and $y_2$, we get

$$w[f_\pi] = w[f_0 e^{i\chi}] = w[f_0] - w[e^{-i\chi}] = w[f_0] - w[v] = 0$$

for $\theta_a \neq \theta_b$.

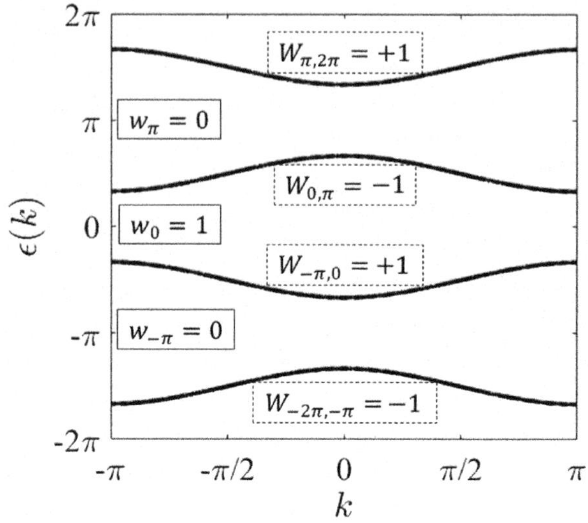


**Fig. 8.5** Quasienergy band diagram of a 1D Floquet microring lattice with $\theta_a < \theta_b$. The band invariants $W_{\xi,\xi'}$ and gap invariants $w_\xi$ are also indicated

$$w_\pi = w[f_\pi] = 0, \quad \theta_a \neq \theta_b \tag{8.47}$$

Thus the 0 bandgap behaves like the bandgap in the SSH lattice, being topologically nontrivial when $\kappa_a < \kappa_b$. On the other hand, the $\pi$ bandgap is always topologically trivial, i.e., it always behaves as a normal insulator bandgap and does not support an edge mode.[4]

The winding numbers of the quasienergy bands (the band invariants) and those of the bandgaps (the gap invariants) are related. For the 1D chiral system, it can be shown [38] that $w_\pi - w_0 = W_{0,\pi}$, where $W_{0,\pi} = w[h_{\text{eff}}]$ is the winding number of the quasienergy band between the 0 and $\pi$ bandgaps. This relationship can be verified for the band diagram of the microring lattice in Fig. 8.5.

## 8.3 Topological Photonic Insulators in 2D Microring Lattices

One-dimensional microring lattices are simple systems that are useful for demonstrating the key concepts of topological insulators. However, since the edge states of a 1D lattice are localized at the two ends of the array, they are not useful for optical transport. On the other hand, 2D topological lattices can support edge modes that propagate along the sample boundaries, making them more useful for device applications. In addition, since these edge modes are topologically protected, they

[4]If we allow the coupling angles $\theta_a$ and $\theta_b$ to exceed $\pi/2$ (in which case $\tau_a < 0$ and $\tau_b < 0$), the winding number $w_\pi$ can become nontrivial for $\theta_a < \theta_b$, indicating that the $\pi$-bandgap can support edge modes in this case. The existence of these edge modes has been confirmed in 1D Floquet TPIs based on periodically coupled waveguide arrays [39].

are immune to lattice imperfections, such as those caused by fabrication variations, suggesting that they can be used to realize fabrication tolerant photonic devices [25]. As in the case of 1D microring arrays, 2D microring lattices can be treated as static systems with time-independent Hamiltonians, or as periodically-driven systems. The latter case can lead to the prediction of new topological behaviors not observed in static systems, such as anomalous Floquet insulators (AFIs).

In this section we will first review various realizations of TPIs using 2D microring lattices as static systems. We will then provide a detailed treatment of a general 2D microring lattice as a Floquet system, covering both its static and dynamical topological behaviors. We will also present our experimental effort in realizing 2D microring lattices on the Silicon-on-Insulator (SOI) platform.

### *8.3.1 2D Microring Lattices as Chern Insulators*

The first TPI based on 2D microring lattices was proposed in 2011 and demonstrated in 2013 by Hafezi et al. [13, 26]. These lattices sought to emulate the quantum Hall effect in a 2D electron gas subject to a perpendicular magnetic field. The system consists of a square lattice of microrings supporting light propagation in the same clockwise or counter-clockwise direction (i.e., single-spin site rings), with adjacent microrings connected via off-resonant link rings, as shown in Fig. 8.6a. The magnetic field is emulated by introducing a synthetic gauge field in the form of a gradient in the hopping phase between site rings in one direction ($y$ direction in Fig. 8.6a), and constant hopping phase in the other direction ($x$ direction). The net phase accumulation around each plaquette (shown by the red dashed line in Fig. 8.6a) is thus $2\pi\alpha$, where the value of $\alpha$ can be set to represent the ratio of the magnetic flux penetrating each plaquette to one flux quantum [40]. The Hamiltonian of the system can be derived using the tight-binding approximation with the effect of the link rings neglected to give

$$-J\left(\psi_{x,y-1} + \psi_{x,y+1} + \psi_{x-1,y}e^{i2\pi\alpha y} + \psi_{x+1,y}e^{-i2\pi\alpha y}\right) = E\psi_{x,y}$$

where $J$ is the hopping rate. The above equation has the same form as the discrete Harper equation for electrons in a 2D lattice subjected to a perpendicular magnetic field [41].

A disadvantage of the above TPI microring lattice is that it is not translationally invariant along the $y$ direction, which could make it less useful for device applications. More recently, it was shown by the same group that nontrivial topological behavior in a 2D microring lattice can also be observed by introducing next-nearest neighbor couplings between single-spin site rings using off-resonant link rings, as shown in Fig. 8.6b. The lattice has translational symmetry in this case. Although the net gauge flux is zero, the combination of next-nearest neighbor (NNN) hopping and directional nearest neighbor (NN) hopping between site rings emulates a staggered local gauge flux that breaks the time-reversal symmetry between the two pseudo-spin states of the lattice and gives rise to nontrivial topological behaviors [42].

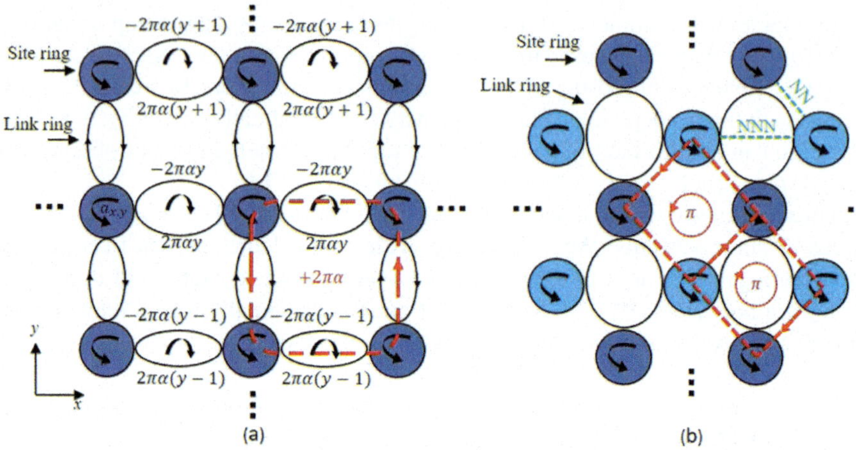


**Fig. 8.6** Realizations of Chern insulators in 2D microring lattices using **a** a synthetic magnetic field, adapted by permission from Nature Publishing Group, Nature Physics [13], copyright (2011) and **b** a staggered local gauge flux, adapted by permission from [42], Copyright (2019) by the American Physical Society

Both types of lattices described above are static 2D systems whose topological behaviors can be characterized by an invariant called the Chern number. Their Chern insulator behaviors are achieved by imposing a synthetic local gauge flux on the lattice via the phase detunings of the link rings. However, a synthetic magnetic field is not necessary for observing nontrivial topological behaviors, as shown by Liang and Chong in [43]. The authors instead showed that microring lattices consisting of strongly coupled single-spin site rings can also support edge modes in a bulk bandgap, even though the Chern numbers of all energy bands are trivial. Pasek and Chong later used Chalker-Coddington network theory [44] to show that these edge states are similar to those in a Floquet insulator [45], although they did not explicitly compute the dynamical invariants characterizing the Floquet TPI behaviors of the lattice.

In the next section, we treat the 2D microring lattice as a Floquet system with a periodically-varying Hamiltonian. In contrast to previous microring lattices, which assume that all site rings support the same pseudo-spin state and require link rings for coupling between site rings, our lattice is formed using direct-coupled microrings and is thus more compact and allows for the natural spin flipping which occurs between adjacent resonators. We compute the topological phase map of the microring lattice and show that the system can exhibit rich topological characteristics, including normal insulator (NI), Chern insulator (CI) and anomalous Floquet insulator (AFI) behaviors.

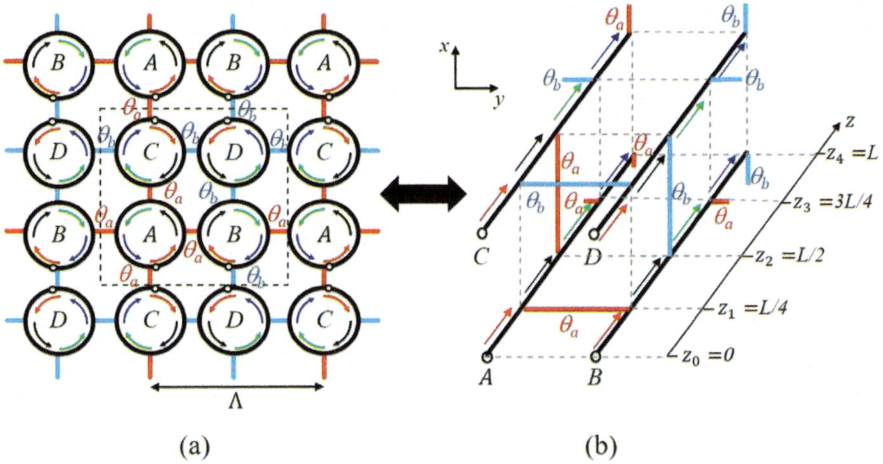


**Fig. 8.7** **a** Schematic of 2D Floquet microring lattice with each unit cell (dashed square) consisting of four resonators and characterized by two coupling angles $\theta_a$ and $\theta_b$. **b** Equivalent coupled waveguide array for one unit cell. Reprinted by permission from [27], Copyright (2020) by the American Physical Society

## *8.3.2 2D Microring Lattices as Floquet Insulators*

Our Floquet lattice is a square lattice of direct-coupled microring resonators with identical resonant frequencies, as shown in Fig. 8.7a. The coupling strength between adjacent microrings is defined by the coupling angle $\theta$, so that $\kappa^2 = \sin^2\theta$ gives the fraction of power coupled between the two resonators. We assume that each microring supports only a clockwise or counter-clockwise propagating wave; however, due to evanescent coupling, neighbor resonators have opposite directions of wave propagation (or pseudo-spin flipping), as depicted in Fig. 8.7a. We define a unit cell of the lattice to consist of four microrings labeled $A$, $B$, $C$, and $D$, as indicated by the dashed square in the figure. Since a square microring lattice with uniform coupling strength is topologically trivial, in order to realize nontrivial topological behaviors, we allow the coupling angle between microring $A$ and its neighbors (denoted by $\theta_a$) to be different from the coupling between microring $D$ and its neighbors (denoted by $\theta_b$). The microring lattice is thus characterized by two coupling angles $(\theta_a, \theta_b)$. The lattice is translationally invariant, with periodicity of $\Lambda \sim 4R$ in the $x$ and $y$ directions.

**Floquet-Bloch Hamiltonian of a 2D Microring Lattice**: To determine the Hamiltonian of the microring lattice, we follow the same approach in Sect. 2.2 and convert the lattice into an equivalent coupled waveguide array. This is accomplished by cutting the microrings at the points indicated by the open circles in Fig. 8.7a and unrolling them to form a 2D array of periodically-coupled waveguides, as shown in Fig. 8.7b. Lines connecting adjacent waveguides indicate direct couplings between

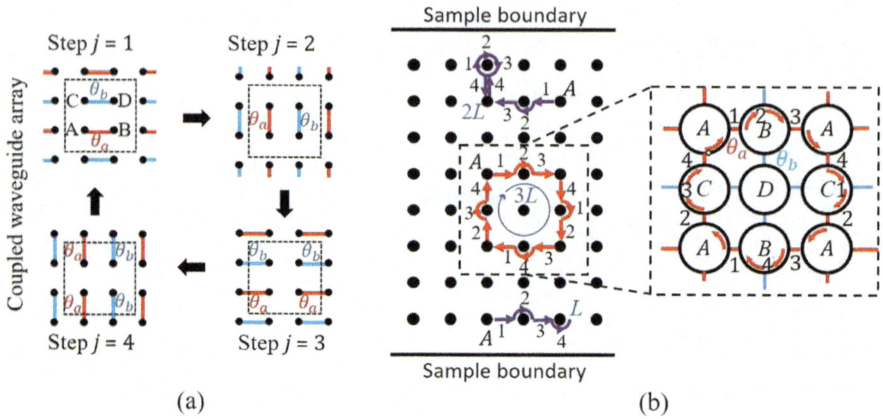


**Fig. 8.8** **a** Coupling sequence in one period ($L = 2\pi R$), reprinted with permission from [46] © The Optical Society. **b** Driving protocol of the 2D microring lattice in the limit of perfect coupling ($\kappa_a \to 1$) and ($\kappa_b \to 0$). The localized bulk mode is shown by the red arrows and edge modes along the top and bottom boundaries are shown by the purple arrows, Reprinted by permission from [27], Copyright (2020) by the American Physical Society. Inset figure shows the corresponding trajectory of a bulk mode in the microring lattice

the corresponding microrings. Within each period $L = 2\pi R$, light in each waveguide undergoes four coupling steps, as shown in Fig. 8.8a. This coupling (or hopping) sequence guarantees that light will partially return to its initial point after three periods. Figure 8.8b illustrates this for the simple case of perfect coupling ($\kappa_a = 1$ and $\kappa_b = 0$). Starting from point $A$ in the bulk of the waveguide array, light follows the trajectory shown by the red arrows and returns to the initial point after $3L$. The corresponding trajectory in the microring lattice is shown by the red arrows in the inset figure. The fact that light propagation forms localized loops in the bulk of the lattice implies that light is expelled from the bulk for certain frequencies and forms propagating edge modes in the presence of boundaries, as shown by the purple arrows in Fig. 8.8b. This behavior suggests that the lattice has the characteristics of a topological insulator.

Following the coupling sequence in Fig. 8.8a, we can write the equations of motion for the fields $[\psi^A_{m,n}, \psi^B_{m+1,n}, \psi^C_{m,n+1}, \psi^D_{m+1,n+1}]$ in each unit cell over one period as:

$$
\begin{aligned}
\frac{\partial \psi^A_{m,n}}{\partial z} &= i\beta\psi^A_{m,n} + ik_a(1)\psi^B_{m+1,n} + k_a(2)\psi^C_{m,n+1} \\
&\quad + ik_a(3)\psi^B_{m-1,n} + ik_a(4)\psi^C_{m,n-1} \\
\frac{\partial \psi^B_{m+1,n}}{\partial z} &= i\beta\psi^B_{m+1,n} + ik_a(1)\psi^A_{m,n} + ik_b(2)\psi^D_{m+1,n+1} \\
&\quad + ik_a(3)\psi^A_{m+2,n} + ik_b(4)\psi^D_{m+1,n-1} \\
\frac{\partial \psi^C_{m,n+1}}{\partial z} &= i\beta\psi^C_{m,n+1} + ik_b(1)\psi^D_{m+1,n+1} + ik_a(2)\psi^A_{m,n} \\
&\quad + ik_b(3)\psi^D_{m-1,n+1} + ik_a(4)\psi^A_{m,n+2} \\
\frac{\partial \psi^D_{m+1,n+1}}{\partial z} &= i\beta\psi^D_{m+1,n+1} + ik_b(1)\psi^C_{m,n+1} + ik_b(2)\psi^B_{m+1,n} \\
&\quad + ik_b(3)\psi^C_{m+2,n+1} + ik_b(4)\psi^B_{m+1,n+2}
\end{aligned}
\tag{8.48}
$$

where $k_{(a,b)}(j) = k_{(a,b)} = 4\theta_{(a,b)}/L$ in step $j$ and equals 0 otherwise. Since the waveguide array is periodic in $x$ and $y$, we apply Bloch's condition to get $\psi^{A,B,C,D}_{m\pm l,n\pm v} = \psi^{A,B,C,D}_{m,n} e^{i(\pm lk_x,\pm vk_y)\Lambda/2}$ where $l$ and $v$ are integers. Equation 8.48 can be written for the state vector $|\psi\rangle = [\psi^A_{m,n}, \psi^B_{m,n}, \psi^C_{m,n}, \psi^D_{m,n}]^T$ of each unit cell in the form

$$
i\frac{\partial}{\partial z}|\psi(\mathbf{k},z)\rangle = [-\beta I + H_{\mathrm{FB}}(\mathbf{k},z)]|\psi(\mathbf{k},z)\rangle \tag{8.49}
$$

where $\mathbf{k} = (k_x, k_y)$ is the momentum vector and $I$ is the identity matrix. The FB Hamiltonian $H_{\mathrm{FB}}$ is given by

$$
H_{\mathrm{FB}}(\mathbf{k},z) = \sum_{j=1}^{4} H(j) \tag{8.50}
$$

where $H(j) = H_j$ in step $j$ and zero otherwise. The Hamiltonian in each step is

$$
H_1 = -\begin{pmatrix} 0 & k_a e^{ik_x\Lambda} & 0 & 0 \\ k_a e^{-ik_x\Lambda} & 0 & 0 & 0 \\ 0 & 0 & 0 & k_b e^{ik_x\Lambda} \\ 0 & 0 & k_b e^{-ik_x\Lambda} & 0 \end{pmatrix}
$$

$$
H_2 = -\begin{pmatrix} 0 & 0 & k_a e^{ik_y\Lambda} & 0 \\ 0 & 0 & 0 & k_b e^{ik_y\Lambda} \\ k_a e^{-ik_y\Lambda} & 0 & 0 & 0 \\ 0 & k_b e^{-ik_y\Lambda} & 0 & 0 \end{pmatrix}
$$

$$H_3 = -\begin{pmatrix} 0 & k_a e^{-ik_x\Lambda} & 0 & 0 \\ k_a e^{ik_x\Lambda} & 0 & 0 & 0 \\ 0 & 0 & 0 & k_b e^{-ik_x\Lambda} \\ 0 & 0 & k_b e^{ik_x\Lambda} & 0 \end{pmatrix}$$

$$H_4 = -\begin{pmatrix} 0 & 0 & k_a e^{-ik_y\Lambda} & 0 \\ 0 & 0 & 0 & k_b e^{-ik_y\Lambda} \\ k_a e^{ik_y\Lambda} & 0 & 0 & 0 \\ 0 & k_b e^{ik_y\Lambda} & 0 & 0 \end{pmatrix}$$

The FB Hamiltonian is periodic in $z$ with a periodicity equal to the microring circumference, i.e., $H_{\text{FB}}(\mathbf{k}, z) = H_{\text{FB}}(\mathbf{k}, z + L)$. We note that unlike the 1D Floquet microring lattice, the FB Hamiltonian in (8.50) does not have chiral nor time-reversal symmetries.

Noting that the Hamiltonian $H_j$ is independent of $z$ in each step, we can obtain the evolution operator using (8.25) as (dropping the $e^{i\beta z}$ term)

$$U(\mathbf{k}, z) = \begin{cases} e^{-iH_1(\mathbf{k})z}, & 0 \le z < L/4 \\ e^{-iH_2(\mathbf{k})(z-L/4)} e^{-iH_1(\mathbf{k})L/4}, & L/4 \le z < L/2 \\ e^{-iH_3(\mathbf{k})(z-L/2)} e^{-iH_2(\mathbf{k})L/4} e^{-iH_1(\mathbf{k})L/4}, & L/2 \le z < 3L/4 \\ e^{-iH_4(\mathbf{k})(z-3L/4)} e^{-iH_3(\mathbf{k})L/4} e^{-iH_2(\mathbf{k})L/4} e^{-iH_1(\mathbf{k})L/4}, & 3L/4 \le z \le L \end{cases} \tag{8.51}$$

The Floquet operator is then

$$U_F(\mathbf{k}) = U(\mathbf{k}, L) = e^{-iH_4(\mathbf{k})(L/4)} e^{-iH_3(\mathbf{k})L/4} e^{-iH_2(\mathbf{k})L/4} e^{-iH_1(\mathbf{k})L/4} = e^{-iH_{\text{eff}}(\mathbf{k})L} \tag{8.52}$$

where $H_{\text{eff}}$ is the effective Hamiltonian. The Floquet operator is a unitary matrix with complex eigenvalues of unit magnitude. Its eigenstates are obtained from

$$U_F(\mathbf{k})|\psi_n(\mathbf{k})\rangle = e^{-i\varepsilon_n(\mathbf{k})L}|\psi_n(\mathbf{k})\rangle \tag{8.53}$$

Over each Floquet Brillouin zone, the Floquet spectrum consists of 4 quasienergy bands $\varepsilon_n(\mathbf{k})$ with 3 bandgaps, which we label I, II and III, as shown in Fig. 8.9a. Bandgaps I and III are symmetric with respect to $\varepsilon L = \pi$.

Figure 8.9a shows the Floquet spectrum of quasienergy bands over one period ($0 \le \varepsilon L \le 2\pi$) of an infinite lattice with $\theta_a = 0.48\pi$ and $\theta_b = 0.1\pi$. Three distinct bandgaps (I, II, and III) can be seen, with bandgap II centered at $\varepsilon L = \pi$ and bandgaps I and III symmetric about this energy. Figure 8.9b shows the regions where bandgaps I and II are open as functions of the coupling angles. Bandgap II ($\pi$ bandgap) is always open except for coupling angle values on the dashed lines.

**Topological Invariants of 2D Microring Lattices**: Since the quasienergy bands of the 2D microring lattice are periodic in $k_x$ and $k_y$, the surface of each band over one Brillouin zone can be wrapped around in $k_x$ and $k_y$ (so that the points at $k_x = 0$

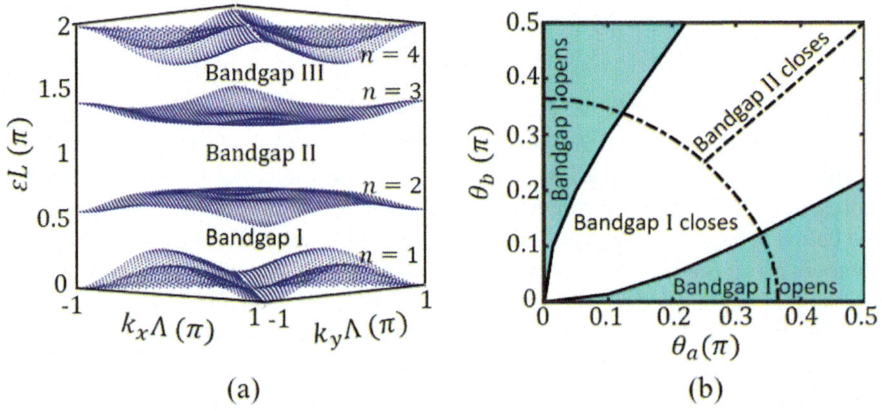


**Fig. 8.9** **a** Band diagram over one Floquet Brillouin zone of an infinite 2D microring lattice with $\theta_a = 0.48\pi$ and $\theta_b = 0.1\pi$. Floquet bands ($n = 1, 2, 3$, and 4) and bulk bandgaps (I, II, and III) are labeled in the figure (there is no gap between bands $n = 1$ and $n = 4$). **b** Map of coupling angles $\theta_a$ and $\theta_b$ showing the regions where bandgaps I and II are open. Bandgap II is always open except for the dashed lines. Bandgap I is open only in the cyan regions

join the points at $k_x = 2\pi/\Lambda$ and so on) to form a torus. The integral of the Berry curvature of the $n$th Bloch mode, $|\psi_n(\mathbf{k})\rangle$, over the torus surface defines the Chern invariant of the quasienergy band $n$ [32]:

$$C_n = \frac{1}{2\pi}\int_{\mathrm{BZ}} (\nabla_{\mathbf{k}} \times \mathbf{A} \cdot \hat{\mathbf{k}}_z)\mathrm{d}k_x\mathrm{d}k_y = \frac{1}{2\pi}\int\int (\partial_{k_x}A_y - \partial_{k_y}A_x)\mathrm{d}k_x\mathrm{d}k_y \tag{8.54}$$

where $\partial_k$ is the partial derivative with respect to $k$, $\nabla_{\mathbf{k}} \times \mathbf{A}$ is the Berry curvature ($\nabla_{\mathbf{k}}$ is the gradient with respect to $\mathbf{k}$) and $\mathbf{A}$ is the Berry phase connection defined as

$$\mathbf{A} = -i\langle\psi_n(\mathbf{k})|\nabla_{\mathbf{k}}|\psi_n(\mathbf{k})\rangle \tag{8.55}$$

Equation 8.54 can also be expressed as [30]

$$C_n = C[P_n] = \frac{1}{2\pi i}\int_{\mathrm{BZ}} \mathrm{Tr}\{P_n[\partial_{k_x}P_n, \partial_{k_y}P_n]\}\mathrm{d}k_x\mathrm{d}k_y \tag{8.56}$$

where [., .] is the commutator and $P_n(\mathbf{k}) = |\psi_n(\mathbf{k})\rangle\langle\psi_n(\mathbf{k})|$ is the projector onto eigenstate $|\psi_n(\mathbf{k})\rangle$ of the Floquet operator.

The topological behavior of a 2D static system is characterized by the Chern number. Specifically, the number of edge states $N_\xi$ in a bandgap at energy $\xi$ is equal to the sum of the Chern numbers of all the energy bands below that bandgap:

$$N_\xi = \sum_{E_n<\xi} C[P_n] \tag{8.57}$$

The lattice is thus a NI if $N_\xi = 0$, otherwise it is a CI. The relation in (8.57), however, applies only to static systems, which have a finite number of energy bands. For a Floquet system, whose quasienergy spectrum is periodic, the Chern number only represents the difference between the number of edge states in the bandgaps above and below that quasienergy band. Notably, edge states can still exist even if the Chern numbers of all the Floquet bands are zero. Thus the Chern number cannot be used to establish the bulk-edge correspondence of a Floquet insulator.

As in the case of periodically-driven 1D systems, to completely characterize the topological behaviors of a 2D Floquet system, we need to determine the dynamical gap invariant, or winding number, which depends on the complete evolution history of the system over each driving period. For a 2D lattice, the winding number associated with a bandgap at quasienergy $\xi$ is given by [30]

$$w[U_\xi] = \frac{1}{8\pi^2}\int_0^L \mathrm{d}z \int_{\mathrm{BZ}} \mathrm{d}k_x \mathrm{d}k_y \mathrm{Tr}\left\{U_\xi^{-1}\partial_z U_\xi [U_\xi^{-1}\partial_{k_x} U_\xi, U_\xi^{-1}\partial_{k_y} U_\xi]\right\} \tag{8.58}$$

where $U_\xi$ is the periodized evolution operator with an open bandgap around $\xi$. As in the 1D case, we can periodize the evolution operator $U(\mathbf{k}, z)$ by multiplying it with an operator $V_\xi$ [30]

$$U_\xi(\mathbf{k}, z) = U(\mathbf{k}, z)V_\xi(\mathbf{k}, z) \tag{8.59}$$

where

$$V_\xi(\mathbf{k}) = e^{iH_{\mathrm{eff},\xi}(\mathbf{k})z} \tag{8.60}$$

with the eigenvalues of the effective Hamiltonian $H_{\mathrm{eff},\xi}$ chosen to be between $\xi$ and $\xi + 2\pi/L$. The winding number is equal to the number of edge modes that can exist in the bandgap. In addition, the Chern number of the $n$th Floquet band is related to the winding number of the lower ($\xi'$) and upper ($\xi$) bandgaps via the equation $C_n = w[U_\xi] - w[U_{\xi'}]$.

To investigate the topological behaviors of the 2D microring lattice, we computed the Chern numbers of the energy bands and winding numbers of the bandgaps for four lattice samples, labeled $M$, $N$, $O$, and $P$, with coupling angles $(\theta_a; \theta_b)$ = $(0.146\pi; 0.019\pi)$, $(0.315\pi; 0.016\pi)$, $(0.397\pi; 0.304\pi)$, and $(0.473\pi; 0.026\pi)$, respectively. Figure 8.10 shows the projected band diagrams of the samples with 10 unit cells in the $y$ direction and infinite extent in the $x$ direction. The Chern numbers and winding numbers are also indicated for the energy bands and bandgaps, respectively. We observe that bandgap II is open in all four samples. The winding number $w_{\mathrm{II}}$ in bandgap II of samples $M$ and $N$ are trivial, indicating that these samples behave as NIs in this bandgap. This is also confirmed by the absence of edge modes in bandgap II of the projected band diagrams of these samples.

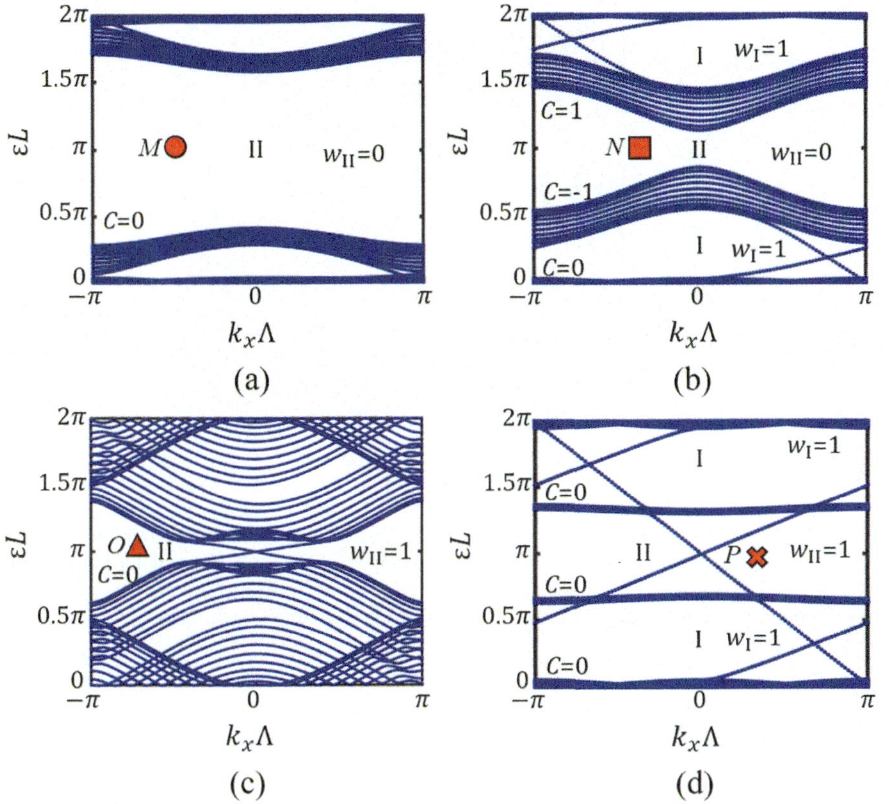


**Fig. 8.10** Projected band diagrams of semi-infinite microring lattices with 10 unit cells in the $y$ direction and infinite extent in the $x$ direction: **a** sample $M$ with coupling angles $\theta_a = 0.146\pi$, $\theta_b = 0.019\pi$; **b** sample $N$ with $\theta_a = 0.315\pi$, $\theta_b = 0.016\pi$; **c** sample $O$ with $\theta_a = 0.397\pi$, $\theta_b = 0.304\pi$; and **d** sample $P$ with $\theta_a = 0.473\pi$, $\theta_b = 0.026\pi$. The Chern numbers $C$ of the Floquet bands and winding numbers $w$ of the bandgaps are also shown [47]

On the other hand, samples $O$ and $P$ behave as topological insulators in bandgap II because they have nontrivial winding numbers ($w_{\text{II}} = 1$). The topological behavior of these samples can be verified by the existence of the edge states that cross bandgap II in the projected band diagrams in Fig. 8.10c and d. The two edge states in the bulk bandgap correspond to the two orthogonal pseudo-spin states with opposite directions of light propagation in each microring. We note that the Chern numbers of all the quasienergy bands of these samples are zero. Thus as static systems, these lattices are topologically trivial. However, their dynamical invariants are non-zero, indicating that their nontrivial topological behavior arises from the periodic nature of the system evolution. These lattices are thus classified as AFIs.

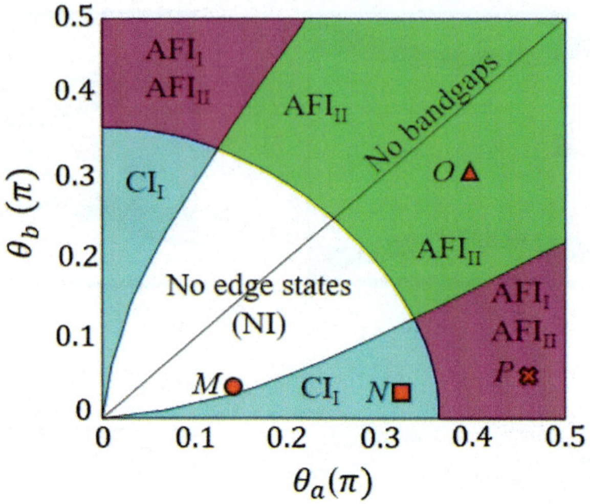


**Fig. 8.11** Topological phase map of 2D microring lattices characterized by coupling angles $\theta_a$ and $\theta_b$ . NI indicates normal insulator behavior in bandgap II; $CI_I$ is Chern insulator in bandgap I; $AFI_I$ and $AFI_{II}$ indicate AFI behavior in bandgaps I and II, respectively. The markers $M, N, O, P$ correspond to the four samples in Fig. 8.10. Reprinted with permission from [46] © The Optical Society

With respect to the topological behaviors of bandgap I (and III), Fig. 8.10b and d shows that these bandgaps are open for lattices $N$ and $P$, both with nontrivial winding numbers $w_I$. These lattices thus behave as topological insulators in these bandgaps. For lattice $N$, the Chern number of band $n = 2$ is also nontrivial and satisfies the relation $C_2 = w_I - w_{II}$. In this case, the lattice is classified as a Floquet CI. However, since lattice $P$ has zero Chern numbers for all the energy bands, its topological behaviors in bandgaps I and III are of the AFI type.

A complete topological characterization of the 2D microring lattice is summarized by the topological phase map in Fig. 8.11, which predicts the various topological behaviors of the lattice over the range of coupling angle values $0 \leq \theta_a, \theta_b \leq \pi/2$. The four samples $M$, $N$, $O$, and $P$ are also marked on the map. In the white region, the lattice has only one bulk bandgap, bandgap II, which is a NI. In the cyan regions, the lattice behaves as a CI in bandgap I and NI in bandgap II. In the green regions, the lattice again only has bandgap II open, which behaves as an AFI. In the purple regions, the lattice behaves as an AFI in all three bandgaps. The topological phase map is symmetric with respect to the line $\theta_a = \theta_b$ since exchanging the two coupling angles does not change the lattice.

For direct verification of the edge modes predicted by the topological phase map, we computed the light intensity distribution in a finite sample of $5 \times 10$ unit cells of lattice $P$ using the field coupling method in [48]. Light enters the sample via an input waveguide coupled to microring $A$ of a unit cell on the left boundary and exits the sample via an output waveguide coupled to microring $B$ of a unit cell on the right

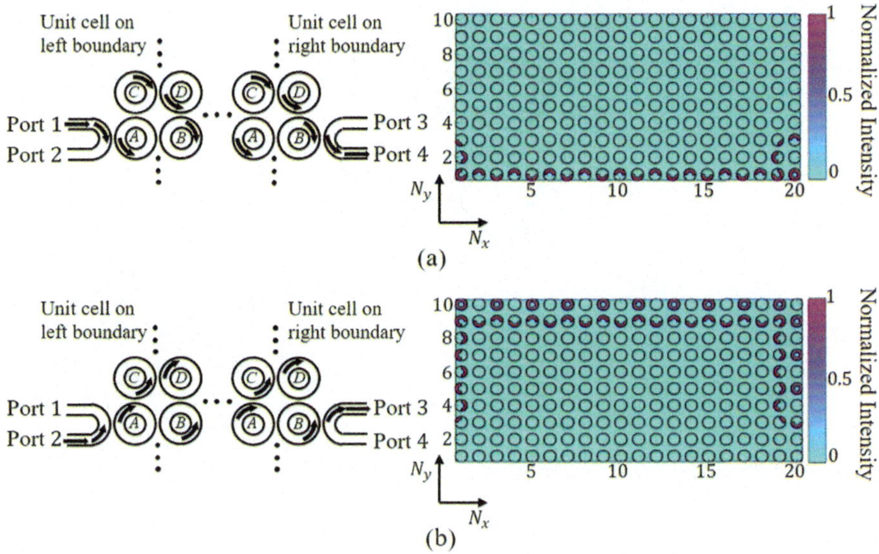


**Fig. 8.12** Schematic for light excitation and intensity distribution of the **a** counter-clockwise and **b** clockwise edge modes at quasienergy $\xi = \pi L$ in an AFI microring lattice with $5 \times 10$ unit cells and coupling angles $\theta_a = 0.473\pi$, $\theta_b = 0.026\pi$

boundary. Figure 8.12a and b shows the excitation schematics and light intensity distributions of the two chiral edge states at the center of bandgap II ($\xi = \pi L$). The counter-clockwise edge mode is excited by injecting light into port 1 of the input waveguide while the clockwise edge mode is excited using port 2. Figure 8.12a shows that the edge mode corresponding to the counter-clockwise pseudo-spin state in microring $A$ travels in the counter-clockwise direction along the sample boundaries and exits from port 4 of the output waveguide. On the other hand, the edge mode with the clockwise pseudo-spin in microring $A$ travels in the opposite direction along the sample boundaries to exit from port 3 of the output waveguide, as seen in Fig. 8.12b. Similar edge mode field patterns are also obtained for light excitation in bandgaps I and III for both CIs and AFIs of samples $N$, $O$, and $P$.

### *8.3.3 Experimental Realization of 2D Floquet Microring Lattices*

In this section, we review our experimental effort in realizing 2D Floquet microring TPIs on a silicon photonics platform [27]. To realize a square lattice of microrings with identical resonance frequencies, the microrings must be identical with identical evanescent coupling gaps between adjacent resonators. However, we also have the requirement that in each unit cell, the coupling strength between microring $A$ and

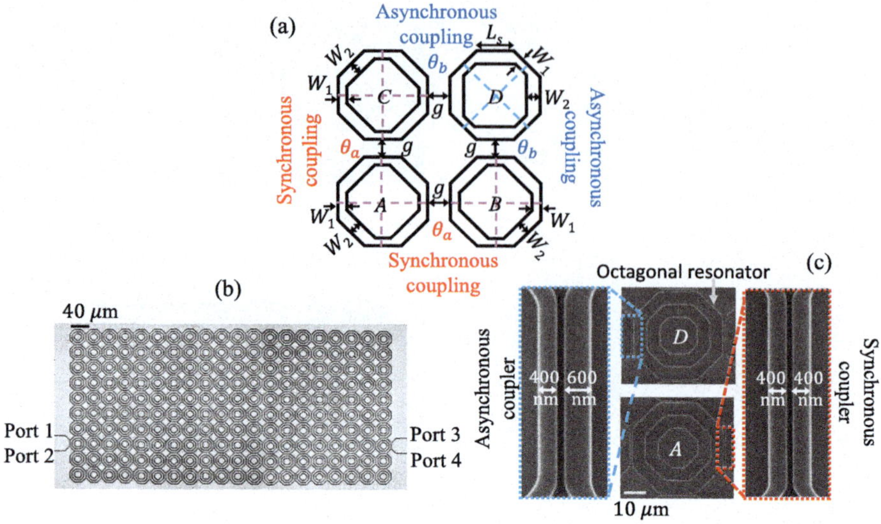


**Fig. 8.13** **a** Schematic of a unit cell of the Floquet microring lattice consisting of octagons with alternating widths $W_1$ and $W_2$. Octagon $D$ is rotated by 45° with respect to the other resonators to achieve asynchronous coupling. **b** Optical microscope image of a fabricated octagon lattice with $5 \times 10$ unit cells. **c** SEM images of octagon resonators $A$ and $D$, with zoomed-in views of the synchronous and asynchronous coupling sections. Reprinted by permission from [27], Copyright (2020) by the American Physical Society

its neighbors must be different from the coupling strength between microring $D$ and its neighbors. To realize such a lattice, we used octagonal resonators with the sides having identical lengths but alternating widths $W_1$ and $W_2$, as shown in Fig. 8.13a. To maintain a square lattice, we set the coupling gaps between adjacent octagons to be identical. Different coupling strengths between adjacent octagons can be achieved by exploiting the difference between synchronous coupling between waveguides of the same widths and asynchronous coupling between waveguides of different widths. In the lattice design, octagons $A$, $B$, and $C$ are oriented in the same way such that coupling between resonator $A$ and its neighbors $B$ and $C$ occurs synchronously between waveguides of the same width $W_1$. By rotating octagon $D$ by 45° with respect to the other three resonators, different coupling strengths between resonator $D$ and its neighbors are obtained due to asynchronous coupling between waveguides of different widths $W_1$ and $W_2$.

We fabricated 2D microring lattices on an SOI substrate with a 220 nm-thick silicon layer on a 2 μm-thick $SiO_2$ layer. The devices were designed to operate with the TE polarization at 1550 nm wavelength. The octagons were designed to have sides of length $L_s = 16.06$ μm and alternating widths $W_1 = 400$ nm and $W_2 = 600$ nm. The coupling gaps between adjacent octagons were $g = 225$ nm. Using Finite Difference Time Domain simulation in the Lumerical software [49], we determined the coupling angles to be $\theta_a = 0.473\pi$ and $\theta_b = 0.026\pi$ 1620 nm wavelength, which corresponds

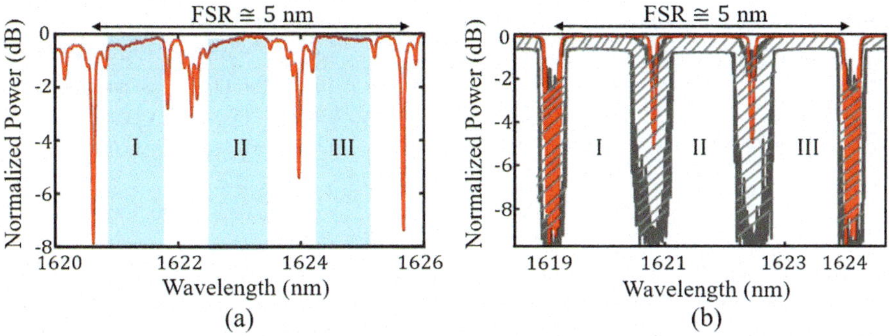


**Fig. 8.14** **a** Measured power transmission of AFI microring lattice. **b** Simulated transmission spectrum of a lattice with 5 × 10 unit cells and coupling angles $\theta_a = 0.473\pi$ and $\theta_b = 0.026\pi$. The red trace is the normalized transmission of the ideal lattice with 3 dB/cm propagation loss in the waveguides. The hatched area shows the possible variations in transmission due to a ±5% deviation in the coupling parameters and round-trip phases of the microrings. Reprinted by permission from [27], Copyright (2020) by the American Physical Society

to lattice $P$ in the topological phase map in Fig. 8.11. We fabricated a lattice sample consisting of 5 × 10 unit cells, with an input waveguide coupled to resonator $A$ of a unit cell located at the left boundary for edge mode excitation, and an output waveguide coupled to resonator $B$ of a unit cell on the right boundary for transmission measurements. The coupling angles between the input/output waveguides and the octagon resonators were set equal to $\theta_a$. Figure 8.13b shows an optical microscope image of the fabricated lattice. Figure 8.13c shows SEM images of octagon resonators $A$ and $D$, with zoomed-in views of the synchronous and asynchronous coupling sections. The two smaller octagons inside each octagon are dummy structures to achieve more uniform pattern density over the lattice.

We measured the transmission spectrum of the lattice by coupling TE-polarized laser light into port 1 of the input waveguide and measured the transmitted light at port 4 of the output waveguide. Figure 8.14a shows the normalized transmitted power over one free spectral range (FSR $\sim$ 5 nm) of the octagon resonators. We observe three regions where the transmission is high and relative flat, which we identify as the three bandgaps I, II and III. The high transmission is evidence of AFI edge modes being excited, as predicted by the projected band diagram for lattice $P$ in Fig. 8.10d, which transported light along the lattice boundaries to the output waveguide. On the other hand, the regions where the transmission is low correspond to the bulk transmission bands (the passbands of the lattice). The low transmission measured at the output port is due to light spreading out into the bulk lattice and being partially reflected back into port 2 of the input waveguide. We also observe sharp transmission dips within these passbands, which are caused by multiple interferences and localized resonances of light in the bulk lattice.

**Table 8.1** Parameters of fabricated octagon lattice samples $M, N$ and $O$ [47]

| Sample | $L_s$ (μm) | $W_1$ (nm) | $W_2$ (nm) | $g$ (nm) | $\theta_a(\pi)$ | $\theta_b(\pi)$ | Wavelength | Edge mode type |
|---|---|---|---|---|---|---|---|---|
| $M$ | 13.14 | 400 | 600 | 275 | 0.146 | 0.019 | 1516 nm | NI (bandgap II) |
| $N$ | 16.06 | 400 | 600 | 225 | 0.315 | 0.016 | 1533 nm | CI (bandgap I) |
| $O$ | 16.06 | 400 | 410 | 225 | 0.397 | 0.304 | 1554 nm | AFI (bandgap II) |

To verify the measured spectrum, we computed the power transmission of the lattice sample using the field coupling method in [48]. The red trace in Fig. 8.14b shows the simulated transmission power assuming 3dB/cm propagation loss in the silicon octagon waveguides. To account for fabrication variations, we also simulated the transmission in the presence of $\pm 5\%$ random variations in the coupling strengths and microring round-trip phases. The range of expected transmission values is shown by the hatched area in the plot. Comparing the simulated and measured transmission spectra shows good qualitative agreement between the two plots, including the characteristic high and flat-band transmission in the bulk bandgaps, and low transmission and sharp dips in the passbands.

To obtain direct evidence of AFI edge modes in the bulk bandgaps, we injected light at 1623 nm, which is located in bandgap II, into port 1 of the input waveguide and imaged the scattered light intensity distribution using a near-infrared (NIR) camera. The camera image is shown in Fig. 8.15a, which clearly shows light being localized along the bottom boundary of the lattice as it propagated from the input waveguide on the left to the output waveguide on the right. For more clarity, the lower left plot of the figure shows the map of scattered light intensity reconstructed from the raw NIR camera data. The lower right plot shows the simulated distribution of light intensity in the resonators, which shows good agreement with the imaged scattered light pattern. When light at the same wavelength was injected into port 2 of the input waveguide, the edge mode corresponding to the opposite pseudo-spin state was excited, which propagated along the top edge of the lattice. This is evident from the NIR camera image and simulated light intensity in Fig. 8.15b. When we tuned the input light wavelength to 1624 nm, which is located in the bulk passband, the camera image in Fig. 8.15c shows light spreading into the bulk of the lattice and no edge mode is observed.

To observe topological behaviors of 2D microring lattices in other regions of the topological phase map, we also designed and fabricated octagonal resonator lattices corresponding to samples $M, N$, and $O$ in Fig. 8.11 [47]. These lattices have different waveguide widths $W_1$, $W_2$, octagon side length $L_s$ and coupling gap $g$ as listed in Table 8.1. The simulated values of the coupling angles $\theta_a$ and $\theta_b$ are also shown in the table along with the operating wavelength and the expected topological behaviors of the samples.

The measured transmission spectra of the fabricated lattices are shown in Fig. 8.16a–c. We observe that these spectra correlate well with the projected band diagrams in Fig. 8.10. In particular, the transmission is high and flat in nontrivial

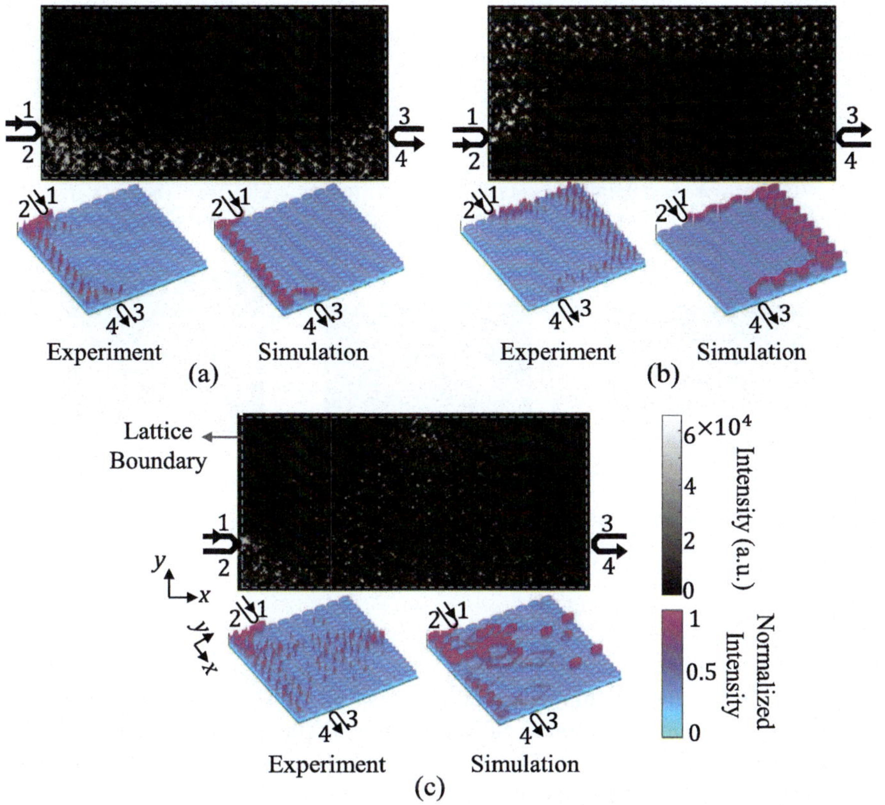


**Fig. 8.15** Direct NIR imaging of scattered light patterns from the AFI microring lattice: **a** counter-clockwise AFI edge mode excited from port 1 of input waveguide at $\lambda = 1623$ nm (in bandgap II), **b** clockwise AFI edge mode excited from port 2 of input waveguide at the same wavelength; **c** bulk mode excitation at $\lambda = 1624$ nm (in a passband). In each figure, the lower left plot shows the map of scattered light constructed from the raw data of the NIR camera. The lower right plot shows the simulated distribution of light intensity in the microrings. Reprinted by permission from [27], Copyright (2020) by the American Physical Society

bandgaps, which support edge modes, and low in trivial bandgaps, where the lattice behaves as a NI. In the bulk passbands, the transmission has an irregular pattern since light is spread out throughout the bulk lattice and partially reflected back. For the nontrivial bandgaps, the types of edge modes (CI or AFI) can be identified based on the computed Chern numbers and winding numbers in the projected band diagrams and indicated in the transmission spectra.

We also performed NIR imaging of the scattered light patterns from the lattices at the wavelengths indicated by the black arrows in the transmission spectra. These images are shown in Fig. 8.16d–f. For lattice $M$, which is a NI, Fig. 8.16d shows that input light at $\lambda = 1516$ nm is reflected and cannot propagate into the lattice bulk. For lattice $N$, which is a CI in bandgaps I and III, the scattered light pattern

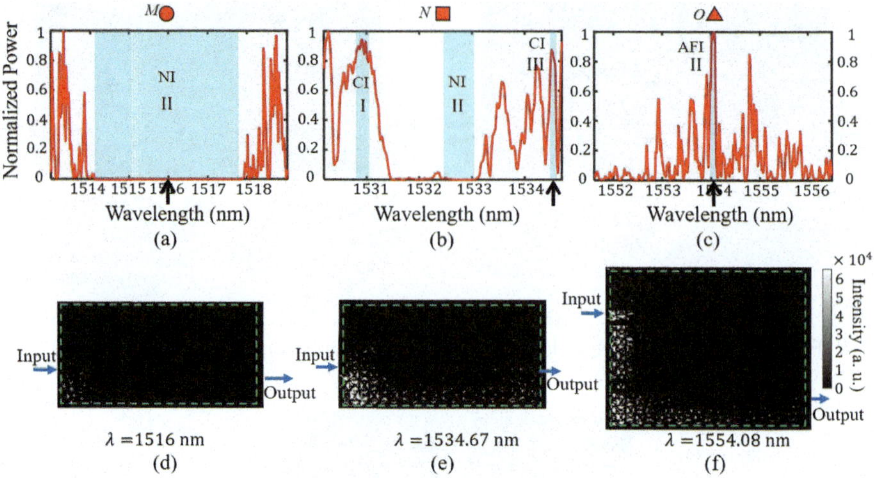


**Fig. 8.16** **a–c** Measured transmission spectra of fabricated samples $M$, $N$ and $O$. **d–f** NIR camera images of scattered light intensity distributions taken at the wavelengths shown by the black arrows in plots (**a–c**) [47]

in Fig. 8.16e clearly shows the formation of an edge mode along the bottom sample boundary when excited by input light at $\lambda = 1534.67$ nm (in bandgap III). For lattice $O$, although bandgap II is almost closed, it can still support an AFI edge mode, as can be verified by the image in Fig. 8.16f for input light at $\lambda = 1554.08$ nm. These results provide additional evidence that our octagonal resonator lattices behave as predicted by the topological phase map. We also note that by exploiting the frequency dispersion of the coupling angles $\theta_a$ and $\theta_b$, it is possible to observe quantum phase transition across a boundary on the topological phase map, e.g., from a NI to an AFI [27].

## 8.4 Conclusion

In this chapter we reviewed the basic concepts of topological photonic insulators and presented methods for analyzing the topological behaviors of 1D and 2D microring lattices. In particular, we focused our attention on the treatment of microring lattices as periodically-driven quantum systems, which are found to exhibit richer topological characteristics than their static counterparts. We also reviewed our recent effort in realizing Floquet insulators based on octagonal resonator lattices in silicon photonics, which provides a versatile nanophotonic platform for exploring potential applications of Floquet TPIs.

Topological photonics is an active area of research with promising potential for breakthroughs in both fundamental science and engineering applications. New topological phenomena such as light trapping in high-Q bulk mode resonance [50] and corner states [51–53] could lead to novel lasers and optical filter devices. Floquet systems provide the flexibility of engineering the driving protocol, either spatially or temporally, to achieve new types of 2D TPIs, such as those protected by time reversal symmetry [54]. Non-Hermitian photonics represents another interesting approach for realizing topological insulators [55–57]. By introducing nonlinear materials in TPIs, novel nonlinear topological devices could be realized, including robust nonlinear optical isolators, switches, and enhanced harmonic generation [20, 58–60]. Finally, there are also many interesting applications of TPIs in quantum photonics, such as topologically-protected transport of single photons and photon entanglements, as well as robust quantum light sources and amplifiers [29, 61, 62].

## References

1. D.C. Tsui, H.L. Stormer, A.C. Gossard, Phys. Rev. Lett. **48**(22), 1559 (1982)
2. D.J. Thouless, M. Kohmoto, M.P. Nightingale, M. den Nijs, Phys. Rev. Lett. **49**(6), 405 (1982)
3. M. Kohmoto, Ann. Phys. **160**(2), 343 (1985)
4. F.D.M. Haldane, S. Raghu, Phys. Rev. Lett. **100**(1), 013904 (2008)
5. S. Raghu, F.D.M. Haldane, Phys. Rev. A **78**(3), 033834 (2008)
6. Z. Wang, Y. Chong, J.D. Joannopoulos, M. Soljačić, Nature **461**(7265), 772 (2009)
7. Z. Yang, F. Gao, X. Shi, X. Lin, Z. Gao, Y. Chong, B. Zhang, Phys. Rev. Lett. **114**(11), 114301 (2015)
8. C. He, X. Ni, H. Ge, X.C. Sun, Y.B. Chen, M.H. Lu, X.P. Liu, Y.F. Chen, Nat. Phys. **12**(12), 1124 (2016)
9. Y.G. Peng, C.Z. Qin, D.G. Zhao, Y.X. Shen, X.Y. Xu, M. Bao, H. Jia, X.F. Zhu, Nat. Commun. **7**(1), 1 (2016)
10. Q. Wei, Y. Tian, S.Y. Zuo, Y. Cheng, X.J. Liu, Phys. Rev. B **95**(9), 094305 (2017)
11. W.J. Chen, S.J. Jiang, X.D. Chen, B. Zhu, L. Zhou, J.W. Dong, C.T. Chan, Nat. Commun. **5**(1), 1 (2014)
12. F. Gao, Z. Gao, X. Shi, Z. Yang, X. Lin, H. Xu, J.D. Joannopoulos, M. Soljačić, H. Chen, L. Lu, Y. Chong, B. Zhang, Nat. Commun. **7**(1), 1 (2016)
13. M. Hafezi, E.A. Demler, M.D. Lukin, J.M. Taylor, Nat. Phys. **7**(11), 907 (2011)
14. L.J. Maczewsky, J.M. Zeuner, S. Nolte, A. Szameit, Nat. Commun. **8**(1), 13756 (2017)
15. M.C. Rechtsman, J.M. Zeuner, Y. Plotnik, Y. Lumer, D. Podolsky, F. Dreisow, S. Nolte, M. Segev, A. Szameit, Nature **496**(7444), 196 (2013)
16. S. Peng, N.J. Schilder, X. Ni, J. van de Groep, M.L. Brongersma, A. Alù, A.B. Khanikaev, H.A. Atwater, A. Polman, Phys. Rev. Lett. **122**(11), 117401 (2019)
17. S. Stützer, Y. Plotnik, Y. Lumer, P. Titum, N.H. Lindner, M. Segev, M.C. Rechtsman, A. Szameit, Nature **560**(7719), 461 (2018)
18. L. Lu, J.D. Joannopoulos, M. Soljačić, Nat. Photonics **8**(11), 821 (2014)
19. T. Ozawa, H.M. Price, A. Amo, N. Goldman, M. Hafezi, L. Lu, M.C. Rechtsman, D. Schuster, J. Simon, O. Zilberberg et al., Rev. Mod. Phys. **91**(1), 015006 (2019)
20. X. Zhou, Y. Wang, D. Leykam, Y.D. Chong, New J. Phys. **19**(9), 095002 (2017)
21. P. St-Jean, V. Goblot, E. Galopin, A. Lemaître, T. Ozawa, L. Le Gratiet, I. Sagnes, J. Bloch, A. Amo, Nat. Photonics **11**(10), 651 (2017)

22. B. Bahari, A. Ndao, F. Vallini, A. El Amili, Y. Fainman, B. Kanté, Science **358**(6363), 636 (2017)
23. M.A. Bandres, S. Wittek, G. Harari, M. Parto, J. Ren, M. Segev, D.N. Christodoulides, M. Khajavikhan, Science **359**, 6381 (2018)
24. S. Klembt, T. Harder, O. Egorov, K. Winkler, R. Ge, M. Bandres, M. Emmerling, L. Worschech, T. Liew, M. Segev et al., Nature **562**(7728), 552 (2018)
25. Y. Wu, C. Li, X. Hu, Y. Ao, Y. Zhao, Q. Gong, Adv. Opt. Mater. **5**(18), 1700357 (2017)
26. M. Hafezi, S. Mittal, J. Fan, A. Migdall, J.M. Taylor, Nat. Photonics **7**(12), 1001 (2013)
27. S. Afzal, T.J. Zimmerling, Y. Ren, D. Perron, V. Van, Phys. Rev. Lett. **124**(25), 253601 (2020)
28. Y.E. Kraus, Y. Lahini, Z. Ringel, M. Verbin, O. Zilberberg, Phys. Rev. Lett. **109**(10), 106402 (2012)
29. S. Barik, A. Karasahin, C. Flower, T. Cai, H. Miyake, W. DeGottardi, M. Hafezi, E. Waks, Science **359**(6376), 666 (2018)
30. M.S. Rudner, N.H. Lindner, E. Berg, M. Levin, Phys. Rev. X **3**(3), 031005 (2013)
31. N.H. Lindner, G. Refael, V. Galitski, Nat. Phys.**7**(6), 490 (2011)
32. J.K. Asbóth, L. Oroszlany, A. Palyi, in *Lecture Notes in Physics*, vol. 919, ed. by W. Beiglbock, J. Ehlers, K. Hepp, H. Weidenmuller (Springer, Switzerland, 2016)
33. J.K. Asbóth, B. Tarasinski, P. Delplace, Phys. Rev. B **90**(12), 125143 (2014)
34. M. Parto, S. Wittek, H. Hodaei, G. Harari, M.A. Bandres, J. Ren, M.C. Rechtsman, M. Segev, D.N. Christodoulides, M. Khajavikhan, Phys. Rev. Lett. **120**(11), 113901 (2018)
35. H. Zhao, P. Miao, M.H. Teimourpour, S. Malzard, R. El-Ganainy, H. Schomerus, L. Feng, Nat. Commun. **9**(1), 1 (2018)
36. T.J. Zimmerling, V. Van, Opt. Lett. **45**(3), 714 (2020)
37. V. Van, *Optical microring resonators: theory, techniques, and applications* (CRC Press, 2016)
38. M. Fruchart, Phys. Rev. B **93**, 115429 (2016)
39. M. Bellec, C. Michel, H. Zhang, S. Tzortzakis, P. Delplace, EPL **119**(1), 14003 (2017)
40. D.R. Hofstadter, Phys. Rev. B **14**(6), 2239 (1976)
41. F. Harper, Ph.D. thesis, University of Oxford, 2015
42. S. Mittal, V.V. Orre, D. Leykam, Y.D. Chong, M. Hafezi, Phys. Rev. Lett. **123**(4), 043201 (2019)
43. G.Q. Liang, Y.D. Chong, Phys. Rev. Lett. **110**(20), 203904 (2013)
44. J.T. Chalker, P.D. Coddington, J. Phys. C Solid State Phys. **21**(14), 2665 (1988)
45. M. Pasek, Y.D. Chong, Phys. Rev. B **89**(7), 075113 (2014)
46. S. Afzal, V. Van, Opt. express **26**(11), 14567 (2018)
47. S. Afzal, T.J. Zimmerling, Y. Ren, D. Perron, V. Van, Experimental Investigation of Topological Phases of Two-Dimensional Floquet Microring Lattices. Paper presented at 14th Pacific Rim Conference on Lasers and Electro-Optics (CLEO PR 2020), p. C9H_2. Optical Society of America, 2020
48. A. Tsay, V. Van, IEEE J. Quantum Electron. **47**(7), 997 (2011)
49. Lumerical Inc. http://www. lumerical. com/products/
50. S. Afzal, V. Van, arXiv preprint arXiv:2102.02903 (2021)
51. W.A. Benalcazar, B.A. Bernevig, T.L. Hughes, Phys. Rev. B **96**(24), 245115 (2017)
52. S. Mittal, V.V. Orre, G. Zhu, M.A. Gorlach, A. Poddubny, M. Hafezi, Nat. Photonics **13**(10), 692 (2019)
53. X.D. Chen, W.M. Deng, F.L. Shi, F.L. Zhao, M. Chen, J.W. Dong, Phys. Rev. Lett. **122**(23), 233902 (2019)
54. B. Höckendorf, A. Alvermann, H. Fehske, Phys. Rev. B **99**(24), 245102 (2019)
55. S. Longhi, D. Gatti, G. Della Valle, Sci. Rep. **5**, 13376 (2015)
56. S. Weimann, M. Kremer, Y. Plotnik, Y. Lumer, S. Nolte, K.G. Makris, M. Segev, M.C. Rechtsman, A. Szameit, Nat. Mater. **16**(4), 433 (2017)
57. K. Mochizuki, D. Kim, N. Kawakami, H. Obuse, Phys. Rev. A **102**(6), 062202 (2020)
58. D. Leykam, Y.D. Chong, Phys. Rev. Lett. **117**(14), 143901 (2016)
59. S. Kruk, A. Poddubny, D. Smirnova, L. Wang, A. Slobozhanyuk, A. Shorokhov, I. Kravchenko, B. Luther-Davies, Y. Kivshar, Nat. Nanotechnol. **14**(2), 126 (2019)

60. D. Smirnova, D. Leykam, Y. Chong, Y. Kivshar, Appl. Phys. Rev. **7**(2), 021306 (2020)
61. S. Mittal, E.A. Goldschmidt, M. Hafezi, Nature **561**(7724), 502 (2018)
62. V. Peano, M. Houde, F. Marquardt, A.A. Clerk, Phys. Rev. X **6**(4), 041026 (2016)

# Chapter 9
# Parallel Digital Gradient Search Technique for Rapid Automated Alignment of Devices on Silicon Photonics Integrated Circuits

**Scott Jordan**

**Abstract** Silicon photonics has emerged as an essential technology for the world moving forward. Colliding needs for escalating data consumption versus environmental sustainability drive the shift from copper-based networking architectures to photon based. New technologies for implementing photonic devices on semiconductor substrates have ignited an explosion of innovation in support of these trends, and while the energy efficiencies of the transition are clear and are rooted in fundamental physics, the economic realities of manufacturing these devices at the necessary scale have posed significant challenges. In particular, the step of aligning photonic elements such as lasers, chips and fibers to maximize transmission has been a time-consuming (and therefore costly) requirement that even more unfortunately must be repeated many times in the test-and-assembly process, starting even before the chips are diced off of the wafer. There is no analog to this step in conventional microelectronic manufacturing. The challenge grows worse as devices grow more complex, for example when adding array channels or additional discrete components, as this introduces physical and geometric dependencies and necessitates optimization across multiple degrees of freedom. Formerly, time-consuming iterative procedures were required to accommodate this, leading to escalating costs. As we discuss here, new technology eliminates this by performing these alignments in parallel across elements, channels, inputs/outputs and degrees of freedom, reducing overall alignment times by typically a factor of 100, resolving these vexing production-economics issues and enabling the industry's development and growth.

## 9.1 An Application-Driven Challenge

Even before the COVID-19 pandemic, the world's appetite for data drove bandwidth consumption at rates that, per some studies, long ago exceeded Moore's law (the semiconductor industry's truism that forecasts a doubling of transistors in integrated

---

S. Jordan (✉)
Physik Instrumente (PI) GmbH & Co., Silicon Valley, California, USA
e-mail: scottj@pi-usa.us

 
D. J. Lockwood and L. Pavesi (eds.), *Silicon Photonics IV*,
Topics in Applied Physics 139, https://doi.org/10.1007/978-3-030-68222-4_9

circuits every two years). Steven Alexander, CTO of networking giant Ciena, documented in 2019 how Internet traffic growth had eclipsed Moore's law [1], and that was before 2020s global viral emergency forced professional and clerical workers and their employers to adapt to working from home, among many other social interactions that suddenly shifted online. This tragic circumstance abruptly turbocharged the world's data consumption, to the point that Telefónica, one of the world's largest multinational network operators, saw a year's worth of traffic growth in the single month of April 2020 [2]. Since at least some of this will stick as the new normal, it is evident that Internet growth has left Moore's law in the dust by now.

## 9.2 Photonics to the Rescue: Energy, Speed, Fidelity, Scalability and Sustainability?

The shift to mobile computing is particularly striking. According to UN data, it only took about twenty years for the number of cell phone subscriptions to approximately match the number of humans on earth. Landline telephones never achieved 25% penetration of global humanity, but mobile broadband exceeded that in just six years [3]. Today's exponentiating demand for data has been driven by applications as diverse as:

- Video conferencing
- Genomics
- Smart cars
- The Internet of Things
- Manufacturing 4.0
- Streaming media
- Selfies
- Personal medicine
- Big data
- Facial recognition, and of course,
- Social networks.

Incidentally, the evident variety of today's applications has an un-obvious consequence: unlike the first photonics industry boom of 1997–2001, which was driven by essentially one application (long-haul telephony, replacing satellite telephone links), the diversity and mushrooming nature of today's applications makes a 2002-style industry collapse hard to imagine. Indeed, the severe economic dislocations of the COVID-19 pandemic came against a background of booming demand for bandwidth, for which photonics is a key enabler.

All these applications reduce to data that is stored, processed and conveyed in bits. Conventionally, to transmit a bit from point A to point B fundamentally requires charging and discharging a wire, even the microscopic ones in CPU and RAM chips. Per Stanford's David A.B. Miller, "Most energy in information processing is in communications, not in logic, even at the gate level. Most energy dissipation in information processing is in charging and discharging wire capacitance, which is ~200 aF/μm" [4, 5].

The promise of photonics in computing and communications is multifold. Scalability, capacity, parallelism, speed: all these benefits are obvious. But, the promise of eliminating wires, especially the short ones in data centers, yields the promise of sustainability compared to the charge/discharge losses of bit-conveyance over copper at all scales.

The problem is practicality. Such devices need to be invented, developed, engineered and manufactured. Fortunately, thanks to clever device engineering and some groundbreaking micro-robotic industrial automation technology, the world is well on its way to mainstream adoption of high-throughput photonic interconnects, from the data center to the home. Soon, we expect this trend to proceed down to the level of on-chip interconnects [6].

Viewed from a zoomed-out perspective, it is clear that the history of photonics is a fascinating tale of progressively shorter link scales, from transoceanic to transcontinental to regional to metro, and now within-and-between data center installations, and progressing farther down to within the chip itself.

## 9.3 Silicon Photonics

In the late 1990s, when photonics technology was first massively deployed to replace satellite links in long-distance telecommunications, the challenge of manufacturing quickly emerged. To couple one of the then-novel laser diodes into a newfangled single-mode fiber required laborious adjustment of those components' mutual position to ensure that the laser's light was efficiently coupled into the fiber. An analog technology called a gradient search was a partial solution, as it allowed the positioning system to quickly reach optimum, at least for the case of a single fiber with a smooth modal profile. But, these positioning systems had significant limitations: They could be fragile, have limited travel ranges, be drifty, and otherwise be not-well-suited to industrial deployment. To address these drawbacks, the *digital* gradient search was developed [7]. This was deployed on industrial-class motion hardware that was already proven in semiconductor manufacturing processes. The combination provided a path forward for repetitive manufacturing of the devices of the day.

Today, the functionalities of photonic data devices have progressed significantly in sophistication and complexity. The key has been the aforementioned integration of photonic technologies onto the same wafers as used for microelectronics. This leverages the world's vast, mature and efficient manufacturing capability. It allows thousands of *photonic* integrated circuits to be minted on a single wafer, alone or alongside microelectronic elements. Silicon photonic (SiPh) is the single most revolutionary development in semiconductors since the invention of the integrated circuit:

The rest of the semiconductor industry's evolution (world-changing as it has been) is basically a tale of different wafer sizes, different coatings, different etching strategies and so on.

Importantly, photonic integrated circuits offer the tantalizing potential for processing multiple channels of light at once. So photonics ICs today routinely integrate multichannel and multi-wavelength structures and offer multiple inputs and outputs arranged in arrays.

The familiar challenge from the late 1990s arises: Those chip-level photonic inputs and outputs need to be coupled to other elements, from optical fibers and fiber arrays and laser diodes to lenses, gratings and similar bulk elements and even other chips. Many of these couplings require meticulous alignment, accurate to tens of nanometers, not only in the sensitive transverse plane (by convention referred to as the "*XY*" plane) but in other degrees of freedom as well. Importantly, any array situation requires precise orientation of the devices not only in *XY* (the plane transverse to the optical axis) but also in theta-*Z* (rotation about the optical axis, fundamental to bringing multiple array channels into correspondence and often required in other degrees of freedom (DOFs) as well. Only measurement of the actual coupling could do the job; machine-vision approaches are too coarse, and passive approaches such as precision V-grooves and interposers have typically required impractical device replication tolerances (though diligent and clever advancements have borne fruit recently, such as Teramount's "Wide Band PhotonicBump" technology).

The 1990s digital gradient search technology could still be utilized for industrial-class manufacturing processes, and in fact, it is still the foundation of successful tool architectures that continue to assemble photonic products today. But when confronted with array elements and other situations requiring multi-DOF optimization, the overall process of global optimization can be slow—often too slow to be practical for testing and packaging those thousands of devices each SiPh wafer represents. This is because of the geometric coupling of the various DOFs. A motion in theta-*Z*, for instance, inevitably causes a de-alignment in *XY* due to the mechanical rotation axis not being precisely coincident with the optical axis. Consequently, a looping process has been needed: Align in *XY*, make a small theta-*Z* improvement, go back and re-align in *XY* and repeat over and over until you converge on a global optimization. The situation is similar for the other DOFs, increasing convergence time in exponential fashion as a function of number of DOFs. Also, many devices (such as the short, multimode waveguide structures common in silicon photonics) exhibit interactions between channels and between inputs and outputs: Optimize the input, optimize the output, but now the input is no longer optimized. Consequently, one must loop *XY* alignments and theta-*Z* (etc.) motions until a consensus optimum is achieved. From there, the same technology can track the alignments, ensuring optimization in the face of drift from thermal changes and even curing stresses.

## 9.4 The Dinosaur Falls: Extinction of the Loops

Most recently, a revolutionary re-thinking of the three-decade-old digital gradient search has allowed industrial positioners to perform *multiple* gradient searches across multiple channels, inputs and DOFs simultaneously, even if they interact optically or geometrically. This reduces or eliminates the loops just mentioned. For instance, instead of iteratively making small theta-$Z$ motions interleaved with $XY$ corrections over a span of minutes, perform the $XY$ and theta-$Z$ alignments *at the same time* using two gradient search processes in parallel and get the job done in a second. This approach can be scaled up to the full 6 degrees of freedom for each element, and multiple positioners can work cooperatively on multiple inputs and outputs of a device even if interactions between the inputs, outputs and channels occur.

This represents time savings, and not a small amount: Process time reductions of 99% or more are common. For example, as mentioned, the multi-minute serial process of looping $XY$ and theta-$Z$ alignments of an array device collapses to a typically 1-second-scale parallel process. Intriguingly, the parallelism of the technology means the overall process time is almost independent of the number of processes performed, so the cost savings multiply with device complexity.

Implemented at the command level entirely in the firmware of industrial-class 6-DOF hexapod microrobots (Fig. 9.1) and other precision mechanisms, this tech-

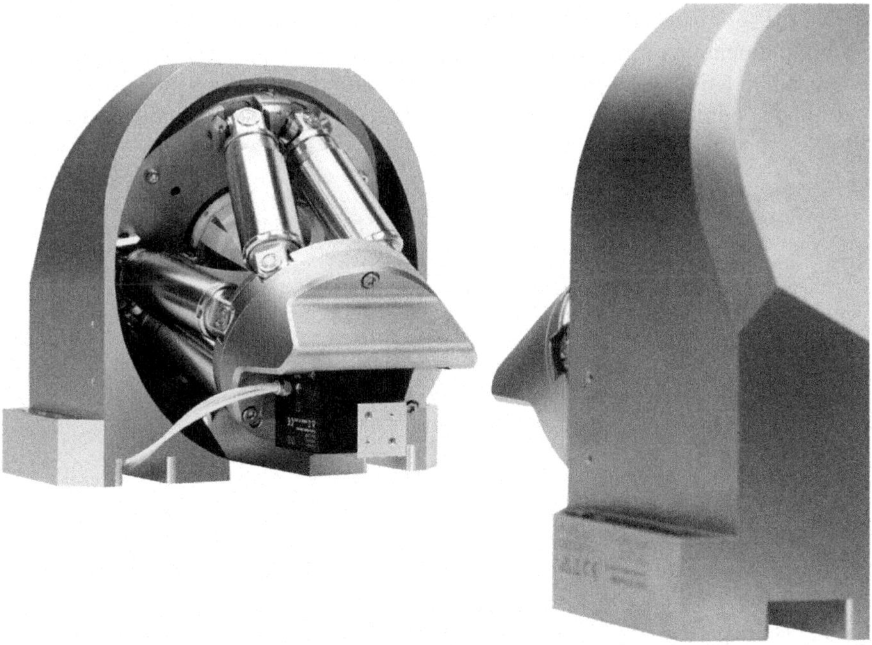

**Fig. 9.1** Alignment micro-robots perform multiple alignment processes simultaneously across degrees of freedom, elements, inputs and outputs, even if they optically or geometrically interact

nology is key to resolving unfavorable economics for the manufacturing of silicon photonics devices (and of course devices based on other wafer technologies such as InGaAs).

## 9.5 Operating Principle

Gradient searches depend on a small cyclic motion ("dither") of one device with respect to another. The position of the dithered element and the coupled optical power (or other figure-of-merit) is monitored synchronously in real time. Per (9.1), this dither motion causes a small, similarly cyclic modulation of the figure-of-merit: $\varepsilon(\theta)$, a function of the instantaneous dither phase angle $\theta$. While it is possible to perform these tasks with all motion and metrology performed by a PC or PLC, it is much faster if a specialized motion controller performs this work, with internal algorithms that deduce the phase and amplitude of the modulation caused by the dither process. This information guides the controller to efficiently drive position uphill to the optimum for the involved axes, where the amplitude of the modulation drops to zero by definition and alignment motion is complete. The dithering process can be continued, however, to provide real-time compensation of drift and disturbances. This is the approach utilized in PI's Fast Multichannel Photonic Alignment (FMPA) micro-robot systems.

Figure 9.2 depicts this process in the simple case of a Gaussian coupling. However, wise implementations are mathematically founded on analytics that do not require or depend on a tidy Gaussian coupling—many SiPh applications yield rather messy, multimode coupling cross sections characterized by local maxima and saddle points.

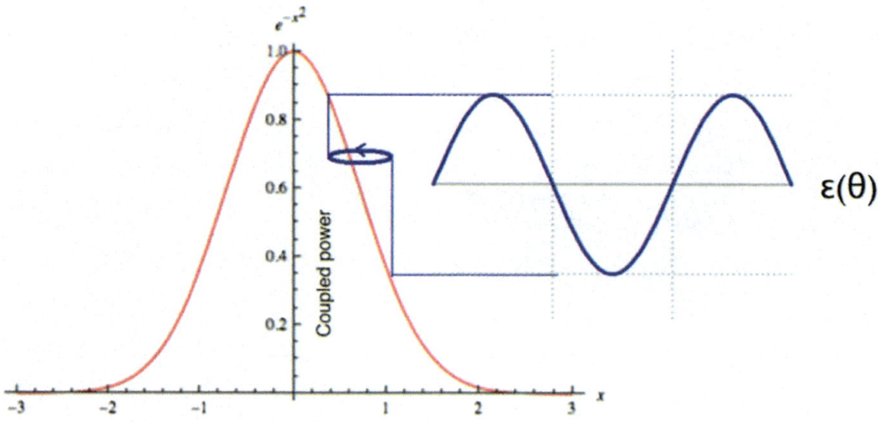


**Fig. 9.2** Graphical depiction of gradient determination via a circular dither, which modulates the observed coupled power (or other figure-of-merit). The phase of the modulation with respect to the dither indicates the direction toward maximum while its amplitude falls to 0 at optimum

The gradient search optimization can be preceded by an efficient, virtually vibrationless single-frequency area scan (either a rectangular sinusoid or a circular spiral) to select the main mode. In the referenced FMPA implementations described in the next section, both the area scan and parallel lock-on each require only a few hundred milliseconds.

$$|\varepsilon(\theta)| = \nabla I = \left(I_{\text{min}} - I_{\text{max}}\right)/I_{\text{min}} \tag{9.1}$$

## 9.6 It is 1985 All Over Again: An Ecosystem Rises

In the 1980s, today's robust and global ecosystem of semiconductor manufacturing toolmakers did not exist. Today, you can construct a fab just by issuing P.O.s, but players in that era had to either build their own production machinery or commission custom manufacturing-automation tooling [8].

Until recently, the situation in the silicon photonics industry has been unfolding in much the same way.

Analogous to the 1980s semiconductor industry, the parallel alignment technology has, since its introduction, been successfully adopted by captive manufacturing-automation integrators internal to some of the industry's largest players. Recently, a global toolmaker ecosystem has started to emerge, again replaying the script of 1980s-era semiconductor manufacturing.

Consider the very first process a SiPh device is subjected to: testing, while it is still on a wafer. The reason is familiar from conventional microelectronics manufacturing: The costs of minting the wafer are a small part of the finished price of a packaged chip. The costs of packaging even a conventional microprocessor can comprise the majority of the final cost of the finished product [9]. So, packaging faulty chips has grave economic consequences. Since yields can be problematic, this means only known-good chips should proceed from wafer into the packaging process. The validation is performed by specialized manufacturing tools: wafer probers. These cycle wafers through, subjecting each chip to contact from precise, needle-like electrodes that electronically stimulate the chip and observe its response, often via racks of sophisticated instruments.

Probing a photonic chip means coupling to its embedded photonic circuitry is also required. Simply making electrical contact is insufficient. Optical connections must be provided, and optical measurement of the chip's performance is required. Worse, while the smallest electrical contacts for probing are typically on the order of 20–30 μm square (meaning probe positioning accuracy must only be on this order), photonic couplings can require alignment accuracies about 1000 times better. And with thousands of SiPh chips on a wafer, this is clearly a costly process, so speeds are paramount.

The parallel gradient search technology provides the answer. It allows rapid optimization of each device's optical I/O across its multiple channels and degrees of

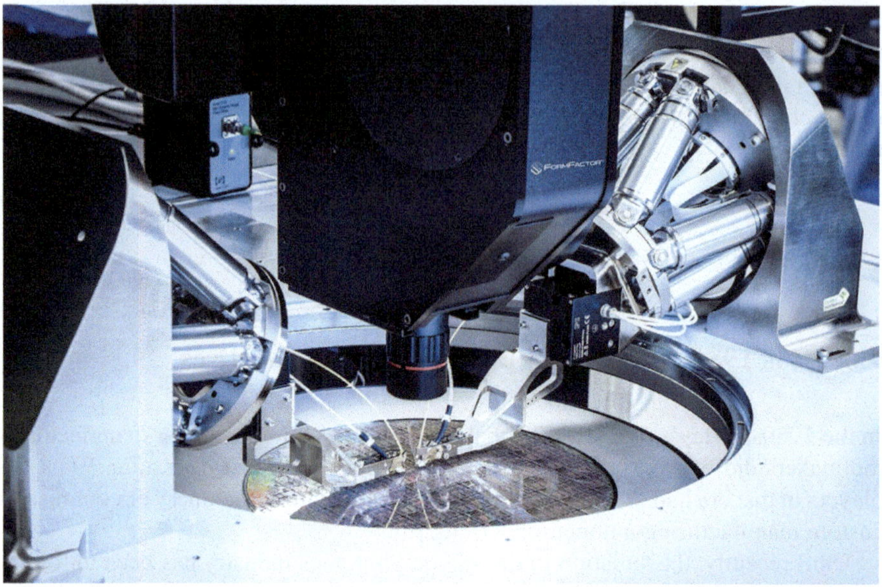

**Fig. 9.3** FormFactor's pioneering cascade CM300 photonics-enabled engineering wafer prober integrates PI's fast multichannel photonics alignment systems for high throughput, wafer-safe, nano-precision optical probing of on-wafer silicon photonics devices. Courtesy FormFactor, Inc.

freedom. It is no surprise that the first OEM implementations of the advanced microrobots with the parallel gradient search were in wafer probers [10] (Fig. 9.3).

Probing chips on-wafer is, of course, just the first step in their birthing process. The chips must next be singulated and then assembled into their final package, often with intermediate testing steps to validate the components' continued good health. At each of these assembly-and-test steps, alignment must again be performed.

(As mentioned, passive-alignment approaches which attempt to eliminate this need by substituting physical-fit techniques provide niche solutions for certain devices and process steps, but these often rely on narrow dimensional tolerances that are, themselves, costly and time-consuming to produce. In any case, they present fresh challenges for the crucial wafer-probing step upon which so much economic feasibility depends.)

The first adoption of the parallel alignment technology in a commercially available assembly-automation tool recently emerged from a micro-assembly toolmaker in The Netherlands, Tegema [11, 12]. Their solution, as illustrated in Fig. 9.4, implements the parallel alignment microrobots in a novel physical layout. The resulting tool is reportedly ten times faster and requires 50% less floor space than competitive entries based on legacy alignment technology.

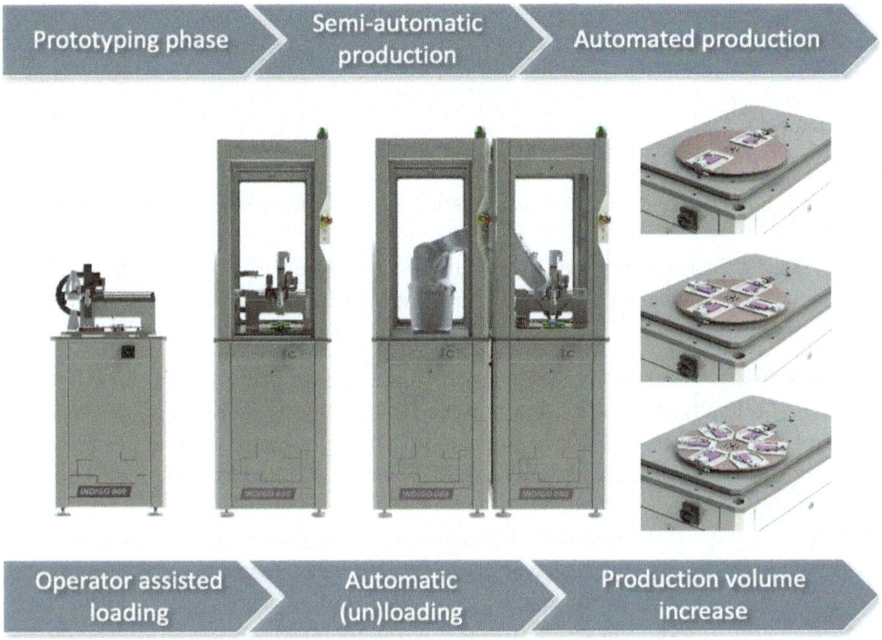


**Fig. 9.4** Fast parallel alignment technology is the enabler behind Tegema's groundbreaking automated-assembly tool for photonic devices, with 10 times higher throughput and 50% of the floorspace of legacy tools

## 9.7 Down Deep: Implementing the Parallel Alignment

Any axes in an FMPA system can perform any of these types of alignments (subject to the physical capabilities of the axes, of course). So, area scans can be performed with long-travel motorized-stage axes, which can be very handy for finding first light and accommodating device and fixturing reproducibilities. Gradient searches are most familiar from transverse optimization, but they can also be performed (for example) in a single linear axis, which is useful for localizing the beam waist in a lensed coupling, or in a gimbaling fashion to optimize an angular orientation, or in combinations of processes operating in parallel. There are many possibilities. These are general-purpose algorithms suitable for all kinds of optimizations, including bulk optic, cavity and pinhole alignments, using figures-of-merit ranging from familiar optical power metrology to photon counts, fast received signal strength indication (RSSI) queries, measures of beam quality in the case of laser-cavity optimization and modulation-transfer function (MTF), 2D Fourier transform and other measures of image quality in imaging optic assembly.

### 9.7.1 Area Scans

Scanning an area to determine the approximate location of the highest coupling peak is useful for a variety of tasks:

- First light seeking
- Profiling for dimensional characterization of a coupling. This can be an important process-control step
- Localizing the main mode of a coupling for subsequent optimization by a gradient search. This hybrid approach helps prevent locking-onto a local maximum and is very powerful.

Traditionally, raster and serpentine scans have been used for this purpose, but they pose several disadvantages. In both, a sweep in one axis is followed by an increment in the orthogonal axis. Each axis must settle at the conclusion of its motion. The stopping and starting can excite resonances in the devices and fixturing, so the settling can take some time.

More recently, scan approaches based on single-frequency sine waves were developed. These do not drive system resonances as raster and serpentine scans do, so they require no settling with every scan line.

In addition to reducing single-frequency rectangular and spiral area scans to a single command, FMPA controllers have automatic curve-fitting capabilities built in, plus a data recorder that can capture the profile on-the-fly for later retrieval, analysis or databasing. These area scans (Fig. 9.5) are very fast, 300 ms or so for typical NanoCube applications (100 × 100 μm area) and loads. The internal curve-fitting capability can fit a Gaussian to a fairly sparse scan (meaning an especially fast scan), allowing good localization of the optimum coupling point without taking time to do a really fine scan. Another capability of the firmware-based FMPA area scan is finding the centroid of a flat-top ("top-hat") coupling, such as seen when probing a deposited photodetector with a single-mode fiber (Fig. 9.6). This allows the scan to terminate with the fiber at the geometric center of a flat or tilted top-hat coupling, maximizing robustness.

A constant-velocity spiral scan may also be selected, allowing data to be acquired with constant spatial density across the spiral.

### 9.7.2 Gradient Search

A new enabling technology for the industry is the gradient search optimizations that can be performed in parallel. This allows (for example) an angular optimization to be performed while an *XY* optimization runs to keep everything locked-on and as a Z optimization also runs to maintain a focus.

Transverse optimizations tend to be the most sensitive and also the most affected by other alignments. Thus, transverse routines tend to be relegated to high-speed,

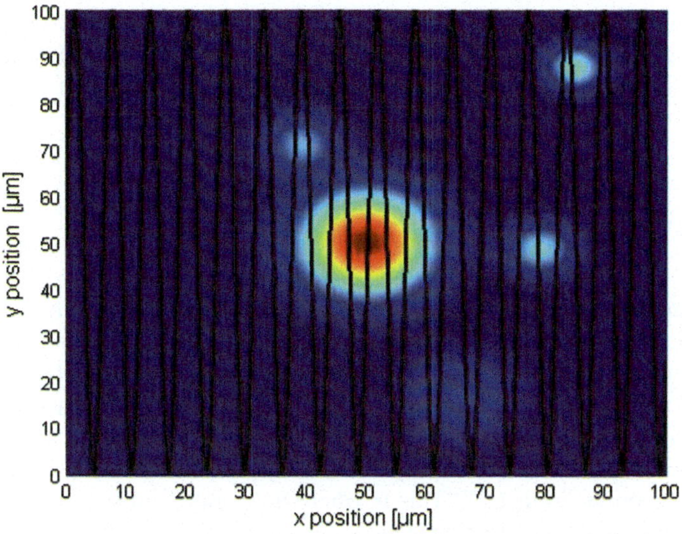


**Fig. 9.5** Simulated optical power distribution showing local maxima. A sinusoidal, rectangular-format area scan or a circular-format spiral scan are both single-frequency and thus do not drive resonances in the fixturing. This allows fast localization of the main mode. Curve fitting, an automatic option, allows accurate localization of peaks that lie between scan lines, allowing good localization with especially fast (typically 300 ms) scans

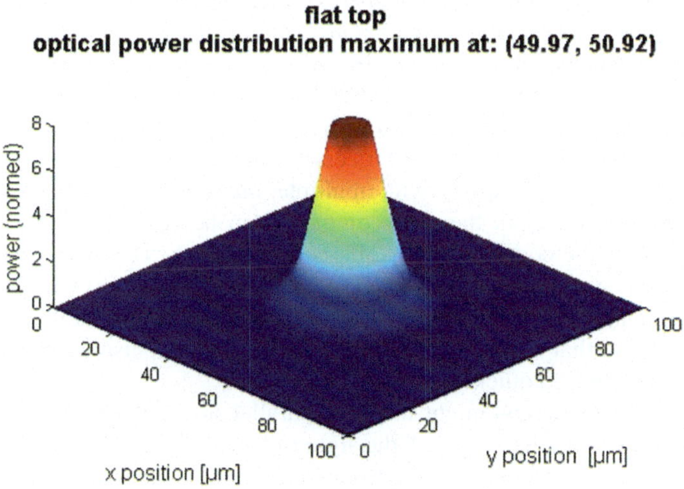


**Fig. 9.6** Flat or tilted top-hat couplings (seen, for example, when a fiber is aligned to a photodiode of larger transverse dimension) can be accommodated by an area scan with automatic centroid fitting. The resulting alignment targets the center of the plateau, the most robust place to terminate

high-resolution piezoelectric flexure stages such as PI's P-616 NanoCube. The speed, wear-free flexure construction and continuous tracking capability allows transverse optimization to be maintained during *Z* and angular optimizations that would ordinarily require the time-consuming, legacy looping approach.

### *9.7.3 Example Case: Array Device Alignment*

In the example just mentioned, an array device must be rotated in theta-*Z* while aligned in *XY*. If the mechanical *Z*-axis were perfectly colinear with the optical axis of one channel, then the job would be done. However, even when a hexapod is used (allowing, as it does, the rotational centerpoint to be placed anywhere in space by software command), as a practical matter chances are 100% that the mechanical and optical axes will not be colinear to the necessary precision to avoid de-alignment in *XY* during motions in theta-*Z*. This is where the parallel processing of (1) the *XY* lock-on and (2) the theta-*Z* optimization pays particular benefits.

In the example applications, we have spotlighted here (FormFactor's wafer prober and Tegema's automated-assembly tool), the low-level implementation provided by the PI controllers is straightforward. In each case, the overall alignment task is divided into processes just as we have done in the previous paragraph: "Process 1" is defined by the system developer as the *XY* transverse lock-on. Various parameters are established in the definition process (including mechanical-axis identifiers, desired frequency, target alignment tolerance, optical input channel, etc.); these may be subsequently changed at any time. The same step is performed for "Process 2" to identify the mechanical axis for theta-*Z* and so on. Later, with the processes having been defined, they can be commanded to operate simultaneously by issuing the *Fast Routine Start* command, specifying processes 1 and 2: *FRS 1 2*.

In the case of applications with multiple micro-robotic stations (hexapods, NanoCubes, stage stacks), the start command can be issued virtually simultaneously to each of the controllers via Ethernet interfacing. These interfaces are fast enough that the commands are received, processed and activated with millisecond-scale determinacy and processes across the stations proceed in parallel as intended, cooperatively, so that overall optimization of elements, inputs/outputs and degrees of freedom is achieved quickly.

Physical considerations might limit the parallelism available in an application. First among these is the speed of the figure-of-merit. For SiPh applications, the figure-of-merit being optimized tends to be optical power, either measured by a power meter or a photodiode/transimpedance amp (perhaps embedded in the device), or via rapid, continuous RSSI queries (perhaps protocol converted via a simple "Arduino"-style microcontroller for compatibility with the micro-robotic controllers). Consider the case of an optical power meter: Unfortunately, the world is rife with low-cost power meters with analog bandwidths on the order of 2 Hz. These are far too slow to drive millisecond-scale fast-automation processes, especially when several processes are desired to run at once! Similarly, for camera- and lens-assembly applications driven by a figure-of-merit based on some measure of image quality such as sharpness or contrast (MTF, 2DFF, etc.), the frame rate and computation speed of the computer performing the analysis is one determinant of how many processes can be achieved in parallel. In the same way, the physical speed capabilities of the motion devices can cap the number of processes possible at one time.

## 9.8 Alignment Enables the Quantum Era

The bizarre physics of the quantum world is the next disruptor for the high tech. It has the attention of government, corporate and venture investors, with major plays by significant companies such as Google and IBM and a venture pace that has quickened from \$7 million for two deals in 2015 to over \$300 million for seven deals in the first four months of 2020 [13, 14]. The bet here is that the indeterminate state of quantum computing elements (qubits) and their ghostly interactions can be leveraged to perform operations impractical in the rigid world of 1s and 0s. In the related field of quantum communications, the notion is that entanglement can convey information instantly without a wire, fiber or stream of electromagnetic radiation required to wend its way from sender to receiver, hindered by intervening obstacles and that pesky speed-of-light limitation.

The field is a Wild West of concepts and approaches, many of them cryogenic in implementation, which would suggest you would not find those in your next iPhone. As the rest of this book demonstrates, it is far too early to declare the victors among these competing ideas, but one thing is certain: Photonics will play a role.

For one thing, information, interrogation and stimulation can be carried in and out of a cryogenic system by photons, adiabatically and without being affected by or generating noise or interference. So, even applications that do not rely on the quantum properties of photons will benefit from utilizing them for connectivity. Then, of course, there are the approaches which do leverage the quantum properties of photons.

We see this emerging already as a signature of the most promising early implementations. For example, in the press releases heralding five-year-old Palo Alto startup PsiQuantum's \$215 million venture round in April, 2020, front-and-center was a FormFactor photonic wafer prober (just like the one discussed here earlier), probing a PsiQuantum wafer [15]. Besides the staggering sums lofting the company, perhaps the most impressive aspect of this news was that wafers were being probed at all! That, after all, is what one does when approaching production status. That in turn suggests that quantum computing is not distant, it is not science fiction, it is not something our grandchildren might see decades hence: Instead, it is imminent.

Generating photons with useful quantum properties is another fertile field of research and development. For example, researchers at Purdue University and the University of Illinois have built workstations integrating inverted microscopes and FMPA-capable piezoelectric nanopositioning stages. These leverage the systems'

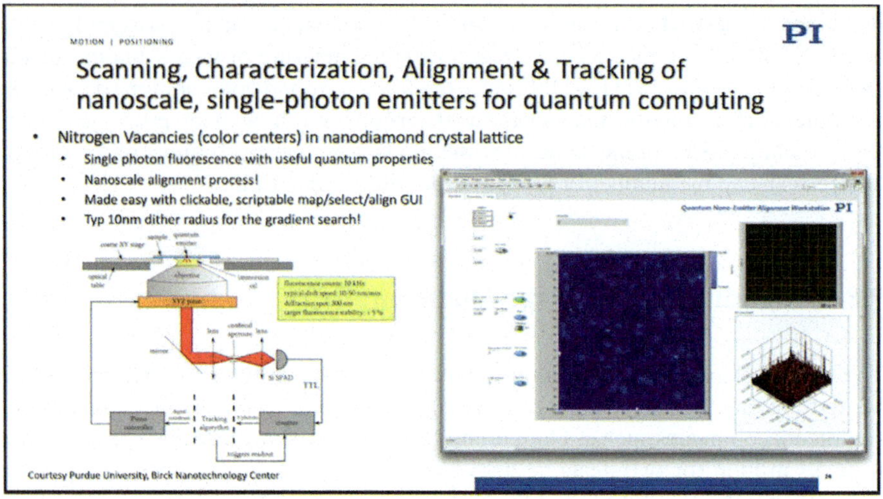


**Fig. 9.7** In this slide from several of the author's presentations, a software workstation utilizes a large/fine area scan to map atomic-scale single-photon emitters, with a scriptable, mouse-able interface to allow selection and tracking of emitters of interest for long-term characterization

fast scanning, optimization and tracking to localize, lock-onto and characterize fluorescing, atomic-scale defects in diamond crystals. Photons emitted by these nanoscale emitters have valuable quantum properties, and the emitters are long-lived, making them a fruitful focus of intensive research aimed at developing manufacturing-friendly elements for future quantum devices.

Figure 9.7 shows a slide featured in several of my industry presentations (e.g., [16]) that describes this workstation and shows a screenshot of the open-source software workstation constructed to support these studies. The large blue box is the output of a large, fine area scan. The dappled dots are the atomic-scale emitters. So what you see here are, essentially, atoms. Further, we constructed the software to allow the user to select an emitter of interest, click it with the mouse, and the system will lock-onto it, compensating for drift and allowing long-term characterization. The workstation is scriptable, for example in Python or MATLAB, allowing remote or unattended operations with the screen serving as a live dashboard. The high speed of the system allows it to churn through a multitude of tasks in coordination with other instrumentation. Recent work by these researchers has yielded not only the world's brightest single-photon emitters but chip-compatible approaches for integration into quantum devices and photonic integrated circuits [17].

## 9.9 Conclusion

From the early days of fiber optics 3–4 decades ago to the latest quantum explorations, a recurring theme is the sheer painfulness of working with photonic elements due to their insistence on being aligned to sub-submicron and even nanoscale accuracies. This is a recurring requirement for research, testing and multiple steps in manufacturing. The time-consuming legacy alignment technologies that served the field for all these years got us this far, but the mainstreaming of photonic connectivity means many thousands of devices, gaudily appointed with multiple functionalities and arrays of inputs and outputs, must be manufactured each day. Today, alignment processes that require minutes are roadblocks to the industry's success.

Fortunately, the industry now benefits from a new generation of alignment micro-robotic technology, a readily implemented approach that reduces the formerly lengthy and repetitive loops of alignment steps into as few as one parallel operation. Embedded in a wide and modular line of mechanisms spanning nanopositioners to hexapods to large stacks of stages to gantries, this has emerged as an essential tool from foundational research to volume production.

## References

1. S. Alexander, World economic forum (2016), https://www.weforum.org/agenda/2016/01/why-the-internet-is-outgrowing-moore-s-law. Accessed 4 July 2020
2. M. Robuck, Fierce telecom (2020), https://www.fiercetelecom.com/telecom/during-one-month-covid-19-telefonica-s-data-traffic-growth-exceeds-yearly-total. Accessed 4 July 2020
3. T. Fernholz, Quartz (2014), https://qz.com/179897/more-people-around-the-world-have-cell-phones-than-ever-had-land-lines/. Accessed 5 July 2020
4. D.A.B. Miller, IEEE/OSA J. Lightwave Technol. **35**, 343 (2017)
5. D. A. B. Miller, in *Optics and Electronics Seminar Series* (Stanford University, 2017)
6. Shin, Compound semiconductor (2019), https://compoundsemiconductor.net/article/107444/Building_Better_Lasers_On_Low-cost_Silicon/feature. Accessed 5 July 2020
7. S. Jordan, US5278934A (11 January 1994)
8. Chip History Center: An Intel Wafer Fab Cleanroom Circa 1980 (1981), https://www.chiphistory.org/128-an-intel-wafer-fab-cleanroom-circa-1980. Accessed 5 July 2020
9. G. Jeong, Amkor technology blog (2017), https://amkor.com/semiconductor-story/tales-semiconductor-material-packaging-saving-cost-vol-2/. Accessed 5 July 2020
10. FormFactor: autonomous silicon photonics measurement assistant (2019), https://www.youtube.com/watch?v=izM33rfVVc4&t=1s. Accessed 5 July 2020
11. Tegema: breakthrough in photonic assembly (2020), https://www.tegema.nl/en/breakthrough-in-photonic-assembly/. Accessed 5 July 5 2020
12. S. Dorrenstein, Assembly and packaging of photonic integrated circuits: a modular concept towards a fully automated process (2020), https://us02web.zoom.us/webinar/register/1115879758161/WN_UY5Esm4jQsmJIV1z_8_XwQ. Accessed 5 July 2020
13. B. Carson, Protocol (2020), https://www.protocol.com/manuals/quantum-computing/vc-investments-bullish-quantum-computing-coronavirus. Accessed 5 July 2020
14. E. Gibney, Nature (2019), https://www.nature.com/articles/d41586-019-02935-4. Accessed 5 July 2020

15. Quantaneo (2020), https://www.quantaneo.com/Quantum-Computing-Startup-Raises-215-Million-for-Faster-Device-%E2%80%8BBuilding-the-world-s-first-useful-quantum-computer_a486.html. Accessed 5 July 2020
16. S. Jordan, M. Price, Photonics (2020), https://www.photonics.com/Webinars/Advancements_in_Precision_Motion_Control_for/w204. Accessed 5 July 2020
17. S. Bogdanov, O. Makarova, Optica (2020), https://www.osapublishing.org/optica/abstract.cfm?uri=optica-7-5-463&origin=search. Accessed 5 July 2020

# Part III
# Advances in Computation Schemes

# Chapter 10
# Neuromorphic Silicon Photonics for Artificial Intelligence

**Bicky A. Marquez, Chaoran Huang, Paul R. Prucnal, and Bhavin J. Shastri**

**Abstract** Recent investigations in neuromorphic photonics, i.e. neuromorphic architectures on photonics platforms, have garnered much interest to enable high-bandwidth, low-latency, low-energy applications of neural networks in machine learning and neuromorphic computing. Although electronics can match biological time scales and exceed them, they eventually reach bandwidth limitations. Neuromorphic photonics exploits the advantages of optical electronics, including the ease of analog processing, and fully parallelism achieved by busing multiple signals on a single waveguide at the speed of light. In this chapter, we summarize silicon photonic on-chip neural network architectures that have been widely investigated from different approaches that can be grouped into three categories: (1) reservoir computing; reconfigurable architectures based on (2) Mach-Zehnder interferometers, and (3) ring-resonators. Our scope is limited to their forward propagation, and includes potential on-chip machine learning tasks and efficiency analyses of the proposed architectures.

## 10.1 Introduction

In the past few years, analog computing has gained important attention due to hardware acceleration-related milestones. Analog hardware has been known as a way to perform efficient operations—since those operations are embedded in the hardware

B. A. Marquez (✉) · B. J. Shastri
Department of Physics, Engineering Physics and Astronomy, Queen's University, Kingston, ON KL7 3N6, Canada
e-mail: bama@queensu.ca

B. J. Shastri
e-mail: bhavin.shastri@queensu.ca

C. Huang · P. R. Prucnal
Department of Electrical Engineering, Princeton University, Princeton, NJ 08544, USA
e-mail: choaranh@princeton.edu

P. R. Prucnal
e-mail: prucnal@princeton.edu

 
D. J. Lockwood and L. Pavesi (eds.), *Silicon Photonics IV*,
Topics in Applied Physics 139, https://doi.org/10.1007/978-3-030-68222-4_10

itself. Each analog system can only implement operations for which it has been built. Therefore, the design of analog machines for specialized task acceleration can be achieved if the task can be broken down into a physical model.

In the field of artificial intelligence (AI), analog computing has been considered as a potential venue to decrease energy and time requirements to run algorithms such as deep neural networks. Analog special-purpose hardware for artificial neural networks (ANNs) would require the construction of a machine that will physically model every single individual component of such networks. This expensive demand should be fulfilled if all neurons are expected to be used in parallel—which is indeed what is required. Considering that current deep networks sizes scale up to thousands or even billions of neurons to solve complex AI related tasks, such a requisite becomes a challenge. For instance, AlexNet requires 650,000 neurons to solve ImagNet [1].

Electronics and photonic platforms are currently the most promising technologies to tackle the expensive calculations performed by deep networks. The analog electronics approach is based on space-efficient topologies such as the resistive crossbar arrays [2]. Despite the fact that passive resistive arrays have been associated with low power consumption, crossbar arrays show fundamental performance flaws when used to model large neural networks. Large crossbar arrays are associated with high energy costs and low bandwidth. Overall, the power consumption, scalability and speed can be greatly affected when working with large networks [3].

The optical platform based on silicon photonics offers high scalability, great bandwidth and less energy consumption for longer distances than its electrical counterpart. This recent expanse in the demonstration of silicon photonic structures for photonic processing belongs to the second wave of optical computing. The first wave occurred in the early 1990s, where photonic processing of optical neural networks were consider slow and bulky. In fact, optical computing never reached the market due to the bulky size of free-space optical systems and to the intensive, low-bandwidth optoelectronic processing of the time. The rejuvenation of the field was possible due to the many advances in optical processing, higher bandwidth achieved and accessible fabrication facilities. As the second wave of analog optical computing comes with important advances for hardware acceleration, this chapter will be focused on the most relevant photonic processing demonstrations. Due to their speed and energy efficiency, photonic neural networks have been widely investigated from different approaches that can be grouped into three categories: (1) reservoir computing [4–7]; reconfigurable architectures based on (2) Mach-Zehnder interferometers [8, 9], and (3) ring-resonators [10–13].

Reservoir computing successfully implements neural networks for fast information processing. Such an advantageous concept is found to be simple and implementable in hardware, however the predefined random weights of their hidden layers cannot be modified [7]. We will describe how to build and utilize a silicon photonic reservoir computer for machine learning applications. This on-chip reservoir will be trained off-line to solve an on-line classification task of isolated spoken digits based on the TI46 corpus [14].

The first reconfigurable architecture that we explore is based on meshes of tunable silicon Mach-Zehnder interferometers (MZIs) that can implement fully connected neural networks. Such architectures are known to demonstrate their versatility to perform unitary matrix operations. In particular, some arrangements of MZIs can be used to implement singular value decomposition on a given layer of an ANN. Such a process is known as a method to reduce the dimensionality of data. Therefore, if applied on each layer of any given ANN, the most computationally expensive parts of AI processing would be alleviated. Here, we will show how to use a $4 \times 4$ MZI-based network to recognize 11 vowel phonemes spoken by 90 different speakers. The training of the network will be performed off-line via backpropagation and the inference stage on-line [15].

Finally, we present an architecture that can implement photonic convolutional neural networks (CNN) for image recognition. The competitive MNIST handwriting dataset [16] is used as a benchmark test for our photonic CNN. We will describe a scalable photonic architecture for parallel processing that can be achieved by using on-chip wavelength division multiplexed (WDM) techniques [11, 17], in conjunction with banks of tunable filters, i.e. photonic synapses, that implement weights on signals encoded onto multiple wavelengths. Silicon microring resonators (MRRs) cascaded in series have demonstrated fan-in and indefinite cascadability [18, 19] which make them ideal as on-chip synaptic weights with small footprint. At first, we train a standard two-layer CNN off-line, after which network parameters are uploaded to the photonic CNN. Then, the on-line inference stage is set to recognize handwritten numbers.

## 10.2 Background: Neuroscience and Computation

Digital computers are typically computing systems that perform logical and mathematical operations with high accuracy. Nowadays, such complex systems significantly outweigh human capabilities for calculation and memory. However, no-one could have imagined the extent that computers were going to reach when they were first envisioned. In 1822, the British mathematician Charles Babbage created the first mechanical computer that could work as an automatic computing machine. At the time, this architecture was known as an analytic engine which could compute several sets of numbers and made hard copies of the results. The core of this first general-purpose computer contained an arithmetic logic unit (ALU), a flow control, punch cards and integrated memory. Unfortunately, many adverse events occurred before the machine could be physically built.

It was not until mid 1930s that a general-purpose computer reached its physical form. Between 1936 and 1938, the German civil engineer Konrad Zuse created the first electromechanical binary programmable computer named Z1. Z1 was capable of executing instructions that were inputted through a punched tape that it could read. This computer also contained a control unit, integrated memory and an ALU that used floating-point logic.

During the same period of time, the British mathematician Alan Turing envisioned an architecture which followed similar principles. However, Turing went further and laid the groundwork for computational science. He defined a computing system as a machine for matching human computing capabilities. He proposed a machine that emulated a human agent following a series of logical instructions. The Turing machine manipulates symbols much as a person manipulates pencil marks on paper during arithmetical operations [20]. Turing motivated his approach by reflecting on how an ideal human computer agent would process information. He argues that human being's information processing principles can be replicated as they are based on symbolic algorithms that are being executed by the brain. Symbolic configurations are executable mechanical procedures that can be mimicked.

In fact, when Turing posed the question "can machines think?" [21], he stated that at least digital computers can follow the same fixed rules that we find in human agents. Digital computers are intended to carry out any operations which could be done by a human computer. Such rules are supplied in a book written following a well defined finite alphabet that consists in discrete strings of elements (digits), see Fig. 10.1. Although the alphabet is fixed, the book is not. The supplied book can be altered whenever it is put on to a new job. As for the previous models of computers, the Turing's digital computer was composed of: a memory component, an executive unit that carries out all the operations involved in calculations, and a control unit to see that those instructions are obeyed correctly in the right order.

If we are to compare a human agent with a Turing machine, we would see that there are many abstractions that should be made to perform one-to-one comparisons. Such abstractions disclose that human cognitive processes are completely procedural and follow any standard logic. Assumptions of this kind attempted to be human-inspired, but in fact they ended up influencing our previous understanding of human cognitive processes. In psychology, the computational theory of intelligence was mainly based

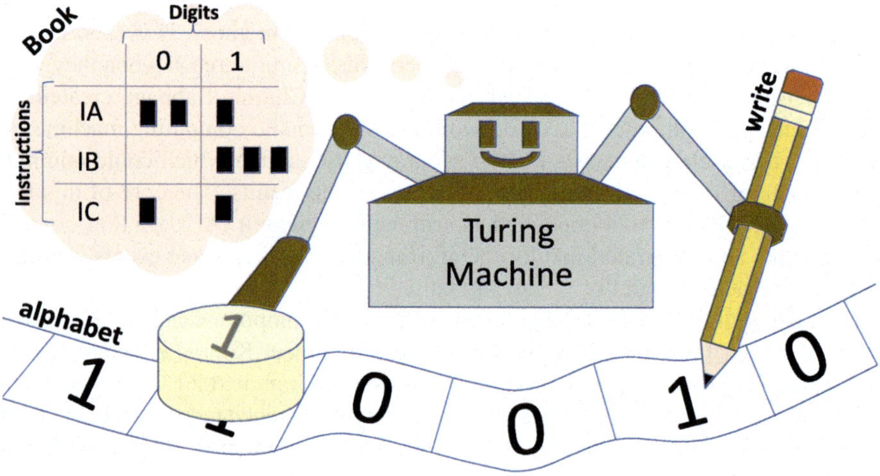


**Fig. 10.1** Turing machine

on the procedural type of abstraction introduced below. The logic theorists Newell and Simon defined intelligence as the process of specifying a goal, assessing the current situation to see how it differs from the goal, and applying a set of operations that reduce the difference [22, 23]. However, this theory has important flaws. The cognitive scientist Steven Pinker pointed out that the vast majority of human acts may not need to crank through a mathematical model or a set of well defined instructions. For instance, cognitive phenomena that include intuition and beliefs cannot be explained and these are frequently used by human agents to forecast and classify [24]. Therefore, a one-to-one mapping between human and digital computers might be a delusion. If so, what can be done from our side to match them?

## *10.2.1 Digital Versus Analog*

Neuromorphic computing approaches that do not use alphabets might be more suited to mimic brain processes. The human brain is one of the most fascinating organic machines in our body composition. The human brain contains around 100 billion neurons, which interact with each other to analyze data coming from an external stimulus. Its 100 trillion set of synaptic interconnections makes the processing of large amounts of information a task that turns out to be fast and well performed. A biological neuron is a cell composed of dendrites, body, axon and synaptic terminals, see the schematic illustration in Fig. 10.2a. The dendrites carry input signals into the cell body, where this incoming information is summed to produce a single reaction. In most cases, the transmission of signals between neurons are represented by action

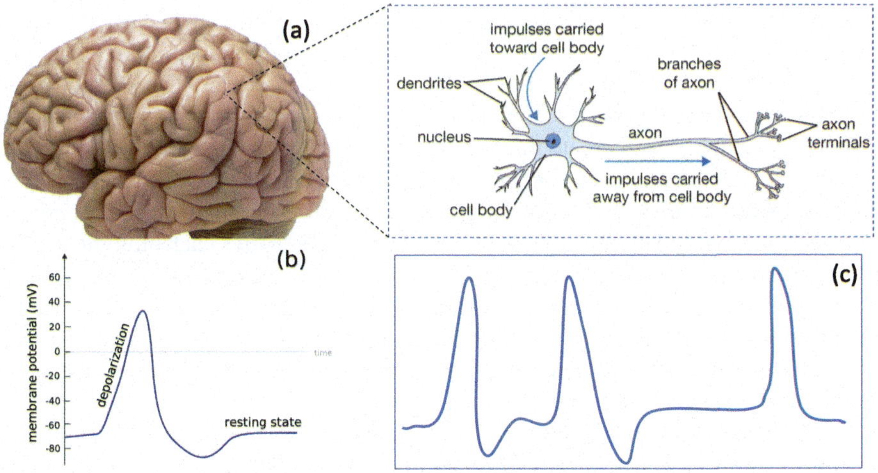


**Fig. 10.2** **a** Illustration of a brain and a biological neuron model, **b** action potential and **c** spiking waveform

potentials at the axon of the cell, which are changes of polarization potential of the cell membrane. The anatomic structure where the neurons communicate with each other is known as a synapse [25]. The cell membrane has a polarization potential of −70 mV at resting state, produced by an imbalanced concentration inside and outside of its charged molecules. This change of the polarization happens when several pulses arrive almost simultaneously at the cell. Then, the potential increases from −70 mV to approximately +40 mV. Some time after the perturbation, the membrane potential becomes negative again but it falls to −80 mV [2, 3]. The cell recovers gradually, and at some point the cell membrane returns to its initial potential (−70 mV), as schematically illustrated in Fig. 10.2b.

The binary nature of digital computers can also be used to mimic neural processes. The digital logic can be compared with the all-or-nothing ("1" or "0") process of the action potential transmission. Nevertheless, the whole story is not being said. This model does not include information about the times between action potential transmissions. The frequency at which neurons spike have functional significance that cannot be dismissed. Therefore, a complete model of neural dynamics would specify the waveform that a series of spikes perform in time (see Fig. 10.2c). This phenomenon has an impact in neural information encoding. For example, the strength of a given stimulus is coded as a frequency value. The stronger the stimulus (information) to the neuron, the smaller the time between spikes [26].

If our nervous system encodes information in such a way, then a digital only representation of the neural functions is inherently incomplete. If we were forced to use it, we would find out that a waveform could be described as well using a binary alphabet, where each spike is a "1", and each resting state is a "0" in time. For this to work, each of the waveforms constituted by a series of spikes and resting states should be identifiable and separable. This will ensure that they can be standardized and used in a "reference book" of operations that describe human cognition. A task of this kind would need an extremely large book which includes each waveform that we detect. Then, each waveform has to be connected to others in a meaningful manner in order to describe different human cognitive processes. This method could be difficult and inefficient to follow if we take into account that some neurons can also spike randomly and create random waveforms that have no meaning.

A more realistic interpretation incorporates a continuous-type model that is typically described by analog systems. This process will create a one-to-one map between the neural system and the analog machine. For this to come true, each biological quantity would be modeled by an analog quantity, i.e. a biological neuron would have its equivalent analog artificial model. For an architecture such as the brain this could be a demanding requirement. As previously introduced, the human brain contains around 100 billion neurons and 100 trillion synaptic interconnections that need to be represented in an artificial machine.

An equivalent analog machine to the brain can possibly be achieved, but we wonder at what cost. The average human brain burns 1300 calories per day in the resting state (54.16 kcal/h = 62.94 watts), which accounts for 20% of the body's energy use. For all the incommensurable amount of operations that the brain needs to perform per day, this amount of energy looks really low. Interestingly, when the brain is thinking, it can burn around 300 calories (12.5 kcal/h = 14.527 watts). As human thoughts can be used for training purposes, a brain that has training activity during e.g. four days can burn around 120 watts in total—for training only. In contrast, a digital machine playing Go such as AlphaGo (which simulates brain training process using binary logic and alphabet) can do the same while burning 50,000 times more energy in one task only [27]. A dedicated (special-purpose) analog machine per task should resolve this problem.

### *10.2.2 Artificial Neural Networks*

Towards the utilization of analog machines to map some of the brain circuitry, we need to define how to model the biological neurons and synapses. Among many others, the most commonly used neural models are spiking artificial neurons and perceptrons. While spiking artificial neurons are significantly more biologically realistic, the field of artificial intelligence (AI) is currently perceptron-based. Nowadays, most significant advances in AI have been achieved using a perceptron as an artificial model of the neuron, therefore in this chapter we will assume that all our artificial neuron models are perceptron-based. A perceptron is shown in its general model by Fig. 10.3a. The output $y$ of the neuron represents the signals coming from the axon of a biological neuron, and it is mathematically described by

$$y = f(\mathbf{W} \cdot \mathbf{x} + b). \tag{10.1}$$

The $x_i$ inputs transmit the information to the neuron through the weights $W_i$, which correspond to the strength of the synapses. The summation of all weighted inputs, and their transformation via activation function $f$, are associated with the physiological role of the neuron's cell body. The bias $b$ represents an extra variable that remains in the system even if the rest of the inputs are absent. The activation function can be linear or nonlinear, and it mimics the firing feature of biological neurons. A nonlinear activation function can be used to set a threshold from which to define activated and

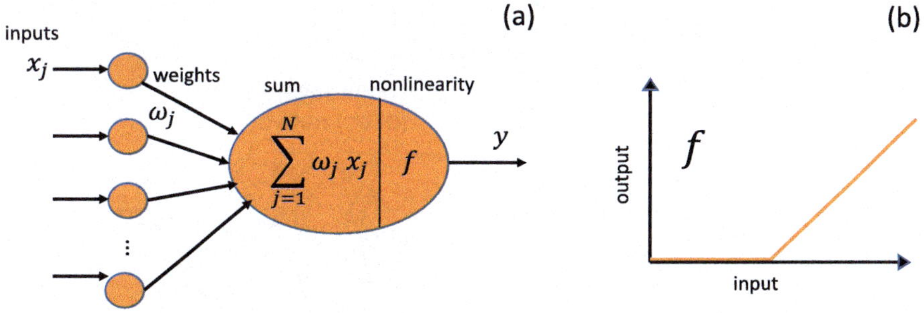


**Fig. 10.3** **a** Schematic diagram of a perceptron and **b** ReLU function

deactivated behaviors in artificial neurons. For instance, a ReLU function $f$ (see Fig. 10.3b) mimics a spiking neuron when its weighted addition $(\mathbf{W} \cdot \mathbf{x} + b) > 0$, otherwise the neuron is considered to be in a resting state.

Indeed, a perceptron can be seen as a compact model of a spiking neuron when data is injected in its input terminals, as the neural response has a one-to-one correspondence with the provided input. It means that different inputs will have different and unique (identifiable and separable) neural responses. Such compact responses have the same functionality of waveforms in biological neural responses.

ANNs are built using perceptrons as neural primitives and the synaptic connections are typically defined as real-valued numbers. Such numbers can be either positive or negative to mimic excitatory and inhibitory neural behavior. Among many other categorization, ANNs can be categorized in the following two main branches: feed forward and recurrent neural networks. Recurrent neural networks (RNNs) are made up of three layers of neurons: input, hidden and output. A RNN has a particular architecture in which outputs of its individual neurons serve as inputs to other neurons on the same hidden layer, see Fig. 10.4a. These feedback connections allow networks' input information to be recycled, transformed and reused. Consequently, RNNs are able to generate internal dynamics that could be advantageous for the development and maintenance of patterns in the networks' high dimensional space (defined by the neurons) [28].

The second branch is led by feed forward neural networks (FFNNs) that represent a conceptually similar configuration to RNNs. FFNNs are also composed from input, internal and output layers. But, different from RNNs, they do not involve any internal feedback between neurons as part of their architectures, see Fig. 10.4b. These networks usually build models that perform smooth function fits to input information. In many cases, FFNNs are designed with more than one layer of neurons to enhance

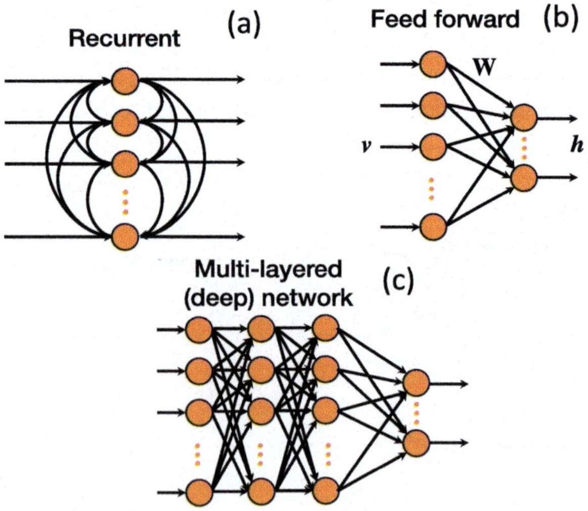


**Fig. 10.4** Schematic diagrams of a **a** recurrent neural network, **b** one layer feedforward neural network and **c** deep network or multilayer neural network

their information processing competencies. Such larger architectures are known as deep networks (Fig. 10.4c), and they are used to solve highly complex problems previously deemed unsolvable by classical methods in an efficient manner, such as pattern classification [1] or human-level control [29].

## 10.3 Electronics and Photonic Platforms

Attempts to build efficient perceptron-based neural networks have been reported through out recent years. Efficiency is expected on any machine that attempts to match or outweigh human computing capabilities. An interesting computing acceleration technique consists in the use of hardware units that perform multiply-accumulate (MAC) operations very fast. A MAC unit performs multiplications and accumulation processes: $a + (w \times x)$. Multiple MAC operations can be run in parallel to perform complex operations such as convolutions and digital filters. By comparing a MAC unit with the concept of a perceptron, we realize that they have a similar mathematical model. A neuron made of $M$ inputs and synapses, one output and a bias term can be therefore written as an array of $M$ MAC operations ($a_i = a_{i-1} + w_i x_i$) [3]. For instance, the weighted addition ($w_1 x_1 + w_2 x_2 + \cdots + w_M x_M + b$) of the neuron:

$$y = f(w_1 x_1 + w_2 x_2 + \cdots + w_M x_M + b), \tag{10.2}$$

can be performed in $M$ blocks as follows: if $a_0 = b$, then the first MAC operation is $a_1 = a_0 + w_1 x_1$. The second MAC operation would be $a_2 = a_1 + w_2 x_2$; and the last one $a_M = a_{M-1} + w_M x_M$. The activation function $f$ can be applied to all the weighted summations at the end of the process. Consequently, a neural network of size $N$ requires $M \times N$ MAC operations per time step. In a fully connected network where $M = N$, the number of MAC operations per time step is $N^2$. MAC operations are typically used in implementations of neural networks in digital electronics. Nevertheless, the serialization of the summands to perform weighted addition makes this process inefficient. As such operations follow a serial processing, the overall computation efficiency will depend on the clock speed of the digital machine. Since 2014, clock rates have saturated at around 8 GHz, and chip designers are looking for alternative solutions such as full parallelism. The most promising technologies used for this purpose are based on specialized analog electronic and photonic platforms.

### *10.3.1 Electronics*

The analog electronics approach is based on space-efficient topologies such as resistive crossbar arrays. In Fig. 10.5a we show the layout of these devices, that consists of tunable resistive elements at each junction that could represent a synaptic weight element. They are typically built as a metal-insulator-metal sandwich, where the insulator can be made of $SiO_x$, with $x < 2$. Tuning is typically performed through the application of input voltages (or currents) to the device in order to change its resistance value. In this case, each column from such a mesh can represent the weighted addition of any neuron. Input values that are injected through voltages $V_i$ are distributed through all the $N$ synaptic weights, which are represented by resistors of conductance $G_j$. The output of the neuron is represented by the resultant current $I_i$ at each column of $N$ elements. Such currents are obtained by the Kirchhoff's current law, where the multiplications and summations ($I_j = \sum_i V_i \cdot G_{i,j}$) are performed.

This architecture is advantageous as the $N^2$ MAC operations can be executed in parallel. If we are to use passive arrays of resistive elements to perform MAC operations, we need to determine how many analog weight values can be represented per resistive device. This decision should be taken for the sake of the entire machine efficiency. It has been shown in [3] that for a maximum of 16 analog values (4 bits) such devices consume fairly low energy. The total energy consumption of an electronic crossbar array is 4.0 aJ/MAC. This number stays almost unchangeable for 8 bits. However, crossbar arrays show fundamental performance flaws when used to model large neural networks. Large crossbar arrays ($L > 100\,\mu$m) are associated with high energy costs and low bandwidth. Overall, the power consumption, scalability and speed can be greatly affected when working with large networks. Nowadays, ANNs have been doubling in size every 3.5 months, therefore platforms with strong limitations to model large neural networks will add to the problem. Accordingly, we continue exploring different platforms in the following.

## *10.3.2 Photonics*

The photonic platform comes as an ideal candidate due to its high scalability, great bandwidth and less energy consumption for longer distances than its electrical counterpart. In particular, silicon photonics can offer analog processing on integrated circuits with high-speed and low power consumption. The silicon mate-

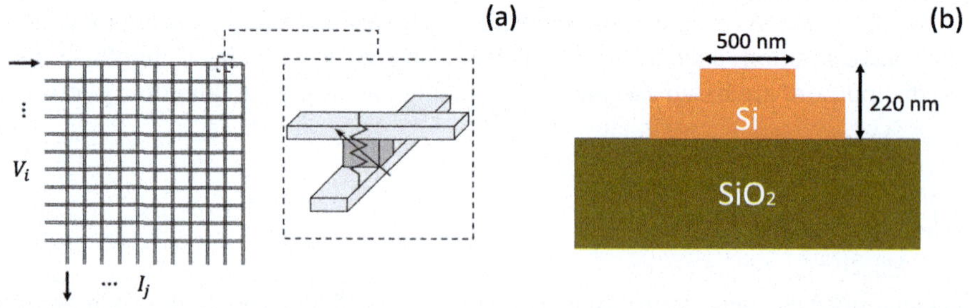


**Fig. 10.5** **a** Electronic crossbar array and **b** schematic diagram of the cross-section of a rib waveguide

rial for integrated photonics offers a manufacturable, low-cost and versatile platform for photonics. In fact, silicon photonic devices can be manufactured with standard silicon foundries and some modified versions of their processing capabilities [30, 31]. Additionally, silicon is well-known for its high refractive index contrast, which allows for submicrometer waveguide dimensions and dense packing of optical functions on the surface of the chip, i.e. small footprint.

In order to make the light propagate in a small micrometer area, a slab waveguide made of three layers can be designed. As shown in Fig. 10.5b, the core layer of the slab waveguide is made of a high index material (silicon, index = 3.48) and the two cladding layers with lower index (air, index = 1.0). In particular, the photonic waveguide illustrated in the same figure is also known as a rib waveguide, which allows for electrical connections to be made to the waveguide [32]. Such a waveguide typically sits on silicon dioxide $SiO_2$ whose refractive index is 1.54. Each on-chip waveguide is designed with a width of 500 nm and a thickness of 220 nm—associated with single mode operation. An outstanding feature of rib waveguides is their capability for signal parallelization, in which hundreds of high speed, multiplexed channels can be independently modulated and detected. Optical channels are defined by different wavelengths, and can be used to transmit independent information with small optical crosstalk.

We introduce a first photonic device which implements weights on signals encoded onto multiple wavelengths. Tuning a given filter on and off resonance changes the transmission of each signal through that filter, effectively multiplying the signal with a desired weight in parallel. Silicon microring resonators (MRRs) have demonstrated to be ideal as on-chip synaptic weights with small footprint [10], see Fig. 10.6a. MRRs can be designed with a ring and one (all-pass) or two (add-drop) adjacent bus waveguides. MRRs are devices capable of trapping light at certain frequencies at which they resonate, according to their physical characteristics. The resonance frequency can be obtained from the wavelength equation $\lambda_R = 2\pi R n_{\text{eff}}/m$, where $R$ is the radius of the ring and $n_{\text{eff}}$ is the effective refractive index. A wide variety of synaptic weights can be represented by an MRR through the tuning of the waveguide refractive index or (as we will see later) by tuning the amount of light that gets trapped in the waveguide. An array of $N$ MRRs can emulate the weighted addition of a single neuron if add-drop MRRs and a photodetector are added to the model as shown in Fig. 10.6b. Each MRR implements a weight value $w_i$, the input values $x_i$ are injected into the neuron through a modulator (a microring [33] or a Mach-Zehnder [34]), and the photodetector adds up all optical signals $[\sum_i w_i \cdot x_i]$.

A second design implements weights encoded onto different phases of Mach Zehnder interferometers (MZI). As shown by Fig. 10.6c, a MZI is designed with a splitter that splits the incoming light into two branches (the upper and lower waveguide), and then a coupler recombines them again. Splitters and couplers are designed with Y-branches or directional couplers [10, 32]. By tuning the amount of phase delay on one of a MZI's arms, a specific weight value can be set. This causes an

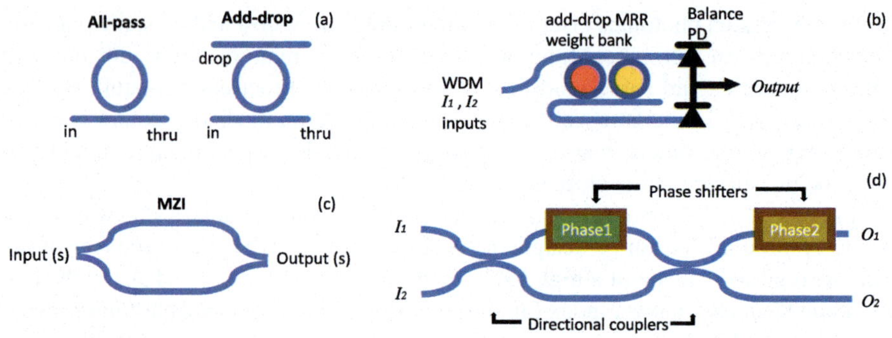


**Fig. 10.6** **a** all-pass and add-drop MRRs, **b** add-drop MRR weightbank with a balance photodetector, **c** the model of a MZI and **d** 2 × 2-MZIs with phase shifters and directional couplers

array of input optical signals **x** to be multiplied by a weight matrix **W**, as **W** · **x**, see Fig. 10.6d. Therefore, each row of the matrix **W** represents the weighted addition part of a neuron.

Meshes composed of $N^2$ MRRs (plus $N$ photodetectors) or $N^2$ MZIs can accomplish $N^2$ MAC operations of the kind $[w_{i,j} \cdot x_i]$ each. Furthermore, both methods can be tuned via the thermo-optic effect, where a voltage (or current) is applied to optimize their weight values. For a maximum of 16 values (4 bits), such devices consume way less energy than their electronic counterparts. The total energy consumption of optical meshes is estimated to be around 2.0 aJ/MAC. However, as reported in [3], the energy required to represent 256 values (8 bits) is 40 times bigger than the 4-bits case. Thankfully, many ANNs still work well when computational precision is low [35, 36].

The total energy consumption estimated in this section does not take into account the amount of energy required to tune these devices to specific weight values. These passive calculations are agnostic to the tuning technology that is used to set the weights, the type of activation function, the control unit and the memory unit used to build a system capable of modeling ANNs. Each of those modules will add to the overall power consumption and speed constrains. Nevertheless, MAC operations have been found to be the most burdensome hardware bottlenecks in ANNs as $N$ grows large [3]. Therefore, such afore mentioned modules should be optimized for the sake of the machine efficiency maintenance. In the following sections, we will describe how to build potentially efficient analog photonic ANNs.

## 10.4 Silicon Photonic Neural Networks

As introduced above, the high speed and parallelism property of the light allows for optical information processing at a high data rate [37]. Furthermore, the high scalability, submicrometer and dense packing of optical functions of silicon integrated photonics makes this a promising technology not only for accelerating AI, but also for ultra fast special-purpose computing. In silicon photonic platforms, there are two main approaches that leverage some (or all) of those afore mentioned features. The first approach harnesses the coherent properties of the light. Under the coherent approach, we present two different experiments. The first experiment is based on MZI-based silicon integrated circuits. This circuit has the property of being fully reconfigurable through thermo-optic tuning. The second coherent experiment differes from the MZI-based circuit in that it consists of a passive non-reconfigurable (non-tunable) circuit for reservoir computing applications only. Finally, we introduce an experiment under the incoherent approach that harnesses the parallelism feature of light and is fully-reconfigurable.

### 10.4.1 MZI-based Processing Unit

As commented in the previous section, weighted additions (MAC operations) are one of the most computationally expensive parts of AI processing. In order to accelerate these operations, a dimensionality reduction method can be applied on any ANN weight matrix $W$. One of the most used methods for this purpose is known as singular value decomposition (SVD). SVD could automatically decompose $W$ as $W = U\Sigma V^{\dagger}$, where $U$ is a unitary matrix, $\Sigma$ is a rectangular diagonal matrix and $V^{\dagger}$ is the complex conjugate of the unitary matrix $V$. Additionally, if we also leverage the high speed property of optical integrated circuits, we can accomplish ultra fast accelerators for AI. Nevertheless, we would need to know how to implement unitary and diagonal matrices in optics.

In fact, theoretical and experimental models have shown that certain arrangements of beam splitter devices can be used to represent unitary matrices [38]. For silicon integrated photonics, MZI have been tested to be equivalent replacements of beam splitters [39]. In Fig. 10.7, we illustrate a MZI-based mesh that can implement $W$ decomposed in $U$, $\Sigma$ and $V^{\dagger}$ [15]. Each $2 \times 2$-MZI (Fig. 10.6(d)) is built with two phase shifters that can be tuned to program the nanophotonic circuit to solve a task. In this case, the Phase term 1, defined as $\theta$, controls the power at the MZI outputs, and the Phase term 2, defined as $\phi$, determines the relative phases of those outputs [40].

A $2 \times 2$-MZI implements the following transformation [40]:

$$R(\theta, \phi)_{MZI} = e^{j\left(\frac{\theta+\pi}{2}\right)} \begin{pmatrix} e^{j\phi} \sin\frac{\theta}{2} & e^{j\phi} \cos\frac{\theta}{2} \\ \cos\frac{\theta}{2} & -\sin\frac{\theta}{2} \end{pmatrix}. \tag{10.3}$$

Such a transformation is associated with the SU(2) rotation group. Unitary matrices $U$ and $V$ are therefore represented by sets of SU(2) transformations that can perform all rotations on the two MZI input signals. For this purpose, the implementation of $U$ and $V$ matrices requires the MZIs' phases $\theta$ and $\phi$ to be tuned. The diagonal matrix $\Sigma$ can be implemented as the change of power to the input signals,

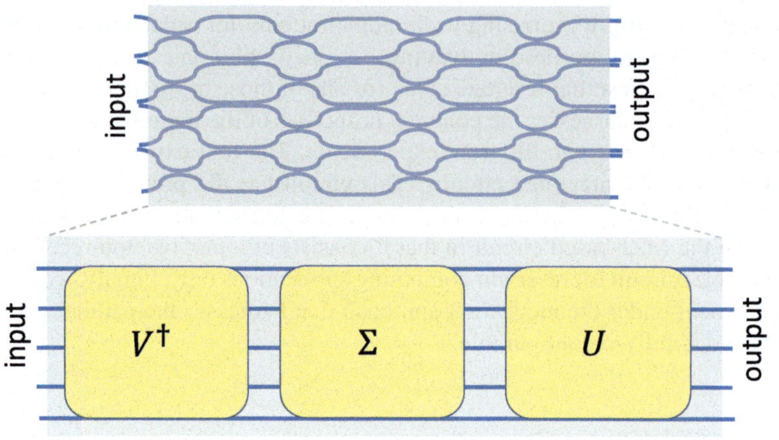


**Fig. 10.7** Illustration of a MZI mesh that performs singular value decomposition

then only $\theta$ should be optimized. The resulting chip will apply a weight matrix $W$ to the amplitude of the input signals [15].

To program the chip, an $N \times N$ unitary matrix $U(N)$ is multiplied by a succession of unitary matrices $R(\theta, \phi)_i$ of the same dimensionality until all its off-diagonal elements are set to zero. This process can be viewed as the result of a rotation of each column of the matrix $U(N)$. In total, $m = N(N-1)/2$ rotations will be performed with $2 \times 2$-MZIs. The resultant diagonal matrix is equal to the identity:

$$U(N)R(\theta_1, \phi_1)_1 R(\theta_2, \phi_2)_2 \ldots R(\theta_m, \phi_m)_m = I, \tag{10.4}$$

such that $U(N)$ can be expressed as a succession of multiplications of the $R(\theta, \phi)_i$ inverses:

$$U(N) = R(\theta_m, \phi_m)_m^{-1} \ldots R(\theta_2, \phi_2)_2^{-1} R(\theta_1, \phi_1)_1^{-1}. \tag{10.5}$$

This representation of $U(N)$ allows for its implementation as a set of phases $\theta$'s and $\phi$'s that will be determined by adjusting the voltage on each internal thermo-optic phase shifter of MZIs from $m = 1$ to $m = N(N-1)/2$. This process is repeated for the unitary matrix $V^{\dagger}$. The diagonal matrix $\Sigma$ contains the effects of photodetection on the uncertainty of the system output [15], and it is implemented by tuning the voltage on each internal thermo-optic phase shifter of a different set of $N$ MZIs.

#### 10.4.1.1 MZI-based Deep Neural Networks

This method can be extended to model large deep neural networks where each layer will be represented by an optical inference unit, followed by a non-linear function that can also be implemented with MZI-based circuits [41], see Fig. 10.8a. As a proof of concept, a $4 \times 4$ weight matrix $W$ (also $4 \times 4$ MAC operations) was implemented on-chip [15], see Fig. 10.8b. This circuit represents matrices $U$ and $\Sigma$. The matrices $U$ is composed of $m = 6$ sets of $2 \times 2$-MZIs, and the matrix $\Sigma$ is composed of $N = 4$ sets of $2 \times 2$-MZIs. $U$ and $\Sigma$ are implemented on a single pass through the chip, and $V^{\dagger}$ is implemented separately in this experiment. The authors of this work suggest that a larger circuit would be required to perform the full matrix decomposition on a single pass. As the unitary matrices $U$ and $V$ are represented by SU(2) transformations that basically perform all rotations of the two MZI input signals, the phases $\theta$ and $\phi$ have to be tuned on MZIs from (1) to (6). The diagonal matrix $\Sigma$ can be implemented as the change of power to the input signals, then only $\theta$ should be optimized on MZIs from (7) to (10).

The first step of this experiment consists in the determination of the weight matrix $W$ for a specific task. The proposed task is called vowel recognition, where 11 vowel phonemes spoken by 90 different speakers have to be recognized. The training of the network was perform via backpropagation off-line (not on-chip but on a computer), and the inference stage was perform on-line (on-chip). Once $W$ is obtained, this matrix is decomposed in $U$, $\Sigma$ and $V^{\dagger}$. The unitary matrix $U$ could be programmed by successively multiplying it by the following array of rotation matrices until all off-diagonal elements are set to zero [40]:

$$U(4)R(\theta,\phi)_1R(\theta,\phi)_2R(\theta,\phi)_4R(\theta,\phi)_3R(\theta,\phi)_5R(\theta,\phi)_6 = I(4), \qquad (10.6)$$

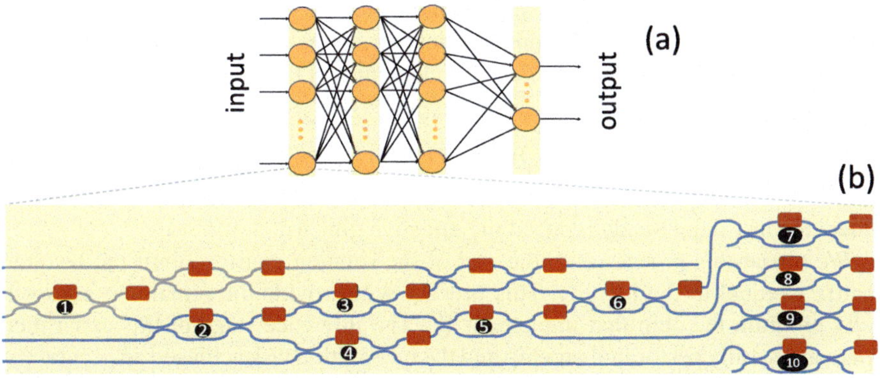


**Fig. 10.8** **a** The model of a deep network with four hidden layers, and **b** a $4 \times 4$ weight matrix represented with sets of $2 \times 2$-MZIs

where,

$$U(4) = \begin{pmatrix} U_{1,1} & U_{1,2} & U_{1,3} & U_{1,4} \\ U_{2,1} & U_{2,2} & U_{2,3} & U_{2,4} \\ U_{3,1} & U_{3,2} & U_{3,3} & U_{3,4} \\ U_{4,1} & U_{4,2} & U_{4,3} & U_{4,4} \end{pmatrix}; \tag{10.7}$$

and

$$R(\theta_i, \phi_i)_i = e^{j\left(\frac{\theta_i+\pi}{2}\right)} \begin{pmatrix} e^{j\phi_i} \sin\frac{\theta_i}{2} & e^{j\phi_i} \cos\frac{\theta_i}{2} & 0 & 0 \\ \cos\frac{\theta_i}{2} & -\sin\frac{\theta_i}{2} & 0 & 0 \\ 0 & 0 & 1 & 0 \\ 0 & 0 & 0 & 1 \end{pmatrix}. \tag{10.8}$$

For instance, when the operation $U(4)R(\theta_1, \phi_1)_1 = A(\theta_1, \phi_1)$ is performed, the element $A_{4,1}$ resultant matrix can be set to zero. Therefore, $\theta_1$ is calculated from:

$$\theta_1 = 2\tan^{-1}\left(-\frac{U_{4,2}}{U_{4,1}}\right). \tag{10.9}$$

We incorporate $A_{4,1} = 0$ to the resultant matrix $A(\theta_1, \phi_1)$, and multiply it by the second rotation $R(\theta, \phi)_2$ to obtain a new matrix $B(\theta_2, \phi_2)$. If the element $B_{4,2}$ is set to zero, then $\theta_2$ can be obtained as well. After we incorporate the element $B_{4,2} = 0$ to the resultant matrix $B$, we proceed to multiply it by $R(\theta, \phi)_4$. This operation allows us to obtain $\theta_4$ once the element $C_{4,3}$ of the resultant matrix $C(\theta_4, \phi_4)$ is set to zero. Once all elements of the last row except the one on the diagonal in $C(\theta_4, \phi_4)$ are set to zero, the next step consists in setting all the elements of its last column to zero—except the diagonal. Since all applied transformations are unitary, the last column will contain only zeros, and the diagonal element will be set to one:

$$C(\theta_4, \phi_4) = U(4)R(\theta, \phi)_1 R(\theta, \phi)_2 R(\theta, \phi)_4 = \begin{pmatrix} U(3) & 0 \\ 0 & 1 \end{pmatrix}, \tag{10.10}$$

where the last step already reduced the effective dimension of $U$ to 3.

We repeat this process with the rest of the rotation matrices until the resultant matrix is equal to the identity. This process will provide the remaining phases to fully program the chip that implement $W$. The inference stage of this experiment was successfully performed on-chip and the recognition task achieved an accuracy of 76.7%. According to the energy calculations performed by the authors of this experiment, a four neuron system would dissipate approximately 8 mW. In the inference stage, this system can perform such calculations quite efficiently, but the heaters used to program each individual MZI burns an important amount of power.

### *10.4.2 Photonic Reservoir Computing*

Reservoir computing is a term that encompasses some types of RNNs that can solve complex tasks with a simplified training methodology. This trend started with echo state networks (ESNs) [42] and Liquid State Machines (LSMs) [43]. These two network architectures come from the fields of machine learning and computational neuroscience, respectively. More recently, a delay echo state network appeared as a novel computational machine. Such an advantageous concept is found to be simple and implementable in hardware [7].

Reservoir computers can also be seen as a random RNN, where the synaptic weights are Gaussian or uniformly [44] distributed. Although the brain's connectivity cannot be assumed to be fully random, there is experimental evidence supporting the assumption that some parts of the brain are described by stochastic architectures. For example, in insects' olfactory systems the odour recognition process is performed by olfactory receptor neurons with structureless (random) synaptic connections [45, 46]. Therefore, reservoir computers might be useful to model a few random (non-trainable) neural dynamics, but it is limited to that.

Reservoirs consist of reservoirs of $m$ neurons in state $\mathbf{x}_n$, internally connected in this case through uniformly randomly distributed internal weights that are defined in a matrix $W$ of dimensionality $m \times m$. The resulting randomly connected network is injected with input data $\{b, y^{in}_{n+1}\}$ according to random offset and teacher/input weights $W^{off}$ and $W^{in}$, respectively. We normalize the largest eigenvalue of $W$ to one. The time-discrete equation that governs the network is [42]

$$\mathbf{x}_{n+1} = f_{NL}(W \cdot \mathbf{x}_n + W^{\text{off}} \cdot b + W^{in} \cdot y^{in}_{n+1}), \tag{10.11}$$

$$y^{\text{out}}_{n+1} = W^{\text{out}} \cdot \mathbf{x}_{n+1} \tag{10.12}$$

where $f_{NL}(\cdot)$ is a nonlinear sigmoid-like activation function and $b$ a constant value which has the role of an offset. The network is trained via a supervised learning rule, where we estimate the output weights $W^{out}$ through which we obtain the network's output $y^{\text{out}}$, see Fig. 10.9a.

The supervised learning rule used to estimate the output weights $W^{out}$ is based on regression. The ridge regression is commonly employed to train reservoirs, and it is executed according to

$$W^{\text{out}}_{\text{op}} = M_{y^T} \cdot M^{\dagger}_x (M_x \cdot M^{\dagger}_x - \lambda \mathbf{I}_m)^{-1}, \tag{10.13}$$

where $M_{y^T}$ and $M_x$ are matrices containing information about target and node responses, respectively. $\lambda$ is the regression parameter.

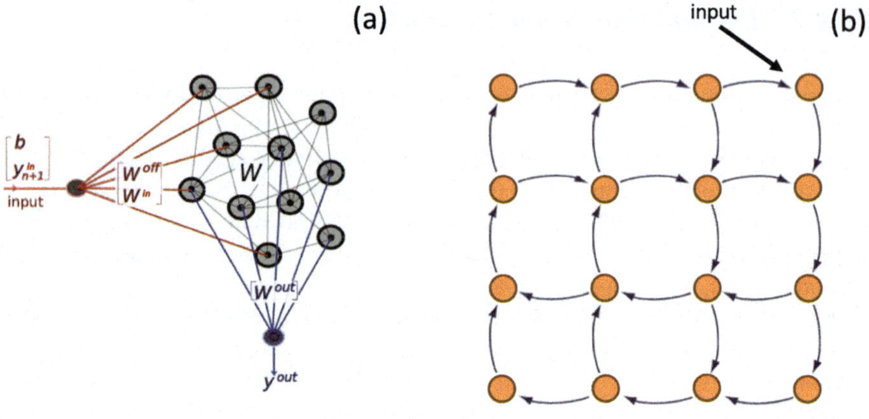


**Fig. 10.9** **a** A schematic diagram of a reservoir computer, and **b** a swirl on chip silicon reservoir

#### 10.4.2.1 Swirl On-Chip Silicon Reservoir Computer

A $6 \times 6$ reservoir computer that can perform speech recognition is shown by Fig. 10.9b [14]. The interconnections between neurons follow a swirl topology, because the connections are oriented as if they were in a whirling motion. Each neuron was designed with 2 cm long waveguides with a square roll shape. The 36 neurons are arranged in a rectangular grid, allowing for nearest neighbor interconnections only. Coupling and splitting between the nodes is done by multimode interferometers with low insertion loss. In order to leverage the full advantages of the maturity of silicon processing technology, this reservoir design does not contain any active element.

The classification task of isolated spoken digits (from 0 to 9) was based on the TI46 speech corpus. This dataset contains 500 samples, where each digit is spoken 10 times by five female speakers. The ridge regression was used to train the network using 75% of the TI46 dataset. Each trained output returns the value $+1$ whenever the corresponding digit is spoken and $-1$ otherwise. The testing stage used the remaining 25% set of digits. This circuit was simulated only with all the parameters of the designed chip and it achieved near 100% performance.

Reservoir computing has demonstrated to be a simple and powerful tool for analog AI. Reservoirs are well suited for hardware implementations as the synaptic weights of each neuron can be set as random—they do not need to be configured to specific values. Nevertheless, its lack of full reconfigurability makes it unable to tackle complex machine learning problems that are being solved by reconfigurable deep feedforward architectures.

### 10.4.3 *Broadcast-and-Weight Architecture*

A scalable photonic architecture for parallel processing can be achieved by using on-chip wavelength division multiplexed (WDM) techniques (see Fig. 10.10), in conjunction with MRRs, i.e. photonic synapses, that implement weights on signals encoded onto multiple wavelengths. Tuning a given MRR on and off resonance changes the transmission of each signal through that filter, effectively multiplying the signal with a desired weight in parallel. Silicon microring resonators (MRRs) cascaded in series have demonstrated fan-in and indefinite cascadability which make them ideal as on-chip synaptic weights with small footprint.

The big advantage of using MRRs to represent weights is that they can be tuned with a wide variety of methods: thermally, electro-optically or through light absorption (phase-change [18] or graphene [47] materials). Those tuning methods can be categorized into two different groups: index and absorption-tuning. In order to illustrate the action of each tuning method, we introduce a simple experiment shown in Fig. 10.11a. In this experiment, an add-drop MRR is fed by a laser at certain

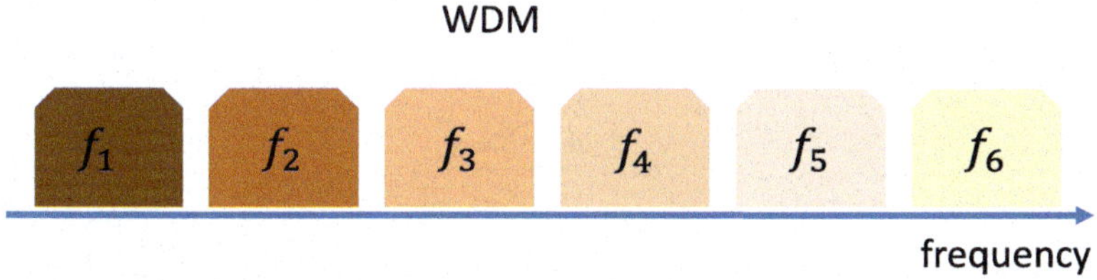


**Fig. 10.10** Illustration of the wave division multiplexing process

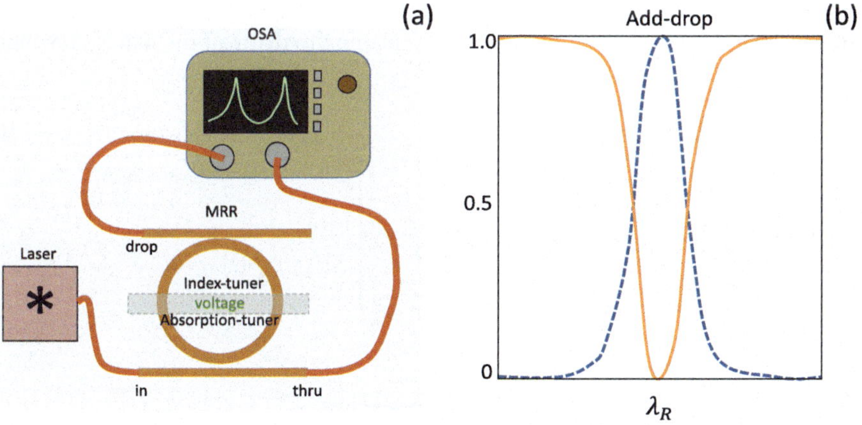


**Fig. 10.11** **a** Illustration of an experiment showing how to test add-drop MRRs that have index or absorption tuners that can be optimized through the application of a voltage. A laser feeds the MRRs and an OSA receives the transmission at the drop port. **b** Transmissions at the drop (dashed line) and the through port (solid curve)

wavelength λ at which this MRR resonates. Under this condition, the MRR begins trapping the incoming light that eventually partially escapes through the drop port via evanescent waves. The output light is captured by an optical spectrum analyzer (OSA). The MRR can perform index or absorption tuning when a certain voltage value is applied. In Fig. 10.11b we show the transmissions at the drop port (dashed curve) and at the through port (solid curve) without any tuning, where the MRR is on resonance at wavelength $\lambda_R$. The transfer function of the through port light intensity with respect to the input light is:

$$T_p(\phi) = \frac{(ar)^2 - 2r^2\cos\phi + r^2}{1 - 2r^2\cos\phi + (r^2a)^2}; \tag{10.14}$$

and the transfer function of the drop port light intensity with respect to the input light is:

$$T_d(\phi) = \frac{(1-r)^2 a}{1 - 2r^2\cos\phi + (r^2a)^2}. \tag{10.15}$$

The parameter $r$ is the self-coupling coefficient, and $a$ defines the propagation loss from the ring and the directional coupler. The phase $\phi$ depends on the wavelength λ of the light and radius $R$ of the MRR [48]:

$$\phi = \frac{4\pi^2 R n_{\text{eff}}}{\lambda}. \tag{10.16}$$

In the case where the coupling losses are negligible, $a \approx 1$, the relationship between the add-drop through and drop transfer functions is $T_p = 1 - T_d$.

In Fig. 10.12 we show the behavior of optical signals detected by the OSA when the voltage is varied. By applying a positive voltage to the index-tuning device, we tune the refractive index of the waveguide, leading to drifts of the natural resonance

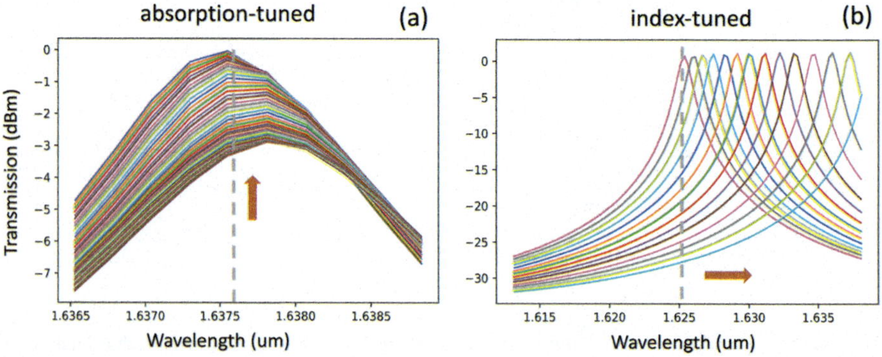


**Fig. 10.12** Transmission versus wavelength for the graphene-based MRR (**a**) and the n-doped MRR, **b** collected from their drop port when different voltage values are applied

frequency. As shown in Fig. 10.12a, we define this type of behavior as a photonic synaptic weight that can be tuned horizontally on the transmission profile, since the drive voltage causes $\lambda_R$ to shift from 1.6250 to 1.6350μ m and beyond. A different weighting-type of methodology is shown in Fig. 10.12b, where we show how weight values can be defined as transmission values that decrease with incrementing negative voltage. We define this type of behavior as a photonic synaptic weight that can be tuned vertically on the transmission profile. The index-tuning approach is so far the most employed tuning method due to the fabrication of index-tuning devices (heaters or NP-modulators) being compatible with standard silicon foundries. Absorption-tuning devices based on graphene have to be manufactured with isolated processes that are not yet standardized by any nanophotonic foundry.

#### 10.4.3.1 Multiwavelength Weighted Additions

The broadcast and weight architecture can perform weighed additions based on WDM techniques as shown by Fig. 10.13. In this illustration we show how to multiply in parallel four vectors contained in set $\mathcal{A}$:

$$\mathcal{A} = \begin{cases} (A_{11}, A_{12}, A_{13}, A_{14}), \\ (A_{21}, A_{22}, A_{23}, A_{24}), \\ (A_{31}, A_{32}, A_{33}, A_{34}), \\ (A_{41}, A_{42}, A_{43}, A_{44}); \end{cases} \tag{10.17}$$

with four vectors contained in set set $\mathcal{B}$:

$$\mathcal{B} = \begin{cases} (B_{11}, B_{12}, B_{13}, B_{14}), \\ (B_{21}, B_{22}, B_{23}, B_{24}), \\ (B_{31}, B_{32}, B_{33}, B_{34}), \\ (B_{41}, B_{42}, B_{43}, B_{44}). \end{cases} \tag{10.18}$$

All vectors are composed of positive real-valued numbers. Vectors from sets $\mathcal{A}$ and $\mathcal{B}$ are implemented as arrays of 8 MRRs cascaded in series. For this implementation we use input and through ports only of add-drop MRRs.

The elements of a given set $\mathcal{A}$ can be mapped to voltage values $Va_{ij}$ that tune each individual MRR($A_{ij}$). Each voltage value has a one-to-one correspondence with a MRR transmission profile $T_{ij}$. The same principle holds for matrix $\mathcal{B}$ with voltage values $Vb_{ij}$. The experimental implementation of this method requires the use of four lasers with different wavelengths $\lambda_i$ (with $i = 1, 2, 3, 4$) that represent four channels. Each channel will represent a vector element from each set through an on resonance MRR. For instance, channel $\lambda_1$ represents $A_{11}$ from $\mathcal{A}$ and $B_{11}$ from $\mathcal{B}$ at the same time. This property allows for optical interactions between the two MRRs that result in the product of the two of them. In Fig. 10.14 we show an illustration

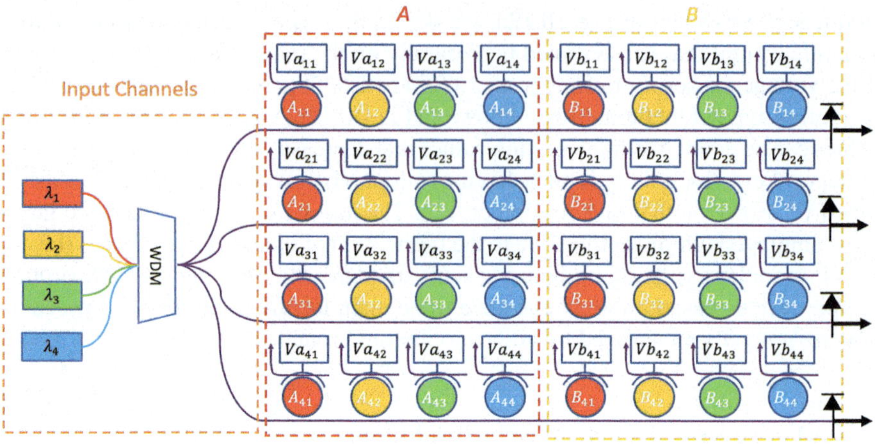


**Fig. 10.13** Photonic architecture for $4 \times 4$ vector-to-vector multiplications between sets of vectors $\mathcal{A}$ and $\mathcal{B}$, containing 4 vectors each

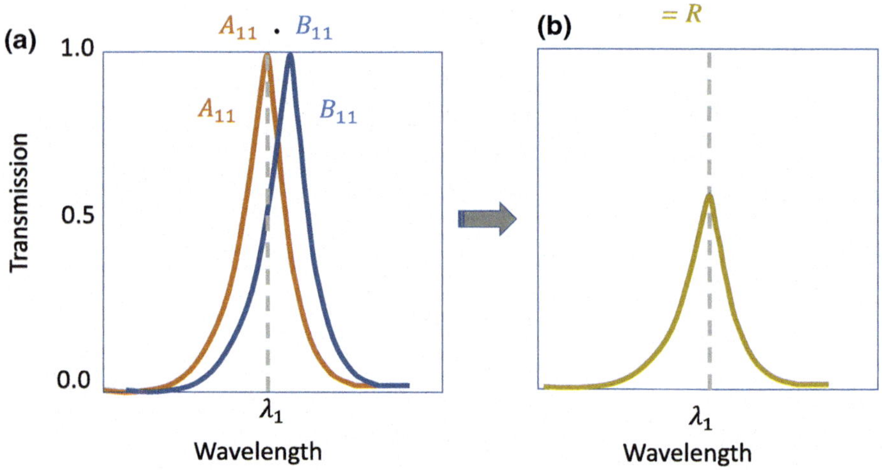


**Fig. 10.14** **a** Transmission versus wavelength curves of two different MRRs [RR($A_{11}$), MRR($B_{11}$)] performing element-to-element optical multiplications, and **b** the product of such multiplication

of the multiplication between two vector elements $A_{11}$ and $B_{11}$, and how the result $R$ is obtained. Let's suppose that the element $A_{11}$ was tuned to have the maximum optical transmission whereas $B_{11}$ was tuned to have half of it. In order to implement $A_{11}$, we leave the MRR($A_{11}$) on-resonance with $\lambda_1$, and we tune MRR($B_{11}$) to be half off-resonance with the same wavelength. They represent real-valued numbers 1 and 0.5, respectively. The result of such multiplication is $R = 0.5$. Two MRRs with different on and off resonance configurations at the same wavelength $\lambda_i$ will therefore perform element-to-element multiplications. A similar process is followed with all other vector elements of sets $\mathcal{A}$ and $\mathcal{B}$.

Furthermore, to perform a full vector-to-vector multiplication of the form:

$$C_1 = \begin{pmatrix} A_{11} & A_{12} & A_{13} & A_{14} \end{pmatrix} \cdot \begin{pmatrix} B_{11} \\ B_{12} \\ B_{13} \\ B_{14} \end{pmatrix}, \tag{10.19}$$

all the sets of multiplications $A_{ij}B_{ij}$, performed per array of eight MRRs, are all summed up by a photodetector located in the end of the array. The photonic circuit shown in Fig. 10.13 performs four weighted additions in parallel by leveraging the signal parallelization property of the light in which hundreds of high speed, multiplexed channels can be independently modulated and detected. Therefore, this architecture can perform four MAC operations in parallel consuming 23.956 Watts of energy—taking into account the tuning process.

#### 10.4.3.2 Convolutions and Convolutional Neural Networks

The efficient parallelism of broadcast and weight architectures for matrix multiplication can be leveraged to perform operatons such as convolutions. A convolution is a weighted summation of two discrete domain functions $f$ and $g$:

$$(f * g)[t] = \sum_{t=-\infty}^{\infty} f[\tau]g[t-\tau], \tag{10.20}$$

where $(f * g)$ represents a weighted average of the function $f[\tau]$ when it is weighting by $g[-\tau]$ shifted by $t$. The weighting function $g[-\tau]$ emphasizes different parts of the input function $f[\tau]$ as $t$ changes.

In this subsection we will learn how to utilize convolutions to do operations on images. Convolutions are well known to perform a highly efficient and parallel matrix multiplication using kernels [49]. Let us introduce the convolution of an image $\mathbb{A}$ with a kernel $\mathbb{B}$ that produces a convolved image $\mathbb{O}$, see Fig. 10.15(a). An image is represented as a matrix of numbers with dimensionality $H \times W \times D$, where $H$ and $W$ are the height and width of the image, respectively; and $D$ refers to the number of channels within the input image. Each element of a matrix $\mathbb{A}$ represents the intensity of a pixel at that particular spatial location. A kernel is a matrix $\mathbb{B}$ of real numbers with dimensionality $R \times R \times D$. The value of a particular convolved pixel is defined by:

$$O_{i,j} = \sum_{h=1}^{D} \sum_{q=0}^{iS+R} \sum_{p=0}^{js+R} B_{q,p,h} A_{iS+q,jS+p,h}. \tag{10.21}$$

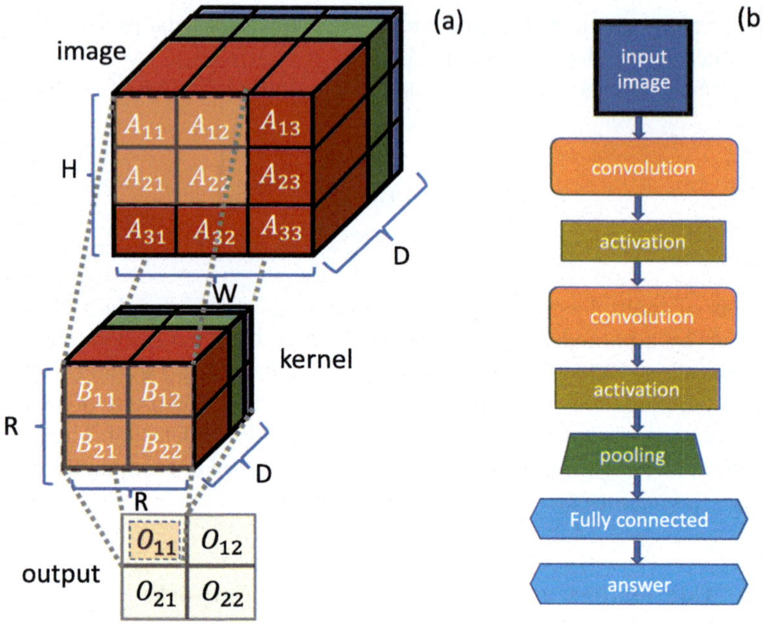


**Fig. 10.15** **a** Illustration of a convolution between and image $\mathbb{A}$ and $\mathbb{B}$, that generates output $\mathbb{O}$, and **b** the block diagram that describes a typical CNN, which contains convolutions, activation functions, pooling and fully connected layers

The additional parameter $S$ is referred to as the "stride" of the convolution. The dimensionality of the output feature is:

$$\left\lceil \frac{H-R}{S} + 1 \right\rceil \times \left\lceil \frac{W-R}{S} + 1 \right\rceil \times K, \tag{10.22}$$

where $K$ is the number of different kernels of dimensionality $R \times R \times D$ applied to an image, and $\lceil \cdot \rceil$ is the ceiling function.

The efficiency of convolutions for image processing is based on the fact that they lower the dimensions of the outputted convolved features. Since kernels are typically smaller than the input images, the feature extraction operation allows efficient edge detection, therefore reducing the amount of memory required to store those features.

As efficiency and parallelism are properties that can be expected from convolution operations, let us incorporate such operations in ANNs for image processing. In particular, we will build convolutional neural networks (CNNs) for image recognition tasks. A CNN consists of some combination of convolutional, nonlinear, pooling and fully connected layers [50]. CNNs are networks suitable to be implemented in photonic hardware since they demand fewer resources to do matrix multiplication and memory usage. The linear operation performed by convolutions allows single feature extraction per kernel. Hence, many kernels are required to extract as many features as possible. For this reason, kernels are usually applied in blocks, allowing the network to extract many different features all at once and in parallel.

In feed-forward networks, it is typical to use a rectified linear unit (ReLU) activation function. Since ReLUs are linear piecewise functions that model an overall nonlinearity, they allow CNNs to be easily optimized during training. The pooling layer introduces a stage where a set of neighbor pixels are encompassed in a single operation. Typically, such an operation consists of the application of a function that determines the maximum value among neighboring values. An average operation can be implemented likewise. Both approaches describe max and average pools, respectively. This statistical operation allows for a direct down-sampling of the image, since the dimensions of the object are reduced by a factor of two. From this step, we aim to make our network invariant and robust to small translations of the detected features.

The triplet, convolution-activation-pooling, is usually repeated several times for different kernels, keeping invariant the pooling and activation functions. Once all possible features are detected, the addition of a fully connected layer is required for the classification stage. This layer prepares and shows the solutions of the task.

CNNs are trained by changing the values of the kernels, analogous to how feed-forward neural networks are trained by changing the weighted connections [51]. The estimated kernel weight values are required in the testing stage. In this work, we trained the CNN to perform image recognition on the MNIST dataset, see Fig. 10.15b. The training stage uses the ADAM optimizer and back-propagation algorithm to compute the gradient function. The optimized parameters to solve MNIST can be categorized in two groups: (i) two $5 \times 5 \times 8$ different kernels and (ii) two fully connected layers of dimensions $800 \times 1$ and $10 \times 1$; and their respective bias terms. These kernels are then defined by eight $5 \times 5$ different filters. In the following we use our photonic CNN simulator to recognize new input images, obtained from a set of 500 images, which are intended to be used for the test step. Our simulator only works at the transfer level and does not simulate noise or distortion from analog components. As it can be seen in the illustration (Fig. 10.15b), a $28 \times 28$ input image from the test dataset is filtered by a first $5 \times 5 \times 8$ kernel, using stride one. The output of this process is a $24 \times 24 \times 8$ convolved feature, with a ReLU activation function already applied. Following the same process, the second group of filters is applied to the convolved feature to generate the second output, i.e. a $20 \times 20 \times 8$ convolved feature.

#### 10.4.3.3 Photonic Deep Convolutional Neural Networks

In general, the elements of the kernel matrix are defined as real-valued numbers, and the elements of the input matrix are typically integer positive numbers. In order to map convolutions onto photonics we need to consider how to represent negative numbers in our photonic circuits. This can be done by using the through and drop ports of the MRRs at once connected to a balance photodetector (BPD). In Fig. 10.16a we show how to represent a single negative number in a photonic circuit. The through and drop ports of the MRR($B_i$) are connected to the BPD, and then

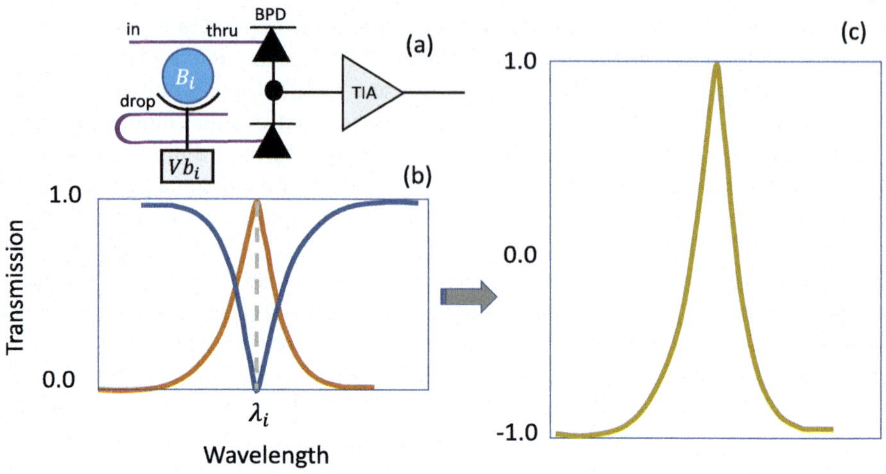


**Fig. 10.16** **a** Add-drop configuration and O/E conversion and amplification. **b** Transmissions of the drop and though ports and **c** output of the balance photodiode

linked to a TIA with gain equal to one. The transmissions of those ports can be seen in Fig. 10.16b, and the output of the BPD can be represented as a waveform shown by Fig. 10.16c. The result of the subtraction of the two optical signals performed by the BPD results in an electrical signal whose range is in $[-1, 1]$. Therefore, any weight value among $[-1, 1]$ can be represented on-chip. The addition of the TIA will allow for the representation of numbers out of this range.

This stage can be performed by an on-chip photonic version of the simulated CNN. Figure 10.17 shows a high-level overview of the proposed testing on-chip architecture. First, we consider an architecture that can produce one convolved pixel at a time. To handle convolutions for kernels with dimensionality up to $R \times R \times D$, we will require $R^2$ lasers with unique wavelengths since a particular convolved pixel can be represented as the dot product of two $1 \times R^2$ vectors. To represent the values of each pixel, we require $DR^2$ add-drop modulators (one per kernel value) where each modulator keeps the intensity of the corresponding carrier wave proportional to the normalized input pixel value. The $R^2$ lasers are multiplexed together using WDM, which is then split into $D$ separate lines. On every line, there are $R^2$ add-drop MRRs (where only input and through ports are being used), resulting in $DR^2$ MRRs in total. Each WDM line will modulate the signals corresponding to a subset of $R^2$ pixels on channel $k$, meaning that the modulated wavelengths on a particular line correspond to the pixel inputs $(A_{m,n,k})_{m\in[i,i+R]n\in[j,j+R]}$ where $k \in [1, D]$, and $S = 1$.

The $D$ WDM lines will then be fed into an array of $D$ photonic weight banks (PWB). Each PWB will contain $R^2$ MRRs with the weights corresponding to the kernel values at a particular channel. Each MRR within a PWB should be tuned to a unique wavelength within the multiplexed signal. The outputs of the weight bank array are electrical signals, each proportional to the dot product $(B_{m,n,k})_{m\in[1,R^2]n\in[1,R^2]} \cdot (A_{p,q,k})_{p\in[i,i+R^2]q\in[j,j+R^2]}$, where $k \in [1, D]$, and $S = 1$. Finally,

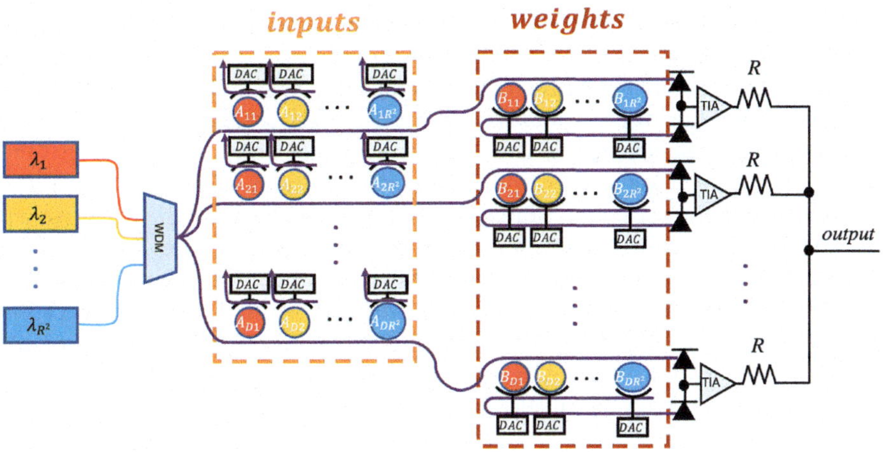


**Fig. 10.17** Photonic architecture for producing a single convolved pixel. We use an array of $R^2$ lasers with different wavelengths $\lambda_i$ to feed the MRRs. The input and kernel values modulate the MRRs via electrical currents proportional to those values. Once the matrix parallel multiplications are performed, the voltage adder has the function to add all signals from weight banks. Here, $R$ are resistance values. Then the output is the convolved feature

the signals from the weight banks need to be added together. This can be achieved using a passive voltage adder. The output from this adder will therefore be the value of a single convolved pixel.

Here, the testing input images modulate the intensities of a group of lasers with identical powers but unique wavelengths. These modulated inputs would be sent into an array of PWBs, which would then perform the convolution for each channel. The kernels obtained in the training step are used to modulate these weight banks. Finally, the outputs of the weight banks would be summed using a voltage adder, which produces the convolved feature. This simulator works using the transfer function of the MRRs, through port and drop port summing equations at the BPDs, and the TIA gain term to simulate a convolution. The simulator assumes that MRRs transfer functions design is based on the averaged transfer function behavior validated experimentally in prior work [34]. The control accuracy of the MRRs is 6-bits as that has been empirically observed [52]. The MRR self-coupling coefficient is equal to the loss, $r = a = 0.99$ [53] in (10.14) and (10.15).

The interfacing of optical components with electronics would be facilitated by the use of digital-to-analog converters (DACs) and analog-to-digital converters (ADCs). The storage of output and retrieving of inputs would be achieved by using memories GDDR SDRAM. The SDRAM is connected to a computer, where the information is already in a digital representation. Then, the implementation of the ReLU non-linearity and the reuse of the convolved feature to perform the next convolution can be performed. The idea is to use the same architecture to implement the triplet convolution-activation-pooling on hardware.

The results of the MNIST task solved by our simulated photonic CNN, for a test set of 500 images, obtained an overall accuracy of 97.6%. This result was found to be 1% better than on-chip MZIs-based theoretical CNNs [54]. These results were obtained with 7 bits of precision. As described in [33], assuming that a maximum of 1024 MRRs can be manufactured in the modulator array, a convolutional unit can support a large kernel size with a limited number of channels, $R = 10$, $D = 12$, or a small kernel size with a large number of channels, $R = 3$, $D = 113$. We will consider both edge cases to get a range of energy consumption values. For the smaller convolution size, we will have $R^2$ lasers, $R^2$ MRRs and DACs in the modulator array, $R^2D$ MRRs and $D$ TIAs in the weight bank array and one ADC to convert back into digital signal. With 100 mW per laser, 19.5 mW per MRR, 26 mW per DAC, 17 mW per TIA [55] and 76 mW per ADC, we get an energy usage of 112 W for the large kernel size and 95 W for the smaller kernel size.

## 10.5 Summary and Concluding Remarks

We have outlined the architectures and motivations behind the use of three photonic integrated circuits that can implement ANNs on-chip. Despite the fact that reservoir computers are simple hardware solutions for machine learning problems that cannot be efficiently solved with classical methods, reconfigurable approaches such as MZI-based meshes and broadcast-and-weight circuits have demonstrated even more versatility. Reconfigurable architectures can be used not only for special-purpose analog tasks, but potentially for general-purpose analog computing as they can be utilized for matrix multiplications, SVD and convolutions. Therefore, it will not come as a surprise if in the future we find a way to create a general-purpose analog machine that is highly optimized for machine learning and neuromorphic computing, but also tackles other more simpler classical tasks.

The reconfigurability of such photonic processing units promise many exciting developments in AI, but there are many challenges that remain with the future implementations of those machines. Such challenges are related to control of the whole processing unit and efficient memory access to the data that needs to be processed. Just like in digital computers, the control unit shall ensure that those data sets are uploaded correctly and in the right order to the processing unit. Since we are working with analog machines, the control unit is particularly important because accurate real-valued numbers only should be inputted to the photonic circuit. The first challenge would be focused on the accurate representation of analog values in photonic processors.

The second critical element is the memory unit. The processes of reading from and writing to memory should not constitute a bottleneck to the processing and control units. This unit has to be optimized in order to receive and deliver information accurately, fast and with low power consumption. Currently, the memory unit is accessed through DACs and ADCs, which slow down the processor considerably. Therefore, the development of analog memories is a crucial step that we need to pursue. For instance, crossbar arrays can be a good solution to this. Crossbar array modules that do not exceed the length limit $L = 100\ \mu$m should help to create an optimized photonic processor. By adding a crossbar-based memory to our machine, we could store analog voltage values on it and tune our MZIs or MRRs with them. The only flaw of this solution is that the cost of transferring information through electric paths has been considered as a bottleneck for power efficiency. In practice, most of the system-energy is lost in data movement between the processor and the memory [56]. However, photonics has been found to efficiently reduce such data movement problems. In fact, optical loss is nearly negligible for intrachip distances [3]. The development of optical memories [57] are a potential solution in the case where analog optical information can be used to tune a device. Such a device should be tuned with light instead of electrons. Unfortunately, this technology has not been developed yet for neuromorphic computing.

In general, the overall system-speed and power remain as challenge points. Many groups and companies have targeted research directions on some of the afore mentioned issues, but it is currently unknown how these problems will be overcome and how a fully working special (or general) purpose photonic processor can be successfully built.

## References

1. Y. Zhang, J. Gao, H. Zhou, Breeds classification with deep convolutional neural network, in *ACM International Conference Proceeding Series*, pp. 145–151 (2020)
2. Q. Xia, J.J. Yang, Memristive crossbar arrays for brain-inspired computing. Nat. Mater. **18**(4), 309–323 (2019)
3. M.A. Nahmias, T.F. De Lima, A.N. Tait, H.T. Peng, B.J. Shastri, P.R. Prucnal, Photonic multiply-accumulate operations for neural networks. IEEE J. Sel. Top. Quantum Electron. **26**(1), 1–18 (2020)
4. F. Duport, A. Smerieri, A. Akrout, M. Haelterman, S. Massar, Fully analogue photonic reservoir computer. Sci. Rep. **3**(6), 22381 (2016)
5. D. Brunner, M.C. Soriano, C.R. Mirasso, I. Fischer, Parallel photonic information processing at gigabyte per second data rates using transient states. Nat. Commun. **4**, 1364 (2013)
6. K. Vandoorne, P. Mechet, T. Van Vaerenbergh, M. Fiers, G. Morthier, D. Verstraeten, B. Schrauwen, J. Dambre, P. Bienstman, Experimental demonstration of reservoir computing on a silicon photonics chip. Nat. Commun. **24**(5), 3541 (2014)
7. L. Larger, M.C. Soriano, D. Brunner, L. Appeltant, J.M. Gutierrez, L. Pesquera, C.R. Mirasso, I. Fischer, Photonic information processing beyond turing: an optoelectronic implementation of reservoir computing. Opt. Express **20**, 3241–3249 (2012)
8. T.W. Hughes, M. Minkov, Y. Shi, S. Fan, Training of photonic neural networks through in situ backpropagation and gradient measurement. Optica **5**, 864–871 (2018)
9. Y. Shen, N.C. Harris, S. Skirlo, M. Prabhu, T. Baehr-Jones, M. Hochberg, X. Sun, S. Zhao, H. Larochelle, D. Englund, M. Soljačić, Deep learning with coherent nanophotonic circuits. Nat. Photonics **11**, 441 (2017)
10. P.R. Prucnal, B.J. Shastri, *Neuromorphic Photonics* (CRC Press, Taylor & Francis Group, Boca Raton, FL, USA, 2017)
11. A.N. Tait, M.A. Nahmias, B.J. Shastri, P.R. Prucnal, Broadcast and weight: an integrated network for scalable photonic spike processing. J. Lightwave Technol. **32**, 4029–4041 (2014)

12. A.N. Tait, A.X. Wu, T.F. de Lima, E. Zhou, B.J. Shastri, M.A. Nahmias, P.R. Prucnal, Microring weight banks. IEEE J. Sel. Top. Quantum Electron. **22**, 312–325 (2016)
13. A.N. Tait, T. Ferreira de Lima, M.A. Nahmias, H.B. Miller, H.-T. Peng, B.J. Shastri, P.R. Prucnal, *A silicon photonic modulator neuron*. arXiv e-prints, p. arXiv:1812.11898 (2018)
14. K. Vandoorne, P. Mechet, T. Van Vaerenbergh, M. Fiers, G. Morthier, D. Verstraeten, B. Schrauwen, J. Dambre, P. Bienstman, Experimental demonstration of reservoir computing on a silicon photonics chip. Nat. Commun. **5**, 1–6 (2014)
15. Y. Shen, N.C. Harris, S. Skirlo, M. Prabhu, T. Baehr-Jones, M. Hochberg, X. Sun, S. Zhao, H. Larochelle, D. Englund, M. Soljacic, Deep learning with coherent nanophotonic circuits. Nat. Photonics **11**(7), 441–446 (2017)
16. Y. LeCun, C. Cortes, in *MNIST Handwritten Digit Database* (2010)
17. A.N. Tait, T.F. De Lima, E. Zhou, A.X. Wu, M.A. Nahmias, B.J. Shastri, P.R. Prucnal, Neuromorphic photonic networks using silicon photonic weight banks. Sci. Rep. **7**(1), 1–10 (2017)
18. J. Feldmann, N. Youngblood, C.D. Wright, H. Bhaskaran, W.H. Pernice, All-optical spiking neurosynaptic networks with self-learning capabilities. Nature **569**(7755), 208–214 (2019)
19. A.N. Tait, T. Ferreira De Lima, M.A. Nahmias, H.B. Miller, H.T. Peng, B.J. Shastri, P.R. Prucnal, Silicon photonic modulator neuron. Phys. Rev. Appl. **11**(6), 1 (2019)
20. L. De Mol, Turing machines, in *The Stanford Encyclopedia of Philosophy*, ed. by E. N. Zalta (Metaphysics Research Lab, Stanford University, winter 2019 ed., 2019)
21. A.M. Turng, Computing machinery and intelligence. Mind **59**(236), 433 (1950)
22. L. Gugerty, Newell and Simon's logic theorist: historical background and impact on cognitive modeling. Proc. Hum. Factors Ergon. Soc. Annu. Meet. **50**(9), 880–884 (2006)
23. A.T. Cianciolo, R.J. Sternberg, *The Nature of Intelligence* (Wiley and Sons Ltd, Hoboken, 2008)
24. S. Pinker, *How the Mind Works (1997/2009)* (W. W. Norton & Company, New York, NY, 2009)
25. I. Khurana, *Medical Physiology for Undergraduate Students*. Elsevier (A Division of Reed Elsevier India Pvt. Limited, 2011)
26. P. Liang, S. Wu, F. Gu, *An Introduction to Neural Information Processing* (Springer, Berlin, 2016)
27. A. Adamatzky, V. Kendon, *From Astrophysics to Unconventional Computation* (Springer, Berlin, 2020)
28. B.A. Marquez, J. Suarez-Vargas, B.J. Shastri, Takens-inspired neuromorphic processor: a downsizing tool for random recurrent neural networks via feature extraction. Phys. Rev. Res. **1**, 33030 (2019)
29. V. Mnih, K. Kavukcuoglu, D. Silver, A.A. Rusu, J. Veness, M.G. Bellemare, A. Graves, M. Riedmiller, A.K. Fidjeland, G. Ostrovski, S. Petersen, C. Beattie, A. Sadik, I. Antonoglou, H. King, D. Kumaran, D. Wierstra, S. Legg, D. Hassabis, Human-level control through deep reinforcement learning. Nature **518**, 529 (2015)
30. W. Bogaerts, R. Baets, P. Dumon, V. Wiaux, S. Beckx, D. Taillaert, B. Luyssaert, J. Van Campenhout, P. Bienstman, D. Van Thourhout, Nanophotonic waveguides in silicon-on-insulator fabricated with cmos technology. J. Lightwave Technol. **23**(1), 401–412 (2005)
31. W. Bogaerts, M. Fiers, P. Dumon, Design challenges in silicon photonics. IEEE J. Sel. Top. Quantum Electron. **20**(4), 1–8 (2014)
32. L. Chrostowski, M. Hochberg, *Silicon Photonics Design: From Devices to Systems* (Cambridge University Press, Cambridge, 2015)
33. V. Bangari, B.A. Marquez, H. Miller, A.N. Tait, M.A. Nahmias, T.F. de Lima, H. Peng, P.R. Prucnal, B.J. Shastri, Digital electronics and analog photonics for convolutional neural networks (deap-cnns). IEEE J. Sel. Top. Quantum Electron. **26**(1), 1–13 (2020)
34. A.N. Tait, H. Jayatilleka, T.F.D. Lima, P.Y. Ma, M.A. Nahmias, B.J. Shastri, S. Shekhar, L. Chrostowski, P.R. Prucnal, Feedback control for microring weight banks. Opt. Express **26**, 26422–26443 (2018)
35. M. Gallus, A. Nannarelli, Handwritten digit classication using 8-bit floating point based convolutional neural networks, in *DTU Compute Technical Report-2018*, vol. 01 (2018)

36. N. Wang, J. Choi, D. Brand, C.-Y. Chen, K. Gopalakrishnan, Training deep neural networks with 8-bit floating point numbers, in *Proceedings of the 32nd International Conference on Neural Information Processing Systems*, NIPS'18, (USA), pp. 7686–7695 (Curran Associates Inc., 2018)
37. P. Ambs, Optical computing: a 60-year adventure. Adv. Opt. Technol. **2010**, 372652 (2010)
38. M. Reck, A. Zeilinger, H.J. Bernstein, P. Bertani, Experimental realization of any discrete unitary operator. Phys. Rev. Lett. **73**, 58–61 (1994)
39. D.A.B. Miller, All linear optical devices are mode converters. Opt. Express **20**, 23985–23993 (2012)
40. F. Shokraneh, M.S. Nezami, O. Liboiron-Ladouceur, Theoretical and experimental analysis of a 4 x 4 reconfigurable mzi-based linear optical processor. J. Lightwave Technol. **38**(6), 1258–1267 (2020)
41. I.A.D. Williamson, T.W. Hughes, M. Minkov, B. Bartlett, S. Pai, S. Fan, Reprogrammable electro-optic nonlinear activation functions for optical neural networks. IEEE J. Sel. Top. Quantum Electron. **26**, 1–12 (2020)
42. H. Jaeger, H. Haas, Harnessing nonlinearity: predicting chaotic systems and saving energy in wireless communication. Science **304**, 78 (2004)
43. W. Maass, T. Natschlaeger, H. Markram, Neural Comput. **14**, 2531 (2002)
44. B.A. Marquez, L. Larger, M. Jacquot, Y.K. Chembo, D. Brunner, Dynamical complexity and computation in recurrent neural networks beyond their fixed point. Sci. Rep. **8**(1), 3319 (2018)
45. A. Gutiérrez, S. Marco, *Biologically Inspired Signal Processing for Chemical Sensing* (Springer, Berlin, 2009)
46. S.J.C. Caron, V. Ruta, L.F. Abbott, R. Axel, Nature **497**, 113 (2013)
47. M. Liu, X. Yin, E. Ulin-Avila, B. Geng, T. Zentgraf, L. Ju, F. Wang, X. Zhang, A graphene-based broadband optical modulator. Nature **474**(7349), 64–67 (2011)
48. W. Bogaerts, P. De Heyn, T. Van Vaerenbergh, K. De Vos, S. Kumar Selvaraja, T. Claes, P. Dumon, P. Bienstman, D. Van Thourhout, R. Baets, Silicon microring resonators. Laser Photonics Rev. **6**(1), 47–73 (2012)
49. I. Goodfellow, Y. Bengio, A. Courville, *Deep Learning* (The MIT Press, Cambridge, 2016)
50. K. O'Shea, R. Nash, *An Introduction to Convolutional Neural Networks*, CoRR, vol. abs/1511.08458 (2015)
51. K. Mehrotra, C.K. Mohan, S. Ranka, *Elements of Artificial Neural Networks* (MIT Press, Cambridge, MA, USA, 1997)
52. P.Y. Ma, A.N. Tait, T.F. de Lima, S. Abbaslou, B.J. Shastri, P.R. Prucnal, Photonic principal component analysis using an on-chip microring weight bank. Opt. Express **27**, 18329–18342 (2019)
53. Y. Tan, D. Dai, Silicon microring resonators. J. Opt. **20**, 054004 (2018)
54. H. Bagherian, S. A. Skirlo, Y. Shen, H. Meng, V. Ceperic, M. Soljacic, *On-Chip Optical Convolutional Neural Networks*. CoRR, vol. abs/1808.03303 (2018)
55. Z. Huang, C. Li, D. Liang, K. Yu, C. Santori, M. Fiorentino, W. Sorin, S. Palermo, R.G. Beausoleil, 25-gbps low voltage waveguide si-ge avalanche photodiode. Optica **3**, 793–798 (2016)
56. S.R. Agrawal, S. Idicula, A. Raghavan, E. Vlachos, V. Govindaraju, V. Varadarajan, C. Balkesen, G. Giannikis, C. Roth, N. Agarwal, E. Sedlar, A many-core architecture for in-memory data processing, in *Proceedings of the 50th Annual IEEE/ACM International Symposium on Microarchitecture*, MICRO-50 '17, New York, NY, USA, pp. 245–258 (Association for Computing Machinery, 2017)
57. T. Alexoudi, G.T. Kanellos, N. Pleros, Optical RAM and integrated optical memories: a survey. Light Sci. Appl. **9**(1), 1–91 (2020)

# Chapter 11
# Quantum Processors in Silicon Photonics

**Stefano Paesani and Anthony Laing**

**Abstract** Machines that can exploit their hardware to process quantum information can solve certain problems exponentially faster than purely classical (conventional) computers. To harness the potentially groundbreaking applications of quantum computers, these machines will be required to control and process quantum systems at a very large scale, potentially involving millions of high-quality quantum information carriers. While a number of challenges must be overcome before silicon photonics can support quantum computing at scale, the capabilities of mature semiconductor fabrication process to integrate large quantum photonic circuits on single devices, means this is a promising and emerging platform for quantum information processing. In this chapter, we will review recent results in developing key building blocks for chip-scale photonic quantum devices and discuss the progress towards useful large-scale quantum computers in the silicon quantum photonics platform.

## 11.1 Introduction

The field of quantum mechanics has provoked deep and philosophical questions about the nature of our universe: Are the outcomes of measurements truly random? Can a particle simultaneously occupy two different states? Can the uncertain states of two separated particles instantaneously become certain and correlated? The field of quantum technologies took hold when, rather than worrying about these questions, scientists instead began to explore how the counter intuitive features of quantum mechanics might be used as resources.

S. Paesani (✉) · A. Laing
Quantum Engineering Technology Labs, H. H. Wills Physics Laboratory, University of Bristol, Bristol BS8 1FD, UK
e-mail: anthony.laing@bristol.ac.uk

Department of Electrical and Electronic Engineering, University of Bristol, Bristol BS8 1FD, UK
e-mail: stefano.paesani@bristol.ac.uk



D. J. Lockwood and L. Pavesi (eds.), *Silicon Photonics IV*,
Topics in Applied Physics 139, https://doi.org/10.1007/978-3-030-68222-4_11

We now know how the combination of particle superpositions and uncertain measurement outcomes enables secret quantum communication. We have discovered how particular quantum states of light can enhance the measurement sensitivity when used as measurement probes. And, perhaps most excitingly, we know how entangled states of particles can be used to run quantum algorithms to solve problems exponentially faster than is possible with classical computers.

Silicon photonics is an appealing platform for quantum information processing. The mature fabrication tools from the microelectronics industry, together with cutting edge techniques from academia, allow the design and lithography of large and complex, yet stable photonic circuits. Reproducible photonic circuitry enables near-identical sources of photons and high-quality interferometers, key elements to the implementation of photonic entangling operations.

The scale of the circuitry required for general purpose photonic quantum computing might seem beyond the capabilities of today's technology. Yet, here we discuss important proof-of-concept demonstrations for photonic quantum processors at significant leaps of complexity over what had been reported only a few years before.

## 11.2 Photonic Quantum Information Processing

We start with a brief background on photonic quantum information processing. The interested readers can find more details, for example, in [1–3].

### *11.2.1 Quantum States of Light*

In the second quantisation formalism, light in a single optical mode is described as an harmonic oscillator with Hamiltonian $\hat{H} = \hbar\omega(\hat{a}^\dagger\hat{a} + 1/2)$, where $\omega$ is the mode frequency and $\hat{a}^\dagger$ ($\hat{a}$) is the bosonic creation (annihilation) operator for the single quantum excitations of the electromagnetic field: photons. If $k$ labels different optical modes, Fock states for photon configurations $\boldsymbol{n}$ describe states with fixed number of photons in the different modes:

$$|\boldsymbol{n}\rangle = |n_1 n_2 \dots n_m\rangle = |n_1\rangle_1 \otimes |n_2\rangle_2 \otimes \dots \otimes |n_m\rangle_m\,, \tag{11.1}$$

where the quantum states $|n_k\rangle$ represent $n_k$ photons in the $k$-th mode and are given by $|n_k\rangle_k = (\hat{a}_k^\dagger)^{n_k}/\sqrt{n_k!}\,|\emptyset\rangle_k$, with $|\emptyset\rangle_k$ the vacuum state on mode $k$. Quantum states can also contain a coherent superposition of different photon numbers, such as the single-mode squeezed vacuum (SMS) and two-mode squeezed vacuum (TMS) states, described in term of Fock states as

$$|\mathrm{SMS}(\xi)\rangle = \frac{1}{\sqrt{\cosh r}} \sum_{n=0}^{\infty} (-e^{i\varphi} \tanh r)^n \frac{\sqrt{(2n)!}}{2^n n!} |2n\rangle , \tag{11.2}$$

$$|\mathrm{TMS}(\xi)\rangle = \frac{1}{\cosh r} \sum_{n=0}^{\infty} (e^{i\varphi} \tanh r)^n |n\rangle_1 |n\rangle_2 , \tag{11.3}$$

where $\xi = re^{i\varphi}$ is the squeezing parameter [1].

### *11.2.2 Encoding Qubits and Qudits in Photons*

While all quantum states of light introduced above can be used to encode and process quantum information [3], in this chapter we will focus on using single photons. A qubit, the quantum equivalent of a classical bit, is a two-level quantum system which can be encoded in the state of a single photon in two optical modes. Although these modes can represent many different photonics degrees of freedom (e.g. polarisation, wavelength, time, orbital angular momentum, etc.), the typical choice in integrated quantum photonics is to use the path of the photon, i.e. the different waveguides the photon can travel in. The qubit is then encoded in a single photon as pictured in Fig. 11.1, where the mapping between the computational state of the qubit and the Fock state is defined as follows: the qubit is in the computation state $|k\rangle$ if the single photon passes through the $k$-th waveguide, with $k \in \{0, 1\}$. Furthermore, this definition can be straightforwardly generalised to go beyond qubits and encode $d$-dimensional quantum systems—qudits—by using $d$ different paths, as also shown in Fig. 11.1 [4, 5].

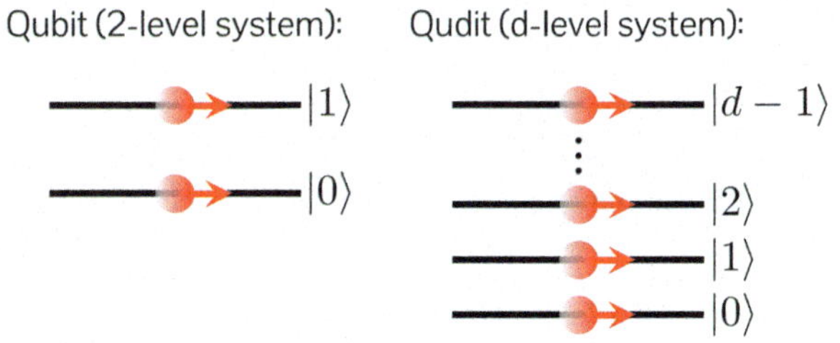


**Fig. 11.1** Encoding of a qubit (left) and a qudit (right) on the spatial modes of a single photon propagating through integrated waveguides

### 11.2.3 Processing Photons with Linear Optics

The typical approach to process photons in integrated quantum circuits is via the use of linear-optical interferometers: linear evolutions conserving the number of photons. Evolutions in a linear-optical system with $m$ modes are described via $m \times m$ unitary matrices, where the unitarity ensures energy conservation (non-unitarity can appear in presence of losses). Linear-optical operations can be constructed using two building blocks: phase shifters and beam-splitters (see Fig. 11.2a), described via the unitaries

$$U_{\rm PS}(\phi) = \begin{bmatrix} 1 & 0 \\ 0 & e^{i\phi} \end{bmatrix}, \qquad U_{\rm BS}(\eta) = \begin{bmatrix} \sqrt{\eta} & i\sqrt{1-\eta} \\ i\sqrt{1-\eta} & \sqrt{\eta} \end{bmatrix}, \tag{11.4}$$

where $\phi$ is the phase applied by the phase shifter, and $\eta$ is the beam-splitter reflectivity. While these components act on single and two modes, they can be combined into Mach–Zehnder interferometers (MZIs, see Fig. 11.2b) and networked to build $m$-mode circuits implementing an arbitrary unitary evolution $U$, known as universal linear-optical schemes. A variety of universal schemes has been developed [6, 7, 9], with two prominent examples by Reck et al. [6] and Clements et al. [7] shown in Figs. 11.2c, d, respectively. Reconfigurable circuits able to implement arbitrary unitary operations on six modes with high fidelity have been recently demonstrated on a silica chip (see Fig. 11.2e) [8]. Such circuits can be used to implement arbitrary single-qubit or single-qudit gates with high precision [5, 10–12].

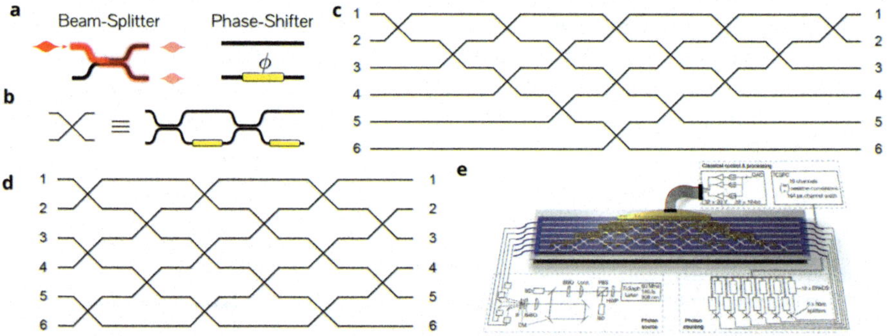


**Fig. 11.2** **a** Building blocks for linear-optical circuits: the beam-splitter mixes amplitudes of different modes, the phase shifter inserts an optical phase $\phi$ in the associated mode. **b** Beam-splitters and phase shifters are combined to form generalised beam-splitters via Mach–Zehnder interferometers and networked to build universal interferometers. Examples of universal linear-optical schemes are shown in **(c)** (Reck scheme [6]) and in **(d)** (Clements scheme [7]). **e** Implementation of a universal scheme via integrated quantum photonics. Image **e** is from [8]

### 11.2.4 Scalable Photonic Quantum Computing Architectures

While single-photon operations can thus be performed deterministically and with high precision in linear optics, the main challenge for photonic quantum processors is to perform entangling operations between different photons. Such operations are required to perform two-qubit gates in universal quantum computing circuits [16]. This difficulty comes from the fact that photons do not directly interact, and therefore photon–photon nonlinearities are highly suppressed (although interesting integrated hybrid approaches are emerging to mediate photon-photon nonlinearities through interfaces with solid-state systems [17]). Nevertheless, universal quantum computing can be scalably implemented in photonics by inducing photon–photon entangling gates through measurements [14, 18]. Due to the probabilistic nature of quantum measurements, such entangling gates are inherently probabilistic, but their success can be heralded via the measurement of auxiliary photons. The heralding enables multiplexing schemes to boost the success probability of such probabilistic entangling operations to near-unity to make them scalable. [14, 18–20].

An example of an heralded entangling gate for the generation of three-photon entanglement, in particular the Greenberger–Horne–Zeilinger (GHZ) state $(|000\rangle + |111\rangle)/\sqrt{2}$, is shown in Fig. 11.3a [13]. Such states play an important role in modern linear-optical quantum computing (LOQC) architectures as they can be fused together to form larger entangled states via fusion gates [14, 19]. An example of a

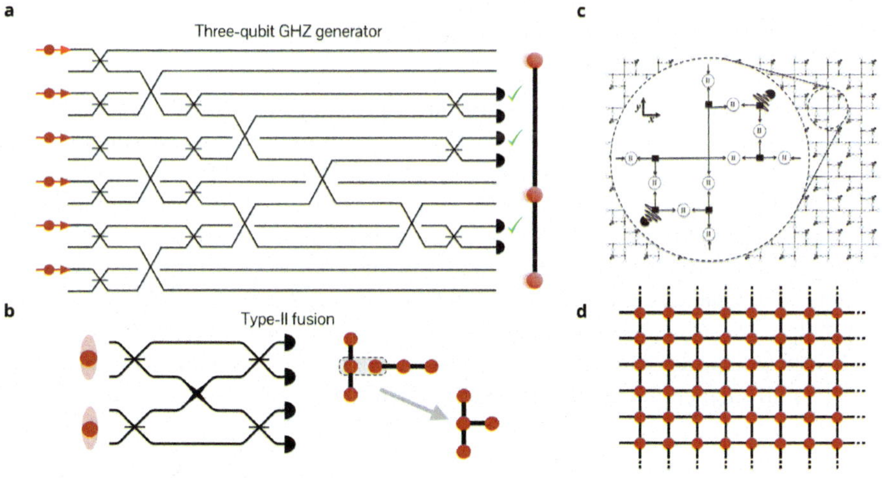


**Fig. 11.3** **a** Optical interferometer for the generation of a three-photon heralded GHZ entangled state $(|000\rangle + |111\rangle)/\sqrt{2}$, with success probability 1/32 [13]. **b** Photonic circuit to perform type-II fusion gates (with success probability 1/2), fusing different photonic entangled states to form larger entanglement resources [14]. **c** Modular LOQC architecture for generating universal resource states for MBQC, e.g. states as shown in **(d)**. Each module includes GHZ generators (black squares), type-II fusion gates (circles), delay lines and single-photon detectors. Image **c** is from [15]

fusion gate (type-II) circuit is reported in Fig. 11.3b, which has a success probability of 1/2 [14]; this success probability can be boosted to 3/4 via the use of additional auxiliary photons [21].

Modular architectures for scalable universal quantum computing with linear optics can be built, for example, by fusing a large number of GHZ states to form entangled resource states for universal measurement-based quantum computing (MBQC), known as cluster states [14, 15, 19, 22, 23]. As shown in Fig. 11.3c, each module in such architecture fuses GHZ states to make a single computational photonic qubit and connects it to other qubits in the lattice of a universal cluster state [14, 19], such as the square lattice of entangled qubits shown in Fig. 11.3d. These types of universal LOQC architectures also enable fault-tolerant photonic quantum computation [15], which is crucial to suppress the exponential amplification of photon loss and computational errors that prevents scaling quantum computation on non-error-corrected quantum devices beyond few tens of qubits [24]. However, the resource costs for such architectures [25] mean that a hardware platform able to integrate millions to billions of components is required to reach a scale where useful fault-tolerant photonic quantum computing applications can become practical. Nearer-term special-purpose photonic quantum devices are possible and will be discussed in Sects. 11.4 and 11.5. Although the prefault-tolerant applications for such devices are significantly more limited, the hardware requirements are reduced to thousands of components [26]; a scale achievable with the current silicon quantum photonic technology.

## 11.3 Silicon Quantum Photonic Technology

To reach the scale required for computationally interesting applications, photonic quantum processors must integrate at least thousands of high-quality components to generate, process and measure quantum states of light. Thanks to its compatibility with mature fabrication facilities, and together with more advanced components, silicon quantum photonics might reach such a scale. Silicon photonics circuits comprising thousands of components are in fact already routinely used for classical applications [27]. Quantum photonic circuits, however, need additional components and generally have stringent performance requirements (for example in term of losses) with respect to purely classical devices. As shown in Fig. 11.4, silicon photonics can integrate all such components into a single platform.

### *11.3.1 Integrated Photon Sources*

Quantum states of light can be generated via spontaneous nonlinear processes arising from the interaction of a pump light with a nonlinear-optical medium [1]. Although silicon does not naturally posses a $\chi^{(2)}$ due to its centro-symmetric crystal structure, preventing the use of spontaneous parametric down-conversion (SPDC) processes

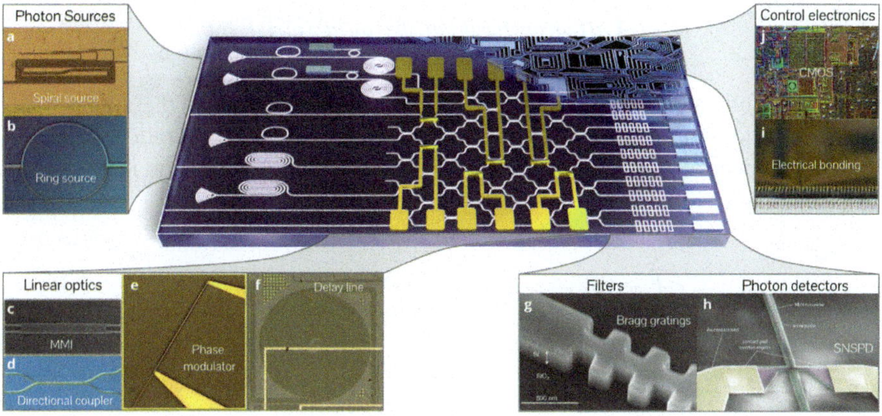


**Fig. 11.4** Blueprint of a typical quantum photonic processor in the silicon platform, which includes waveguide **(a)** and ring **(b)** SFWM sources. Beam-splitters based on MMIs **(c)** or directional couplers **(d)**, phase shifters **(e)** and delay lines **(f)** are networked to form reconfigurable linear-optical circuits. Filters **(g)** are used to suppress the pump light and on-chip detectors **(h)** are used to measure the processed single photons. Classical control electronics can be interfaced with the device via wire-bonding to a PCB **(i)** or via a homogeneous integration or direct chip-to-chip bonding with CMOS electronics **(j)**

for photon generation, it posses a strong $\chi^{(3)}$ nonlinearity. Pairs of photons can thus be generated via the spontaneous four-wave mixing (SFWM) $\chi^{(3)}$ process that naturally arise when a bright laser pump field propagates through silicon [28–30]. In order to be suitable for scalable quantum photonic architectures, the key qualities to be considered for spontaneous photon-pair sources are [30]:

- **Indistinguishability**: Photons emitted from different sources need to be indistinguishable in order to interfere.
- **Purity**: The state in each degree of freedom (frequency, polarisation, etc.) of the individual photons emitted needs to be pure in order to ensure high-quality quantum interference.
- **Heralding efficiency**: the intrinsic losses in the source have to be low to avoid photon loss. In other words, for a spontaneous photon-pair source, if one of the photons (the idler) is detected to herald the emission of the other photon (the signal), the probability that the signal photon is actually present needs to be high.
- **Brightness**: photon emission with low pump power requirements is desirable to limit the power consumption when scaling to large arrays of sources, as well as for facilitating pump-rejection filtering.

#### 11.3.1.1 Waveguide Sources

In the bright-pump approximation, the SFWM process in a silicon waveguide can be described via the Hamiltonian [1]

$$\hat{H} = \int d\omega_i d\omega_s\, F(\omega_i, \omega_s)\hat{a}^\dagger(\omega_i)\hat{a}^\dagger(\omega_s) + h.c., \tag{11.5}$$

where $h.c.$ indicates the Hermitian conjugate, $\omega_s$ ($\omega_i$) represent the signal (idler) frequency and $\hat{a}^\dagger(\omega_s)$ ($\hat{a}^\dagger(\omega_i)$) the associated photon creation operation. $F(\omega_i, \omega_2)$ is the joint spectral amplitude (JSA) of the emitted pairs of signal and idler photons, given by

$$F(\omega_i, \omega_s) = \mathcal{N} \int_0^L dz \int d\omega_{p_1} d\omega_{p_2}$$
$$\alpha_{p_1}(\omega_{p_1})\alpha_{p_2}(\omega_{p_2})\delta(\omega_i + \omega_s - \omega_{p_1} - \omega_{p_2}) \exp(-\mathrm{i}\Delta kz). \tag{11.6}$$

Here, $\mathcal{N}$ is a normalisation factor, $L$ is the waveguide length, $\alpha_{p_1}$ and $\alpha_{p_2}$ are the spectral envelops for the two pump fields, and $\Delta k(\omega_i, \omega_s, \omega_{p_1}, \omega_{p_2})$ is the phase mismatch between the four fields. The factor $\delta(\omega_i + \omega_s - \omega_{p_1} - \omega_{p_2})$ ensures energy conservation, while the phase-matching term $\exp(-\mathrm{i}\Delta kz)$ ensures momentum conservation.

The Hamiltonian in 11.5 describes a squeezing process where two photons are absorbed from the pump light at frequencies $\omega_{p_1}$ and $\omega_{p_2}$, generating a photon pair at frequencies $\omega_i$ and $\omega_s$ [1]. Depending on the choice of frequencies for the pump field and for the emitted photons, SFWM can be operated in two different regimes, illustrated in Fig. 11.5a:

- **Non-degenerate SFWM:** $\omega_{p_1} = \omega_{p_2}$ and $\omega_i \neq \omega_s$. In this case, the photons are generated in two separate spectral modes, and two-mode squeezing is obtained (see 11.3).
- **Degenerate SFWM:** $\omega_{p_1} \neq \omega_{p_2}$ and $\omega_i = \omega_s$. In this case, photons are generated in the same spectral mode, and single-mode squeezing is obtained (see 11.2).

In the low-squeezing regime (i.e. low pump energy), the TMS state generated with non-degerate SFWM approximates the state

$$|\psi\rangle_{i,s} \simeq |0\rangle_i |0\rangle_s - i\gamma LP \int d\omega_i d\omega_s F(\omega_i, \omega_s) |1\rangle_{i,\omega_i} |1\rangle_{s,\omega_s} \tag{11.7}$$

which describes the probabilistic generation of a photon-pair with an approximate probability $p = \gamma^2 L^2 P^2$, where $\gamma$ is the nonlinear parameter of the silicon waveguide and $P$ is the pump power [31].

Both degenerate and non-degenerate SFWM regimes have been demonstrated in silicon single-mode waveguide sources [32, 33]. The typical length for standard single-mode 220 nm SOI waveguide sources is approximately 1 cm, usually wrapped in a spiral shape to reduce the component footprint, as shown in Fig. 11.4a. Being fully passive devices and easily reproducible with standard semiconductor fabrication facilities, waveguide sources are very practical and with high indistinguishability, resulting in a wide usage in silicon quantum photonic processors [5, 32–34]. The

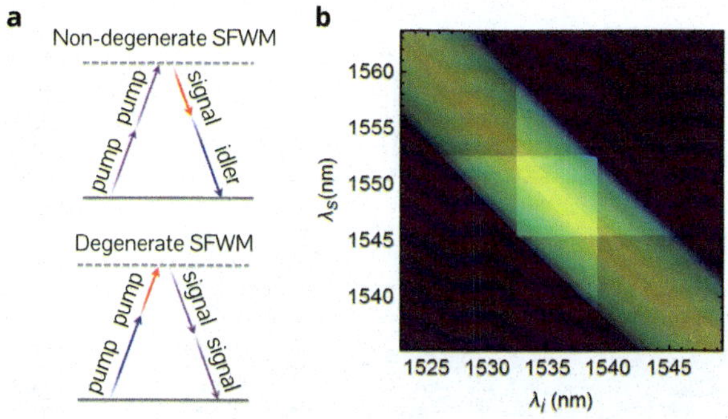


**Fig. 11.5** **a** Spontaneous four-wave mixing schemes. Non-degenerate SFWM (top) generates TMS with photons at difference frequencies, while degenerate SFWM generates SMS with pairs of degenerate photons. **b** Exemplary joint spectra produced in waveguide sources via non-degenerate SFWM. Using spectral filters (shaded areas), the photon spectral purity is increased from 30 to 96%

intrinsic heralding efficiency in waveguide sources is only limited by transmission losses in the waveguide. However, phase-matching conditions in waveguides imply very strong spectral correlations between the photons in each pair, as manifested in the typical JSA shown in Fig. 11.5b. Such spectral correlations imply that the spectral state of the individual photons is highly impure [35]. The single-photon purity can be improved to values >90% via spectrally filtering the photons (see Fig. 11.5b), which comes at the cost of significantly reducing the heralding efficiency of the source [36].

#### 11.3.1.2 Multimodal Waveguide Sources

Multimodal SFWM (mmSFWM) approaches have been recently demonstrated to be capable of solving the limitations of standard waveguide sources in silicon [37, 38]. As shown in the schematic in Fig. 11.5a, in mmSFWM the nonlinear interaction is performed between different transverse modes propagating through a multimode waveguide. Via the use of beam-splitters and mode converters, the pump light is injected into the multimode waveguide in a superposition between different transverse modes, e.g. the TM0 and TM1 modes in the device in Fig. 11.5, with the signal and idler photons also emitted in different transverse modes. By tailoring the group velocities dispersion of the different modes through the design of the multimode waveguide cross section, the phase-matching conditions can be engineered to reduce the spectral correlations between the emitted photons [38]. Moreover, a time delay can be inserted between the pump modes so that, due to the different group velocities of the modes, the temporal overlap between the pump modes gradually increases and decreases along the waveguide, as shown in Fig. 11.5a. This adiabatic switching of the nonlinear interaction further suppress the spectral correlations enabling a near-unit spectral purity [37, 39, 40]. Moreover, due to the lower losses in

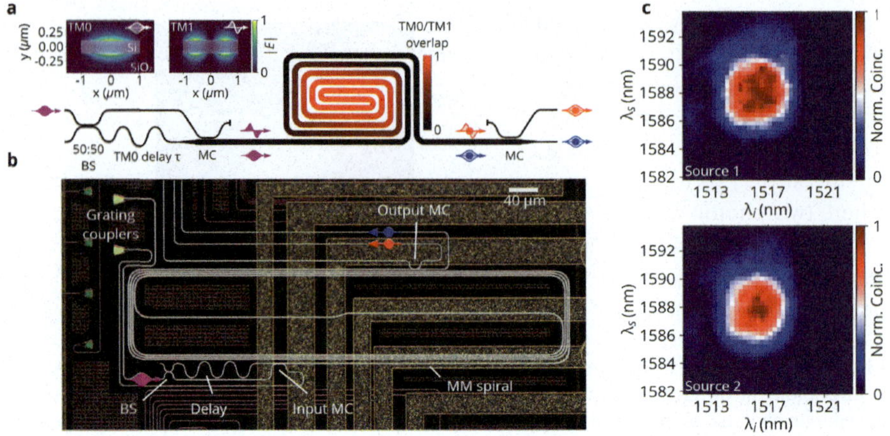


**Fig. 11.6** **a** Schematic of a multimodal source in silicon, where the input pump is prepared in a superposition between the TM0 and TM1 transverse modes via a beam-splitter and a mode converter. A time delay $\tau$ is inserted to increase the source purity. Photons are emitted in the TM0 and TM1 modes via intermodal SFWM along the waveguide and separated at the output by a second mode converter. **b** Microscope image of a multimodal SFWM source in silicon. **c** Joint spectra of the emission from two different multimodal silicon sources show highly uncorrelated spectra (purity of 99%) and high overlap between the sources (indistinguishability > 98%). Images from [37]

multimode waveguides, the heralding efficiency is also significantly improved compared to single-mode waveguide sources. Sources based on mmSFWM have been recently demonstrated on commercial silicon-on-insulator (SOI) photonic devices (see Fig. 11.5b), exhibiting a near-ideal spontaneous photon source performance: spectral purities of 99% without any spectral filtering of the photons (see Fig. 11.5c), source indistinguishability >98%, and intrinsic heralding efficiency >90%. All such performances resulted in the demonstration of a record-high on-chip heralded quantum interference visibility of 96% without spectral filtering of the photons, making this type of sources excellent candidates for photonic quantum information processing [37]. Moreover, mmSFWM has been also recently used to generate photonic entanglement between the transverse modes of photons in silicon multimode waveguides [41].

#### 11.3.1.3 Ring Cavity Sources

Waveguide sources are cavity-free elements. To improve the source brightness, resonant structures such as ring resonators [42–52] or photonic crystal waveguides [53–55] can be used. Ring sources (see Fig. 11.4b) are particularly appealing due to their high miniaturisation and the capability to achieve high Q factors (up to the $10^4$–$10^5$ range for standard single-mode waveguides). Significant benefits to the source brightness are possible when operating SFWM in a cavity: in principle, the photon-pair emission probability is enhanced by a factor $F^6$, with $F$ being the field enhancement in the cavity [56]. However, when using telecom wavelengths in silicon waveguides, the strong pump field inside the ring can induce large nonlinear losses due to effects such as two-photon absorption, which can significantly limit the ring source [57, 58]. These limitations can be avoided by moving to mid-infrared ($\geq$2 $\mu$m) wavelengths in silicon [59] or using wider band-gap materials, such as silicon nitride [60–63], where two-photon absorption is suppressed.

Because the photon emission, as well as the pump fields, need to be resonant with the ring cavity, the narrow-band photon emission reduces spectral correlations between the photons, and thus purities above 80% are readily achieved [49, 51, 56]. Although for standard Gaussian pump spectral envelopes the photon purity from ring resonator sources is limited to be below 93% [64], approaches where more complex pump spectral shapes [65] or dispersive coupling to the resonator, via auxiliary rings or asymmetric MZI structures [64], can boost the purity to near-unity. Such schemes have been recently implemented in silicon devices [52]. The active control of ring resonances via thermal phase shifters has also been demonstrated to ensure a good indistinguishability between multiple ring sources [50, 51]. Moreover, ring structures have also been proposed and demonstrated to mitigate noises when operating SFWM in the degenerate regime for the generation of single-mode squeezed states [63, 66].

### *11.3.2 Linear-Optical Components*

Identically to classical laser light, single photons can be guided in silicon photonic chips through waveguides and manipulated through linear-optical components such as multimode interference (MMI) couplers and evanescent directional couplers (DC) (see Fig. 11.4c, d). In the quantum regime, however, it is crucial to minimise the insertion loss of all linear-optical components in order to reduce the probability of losing single photons—one of the main sources of errors for photonic quantum computing. Linear losses in waveguides have been shown to reach values as low as 0.3 dB/cm at telecom wavelengths, with interesting prospects for building delay lines (see Fig. 11.4f) in silicon, required for multiplexing schemes in linear-optical quantum computing architectures [67, 68]. Beam-splitters implemented by evanescent coupling can offer minimal additional loss, but exhibit less robust splitting ratios, being more sensitive to waveguide fabrication variations. MMI structures are much more robust to fabrication imperfections, but with slightly higher losses (typically $\geq$ 0.1 dB insertion loss) [69].

Phase shifters are commonly realised as thermal heaters (see Fig. 11.4e), which present extremely low losses but are limited to KHz bandwidths [70–74]. High-speed modulation of quantum states can be achieved with carrier-injection or carrier-depletion modulators, showing typical 10 GHz bandwidths [75]. However, their characteristic high and phase-dependent loss can severely reduce their functionality for quantum circuits. Moreover, they have limited functionality to operate at cryogenic temperatures due to carrier freeze-out. Recent developments in thin-film barium titanate on silicon show great potential for the development of on-chip fast modulators, with demonstrated losses below 0.5 dB, speed of tens of GHz [76], and compatible with cryogenic temperatures for full-system integration with superconducting single-photon detectors [77].

Beam-splitters with tunable reflectivities can be built by cascading MMIs and phase shifters into Mach–Zehnder-type structures [78]. Extremely high-precision control of such linear-optical two-mode operations has been demonstrated with extinctions exceeding 60 dB, corresponding to single-qubit computational errors below $10^{-5}$ [79, 80].

### *11.3.3 Detection Systems*

The integration of high-efficiency photon detectors is important to reduce losses due to off-chip coupling, as well as for reducing the system latency (implying less losses in delay lines) and enabling scalable detection of large quantum systems. Superconducting nanowire single-photon detectors (SNSPDs, see Fig. 11.4h) can perform near-ideal single-photon detection with extremely low jitter, death time and dark counts [81–83] and have been integrated in a variety of integrated quantum photonic platforms [84]. The high-yield fabrication of SNSPDs in the NbN material in silicon photonic devices has been demonstrated [85], as well as arrays of up to 240 detectors suitable for the detection of large-scale circuits [86]. Furthermore, the integration of superconducting transition edge sensors (TES), capable of performing photon-number-resolving detection, has also been recently reported in Ti:$LiNbO_3$ waveguides [87], with prospects also for silicon integration. However, superconducting detection systems require the photonic device to operate at cryogenic temperature, typically at $<2$ K temperature for standard SNSPDs, and few tens of mK for TES detectors. Recent engineering efforts have also showed that few-photon number resolution, sufficient for a large part of LOQC applications, is achievable with impedance-matched single standard nanowire superconducting detectors [88].

Apart from single-photon detectors, integrated homodyne detectors have also been recently demonstrated in the silicon quantum photonic platform [89]. In contrast to SNSPDs, these systems can operate at room temperature. A record-high detection bandwidth of up to 9 GHz has been recently reported for on-chip homodyne measurements in silicon [90]. These systems are promising for on-chip continuous-variable photonic quantum applications such as on-chip quantum metrology, quantum random number generation and continuous-variable photonic quantum computing [89, 91, 92].

### *11.3.4 Single-Photon Filters*

The simultaneous integration of single-photon sources and single-photon detectors necessarily requires the integration of filtering structures for high-extinction pump suppression. Such components must suppress the pump below the typical noise level of SNSPDs (less than kHz typical dark counts values), thus providing extinctions above 100 dB. While in most current silicon quantum photonic experiments pump rejection is performed off-chip using standard telecommunication components, devices showing on-chip pump filtering have been realised combining Bragg grating structures and add-drop filters (see Fig. 11.4g), or coherently cascaded unbalanced Mach–Zehnder interferometers. Currently, such filters can achieve up to 150 dB pump suppression when implemented between two silicon chips [44, 93]. For a monolithic integration of pump filters with sources and detectors on a single device, a challenge for filtering comes from the pump light scattered off the waveguides, which might limit the total extinction [93]. The full monolithic system integration of SFWM sources, filters and detectors would represent a major technological milestone for silicon quantum devices, but has yet to be demonstrated.

### *11.3.5 Optical and Electronic Packaging*

When expanding the complexity of photonic quantum processors in microscale silicon devices, a scalable optical and electronic packaging of the system becomes increasingly important. Fortunately, chip packaging methods developed for classical applications can in principle be readily adapted to quantum photonic processors [94]. Optical access to silicon quantum photonic devices is typically achieved via coupling single-mode fibres and fibre arrays via edge-couplers or grating couplers, with sub-1dB coupling losses demonstrated in both approaches [95–99]. For the electronic packaging, most experiments in silicon quantum photonics currently use direct electrical probing [32, 48] or wire-bonding to an external PCB (see Fig. 11.4i) [5, 11, 33, 34, 51, 100, 101]. To decrease the latency of the electronic classical control and read-out systems, crucial to decrease photon losses in feed-forward operations, impressive improvements have been recently shown via direct wire-bonding of quantum photonic chips to microelectronics [90]. Hybrid integration of photonic and electronic CMOS systems [102] as well as flip-chip bonding techniques [103, 104] hold great promise for further speed and scalability enhancements for the classical control of quantum photonic devices (see Fig. 11.4j).

### *11.3.6 Scaling Silicon Quantum Photonic Circuits*

The silicon photonics platform enables to scale quantum photonic circuits where the large number of the components described above can be interconnected on single silicon devices. As shown in Fig. 11.7, this has allowed a rapid scaling of quantum photonic experiments in silicon quantum photonics in recent years, both in terms of circuit complexity [5, 11, 34, 107] and number of photons and qubits generated and processed on-chip [33, 34, 101], which shows a good potential for developing large-scale quantum processors on this platform.

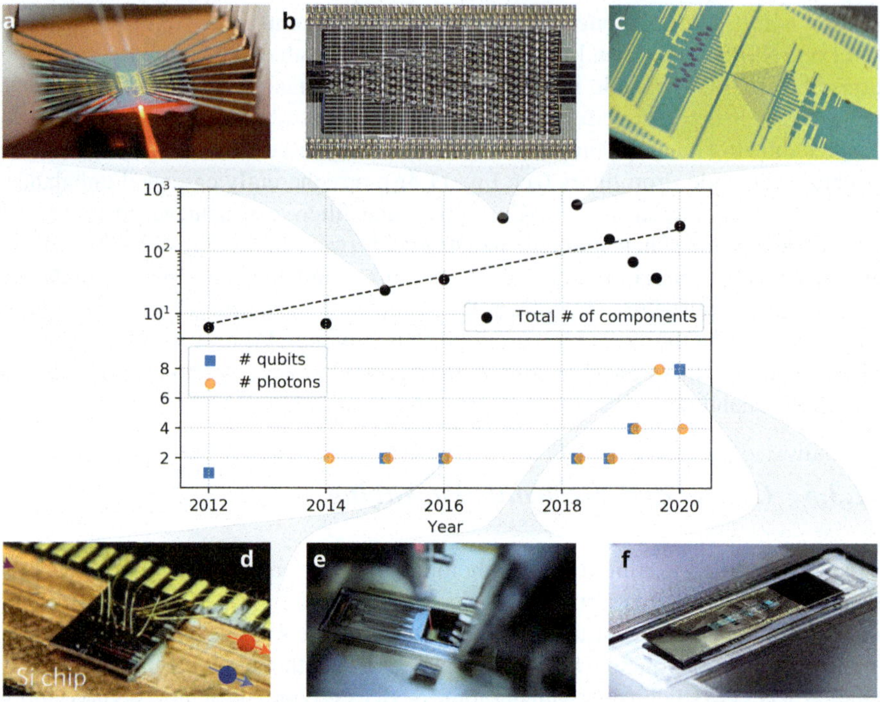


**Fig. 11.7** Progress in the number of components (top) and number of photons and qubits (bottom) for silicon quantum photonic processors. Exemplary devices from some of the reported experiments are shown. Images are from the following references: **a** [48], **b** [11], **c** [5], **d** [100, 105, 106], **e** [33], **f** [34]

## 11.4 Silicon Photonic Quantum Processors

In the past five years, remarkable improvements for quantum photonic processors in silicon have been reported: starting from the first on-chip generation and processing of qubit entanglement using two photons and approximately 10 components in 2015 (see Fig. 11.7), we are now at the stage where experiments with ten or more photons and qubits processed via thousand of optical components are conceivable. While such experiments are still far from the scale required for general purpose quantum computers, they are a good indication of the potential scaling capability of the silicon quantum photonics platform. In this section, the aforementioned recent silicon quantum photonic results are reviewed.

### *11.4.1 Entanglement Generation and Processing in Silicon Photonics*

The first demonstration of path-encoded entanglement between two photonic qubits in silicon photonics was achieved in 2015 using the device shown in Fig. 11.7a. In this integrated photonic circuit, schematised in Fig. 11.8a, a pair of ring sources are coherently coupled to generate a photon pair in superposition between the two sources via non-degenerate SFWM. Such superposition corresponds to a maximally entangled path-encoded state of two qubits (i.e. a Bell pair), obtained after the two emitted photons, which possess different frequencies, are separated and grouped via ring-based filters and a crosser. The generated entanglement was characterised via the use of two integrated MZIs to perform arbitrary local measurements (detection was performed via off-chip SNSPDs) and verified through the violation of a Bell inequality (see Fig. 11.8b). Note that with this approach the two-photon entanglement

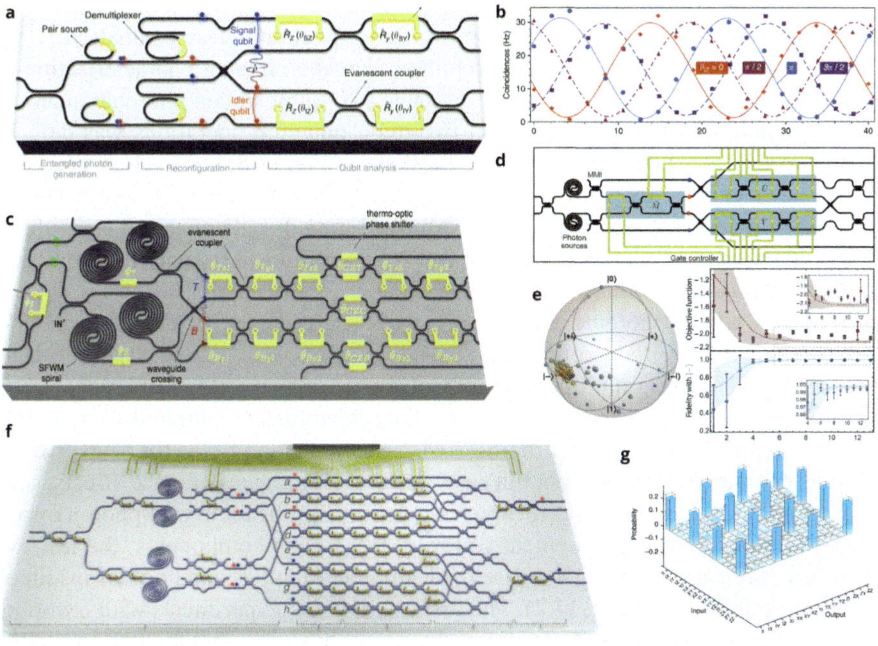


**Fig. 11.8** **a** Silicon quantum device for the generation and measurement of two-qubit path-encoded entanglement via ring resonators sources. **b** Characterisation of the generated on-chip two-qubit entanglement via Bell inequality violation. **c** Silicon device for the generation of two-qubit entanglement and the operation of a post-selected CNOT entangling gate. **d** Silicon device for the operation of arbitrary two-qubit controlled operations. **e** Implementation of a witness-assisted variational quantum simulation algorithm on the device. **f** Large-scale silicon quantum photonic processor for the implementation of arbitrary two-qubit gates. **g** Reconstruction of an exemplary two-qubit gate on the device using quantum process tomography, showing a fidelity of 95%. Images **a** and **b** are from [48], **c** is from [108], **d** is from [106], **e** is from [105] **f** and **g** are from [107]

is generated directly at the sources, where the nonlinearity of SFWM converts the coherent pumping of different sources to the correlated superposition of the photon pair, without the need of entangling gates. Similar ideas have also been used to obtain two-photon entanglement in other degrees of freedom using on-chip sources in silicon, such as frequency entanglement [45, 109–113], time-bin entanglement [46, 114], polarisation entanglement [115] and transverse-mode entanglement [41].

The increase in the complexity of linear-optical circuits has opened the possibility to perform on-chip probabilistic post-selected entangling gates in silicon quantum photonics. A six-mode circuit embedding a post-selected controlled-NOT entangling operation was demonstrated using the device in Fig. 11.8c to add arbitrary tunability to the two-photon entanglement generated and processed on-chip [108]. A more general circuit was reported in the silicon device used in [100, 105, 106] (see Fig. 11.7a for a photo of the chip, and Fig. 11.8d for a circuit schematic), which was able to perform on-chip arbitrary and reconfigurable controlled operations between two qubits. Using such a circuit, a number of proof-of-principle implementations of novel quantum algorithms were demonstrated on this device, such as the quantum simulation of small chemical systems via witness-assisted variational quantum algorithms (see Fig. 11.8e) [105] and via Bayesian approaches to quantum phase estimation [100], and quantum Hamiltonian learning techniques for the efficient characterisation of quantum systems [106]. A fully programmable two-qubit silicon photonic quantum processor able to implement universal two-qubit quantum operations was demonstrated in 2018 using the large-scale circuit shown in Fig. 11.8f [107]. High-precision performance of arbitrary two-qubit operations was demonstrated with this device (see Fig. 11.8g), which was also used to perform some small-scale quantum algorithms, such as the quantum approximate optimisation algorithms [116] and the simulation of Szegedy quantum walks [117].

While the silicon photonic quantum processors described so far display a remarkable increase in the circuit complexity, they are limited to two-qubit generation and processing. Generating multiphoton entanglement adds important difficulties compared to the two-photon case due to the need of probabilistic multiphoton entangling gates, a significant decrease in the SFWM-based generation rate for four or more photons compared to the two-photon case, and the appearances of noise effects, such as single-photon spectral impurities, which are not significant in two-photon experiments. Thanks to the development of lower-loss silicon components, the last two years have seen a remarkable increase in the number of qubits and photons in silicon quantum photonics (see Fig. 11.7). The first multiphoton experiments with four photons generated and processed on silicon chips have been reported in 2018 [49, 101]. The current state of the art is silicon quantum photonics experiments with up to eight photons [33] (see Fig. 11.7e) and eight qubits (see Fig. 11.7f) [34], as described in the following sections.

### 11.4.2 High-Dimensional Quantum Entanglement in Silicon

Recent experiments have exploited the capability of silicon photonics to integrate extremely complex circuits to enlarge the quantum information processing capability for quantum devices with limited number of photons, via the use of high-dimensional quantum systems. In particular, the device in Fig. 11.7c was used to demonstrate the on-chip generation and processing of high-dimensional entanglement between a pair of photons [5]. Qudits, with local dimensionality up to 16, were encoded in the paths of each photon, as described in Fig. 11.1, and maximally entangled states were generated via the coherent pumping of multiple SFWM waveguide sources. The large-scale reconfigurable silicon circuit used, which embedded more than 500 optical components (see Fig. 11.9a), was demonstrated to be able to process the high-dimensional photonic entanglement with very high precision, with the reconstructed entangled states showing unprecedented quality for high-dimensional entangled systems. The reconstructed entangled state for a pair of two qudits in dimension 12 is shown Fig. 11.9b.

Although silicon quantum photonics can enhance the complexity of current photonic quantum processors, increasing the photon number is ultimately required to build up scalable quantum photonics devices. Nevertheless, the resource savings it enables are of potential interest for both near-term applications on prefault-tolerant quantum devices and more efficient large-scale architectures. In a recent silicon quantum photonic experiment, Vigliar et al. have demonstrated such resource savings for the generation of quantum states of up to eight qubits by using four photons, each one encoding two qubits in its four-dimensional Hilbert space [34]. The silicon device, shown in Fig. 11.10a, used a circuit where eight SFWM sources were pumped to generate four photons in two pairs of maximally entangled ququarts (four-dimensional qudits). Using the circuit schematised in Fig. 11.10b, post-selected fusion operations were used two perform entangling gates between the two pairs of qudits, generating a multiphoton high-dimensional entangled states (see Fig. 11.10c). Such four-photon state was able to encode the equivalent of up to eight qubits. By reconfiguring the

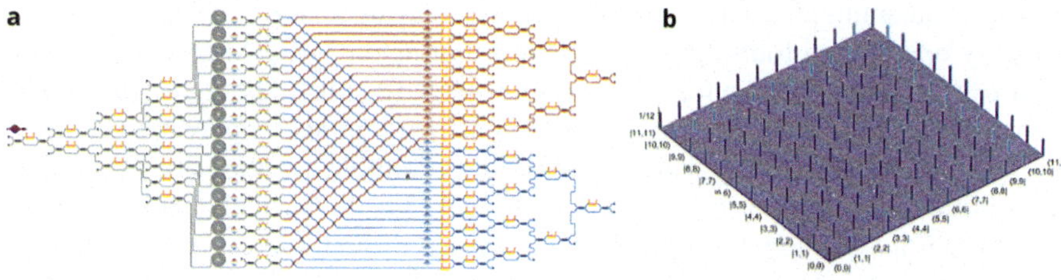


**Fig. 11.9** **a** Circuit schematic for the generation and measurement of high-dimensional photonic entanglement on a reconfigurable large-scale silicon quantum photonic device. **b** Entangled states of two qubits of local dimension 12 reconstructed on the silicon chip via quantum state tomography. Images from [5]

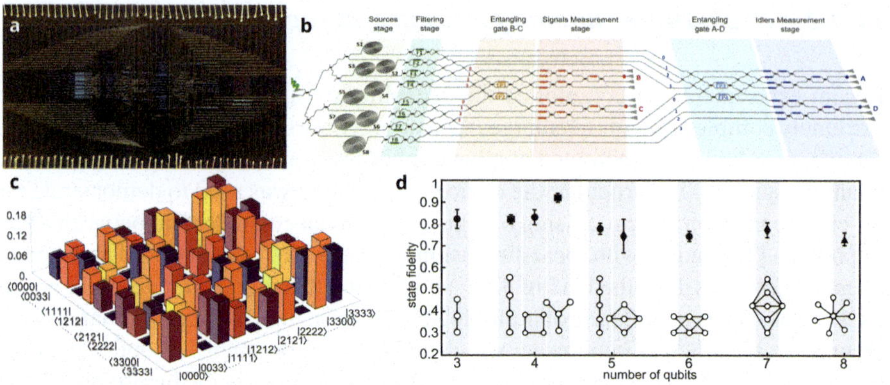


**Fig. 11.10** **a** Silicon chip embedding the photonic circuit in **b** for the generation and processing of eight-qubit entangled states via the use of multiphoton high-dimensional entanglement. **c** Reconstructed entangled state of the four photons, with each photon representing a four-dimensional qudit. **d** Fidelities for different classes of entangled graph states of up to eight qubits that can be generated reconfiguring the circuit in **(b)**

photonic circuit, the device was able to explore a wide range of entanglement classes, so-called graph states, as shown in Fig. 11.10d.

### *11.4.3 Measurement-Based Quantum Computing in Silicon Quantum Photonics*

The generation of highly entangled multiphoton states represents the fundamental resource for measurement-based quantum computation (MBQC), the most relevant quantum computing paradigm for quantum photonics [14, 15, 23]. The scale reached with current silicon quantum photonic devices enables the implementation of simple MBQC protocols fully on-chip. A proof-of-principle MBQC implementation in integrated quantum photonics was firstly reported using a laser-written glass device passively operating a four-qubit state, demonstrating an MBQC implementation of the Grover search algorithm for a four-bin register [118, 119]. The eight-qubit silicon quantum processor shown in Fig. 11.10 demonstrated more complex MBQC protocols on-chip, demonstrating also the embedding of simple quantum error-correcting codes in the measurement-based quantum model [34]. In particular, MBQC operations were implemented using graph states formed of both logical (error-corrected) qubits as well as physical (non-error-corrected). For various single-qubit MBQC operations, increased fidelities were observed when adopting error-protected codes. Such improvements resulted in a significant performance enhancement in the MBQC implementation of the quantum phase estimation algorithm.

### *11.4.4 Networking Silicon Quantum Devices*

One of the most important advantages of quantum photonics compared to other quantum platforms is the capability of photons to travel long distances and efficiently interconnect different and distant processors. In silicon quantum photonics, this capability has enabled a number of demonstrations for integrated systems for quantum key distribution with weak-coherent states of light, which can greatly leverage on silicon components developed for classical communications [120–122]. The chip-to-chip distribution of quantum entanglement is a key task in networking quantum processors, with prospects for the development of modular quantum computing architectures [123, 124] and the quantum internet [125]. The first entanglement distribution between two silicon quantum photonic devices was achieved in 2016, where a path-encoded qubit from an entangled pair was converted into a polarisation qubit via a two-dimensional grating coupler, and sent via a fibre to a second chip to convert it back to path and measure the entanglement [126]. Quantum teleportation between two silicon chips has also been recently demonstrated using the devices schematised in Fig. 11.11a [51]. A qubit from an entangled photon pair was distributed from a silicon Chip A to a second chip B via a path-polarisation interconversion and a fibre link. The state of a third heralded photon in chip A was then teleported to the photon in Chip B via a Bell measurement on the two photons left on Chip A. The teleported states, shown in Fig. 11.11b for few exemplary states, achieved fidelities of approximately 90%.

## 11.5 Applications for Near-Term Photonic Quantum Processors

In previous sections, we discussed how silicon quantum photonics enables a scalable approach to build large-scale photonic quantum processors, with hundreds of quantum optical components linked together in optically stable interferometers. However, a crucial challenge in photonics is to increase the number of quantum information carries, i.e. photons, that can be processed in such large circuits. In fact, the number of photons is a key parameter to determine the computational complexity of the photonic quantum computation performed. However, optical loss in integrated components, as well as non-deterministic photon generation and entangling operations, suppress the computation rate exponentially when increasing the photon number on preloss-tolerant devices. This renders the efficient scaling of photonic architectures challenging. Note that these errors are unique to quantum photonics, as in other quantum platforms (solid-state qubits, superconducting circuits, trapped ions, etc.) the possibility that a quantum information carrier disappears is typically negligible. Fault-tolerant architectures, as the ones mentioned in Sect. 11.2, are tolerant to photon loss and key to enable scalable photonic quantum hardware [15, 19]. However,

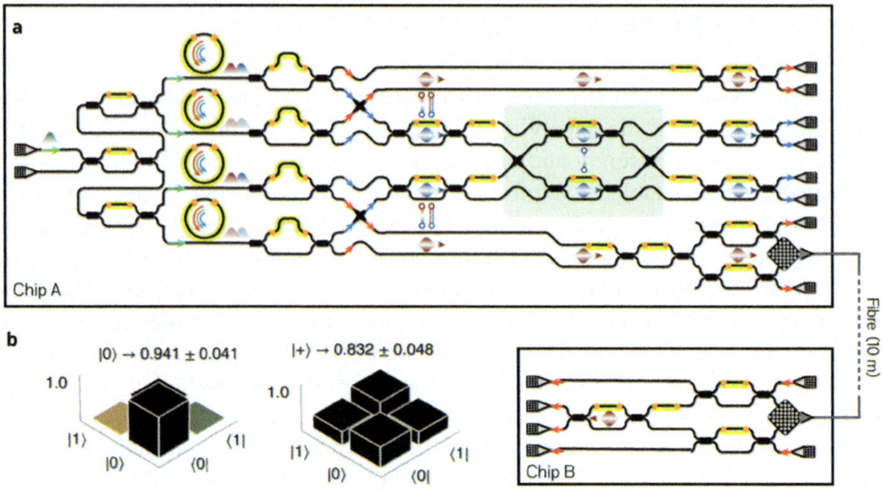


**Fig. 11.11** **a** Schematic of silicon photonic devices used to perform chip-to-chip quantum teleportation. Two pairs of photons are generated on Chip A. A photon from an entangled pair is sent to Chip B via path-polarisation interconversion which allows to transmit the qubit via a 10m-long optical fibre. A post-selected Bell measurement between two photons on Chip A performs the single-qubit quantum state teleportation from Chip A to Chip B, and the circuit on chip is used to characterise the teleported state. **b** Reconstruction of the teleported states on Chip B for some exemplary one-qubit states. Images from [51]

the highly demanding overheads required for fault-tolerance will necessitate considerable technological progress before making such scalable and universal photonic quantum machines accessible.

A nearer-term goal is the development of non-universal machines that perform specialised algorithms on non-fault-tolerant quantum machines [24, 26]. We will here focus on boson sampling [26, 127], a promising approach for such special purpose quantum photonic devices, describing different boson sampling protocols and reviewing recent implementations in silicon photonics and possible applications.

## *11.5.1 Boson Sampling Machines*

Fault-tolerant quantum computation is well beyond current quantum technologies. On the other hand, there are already quantum systems currently accessible, e.g. ultra-cold atomic systems, that allow some degree of control and whose behaviour seems to be intractable to simulate on classical machines [128–130]. It is, however, difficult to interpret these systems in terms of computational machines, i.e. with some well-specified inputs and outputs, as well as to theoretically assess the classical computational complexity for their simulation.

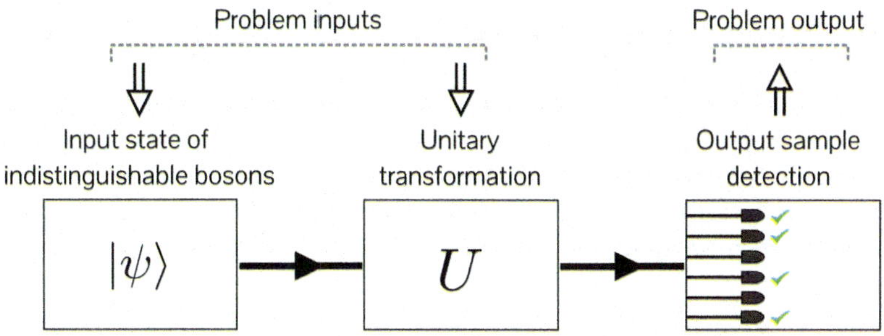


**Fig. 11.12** Schematic representation of the boson sampling computational problem

Boson sampling has been proposed by Aaronson and Arkiphov as an intermediate situation: a well-defined computational model that is provably intractable on classical machines but realistic using near-term experimental capabilities [26]. The use of sampling protocols inspired by boson sampling has recently enabled the ground-breaking demonstration of the first quantum computational advantage, obtained by Google in the superconducting qubits platform [130]. The boson sampling computational model is pictured in Fig. 11.12 and can be schematised as follows.

- *Inputs*: an initial multimode photonic state $|\psi\rangle$ and a linear-optical network with $m$ input modes and $m$ output modes, described by a $m \times m$ unitary transformation $U$ randomly sampled from the Haar random distribution.
- *Experiment*: The input state $|\psi\rangle$ is propagated through the linear-optical circuit $U$ and single-photon detection is performed on the evolved state, using $m$ detectors on the output modes. (Here, we will focus on single-photon detection, although the protocol can be further generalised to include Gaussian measurements as well [131]).
- *Outputs*: the observed $n$-photon coincidence pattern at the single-photon detectors.

The computational model is a *sampling* problem: given the input state and the unitary transformation, the experimenter is required to sample measurement outcomes from the resulting output distribution of the detection patterns. Clearly, this model is quite simple from a technological point of view; no feed-forward, optical nonlinearity or adaptivity is required, making boson sampling machines much more realistic than fault-tolerant quantum hardware for near-term experiments. Still, it has been demonstrated to be intractable to simulate, even approximately, on classical computers if the number of photons is large enough ($n \gtrsim 50$ photons) [26, 132, 133]. The price to pay is universality: the boson sampling computational model is strongly believed to be non-universal.

In recent years, a range of boson sampling variants has been developed and implemented [8, 134–143] to enhance the scalability in practical devices and open a wider set of applications, with the main difference being the input photonic state $|\psi\rangle$. Here, we discuss three of the main protocols.

#### 11.5.1.1 Standard Boson Sampling

In its original proposal [26], boson sampling assumed as input state $|\psi\rangle$ a Fock state of $n$ photons in $n$ different modes, as schematised in Fig. 11.13a. Considering the case where the $n$ photons are injected in modes $\mathbf{j} = \{j_1, j_2, \ldots, j_n\}$ of a $m \times m$ linear interferometer described by a transfer unitary matrix $U$, the probability of obtaining an $n$-photon detection pattern $\mathbf{k} = \{k_1, k_2, \ldots, k_n\}$ is given by [144]

$$p_U(\mathbf{k}|\mathbf{j}) = \frac{1}{\left[\prod_{i=1}^{m} s(\boldsymbol{j})_i!\right]\left[\prod_{i=1}^{m} s(\boldsymbol{k})_i!\right]} \left|\text{Perm}\left(U_{j,k}\right)\right|^2. \tag{11.8}$$

Here, $s(\boldsymbol{j})$ and $s(\boldsymbol{k})$ represent the mode occupancies for the two configurations, i.e. $s(\boldsymbol{k})_i$ indicated how many photons are present in the $i$-th mode for the configuration $\boldsymbol{k}$. The matrix $U_{\mathbf{j},\mathbf{k}}$ is the submatrix of $U$ obtained by taking its rows and columns associated with $\mathbf{k}$ and $\mathbf{j}$, respectively [144], and Perm$(\cdot)$ is the matrix permanent function.

Aaronson and Arkiphov showed that, under mild conjectures, approximate sampling of output states $\mathbf{k}$ from the distribution $p_U(\mathbf{k}|\mathbf{j})$ is intractable on classical machines for large values of $n$ [26]. Current estimates predict that $n \approx 50$ are required to enter a regime where a classical simulation of boson sampling would no longer be possible on supercomputers [132]. However, non-deterministic photon sources and losses limit the scalability of this approach. In fact, if $n$ sources with an efficiency $\epsilon$ are used to produce the $n$-photon input Fock state in the standard boson sampling configuration, the probability of generating such state is thus $p_{BS}(n) = \epsilon^n$. Therefore, if the sources are non-deterministic ($0 \leq \epsilon < 1$), the experimenter would need to wait an exponentially long time before observing an $n$-photon event. Although significant progress has been achieved in high-efficiency solid-state single-photon emitters, this issue has so far limited current implementations of standard boson sampling to systems with less than 15 photons [143].

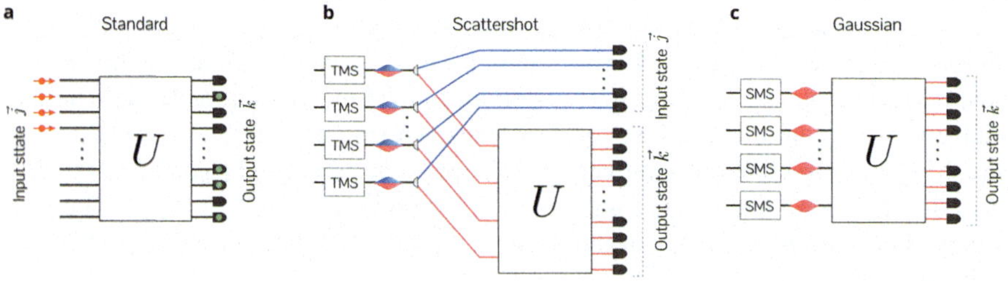


**Fig. 11.13** **a** Schematic representation of the standard boson sampling protocol, where $n$ photons are prepared in the configuration $\boldsymbol{j}$ and injected into a $m \times m$ interferometer described by the unitary transformation $U$. The detection pattern $\boldsymbol{k}$ is recorded at the output of the interferometer. **b** Schematic of scattershot boson sampling. Weak TMS states are generated and separated. The $n$-photon configuration $\boldsymbol{j}$ injected into the interferometer is now random but heralded upon detection of the idlers modes (blue modes). **c** Schematic of Gaussian boson sampling, where the single photons at the input are replaced with SMS states

#### 11.5.1.2 Scattershot Boson Sampling

The scattershot approach to boson sampling, developed by Lund et al. few years after the original boson sampling proposal [145], represents a way to increase the complexity of photonic experiments with probabilistic sources based on spontaneous parametric processes, such as SFWM-based sources in silicon waveguides. The protocol is represented in Fig. 11.13b. To perform $n$-photon boson sampling, $m_0 \geq n$ parametric sources are used, each generating a TMS state (see 11.3). The total state is then a product of TMS states produced from the array of sources:

$$|\psi\rangle = \bigotimes_{j=1}^{m_0} |\mathrm{TMS}(\xi)\rangle_j \,, \tag{11.9}$$

where the two-mode squeezing parameter $\xi$ is for simplicity considered uniform across the sources. Half of each TMS, i.e. the idler modes, are sent directly to a photon counter, while the other $m_0 \leq m$ signal modes are injected into an $m \times m$ interferometer described by $U$. Due to the perfect correlations of the photon number in TMS states, the photon state $\mathbf{j}$ injected into the interferometer corresponds directly to the pattern measured at the idler modes, and a standard boson sampling scenario is thus recovered. Note, however, that now the input state is random but, crucially, heralded. In particular, an $n$-photon input state is generated whenever $n$ out of the $m_0$ probabilistic sources fire, which provides a combinatorial enhancement with respect to the standard boson sampling case. In fact, the probability to generate an $n$-photon input state is now given by [145]:

$$p_{SBS}(n) = \binom{m_0}{n} \epsilon^n (1-\epsilon)^{m_0} \tag{11.10}$$

where $\epsilon = |\tanh(\xi)|^2$ is the photon-pair emission probabilities for each TMS source (in the low-squeezing regime). Considering $m_0 = n^2$, such combinatorial enhancement can now provide a polynomial $n$-photon generation probability scaling as $p_{SBS} \propto 1/\sqrt{n}$, even if the efficiency $\epsilon$ of each source is very low. In contrast to standard boson sampling, the scattershot approach enables an efficient scaling also in presence of probabilistic sources, at the cost of increasing the number of sources and detectors used [138]. It is therefore well-suited for integrated photonics approaches, and in particular silicon quantum photonics, where large arrays of probabilistic sources and detectors are accessible [5, 86].

#### 11.5.1.3 Gaussian Boson Sampling

In scattershot boson sampling, the squeezed states from the sources is collapsed to single-photon states before the interference by measuring the idler modes. Gaussian boson sampling is instead another variant of boson sampling where such collapse happens only after the interference [146, 147]. As schematised in Fig. 11.13c, $m_0$ SMS states, generated for example, via degenerate SFWM in silicon waveguides (see 11.2), are directly injected in the interferometer, followed by single-photon detection at the output. In this situation, the input photon configuration is a coherent superposition of many different possible Fock states, and the probability to detect an output pattern $\boldsymbol{k}$ is given by [147]:

$$p_{\boldsymbol{\xi}}(\mathbf{k}) = \frac{1}{\left[\prod_{i=1}^{m} s(\boldsymbol{k})_i!\right]\sqrt{\det(\boldsymbol{\sigma}_Q)}} \left|\mathrm{Haf}\left[B_{\mathbf{k}}(\boldsymbol{\xi})\right]\right|^2, \tag{11.11}$$

where

$$B(\boldsymbol{\xi}) = U\left(\bigoplus_{i=1}^{m} \tanh(\xi_i)\right)U^{\top}, \tag{11.12}$$

with $\xi_i$ the SMS parameter of the $i$th source, and $B_{\mathbf{k}}$ the submatrix obtained from the rows and columns $\{k_1, k_2, \ldots, k_n\}$ of $B$ [147]. The functions $\det(\cdot)$ and $\mathrm{Haf}(\cdot)$ are, respectively, the determinant and the Hafnian of a matrix [148], while $\sigma_Q = \sigma + \mathbb{1}$ and $\sigma$ is the total covariance matrix of the input SMS states [149, 150]. Also, for Gaussian boson sampling, the probability to detect $n$ photons at the output is combinatorially enhanced [147]:

$$p_{GBS}(n) = \binom{n + m_0/2 - 1}{n} \epsilon^n (1-\epsilon)^{m_0/2}, \tag{11.13}$$

which, for $m_0 = n^2$ scales as $p_{GBS} \propto 1/\sqrt{n}$ similarly to scattershot (with an asymptotic additional speed-up by a constant factor of $e \simeq 2.71$) [147]. In addition, Gaussian boson sampling is also more resource efficient compared to scattershot, for instance, because no auxiliary detectors are required for heralding. However, as the computation time for the calculation of a permanent and a Hafnian of a $n \times n$ matrix are $\mathcal{O}(n2^n)$ and $\mathcal{O}(n2^{n/2})$ respectively [151], it is expected that $2n$ photons are required for a Gaussian boson sampling protocol to achieve a classical run time comparable with a $n$ photon standard boson sampling experiment [146, 147].

### *11.5.2 Scaling Boson Sampling with Silicon Quantum Photonics*

Achieving a regime of quantum computational advantage with boson sampling requires a photonic platform able to generate and process states with tens of photons in hundreds of modes. While significant improvements in quantum dot sources interfaced with low-loss passive interferometers (generally based on bulk optics) have enabled experiments with up to 14 detected photons [139–141, 143], reconfigurable integrated quantum photonics circuitry will be required for practical applications at scale. Although a number of the initial boson sampling demonstrations have been performed via integrated laser-written interferometers in glass chips, they relied on bulk photon sources and were limited to less than 6 photons, presenting significant scaling challenges [8, 135–138, 142, 152–154].

#### 11.5.2.1 Implementing Boson Sampling in Silicon Quantum Photonics

Silicon quantum photonics, enabling the integration of large arrays of high-quality sources [5, 37] and large-scale reconfigurable interferometers [5, 11, 34, 107], offers a promising photonic platform to scale up boson sampling experiments. Recently, the first boson sampling experiment with fully on-chip photon generation and processing (off-chip detection) was reported in a silicon quantum photonic device, where up to 8-photon states were operated to perform scattershot and Gaussian boson sampling [33]. The device used is schematised in Fig. 11.14a and the silicon circuit is shown in Fig. 11.14b. The device consisted of four SFWM sources, reconfigurable asymmetric-MZI-based filters to reject the pump and separate the idler and signal photons, a 12-mode low-loss random walk interferometer and low-loss grating couplers [95] to couple photons off-chip and send them to high-efficiency (>90%) off-chip SNSPDs. To implement different boson sampling protocols, the sources were operated in two different regimes, as shown in Fig. 11.14b. In the first regime, they were pumped using a single-wavelength laser in order to generate weak TMS states via non-degenerate

a
b
c
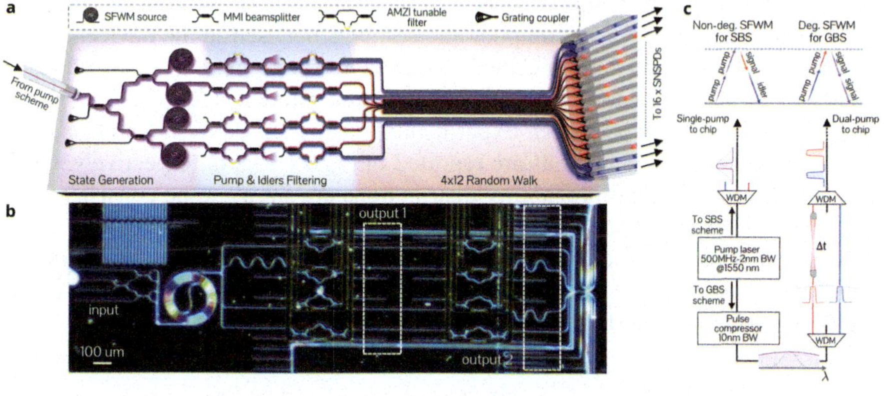


**Fig. 11.14** **a** Schematic of the silicon boson sampling device. Four integrated SFWM sources are used for photon generation, and two layers of asymmetric MZI interferometers are used to filter out the pump light from the photons and to separate the idler and signal photons. A 12-mode unitary transformation is implemented via a low-loss random walk, obtained by coupling together 12 waveguides of length $\sim$ 110 $\mu$m, to interfere the signal photons. Finally, photons are fibre-coupled off-chip via low-loss grating couplers [95] ($\sim$ 1dB loss) and detected off-chip. **b** Optical microscope image of the silicon photonic circuit. **c** Pumping schemes used to generate SMS and TMS states via degenerate and non-degenerate SFWM, respectively. The generated TMS (SMS) states are used to implement scattershot (Gaussian) boson sampling within the same silicon quantum photonic device

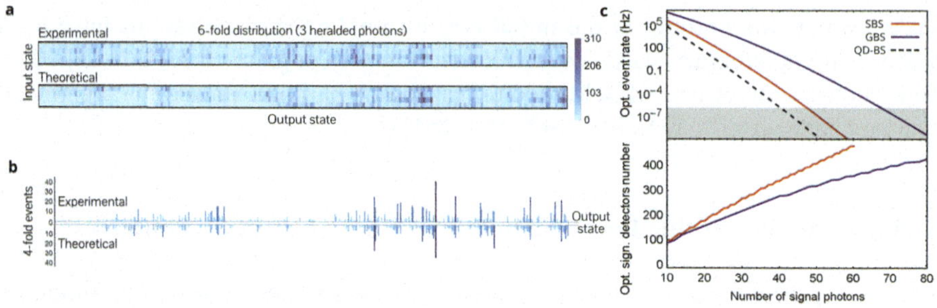


**Fig. 11.15** **a** Input/output scattershot boson sampling experimental and theoretical distributions for six-photon events (three heralded signal photons). **b** Output Gaussian boson sampling experimental and theoretical distributions for four-photon events. **c** Optimal event rate (top) and associated circuit size (bottom) estimated for different numbers of signal photons in the scattershot (red) and Gaussian (blue) boson sampling regimes. Shaded areas represent impractical experiments, where the threshold is set to be 1 event/week

SFWM. In this case, signal and idler photons are emitted at different wavelengths. In the second regime, a dual-wavelength pumping scheme was used to generate weak SMS states via degenerate SFWM. In this second case, all photons are emitted at the signal frequency. The first pumping regime allows scattershot boson sampling, based on TMS states, while the second implements Gaussian boson sampling, based on SMS states. The switching between the two different regimes can be performed within the same chip, by reprogramming the nonlinear effect operated in the sources (degenerate or non-degenerate SFWM) via the choice of the pumping scheme.

Thanks to the low-loss silicon photonic components used in the device, up to 8 photons were generated and processed in the scattershot regime (4 heralded signal photons), the current record in integrated quantum photonics, although at low 8-photon event rates (few per hour). Experimental results for scattershot and Gaussian boson sampling implementations are reported in Fig. 11.15a, b, respectively, where the observed input/output probability distributions are shown to be consistent with the theoretical expectations.

Recently, another boson sampling experiment was reported in the silicon platform, where a bright off-chip source of squeezing was coupled into a passive silicon circuit to perform quantum interference of up to 5 photons [155].

#### 11.5.2.2 Scaling with Near-Term Silicon Devices

While the device discussed above provides an architecture for implementing various boson sampling protocols in silicon quantum photonics, the scale of the protocols implemented is still small, meaning that they can be easily simulated on classical computers. However, we can investigate how the computational complexity of the protocols implementable with such architecture scale when increasing the size of the silicon photon circuits. To represent the capability of near-term silicon quantum photonic devices, we consider low-loss components as those implemented and characterised in the chip of Fig. 11.14a [33], high-quality SFWM sources as in Fig. 11.6 [37], and arrays of integrated SNSPDs [86]. The protocols we consider are scattershot and Gaussian boson sampling. In terms of $N$-photon event rates, the combinatorial enhancement obtained in both protocols when increasing the number of sources (and modes) is compensated by the additional losses obtained when increasing the depth of the interferometer [33]. A trade-off between these two effects is required to achieve high rates. The optimal circuit size and rates with the device parameters described above are shown in Fig. 11.15c for different photon numbers and for both scattershot and Gaussian boson sampling. It can be observed that experiments at a scale of up to $\approx$70 signal photons are estimated to be possible with Gaussian boson sampling, and of $\approx$50 signal photons for scattershot. These values are expected to be at the limit of what is tractable with classical supercomputers [132]. Note that the circuits require hundreds of sources, modes in the interferometer, and detectors; a scale impractical for bulk optical experiments.

To further increase the complexity, e.g. to target experiments with hundreds of photons, significant technological progress is required. A first improvement would be to develop materials with lower transmission losses and more efficient sources of squeezed light. In this direction, promising alternatives are being investigated in the SiN, LiN, LNOI and $LiNbO_3$ integrated photonic platforms [60–63, 66, 156, 157].

## *11.5.3 Quantum Simulation via Boson Sampling*

While boson sampling is an interesting quantum computing model to reach quantum advantages with photonics, the task of sampling the output photons from a random interferometer is a specialised problem with no direct application. However, in recent years, a wide number of applications, from quantum chemistry to graph theory problems, have been shown to be solvable on boson sampling machines, with potential prospects to achieve quantum speed ups in industrially relevant applications with boson sampling. We describe some of these applications, with special focus on molecular quantum chemistry simulations.

### 11.5.3.1 Simulation of Molecular Quantum Dynamics

The reason why complex quantum chemical systems are intractable on classical machines was described by Dirac already in the early days of quantum mechanics [159], namely he noted that the wavefunction of a quantum system grows exponentially with the number of particles, making classical computers unable to exactly simulate quantum systems in an efficient way. This problem led Feynman in 1982 to propose the development of controllable quantum hardware for efficient simulation of complex chemical systems [160]. Quantum chemistry can thus be considered as the original motivation that led to the field of quantum computing in the first place.

The particular system we here focus on are vibrational quantum molecular dynamics. Quantum vibrations in molecules, due to atomic oscillations perturbing a stable molecular configuration, are important, for example, in the study and design of efficient molecular dissociation pathways [161–163]. However, evolving a multiexcitation state across many vibrational modes is computationally inefficient on classical computers even for the basic models based on independent quantum harmonic oscillators. The general idea to map such systems into boson sampling machines is to map the evolution of the bosonic vibrational modes (phonons) to the evolution of bosonic excitations of the electromagnetic field (photons) in optical interferometers.

In more detail, vibrational dynamics in a molecule are essentially small oscillations of the nuclei around a local minimum in the potential energy surface of the molecule. The energy surface depends on the electronic structure of the molecule and the nuclear positions, but, in the Born–Oppenheimer approximation, it is independent from the vibrational state of the molecule. For a general molecule with $N$ atoms the energy surface lives in a $3N-6$ space ($3N-5$ for linear molecules), so that $3N-6$ vibrational modes are possible. In the harmonic approximation, a quadratic form is assumed for the potential energy surface near the stable configuration (see Fig. 11.16a), and $3N-6$ independent normal vibrational modes can be defined (see Fig. 11.16b, top panel), with associated bosonic creation operations $\hat{a}_i^\dagger$ and normal frequencies $\{\omega_i\}$. Another set of vibrational modes that are of practical interest are the so-called *localised modes*, described by bosonic creation operations $\hat{b}_i^\dagger$. These

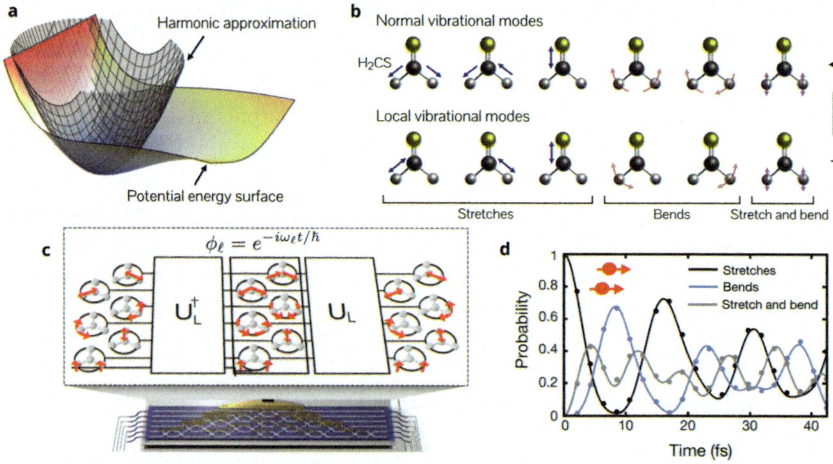


**Fig. 11.16** **a** Example of a molecular potential energy surface and its harmonic approximation. **b** Representation of the normal (top) and localised (bottom) vibrational modes in the $H_2CS$ (Thioformaldehyde) molecule. The change of basis between the localised and normal modes is performed via the unitary matrix $U^L$. **c** Mapping of the quantum evolution of the localised vibrational modes into a reconfigurable boson sampling machines. Such a machine can be implemented and scaled up using universal integrated quantum photonics circuits. **d** Example of experimentally simulated quantum dynamics in the $H_2CS$ (Thioformaldehyde) molecule, reported in [158]

are modes where the vibrational energy is spatially localised on single atoms of the molecule, as shown in Fig. 11.16b bottom panel. Localised modes are of practical importance for understanding many molecular phenomena, such as energy transport and dissociation, and single excitations of these modes can be prepared in quantum chemistry experiments [164]. We therefore focus on simulating quantum dynamics of molecules prepared and measured in localised modes. To describe the dynamics of localised modes, it is convenient to define the basis transformation between the normal and localised modes, given by unitary matrix $U_L$ such that

$$\hat{a}_\ell^\dagger \mapsto \sum_{k=1}^{3N-6} U_{L_{k,\ell}} \hat{b}_k^\dagger. \tag{11.14}$$

The unitary evolution $U(t)$ of the localised modes can then be obtained by converting them into the normal modes, which in the harmonic approximation are independent and evolve according to $\oplus_\ell e^{-iw_\ell t/\hbar}$, and then convert back into the localised modes basis (see Fig. 11.16c):

$$U(t) = U_L \left( \bigoplus_\ell e^{-iw_\ell t/\hbar} \right) {U_L}^\dagger. \tag{11.15}$$

Due to the analogy between bosonic vibrational modes and photons, the ingredients described so far can be mapped into a photonic scenario in the following way [158]:

- Vibrational modes ↔ Optical modes.
- Vibrational mode $\ell$ initialised with $n$ excitations ↔ Optical mode $\ell$ initialised with $n$ photons.
- Evolution of the molecular vibrations described by $U(t)$ ↔ Evolution of the photons in a linear interferometer described by $U(t)$.
- Measurement of the final vibrational configuration ↔ Photon number detection at the output of the interferometer.

The experimental scenarios described above can be directly mapped into boson sampling machines: photons need to be prepared in an input state that matches the initial vibrational state of the molecule, and output configurations are sampled after the evolution according to $U(t)$ implemented via a reconfigurable integrated interferometer (see Fig. 11.16c). Standard boson sampling corresponds to the case where the molecular state is initialised in Fock vibrational states, i.e. with a fixed number of excitations, while if molecules are prepared in squeezed states (or other Gaussian states) the simulation is mapped into Gaussian boson sampling.

Integrated quantum photonics is a very promising platform to implement these simulations and scale them up into computationally interesting regimes. The first photonic quantum simulation of molecular quantum dynamics was implemented using a fully reconfigurable 6-mode integrated interferometer on a silica chip [8, 158]. In this experiment, the molecular quantum dynamics for a wide range of molecules were simulated with up to four photons (see Fig. 11.16d for an example), both in the harmonic approximation and in the anharmonic regime. A proof-of-principle demonstration of how these simulators could be used for the design of more efficient molecular dissociation processes was also reported [158]. The scalability of the silica photonic platform used was, however, limited due to the large footprint of the silica device and the use of bulk off-chip sources. As discussed above, silicon quantum photonics has huge potential to overcome such limitations.

#### 11.5.3.2 Calculation of Molecular Franck-Condon Profiles

A different quantum chemistry problem that can be mapped to boson sampling is the calculation of molecular vibronic (vibrational and electronic) spectra, known as Franck–Condon profiles. While the molecular vibrations discussed in the previous section considered fixed potential energy surfaces, vibronic transitions represent the transition from an initial set of vibrational modes to a new set of vibrational modes that arise when the potential energy surface is modified following a modification in the electronic structure (see Fig. 11.17a). While the spectra for such transitions, i.e. the Franck–Condon profiles, are useful to investigate chemical properties of the molecules, such as their performance as solar cells [165] or as dyes [166], their prediction using classical approaches is computationally challenging already for molecules of modest size [167–169]. On the other hand, Huh et al. have shown how such calculations can be implemented on a variant of Gaussian boson sampling [170].

The mapping makes use of the Doktorov transformation to describe the transition of the molecular vibrational operators $\hat{a}_i^\dagger$ [171] in vibronic processes, given by the operator

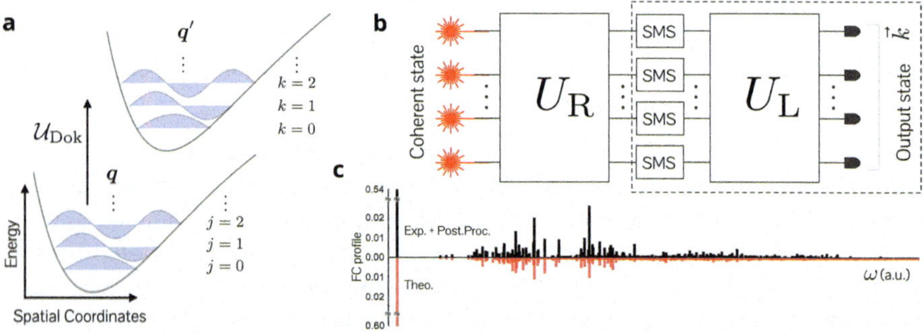


**Fig. 11.17** **a** Simplified schematic of a vibronic transition. The molecule is initially in an electronic configuration with normal vibrational modes (in the harmonic approximation) $\boldsymbol{q}$ associated with the energy surface. When a process, e.g. photon absorption, induces a change in the electronic structure, the energy surface is modified, defining a new set of normal vibrational modes $\boldsymbol{q}'$. In the Franck-Condon approximation, the transformation between the two sets of vibrational modes is given by a linear mapping $\mathcal{U}_{\rm Dok}$. **b** Schematic of the photonic circuit required for the calculation of Franck-Condon profiles. c Reconstructed Franck-Condon profile for a synthetic molecule with the silicon quantum photonic boson sampling device in Fig. 11.14

$$\mathcal{U}_{\mathrm{Dok}} = \mathcal{U}_{\mathrm{L}} \left[ \otimes_{i=1}^{m} \mathcal{S}(\xi_i) \right] \mathcal{U}_{\mathrm{R}}^{\dagger} \left[ \otimes_{i=1}^{m} \mathcal{D}(\alpha_i) \right], \tag{11.16}$$

where $\mathcal{D}(\alpha_i)$ represents single-mode displacement operators with amplitudes $\alpha_i$, $\mathcal{S}(\xi_i)$ represents a single-mode squeezing operator with squeezing parameter $\xi_i$, and $\mathcal{U}_{\mathrm{R}}$ and $\mathcal{U}_{\mathrm{L}}$ are unitary evolutions. If we consider for simplicity the molecule to be initially at zero-temperature (i.e. the initial vibrational state of the molecule in the vacuum state), the Franck–Condon probability for a vibronic transition to a final vibrational configuration $|\boldsymbol{k}\rangle$ is then given by [172, 173]

$$p_{\mathrm{FC}}(\boldsymbol{k}) = |\langle \boldsymbol{k} | \mathcal{U}_{\mathrm{Dok}} | 0 \ldots 0 \rangle|^2. \tag{11.17}$$

Because, as for the simulation of molecular quantum dynamics, the vibrational operators $\hat{a}_i^{\dagger}$ have direct analogy to photonic operators, the vibronic transformation in 11.16 is analogous to the optical circuit shown in Fig. 11.17b, with the Franck–Condon probabilities $p_{\mathrm{FC}}(\boldsymbol{k})$ corresponding to the probability to detect the photon configuration $|\boldsymbol{k}\rangle$ at the output.

Note that the circuit required to calculate Franck–Condon probabilities is a Gaussian boson sampling circuit (boxed in Fig. 11.17b), with the addition of coherent states and a first unitary transformation $U_{\mathrm{R}}$. In practice, such additional resource implies weak laser light injected in the signal modes of the interferometer, which is easy to implement. If the particular molecule under study does not require displacement, the quantum simulation circuit reduces exactly to a Gaussian boson sampling machine [174].

Proof-of-principle demonstrations of the quantum simulation of Franck–Condon profiles via Gaussian boson sampling have been recently reported in a variety of platforms, including fibre-optical set-ups [174], trapped ions [175], superconducting cavities [176], as well as in the silicon quantum photonic processor discussed in Sect. 11.5.2.1 [33]. A Franck–Condon profile experimentally reconstructed on the silicon quantum photonic processor is shown in Fig. 11.17c. In this case, because the interferometer in the chip is not reconfigurable (see schematic in Fig. 11.14a), the calculation was performed for a synthetic molecule associated with the passive circuit used, and with no displacement. Reconfigurable large-scale silicon quantum photonic circuits, such as those reported in Sect. 11.4, are promising to scale this implementation to larger applications, and recent ring-based schemes can be used to reduce computational errors arising from noises in the on-chip SFWM single-mode squeezers [63, 66].

#### 11.5.3.3 Other Boson Sampling Applications

Apart from the quantum simulations applications already discussed, boson sampling has been recently mapped also to other types of problems, including the simulation of spin Hamiltonians [177], molecular docking [178], the enhancement of classical optimisation heuristics [179], certain graph theory calculations [180, 181] and quantum identification and cryptography protocols [182]. This wide portfolio of different near-term applications for reconfigurable photonic devices represents a valuable motivation for the development of large-scale integrated boson sampling photonic hardware.

## 11.6 Outlook

While the mature fabrication tools of the silicon industry have enabled a series of demonstrations of photonic quantum processors with successively more complex circuitry, a powerful general purpose quantum computer remains a highly ambitious goal. As with all current proposals for quantum computing hardware, a number of challenges must be overcome before silicon photonics can support quantum computing at scale. These challenges include filtering strong pump light and full system integration with SNSPDs, developing fast low-loss switches for photonic feed-forward operations, improving the quality of spontaneous photon sources, decreasing photon loss, and enhancing the success probability for probabilistic entangling gates.

The technological journey to general purpose quantum computing is perhaps foreshadowed by that of experimentally demonstrating a Bose–Einstein condensate. Also, a quantum state of matter, seventy years elapsed between its prediction, made in 1925, and its experimental demonstration in 1995. Feynman famously proposed quantum computers in 1982, and a similar 70 year development would mean we have just over thirty years to wait for a general purpose quantum computer, at time of writing.

## References

1. C. Gerry, P. Knight, P.L. Knight, *Introductory Quantum Optics* (Cambridge University Press, Cambridge, 2005)
2. P. Kok, W.J. Munro, K. Nemoto, T.C. Ralph, J.P. Dowling, G.J. Milburn, Linear optical quantum computing with photonic qubits. Rev. Mod. Phys. **79**(1), 135 (2007)
3. J.L. O'Brien, A. Furusawa, J. Vučković, Photonic quantum technologies. Nat. Photonics **3**(12), 687–695 (2009)
4. C. Schaeff, R. Polster, R. Lapkiewicz, R. Fickler, S. Ramelow, A. Zeilinger, Scalable fiber integrated source for higher-dimensional path-entangled photonic qunits. Opt. Express **20**(15), 16145–16153 (2012)
5. J. Wang, St Paesani, Y. Ding, R. Santagati, P. Skrzypczyk, A. Salavrakos, J. Tura, R. Augusiak, L. Mančinska, D. Bacco et al., Multidimensional quantum entanglement with large-scale integrated optics. Science **350**(6386), 285 (2018)
6. M. Reck, A. Zeilinger, H.J. Bernstein, P. Bertani, Experimental realization of any discrete unitary operator. Phys. Rev. Lett. **73**, 58–61 (1994)
7. W.R. Clements, P.C. Humphreys, B.J. Metcalf, W.S. Kolthammer, I.A. Walmsley, Optimal design for universal multiport interferometers. Optica **3**(12), 1460–1465 (2016)
8. J. Carolan, C. Harrold, C. Sparrow, E. Martín-López, N.J. Russell, J.W. Silverstone, P.J. Shadbolt, N. Matsuda, M. Oguma, M. Itoh, G.D. Marshall, M.G. Thompson, J.C.F. Matthews, T. Hashimoto, J.L. O'Brien, A. Laing, Universal linear optics. Science **349**, 711–716 (2015)

9. MYu. Saygin, I.V. Kondratyev, I.V. Dyakonov, S.A. Mironov, S.S. Straupe, S.P. Kulik, Robust architecture for programmable universal unitaries. Phys. Rev. Lett. **124**, 010501 (2020)
10. C.M. Wilkes, X. Qiang, J. Wang, R. Santagati, S. Paesani, X. Zhou, D.A. Miller, G.D. Marshall, M.G. Thompson, J.L. O'Brien, 60 db high-extinction auto-configured mach-zehnder interferometer. Opt. Lett. **41**(22), 5318–5321 (2016)
11. N.C. Harris, G.R. Steinbrecher, M. Prabhu, Y. Lahini, J. Mower, D. Bunandar, C. Chen, F.N. Wong, T. Baehr-Jones, M. Hochberg, S. Lloyd, Quantum transport simulations in a programmable nanophotonic processor. Nat. Photonics **11**(7), 447–452 (2017)
12. C. Taballione, T.A. Wolterink, J. Lugani, A. Eckstein, B.A. Bell, R. Grootjans, I. Visscher, J.J. Renema, et al., 8x8 programmable quantum photonic processor based on silicon nitride waveguides. arXiv:1805.10999 (2018)
13. Michael Varnava, Daniel E. Browne, Terry Rudolph, How good must single photon sources and detectors be for efficient linear optical quantum computation? Phys. Rev. Lett. **100**, 060502 (2008)
14. Daniel E. Browne, Terry Rudolph, Resource-efficient linear optical quantum computation. Phys. Rev. Lett. **95**, 010501 (2005)
15. Terry Rudolph, Why i am optimistic about the silicon-photonic route to quantum computing. APL Photonics **2**(3), 030901 (2017)
16. M.A. Nielsen, I. Chuang, *Quantum Computation and Quantum Information* (Cambridge University Press, Cambridge, 2002)
17. Je-Hyung Kim, Shahriar Aghaeimeibodi, Jacques Carolan, Dirk Englund, Edo Waks, Hybrid integration methods for on-chip quantum photonics. Optica **7**(4), 291–308 (2020)
18. E. Knill, R. Laflamme, G.J. Milburn, A scheme for efficient quantum computation with linear optics. Nature **409**(6816), 46–52 (2001)
19. Mercedes Gimeno-Segovia, Pete Shadbolt, Dan E. Browne, Terry Rudolph, From three-photon greenberger-horne-zeilinger states to ballistic universal quantum computation. Phys. Rev. Lett. **115**, 020502 (2015)
20. M. Gimeno-Segovia, H. Cable, G.J. Mendoza, P. Shadbolt, J.W. Silverstone, J. Carolan, M.G. Thompson, J.L. O'Brien, T. Rudolph, Relative multiplexing for minimising switching in linear-optical quantum computing. New J. Phys. **19**(6), 063013 (2017)
21. Fabian Ewert, Peter van Loock, 3/4-efficient bell measurement with passive linear optics and unentangled ancillae. Phys. Rev. Lett. **113**, 140403 (2014)
22. Hans J. Briegel, Robert Raussendorf, Persistent entanglement in arrays of interacting particles. Phys. Rev. Lett. **86**, 910–913 (2001)
23. R. Raussendorf, H.J. Briegel, A one-way quantum computer. Phys. Rev. Lett. **86**(22), 5188 (2001)
24. John Preskill, Quantum computing in the nisq era and beyond. Quantum **2**, 79 (2018)
25. Ying Li, Peter C. Humphreys, Gabriel J. Mendoza, Simon C. Benjamin, Resource costs for fault-tolerant linear optical quantum computing. Phys. Rev. X **5**, 041007 (2015)
26. S. Aaronson, A. Arkhipov, The computational complexity of linear optics. Quant. Inf. Comp. **13**, 1383 (2014)
27. D. Thomson, A. Zilkie, J.E. Bowers, T. Komljenovic, G.T. Reed, L. Vivien, D. Marris-Morini, E. Cassan, L. Virot, J.-M. Fédéli et al., Roadmap on silicon photonics. J. Opt. **18**(7), 073003 (2016)
28. J.E. Sharping, K.F. Lee, M.A. Foster, A.C. Turner, B.S. Schmidt, M. Lipson, A.L. Gaeta, P. Kumar, Generation of correlated photons in nanoscale silicon waveguides. Opt. Express **14**(25), 12388–12393 (2006)
29. Q. Lin, O.J. Painter, G.P. Agrawal, Nonlinear optical phenomena in silicon waveguides: modeling and applications. Opt. Express **15**(25), 16604–16644 (2007)
30. L. Caspani, C. Xiong, B.J. Eggleton, D. Bajoni, M. Liscidini, M. Galli, R. Morandotti, D.J. Moss, Integrated sources of photon quantum states based on nonlinear optics. Light Sci. Appl. **6**(11), e17100 (2017)
31. D. Bonneau, *Integrated Quantum Photonics at Telecommunication Wavelength in Silicon-On-Insulator and Lithium Niobate Platforms*. Ph.D. thesis (University of Bristol, UK, 2013)

32. J.W. Silverstone, D. Bonneau, K. Ohira, N. Suzuki, H. Yoshida, N. Iizuka, M. Ezaki, C.M. Natarajan, M.G. Tanner, R.H. Hadfield, V. Zwiller, On-chip quantum interference between silicon photon-pair sources. Nat. Photonics **8**(2), 104–108 (2014)
33. S. Paesani, Y. Ding, R. Santagati, L. Chakhmakhchyan, C. Vigliar, K. Rottwitt, L.K. Oxenløwe, J. Wang, M.G. Thompson, A. Laing, Generation and sampling of quantum states of light in a silicon chip. Nat. Phys. **15**(9), 925–929 (2019)
34. C. Vigliar, S. Paesani, Y. Ding, J.C. Adcock, J. Wang, S. Morley-Short, D. Bacco, L.K. Oxenløwe, M.G. Thompson, J.G. Rarity, et al., *Error Protected Qubits in a Silicon Photonic Chip*. arXiv preprint arXiv:2009.08339 (2020)
35. I. Jizan, L.G. Helt, C. Xiong, M.J. Collins, D.Y. Choi, C.J. Chae, M. Liscidini, M.J. Steel, B.J. Eggleton, A.S. Clark, Bi-photon spectral correlation measurements from a silicon nanowire in the quantum and classical regimes. Sci. Rep. **5**(1), 1–9 (2015)
36. Evan Meyer-Scott, Nicola Montaut, Johannes Tiedau, Linda Sansoni, Harald Herrmann, Tim J. Bartley, Christine Silberhorn, Limits on the heralding efficiencies and spectral purities of spectrally filtered single photons from photon-pair sources. Phys. Rev. A **95**, 061803 (2017)
37. Stefano Paesani, Massimo Borghi, Stefano Signorini, Alexandre Maïnos, Lorenzo Pavesi, Anthony Laing, Near-ideal spontaneous photon sources in silicon quantum photonics. Nat. Commun. **11**(1), 1–6 (2020)
38. Stefano Signorini, Mattia Mancinelli, Massimo Borghi, Martino Bernard, Mher Ghulinyan, Georg Pucker, Lorenzo Pavesi, Intermodal four-wave mixing in silicon waveguides. Photonics Res. **6**(8), 805–814 (2018)
39. B. Fang, O. Cohen, J.B. Moreno, V.O. Lorenz, State engineering of photon pairs produced through dual-pump spontaneous four-wave mixing. Opt. Express **21**(3), 2707–2717 (2013)
40. Y. Zhang, R. Spiniolas, K. Shinbrough, B. Fang, O. Cohen, V.O. Lorenz, Dual-pump approach to photon-pair generation: demonstration of enhanced characterization and engineering capabilities. Opt. Express **27**(13), 19050–19061 (2019)
41. L.T. Feng, M. Zhang, X. Xiong, Y. Chen, H. Wu, M. Li, G.P. Guo, G.C. Guo, D.X. Dai, X.F. Ren, On-chip transverse-mode entangled photon pair source. npj Quant. Inf. **5**(1), 1–7 (2019)
42. S. Azzini, D. Grassani, M.J. Strain, M. Sorel, L.G. Helt, J.E. Sipe, M. Liscidini, M. Galli, D. Bajoni, Ultra-low power generation of twin photons in a compact silicon ring resonator. Opt. Express **20**(21), 23100–23107 (2012)
43. E. Engin, D. Bonneau, C.M. Natarajan, A.S. Clark, M.G. Tanner, R.H. Hadfield, S.N. Dorenbos, V. Zwiller, K. Ohira, N. Suzuki, H. Yoshida, Photon pair generation in a silicon micro-ring resonator with reverse bias enhancement. Opt. Express **21**(23), 27826–27834 (2013)
44. N.C. Harris, D. Grassani, A. Simbula, M. Pant, M. Galli, T. Baehr-Jones, M. Hochberg, D. Englund, D. Bajoni, C. Galland, Integrated source of spectrally filtered correlated photons for large-scale quantum photonic systems. Phys. Rev. X. **4**(4), 041047 (2014)
45. D. Grassani, S. Azzini, M. Liscidini, M. Galli, M.J. Strain, M. Sorel, J.E. Sipe, D. Bajoni, Micrometer-scale integrated silicon source of time-energy entangled photons. Optica **2**(2), 88–94 (2015)
46. Ryota Wakabayashi, Mikio Fujiwara, Ken-ichiro Yoshino, Yoshihiro Nambu, Masahide Sasaki, Takao Aoki, Time-bin entangled photon pair generation from si micro-ring resonator. Opt. Express **23**(2), 1103–1113 (2015)
47. S.F. Preble, M.L. Fanto, J.A. Steidle, C.C. Tison, G.A. Howland, Z. Wang, P.M. Alsing, On-chip quantum interference from a single silicon ring-resonator source. Phys. Rev. Appl. **4**(2), 021001 (2015)
48. J.W. Silverstone, R. Santagati, D. Bonneau, M.J. Strain, M. Sorel, J.L. O'Brien, M.G. Thompson, Qubit entanglement between ring-resonator photon-pair sources on a silicon chip. Nat. Commun. **6**(1), 1–7 (2015)
49. I.I. Faruque, G.F. Sinclair, D. Bonneau, J.G. Rarity, M.G. Thompson, On-chip quantum interference with heralded photons from two independent micro-ring resonator sources in silicon photonics. Opt. Express **26**(16), 20379–20395 (2018)
50. J. Carolan, U. Chakraborty, N.C. Harris, M. Pant, T. Baehr-Jones, M. Hochberg, D. Englund, Scalable feedback control of single photon sources for photonic quantum technologies. Optica. **6**(3), 335–340 (2019)

51. D. Llewellyn, Y. Ding, I.I. Faruque, S. Paesani, D. Bacco, R. Santagati, Y.J. Qian, Y. Li, Y.F. Xiao, M. Huber, M. Malik, Chip-to-chip quantum teleportation and multi-photon entanglement in silicon. Nat. Phys. **16**(2), 148–153 (2020)
52. Yingwen Liu, Wu Chao, Gu Xiaowen, Yuechan Kong, Yu. Xinxin, Renyou Ge, Xinlun Cai, Xiaogang Qiang, Wu Junjie, Xuejun Yang et al., High-spectral-purity photon generation from a dual-interferometer-coupled silicon microring. Opt. Lett. **45**(1), 73–76 (2020)
53. C. Xiong, C. Monat, A.S. Clark, C. Grillet, G.D. Marshall, M.J. Steel, J. Li, L. O'Faolain, T.F. Krauss, J.G. Rarity et al., Slow-light enhanced correlated photon pair generation in a silicon photonic crystal waveguide. Opt. Lett. **36**(17), 3413–3415 (2011)
54. Stefano Azzini, Davide Grassani, Matteo Galli, Dario Gerace, Maddalena Patrini, Marco Liscidini, Philippe Velha, Daniele Bajoni, Stimulated and spontaneous four-wave mixing in silicon-on-insulator coupled photonic wire nano-cavities. Appl. Phys. Lett. **103**(3), 031117 (2013)
55. M.J. Collins, C. Xiong, I.H. Rey, T.D. Vo, J. He, S. Shahnia, C. Reardon, T.F. Krauss, M.J. Steel, A.S. Clark et al., Integrated spatial multiplexing of heralded single-photon sources. Nat. Commun. **4**(1), 1–7 (2013)
56. L.G. Helt, Z. Yang, M. Liscidini, J.E. Sipe, Spontaneous four-wave mixing in microring resonators. Opt. Lett. **35**(18), 3006–3008 (2010)
57. C.A. Husko, A.S. Clark, M.J. Collins, A. De Rossi, S. Combrié, G. Lehoucq, I.H. Rey, T.F. Krauss, C. Xiong, B.J. Eggleton, Multi-photon absorption limits to heralded single photon sources. Sci. Rep. **4**, 33087 (2013)
58. Yuan Guo, Wei Zhang, Ning Lv, Qiang Zhou, Yidong Huang, Jiangde Peng, The impact of nonlinear losses in the silicon micro-ring cavities on cw pumping correlated photon pair generation. Opt. Express **22**(3), 2620–2631 (2014)
59. Lawrence M Rosenfeld, Dominic A Sulway, Gary F Sinclair, Vikas Anant, Mark G Thompson, John G Rarity, and Joshua W Silverstone. Mid-infrared quantum optics in silicon. *arXiv preprint* arXiv:1906.10158, 2019
60. Avik Dutt, Kevin Luke, Sasikanth Manipatruni, Alexander L. Gaeta, Paulo Nussenzveig, Michal Lipson, On-chip optical squeezing. Phys. Rev. Appl. **3**, 044005 (2015)
61. R. Cernansky, A. Politi, *Nanophotonic Source of Broadband Quadrature Squeezing*. arXiv preprint arXiv:1904.07283 (2019)
62. V.D. Vaidya, B. Morrison, L.G. Helt, R. Shahrokhshahi, D.H. Mahler, M.J. Collins, K. Tan, J. Lavoie, A. Repingon, M. Menotti, et al., *Broadband Quadrature-Squeezed Vacuum and Nonclassical Photon Number Correlations from a Nanophotonic Device*. arXiv preprint arXiv:1904.07833 (2019)
63. Y. Zhang, M. Menotti, K. Tan, V.D. Vaidya, D.H. Mahler, L. Zatti, M. Liscidini, B. Morrison, Z. Vernon, *Single-Mode Quadrature Squeezing Using Dual-Pump Four-Wave Mixing in an Integrated Nanophotonic Device*. arXiv preprint arXiv:2001.09474 (2020)
64. Z. Vernon, M. Menotti, C.C. Tison, J.A. Steidle, M.L. Fanto, P.M. Thomas, S.F. Preble, A.M. Smith, P.M. Alsing, M. Liscidini et al., Truly unentangled photon pairs without spectral filtering. Opt. Lett. **42**(18), 3638–3641 (2017)
65. J.B. Christensen, J.G. Koefoed, K. Rottwitt, C.J. McKinstrie, Engineering spectrally unentangled photon pairs from nonlinear microring resonators by pump manipulation. Opt. Lett. **43**(4), 859–862 (2018)
66. Z. Vernon, N. Quesada, M. Liscidini, B. Morrison, M. Menotti, K. Tan, J.E. Sipe, Scalable squeezed-light source for continuous-variable quantum sampling. Phys. Rev. Appl. **12**, 064024 (2019)
67. Jaime Cardenas, Carl B. Poitras, Jacob T. Robinson, Kyle Preston, Long Chen, Michal Lipson, Low loss etchless silicon photonic waveguides. Opt. Express **17**(6), 4752–4757 (2009)
68. M.P. Nezhad, O. Bondarenko, M. Khajavikhan, A. Simic, Y. Fainman, Etch-free low loss silicon waveguides using hydrogen silsesquioxane oxidation masks. Opt. Express **19**(20), 18827–18832 (2011)
69. J.C. Mikkelsen, W.D. Sacher, J.K. Poon, Dimensional variation tolerant silicon-on-insulator directional couplers. Opt. Express **22**(3), 3145–3150 (2014)

70. J. Song, Q. Fang, S.H. Tao, T.Y. Liow, M.B. Yu, G.Q. Lo, D.L. Kwong, Fast and low power Michelson interferometer thermo-optical switch on SOI. Opt. Express **16**(20), 15304–15311 (2008)
71. J. Van Campenhout, W.M. Green, S. Assefa, Y.A. Vlasov, Integrated NiSi waveguide heaters for CMOS-compatible silicon thermo-optic devices. Opt. Lett. **35**(7), 1013–1015 (2010)
72. Q. Fang, J.F. Song, T.Y. Liow, H. Cai, M.B. Yu, G.Q. Lo, D.L. Kwong, Ultralow power silicon photonics thermo-optic switch with suspended phase arms. IEEE Photonics Technol. Lett. **23**(8), 525–527 (2011)
73. M.R. Watts, J. Sun, C. DeRose, D.C. Trotter, R.W. Young, G.N. Nielson, Adiabatic thermo-optic Mach-Zehnder switch. Opt. Lett. **38**(5), 733–735 (2013)
74. N.C. Harris, Y. Ma, J. Mower, T. Baehr-Jones, D. Englund, M. Hochberg, C. Galland, Efficient, compact and low loss thermo-optic phase shifter in silicon. Opt. Express **22**(9), 10487–10493 (2014)
75. G.T. Reed, G. Mashanovich, F.Y. Gardes, D.J. Thomson, Silicon optical modulators. Nat. Photonics **4**(8), 518–526 (2010)
76. S. Abel, F. Eltes, J.E. Ortmann, A. Messner, P. Castera, T. Wagner, D. Urbonas, A. Rosa, A.M. Gutierrez, D. Tulli et al., Large pockels effect in micro-and nanostructured barium titanate integrated on silicon. Nat. Mater. **18**(1), 42–47 (2019)
77. F. Eltes, G.E. Villarreal-Garcia, D. Caimi, H. Siegwart, A.A. Gentile, A. Hart, P. Stark, G.D. Marshall, M.G. Thompson, J. Barreto, J. Fompeyrine, An integrated optical modulator operating at cryogenic temperatures. Nat. Mater. **19**(11), 1164–1168 (2020)
78. D.A.B. Miller, Perfect optics with imperfect components. Optica **2**(8), 747–750 (2015)
79. N.C. Harris, D. Bunandar, M. Pant, G.R. Steinbrecher, J. Mower, M. Prabhu, T. Baehr-Jones, M. Hochberg, D. Englund, Large-scale quantum photonic circuits in silicon. Nanophotonics **5**(3), 456–468 (2016)
80. C.M. Wilkes, X. Qiang, J. Wang, R. Santagati, S. Paesani, X. Zhou, D.A. Miller, G.D. Marshall, M.G. Thompson, J.L. O'Brien, 60 dB high-extinction auto-configured Mach-Zehnder interferometer. Opt. Lett. **41**(22), 5318–5321 (2016)
81. C.M. Natarajan, M.G. Tanner, R.H. Hadfield, Superconducting nanowire single-photon detectors: physics and applications. Superconductor Sci. Technol. **25**(6), 063001 (2012)
82. F. Marsili, V.B. Verma, J.A. Stern, S. Harrington, A.E. Lita, T. Gerrits, I. Vayshenker, B. Baek, M.D. Shaw, R.P. Mirin et al., Detecting single infrared photons with 93% system efficiency. Nat. Photonics **7**(3), 210–214 (2013)
83. B. Korzh, Q.Y. Zhao, J.P. Allmaras, S. Frasca, T.M. Autry, E.A. Bersin, A.D. Beyer, R.M. Briggs, B. Bumble, M. Colangelo, G.M. Crouch, Demonstration of sub-3 ps temporal resolution with a superconducting nanowire single-photon detector. Nat. Photonics **14**(4), 250–255 (2020)
84. Simone Ferrari, Carsten Schuck, Wolfram Pernice, Waveguide-integrated superconducting nanowire single-photon detectors. Nanophotonics **7**(11), 1725–1758 (2018)
85. F. Najafi, J. Mower, N.C. Harris, F. Bellei, A. Dane, C. Lee, X. Hu, P. Kharel, F. Marsili, S. Assefa, K.K. Berggren, On-chip detection of non-classical light by scalable integration of single-photon detectors. Nat. commun. **6**(1), 1–8 (2015)
86. C. Schuck, W.H. Pernice, O. Minaeva, M. Li, G. Gol'Tsman, A.V. Sergienko, H.X. Tang, Matrix of integrated superconducting single-photon detectors with high timing resolution. IEEE Ttrans. Appl. Superconductivity **23**(3), 2201007 (2013)
87. J.P. Höpker, T. Gerrits, A. Lita, S. Krapick, H. Herrmann, R. Ricken, V. Quiring, R. Mirin, S.W. Nam, C. Silberhorn, et al., *Integrated Transition Edge Sensors on Lithium Niobate Waveguides*. arXiv preprint arXiv:1812.08483 (2018)
88. D. Zhu, M. Colangelo, C. Chen, B.A. Korzh, F.N. Wong, M.D. Shaw, K.K. Berggren, Resolving photon numbers using a superconducting nanowire with impedance-matching taper. Nano Lett. **20**(5), 3858–3863 (2020)
89. F. Raffaelli, G. Ferranti, D.H. Mahler, P. Sibson, J.E. Kennard, A. Santamato, G. Sinclair, D. Bonneau, M.G. Thompson, J.C. Matthews, A homodyne detector integrated onto a photonic chip for measuring quantum states and generating random numbers. Quant. Sci. Technol. **3**(2), 025003 (2018)

90. J.F. Tasker, J. Frazer, G. Ferranti, E.J. Allen, L.F. Brunel, S. Tanzilli, V. D'Auria, J.C.F. Matthews, *9 ghz Measurement of Squeezed Light by Interfacing Silicon Photonics and Integrated Electronics*. arXiv preprint arXiv:2009.14318 (2020)
91. F. Raffaelli, P. Sibson, J.E. Kennard, D.H. Mahler, M.G. Thompson, J.C. Matthews, Generation of random numbers by measuring phase fluctuations from a laser diode with a silicon-on-insulator chip. Opt. Express **26**(16), 19730–19741 (2018)
92. F. Lenzini, J. Janousek, O. Thearle, M. Villa, B. Haylock, S. Kasture, L. Cui, H.P. Phan, D.V. Dao, H. Yonezawa, P.K. Lam, Integrated photonic platform for quantum information with continuous variables. Sci. Adv. **4**(12), eaat9331,eaat9331 (2018)
93. Mateusz Piekarek, Damien Bonneau, Shigehito Miki, Taro Yamashita, Mikio Fujiwara, Masahide Sasaki, Hirotaka Terai, Michael G. Tanner, Chandra M. Natarajan, Robert H. Hadfield et al., High-extinction ratio integrated photonic filters for silicon quantum photonics. Opt. Lett. **42**(4), 815–818 (2017)
94. Lee Carroll, Jun-Su Lee, Carmelo Scarcella, Kamil Gradkowski, Matthieu Duperron, Lu Huihui, Yan Zhao, Cormac Eason, Padraic Morrissey, Marc Rensing et al., Photonic packaging: transforming silicon photonic integrated circuits into photonic devices. Appl. Sci. **6**(12), 426 (2016)
95. Y. Ding, H. Ou, C. Peucheret, Ultra-high-efficiency apodized grating coupler using fully etched photonic crystals. Opt. Lett. **38**, 2732–2734 (2013)
96. Y. Ding, C. Peucheret, H. Ou, K. Yvind, Fully etched apodized grating coupler on the SOI platform with- 05.8 dB coupling efficiency. Opt. Lett. **39**(18), 5348–50 (2014)
97. S.J. McNab, N. Moll, Y.A. Vlasov, Ultra-low loss photonic integrated circuit with membrane-type photonic crystal waveguides. Opt. Express **11**(22), 2927–2939 (2003)
98. M. Pu, L. Liu, H. Ou, K. Yvind, J.M. Hvam, Ultra-low-loss inverted taper coupler for silicon-on-insulator ridge waveguide. Opt. Commun. **283**(19), 3678–3682 (2010)
99. W.S. Zaoui, A. Kunze, W. Vogel, M. Berroth, J. Butschke, F. Letzkus, J. Burghartz, Bridging the gap between optical fibers and silicon photonic integrated circuits. Opt. Express **22**(2), 1277–1286 (2014)
100. S. Paesani, A.A. Gentile, R. Santagati, J. Wang, N. Wiebe, D.P. Tew, J.L. O'Brien, M.G. Thompson, Experimental bayesian quantum phase estimation on a silicon photonic chip. Phys. Rev. Lett. **118**, 100503 (2017)
101. C. Ostle, R.C. Thompson, D. Broughton, L. Gregory, M. Wootton, D.G. Johns, The rise in ocean plastics evidenced from a 60-year time series. Nat. Commun. **10**(1), 1–6 (2019)
102. A.H. Atabaki, S. Moazeni, F. Pavanello, H. Gevorgyan, J. Notaros, L. Alloatti, M.T. Wade, C. Sun, S.A. Kruger, H. Meng, Qubaisi K. Al, Integrating photonics with silicon nanoelectronics for the next generation of systems on a chip. Nature **556**(7701), 349–54 (2018)
103. J.M. Fedeli, L. Di Cioccio, D. Marris-Morini, L. Vivien, R. Orobtchouk, P. Rojo-Romeo, C. Seassal, F. Mandorlo, Development of silicon photonics devices using microelectronic tools for the integration on top of a cmos wafer, in *Advances in Optical Technologies* (2008)
104. Y. Arakawa, T. Nakamura, Y. Urino, T. Fujita, Silicon photonics for next generation system integration platform. IEEE Commun. Mag. **51**(3), 72–77 (2013)
105. R. Santagati, J. Wang, A.A. Gentile, S. Paesani, N. Wiebe, J.R. McClean, S. Morley-Short, P.J. Shadbolt, D. Bonneau, J.W. Silverstone, D.P. Tew, Witnessing eigenstates for quantum simulation of Hamiltonian spectra. Sci. Adv. **4**(1), eaap9646 (2018)
106. J. Wang, S. Paesani, R. Santagati, S. Knauer, A.A. Gentile, N. Wiebe, M. Petruzzella, J.L. O'Brien, J.G. Rarity, A. Laing, M.G. Thompson, Experimental quantum Hamiltonian learning. Nat. Phys. **13**(6), 551–555 (2017)
107. X. Qiang, X. Zhou, J. Wang, C.M. Wilkes, T. Loke, S. O'Gara, L. Kling, G.D. Marshall, R. Santagati, T.C. Ralph et al., Large-scale silicon quantum photonics implementing arbitrary two-qubit processing. Nat. Photonics **12**(9), 534–539 (2018)
108. R. Santagati, J.W. Silverstone, M.J. Strain, M. Sorel, S. Miki, T. Yamashita, M. Fujiwara, M. Sasaki, H. Terai, M.G. Tanner et al., Silicon photonic processor of two-qubit entangling quantum logic. J. Opt. **19**(11), 114006 (2017)

109. Ken-ichi Harada, Hiroki Takesue, Hiroshi Fukuda, Tai Tsuchizawa, Toshifumi Watanabe, Koji Yamada, Yasuhiro Tokura, Sei-ichi Itabashi, Generation of high-purity entangled photon pairs using silicon wire waveguide. Opt. Express **16**(25), 20368–20373 (2008)
110. J. Chen, Z.H. Levine, J. Fan, A.L. Migdall, Frequency-bin entangled comb of photon pairs from a silicon-on-insulator micro-resonator. Opt. Express **19**(2), 1470–1483 (2011)
111. C.M. Gentry, J.M. Shainline, M.T. Wade, M.J. Stevens, S.D. Dyer, X. Zeng, F. Pavanello, T. Gerrits, S.W. Nam, R.P. Mirin et al., Quantum-correlated photon pairs generated in a commercial 45 nm complementary metal-oxide semiconductor microelectronic chip. Optica **2**(12), 1065–1071 (2015)
112. F. Mazeas, M. Traetta, M. Bentivegna, F. Kaiser, D. Aktas, W. Zhang, C.A. Ramos, L.A. Ngah, T. Lunghi, E. Picholle, N. Belabas-Plougonven, High-quality photonic entanglement for wavelength-multiplexed quantum communication based on a silicon chip. Opt. Express **24**(25), 28731–28738 (2016)
113. D. Oser, S. Tanzilli, F. Mazeas, C. Alonso-Ramos, X. Le Roux, G. Sauder, X. Hua, O. Alibart, L. Vivien, É. Cassan et al., High-quality photonic entanglement out of a stand-alone silicon chip. npj Quant. Inf. **6**(1), 1–6 (2020)
114. M. Fujiwara, R. Wakabayashi, M. Sasaki, M. Takeoka, Wavelength division multiplexed and double-port pumped time-bin entangled photon pair generation using si ring resonator. Opt. Express **25**(4), 3445–3453 (2017)
115. J. Suo, S. Dong, W. Zhang, Y. Huang, J. Peng, Generation of hyper-entanglement on polarization and energy-time based on a silicon micro-ring cavity. Opt. Express **23**(4), 3985–3995 (2015)
116. E. Farhi, J. Goldstone, S. Gutmann, *A Quantum Approximate Optimization Algorithm*. arXiv preprint arXiv:1411.4028 (2014)
117. M. Szegedy, *Spectra of Quantized Walks and a* $\sqrt{\delta\varepsilon}$ *Rule*. arXiv preprint quant-ph/0401053 (2004)
118. L.K. Grover, A fast quantum mechanical algorithm for database search, in *Proceedings of the Twenty-Eighth Annual ACM Symposium on Theory of Computing*, pp. 212–219 (1996)
119. M.A. Ciampini, A. Orieux, S. Paesani, F. Sciarrino, G. Corrielli, A. Crespi, R. Ramponi, R. Osellame, P. Mataloni, Path-polarization hyperentangled and cluster states of photons on a chip. Light Sci. Appl. **5**(4), e16064 (2016)
120. P. Sibson, J.E. Kennard, S. Stanisic, C. Erven, J.L. O'Brien, M.G. Thompson, Integrated silicon photonics for high-speed quantum key distribution. Optica **4**(2), 172–177 (2017)
121. Y. Ding, D. Bacco, K. Dalgaard, X. Cai, X. Zhou, K. Rottwitt, L.K. Oxenløwe, High-dimensional quantum key distribution based on multicore fiber using silicon photonic integrated circuits. npj Quant. Inf. **3**(1), 1–7 (2017)
122. D. Bunandar, A. Lentine, C. Lee, H. Cai, C.M. Long, N. Boynton, N. Martinez, C. DeRose, C. Chen, M. Grein, D. Trotter, Metropolitan quantum key distribution with silicon photonics. Phys. Rev. X **8**(2), 021009 (2018)
123. C. Monroe, J. Kim, Scaling the ion trap quantum processor. Science **339**(6124), 1164–1169 (2013)
124. C. Monroe, R. Raussendorf, A. Ruthven, K.R. Brown, P. Maunz, L.-M. Duan, J. Kim, Large-scale modular quantum-computer architecture with atomic memory and photonic interconnects. Phys. Rev. A **89**(2), 022317 (2014)
125. H.J. Kimble, The quantum internet. Nature **453**(7198), 1023–1030 (2008)
126. J. Wang, D. Bonneau, M. Villa, J.W. Silverstone, R. Santagati, S. Miki, T. Yamashita, M. Fujiwara, M. Sasaki, H. Terai, M.G. Tanner, Chip-to-chip quantum photonic interconnect by path-polarization interconversion. Optica **3**(4), 407–413 (2016)
127. D.J. Brod, E.F. Galvão, A. Crespi, R. Osellame, N. Spagnolo, F. Sciarrino, Photonic implementation of boson sampling: a review. Adv. Photonics **1**(3), 034001 (2019)
128. S. Trotzky, Y.A. Chen, A. Flesch, I.P. McCulloch, U. Schollwöck, J. Eisert, I. Bloch, Probing the relaxation towards equilibrium in an isolated strongly correlated one-dimensional Bose gas. Nat. Phys. **8**(4), 325–30 (2012)

129. H. Bernien, S. Schwartz, A. Keesling, H. Levine, A. Omran, H. Pichler, S. Choi, A.S. Zibrov, M. Endres, M. Greiner et al., Probing many-body dynamics on a 51-atom quantum simulator. Nature **551**(7682), 579–584 (2017)
130. F. Arute, K. Arya, R. Babbush, D. Bacon, J.C. Bardin, R. Barends, R. Biswas, S. Boixo, F.G. Brandao, D.A. Buell, B. Burkett, Quantum supremacy using a programmable superconducting processor. Nature **574**(7779), 505–510 (2019)
131. L. Chakhmakhchyan, N.J. Cerf, Boson sampling with Gaussian measurements. Phys. Rev. A. **96**(3), 032326 (2017)
132. A. Neville, C. Sparrow, R. Clifford, E. Johnston, P.M. Birchall, A. Montanaro, A. Laing, Classical boson sampling algorithms with superior performance to near-term experiments. Nat. Phys. **13**(12), 1153 (2017)
133. A.M. Dalzell, A.W. Harrow, D.E. Koh, R.L. La Placa, *How Many Qubits are Needed for Quantum Computational Supremacy?* arXiv:1805.05224 (2018)
134. A. Crespi, R. Osellame, R. Ramponi, D.J. Brod, E.F. Galvao, N. Spagnolo, C. Vitelli, E. Maiorino, P. Mataloni, F. Sciarrino, Integrated multimode interferometers with arbitrary designs for photonic boson sampling. Nat. Photonics **7**(7), 545 (2013)
135. M. Tillmann, B. Dakić, R. Heilmann, S. Nolte, Alexander Szameit, Philip Walther, Experimental boson sampling. Nat. Photonics **7**(7), 540 (2013)
136. M.A. Broome, A. Fedrizzi, S. Rahimi-Keshari, J. Dove, S. Aaronson, T.C. Ralph, A.G. White, Photonic boson sampling in a tunable circuit. Science **339**(6121), 794–798 (2013)
137. J.B. Spring, B.J. Metcalf, P.C. Humphreys, W.S. Kolthammer, X.M. Jin, M. Barbieri, A. Datta, N. Thomas-Peter, N.K. Langford, D. Kundys, J.C. Gates, Boson sampling on a photonic chip. Science **339**(6121), 798–801 (2013)
138. M. Bentivegna, N. Spagnolo, C. Vitelli, F. Flamini, N. Viggianiello, L. Latmiral, P. Mataloni, D.J. Brod, E.F. Galvão, A. Crespi et al., Experimental scattershot boson sampling. Sci. Adv. **1**(3), e1400255 (2015)
139. Yu. Hui Wang, Yu-Huai Li He, Su Zu-En, Bo Li, He-Liang Huang, Xing Ding, Ming-Cheng Chen, Chang Liu, Jian Qin et al., High-efficiency multiphoton boson sampling. Nat. Photonics **11**(6), 361 (2017)
140. J.C. Loredo, M.A. Broome, P. Hilaire, O. Gazzano, I. Sagnes, A. Lemaitre, M.P. Almeida, P. Senellart, A.G. White, Boson sampling with single-photon fock states from a bright solid-state source. Phys. Rev. Lett. **118**, 130503 (2017)
141. H. Wang, W. Li, X. Jiang, Y.M. He, Y.H. Li, X. Ding, M.C. Chen, J. Qin, C.Z. Peng, C. Schneider, M. Kamp, W.-J. Zhang, H. Li, L.-X. You, Z. Wang, J.P. Dowling, S. Höfling, C.-Y. Lu, J.-W. Pan, Toward scalable boson sampling with photon loss. Phys. Rev. Lett. **120**(23), 230502 (2018)
142. H.S. Zhong, Y. Li, W. Li, L.C. Peng, Z.E. Su, Y. Hu, Y.M. He, X. Ding, W. Zhang, H. Li, L. Zhang, Z. Wang, L.-X. You, X.-L. Wang, X. Jiang, L. Li, Y.-A. Chen, N.-L. Liu, C.-Y. Lu, J.-W. Pan, *12-Photon Entanglement and Scalable Scattershot Boson Sampling with Optimal Entangled-Photon Pairs from Parametric Down-Conversion*. arXiv:1810.04823 (2018)
143. H. Wang, J. Qin, X. Ding, M.C. Chen, S. Chen, X. You, Y.M. He, X. Jiang, L. You, Z. Wang, C. Schneider, J.J. Renema, S. Höfling, C.-Y. Lu, J.-W. Pan, Boson sampling with 20 input photons and a 60-mode interferometer in a $10^{14}$-dimensional hilbert space. Phys. Rev. Lett. **123**, 250503 (2019)
144. S. Scheel, *Permanents in Linear Optical Networks*. arXiv:quant-ph/0406127 (2004)
145. A.P. Lund, A. Laing, S. Rahimi-Keshari, T. Rudolph, J.L. O'Brien, T.C. Ralph, Boson sampling from a gaussian state. Phys. Rev. Lett. **113**, 100502 (2014)
146. C.S. Hamilton, R. Kruse, L. Sansoni, S. Barkhofen, C. Silberhorn, I. Jex, Gaussian boson sampling. Phys. Rev. Lett. **119**, 170501 (2017)
147. R. Kruse, C.S. Hamilton, L. Sansoni, S. Barkhofen, C. Silberhorn, I. Jex, *A Detailed Study of Gaussian Boson Sampling*. arXiv:1801.07488 (2018)
148. E.R. Caianiello, *Combinatorics and Renormalization in Quantum Field Theory* (Benjamin Inc, W. A, 1973)
149. S. Olivares, Quantum optics in the phase space. Eur. Phys. J. **203**(1), 3–24 (2012)

150. G. Adesso, S. Ragy, A.R. Lee, Continuous variable quantum information: Gaussian states and beyond. Open Syst. Inf. Dyn. **21**(01n02), 1440001 (2014)
151. A. Björklund, Counting perfect matchings as fast as ryser, in *Proceedings of the Twenty-Third Annual ACM-SIAM Symposium on Discrete Algorithms*, pp. 914–921 (Society for Industrial and Applied Mathematics, 2012)
152. N. Spagnolo, C. Vitelli, M. Bentivegna, D.J. Brod, A. Crespi, F. Flamini, S. Giacomini, G. Milani, R. Ramponi, P. Mataloni et al., Experimental validation of photonic boson sampling. Nat. Photonics **8**(8), 615–620 (2014)
153. J. Carolan, J.D. Meinecke, P.J. Shadbolt, N.J. Russell, N. Ismail, K. Wörhoff, T. Rudolph, M.G. Thompson, J.L. O'brien, J.C. Matthews et al., On the experimental verification of quantum complexity in linear optics. Nat. Photonics **8**(8), 621 (2014)
154. Y. He, X. Ding, Z.E. Su, H.L. Huang, J. Qin, C. Wang, S. Unsleber, C. Chen, H. Wang, Y.M. He, X.L. Wang, W.-J. Zhang, S.-J. Chen, C. Schneider, M. Kamp, L.-X. You, Z. Wang, S. Höfling, C.-Y. Lu, J.-W. Pan, Time-bin-encoded boson sampling with a single-photon device. Phys. Rev. Lett. **118**(19), 190501 (2017)
155. B.A. Bell, G.S. Thekkadath, R. Ge, X. Cai, I.A. Walmsley, Testing multi-photon interference on a silicon chip. Opt. Express **27**(24), 35646–35658 (2019)
156. M. Zhang, C. Wang, R. Cheng, A. Shams-Ansari, M. Lončar, Monolithic ultra-high-q lithium niobate microring resonator. Optica **4**(12), 1536–1537 (2017)
157. J. Zhao, C. Ma, M. Rüsing, S. Mookherjea, High quality entangled photon pair generation in periodically poled thin-film lithium niobate waveguides. Phys. Rev. Lett. **124**, 163603 (2020)
158. C. Sparrow, E. Martín-López, N. Maraviglia, A. Neville, C. Harrold, J. Carolan, Y.N. Joglekar, T. Hashimoto, N. Matsuda, J.L. O'Brien et al., Simulating the vibrational quantum dynamics of molecules using photonics. Nature **557**(7707), 660 (2018)
159. P.A.M. Dirac, *The Principles of Quantum Mechanics* (Oxford University Press, Oxford, 1981)
160. R. Feynman, Simulating physics with computers. Int. J. Theor. Phys. **21**, 467 (1982)
161. A. Assion, T. Baumert, M. Bergt, T. Brixner, B. Kiefer, V. Seyfried, M. Strehle, G. Gerber, Control of chemical reactions by feedback-optimized phase-shaped femtosecond laser pulses. Science **282**(5390), 919–922 (1998)
162. F. Gatti, *Molecular Quantum Dynamics: From Theory to Applications* (Springer, Berlin, 2014)
163. F. Gatti, B. Lasorne, H.-D. Meyer, A. Nauts, *Applications of Quantum Dynamics in Chemistry*, vol. 98 (Springer, Berlin, 2017)
164. P.M. Hundt, B. Jiang, M.E. van Reijzen, H. Guo, R.D. Beck, Vibrationally promoted dissociation of water on Ni (111). Science **344**(6183), 504–507 (2014)
165. J. Hachmann, R. Olivares-Amaya, S. Atahan-Evrenk, C. Amador-Bedolla, R.S. Sánchez-Carrera, A. Gold-Parker, L. Vogt, A.M. Brockway, A. Aspuru-Guzik, The harvard clean energy project: large-scale computational screening and design of organic photovoltaics on the world community grid. J. Phys. Chem. Lett. **2**(17), 2241–2251 (2011)
166. M. Gross, D.C. Müller, H.G. Nothofer, U. Scherf, D. Neher, C. Bräuchle, K. Meerholz, Improving the performance of doped $\pi$-conjugated polymers for use in organic light-emitting diodes. Nature **405**(6787), 661 (2000)
167. F. Santoro, A. Lami, R. Improta, V. Barone, Effective method to compute vibrationally resolved optical spectra of large molecules at finite temperature in the gas phase and in solution. J. Chem. Phys. **126**(18), 184102 (2007)
168. H.-C. Jankowiak, J.L. Stuber, R. Berger, Vibronic transitions in large molecular systems: Rigorous prescreening conditions for franck-condon factors. J. Chem. Phys. **127**(23), 234101 (2007)
169. D. Hayes, J. Wen, G. Panitchayangkoon, R.E. Blankenship, G.S. Engel, Robustness of electronic coherence in the fenna-matthews-olson complex to vibronic and structural modifications. Faraday Discuss. **150**, 459–469 (2011)
170. J. Huh, G.G. Guerreschi, B. Peropadre, J.R. McClean, A. Aspuru-Guzik, Boson sampling for molecular vibronic spectra. Nat. Photonics **9**(9), 615 (2015)
171. E.V. Doktorov, I.A. Malkin, V.I. Man'ko, Dynamical symmetry of vibronic transitions in polyatomic molecules and the franck-condon principle. J. Mol. Spectrosc. **64**, 302 (1977)

172. J. Franck, E.G. Dymond, Elementary processes of photochemical reactions. Trans. Faraday Soc. **21**, 536 (1926)
173. E.U. Condon, Nuclear motions associated with electron transitions in diatomic molecules. Phys. Rev. **32**, 858 (1928)
174. W.R. Clements, J.J. Renema, A. Eckstein, A.A. Valido, A. Lita, T. Gerrits, S.W. Nam, W.S. Kolthammer, J. Huh, I.A. Walmsley, *Experimental Quantum Optical Approximation of Vibronic Spectroscopy*. arXiv:1710.08655 (2017)
175. Y. Shen, L. Yao, K. Zhang, J. Zhang, S. Zhang, J. Huh, K. Kim, Quantum optical emulation of molecular vibronic spectroscopy using a trapped-ion device. Chem. Sci. **9**(4), 836–840 (2018)
176. C.S. Wang, J.C. Curtis, B.J. Lester, Y. Zhang, Y.Y. Gao, J. Freeze, V.S. Batista, P.H. Vaccaro, I.L. Chuang, L. Frunzio, L. Jiang, Efficient multiphoton sampling of molecular vibronic spectra on a superconducting bosonic processor. Phys. Rev. X. **10**(2), 021060 (2020)
177. D.G. Olivares, B. Peropadre, A. Aspuru-Guzik, J.J. García-Ripoll, Quantum simulation with a boson sampling circuit. Phys. Rev. A. **94**(2), 022319 (2016)
178. L. Banchi, M. Fingerhuth, T. Babej, C. Ing, J.M. Arrazola, Molecular docking with Gaussian boson sampling. Sci Adv. **6**(23), eaax1950 (2020)
179. J.M. Arrazola, T.R. Bromley, Using Gaussian boson sampling to find dense subgraphs. Phys. Rev. Lett. **121**(3), 030503 (2018)
180. K. Brádler , P.L. Dallaire-Demers, P. Rebentrost, D. Su, C. Weedbrook, Gaussian boson sampling for perfect matchings of arbitrary graphs. arXiv:1712.06729 (2017)
181. K. Brádler, S. Friedland, J. Izaaca, N. Killoran, D. Su, *Graph Isomorphism and Gaussian Boson Sampling*. arXiv:1810.10644 (2018)
182. Z. Huang, P.P. Rohde, D.W. Berry, P. Kok, J.P. Dowling, C. Lupo, *Boson Sampling Private-Key Quantum Cryptography. arXiv preprint* arXiv:1905.03013 (2019)

# Chapter 12
# An Open Silicon Photonics Ecosystem for Computercom Applications

**Marco Fiorentino, Zhihong Huang, Di Liang, Sagi Mathai, M. Ashkan Seyedi, and Raymond G. Beausoleil**

**Abstract** We present progress toward an open Silicon photonics ecosystem targeted at computercom applications. The ecosystem is centered around a development kit comprising verified devices that can be laid out and simulated using industry-standard tools and fabricated in a commercial foundry. The ecosystem includes partners for testing, attaching fiber, and packaging of the finished photonics integrated circuits.

## 12.1 Introduction

The push to build bigger, better, and faster computers has accelerated in recent times thanks to the advent of technologies like big data and machine learning. In high-performance computing (HPC), the push to exascale performance (i.e., building a computer with $10^{18}$ floating-point operations per second) has put a strain on existing technologies. Systems with up to 100,000 sockets are now being built as the other avenues to improve performance are being limited by other factors. To achieve such scales, HPC systems will require affordable interconnects. Photonics networks are already widely deployed because they provide the bandwidth and reach that is required to build modern HPC systems. However, in its current implementation, HPC photonics relies on very expensive active optical cables (AOCs) and therefore

M. Fiorentino (✉) · Z. Huang · D. Liang · S. Mathai · M. Ashkan Seyedi · R. G. Beausoleil
Hewlett Packard Laboratories, Milpitas, USA
e-mail: marco.fiorentino@hpe.com

Z. Huang
e-mail: zhihong.huang@hpe.com

S. Mathai
e-mail: sagi.mathai@hpe.com

M. Ashkan Seyedi
e-mail: ashkan.seyedi@hpe.com

R. G. Beausoleil
e-mail: ray.beausoleil@hpe.com



D. J. Lockwood and L. Pavesi (eds.), *Silicon Photonics IV*,
Topics in Applied Physics 139, https://doi.org/10.1007/978-3-030-68222-4_12

its reach is limited by overall system cost considerations. In addition to lowering the system costs, low-cost photonic links enable new, high-performance architectures [1]. Silicon photonics (SiPh) is considered one of the most promising technologies to enable low-cost optical links. The reason for this is twofold. Firstly because SiPh allows the integration of multiple components in a single photonic integrated circuit (PIC), the fabrication and assembly costs are much lower compared to a solution that uses discrete optical components. Secondly SiPh uses a material platform that has been widely developed by the semiconductor industry and can therefore leverage large investments made in developing materials and fabrication tools for tight control in high-resolution fabrication and can adopt successful model of the highly standardized design flow in EDA tools. With these advantages SiPh has shown, in the past few years, a significant market penetration in data centers and HPCs. However, all the companies that have successfully entered the SiPh business have done so using a vertically integrated model whereas they exclusively own much of the intellectual property involved in the design and assembly of SiPh components. This vertical model is significantly different to the foundry and outsourced assembly and test (OSAT) model that has been common in the semiconductor industry. We believe that the vertical model has stifled innovation by creating a steep entrance barrier for innovative companies and reduced the impact of SiPh for application where the total addressable market (TAM) is too small to justify the investment for developing a design kit from scratch. A more open and diverse range of solutions would stimulate competition and innovation and encourage system integrators to adopt SiPh more widely. For this reason, HPE in collaboration with a number of partners has started building an open SiPh industrial ecosystem targeted at computercom applications in datacenters and HPCs. The open SiPh ecosystem would make available to the design community a new way to design, build, test, and assemble SiPh PICs using verified IP, industry-standard tools, and verified fabrication and assembly processes. In the following sections, we will outline the progress we have made toward an open-source SiPh ecosystem as well as point out open questions and issues. In Sect. 12.2, we discuss the process requirements for an open-source SiPh foundry. We will outline a list of required components and their process requirements and also review the status of process verification and control. In Sect. 12.3, we describe the process toward a set of tools designed for a SiPh open ecosystem. These include device design tools, place and route tools, and device models that allow for co-simulation of photonic and electronic components. In Sect. 12.4, we share a list of devices currently available in the design kit and some of the directions of development that our partners and us are considering. Before concluding, in Sect. 12.4.5, we describe our progress toward an innovative fiber attach solution as well as the prospects for SiPh interposer integration.

## 12.2 Process

The main requirement for the SiPh fabrication process is that it has to be compatible with the required Product Design Kit (PDK) components while having a high yield in order to keep fabrication costs low. With a view toward standardizing devices and processes, changes in the fabrication parameters, requiring requalifications of the components, are discouraged as they are both expensive and can affect the stability of the standard process. It should be noted that the choice of some of the parameters is somewhat arbitrary and is sometimes dictated by compatibility with previous devices. Here, we present a list of process components and steps and some indicative parameters.

**Silicon wafer** Production-level SiPh is currently almost exclusively based on 300 mm diameter wafers. The larger wafers allow for a cost reduction compared to the smaller formats as many of the fabrication operations (e.g., spin coating, depositions, implants, and etching) happen in parallel for all the dies on a wafer. Also, fabrication equipment for the 65 and 45 nm nodes is fully depreciated and increasingly available for processes like SiPh. High-quality 300 mm silicon on oxide (SOI) wafers are available from several supplier thus reducing supply chain issues. Top silicon thicknesses of 220–270 nm provide a good support for a range of high-quality devices. The thickness of the buried oxide is subject to a number of constraints. On one side thicknesses in excess of 1 $\mu$m are required to insure low waveguide losses, especially for the multimode waveguides used for long-range transport. But very thick oxides in excess of 3 $\mu$m are difficult to fabricate and create thermal effects during fabrication that can affect yields. Edge coupling requires thick oxides in order to avoid coupling the expanded beams at the coupler into the silicon substrate. Based on these requirements, oxide thicknesses between 1 and 3 $\mu$m are typically used.

**Etches** A number of silicon etches are required for a functional PIC. There are typically three waveguide etches to that allow one to fabricate a shallow multimode waveguide (25–30% of the top silicon), a shallow single-mode waveguide ($\simeq$50% of the top silicon) and a deep single-mode waveguide ($\simeq$75% of the top silicon). A full etch of the top silicon, sometimes called a trench etch, is used for electrical an thermal insulation of the devices. A grating etch is also needed albeit sometimes one of the waveguide etches can be used for the gratings as well.

**Implants** To implement plasma-effect modulators at least four implants are required: n++ and p++ for the contacts n+ and p+ for the PN and PiN junctions. The highly doped layers are also used to build resistive heaters embedded in the top silicon.

**Ge block** Germanium photodetectors (PDs) are key components for SiPh PICs. Foundries selectively grow silicon–germanium in cavities etched in the top silicon and then do p and n implants in the germanium to create the PN detector junction. To contact the photodiodes, two strategies can be adopted. Directly contacting the germanium reduces the number of interfaces where charges can be trapped but requires the development of a new metal–semiconductor contact. Contacting the Ge p and n interfaces through highly doped silicon is more prone to trapping but leverages high-quality contacts available on silicon. As detailed below we are also exploring avalanche photodiodes (APD) as means to improve sensitivity and link budgets [2]. The APD process comprises several additional implant steps and silicon epitaxy to create the absorption, multiplication, and charge layers. Compatibility of these steps with high-volume high-yield foundry platforms is a current subject of study.

**Laser integration** Integrating III-V lasers in a SiPh PIC platform is very attractive because of the expected reduction in the cost, complexity, integration density, and power that such integration promises. This integration is fairly complex from the process point of view because of the fundamental incompatibility of the silicon and III-V material systems. We have focussed on a selective bonding of III-V onto Silicon as this heterogeneous technology has given very good performance [3, 4], while initial tests have proved encouraging the integration of III-V bonding with the SiPh process at a volume foundry is still under investigation. It is worth noticing that establishing a heterogeneous integration platform would open the way to a great increase of the SiPh platform capabilities. If the integration push is successful, in addition to lasers we would have access to integrated amplifiers, photodetectors [5], and various types of modulators [4].

**Process verification** A key component of a successful CMOS microelectronics has been a tight control of the relevant process parameters and SiPh should follow the same path. In addition to the tools available for standard semiconductor fabrication, there are new tools that can specifically address some of the requirements of SiPh process verification, for example, control of the etch depth greatly affects the performance of devices especially in gratings and microring-based dense wavelength division multiplexing (DWDM) links [6]. Variations as small as 1 nm can cause significant changes in the microring resonant wavelength and therefore affect power consumption and yields. Quantifying and controlling theses variations therefore is a key task in the SiPh process deployment. As shown in Fig. 12.1, we have developed some innovative measurement and analysis techniques that allow us to carefully quantify foundries' parameter control and give feedback on areas of improvement.

## 12.3 Design Tools

A robust set of design tools is a key ingredient for a successful SiPh open ecosystem. The goal of the design tools is to enable different technologist to access the underlying technologies easily and transparently. The first technologists involved in the design tools are the SiPh device designers. This group is tasked with designing, parameterizing, testing, and modeling devices as well as with defining the devices PDK cells. The cells are collected in a foundry-specific design manual and are used

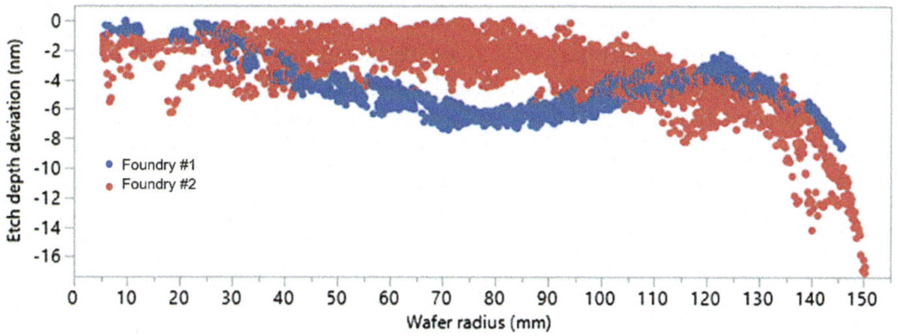


**Fig. 12.1** Comparison of two foundries etch depth control for microring resonators. By using optical testing, we were able to determine the etch depth variations across wafers from two foundries. Foundry 1 has more random variation whereas foundry 2 has a systematic "donut" shaped variation but lower random variations

by SiPh system designers to build schematics and layouts for SiPh PICs. These designers design functional PICs based on system requirements and use verification and simulation tools to insure correctness and performance. SiPh designers also need to interface with designers of the driver and receiver electronics that interface with the photonic PICs. The models of the optoelectronic response of SiPh devices created by the device designers need to be compatible with commonly used electronic simulation tools in order to allow an end-to-end simulation of photonic links. Finally, if complex packaging is required for the optoelectronic devices, thermal and mechanical simulations will be required to correctly design the packages in order to guarantee performance across all the required environmental conditions. Pieces of this ecosystem exist in the industry and others are being designed but a fully integrated workflow is still being developed. Here, we describe some of the existing tools and the work we have been doing to fill some of the gaps we have observed.

**Device design tools** In the past few years, there has been an explosion of tools that allow the simulation and optimization of photonic devices. Many of these tools are especially suited for a particular type of device or analysis. From the point of view of the SiPh ecosystem, we are mostly interested in a collection of tools that are integrated and help streamlining the device designer workflow from device conception and layout to the initial simulation through to test, verification, and model build. We currently use the Ansys/Lumerical DeviceSuite [7] because it combines a set of photonic device simulation tools with the ability to simulate optoelectronic response with the possibility to build a compact model library (CML) of the devices that can be widely distributed. These models can be then used in a system simulation tool to ensure end-to-end performance.

**Device optimization and models** Our group has been active in the design and optimization of individual devices, for example, we have used adjoint methods [8] as an effective and powerful tool for optimizing photonic devices such as grating couplers that comprise a large number of parameters. The adjoint method simplifies the exploration of large parameter spaces by calculating the gradients of the problem figure-of-merit (FoM) in an efficient manner. We have also worked on models for devices we designed such as APDs [9]. By building APD models that are valid for a large range of gains, we can include APDs in our link-level simulations and optimize performance across various scenarios. We also used data mining techniques to classify and parameterize the resonance of multiring transceivers [10]. The data thus obtained allowed us to construct functional models for microring modulators and filters/switches.

**System design** Design automation is one of the key requirements for the system design tools. Ideally the SiPh system design workflow should be similar to the electronic design workflow. The designer starts with a schematic of the circuit, verifies the functionality of the circuit using compact models, and then is guided through the layout by an automated layout tool. Once the layout is complete, a layout-versus-schematic tool verifies that the final layout reflects the original design intent. The final step before tapeout is the verification that the layout complies with the process design rules. This workflow has been at least partially realized thanks to tools like Mentor Graphics Lightsuite [11] that allows schematic driven layout of complex photonic circuits. The integration of LightSuite with the Caliber verification tool also simplifies compliance with the design rules.

## 12.4 Devices

Devices are the heart of the open ecosystem as they enable users to create functional circuits. In the early versions, our device library will be primarily intended to support WDM-based links for multiple wavelengths on a fiber, using a grating coupler for top-down light input and output. The block diagram in Fig. 12.2 shows a typical link implementation. There are also cells for Mach–Zehnder modulators and add-drop filters. Numerous extensions are expected including nitride waveguides for low-loss transport and edge couplers. Further extensions are discussed below.

### *12.4.1 Passives*

#### 12.4.1.1 Waveguides

**Shallow single-mode waveguide** This TE (transverse electric) waveguide is used for optical routing between devices using shallow single-mode waveguides to reduce system losses. Propagation losses are expected to be around 1.5 dB/cm.

**Deep single-mode waveguide** This TE waveguide is used for coupling in and out of resonators to optimize mode matching. Given the high losses of 3 dB/cm it is not recommended for routing.

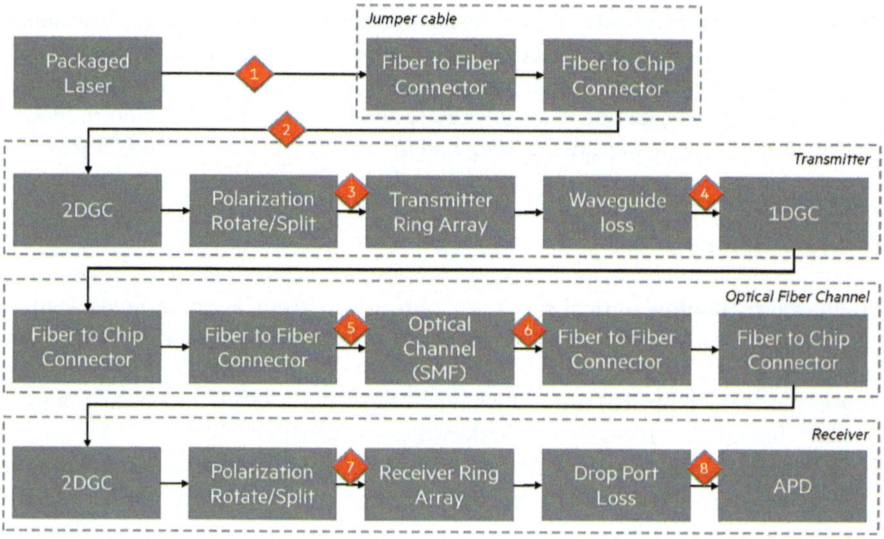


**Fig. 12.2** Typical DWDM ring-based link architecture with various test points throughout (red diamonds)

**Shallow multimode waveguide** This TE waveguide is designed for optical routing over longer distances, typically more than 1 mm. Losses of 0.5 dB/cm are expected.

**Nitride waveguide** This waveguide is being considered for very low-loss transport and as a transition waveguide for edge couplers. Loss and other parameters are not defined at this time.

#### 12.4.1.2 Waveguide Transitions

In addition to the various types of waveguides listed above, we require low-loss adiabatic transitions and bends for the waveguides.

**Single-mode Shallow to Multimode Shallow Taper** This taper is used to transition between the two transport waveguides. Losses are expected to be less than 0.1 dB per transition with reflections of less than −40 dB.

**Single-mode Shallow to Single-Mode Deep Taper** This taper is used mainly in conjunction with ring resonator and to introduce sharp bends. Losses are expected to be less than 0.1 dB with reflections of less than −25 dB.

**Silicon to Nitride Taper** This taper is used for vertical coupling of light from top-silicon to the nitride waveguide. Losses are expected to be less than 0.02 dB with reflections of less than −25 dB.

**Single-mode Shallow Bend** A 90° bend in the single-mode shallow waveguide. Propagation losses are 1.5 dB/cm and the minimum bend radius is 20 μm.

**Multimode Shallow Bend** A 90° bend in the multimode shallow waveguide. Propagation losses are 1.5 dB/cm and the minimum bend radius is 50 μm.

**Single-mode Deep Bend** A 90° bend in the single-mode deep waveguide. This bend is used to create ring resonator cavity and is not recommended for individual use. Propagation losses are 3 dB/cm and the minimum bend radius is 4 μm.

#### 12.4.1.3 Splitters

**Single-mode Shallow optical tap** An optical tap from a single-mode shallow waveguide. This directional tap is available in tap ratios −10/−13/−16 dBs with reflections ≤−30 dB.

**Single-mode deep optical tap** An optical tap from a single-mode deep waveguide. Tap ratios −10/−13/−16; reflections ≤−30 dB.

**1 × 2 MMI** Multimode interferometer in a 1 × 2 configuration using single-mode shallow input/output waveguides. Splitting ratio −3 dB; insertion losses 0.2 dB; reflections ≤−30 dB.

**Polarization Beam Rotator and Splitter** This device is under investigation for inclusion. The rotator is intended for use with nitride-based edge coupler, which is currently not the primary optical I/O scheme.

#### 12.4.1.4 Fiber Couplers and Crossings

**1D DWDM Grating Coupler** Single-polarization grating coupler (GC) for off-surface emission of light for coupling into a single-mode fiber for use with DWDM links. Expected: Insertion loss 1.5 dB; optical bandwidth (1 dB) 21 nm; angle of emission 8 degrees; center wavelength 1300–1320; reflection −15 dB.

**1D CWDM Grating Coupler** Single-polarization grating coupler for off-surface emission of light for coupling into a single-mode fiber for use with CWDM (coarse wavelength division multiplexing) links for wider channel spacing. Expected: Insertion loss 3 dB; optical bandwidth (1 dB) 80 nm; angle of emission 8 degrees; center wavelength 1300–1320 nm; reflection −20 dB.

**2D DWDM Grating Coupler** Double-polarization grating coupler for off-surface emission of light for coupling into a single-mode fiber for use with DWDM links. Expected: Insertion loss 2.5 dB; optical bandwidth (1dB) 20 nm; angle of emission 8 degrees; center wavelength 1300–1320 nm; reflection −40 dB.

**Nitride Edge Coupler** Dual-polarization edge coupler using a nitride-based mode converter and taper. Expected: Insertion loss 1 dB; optical bandwidth (1dB) 100 nm; angle of emission 90°; center wavelength 1310 nm; reflection −20 dB.

**4-way optical crossing** Four-port optical crossing using shallow single-mode waveguides. Expected: Insertion loss 0.2 dB; optical crosstalk −30 dB; reflection −25 dB.

## 12.4.2 Active Components

**Ring-Based Filter** Here, a single ring is used as a spectral filter. The parametric device cell allows for user-input FSR and spectral linewidth and generates the shape with a correct radius. This filter is used to statically route wavelengths inside of a circuit. The filters are not intended for use as demux on the receiver (where the PIN ring modulator described below is the recommended device) but, e.g., for wavelength routing in an optical crossbar switch implementation. Expected specifications: insertion loss 0.05 dB; extinction ratio >3 dB; tuning range 15 nm; spectral linewidth 0.3–1 nm (user-defined); drop port loss 1.5 dB; free spectral range (FSR) 1–20 nm (user-defined); thermal wavelength control power 3 mW/nm; thermal control bias range <4 V; thermal control current range < 25 mA; carrier-injection wavelength control power 10 μW/nm; carrier-injection control bias range <1 V; carrier-injection control current range <5 mA.

**On-chip SOI Arrayed Waveguide Grating** An on-chip AWG for CWDM application designed with a 20 nm channel spacing. Expected specifications: insertion loss 4 dB; footprint 0.7 $mm^2$; drop port loss 1.5 dB. Design for 25 nm channel spacing also on-going.

**Mach–Zehnder Interferometer (MZI) lattice for CWDM** We also plan to include a nested MZI lattice for CWDM filtering. Expected specifications: channel spacing 20 nm; footprint 0.01 $mm^2$. Other specifications like insertion loss and tuning power are still under definition.

**Quadrature bias diode** PIN junction integrated into waveguide for plasma-effect phase shift. Expected specifications: insertion loss 1 dB; plasma wavelength control power 10 μW/GHz; plasma control bias range <1 V; plasma control current range < 5 mA.

**Resistive Heater** Resistive heating element integrated into the top silicon and intended for use with ring resonator cavities. Expected specifications: thermal wavelength control power 3 mW/nm; bias range <4 V; current range <25 mA

**PIN Ring Modulator** Single ring used as a modulator for carrier-injection diode. Figure 12.3 shows a typical ring modulator with corresponding pads for driving the RF diode for modulation, receiving photocurrent from the integrated PD/APD and two pads for controlling the integrated resistive heater. Optical insertion loss 0.1 dB; extinction ratio 12 dB; tuning range 15 nm; linewidth (FWHM) 0.11 nm; drop port loss 1.5 dB; FSR 13 nm; thermal wavelength control power 3 mW/nm; thermal control bias range <4 V; thermal control current range <25 mA; thermal time constant 4–5 μs; plasma wavelength control power 10 μW/GHz; plasma control bias range <1 V; plasma control current range <5 mA; RF drive voltage Vdc 0.4 V; RF drive voltage $V_{PP}$ 2 0.5 V; electrical $S_{11}$ 3 dB bandwidth 1.5 GHz; optical $S_{21}$ 3 dB bandwidth 16 GHz.

**PN Ring Modulator** This device comprises a single ring used as a modulator for carrier-depletion diode. This ring has a similar configuration and PIN ring but has a PN junction inside the ring waveguide. Expected specifications: optical insertion loss 2.5 db; extinction ratio 4 dB; tuning range 15 nm; linewidth 0.3 nm;

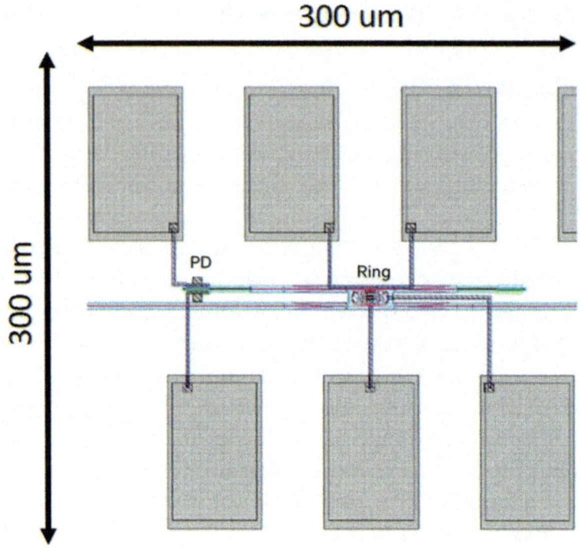


**Fig. 12.3** Layout of a ring cell. The pads are intended for probing or wire-bonding

drop port loss 2 dB; FSR 13 nm; thermal wavelength control power 3 mW/nm; thermal control bias range <4 V; thermal control current range < 25 mA; thermal time constant 4–5 μs; plasma wavelength control 10 μW/nm; plasma control bias range <1 V; plasma control current range <5 mA; RF drive voltage Vdc 1 V; RF drive voltage $V_{PP}$ 2 V; electrical $S_{11}$ 3 dB bandwidth 15 GHz; optical $S_{21}$ 3 dB bandwidth 30 GHz.

**PN MZI Modulator** An integrated MZI with segmented PN junction for PAM-4 modulation is also planned for inclusion but the devices specifications are not available at the time of writing.

### *12.4.3 Lasers*

In a SiPh link, the laser represents one of the most expensive components thanks to the fact that most other components are integrated in the SiPh PIC and therefore benefit from the economies of scale associated with the silicon industry. Possible laser configurations are sketched in Fig. 12.4. Figure 12.4a shows a common off-chip configuration where the laser is independently fabricated and packaged and is then coupled through a fiber to the PIC chip. This configuration has the advantage of being able to leverage existing laser packages and fabrication technologies as well as being able to isolate the laser in an environment that has smaller temperature variations. However, it is the most expensive solution and has the highest losses. In the second on-chip configuration Fig. 12.4b, the laser is independently fabricated and then it is attached to the SiPh PIC. Light from the laser is coupled to the PIC

**Fig. 12.4** Laser integration evolution

via either a vertical or an edge coupler. This configuration has the advantage of simplifying the packaging and reducing losses (because there are fewer interfaces between the laser and the PIC) while still keeping the laser fabrication separated from the PIC fabrication. In the third integrated configuration Fig. 12.4c, the laser is directly fabricated on the PIC and light is usually coupled between the laser and the PIC though evanescent coupling. This configuration is the most efficient for cost and power but requires the processing of III-V compound semiconductors that are in general incompatible with CMOS silicon fabrication lines. Our group has explored several options for the lasers and we expect that these solution will be available to users through ecosystem partners. In particular, here we focus on multiline comb lasers engineered for DWDM applications.

**Off-chip lasers** Multiline comb lasers based on quantum dots have been demonstrated in the past [12, 13]. These lasers are packaged separately and are procured as a qualified part. Qualified lasers are included in the development kit as models to allow full link simulations. In our first implementation, lasers will be available as a butterfly-packaged test component. Other packages may be possible. Expected specifications: bias current 175 mA; bias voltage typical 2 V; saturable absorber reverse bias 1 V; operating temperature 10–80° C; spectral channel spacing 50 GHz; number of channels 16; relative intensity noise $<-140$ dB/$\sqrt{\text{Hz}}$; typical optical power per line at the fiber facet 4 dBm; typical wall plug efficiency 7%.

**On-chip** Single-line laser chips that can be directly attached to a PIC have been demonstrated in the past [14]. We foresee a solution where a multiline comb quantum dot laser can be directly attached to a PIC using solder ball or copper pillars that are used both for mechanical attachment and electrical connection. Light from the laser can be coupled into the PIC through an off-surface grating. Because the laser chip is in many ways similar to the one used in the off-chip laser solution, we expect similar properties with a considerable increase in the efficiency of power delivery to the chip thanks to the reduction in the number of optical interfaces in the optical path.

**Integrated** The ultimate goal of laser integration is to build the laser in the same platform as the rest of the PIC. Our group [3, 4] has worked extensively to develop this solution. As shown schematically in Fig. 12.5, such a system would include a comb laser evanescently coupled to a SiPh PIC. The lasers are built using a bonding process to attach III-V coupons onto a SiPh wafer followed by lithography and etch steps to define the laser structures. In addition to CW (continuous wave) laser combs, this platform supports photodiodes [5] (both PIN and APD) as well as high-speed modulators based on a MOS capacitor built at the interface between the top

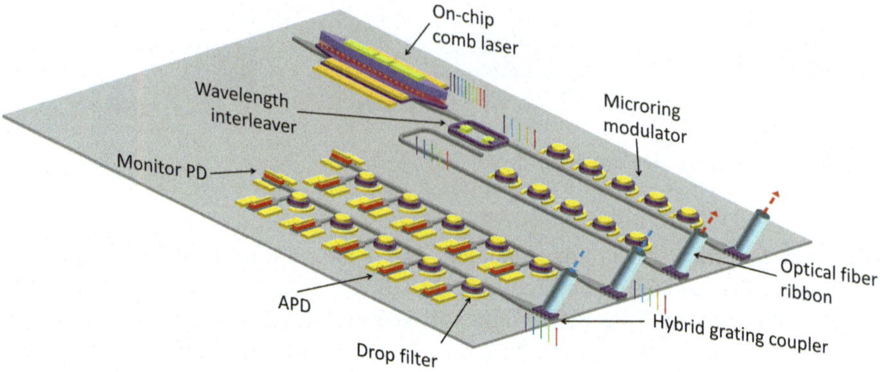


**Fig. 12.5** Vision for a fully integrated PIC with on-chip comb laser

silicon layer and the III-V bonded material [15]. This platform further reduces the laser interface losses and potentially reduces individual laser costs as thousands of lasers can be built in parallel using standard photolithographic techniques.

### *12.4.4 Detectors*

Several foundries have developed Ge PIN detectors that can be integrated in the SiPh platform and are efficient at both the conventional (1550 nm) and the short (1300 nm) telecom window. Such detectors are usually built using silicon–germanium selective growth and implants. The main processing issue is to overcome the mismatch between the Si and Ge lattice spacing using appropriate SiGe buffer layers. Once the detector material is defect-free and the detector dark noise is low it is easy to build detectors with high efficiency and bandwidth. This is an effect of the detectors' waveguide configuration where the responsivity is primarily dictated by the detector length along the waveguide axis and the bandwidth is dictated by the detector width. Our group has researched detectors beyond the standard SiGe PINs. In particular, we have demonstrated low-voltage SiGe avalanche photodiodes that are compatible with both the SiPh fabrication process and the power supplies typically present in computing environments. We have also explored quantum dot APDs as an alternative to SiGe APDs that are compatible with our integrated laser platform.

**PIN** This is the standard waveguide-based germanium PD for 1300 nm. This photodiode will be available to users as a p-cell. The expected typical responsivity is 0.8 A/W at an operating bias of −1 V with a sensitivity of at least −10 dBm and optical 3 dB bandwidth of 16 GHz.

**Silicon–Germanium Avalanche PD** Our group has extensively studied avalanche photodiodes ([2, 16–18]) because of their ability to improve link sensitivity without a corresponding increase in the power consumption. We are actively pursuing the inclusion of SiGe waveguide APDs in the standard SiGe platform. To ensure the compatibility of the APD bias voltage with computer power supplies, we have designed three-terminal APDs with breakdown voltages as low as 6 V [2]. In addition to the low bias voltage, these devices have a bandwidth of 18.9 GHz, a DC photocurrent gain of 15, and sensitivity of −11.4 dBm at a data rate of 25 Gb/s.

**Quantum dot PD and APD** The rationale behind the development of quantum dot detectors is that, if the SiPh platform is expanded to include quantum dot materials for lasers, the same materials can be used for PD and APD that have excellent properties, for example, our group has demonstrated [5] an avalanche photodiode by bonding indium arsenide (InAs) QDs on an SOI platform. Our demonstration has shown encouraging performance with dark currents of 1 $\mu$A/cm$^2$, responsivities of 0.34 A/W at 1310 nm and 0.9 A/W at 1280 nm, and a 3 dB bandwidth of 15 GHz. These device are also able to function at temperatures up to 60 °C. We are considering the inclusion of these devices in the platform in later releases.

### *12.4.5 Packaging*

The main goal of our packaging effort is to enable close integration of the SiPh devices with network switches, network interface cards (NICs) and other networking, compute, and memory components. The integration improves signal integrity and power consumption. As shown in Fig. 12.6, the industry has been moving from front-panel active optical cables to on-board optics with the final goal of integrating photonics in the same package with the logic components. There are two main elements to achieve this integration: the fiber attach and the co-packaging of SiPh and CMOS circuits. For the fiber attach in addition to the requirements related to optical performance, yield, and reliability, there are additional requirements, for example, it is important to make the fiber attach compatible with reflowing complete optoelectronic assemblies in a workflow that is compatible with the existing infrastructure. It is also important to allow components to be separated on the field for service. This requirement excludes the option of pig-tailing the fiber to the assembly, as a pig-tailed assembly with tens or hundreds of fibers would be impractical. The main requirement for the co-packaging is the compatibility with the fiber attach process. This is a particularly stringent requirement if 3D integration of the SiPh and CMOS is the goal. Our group, in collaboration with ecosystem partners, has started addressing these requirements.

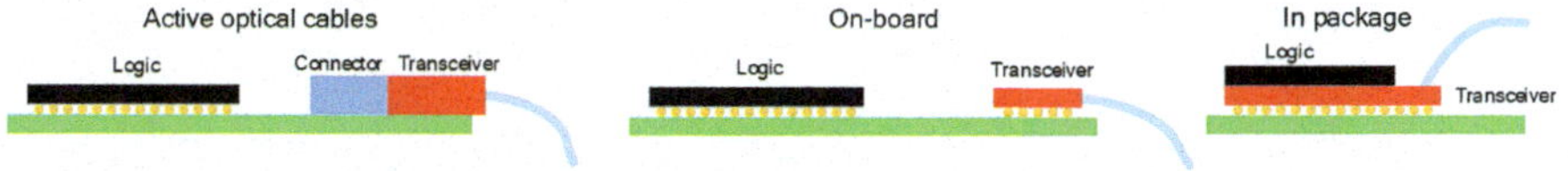


**Fig. 12.6** Evolution of photonic packaging

**Fiber attach** Our approach to fiber attach is delineated in [19]. A common approach to fiber attach for SiPh PICs is to pigtail the circuit, i.e., bonding fiber/v-groove arrays to on-chip grating couplers. This approach precludes the standard approach to the assembly of transceivers and other optoelectronic devices, i.e., the surface mount and reflow of the devices on standard system printed boards. To obviate this, we have developed a $1 \times 8$ single-mode (SM) optical interface for DWDM SiPh PICs. Our approach is based on a low profile, small form factor, detachable, and expanded-beam optical connector. The connector consists of four parts: a GC array, a surface-mount glass microlens array chip, an injection-molded solder-reflowable optical socket, and an injection-molded single mode light-turn ferrule. On-wafer measurements of the GC coupling loss to SMF28 fiber at 1310 nm is 1.4 dB with a 1 dB bandwidth of 22 nm. The fiber-to-fiber insertion loss through the light-turn ferrule, two microlenses and GCs, and a looped back SOI waveguide ranged from 4.1–6.3 dB, with insertion loss repeatability of 0.2 dB after multiple mating cycles. Users of the open ecosystem will have access to this solution through a design cell for the coupler array that includes the GCs as well as the structures necessary for connector alignment. Partner OSATs will then be able to attach the other on-chip components. Jumpers and cables that can connect the PICs to other optoelectronics devices will be available through other ecosystem partners.

**Packaging** A small transceiver chip like the one depicted in Fig. 12.7 can be in principle realized by wire-bonding the SiPh interposer to the organic substrate. In contrast this solution is not feasible for assemblies involving large logic chips and multiple connectors. The electronic industry has developed systems to co-package multiple chips on a silicon through-silicon-via (TSV) interposer (see e.g. [20]). The fabrication techniques used in those processes can be adapted to a SiPh TSV

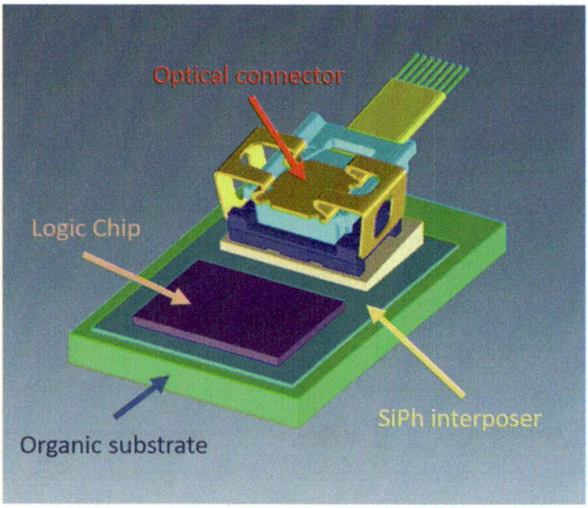


**Fig. 12.7** Fiber attach and package. We thank USConec for permission to use the image of their connector [21]

interposer although the assembly process will need changes in order to ensure that the optical surfaces are not damaged. Our group is working with partners in the ecosystem to enable a solution for large-size TSV-enabled interposers.

## 12.5 Conclusions and Future Work

In this chapter, we have outlined an open SiPh ecosystem for computercom applications. This ecosystem centers around a design kit comprising a number of functional devices that have been fabricated and verified at a major foundry. The device catalog is expanding but in its first release would allow to design, verify, and build functional high-performance circuits for DWDM and CWDM links. Ecosystem partners would also allow one to test the PIC, assemble the known good PICs with logic circuit fabricated in a different process and attach a fiber array using a detachable, reflowable connector. This subassembly could be then handed off to an electronic assembly house for final assembly on an organic substrate. Substantial progress in all the areas of this process has been demonstrated and we expect to continue to share additional progress toward our goal.

## References


1. J. H. Ahn, N. Binkert, A. Davis, M. McLaren, and R. S. Schreiber. Hyperx: topology, routing, and packaging of efficient large-scale networks. In *Proceedings of the Conference on High Performance Computing Networking, Storage and Analysis*, pages 1–11, 2009
2. Xiaoge Zeng, Zhihong Huang, Binhao Wang, Di Liang, Marco Fiorentino, and Raymond G Beausoleil. Silicon–germanium avalanche photodiodes with direct control of electric field in charge multiplication region. *Optica*, 6(6):772–777, 2019
3. Di Liang, Erwen Li, Sudharsanan Srinivasan, Antoine Descos, Zhixin Liu, Geza Kurczveil, Zhihong Huang, and Raymond G. Beausoleil. Heterogeneous multi-wavelength optical injection locked system-on-chip: a proposal & proof-of-concept experiment. In *Asia Communications and Photonics Conference (ACPC) 2019*, page M4D.4. Optical Society of America, 2019
4. Di Liang, Ashkan Roshan-Zamir, Yang-Hang Fan, Chong Zhang, Binhao Wang, Antoine Descos, Wenqing Shen, Kunzhi Yu, Cheng Li, Gaofeng Fan, et al. Fully-integrated heterogeneous dml transmitters for high-performance computing. *Journal of Lightwave Technology*, 2019
5. Bassem Tossoun, Géza Kurczveil, Chong Zhang, Antoine Descos, Zhihong Huang, Andreas Beling, Joe C. Campbell, Di Liang, Raymond G. Beausoleil, Indium arsenide quantum dot waveguide photodiodes heterogeneously integrated on silicon. Optica **6**(10), 1277–1281 (2019). Oct
6. P. Sun, J. Hulme, T. Van Vaerenbergh, J. Rhim, C. Baudot, F. Boeuf, N. Vulliet, A. Seyedi, M. Fiorentino, R.G. Beausoleil, Statistical behavioral models of silicon ring resonators at a commercial cmos foundry. IEEE Journal of Selected Topics in Quantum Electronics **26**(2), 1–10 (2020)
7. Lumerical. Lumerical Website. See https://www.lumerical.com/products/device

8. Peng Sun, Thomas Van Vaerenbergh, Marco Fiorentino, Raymond Beausoleil, Adjoint-method-inspired grating couplers for cwdm o-band applications. Optics Express **28**(3), 3756–3767 (2020)
9. Binhao Wang, Zhihong Huang, Xiaoge Zeng, Wayne V Sorin, Di Liang, Marco Fiorentino, and Raymond G Beausoleil. A compact model for si-ge avalanche photodiodes over a wide range of multiplication gain. *Journal of Lightwave Technology*, 37(13):3229–3235, 2019
10. Peng Sun, Jared Hulme, Ashkan Seyedi, Marco Fiorentino, and Raymond Beausoleil. Data-mining-assisted resonance labeling in ring-based dwdm transceivers. In *Optical Fiber Communication Conference*, pages M3A–3. Optical Society of America, 2020
11. MentorGraphics. Mentor Website. See https://www.mentor.com/products
12. L. Gregory, Wojcik, Dongliang Yin, Alexey R. Kovsh, Alexey E. Gubenko, Igor L. Krestnikov, Sergey S. Mikhrin, Daniil A. Livshits, David A. Fattal, Marco Fiorentino, and Raymond G. Beausoleil. A single comb laser source for short reach WDM interconnects, in *Novel In-Plane Semiconductor Lasers VIII*, volume 7230, ed. by Alexey A. Belyanin, Peter M. Smowton (International Society for Optics and Photonics, SPIE, 2009), pp. 89–100
13. Ashkan Seyedi, Marco Fiorentino, and Ray Beausoleil. True Concurrent Modulation of a Multi-Channel Ring Modulator Transmitter Driven by a Comb Laser. In *International Technical Conference and Exhibition on Packaging and Integration of Electronic and Photonic Microsystems*. ASME, 08 2017
14. Peter De Dobbelaere. Silicon photonics technology platform for integration of optical ios with asics. In *2013 IEEE Hot Chips 25 Symposium (HCS)*, pages 1–18. IEEE, 2013
15. Di Liang, Geza Kurczveil, Zhihong Huang, Binhao Wang, Antoine Descos, Sudharsanan Srinivasan, Yingtao Hu, Xiaoge Zeng, Wayne V Sorin, Stanley Cheung, et al. Integrated green dwdm photonics for next-gen high-performance computing. In *Optical Fiber Communication Conference*, pages Th1E–2. Optical Society of America, 2020
16. Binhao Wang, Zhihong Huang, Xiaoge Zeng, Di Liang, Marco Fiorentino, Wayne V Sorin, and Raymond G Beausoleil. 50 gb/s pam4 low-voltage si-ge avalanche photodiode. In *CLEO: Science and Innovations*, pages SM4J–7. Optical Society of America, 2019
17. Binhao Wang, Zhihong Huang, Xiaoge Zeng, Di Liang, Wayne V Sorin, Marco Fiorentino, and Raymond G Beausoleil. Low-voltage si-ge avalanche photodiodes for datacom. In *2019 Photonics & Electromagnetics Research Symposium-Fall (PIERS-Fall)*, pages 2526–2531. IEEE, 2019
18. Zhihong Huang, Binhao Wang, Yuan Yuan, Di Liang, Marco Fiorentino, and Raymond G Beausoleil. 64gbps pam4 modulation for a low energy si-ge waveguide apd with distributed bragg reflectors. In *Optical Fiber Communication Conference*, pages W4G–8. Optical Society of America, 2020
19. Sagi Mathai, Paul Rosenberg, George Panotopoulos, Dan Kurtz, Darrell Childers, Thomas Van Vaerenbergh, Peng Sun, Jared Hulme, Jinsoo Rhim, Ashkan Seyedi, et al. Detachable $1 \times 8$ single mode optical interface for dwdm microring silicon photonic transceivers. In *Optical Interconnects XX*, volume 11286, page 112860A. International Society for Optics and Photonics, 2020
20. TSMC. TSMC Website. See https://www.tsmc.com/english/dedicatedFoundry/services/cowos.htm
21. Mike Hughes, Darrell Childers, Dan Kurtz, Dirk Schoellner, Shubhrangshu Sengupta, and Ke Wang. A single-mode expanded beam separable fiber optic interconnect for silicon photonics. In *Optical Fiber Communication Conference (OFC) 2019*, page Tu2A.6. Optical Society of America, 2019

# Index




D. J. Lockwood and L. Pavesi (eds.), *Silicon Photonics IV*,
Topics in Applied Physics 139, https://doi.org/10.1007/978-3-030-68222-4

The manufacturer's authorised representative in the EU is Springer Nature Customer Service Centre GmbH, Europaplatz 3, 69115 Heidelberg, Germany. If you have any concerns regarding our products, please contact ProductSafety@springernature.com

Printed and bound by CPI Group (UK) Ltd, Croydon, CR0 4YY
30/09/2026
02225594-0004